Self-similar and Self-affine Sets and Measures

Mathematical
Surveys
and
Monographs

Volume 276

Self-similar and Self-affine Sets and Measures

Balázs Bárány
Károly Simon
Boris Solomyak

AMERICAN MATHEMATICAL SOCIETY
Providence, Rhode Island

2020 *Mathematics Subject Classification.* Primary 28A80; Secondary 28A78, 28A75, 37D35, 42A38.

For additional information and updates on this book, visit
www.ams.org/bookpages/surv-276

Library of Congress Cataloging-in-Publication Data

Names: Bárány, Balázs, 1984- author. | Simon, Károly, 1961- author.
Title: Self-similar and self-affine sets and measures/ Balázs Bárány, Károly Simon, Boris Solomyak.
Description: Providence, Rhode Island: American Mathematical Society, [2023] | Series: Mathematical surveys and monographs, 0076-5376; Volume 276 | Includes bibliographical references and index.
Identifiers: LCCN 2023030385 | ISBN 9781470470463 (paperback) | 9781470475505 (ebook)
Subjects: LCSH: Self-similar processes. | Fractals. | Hausdorff measures. | AMS: Measure and integration – Classical measure theory – Fractals. | Measure and integration – Classical measure theory – Hausdorff and packing measures. | Measure and integration – Classical measure theory – Length, area, volume, other geometric measure theory. | Dynamical systems and ergodic theory – Dynamical systems with hyperbolic behavior – Thermodynamic formalism, variational principles, equilibrium states. | Harmonic analysis on Euclidean spaces – Harmonic analysis in one variable – Fourier and Fourier-Stieltjes transforms and other transforms of Fourier type.
Classification: LCC QA274.9 .B37 2023 | DDC 519.2/3–dc23/eng/20231002
LC record available at https://lccn.loc.gov/2023030385

♾ The paper used in this book is acid-free and falls within the guidelines
established to ensure permanence and durability.
Visit the AMS home page at `https://www.ams.org/`

10 9 8 7 6 5 4 3 2 1   28 27 26 25 24 23

Contents

Preface

This book is devoted to a field at the intersection of fractal geometry and dynamical systems theory. Although certain parts of the theory have been around for a long time, in particular, the classical Cantor set and its relatives, as well as natural measures on them, in the last few decades it has seen an explosive growth. On the one hand, it was the insight of Mandelbrot who realized that diverse phenomena in many areas of Mathematics, Physics, Biology, and other Sciences, as well as in Nature, which have previously been viewed as "anomalous" or "weird" are actually widespread and have a common feature which he called "fractal". We will not attempt to give a definition of a fractal, but it has now become well established that introducing this notion was crucial and transformed the way we view many things in mathematics and far beyond. It has, of course, been tremendously aided by the advance of computers and the beautiful pictures which are easy to produce and bring the beauty of the subject to the general public. On the other hand, Dynamical Systems theory, in particular, of what is sometimes called the "chaotic" behavior, naturally lead to fractals, as many of the objects appearing there, such as attractors, repellers, basin boundaries, sets of parameters with distinguished properties, etc., are fractal.

There are many excellent books on fractals, among them we should mention the books by Rogers [**Rog70**], Falconer [**Fal85, Fal03, Fal97**], Barnsley [**Bar93**], Mattila [**Mat95, Mat15**], Edgar [**Edg98**], Bishop–Peres [**BP17**], and Fraser [**Fra20**]. Our goal is different, first of all, in that we focus almost exclusively on the class of *self-similar* and *self-affine* fractals. Roughly speaking, if we say that a "fractal" is a set where "zooming in" we see a picture that resembles the entire set, which is the general (imprecise) concept of self-similarity, then self-similar and self-affine sets are those for which this resemblance is precise and is given by a linear contracting similarity map or a general linear affine contracting map, respectively. These are the "model" and "test" cases, which have always been the first under investigation, and surprisingly, even simple-minded questions about them turn out to be very difficult. In the last decades, there have been many breakthroughs, and our aim is to give an introduction to some of them. Specific questions are usually about *"fractal" dimension*, the Hausdorff and box-counting dimensions being the most widely used, and on the topological nature of the fractals, such as: when do they have nonempty interior? In order to address these questions, it is unavoidable that we have to study *measures* on the fractal sets, so we are naturally led to self-similar and self-affine measures. We consistently use the framework of *iterated function systems*, first developed by Hutchinson, Diaconis, and Barnsley, and the intimately connected *symbolic spaces*, which provide the "coding", or "address" structure and facilitate the application of symbolic dynamics.

We concentrate on a few directions in the (relatively) recent developments. We try to choose the simplest special cases, where the proofs can be presented in a precise but less technical way, in order to introduce a range of important techniques for those interested to learn the subject in depth. It is unavoidable that many results that we rely on, are stated without proofs. Moreover, the subject has grown so much in recent years, that is impossible to give a comprehensive account in a single volume.

One of the key areas that we focus on is the investigation of self-similar sets with "overlaps". Such sets arise naturally, in particular, from orthogonal projections of nonoverlapping Cantor-like sets in higher dimensions. A closely related direction is the study of self-similar measures with overlaps, of which one of the best known is the family of so-called *infinite Bernoulli convolutions*, or simply Bernoulli convolutions. They are easy to define: for a parameter $0 < \lambda < 1$ consider ν_λ, the distribution of the random series $\sum_{n=0}^{\infty} \pm\lambda^n$, where the signs are chosen independently with probabilities $(1/2, 1/2)$. In the case $\lambda < 1/2$ we get the Cantor–Lebesgue measure on a homogeneous self-similar Cantor set of dimension less than 1, for $\lambda = 1/2$ this is a uniform measure on $[-2, 2]$, but the "overlapping" case $\lambda \in (1/2, 1)$ presents a major challenge. These measures and their close relatives come up in many areas, both pure and applied, and they have been studied since the 1930's by many authors, including Erdős, Salem, Kahane, and Garsia, by a variety of methods, with the help of harmonic analysis, number theory, and probability.

In the 1990s and early 2000s, the method of "transversality" has been developed, which led to a first breakthrough in the understanding of the overlapping case, both for self-similar sets and measures. The authors of this book, with co-authors, have been among the main contributors in this development, and it constitutes a large and important part of the book. More recently, a further breakthrough occurred, starting with the work of Hochman, followed by Shmerkin and Varjú, with various co-authors. The details of this latest stage, which is currently very "hot" and is still being developed, remains beyond the scope of our book; however we do summarize the main results obtained as of this writing.

Another great challenge is computing the dimension of general self-affine sets, even in the plane when there are no overlaps. Here the difficulty, compared with the self-similar case, is "nonconformality". In the book we present first the results obtained using the transversality method, which generalize a classical theorem of Falconer. In the study of self-affine sets, dynamical systems and ergodic theory tools appear even more prominently. Here also there were recent breakthroughs; an important one was started by Feng and Hu, who adapted the approach of Ledrappier and Young from dynamical systems theory to the self-affine case. A part of this theory is presented in the book. In this area as well, the most recent developments, e.g., those by Käenmäki, Morris, Bárány, Hochman, and Rapaport, are only summarized.

Once again, we emphasize that in such an active and broad field it is impossible to be comprehensive; we focus on a few specific directions which we consider interesting and important. Among the many topics which are completely left out or only mentioned briefly, are graph-directed systems, random fractals, iterated function systems that are contracting on-average, iterated function systems with infinitely many maps (an excellent book by Mauldin and Urbański [**MU03**] is devoted to the latter), and there is only one chapter about nonlinear systems.

On the other hand, in a few places, for the sake of completeness, our presentation bears similarity to the existing books; notably, to Falconer [**Fal97**, Chapter 11] in parts of our chapter on multifractals, to Mattila's books: [**Mat95**, Chapter 9] in the proof of Marstrand's theorem and [**Mat15**, Chapters 8–9] in the treatment of Cantor measures with Pisot parameters and Bernoulli convolutions, as well as to Bishop–Peres [**BP17**, Chapter 9] in the discussion of projections of self-similar sets and the Bandt–Graf theorem.

A few words about the contents and organization of the book. The long introductory chapter is aimed at a newcomer to the subject; there we present many examples, introduce the basic notions, and outline the main directions of what follows. A more experienced reader can skip it after a quick perusal and only return to it as necessary. The second chapter contains a few results from geometric measure theory that we need, among them results about densities of Radon measures, the Marstrand projection theorem, whose proof by Kaufman has been a "precursor" of the transversality method, and slicing theorems, with proofs or partial proofs, as well as additional background material on dimension of measures. In Chapter 3 we begin an in-depth investigation of self-similar sets and measures. We show that some results hold without any assumptions of nonoverlapping, such as equality of the Hausdorff and box-counting dimensions and lower semicontinuity in parameter dependence. We also prove "Laws of Pure Types" for self-similar measures and show the existence of L^q dimensions and the entropy dimensions for them. In Chapter 4 we turn to various conditions of being overlapping or nonoverlapping, the classical one being the Open Set Condition. The main part of this chapter is devoted to a large class of systems for which weaker forms of separation holds (namely, some of the "cylinders" are allowed to coincide). This turns out to be remarkably subtle, and we present a few featured special systems in which the methods can be understood more clearly. Some of these more specialized sections may be skipped on first reading. A short Chapter 5 gives a glimpse into the Multifractal Analysis for self-similar measures, which is mainly beyond the scope of the book. Chapter 6 contains the key parts of the transversality method for self-similar sets and measures and includes a discussion of more recent developments. In Chapter 7 we mainly focus on the questions of positivity of the Hausdorff and packing measure of self-similar sets in their dimension. This includes connections with the weak separation property and the Assouad dimension. The chapter also contains sections on important examples, such as projections of planar self-similar sets (without rotations) and the $\{0, 1, 3\}$-problem. Chapter 8 contains a mixture of old and more recent results, with a focus on Number Theory connections, often with the help of Fourier Analysis. It includes classical results of Erdős and Garsia on Bernoulli convolutions with a modern exposition, connections with β-expansions, and a discussion of recent breakthrough results on dimensions of Bernoulli convolutions. Chapter 9 presents some elements of Ergodic Theory, mostly without proof, and it opens the informal second part of the book devoted to self-affine sets and measures. It plays a role similar to Chapter 2 on elements of geometric measure theory. Ergodic Theory is used throughout the book, including the earlier chapters, in particular, the Ergodic and Shannon–McMillan–Breiman Theorem. In order to help a reader unfamiliar with the basics, some of the notions and results of Ergodic Theory are presented already in Chapter 1. In Chapters 10–13 the use of Ergodic Theory becomes more sophisticated, which includes Rokhlin's Theorem on disintegration of measures, Multiplicative Ergodic Theorem of Oseledeč, and Thermodynamic Formalism. Chapters 10–13 constitute the second "main core" of the book; they are devoted to selected results on self-affine sets and measures. In the first part of Chapter 10 we present the self-affine transversality condition and its consequences, following Jordan, Pollicott, and Simon. After that we introduce the sub-additive thermodynamic formalism for self-affine IFS's, including a presentation of Käenmäki's results and the notion of domination, with proofs in special cases. Chapter 11 is devoted to the Feng–Hu theory, which may be thought of as the adaptation of the Ledrappier–Young theory to the context of diagonally-self-affine IFS. We first present it the simplest nontrivial case, in order to motivate the notion of projection entropy. As an application, we consider a large variety of self-affine families, starting with the classical Bedford–McMullen carpets. In Chapter 12 we present a selection of results on dimension conservation and exact dimensionality for self-affine measures, highlighting the role of the Furstenberg measure and the Ledrappier–Young formula, extended to the general planar case. Chapter 13 studies projections of self-affine sets and measures, using the method of entropy averages introduced by Hochman and Shmerkin. In most of the results in Chapters 10–13 we restrict ourselves to the planar case, in order to make the presentation more accessible. Chapter 14 is somewhat special, in that it treats nonlinear conformal IFS on the line. We demonstrate how the transversality methods presented in Chapter 6 may be adapted to a nonlinear setting. Finally, several appendices supply the needed background material from diverse fields.

Parts of this book have been used by the authors in graduate and advanced undergraduate courses over the years: at the Budapest University of Technology and Economics (B.B. & K.S.), at the University of Washington and the Bar-Ilan University (B.S.), as well as lecture series and mini-courses at various schools and workshops. In particular, we think that Chapter 1 (Sections 1.1–1.6, 1.8–1.12); Chapter 2 (Sections 2.1–2.4); Chapter 3 (Sections 3.1–3.4); Chapter 4 (Section 4.1); Chapter 5 (Sections 5.1 and 5.3); Chapter 6 (Sections 6.1 and 6.2), and Chapter 8 (Section 8.1), are suitable material for an advanced undergraduate course. Much of the rest can be used for a graduate course, according to the Lecturer's taste.

In spite of our best efforts, errors (hopefully, minor) undoubtedly remain. We would appreciate receiving email about errors found by the readers. Corrections will be posted on the authors' personal websites.

Acknowledgment. The authors would like to express their gratitude to Simon Baker, Christoph Bandt, Jonathan M. Fraser, Thomas Jordan, Antti Käenmäki, Vilma Orgoványi, Yuval Peres, Mark Pollicott, Dániel Rudolf Prokaj, Feliks Przytycki, Ariel Rapaport, Alex Rutar, Adam Śpiewak, Mariusz Urbański, and the anonymous referees for their careful proofreading, useful comments and valuable suggestions. We are grateful to Pablo Shmerkin for his kind permission to use some excerpts of his unpublished Lecture Notes.

CHAPTER 1

Introduction

This chapter is intended as a gentle introduction to the concepts which form the core of the book. We present a large number of examples, some standard, some less so, which are defined from "scratch". We then move to the definitions of dimension and discuss some of the basic results, mostly without complete proofs, but with some hints for the proofs or proofs in important special cases.

The Sierpiński gasket (or Sierpiński triangle) is one of the best known fractals. To construct it we start with the equilateral triangle ΔABC of unit side length (see Figure 1.1). In the first step we remove the open triangle formed by the

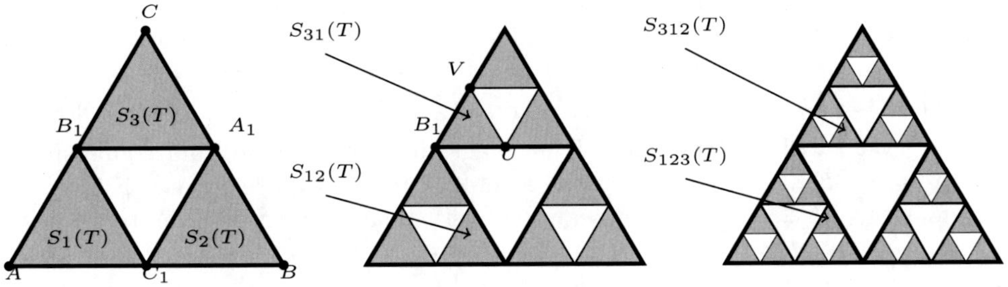

FIGURE 1.1. T is the "big" equilateral triangle. The first three approximations of the Sierpiński gasket.

midpoints $\Delta A_1B_1C_1$. Each of the remaining three triangles ΔAC_1B_1, ΔBA_1C_1 and ΔCB_1A_1 are similar copies (ratio 1/2) of the original triangle. We repeat the process of removing the open midpoints triangles in each remaining triangle ad infinitum. The first three steps of our infinite iterative process are shown in Figure 1.1. The collection of those points which are not removed in any step of this process from the set is called the Sierpiński gasket.

We can look at the construction of the Sierpiński gasket from the point of view of the triangles that are retained after each step rather than the ones that are removed. Namely, first observe (see Figure 1.1) that the so-called level-1 triangles ΔAC_1B_1, ΔBA_1C_1 and ΔCB_1A_1 are the images of $T := \Delta ABC$ under the maps defined for $\mathbf{x} = \begin{bmatrix} x_1 \\ x_2 \end{bmatrix} \in \mathbb{R}^2$ by

$$(1.1) \qquad S_1(\mathbf{x}) = \frac{1}{2} \cdot \mathbf{x},\ S_2(\mathbf{x}) = \frac{1}{2} \cdot \mathbf{x} + \begin{bmatrix} \frac{1}{2} \\ 0 \end{bmatrix} \text{ and } S_3(\mathbf{x}) = \frac{1}{2} \cdot \mathbf{x} + \begin{bmatrix} \frac{1}{4} \\ \frac{\sqrt{3}}{4} \end{bmatrix}.$$

The level-2 rectangles (see the Figure 1.1) can be obtained by the composition $S_{i_1i_2}(T)$, where we write

$$S_{i_1\dots i_n} := S_{i_1} \circ \cdots \circ S_{i_n}.$$

A nonempty bounded set which can be represented as a finite (not necessarily disjoint) union of its own similar copies is called a *self-similar set*. Usually we obtain such a set as a result of an infinite iterative process.

1.1. Self-similar sets

For integers $m \geqslant 2$ and $d \geqslant 1$, we call the tuple $\mathcal{S} = (S_1, \dots, S_m)$ of m contracting similarity transformations acting on $\mathbb{R}^d$ a *self-similar IFS* on $\mathbb{R}^d$ with contraction ratios $0 < r_i < 1$, $i = 1, \dots, m$, if

$$\forall i \leqslant m,\ \forall \mathbf{x}, \mathbf{y} \in \mathbb{R}^d,\ \|S_i(\mathbf{x}) - S_i(\mathbf{y})\| = r_i \|\mathbf{x} - \mathbf{y}\|. \tag{1.2}$$

We always assume that not all of the fixed points of the maps in $\mathcal{S}$ are identical.

NOTATION 1. Denote by $\mathscr{S}_m(d)$ the collection of ordered tuples $\mathcal{S} := (S_1, \dots, S_m)$ of (strictly) contracting similarities on $\mathbb{R}^d$ satisfying the following property: there exist $i, j \in \{1, \dots, m\}$ such that the fixed points of S_i and S_j are distinct. We emphasize that we allow $S_i \equiv S_j$ for some distinct $i, j \in \{1, \dots, m\}$. That is, repetition in the tuple $(S_1, \dots, S_m)$ is allowed.

EXAMPLE 1.1.1 (Sierpiński gasket revisited). The transformations in (1.1) provide a self-similar IFS in $\mathbb{R}^2$ which we denote $\mathcal{S}_{SG}$. One can easily check that all of the maps $S_i \in \mathcal{S}_{SG}$ send the equilateral triangle T with vertices $A = (0,0)$, $B = (1,0)$ and $C = \left(\frac{1}{2}, \frac{\sqrt{3}}{2}\right)$ into itself (see Figure 1.1). The sets $S_1(T)$, $S_2(T)$ and $S_3(T)$ are the level-1 cylinder triangles. The images of T by the two-fold compositions $S_{ij}(T) := S_i \circ S_j$, $1 \leqslant i, j \leqslant 3$, are the level-2 cylinder triangles. Since $S_j(T) \subset T$, therefore $S_{ij}(T) \subset S_i(T)$. This way for every n and for every infinite sequence $(i_1, i_2, \dots) \in \{1,2,3\}^{\mathbb{N}}$ we have $S_{i_1 \dots i_{n+1}}(T) \subset S_{i_1 \dots i_n}(T)$. The set of points Λ that remain after infinitely many iterations:

$$\Lambda := \bigcap_{n=1}^{\infty} \bigcup_{(i_1 \dots i_n) \in \{1,2,3\}^n} S_{i_1 \dots i_n}(T)$$

is the Sierpiński gasket. Clearly, $\Lambda = \bigcup_{i=1}^{3} S_i(\Lambda)$. Therefore, the Sierpiński gasket is self-similar.

Another well known self-similar fractal set is the Sierpiński carpet.

EXAMPLE 1.1.2 (Sierpiński carpet). Let

$$\mathcal{S}_{SC} := \left(S_i(\mathbf{x}) = \frac{1}{3} \cdot (\mathbf{x} + \mathbf{t}_i)\right)_{i=1}^{8}, \text{ where } \mathbf{t}_i \in \{(0,0), \dots, (2,2)\} \setminus \{(1,1)\}. \tag{1.3}$$

For the unit square $Q = [0,1]^2$ we have $S_i(Q) \subset Q$ for all $i = 1, \dots, 8$. The attractor of the IFS $\mathcal{S}_{SC}$ is the set

$$\Lambda = \bigcap_{n=1}^{\infty} \bigcup_{(i_1 \dots i_n) \in \{1, \dots, 8\}^n} S_{i_1 \dots i_n}(Q).$$

The level-1, level-2 and level-3 cylinder squares of the Sierpiński carpet are shown in Figure 1.2.

Finally, the simplest self-similar fractal is the middle-third Cantor set:

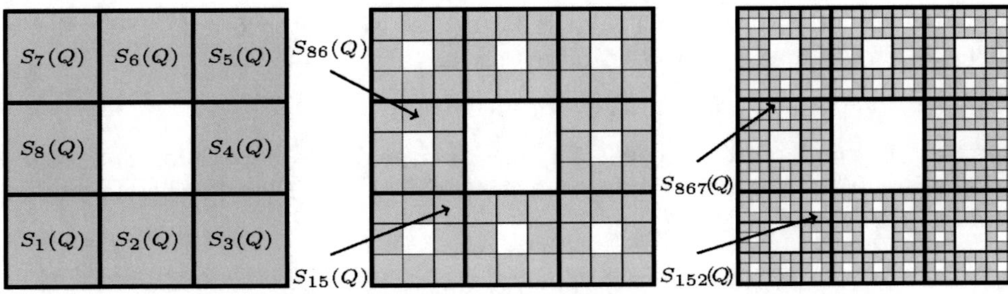

FIGURE 1.2. Q is the "large" square. The first three approximations of the Sierpiński carpet.

EXAMPLE 1.1.3 (Middle-third Cantor set). Consider the IFS

$$\mathcal{S}_C := \left(S_1(x) = \frac{x}{3}, \quad S_2(x) = \frac{x}{3} + \frac{2}{3}\right), \qquad x \in \mathbb{R}. \tag{1.4}$$

The middle-third Cantor set (see [**Rud76**]) is the attractor Λ of $\mathcal{S}_C$, see Figure 1.3:

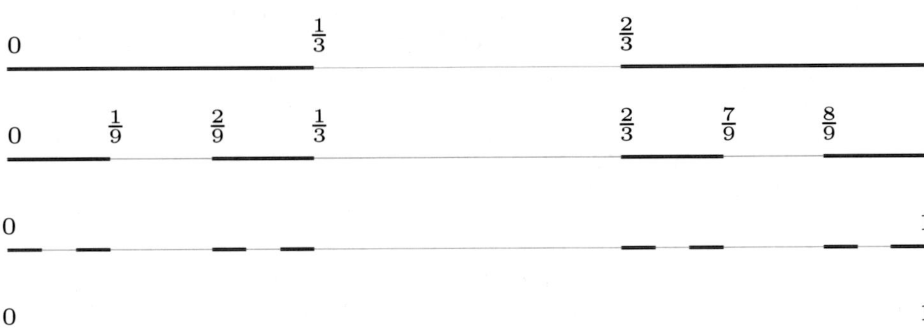

FIGURE 1.3. The first four approximations of the middle-third Cantor set.

$$\Lambda = \bigcap_{n=1}^{\infty} \bigcup_{(i_1,\dots,i_n)\in\{1,2\}^{\mathbb{N}}} S_{i_1\dots i_n}([0,1]). \tag{1.5}$$

A natural generalisation of the middle-third Cantor set is the following:

EXAMPLE 1.1.4 (Middle-α Cantor set). Fix $\alpha \in (0,1)$. We remove the (open) middle-α portion from the interval $[0,1]$. This way we obtain two intervals each of which has length $(1-\alpha)/2$ and repeat the same procedure for these smaller intervals. We iterate this ad infinitum to get the middle-α Cantor set. More precisely, let

$$\mathcal{S}_\alpha := \left(S_1(x) = \frac{1-\alpha}{2}\cdot x, \ S_2(x) = \frac{1-\alpha}{2}\cdot x + \frac{1+\alpha}{2}\right).$$

Then the middle-α Cantor set $\mathcal{C}_\alpha$ is defined by (1.5).

In the general case when our self-similar IFS is given in the form (1.2) we choose our favorite nonempty compact set H satisfying

$$S_i(H) \subset H \quad \text{for all} \quad i = 1, \dots, m. \tag{1.6}$$

This was the equilateral triangle T for $\mathcal{S}_{SG}$ and the unit square Q for $\mathcal{S}_{SC}$. If we have no better geometric intuition, we can always choose H to be the closed ball of radius R centered at the origin:

$$H := \overline{B}(\mathbf{0}, R) \quad \text{for} \quad R := \max_i \left\{ \frac{\|S_i(\mathbf{0})\|}{1 - r_i} \right\}. \tag{1.7}$$

The sets $S_i(H)$, $i = 1, \dots, m$, are the level-1 cylinders. The level n cylinders are the images of H by the n-fold compositions $S_{i_1 \dots i_n}$. (See Figure 1.1 and Figure 1.2.) Using that S_i is a similarity map with a contraction ratio $r_i < 1$, we obtain that

$$|S_{i_1, \dots, i_n}(H)| \leqslant r_{\max}^n \cdot |H|, \tag{1.8}$$

where we write $r_{\max} := \max_i r_i < 1$ and for an arbitrary set E its diameter is denoted by $|E|$. Further, (1.6) yields that for all $(i_1, \dots, i_{n+1})$ we have:

$$S_{i_1, \dots, i_{n+1}}(H) \subset S_{i_1, \dots, i_n}(H). \tag{1.9}$$

This means that the set of points $\bigcup_{i_1 \dots i_n} S_{i_1 \dots i_n}(H)$ which remain after all possible n-fold iterations is a decreasing sequence of nonempty compact sets. So, their intersection Λ, which can be interpreted as the set of all points that remain after infinitely many iterations:

$$\Lambda := \bigcap_{n=1}^{\infty} \bigcup_{(i_1 \dots i_n) \in \{1, \dots, m\}^n} S_{i_1 \dots i_n}(H), \tag{1.10}$$

is also a nonempty compact set. We say that Λ is the *attractor* or *invariant set* of the IFS $\mathcal{S}$. Observe that by (1.8) and (1.9) for every $\mathbf{i} = (i_1, i_2, \dots) \in \{1, \dots, m\}^{\mathbb{N}}$ the set $\bigcap_{n=1}^{\infty} S_{i_1 \dots i_n}(H)$ consists of a single point. This point can be represented as $\lim_{n \to \infty} S_{i_1 \dots i_n}(\mathbf{0})$, where we could have taken any $\mathbf{h} \in \mathbb{R}^d$ instead of $\mathbf{0}$. Indeed, $S_{i_1 \dots i_n}$ is a similarity map with a contraction ration $r_{i_1} \cdots r_{i_n} \leqslant r_{\max}^n < 1$. Thus

$$\lim_{n \to \infty} |S_{i_1 \dots i_n}(\mathbf{h}) - S_{i_1 \dots i_n}(\mathbf{0})| \leqslant r_{\max}^n \cdot |\mathbf{h} - \mathbf{0}| \longrightarrow 0$$

holds for all $\mathbf{h} \in \mathbb{R}^d$. So, for every $\mathbf{i} = (i_1, i_2, \dots) \in \{1, \dots, m\}^{\mathbb{N}}$ we can write

$$\Pi(\mathbf{i}) = \lim_{n \to \infty} S_{i_1 \dots i_n}(\mathbf{0}), \quad \text{or alternatively,} \quad \{\Pi(\mathbf{i})\} = \bigcap_{n=1}^{\infty} S_{i_1 \dots i_n}(H). \tag{1.11}$$

Hence, (1.11) provides a symbolic representation of the points of the attractor by the elements of the symbolic space

$$\Sigma := \{1, \dots, m\}^{\mathbb{N}}.$$

By the definition of Λ,

$$\Lambda = \bigcup_{\mathbf{i} \in \Sigma} \Pi(\mathbf{i}) = \bigcup_{\mathbf{i} \in \Sigma} \lim_{n \to \infty} S_{i_1 \dots i_n}(\mathbf{0}). \tag{1.12}$$

This shows that the definition of Λ remains the same if we replace H with any other set for which (1.6) holds.

In some cases the *natural projection* $\Pi\colon \Sigma \to \Lambda$ (see (1.11)) is bijective (like in the case of the middle-α Cantor set), whereas in many other cases it is only surjective but not injective. For example, in the case of the Sierpiński gasket (see Figure 1.1) we have $C_1 = \Pi(1\overline{2}) = \Pi(2\overline{1})$, where $\overline{i} = (i, i, \dots)$ for $i = 1, 2$. It is immediate from the definition of the natural projection Π that

$$\Pi(\mathbf{i}) = S_{i_1}(\Pi(\sigma\mathbf{i})), \quad \text{for all } \mathbf{i} \in \Sigma, \tag{1.13}$$

where σ is the left shift on the symbolic space Σ, that is,

$$\sigma(i_1, i_2, i_3, \dots) := (i_2, i_3, \dots). \tag{1.14}$$

Using this, one can see that

$$\Lambda = \bigcup_{k=1}^{m} S_k(\Lambda), \tag{1.15}$$

so Λ is really a self-similar set in the sense that it is the union of its own similar copies. On the other hand, we have the following

FACT 1.1.5. *The set Λ is the unique nonempty compact set for which* (1.15) *holds.*

More generally, a tuple of strict contractions $\mathcal{F} = (f_1, \dots, f_m)$ on $\mathbb{R}^d$, satisfying

$$\|f_i(\mathbf{x}) - f_i(\mathbf{y})\| \leqslant r_i\|\mathbf{x} - \mathbf{y}\|, \quad \text{with} \quad 0 < r_i < 1 \quad \text{for all } i \leqslant m, \ \mathbf{x}, \mathbf{y} \in \mathbb{R}^d, \tag{1.16}$$

is called an *iterated function system* (IFS) on $\mathbb{R}^d$. We will show in Theorem 1.3.6 that, given an IFS $\mathcal{F}$, there is a unique nonempty compact set

$$\Lambda_{\mathcal{F}} = \bigcup_{i=1}^{m} f_i(\Lambda_{\mathcal{F}}) \subset \mathbb{R}^d, \tag{1.17}$$

called the *attractor* of the IFS.

For now, we return to our self-similar IFS, the attractor Λ satisfying (1.15) and the natural projection Π defined by (1.11). When $\Pi(\mathbf{i}) = x$ we say that $\mathbf{i} \in \Sigma$ is a *code* (or address) of $x \in \Lambda$. By (1.12) every element of Λ has at least one code. We denote the set of points having a code beginning with $i_1 \dots i_n$ by $\Lambda_{i_1\dots i_n} := S_{i_1\dots i_n}(\Lambda)$ and call these sets the *level-n cylinders* of the attractor Λ. For the middle-α Cantor set (Example 1.1.4) every point has a unique code. For the Sierpiński gasket (Example 1.1.1) all points have at most two codes, and for the Sierpiński carpet (Example 1.1.2) all points have at most four codes (for example, $(1/9, 1/3)$ has four codes). However, in some cases there are infinitely many codes of some elements of the attractor. This happens for the "golden gasket" investigated by Broomhead, Montaldi, and Sidorov [**BMS04**], see Figure 1.4.

EXAMPLE 1.1.6 (Golden gasket). Let $\lambda = \frac{\sqrt{5}-1}{2}$ be the reciprocal of the golden ratio. Let Λ be the attractor of the IFS

$$\mathcal{S}_{\text{gold}} := \left(S_i(\mathbf{x}) = \lambda \cdot \mathbf{x} + \mathbf{t}_i\right)_{i=1}^{3}, \quad \mathbf{t}_1 = \begin{bmatrix} 0 \\ 0 \end{bmatrix}, \mathbf{t}_2 = \begin{bmatrix} 1-\lambda \\ 0 \end{bmatrix}, \mathbf{t}_3 = \begin{bmatrix} \frac{1-\lambda}{2} \\ \frac{(1-\lambda)\sqrt{3}}{2} \end{bmatrix} \tag{1.18}$$

One can see in the second triangle of Figure 1.4 that

$$\triangle C_2C_3R = S_1(S_2(S_2(T))) = S_2(S_1(S_1(T))), \tag{1.19}$$

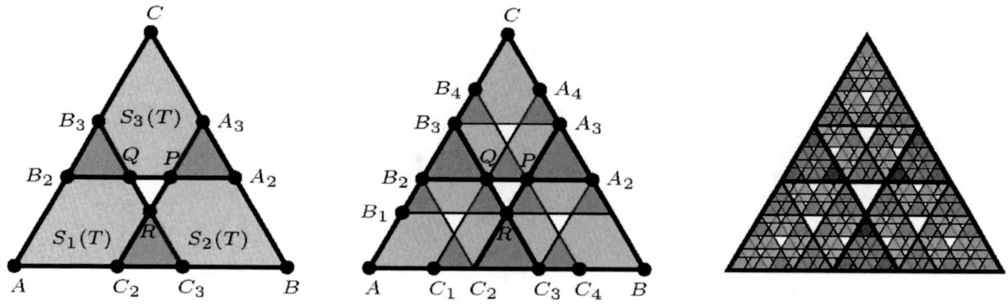

FIGURE 1.4. The first, second and fourth approximations of the golden gasket.

where $T := \triangle ABC$. That is, $\Lambda_{122} = \Lambda_{211}$ and similarly, $\Lambda_{233} = \Lambda_{322}$ and $\Lambda_{133} = \Lambda_{311}$. The authors of [**BMS04**], in fact, investigate the entire family of "fat gaskets", which are attractors of the IFS in (1.18) with $\lambda \in (1/2, 1)$, and especially those which correspond to $\lambda = \omega_m$, where ω_m is the positive solution of the equation $1 = x + x^2 + \cdots + x^m$, for $m \geqslant 2$. (Note that ω_2 is the reciprocal of the golden ratio.)

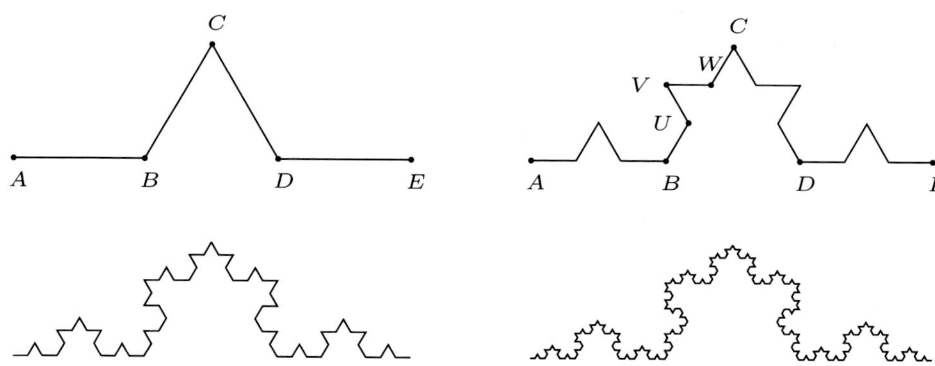

FIGURE 1.5. The first four approximations of the von Koch curve.

Using (1.15) we construct self-similar sets of somewhat different nature.

EXAMPLE 1.1.7 (The von Koch curve). The "generator" of the von Koch curve is the broken line $ABCDE$ at the top left of Figure 1.5. We consider the self-similar IFS of four similarity transformations taking the line segment AE into the line segments AB, BC, CD, DE:

$$
(1.20)\quad \mathcal{S}_K := \Bigg(S_1(\mathbf{x}) = \frac{1}{3}\mathbf{x}, \qquad S_2(\mathbf{x}) = \frac{1}{3}\begin{bmatrix} \frac{1}{2} & -\frac{\sqrt{3}}{2} \\ \frac{\sqrt{3}}{2} & \frac{1}{2} \end{bmatrix} \cdot \mathbf{x} + \begin{bmatrix} \frac{1}{3} \\ 0 \end{bmatrix},
$$

$$
S_3(\mathbf{x}) = \frac{1}{3}\begin{bmatrix} \frac{1}{2} & \frac{\sqrt{3}}{2} \\ -\frac{\sqrt{3}}{2} & \frac{1}{2} \end{bmatrix} \cdot \mathbf{x} + \begin{bmatrix} \frac{1}{2} \\ \frac{\sqrt{3}}{6} \end{bmatrix}, \quad S_4(\mathbf{x}) = \frac{1}{3}\mathbf{x} + \begin{bmatrix} \frac{2}{3} \\ 0 \end{bmatrix} \Bigg)
$$

The sets of $\{\cup_{i_1\dots i_n} S_{i_1\dots i_n}(AE)\}_{n=1}^4$ are the four broken lines of Figure 1.5. Continuing this iteration ad infinitum we obtain the von Koch curve γ_K, which is the unique nonempty compact set satisfying $\gamma_K = \bigcup_{i=1}^4 S_i(\gamma_K)$.

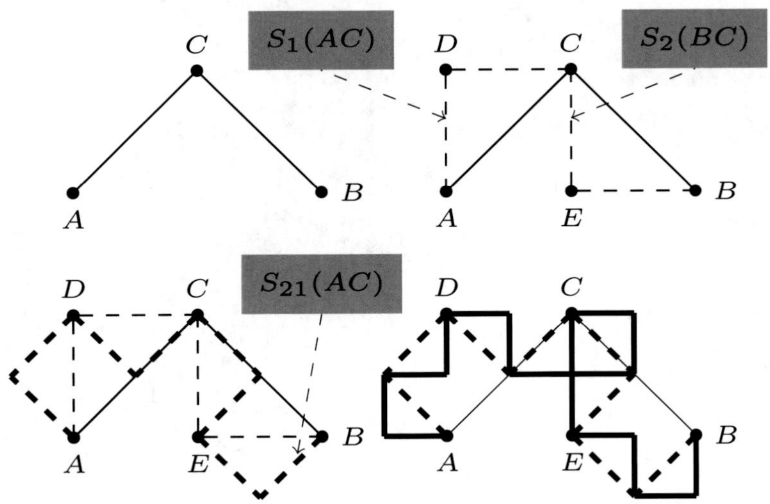

FIGURE 1.6. The first four approximations of the Heighway dragon.

The next example is the "Heighway dragon". John Heighway constructed it originally as follows: take a sheet of cardboard and fold it in half. Repeat the same process n times always folding in the same direction. Finally, open the sheet in such a way that there is a right angle at every fold. This is the n-th approximation to the Heighway dragon. More precisely:

EXAMPLE 1.1.8 (Heighway dragon). Consider the self-similar IFS in the plane:

$$\mathcal{S}_H := \left(S_1(\mathbf{x}) = \begin{bmatrix} \frac{1}{2} & -\frac{1}{2} \\ \frac{1}{2} & \frac{1}{2} \end{bmatrix} \cdot \mathbf{x}, \quad S_2(\mathbf{x}) = \begin{bmatrix} -\frac{1}{2} & -\frac{1}{2} \\ \frac{1}{2} & -\frac{1}{2} \end{bmatrix} \cdot \mathbf{x} + \begin{bmatrix} 1 \\ 0 \end{bmatrix}\right). \tag{1.21}$$

The contraction ratio for both of these similarity transformations is $1/\sqrt{2}$, and S_1, S_2 rotate by 45°, 135° respectively. The attractor of the IFS $\mathcal{S}_H$ is called the *Heighway dragon*. It is, in fact, a curve. Indeed, the nth approximation of the Heighway dragon is a union of 2^n segments each having length $2^{-n/2}$. This gives a natural parametrization of the nth approximation on the interval $[0,1]$. These parameterizations converge uniformly on $[0,1]$ to the parametrization of the Heighway dragon curve. We remark here that the Heighway dragon is a space filling (Peano) curve in the sense that it has nonempty interior! Moreover, the entire plane can be tiled by isometric copies of the Heighway dragon, and the interior of Heighway dragon is a countable union of disjoint open disk-like planar sets (see [**NN03**]).

The most trivial example of a self-similar set is the unit interval $[0,1]$.

EXAMPLE 1.1.9. Let $S_1(x) = x/2$, $S_2(x) = (x+1)/2$. Then $[0,1] = S_1([0,1]) \cup S_2([0,1])$.

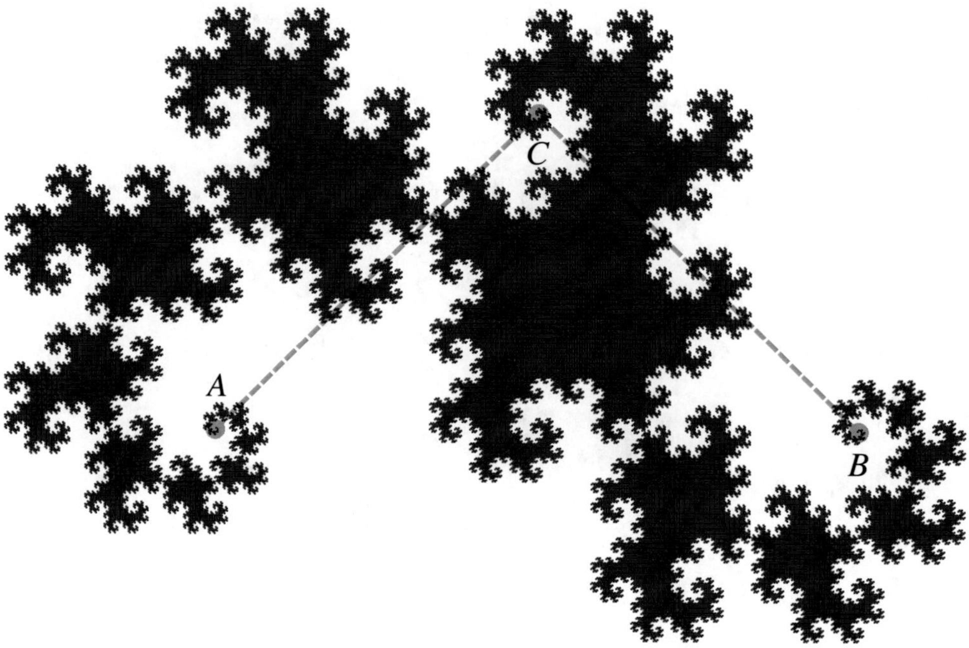

FIGURE 1.7. The 18-th approximation of the Heighway dragon.

In all of the examples we have seen above the contraction ratio of all of the maps of the IFS was identical. We say that these self-similar IFS are *homogeneous.* A nonhomogeneous self-similar IFS is e.g.

EXAMPLE 1.1.10. Let

$$S_1(x) = \frac{1}{2} \cdot x \text{ and } S_2(x) = \frac{1}{4} \cdot x + \frac{3}{4}. \tag{1.22}$$

Then the nth approximation of the attractor Λ is a union of 2^n intervals among which there are $\binom{n}{i}$ of length $4^{-i} \cdot 2^{i-n}$ for every $0 \leqslant i \leqslant n$.

A fairly general example of not necessarily homogeneous self-similar IFS is as follows:

EXAMPLE 1.1.11. Let $\mathcal{S}_{\text{gen}} := (S_i(x) = r_i x + t_i)_{i=1}^m$ be an IFS on $\mathbb{R}$, where r_i, t_i are chosen so that:

(1) $r_i > 0$,
(2) $t_1 = 0$ and $r_m + t_m = 1$,
(3) $r_i + t_i < t_{i+1}$ for all $i = 1, \ldots, m-1$,

Then for $I_i := S_i([0,1])$ we have:

(a) $I_i < I_{i+1}$ (that is, the right endpoint of I_i is smaller than the left endpoint of I_{i+1}) for $1 \leqslant i \leqslant m-1$.
(b) The left endpoint of I_1 is 0.
(c) The right endpoint of I_m is 1.

1.1.1. The size of the attractors. A self-similar set can be relatively "large" like the ones in Examples 1.1.8 and 1.1.9. However, in many cases it is a set of zero Lebesgue measure:

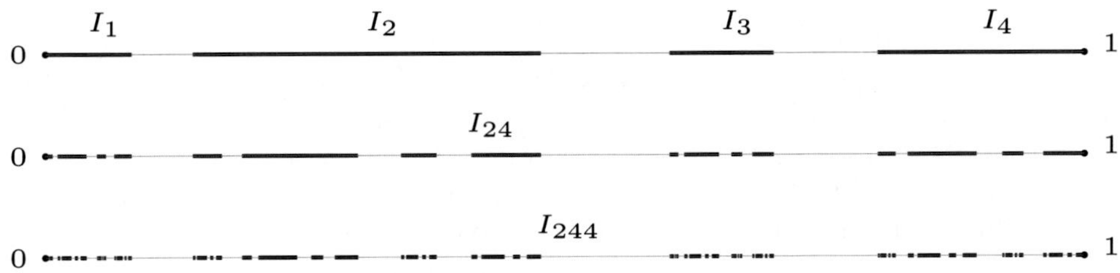

FIGURE 1.8. The first three approximations of a nonhomogeneous self-similar IFS.

LEMMA 1.1.12. *Let $\Lambda_{\mathcal{F}}$ be the attractor* (1.17) *of the IFS $\mathcal{F}$ from* (1.16). *If*

$$\widehat{r} := \sum_{i=1}^{m} r_i^d < 1, \tag{1.23}$$

then the d-dimensional volume (the $\mathcal{L}^d$ measure) of Λ is zero.

PROOF. It follows from (1.17) that

$$\forall n, \quad \mathcal{L}^d(\Lambda) \leqslant \sum_{i=1}^{m} \mathcal{L}^d(f_i(\Lambda)) \leqslant \sum_{i=1}^{m} r_i^d \mathcal{L}^d(\Lambda) = \widehat{r}\mathcal{L}^d(\Lambda). \tag{1.24}$$

So, by (1.23) we obtain that $\mathcal{L}^d(\Lambda) = 0$, since Λ is a bounded set. □

Note that (1.23) holds for all the examples of self-similar sets above but the golden gasket, the Heighway dragon, and the unit interval. In the case of the unit interval and the Heighway dragon we have $\widehat{r} = 1$ and both of these sets contain interior points. However, as it was proved in [**BMS04**], for the golden gasket the $\mathcal{L}^2$-measure is zero, whereas $\widehat{r} = 3 \cdot \left(\frac{\sqrt{5}-1}{2}\right)^2 = 1.145\cdots > 1$. Thus the assertion of Lemma 1.1.12 is sharp and its converse does not hold.

1.1.2. Heuristic introduction to the fractal dimension for homogeneous self-similar sets. The "size" of most self-similar or self-affine sets cannot be quantified by its Lebesgue measure properly, therefore we need to introduce some concepts of fractal dimension. The idea comes from the following observation: consider for instance the middle-third Cantor set (Example 1.1.3). The nth approximation $\mathcal{C}_{1/3}^n$ consists of 2^n intervals of length $1/3^n$. Clearly, $\mathcal{L}^1(\mathcal{C}_{1/3}^n) = 2^n \cdot (1/3^n) \to 0$. However, if we raise the length of these cylinder intervals to a power s, then the smaller s we choose the greater the number obtained will be. If we do this for all the level-n cylinders (for some $n \in \mathbb{N}$), then the total effect is:

$$S_n(s) := \sum_{\substack{I \subset C^n \\ \text{level-}n\text{ cylinders}}} |I|^s = 2^n((1/3)^s)^n = (2 \cdot (1/3^s))^n. \tag{1.25}$$

So if we choose $s_0 = \log 2/\log 3$, then $S_n(s_0) = 1$ for all n. Observe that the choice of s_0 is critical in the sense that

$$\text{if } s' < s_0 < s'', \quad \text{then } S_n(s') \to \infty \text{ and } S_n(s'') \to 0. \tag{1.26}$$

If we repeat the same for the middle-α Cantor set $\mathcal{C}_\alpha$ (Example 1.1.4), where the length of each of the 2^n level-n cylinder intervals is $\left(\frac{1-\alpha}{2}\right)^n$, then we get $S_n^{(\alpha)}(s) =$

$\left(2 \cdot \left(\frac{1-\alpha}{2}\right)^s\right)^n$. The critical value s_c, for which $S_n^\alpha(s_c) = 1$, is $s_c^{(\alpha)} := \frac{\log 2}{-\log((1-\alpha)/2)} \in (0,1)$. Clearly, the assertion that corresponds to (1.26) also holds. Note that the smaller $\alpha \in (0, 1/2)$ results in the greater critical value $s_c^{(\alpha)}$. On the other hand, the smaller value of α results in larger portions of the cylinder intervals retained at each step, which yields the larger set $\mathcal{C}_\alpha$. It appears to be adequate to quantify the size of a self-similar set with this critical exponent. Observe that if we consider the unit interval, the unit square, or the unit cube as self-similar sets and compute the critical exponent for these sets, then we get $s_c = 1,\ 2$, or 3 respectively. We can think of this critical exponent as a kind of "fractal dimension" of a homogeneous self-similar IFS with cylinders without significant overlaps. In order to extend this idea and make it precise, we will introduce various notions of dimension in Section 1.4.

1.2. Self-affine sets

A finite tuple $\mathcal{F}$ of contracting affine maps acting on $\mathbb{R}^d$

$$\mathcal{F} = (f_i(\mathbf{x}) = A_i\mathbf{x} + \mathbf{t}_i)_{i=1}^m, \qquad \mathbf{x}, \mathbf{t}_i \in \mathbb{R}^d, \tag{1.27}$$

is called a *self-affine* IFS on $\mathbb{R}^d$. We always assume that the matrices A_i are nonsingular.

NOTATION 2. Similarly to the case of self-similar systems, denote by $\mathscr{A}_m(d)$ the collection of ordered tuples $\mathcal{F} := (f_1, \dots, f_m)$ of (strictly) contracting affinities on $\mathbb{R}^d$ satisfying the following property: there exist $i, j \in \{1, \dots, m\}$ such that the fixed points of f_i and f_j are distinct.

The attractor $\Lambda = \Lambda_\mathcal{F}$ of $\mathcal{F}$ is the unique compact set satisfying (1.17). If all of the matrices are of the form $A_i = r_i \cdot Q_i$, where Q_i is an orthogonal matrix, then our self-affine system is in fact self-similar. A typical example of self-affine IFS is as follows:

EXAMPLE 1.2.1. Consider the IFS $\mathcal{F} = (f_1(\mathbf{x}), f_2(\mathbf{x}))$ acting in the plane, where

$$f_1\begin{bmatrix} x_1 \\ x_2 \end{bmatrix} = \begin{bmatrix} -\frac{1}{5} & -\frac{\sqrt{3}}{8} \\ \frac{\sqrt{3}}{5} & -\frac{1}{8} \end{bmatrix}\begin{bmatrix} x_1 \\ x_2 \end{bmatrix} + \begin{bmatrix} -\frac{3}{10} \\ \frac{3\cdot\sqrt{3}}{10} \end{bmatrix},$$

$$f_2\begin{bmatrix} x_1 \\ x_2 \end{bmatrix} = \begin{bmatrix} \frac{\sqrt{3}}{5} & -\frac{1}{8} \\ \frac{1}{5} & \frac{\sqrt{3}}{8} \end{bmatrix}\begin{bmatrix} x_1 \\ x_2 \end{bmatrix} + \begin{bmatrix} \frac{3\sqrt{3}}{10} \\ \frac{3}{10} \end{bmatrix},$$

see Figure 1.9. Note that in the self-affine case the shape of the cylinders may be significantly different.

A somewhat special but very well known example is the Hironaka curve.

EXAMPLE 1.2.2. The Hironaka curve is the attractor $\Lambda_{\mathrm{Hir}} \subset \mathbb{R}^2$ (see Figure 1.10) of the self-affine IFS:

$$\mathcal{F}_{\mathrm{Hir}} := \left(f_i(\mathbf{x}) = \begin{bmatrix} \frac{1}{3} & 0 \\ 0 & \frac{1}{2} \end{bmatrix} \cdot \mathbf{x} + \mathbf{t}_i\right)_{i=1}^3, \quad \mathbf{t}_1 = \begin{bmatrix} 0 \\ 0 \end{bmatrix}, \mathbf{t}_2 = \begin{bmatrix} \frac{1}{3} \\ \frac{1}{2} \end{bmatrix}, \mathbf{t}_3 = \begin{bmatrix} \frac{2}{3} \\ 0 \end{bmatrix}. \tag{1.28}$$

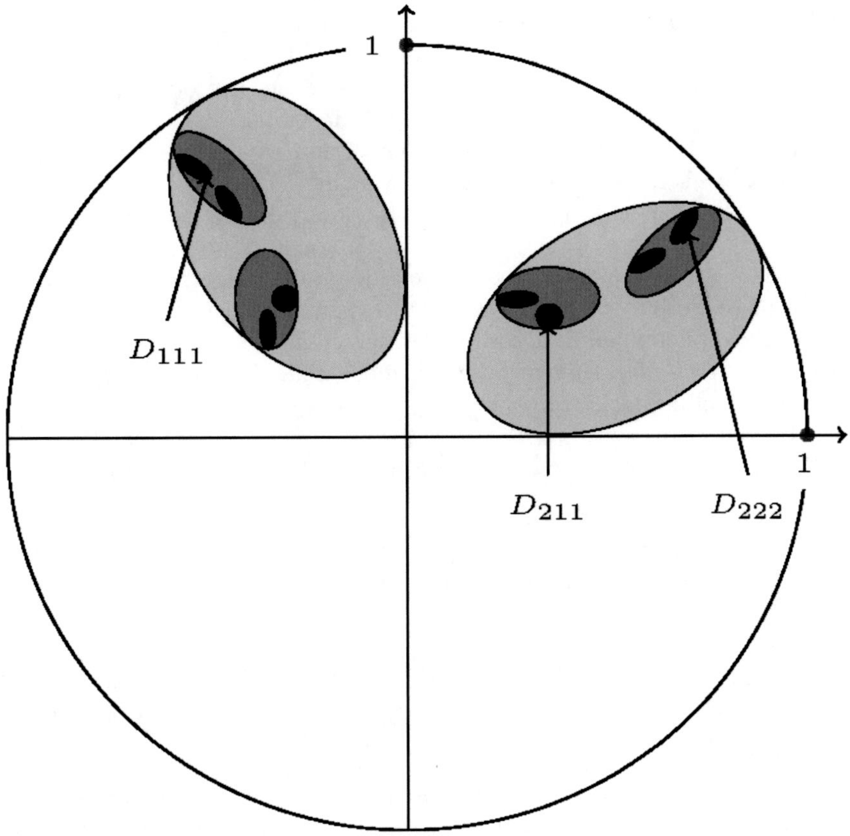

FIGURE 1.9. The level one, two, and three cylinders for Example 1.2.1.

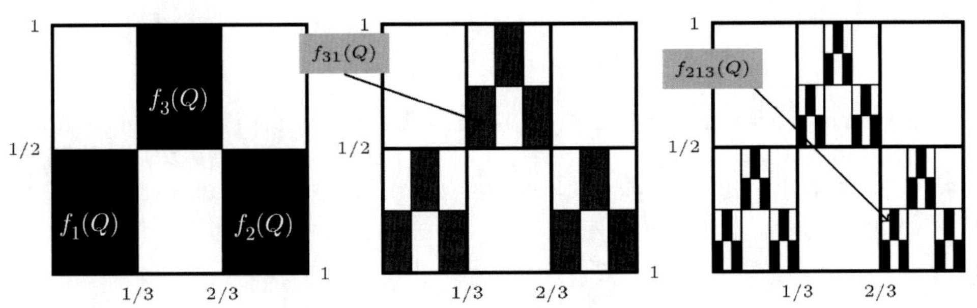

FIGURE 1.10. The Hironaka curve.

One can notice that the attractor Λ_{Hir} is actually not a curve, because it is disconnected. Hironaka's curve, the way he defined it, is obtained from Λ_{Hir} by adding a countable set of vertical line segments to it, which connect cylinders of different levels.

Having a look at the first three approximations of these attractors we can see that some of the cylinders are relatively long and thin. This phenomenon is due to the fact that the contraction ratio is not the same in different directions. We can best see this in Figure 1.10. The cylinder rectangles of the third approximation have the size $3^{-3} \times 2^{-3}$. For the nth approximation the ratio of the lengths of the sides of the cylinders is $2^{-n}/3^{-n} \to \infty$ as $n \to \infty$. On the other hand, for some IFS, as we can see in Figure 1.9, for some cylinders (like D_{211} which has a shape of a filled circle) the effects of different contraction ratios in different directions counterbalance each other. Nevertheless, even in this case one can observe the tendency that some of the higher order cylinders become relatively longer and thinner than the cylinders of lower levels. Then it can happen that on some very high level there are exponentially many cylinders of diameter, say r, which are disjoint but lie so close to each other that even their union can be covered by a disk of radius $2r$. The phenomenon described (which cannot happen for the self-similar IFS *with disjoint cylinders*) makes it much more difficult to obtain even the most basic results for self-affine systems.

1.3. Iterated function systems and the symbolic space

Here we prove an extension of Fact 1.1.5 to the case of a general contracting IFS on $\mathbb{R}^d$. The notion of IFS was popularized by Hutchinson [**Hut81**]. Recall that $f : \mathbb{R}^d \to \mathbb{R}^d$ is a *Lipschitz function* with a Lipschitz constant $r > 0$ if

$$\|f(\mathbf{x}) - f(\mathbf{y})\| \leqslant r \|\mathbf{x} - \mathbf{y}\| \quad \text{for all } \mathbf{x}, \mathbf{y} \in \mathbb{R}^d. \tag{1.29}$$

DEFINITION 1.3.1. A tuple of functions $\mathcal{F} = (f_1, \dots, f_m)$, where every $f_i \colon \mathbb{R}^d \to \mathbb{R}^d$ is Lipschitz with a Lipschitz constants $r_i < 1$, is called a *contracting iterated function system* (IFS).

The construction of the attractor proceeds exactly as in the self-similar case described in Section 1.1. The symbolic representation of the attractor will be important for us; thus we start with some background material on the properties of symbolic spaces.

1.3.1. Symbolic space.

The *symbolic space* corresponding to the IFS $\mathcal{F}$ is

$$\Sigma := \{1, \dots, m\}^{\mathbb{N}}.$$

The elements of Σ are denoted by $\mathbf{i} = (i_1, i_2, \dots)$, $\mathbf{j} = (j_1, j_2, \dots) \in \Sigma$. The common prefix of $\mathbf{i}$ and $\mathbf{j}$ is defined by $\mathbf{i} \wedge \mathbf{j} = (i_1, \dots, i_n)$, where $n := \sup\{k \geqslant 1 : i_k = j_k\}$. The length of the common prefix is $|\mathbf{i} \wedge \mathbf{j}| := n$. We introduce a metric ϱ on Σ as follows:

$$\varrho(\mathbf{i}, \mathbf{j}) := m^{-|\mathbf{i} \wedge \mathbf{j}|}. \tag{1.30}$$

This metric defines a topology on Σ, which agrees with the the (infinite) product topology. By Tychonoff's Theorem, Σ is compact. (Sometimes, a different metric on Σ, defining the same topology, is more convenient.) The collection of all finite words over the alphabet $\{1, \dots, m\}$ is $\Sigma^* := \bigcup_{n=0}^{\infty} \Sigma_n$, where $\Sigma_n := \{1, \dots, m\}^n$. The most important subsets of Σ are the symbolic cylinders:

$$[i_1, \dots, i_n] := \{\mathbf{j} \in \Sigma : j_k = i_k, \quad \text{for all } k \leqslant n\}. \tag{1.31}$$

The symbolic cylinders are clopen (closed and open) sets, and they generate the topology. Denote by $\mathscr{S}$ the collection of all symbolic cylinders, together with the empty set $\varnothing$.

The Borel σ-algebra $\mathcal{B}$ in Σ is generated by $\mathscr{S}$. We will frequently need to consider finite measures on $(\Sigma, \mathcal{B})$. Observe that $\mathscr{S}$ forms a *semi-algebra* in Σ (see Appendix B.1 for definition).

There is a simple intuitive way to define a finite measure on $(\Sigma, \mathcal{B})$, using the idea of "splitting the mass". Namely, suppose we have a function $\phi : \Sigma^* \to \mathbb{R}_+$ satisfying the property

$$\phi(\mathbf{i}) = \sum_{j=1}^{m} \phi(\mathbf{i}j), \quad \text{for all } \mathbf{i} \in \Sigma^*, \tag{1.32}$$

where $\mathbf{i}j$ denotes the word obtained by the concatenation of $\mathbf{i}$ and j.

LEMMA 1.3.2. *Let ϕ be a function on Σ^* satisfying* (1.32). *Then the set-function defined on $\mathscr{S}$ by*

$$\mu(\varnothing) = 0, \quad \mu([\mathbf{i}]) := \phi(\mathbf{i}), \quad \textit{for all } \mathbf{i} \in \Sigma^*, \tag{1.33}$$

extends uniquely to a finite Borel measure on Σ.

PROOF. By the Measure Extension Theorem (see Theorem B.1.3), it is enough to show that μ is countably additive on $\mathscr{S}$, that is, if $A = \cup_{n \geqslant 1} A_n$ is a finite or countable disjoint union and $A, A_n \in \mathcal{A}$, then $\mu(A) = \sum_{n \geqslant 1} \mu(A_n)$. Note, however, that A is compact and A_n are open, hence the disjoint union is necessarily finite. Finite additivity is easy to check, as we explain next. □

It is convenient to visualize the set of words Σ^* as an infinite tree, as, e.g., in [**BP17**, Section 3.1]. This tree is uniquely (up to isomorphism) defined by the property that the vertices are arranged in levels enumerated by $n = 0, 1, 2, \ldots$, with every vertex of level n having exactly m edges leading to the vertices of the next level. Thus, every vertex has m "children" and a unique "parent" (except the unique root vertex at level 0). Such a tree is called the *m-regular tree.*

A word $\mathbf{i} = i_1 \ldots i_n$ is identified with a vertex at level n. The empty word is the root vertex. The set of infinite (nonbacktracking) paths in the tree from the root may be identified with the symbolic space Σ. For a vertex $\mathbf{i} = i_1 \ldots i_n$, with $n \geqslant 1$, we denote by $\mathbf{i}^- = i_1 \ldots i_{n-1}$ its "parent" vertex. A *flow* on the tree is a nonnegative function f on the set of edges, satisfying the "inflow=outflow" property: $f(\mathbf{i}^-, \mathbf{i}) = \sum_{\mathbf{j}:\, \mathbf{i}=\mathbf{j}^-} f(\mathbf{i}, \mathbf{j})$. It is immediate that there is a one-to-one correspondence between the set of flows and the set of functions ϕ satisfying (1.32) via the formula

$$\phi(\mathbf{i}) = f(\mathbf{i}^-, \mathbf{i}).$$

A *cut-set* is a set of edges that intersects every infinite path from the root, and a *minimal cut-set* is a cut-set which is minimal under the inclusion relation. Minimal cut-sets are finite by compactness; they correspond to disjoint open covers of Σ. The *norm* of the flow $\|f\|$ is defined as the sum of the flows from the root vertex to its "children".

LEMMA 1.3.3. *Let f be a flow on the m-regular tree identified with Σ^*. Then for any minimal cut-set $\mathscr{P}$ holds*

$$\|f\| = \sum_{\mathbf{e} \in \mathscr{P}} f(\mathbf{e}).$$

Equivalently, every function ϕ satisfying (1.32) *is finitely additive (hence countably additive) on the set of symbolic cylinders $\mathscr{S}$.*

The proof is by induction on the maximal length of the word corresponding to $\mathbf{e} \in \mathscr{P}$, see [**BP17**, Lemma 2.2.4].

EXAMPLE 1.3.4 (Bernoulli measure on Σ). Let $\mathbf{p} = (p_1, \ldots, p_m)$ be a probability vector. Define the function $\phi : \mathcal{S} \to \mathbb{R}_+$ by

$$\phi([\mathbf{i}]) := p_{\mathbf{i}}, \quad \text{where} \quad p_{\mathbf{i}} := p_{i_1} \cdots p_{i_n} \quad \text{for} \quad \mathbf{i} = i_1 \ldots i_n \in \Sigma_n.$$

Then ϕ satisfies (1.32) and hence extends uniquely to a measure on $(\Sigma, \mathcal{B})$ denoted by $\mu_{\mathbf{p}}$. Alternatively, the measure $\mu_{\mathbf{p}}$ may be defined as an infinite product (Bernoulli) measure $\mathbf{p}^{\mathbb{N}}$, whose existence and uniqueness follows from the Kolmogorov Extension Theorem. From the point of view of Probability, $\mu_{\mathbf{p}}$-a.e. $\mathbf{i} \in \Sigma$ produces a sequence of i.i.d. random variables $\{i_n\}_{n \geqslant 1}$ distributed according to $\mathbf{p}$.

1.3.1.1. *Shift-invariant measures.* Recall that we defined in (1.14) the *left shift* transformation on Σ:

$$\sigma(i_1, i_2, i_3, \ldots) := (i_2, i_3, \ldots).$$

The left shift is a continuous (in fact, it is m-Lipschitz in the metric (1.30)) m-to-1 onto map on the symbolic space Σ. An important role will be played by *σ-invariant*, or simply *invariant* measures. A Borel measure μ on Σ is said to be σ-invariant if

$$\mu(\sigma^{-1}A) = \mu(A) \quad \text{for all Borel } A \subset \Sigma. \tag{1.34}$$

By Theorem B.1.7, in order to show that μ is σ-invariant it is enough to verify (1.34) for symbolic cylinder sets, which form a semi-algebra generating the Borel σ-algebra on Σ.

LEMMA 1.3.5. *Let $\mathbf{p} = (p_1, \ldots, p_m)$ be a probability vector and $\mu_{\mathbf{p}} = \mathbf{p}^{\mathbb{N}}$ the corresponding measure on Σ, defined in Example 1.3.4. Then $\mu_{\mathbf{p}}$ is σ-invariant.*

PROOF. Let $\mathbf{i} \in \Sigma^*$. Then

$$\mu_{\mathbf{p}}(\sigma^{-1}[\mathbf{i}]) = \mu_{\mathbf{p}}\Big(\bigsqcup_{j=1}^{m} [j\mathbf{i}]\Big) = \sum_{j=1}^{m} \mu_{\mathbf{p}}([j\mathbf{i}]) = \sum_{j=1}^{m} p_j \mu_{\mathbf{p}}([\mathbf{i}]) = \mu_{\mathbf{p}}([\mathbf{i}]),$$

by the definition of the product measure $\mu_{\mathbf{p}}$. Thus (1.34) holds for symbolic cylinder sets, hence $\mu_{\mathbf{p}}$ is σ-invariant. □

1.3.2. Existence and uniqueness of the attractor. Assume that the IFS $\mathcal{F} = (f_1, \ldots, f_m)$ consists of functions $f_i : \mathbb{R}^d \to \mathbb{R}^d$ with Lipschitz constants $r_i < 1$. Let $\mathbf{b}_i$ be the fixed point of f_i. Then for the closed ball

$$\overline{B} := \overline{B}(\mathbf{0}, R) \text{ where } R := \max_i \left\{ \frac{\|\mathbf{b}_i\|}{1 - r_i} \right\} \tag{1.35}$$

we have $f_i(\overline{B}) \subset \overline{B}$ for all $i = 1, \ldots, m$. The nonempty compact set

$$\Lambda_{\mathcal{F}} = \bigcap_{n=1}^{\infty} \bigcup_{(i_1, \ldots, i_n) \in \Sigma_n} f_{i_1 \ldots i_n}(\overline{B}) \tag{1.36}$$

is called the *attractor* or *invariant set* of the IFS $\mathcal{F}$.

Denote by $\mathfrak{C}$ the space of all nonempty compact sets contained in $\overline{B}$, the ball from (1.35). The space $\mathfrak{C}$ is equipped with the *Hausdorff metric*, defined for $K_1, K_2 \in \mathfrak{C}$ as follows:

$$\text{dist}_H(K_1, K_2) := \inf\{\delta : K_1 \subset B(K_2, \delta) \text{ and } K_2 \subset B(K_1, \delta)\}, \tag{1.37}$$

where $B(K, \delta)$ denotes the open δ-neighborhood of a set K. Then $(\mathfrak{C}, \operatorname{dist}_H)$ is a complete metric space (see e.g. [**Fal85**, Theorem 3.6] for a proof of this well-known fact).

THEOREM 1.3.6 (Hutchinson [**Hut81**]). *The attractor $\Lambda_{\mathcal{F}}$, defined in (1.36), is the unique nonempty compact set satisfying*

$$\Lambda_{\mathcal{F}} = \bigcup_{i=1}^{m} f_i(\Lambda_{\mathcal{F}}). \tag{1.38}$$

Moreover, for any nonempty compact set $K \subset \mathbb{R}^d$, the iterates $\Phi^n(K)$ of the map $\Phi : K \to \bigcup_{i=1}^{m} f_i(K)$ converge to $\Lambda_{\mathcal{F}}$ in the Hausdorff metric as $n \to \infty$.

PROOF. The map $\Phi : \mathfrak{C} \to \mathfrak{C}$

$$\Phi : K \to \bigcup_{i=1}^{m} f_i(K) \tag{1.39}$$

is a contraction on $(\mathfrak{C}, \operatorname{dist}_H)$ since the maps f_i, $i = 1, \ldots, m$, are contractions.

Indeed, for every compact sets $A, B, C, D \in \mathfrak{C}$ and for every Lipschitz map $S \colon \mathbb{R}^d \to \mathbb{R}^d$ with a Lipschitz constant $r > 0$,

$$\operatorname{dist}_H(S(A), S(B)) \leqslant r \cdot \operatorname{dist}_H(A, B) \text{ and} \tag{1.40}$$

$$\operatorname{dist}_H(A \cup B, C \cup D) \leqslant \max\{\operatorname{dist}_H(A, C), \operatorname{dist}_H(B, D)\}. \tag{1.41}$$

Hence, $\operatorname{dist}_H(\Phi(A), \Phi(B)) \leqslant (\max_i r_i) \cdot \operatorname{dist}_H(A, B)$ for any compact sets $A, B \in \mathfrak{C}$.

Observe that equation (1.15) means that the set $\Lambda_{\mathcal{F}}$ is a fixed point of Φ, and by the Banach fixed point theorem the fixed point of Φ is unique. Moreover, it follows that for any $K \in \mathfrak{C}$ we have $\Phi^n(K) \to \Lambda_{\mathcal{F}}$ in the Hausdorff metric. □

A simple corollary of Theorem 1.3.6 is continuous dependence of $\Lambda_{\mathcal{F}}$ on $\mathcal{F}$ in the Hausdorff metric.

PROPOSITION 1.3.7. *Let $\mathcal{F} = (f_1, \ldots, f_m)$ and $\mathcal{G} = (g_1, \ldots, g_m)$ be IFS's with attractors $\Lambda_{\mathcal{F}}$ and $\Lambda_{\mathcal{G}}$ respectively. Let $\overline{B}$ be a closed set such that $f_i(\overline{B}), g_i(\overline{B}) \subset \overline{B}$ for every $i = 1, \ldots, m$, and let $r_{\max}$ be an upper bound for the contraction ratios of f_i and g_i. Then*

$$\operatorname{dist}_H(\Lambda_{\mathcal{F}}, \Lambda_{\mathcal{G}}) \leqslant \frac{\max_{i=1,\ldots,m} \max_{\mathbf{x} \in \overline{B}} \|f_i(\mathbf{x}) - g_i(\mathbf{x})\|}{1 - r_{\max}}.$$

PROOF. Using the definition of the attractor (1.38), and applying (1.40) and (1.41), we get

$$\operatorname{dist}_H(\Lambda_{\mathcal{F}}, \Lambda_{\mathcal{G}}) \leqslant \max_{i=1,\ldots,m} \operatorname{dist}_H(f_i(\Lambda_{\mathcal{F}}), g_i(\Lambda_{\mathcal{G}})).$$

For every $i = 1, \ldots, m$,

$$\begin{aligned} \operatorname{dist}_H(f_i(\Lambda_{\mathcal{F}}), g_i(\Lambda_{\mathcal{G}})) &\leqslant \operatorname{dist}_H(f_i(\Lambda_{\mathcal{F}}), g_i(\Lambda_{\mathcal{F}})) + \operatorname{dist}_H(g_i(\Lambda_{\mathcal{F}}), g_i(\Lambda_{\mathcal{G}})) \\ &\leqslant \max_{\mathbf{x} \in \overline{B}} \|f_i(\mathbf{x}) - g_i(\mathbf{x})\| + r_i \operatorname{dist}_H(\Lambda_{\mathcal{F}}, \Lambda_{\mathcal{G}}), \end{aligned}$$

where r_i is the contraction ratio of g_i. Hence, the claim follows. □

In the sequel we frequently use the following:

NOTATION 3. Given an IFS $\mathcal{F} = (f_1, \ldots, f_m)$ let $\Lambda = \Lambda_{\mathcal{F}}$ be the attractor of $\mathcal{F}$. The *natural projection* $\Pi : \Sigma \to \Lambda$ is defined by

$$\Pi(\mathbf{i}) := \lim_{n\to\infty} f_{i_1\ldots i_n}(\mathbf{0}), \text{ or alternatively } \{\Pi(\mathbf{i})\} := \bigcap_{n=1}^{\infty} f_{i_1\ldots i_n}(\overline{B}), \tag{1.42}$$

for the closed ball $\overline{B}$ defined in (1.35). We have

$$\Lambda = \bigcup_{\mathbf{i}\in\Sigma} \Pi(\mathbf{i}). \tag{1.43}$$

We denote

$$\Lambda_{i_1\ldots i_n} := f_{i_1\ldots i_n}(\Lambda) \tag{1.44}$$

and we call these sets the level-n cylinders of the attractor Λ.

An interesting consequence of the formula (1.36) and the definition of the natural projection is that $x \in \Lambda_{\mathcal{F}}$ if and only if there exists $(i_1, i_2, \ldots) \in \Sigma$ such that the sequence $\{y_n\}_{n=0}^{\infty}$ of points in $\mathbb{R}^d$ is bounded, where $y_0 := x$ and $y_n := f_{i_n}^{-1}(y_{n-1})$ for $n \geqslant 1$.

1.3.3. Conjugation of systems.

Sometimes it will be convenient to rescale or translate the self-similar or self-affine set under consideration. This is possible by applying *conjugation.* A function is called *bi-Lipschitz* if it is Lipschitz, invertible, and the inverse is Lipschitz as well.

DEFINITION 1.3.8. Let $\mathcal{F} = (f_1, \ldots, f_m)$ and $\widetilde{\mathcal{F}} = (\widetilde{f}_1, \ldots, \widetilde{f}_n)$ be two iterated function systems on $\mathbb{R}^d$. We say that $\mathcal{F}$ is *conjugate* to $\widetilde{\mathcal{F}}$ if $n = m$ and there exist a permutation $\tau\colon \{1, \ldots, m\} \to \{1, \ldots, m\}$ and a homeomorphism $h\colon \mathbb{R}^d \to \mathbb{R}^d$, such that $\widetilde{f}_{\tau(i)} = h \circ f_i \circ h^{-1}$. If h is a bi-Lipschitz function, then $\mathcal{F}$ and $\widetilde{\mathcal{F}}$ are said to be *bi-Lipschitz conjugate.*

REMARK 1.3.9. It may happen that a tuple of functions $\mathcal{F}$ is conjugate to $\widetilde{\mathcal{F}}$, where $\widetilde{\mathcal{F}}$ is a contracting IFS, even though the original tuple $\mathcal{F}$ is not.

Redefining the indices of $\mathcal{F}'$ we may assume without loss of generality that τ is the identity. If $\Pi_{\mathcal{F}}$ and $\Pi_{\widetilde{\mathcal{F}}}$ denote the natural projection of $\mathcal{F}$ and $\widetilde{\mathcal{F}}$ respectively, then

$$\begin{aligned}
\Pi_{\widetilde{\mathcal{F}}}(\mathbf{i}) &= \lim_{n\to\infty} \widetilde{f}_{i_1} \circ \cdots \circ \widetilde{f}_{i_n}(\mathbf{0}) \\
&= \lim_{n\to\infty} h \circ f_{i_1} \circ h^{-1} \circ h \circ f_{i_2} \circ h^{-1} \cdots \circ h \circ f_{i_n} \circ h^{-1}(\mathbf{0}) \\
&= \lim_{n\to\infty} h \circ f_{i_1} \circ \cdots \circ f_{i_n}(h^{-1}(\mathbf{0})) = h(\Pi_{\mathcal{F}}(\mathbf{i})),
\end{aligned}$$

and so by (1.43), $\Lambda_{\widetilde{\mathcal{F}}} = h(\Lambda_{\mathcal{F}})$, where $\Lambda_{\widetilde{\mathcal{F}}}$ and $\Lambda_{\mathcal{F}}$ are the attractors of $\widetilde{\mathcal{F}}$ and $\mathcal{F}$ respectively.

It is clear that if $\mathcal{S} = (S_1, \ldots, S_m)$ is a self-similar IFS and h is a similarity map, then $\mathcal{S}' = (h \circ S_1 \circ h^{-1}, \ldots, h \circ S_m \circ h^{-1})$ is a self-similar IFS as well, with the contracting ratios unchanged.

1.4. Preliminaries on fractal geometry

In this section we introduce the box (or box-counting), Hausdorff, and packing dimensions. Although the iterated function systems and fractals that we study in this book will be in $\mathbb{R}^d$, we will occasionally need to consider dimensions in other spaces, particularly the symbolic space. Some of the theory applies to subsets of a complete metric space, but often the structure of Euclidean space gives an advantage. We give almost no proofs, referring the reader to classical sources instead.

1.4.1. The box dimension.

DEFINITION 1.4.1. Let (X, ϱ) be a metric space. A subset $E \subset X$ is called *totally bounded* if for any $\delta > 0$ it can be covered by finitely many balls of radius δ. Given $E \subset X$, a totally bounded nonempty set, let $N_\delta(E)$ be the minimal number of sets of diameter δ needed to cover E. The *lower and upper box dimensions* of E are defined by

$$\underline{\dim}_B(E) := \liminf_{\delta \to 0} \frac{\log N_\delta(E)}{-\log \delta}, \tag{1.45}$$

$$\overline{\dim}_B(E) := \limsup_{\delta \to 0} \frac{\log N_\delta(E)}{-\log \delta}. \tag{1.46}$$

If the limit exists, then we call it the box dimension of E and denote it by $\dim_B(E)$.

The following lemma is left as an exercise (see [**Fal03**, Section 3.1]).

LEMMA 1.4.2. *Let E be a bounded subset of $\mathbb{R}^d$.*

(i) *The value of the box dimension does not change if $N_\delta(E)$ in Definition* 1.4.1 *in any of the following:*
   (a) *the minimal number of closed balls of radius δ that cover E;*
   (b) *the minimal number of cubes of side δ that cover E;*
   (c) *the number of δ-mesh cubes that intersect E, i.e.,*

$$\#\left\{(j_1, \ldots, j_d) \in \mathbb{Z}^d : \prod_{k=1}^{d} [j_k \cdot \delta, (j_k + 1) \cdot \delta) \cap E \neq \varnothing\right\},$$

   where $\#(A)$ denotes the cardinality of A,
   (d) *the minimal number of sets of diameter at most δ that cover E;*
   (e) *the maximal number of disjoint balls of radius δ with centers in E.*

(ii) *Suppose that $E \subset \mathbb{R}^d$, $E \neq \varnothing$, is bounded. Then*

$$\underline{\dim}_B(E) = d - \limsup_{r \to 0} \frac{\log \mathcal{L}^d(B(E, \delta))}{\log \delta}$$

and

$$\overline{\dim}_B(E) = d - \liminf_{r \to 0} \frac{\log \mathcal{L}^d(B(E, \delta))}{\log \delta},$$

where $B(E, \delta)$ is the open δ-neighborhood of E.

The equations in part (ii) were used by Minkowski as the definition, and so the box dimension is sometimes called Minkowski dimension. The definition in (i)(c) is especially well suited for numerical computations.

In general, the box dimension need not exist, as the following example shows.

EXAMPLE 1.4.3. We construct a set $\Lambda \subset [0,1]$ such that

$$\underline{\dim}_B(\Lambda) = \frac{\log 2}{\log 9} \quad \text{and} \quad \overline{\dim}_B(\Lambda) = \frac{\log 2}{\log 3}. \tag{1.47}$$

Let $\{n_k\}_{k=1}^{\infty}$ be a sequence satisfying

$$n_k \in \mathbb{N} \quad \text{and} \quad \frac{n_1 + \cdots + n_k}{n_{k+1}} \to 0. \tag{1.48}$$

For $\ell = 1, 2$ we define

$$u_1^{(\ell)}(x) := \frac{x}{3^\ell}, \qquad u_2^{(\ell)}(x) := \frac{x}{3^\ell} + 1 - \frac{1}{3^\ell}.$$

Note that the attractor of the IFS $\left(u_1^{(1)}(x), u_2^{(1)}(x)\right)$ is the middle-third Cantor set. Its dimension (the critical exponent discussed in Section 1.1.2) is $\log 2/\log 3$. The attractor of $\left(u_1^{(2)}(x), u_2^{(2)}(x)\right)$ is a middle-7/9 Cantor set and its dimension is $\log 2/\log 9$. Now we build a compact (not self-similar) set by combining these two IFS's in such a way that for a long time (n_k steps) we apply only one of them and "forget" about the other. Then we switch to the other IFS for a much longer time (n_{k+1} steps) and then go back to the original one for a still much longer time (n_{k+2} steps), and iterate this ad infinitum. Namely, for $i = 1, 2$ let

$$f_{n,i}(x) := \begin{cases} u_i^{(1)}(x), & \text{if } n_k \leqslant n \leqslant n_{k+1} - 1 \text{ and } k \text{ is odd;} \\ u_i^{(2)}(x), & \text{if } n_k \leqslant n \leqslant n_{k+1} - 1 \text{ and } k \text{ is even.} \end{cases}$$

Then the set

$$\Lambda := \bigcap_{n=1}^{\infty} \bigcup_{i_1 \ldots i_n \in \{1,2\}^n} f_{1,i_1} \circ \cdots \circ f_{n,i_n}[0,1] \tag{1.49}$$

satisfies (1.47). The details are left as an exercise.

Although in many cases the box dimension can be easily estimated, it has the disadvantage that every countable dense set of a d-dimensional open set has dimension d, contradicting our intuition that a countable set should be of zero dimension. The same can never happen if we use the *Hausdorff dimension.*

1.4.2. The Hausdorff dimension. First we define the t-dimensional Hausdorff measure for $t \geqslant 0$.

Let (X, ϱ) be a metric space, $E \subset X$, and $t \geqslant 0$. We say that $\{A_i\}_{i=1}^{\infty}$ is a δ-cover of E for $\delta > 0$ if $E \subset \bigcup_{i=1}^{\infty} A_i$ and $|A_i| \leqslant \delta$ for all i (the δ-cover may actually be finite or countable, but we will still denote it $\{A_i\}_{i=1}^{\infty}$). For $\delta > 0$ we consider the Hausdorff pre-measure which is the following set function:

$$\mathcal{H}_\delta^t(E) = \inf\Big\{\sum_{i=1}^{\infty} |A_i|^t : \ \{A_i\}_{i=1}^{\infty} \text{ is a } \delta\text{-cover of } E\Big\}. \tag{1.50}$$

To make it a measure, we take the limit as $\delta \to 0$ (note that the limit exists, because the supply of δ-covers increases with δ, so the function $\delta \mapsto \mathcal{H}_\delta^t(E)$ is nonincreasing).

More precisely, define

$$\mathcal{H}^t(E) = \lim_{\delta \to 0} \left\{ \underbrace{\inf \left\{ \sum_{i=1}^{\infty} |A_i|^t : E \subset \bigcup_{i=1}^{\infty} A_i, |A_i| \leqslant \delta \right\}}_{\mathcal{H}^t_\delta(E)} \right\}. \tag{1.51}$$

Then $\mathcal{H}^t$ is a metric outer measure (see Appendix B.1). Therefore, the restriction of $\mathcal{H}^t$ to the σ-algebra of $\mathcal{H}^t$-measurable sets is a Borel measure, called the t-dimensional Hausdorff measure. The d-dimensional Hausdorff measure in $\mathbb{R}^d$ is a constant multiple of the d-dimensional volume (the Lebesgue measure) $\mathcal{L}^d$ (see [**Fal85**, p.11] for the proof):

LEMMA 1.4.4. *For any Borel set $E \subset \mathbb{R}^d$ we have $\mathcal{H}^d(E) = \frac{2^d (d/2)!}{\pi^{d/2}} \cdot \mathcal{L}^d(E)$.*

Thus the measure $\mathcal{H}^t$ for a noninteger $t > 0$ can be considered as an extension of the volume to the noninteger dimension t. It follows from the definition of the Hausdorff measure that for any $E \subset X$ and $0 \leqslant \alpha < \beta$ we have $\mathcal{H}^\beta_\delta(E) \leqslant \delta^{\beta-\alpha}\mathcal{H}^\alpha_\delta(E)$. Thus we obtain the following implications:

$$\begin{aligned} \mathcal{H}^\alpha(E) < \infty &\implies \mathcal{H}^\beta(E) = 0; \\ 0 < \mathcal{H}^\beta(E) &\implies \mathcal{H}^\alpha(E) = \infty. \end{aligned}$$

So, for a given set E, if we choose t which is "too small", then $\mathcal{H}^t(E) = \infty$. On the other hand, if t is "too large", then $\mathcal{H}^t(E) = 0$.

The value of t where the t-dimensional Hausdorff measure drops from infinity to zero is called the Hausdorff dimension of the set E, see Figure 1.11. That is,

$$\dim_{\mathrm{H}}(E) = \inf \{ t : \mathcal{H}^t(E) = 0 \} = \sup \{ t : \mathcal{H}^t(E) = \infty \}. \tag{1.52}$$

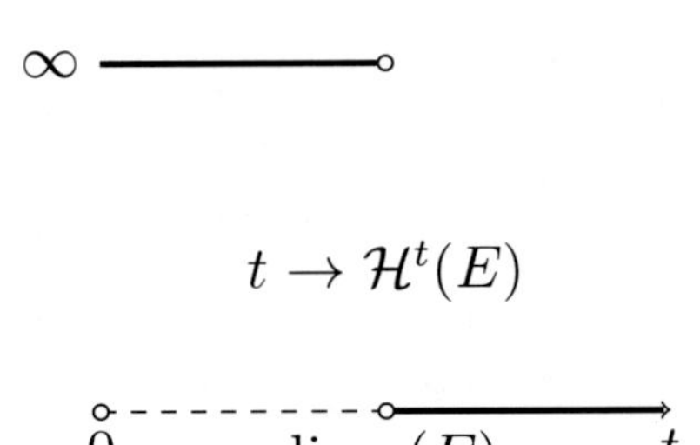

FIGURE 1.11. For $t_0 = \dim_{\mathrm{H}}(E)$ the t_0-dimensional Hausdorff measure $\mathcal{H}^{t_0}(E)$ can be zero, infinite, or a (finite) positive number.

We can compare the definition of the Hausdorff measure with the characterization of the box dimension given next.

LEMMA 1.4.5. *Let $E \subset X$ be totally bounded. For every $\delta > 0$ and $t \geqslant 0$ consider*

$$U_\delta^t(E) := \inf\left\{\sum_{i=1}^{K} |A_i|^t : K \in \mathbb{N},\ E \subset \bigcup_{i=1}^{K} A_i,\ |A_i| = \delta\right\}.$$

Further, let

$$\underline{U}^t(E) := \liminf_{\delta\to 0} U_\delta^t(E), \qquad \overline{U}^t(E) := \limsup_{\delta\to 0} U_\delta^t(E). \tag{1.53}$$

Then

$$\begin{aligned}\underline{\dim}_B(E) &= \inf\left\{t : \underline{U}^t(E) = 0\right\} = \sup\left\{t : \underline{U}^t(E) = \infty\right\}\\ \overline{\dim}_B(E) &= \inf\left\{t : \overline{U}^t(E) = 0\right\} = \sup\left\{t : \overline{U}^t(E) = \infty\right\}.\end{aligned}$$

In other words, we get the (upper and lower) box dimension by restricting the covers to sets of diameter exactly δ, whereas in the definition of the Hausdorff dimension we consider covers by sets of diameter less than or equal to δ. The proof of Lemma 1.4.5 is left as an exercise. From Lemma 1.4.5 we see immediately that

LEMMA 1.4.6. *For every totally bounded $E \subset X$ we have*

$$\dim_{\mathrm{H}}(E) \leqslant \underline{\dim}_{\mathrm{B}}(E) \leqslant \overline{\dim}_{\mathrm{B}}(E).$$

It follows from the definition and Lemma 1.4.4 that the Hausdorff dimension has the following basic properties, where X and Y are two arbitrary metric spaces:

LEMMA 1.4.7 (Elementary properties of the Hausdorff dimension). *The following properties hold:*

EP0: *Every countable set has zero Hausdorff dimension.*
EP1: *For every $F \subset \mathbb{R}^d$ we have $\dim_{\mathrm{H}}(F) \leqslant d$.*
EP2: *If $\mathcal{L}^d(E) > 0$ for a Borel set $E \subseteq \mathbb{R}^d$ then $\dim_{\mathrm{H}}(E) = d$.*
EP3: *For any $k < d$, a k-dimensional smooth surface in $\mathbb{R}^d$ has Hausdorff dimension k.*
EP4: *For a Lipschitz map $f : X \to Y$ and a set $E \subset X$ we have $\dim_{\mathrm{H}}(f(E)) \leqslant \dim_{\mathrm{H}}(E)$.*
EP5: *For a bi-Lipshitz map $f : X \to Y$ and a set $E \subset X$ we have $\dim_{\mathrm{H}}(E) = \dim_{\mathrm{H}}(f(E))$.*
EP6: *Let $\{E_i\}_{i=1}^{\infty}$ be a sequence of sets in X. Then*

$$\dim_{\mathrm{H}}\Big(\bigcup_{i=1}^{\infty} E_i\Big) = \sup_i \dim_{\mathrm{H}}(E_i).$$

DEFINITION 1.4.8. Let (X, ϱ_X) and (Y, ϱ_Y) be two metric spaces. A map $f : X \to Y$ is said to be *Hölder continuous*, or simply *Hölder*, with exponent α and constant $c > 0$, if

$$\varrho_Y(f(x), f(y)) \leqslant c\varrho_X(x,y)^\alpha \quad \text{for all } x, y \in X. \tag{1.54}$$

LEMMA 1.4.9. *Suppose that $f : X \to Y$ is* Hölder, *with exponent α and constant $c > 0$. Then for any set $E \subset X$ and $t \geqslant 0$,*

$$\mathcal{H}^t(f(E)) \leqslant c^t \mathcal{H}^{t\alpha}(E), \tag{1.55}$$

and hence

$$\dim_{\mathrm{H}}(f(E)) \leqslant \frac{\dim_{\mathrm{H}}(E)}{\alpha}. \tag{1.56}$$

PROOF. For every $\delta > 0$ and $\varepsilon > 0$ one can find a δ-cover $\{A_i\}$ of E such that $\sum_i |A_i|^{t\alpha} \leqslant \mathcal{H}_\delta^{t\alpha}(E) + \varepsilon$ and $|f(A_i)| \leqslant c|A_i|^\alpha \leqslant c\delta^\alpha$. Then $\{f(A_i)\}$ is a $c\delta^\alpha$-cover of $f(E)$ and hence

$$\mathcal{H}_{c\delta^\alpha}^t(f(E)) \leqslant \sum_i |f(A_i)|^t \leqslant c^t \sum_i |A_i|^{\alpha t} \leqslant c^t \mathcal{H}_\delta^{t\alpha}(E) + \varepsilon.$$

The inequality (1.55) follows by letting $\delta \to 0$ and then $\varepsilon \to 0$. □

A simple consequence of (1.55) is that for any linear map $S\colon \mathbb{R}^d \to \mathbb{R}^d$ with $\|S(x) - S(y)\| = \rho\|x - y\|$ we have

$$\mathcal{H}^t(S(E)) = \rho^t \mathcal{H}^t(E). \tag{1.57}$$

Observe that the natural projection from the symbolic space onto the attractor is Hölder continuous.

LEMMA 1.4.10. *Let $\mathcal{F} = (f_1, \dots, f_m)$ be an IFS of strict contractions on $\mathbb{R}^d$, with the Lipschitz constants $0 < r_i < 1$ for $i \leqslant m$. Let $\Pi : \Sigma \to \Lambda$ be the natural projection corresponding to $\mathcal{F}$, and let $r_{\max} = \max_i r_i$. Then Π is Hölder continuous with exponent $\alpha = \frac{\log(r_{\max}^{-1})}{\log m}$, where Σ is equipped with the metric defined in* (1.30).

PROOF. Let $\mathbf{i}, \mathbf{j} \in \Sigma$ and let $\mathbf{w} = \mathbf{i} \wedge \mathbf{j}$ be the common initial segment of the sequences $\mathbf{i}, \mathbf{j}$. Suppose that $|\mathbf{i} \wedge \mathbf{j}| = n \geqslant 0$. Then we have by the definition of the natural projection (1.42):

$$\begin{aligned} |\Pi(\mathbf{i}) - \Pi(\mathbf{j})| &= |\Pi(\mathbf{w}\,\sigma^n\mathbf{i}) - \Pi(\mathbf{w}\,\sigma^n\mathbf{j})| \\ &= |f_{\mathbf{w}} \circ \Pi(\sigma^n\mathbf{i}) - f_{\mathbf{w}} \circ \Pi(\sigma^n\mathbf{j})| \\ &\leqslant r_{\max}^{|\mathbf{i}\wedge\mathbf{j}|} \cdot |\Lambda| = |\Lambda| \cdot \varrho(\mathbf{i}, \mathbf{j})^\alpha, \end{aligned}$$

where we recall that $\varrho(\mathbf{i}, \mathbf{j}) = m^{-|\mathbf{i}\wedge\mathbf{j}|}$. □

1.4.3. Hausdorff content. Now we again consider a general metric space (X, ϱ). A useful tool for estimating the Hausdorff measure is the *s-dimensional Hausdorff content*, which is the following set function:

$$\mathcal{H}_\infty^s(E) = \inf\Big\{\sum_i |A_i|^s : \{A_i\}_{i=1}^\infty \text{ is a cover of } E\Big\}. \tag{1.58}$$

Compared to the Hausdorff measure $\mathcal{H}^s$, the Hausdorff content $\mathcal{H}_\infty^s$ is an outer measure, but not *metric* outer measure. Clearly, for every set $E \subset X$,

$$\mathcal{H}_\infty^s(E) \leqslant \mathcal{H}^s(E).$$

Notice that $\mathcal{H}_\infty^s(E)$ is finite for every bounded subset $E \subset X$. On the other hand, by the definition of the Hausdorff content, if $\mathcal{H}_\infty^s(E) = 0$ then for every $\delta > 0$ there exists a cover $\{A_i\}$ of E for which $\sum_i |A_i|^s < \delta$ and consequently, $|A_i| < \delta^{1/s}$ for all i. Hence, $\mathcal{H}_{\delta^{1/s}}^s(E) < \delta$ and so $\mathcal{H}^s(E) = 0$. Thus, the Hausdorff content can be used for the characterization of the Hausdorff dimension, namely,

$$\dim_{\mathrm{H}}(E) = \inf\{s \geqslant 0 : \mathcal{H}_\infty^s(E) = 0\}.$$

1.4.4. Mass distribution principle and Frostman measures. The following lemma is the simplest way to estimate the Hausdorff dimension from below.

LEMMA 1.4.11 (Mass distribution principle). *Let E be a Borel set in a metric space, and let μ be a positive finite Borel measure such that $\mu(E) > 0$. Assume that for some $t > 0$ there exist $c > 0$ and $\delta > 0$ such that*

$$|A| < \delta \Longrightarrow \mu(A) < c \cdot |A|^t. \tag{1.59}$$

Then we have

$$\mathcal{H}^t(E) \geqslant \frac{\mu(E)}{c} \quad \textit{and} \quad \dim_{\mathrm{H}}(E) \geqslant t. \tag{1.60}$$

PROOF. Let $\{A_i\}_{i=1}^{\infty}$ be an arbitrary cover of the set E with $|A_i| \leqslant \delta$. Then the sum in (1.51) (in the definition of the Hausdorff measure) is

$$\sum_{i=1}^{\infty} |A_i|^t \geqslant \frac{1}{c} \sum_{i=1}^{\infty} \mu(A_i) \geqslant \frac{1}{c} \mu\Big(\bigcup_{i=1}^{\infty} A_i\Big) \geqslant \frac{1}{c} \mu(E).$$

□

The property (1.59) (in a slightly different form) is very important; such measures have a special name.

DEFINITION 1.4.12. A compactly supported positive Borel measure μ on a metric space X is called a *Frostman measure* (with exponent t) if $0 < \mu(X) < \infty$ and

$$\mu(B(x,r)) \leqslant Cr^t \quad \text{for all } x \in X, \ r > 0. \tag{1.61}$$

In fact, every measure satisfying (1.59) is a Frostman measure with the same exponent, since for a ball of diameter greater that δ, that is, with $r > \delta/2$, we can estimate

$$\mu(B(x,r)) \leqslant \mu(X) \leqslant \frac{\mu(X)}{\delta^t} \cdot r^t.$$

And conversely, if (1.61) holds, then $\mu(A) \leqslant (C+\varepsilon)|A|^t$ for *any* Borel set A and any $\varepsilon > 0$, since the set A is contained in any open ball with a center in A, of any radius strictly greater than $|A|$. Thus we can repeat the proof of Lemma 1.4.11 for an arbitrary cover to obtain the following: if μ is a Borel measure on X, $E \subset X$ is Borel,

$$0 < \mu(E) < \infty \text{ and } \forall x \in X, \ \forall r > 0, \ \mu(B(x,r)) \leqslant Cr^t \quad \Longrightarrow \quad \mathcal{H}^t_\infty(E) \geqslant \frac{\mu(E)}{C}. \tag{1.62}$$

It turns out that a kind of "converse statement" to the last assertion holds as well; it is often called "Frostman's Lemma". We state it in two different versions.

THEOREM 1.4.13. *Let $t > 0$.*

(i) *If $E \subset \mathbb{R}^d$ is a Borel set, such that $\mathcal{H}^t(E) = \infty$, then there exists a compact subset F of E satisfying $0 < \mathcal{H}^t(F) < \infty$ and*

$$\mathcal{H}^t(B(x,r) \cap F) \leqslant Cr^t \quad \textit{for all } x \in \mathbb{R}^d, \ r > 0,$$

for some $C > 0$, so that $\mathcal{H}_t|_F$ is a Frostman measure with exponent t for E.

(ii) *If X is a compact metric space with $\mathcal{H}^t(X) > 0$, then there exists a finite positive Borel measure μ on X such that $\mu(X) > 0$ and for some $C > 0$,*

$$\mu(B(x,r)) \leqslant Cr^t \quad \text{for all } x \in X.$$

For the proof of part (i) see Falconer [**Fal85**, Theorems 5.4 and 5.5.]; for the proof of part (ii) see Mattila [**Mat95**, Theorem 8.17]. For an alternative proof in the Euclidean space see Bishop and Peres [**BP17**, Lemma 3.1.1 and Appendix B].

Now we turn to another frequently used method to give a lower bound on the Hausdorff dimension for subsets of $\mathbb{R}^d$.

1.4.5. Potential-theoretic characterization of the Hausdorff dimension. Following Falconer [**Fal03**], we call a measure μ on $\mathbb{R}^d$ a *mass distribution* if the support of μ is compact and $0 < \mu(\mathbb{R}^d) < \infty$.

Let μ be a mass distribution on $\mathbb{R}^d$ and $t > 0$. The t-energy of μ is defined by

$$\mathcal{E}_t(\mu) := \iint |x-y|^{-t}\, d\mu(x)\, d\mu(y). \tag{1.63}$$

The condition (1.61) and $\mathcal{E}_t(\mu) < \infty$ are closely related to each other.

LEMMA 1.4.14. *Let μ be a mass distribution on $\mathbb{R}^d$.*

(i) *If there exists $c > 0$ such that*

$$\mu(B(x,r)) \leqslant cr^t \quad \text{for all } x \in \mathbb{R}^d,\ r > 0, \tag{1.64}$$

then $\mathcal{E}_s(\mu) < \infty$ for $s < t$.

(ii) *If $\mathcal{E}_t(\mu) < \infty$, then* (1.64) *holds for a restriction of μ to a set of positive measure.*

We provide a quick proof following Mattila [**Mat95**, pp. 109-110]. Note that the proof extends verbatim to the setting of a finite Borel measure ν on a compact metric space (X, ϱ), with the t-energy of ν defined by $\mathcal{E}_t(\nu) = \iint \rho(x,y)^{-t}\, d\nu(x) d\nu(y)$.

PROOF OF (I). Define the s-potential of the measure μ at the point x by

$$\Phi_s(\mu, x) := \int \frac{d\mu(y)}{|x-y|^s}.$$

Using Theorem B.5.1 and then (1.64), we have

$$\begin{aligned}
\Phi_s(\mu,x) &= \int_0^\infty \mu\{y : |x-y|^{-s} \geqslant u\}\, du \\
&= \int_0^\infty \mu(B(x, u^{-1/s}))\, du \leqslant \int_0^\infty \min\{\mu(\mathbb{R}^d), Cu^{-t/s}\}\, du < \infty,
\end{aligned}$$

since $t > s$. Since $\mathcal{E}_s(\mu) = \int \Phi_s(\mu,x)\, d\mu(x)$, the claim follows. □

The proof of part (ii) of Lemma 1.4.14 is contained in the proof of the next lemma.

LEMMA 1.4.15. *For a Borel set $E \subset \mathbb{R}^d$ and for a mass distribution μ such that $\mu(E^c) = 0$ we have*

$$\mathcal{E}_t(\mu) < \infty \Longrightarrow \dim_{\mathrm{H}}(E) \geqslant t.$$

It follows that in order to prove that the Hausdorff dimension of a given Borel set E is at least t, it is enough to find a mass distribution μ supported on E with finite t-energy.

PROOF OF LEMMA 1.4.15. We follow Mattila [**Mat95**, p. 109]. Using that $\mathcal{E}_t(\mu) = \int \Phi_t(\mu, x)d\mu(x) < \infty$ we can find a positive number M such that

$$E_M := \{x \in E : \Phi_t(\mu, x) \leqslant M\}$$

is a set of positive measure μ. Fix such an M and let $\nu := \mu|_{E_M}$. Observe that the assertion of the lemma would follow from the mass distribution principle if we can verify (1.59). To this end, choose an arbitrary bounded Borel set A and fix an arbitrary $x \in A \cap E_M$. (If there is no such x then (1.59) obviously holds for the set A.) Then

$$M \geqslant \int \frac{d\mu(y)}{|x-y|^t} \geqslant \int \frac{d\nu(y)}{|x-y|^t} \geqslant \int_A \frac{d\nu(y)}{|x-y|^t} \geqslant \nu(A) \cdot |A|^{-t},$$

as desired. □

The next theorem states that, in a sense, the converse of the previous lemma holds. We refer to it as the *potential-theoretic characterization of the Hausdorff dimension.*

THEOREM 1.4.16. *For a Borel set $E \subset \mathbb{R}^d$ and every $0 < t_1 < \dim_{\mathrm{H}}(E) < t_2$, hold:*

(i) *there exists a mass distribution μ supported on E such that $\mathcal{E}_{t_1}(\mu) < \infty$;*

(ii) *for every mass distribution μ on E we have $\mathcal{E}_{t_2}(\mu) = \infty$.*

Part (i) follows from Theorem 1.4.13 and Lemma 1.4.14(i). Part (ii) follows from Lemma 1.4.15.

1.4.6. Packing measure and dimension of a set. Packing measures are in a sense "dual" to Hausdorff measures. They are defined in terms of the most economical "packings" by balls. However, unlike the Hausdorff measure, this is a 2-step process, via a "pre-measure".

Let (X, ϱ) be a metric space. For $E \subset X$ a *δ-packing of E* is a collection of disjoint closed balls $\{\overline{B}(x_i, r_i)\}$ such that $x_i \in E$ and $0 < r_i \leqslant \delta$ for all i. For $s > 0$ let

$$\mathcal{P}^s_\delta(E) = \sup\Big\{\sum_i |\overline{B}_i|^s : \ \{\overline{B}_i\} \text{ is a } \delta\text{-packing of } E\Big\}.$$

Since $\mathcal{P}^s_\delta(E)$ decreases with δ, there is a limit

$$\widetilde{\mathcal{P}}^s(E) = \lim_{\delta \to 0} \mathcal{P}^s_\delta(E).$$

This is the *s-dimensional packing pre-measure.* The problem with $\widetilde{\mathcal{P}}^s(E)$ is that, although it is finitely sub-additive, it is not countably sub-additive.

DEFINITION 1.4.17. The s-dimensional *packing measure* of a set E is given by

$$\mathcal{P}^s(E) = \inf\Big\{\sum_{i=1}^\infty \widetilde{\mathcal{P}}^s(E_i) : \ E \subset \bigcup_{i=1}^\infty E_i\Big\}. \tag{1.65}$$

The *packing dimension* is defined by

$$\dim_{\mathrm{P}}(E) = \inf\{s > 0 : \ \mathcal{P}^s(E) = 0\}.$$

Packing measures and dimension were introduced by Tricot [**Tri82**] and investigated by Taylor and Tricot [**TT85**]. One can show that $\mathcal{P}^s$ is a metric outer measure, hence all Borel sets are $\mathcal{P}^s$-measurable. In fact, $\mathcal{P}^s$ (like $\mathcal{H}^s$) is a Borel regular measure (see Definition B.2.1). The following inequalities always hold:

$$\dim_{\mathrm{H}}(E) \leqslant \dim_{\mathrm{P}}(E) \leqslant \overline{\dim}_{\mathrm{B}}(E).$$

It turns out that in $\mathbb{R}^d$ one can express the packing dimension directly, in terms of the upper box dimension:

PROPOSITION 1.4.18. *For any $F \subset \mathbb{R}^d$ holds*

$$\dim_{\mathrm{P}}(F) = \inf\Big\{\sup_{i\in I}\overline{\dim}_{\mathrm{B}}(F_i) : \ F \subseteq \bigcup_{i\in I} F_i,\ I \text{ is countable and } F_i \text{ are bounded}\Big\}.$$

For the proof, see, e.g., [**BP17**, Prop. 2.7.1]. A useful corollary is as follows:

COROLLARY 1.4.19 (see [**BP17**, Cor. 2.8.2]). *Let $F \subset \mathbb{R}^d$ be a compact set such that for all $U \subset \mathbb{R}^d$ open for which $U \cap F \neq \varnothing$ we have $\overline{\dim}_{\mathrm{B}}(F \cap U) = \overline{\dim}_{\mathrm{B}}(F)$. Then* $\dim_{\mathrm{P}}(F) = \overline{\dim}_{\mathrm{B}}(F)$.

1.4.7. Frostman-type lemmas. We borrowed the term from the book [**PU10**, Section 7.6]. In fact, the next theorem is combined from several sources; it demonstrates how the knowledge about the decay of the measure of small balls yields information about dimensions of the support of the measure. It is also a clear manifestation of the Hausdorff vs. packing dimension duality.

THEOREM 1.4.20. *Let μ be a finite Borel measure on $\mathbb{R}^d$, and let $A \subset \mathbb{R}^d$ be a bounded Borel set.*

(i) *If $\mu(A) > 0$ and there exists $\theta_1 > 0$ such that for every $x \in A$,*
$$\liminf_{r\to 0} \frac{\log \mu(B(x,r))}{\log r} \geqslant \theta_1,$$
then $\dim_{\mathrm{H}}(A) \geqslant \theta_1$.

(ii) *If there exists $\theta_2 > 0$ such that for every $x \in A$,*
$$\liminf_{r\to 0} \frac{\log \mu(B(x,r))}{\log r} \leqslant \theta_2,$$
then $\dim_{\mathrm{H}}(A) \leqslant \theta_2$.

(iii) *If $\mu(A) > 0$ and there exists $\theta_1 > 0$ such that for every $x \in A$,*
$$\limsup_{r\to 0} \frac{\log \mu(B(x,r))}{\log r} \geqslant \theta_1,$$
then $\dim_{\mathrm{P}}(A) \geqslant \theta_1$.

(iv) *If there exists $\theta_2 > 0$ such that for every $x \in A$,*
$$\liminf_{r\to 0} \frac{\log \mu(B(x,r))}{\log r} \leqslant \theta_2,$$
then $\dim_{\mathrm{P}}(A) \leqslant \theta_2$.

All the claims (i)–(iv) will follow from the corresponding assertions in Theorem 2.1.2 (i)–(iv) about densities of measures, which is partially proven in the next chapter. In fact, part (i) is a more sophisticated version of the Mass Distribution Principle, and part (ii) may be deduced using the Besicovitch covering theorem; for the proof of these parts we refer the reader to [**Fal03**, Proposition 4.9]. Parts (iii) and (iv) follow from Theorem 2.1.2 (iii) and (iv), which are less well known and are proved completely in Section 2.1.

1.4.8. Basic properties of dimensions. Here we borrow an idea from Jonathan Fraser's PhD thesis [**Fra13**, p. 7] and collect in a table some of the most important properties of the dimensions that we have introduced so far. In the table below we assume that the sets $E, F, E_1, E_2, \dots,$ are subsets of a complete metric space (X, ϱ). We write $\mathfrak{C}(X)$ for the set of all nonempty compact subsets of X, and dist_H is the Hausdorff metric on $\mathfrak{C}(X)$, that is,

$$\mathrm{dist}_H(E, F) = \inf\{\varepsilon > 0 : E \subseteq B(F, \varepsilon) \text{ and } F \subseteq B(E, \varepsilon)\}.$$

TABLE 1. Basic properties of the dimensions of sets

Property	$\dim_{\mathrm{H}}$	$\dim_{\mathrm{P}}$	$\underline{\dim}_{\mathrm{B}}$	$\overline{\dim}_{\mathrm{B}}$
$E \subseteq F \Rightarrow \dim(E) \leqslant \dim(F)$,	Yes	Yes	Yes	Yes
$\dim(E \cup F) = \max\{\dim(E), \dim(F)\}$	Yes	Yes	No	Yes
$\dim(\bigcup_{i=1}^{\infty} E_i) = \sup_{1 \leqslant i \leqslant \infty} \dim(E_i)$	Yes	Yes	No	No
$\dim(f(E)) \leqslant \dim(E)$, if f is Lipschitz	Yes	Yes	Yes	Yes
$\dim(E) = \dim(\overline{E}\ )$	No	No	Yes	Yes
$\dim(U) = n$ if $X = \mathbb{R}^n$, $U \neq \varnothing$, $U \subset \mathbb{R}^n$ open	Yes	Yes	Yes	Yes
$K \mapsto \dim(K)$ is measurable on $(\mathfrak{C}(X), \mathrm{dist}_H)$	Yes	No	Yes	Yes

The verification of the last property may be found in [**MM97**]. All the other properties are proved in [**Fal03**, Chapter 2,3].

1.5. Similarity dimension; the dimension of self-similar sets with well separated cylinders

The definition of the Hausdorff measure offers a simple way to give a natural upper bound on the Hausdorff dimension of a self-similar set Λ, which in many cases is the Hausdorff dimension itself. In order to do that, we need to estimate $\mathcal{H}^t(\Lambda)$. We compute $\sum_{i=1}^{\infty} |A_i|^t$ (which is the sum highlighted in (1.51)) in the particular case when $\{A_i\}$ is the collection of level-n cylinders for some large n. If Λ is the attractor of the self-similar IFS $\mathcal{S} = (S_1, \dots, S_m)$ on $\mathbb{R}^d$ and the contraction ratio of S_i is $0 < r_i < 1$, then the diameter of the level-n cylinder $S_{i_1 \dots i_n}(\Lambda)$ is equal to $|\Lambda|\, r_{i_1} \cdots r_{i_n}$. So, the sum in (1.51) can be computed as follows:

$$\sum_{i=1}^{\infty} |A_i|^t = |\Lambda| \cdot \sum_{i_1 \dots i_n} r_{i_1}^t \cdots r_{i_n}^t = |\Lambda| \cdot \left(r_1^t + \dots + r_m^t\right)^n \tag{1.66}$$

When the infimum in formula (1.51) is attained for this natural system of covers by level-n cylinder sets, the Hausdorff dimension of Λ is the solution s of the equation:

$$\boxed{r_1^s + \dots + r_m^s = 1.} \tag{1.67}$$

In fact, as $n \to \infty$ the sum in (1.66) tends to infinity if $t < s$, to zero if $t > s$, and it is a constant $|\Lambda|$ if $t = s$. In all cases the solution of the equation (1.67) is called the *similarity dimension* of the self-similar IFS $\mathcal{S}$. The calculation above implies

LEMMA 1.5.1. *Let Λ be the attractor of a self-similar IFS in $\mathbb{R}^d$. Then $\dim_{\mathrm{H}}(\Lambda) \leqslant \min\{d, s\}$, where s is the similarity dimension of the IFS. Moreover, $\mathcal{H}^s(\Lambda) < \infty$.*

Observe that a very similar calculation was carried out in the case of middle-α Cantor sets in Section 1.1.2. The critical exponent obtained there is exactly the similarity dimension. The intuition is that if the cylinders of the self-similar set are disjoint or if their intersection is not too large (when the Open Set Condition holds, see below), then the similarity dimension, the box dimension, and the Hausdorff dimension should be the same.

On the other hand, if there is a "heavy overlap" between the cylinders of the attractor, then it may be possible to find a system of covers of the attractor for which the sum in the definition of the Hausdorff dimension (1.51) is much smaller than that for the natural cover by level-n cylinder sets. Then it can happen that the Hausdorff dimension of the attractor is less than the similarity dimension. For instance, for the variant of the golden gasket from Example 1.1.6 with $\lambda = \omega_3 \approx 0.5368$, the similarity dimension is $\frac{\log 3}{-\log \omega_3} \approx 1.80284$, whereas the Hausdorff dimension of the attractor is approximately 1.7319, as shown in [**BMS04**]. (Such a "dimension drop" also occurs for the golden gasket itself, but then it is not surprising that the "natural cover" by cylinder sets is not optimal, since its similarity dimension is $2.283\ldots > 2$.) To avoid such a phenomenon, the following conditions have been introduced:

DEFINITION 1.5.2. Let $\mathcal{F} = (f_1, \ldots, f_m)$ be a contracting IFS in $\mathbb{R}^d$, and Λ is its attractor.

(a) The **Strong Separation Property (SSP)** holds for $\mathcal{F}$ if

$$f_i(\Lambda) \cap f_j(\Lambda) = \varnothing \text{ for all } i \neq j. \tag{1.68}$$

(b) The **Open Set Condition (OSC)** holds for $\mathcal{F}$ if there exists a nonempty bounded open set $V \subset \mathbb{R}^d$ such that
- $f_i(V) \subset V$ for all $i = 1, \ldots, m$;
- $f_i(V) \cap f_j(V) = \varnothing$ for all $i \neq j$.

(c) The **Strong Open Set Condition (SOSC)** holds for $\mathcal{F}$ if the set V in the definition of the OSC can be chosen such that $V \cap \Lambda \neq \varnothing$.

We note that the SSP is sometimes called the Strong Separation Condition.

The SSP is a very strong assumption. Usually, we can obtain the same results if we assume only the OSC, but the proofs are more technical. Observe that the IFS of the middle-α Cantor set satisfies the SSP, whereas those of Sierpiński gasket, Sierpiński carpet, and the Von Koch curve, satisfy the OSC but not the SSP.

The OSC was introduced by P. A. P. Moran in [**Mor46**] in 1946, and became widely known after the work of J. Hutchinson [**Hut81**] in 1981.

THEOREM 1.5.3 (Moran [**Mor46**], Hutchinson [**Hut81**]). *Assume that the self-similar IFS $\mathcal{S} = (S_1, \ldots, S_m)$ acts on $\mathbb{R}^d$ and satisfies the OSC. The similarity ratio of S_i is $0 < r_i < 1$, $i = 1, \ldots, m$. Let s be the similarity dimension, that is, $r_1^s + \cdots + r_m^s = 1$. Then for the attractor Λ of the IFS $\mathcal{S}$ we have*

$$0 < \mathcal{H}^s(\Lambda) < \infty. \tag{1.69}$$

Furthermore,

$$\dim_{\mathrm{H}}(\Lambda) = \dim_{\mathrm{B}}(\Lambda) = s, \tag{1.70}$$

in particular, the box dimension exists.

The complete proof of the theorem may be found in [**Hut81**], and we will discuss it later in detail in Section 4.1. Here we prove it in the special case when the Strong Separation Property holds. The proof consists of two parts. First we show (1.69), assuming the SSP, which implies that $\dim_{\mathrm{H}}(\Lambda) = s$. Then we prove that $\overline{\dim}_{\mathrm{B}}(\Lambda) \leqslant s$, which yields (1.70), in view of Lemma 1.4.6. This second part will be stated as an independent Lemma 1.5.4, since it holds without any separation conditions.

PROOF OF (1.69). We already know that $\mathcal{H}^s(\Lambda) < \infty$ from Lemma 1.5.1, so it remains to prove positivity of the Hausdorff measure. For this purpose, we will use the Mass Distribution Principle (Lemma 1.4.11). First we define a measure on the symbolic space $\Sigma = \{1, \ldots, m\}^{\mathbb{N}}$. For a symbolic cylinder $[i_1, \ldots, i_n] \subset \Sigma$ (see (1.31) for the definition), we let

$$\mu([i_1 \ldots i_n]) = r^s_{i_1 \ldots i_n} := r^s_{i_1} \cdot \ldots \cdot r^s_{i_n}.$$

Using that $\sum\limits_{i_1 \ldots i_k} r^s_{i_1 \ldots i_k} = (r^s_1 + \cdots + r^s_m)^k = 1$, by Kolmogorov's extension principle (or by Lemma 1.3.2), one can check that μ may be extended to a Borel probability measure on the symbolic space Σ. The extended measure will also be denoted by μ. (In fact, $\mu = \mu_{\mathbf{p}} = \mathbf{p}^{\mathbb{N}}$ is the infinite product (Bernoulli) measure, with $\mathbf{p} = (r^s_1, \ldots, r^s_m)$, defined in Section 1.3.1.) Then we consider the push-forward measure $\nu = \Pi_*(\mu)$, that is, $\nu(A) = \mu(\Pi^{-1}(A))$ for all Borel $A \subset \mathbb{R}^d$. It is clear that ν is a probability measure supported on the attractor Λ. Recall that we assumed the SSP, which is equivalent to the natural projection Π being bijective. Thus we have

$$\nu(\Lambda_{i_1 \ldots i_n}) = \mu([i_1 \ldots i_n]) = r^s_{i_1 \ldots i_n}. \tag{1.71}$$

Furthermore, the Strong Separation Property implies that

$$\widetilde{d} := \min\{\mathrm{dist}(\Lambda_i, \Lambda_j) : \ i \neq j\} > 0.$$

Then, using that the IFS consists of similarities, we obtain for all $(i_1, \ldots, i_n) \in \{1, \ldots, m\}^n$:

$$\min\{\mathrm{dist}(\Lambda_{i_1 \ldots i_n i}, \Lambda_{i_1 \ldots i_n j}) : \ i \neq j\} = r_{i_1 \ldots i_n} \widetilde{d}. \tag{1.72}$$

In order to verify (1.59), it is enough to show that there exists $C > 0$ such that

$$\nu(B(x, r)) \leqslant C \cdot r^s$$

for all $x \in \Lambda$ and r sufficiently small, since every set A of diameter r intersecting the attractor is covered by a closed ball of radius r centered at a point of the attractor. Fix $x \in \Lambda$ and $0 < r < \widetilde{d}$. Let $\mathbf{i} = (i_1, i_2, \ldots)$ be the symbolic code of the point x, that is, $x = \Pi(\mathbf{i})$. We can find $n \geqslant 1$ so that

$$r_{i_1 \ldots i_n} \widetilde{d} \leqslant r < r_{i_1 \ldots i_{n-1}} \widetilde{d}.$$

Then (1.72) implies

$$B(x, r) \cap \Lambda = B(x, r) \cap \Lambda_{i_1 \ldots i_{n-1}}.$$

Thus

$$\nu(B(x,r)) \leqslant \nu(\Lambda_{i_1\dots i_{n-1}}) = r^s_{i_1\dots i_{n-1}} \leqslant r^{-s}_{\min} r^s_{i_1\dots i_n} \leqslant (\widetilde{d})^{-s} r^{-s}_{\min} r^s.$$

Here we denoted $r_{\min} = \min\{r_i : i = 1, \dots, m\}$. This concludes the proof of (1.59), so by the Mass Distribution Principle we obtain (1.60), which gives

$$\mathcal{H}^s(\Lambda) \geqslant (\widetilde{d})^s r^s_{\min} \nu(\Lambda) = (\widetilde{d})^s r^s_{\min} > 0,$$

as desired. □

LEMMA 1.5.4. *Let Λ be the attractor of a self-similar IFS $(S_1, \dots, S_m)$ where S_i has contraction ratio $0 < r_i < 1$, for $i = 1, \dots, m$. Let s be the similarity dimension. Then*

$$\overline{\dim}_{\mathrm{B}}(\Lambda) \leqslant s.$$

PROOF. We introduce the important notion of *Moran cut-set.* For $0 < r \leqslant 1$ the Moran cut-set (see Section 1.3.1 for the definition of a cut-set) is defined by

$$\widetilde{\mathcal{M}}_r := \{\mathbf{i} \in \Sigma^* : r_{\mathbf{i}} \leqslant r < r_{\mathbf{i}^-}\}, \tag{1.73}$$

where we obtain the word $\mathbf{i}^-$ from $\mathbf{i}$ by deleting its last letter, that is,

$$\mathbf{i}^- := (i_1, \dots, i_{n-1}) \text{ if } \mathbf{i} = (i_1, \dots, i_n). \tag{1.74}$$

We use the notation $r_{\mathbf{i}} = r_{i_1\dots i_n}$. By definition,

$$\mathbf{i} \in \widetilde{\mathcal{M}}_r \implies r r_{\min} < r_{\mathbf{i}} \leqslant r. \tag{1.75}$$

The set $\widetilde{\mathcal{M}}_r$ is a *minimal* cut-set, because the corresponding cylinder sets form a partition of Σ. In fact, for any $0 < r \leqslant 1$ we have

$$\Sigma = \bigcup \{[\mathbf{i}] : \mathbf{i} \in \widetilde{\mathcal{M}}_r\}, \quad [\mathbf{i}] \cap [\mathbf{j}] = \varnothing \text{ for } \mathbf{i} \neq \mathbf{j} \text{ in } \widetilde{\mathcal{M}}_r.$$

We will also need the *Moran cover*

$$\mathcal{M}_r := \{\Lambda_{\mathbf{i}} : \mathbf{i} \in \widetilde{\mathcal{M}}_r\}, \tag{1.76}$$

which is a cover of the attractor Λ, but its elements are disjoint only if the SSP holds.

We can use Moran covers to estimate $\overline{\dim}_{\mathrm{B}}(\Lambda)$ because all elements of $\mathcal{M}_r$ have roughly the same diameter:

$$\Lambda_{\mathbf{i}} = \Lambda_{i_1\dots i_n} \in \mathcal{M}_r \implies r r_{\min}|\Lambda| \leqslant |\Lambda_{\mathbf{i}}| = r_{\mathbf{i}}|\Lambda| \leqslant r|\Lambda|. \tag{1.77}$$

Since every set of diameter less than $r|\Lambda|$ can be covered by a set of diameter equal to $r|\Lambda|$, we obtain

$$N_{r|\Lambda|}(\Lambda) \leqslant \#\mathcal{M}_r \tag{1.78}$$

(see Definition 1.46). So it remains to estimate the cardinality of the Moran cover $\#\mathcal{M}_r = \#\widetilde{\mathcal{M}}_r$. Recall that $\sum_{i=1}^m r_i^s = 1$, hence for any $0 < r \leqslant 1$, by Lemma 1.3.3,

$$\sum_{\mathbf{i} \in \widetilde{\mathcal{M}}_r} r^s_{\mathbf{i}} = 1, \tag{1.79}$$

since $\widetilde{\mathcal{M}}_r$ is a minimal cut-set. Equivalently, this follows from the fact that μ is a probability measure on Σ, where $\mu = \mathbf{p}^{\mathbb{N}}$, with $\mathbf{p} = (r_1^s, \dots, r_m^s)$, was introduced in the previous proof. Now, (1.75) and (1.79) imply $r^s_{\min} r^s \cdot \#\mathcal{M}_r < 1$, hence

$$\#\mathcal{M}_r \leqslant r^{-s}_{\min} r^{-s}.$$

Combining this with (1.78) yields

$$\frac{\log N_{r|\Lambda|}(\Lambda)}{-\log(r|\Lambda|)} \leqslant \frac{-s\log r - s\log r_{\min}}{-\log r - \log|\Lambda|} \to s, \quad \text{as } r \to 0,$$

which implies $\overline{\dim}_{\mathrm{B}}(\Lambda) \leqslant s$ by (1.46). □

1.6. Self-similar sets with overlaps

One of the main topics of this book is the study of *self-similar sets with overlaps.* Such are the self-similar sets for which we do not have any separation conditions. We will consider some parametrized families of self-similar IFS in $\mathbb{R}^d$ with possibly "heavy" overlaps between the cylinders. We will show that if the family satisfies a technical *transversality condition*, then the dimension of the attractor is equal to the minimum of d and the similarity dimension, at least for Lebesgue almost every parameter. Here we only demonstrate this on two examples.

Consider the self-similar IFS on the real line:

$$\mathcal{F}_{\mathbf{r},\mathbf{t}} = (f_1, \ldots, f_m)_{j=1}^m,$$

where

$$f_j(x) = r_j x + t_j, \quad 0 < |r_j| < 1 \ \text{ and } t_j \in \mathbb{R}, \ j = 1, \ldots, m.$$

Denote its attractor by $\Lambda_{\mathbf{r},\mathbf{t}}$ where $\mathbf{r} = (r_1, \ldots, r_m)$ and $\mathbf{t} = (t_1, \ldots, t_m)$. Let $s = s(\mathbf{r})$ be the similarity dimension, that is, $\sum_{j=1}^m r_j^s = 1$.

THEOREM 1.6.1.

(i) *We have*

$$\dim_{\mathrm{H}}(\Lambda_{\mathbf{r},\mathbf{t}}) = \min\{1, s(\mathbf{r})\} \ \textit{for a.e. } \mathbf{t} \in \mathbb{R}^m. \tag{1.80}$$

(ii) *If* $s(\mathbf{r}) > 1$ *then* $\mathcal{L}^1(\Lambda_{\mathbf{r},\mathbf{t}}) > 0$ *for a.e.* $\mathbf{t} \in \mathbb{R}^m$.

Part (i) was proved by Falconer [**Fal87**] under the assumption that $\sum_{j=1}^m |r_j| < 1$. The scope of part (i) was extended and part (ii) was proved by Simon and Solomyak [**SS02**]. The proof of the theorem will be given in Chapter 6.

One of the ideas of the proof is to consider the attractor $\Lambda_{\mathbf{r},\mathbf{t}}$ as an *orthogonal projection* of a self-similar set in a higher-dimensional space, which in many cases will have no overlaps. If a self-similar IFS in $\mathbb{R}^d$ for $d > 1$ consists of maps whose linear parts are of the form $\mathbf{x} \mapsto r_j\mathbf{x}$ (without rotations), then all projections of its attractor to linear subspaces are self-similar. This often yields self-similar sets with overlaps, even when the original IFS has strong separation, and it is a rich source of examples.

1.6.1. The $\{0, 1, 3\}$-problem and expansions with deleted digits. This problem was first studied by Keane and Smorodinsky in the early 1990's. It concerns the IFS

$$\mathcal{S}_{013}^{\lambda} := (S_0(x) = \lambda x, \ \ S_1(x) = \lambda x + 1, \ \ S_3(x) = \lambda x + 3), \qquad x \in \mathbb{R},$$

where $\lambda \in (0, 1)$. We will denote its attractor by $\mathcal{K}^{\lambda}$. One can check that

$$\mathcal{K}^{\lambda} = \Big\{x = \sum_{n=0}^{\infty} a_n \lambda^n : \ a_n \in \{0, 1, 3\}\Big\}. \tag{1.81}$$

Thus $\mathcal{K}^\lambda$ is the set of base-λ expansions with digits $\{0,1,3\}$ (2 is a "deleted digit"). The motivation came from a problem of Palis who asked whether the arithmetic difference $\mathcal{C}_\alpha - \mathcal{C}_\beta$ contains an interval if and only if it has positive Lebesgue measure. (Here $\mathcal{C}_\alpha$ and $\mathcal{C}_\beta$ are the middle-α and middle-β Cantor sets. We do not address this problem in this book, but refer the reader to [**PT93**].) One can see that if, for instance, $\gamma = \lambda^2$, where $\lambda = \frac{1-\alpha}{2}$ and $\gamma = \frac{1-\beta}{2}$ are the contraction ratios of the Cantor sets, then the arithmetic difference $\mathcal{C}_\alpha - \mathcal{C}_\beta$ may be expressed in terms of base-λ expansions, with an "unusual" set of digits. The $\{0,1,3\}$-problem was chosen as one of the "simplest nontrivial examples of this kind", according to Meir Smorodinsky.

We collect elementary properties of the sets $\mathcal{K}^\lambda$ in the next lemma, illustrated in Figure 1.12.

LEMMA 1.6.2. *(i) The smallest closed interval containing* $\mathcal{K}^\lambda$ *is* $I_\lambda := [0, \frac{3}{1-\lambda}]$.
(ii) For all $\lambda \in (0,1)$, $\dim_H(\mathcal{K}^\lambda) \leqslant \min\{1, \frac{\log 3}{|\log \lambda|}\}$.
(iii) the IFS $\mathcal{S}^\lambda_{013}$ *satisfies the Open Set Condition for all* $\lambda \in (0, \frac{1}{4}]$, *hence*

$$\dim_H(\mathcal{K}^\lambda) = s_\lambda := \frac{\log 3}{|\log \lambda|} \quad \textit{and} \quad 0 < \mathcal{H}^{s_\lambda}(\mathcal{K}^\lambda) < \infty, \quad \lambda \in (0, 1/4];$$

(iv) $\mathcal{K}^\lambda = I_\lambda$ *(the line segment) for all* $\lambda \in [\frac{2}{5}, 1]$.

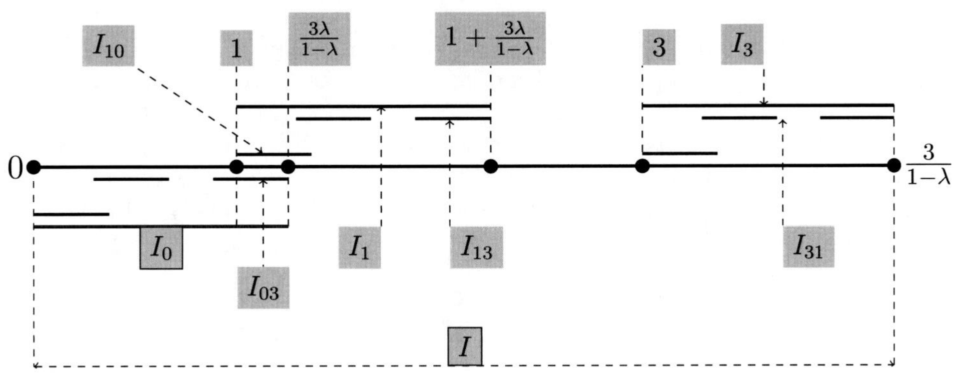

FIGURE 1.12. The set $\mathcal{K}^\lambda$ for $\lambda = 0.295$

The proof is left as an exercise. The lemma implies that the nontrivial parameter range is $\lambda \in (\frac{1}{4}, \frac{2}{5})$, when two first cylinder intervals of level one intersect, but there is a gap between the second and the third interval, hence the attractor $\mathcal{K}^\lambda$ is disconnected. What can we say about the measure-theoretic, fractal-dimensional, and topological properties of $\mathcal{K}^\lambda$? It is natural to distinguish between the *subcritical case*: $\lambda \in (\frac{1}{4}, \frac{1}{3})$, *critical case*: $\lambda = \frac{1}{3}$, and *supercritical case:* $\lambda \in (\frac{1}{3}, \frac{2}{5})$, depending on whether the similarity dimension s_λ is smaller, equal, or greater than one. The critical case will be considered in Example 4.3.1; this is a classical example of an IFS satisfying the *Weak Separation Condition.*

For $\lambda < \frac{1}{3}$, we have by Lemma 1.6.2(i) that $\mathcal{K}^\lambda$ is a Cantor set of dimension less than one. Indeed, $\mathcal{K}^\lambda$ has zero Lebesgue measure, hence totally disconnected, and an attractor of a nondegenerate IFS cannot have isolated points. M. Keane raised a question whether $\lambda \mapsto \dim_H(\mathcal{K}^\lambda)$ is continuous. This was answered negatively by Pollicott and Simon [**PS95**].

THEOREM 1.6.3 (Pollicott and Simon 1995 [**PS95**]). *Let $\mathcal{K}^\lambda$ be the attractor of the IFS $\mathcal{S}_{013}^\lambda$ with parameter λ. Then*

(i) $\dim_H(\mathcal{K}^\lambda) = s_\lambda = \frac{\log 3}{|\log \lambda|}$ for Lebesgue-a.e. $\lambda \in (\frac{1}{4}, \frac{1}{3})$;

(ii) there is a dense subset $\mathcal{E} \subset (\frac{1}{4}, \frac{1}{3})$ such that $\dim_H(\mathcal{K}^\lambda) < s_\lambda$ for $\lambda \in \mathcal{E}$.

In fact, [**PS95**] considered a more general case of finitely many integer digits. The paper [**PS95**] was quite important, because it introduced the *transversality method.* We will treat this method in depth in Chapter 6 and obtain the results on the sets $\mathcal{K}^\lambda$ as an application in Section 6.5. In the supercritical case, after some partial results in [**PS95, KSS95**], the following was shown in [**Sol95**].

THEOREM 1.6.4. *For Lebesgue-a.e. $\lambda \in (\frac{1}{3}, \frac{2}{5})$ the attractor $\mathcal{K}^\lambda$ has positive Lebesgue measure.*

This result was also proved by the transversality method, see Section 6.5. Now much stronger results are known. In particular, it follows from the work of Rapaport and Varjú [**RV20**, Theorem A.1] that the only reason for the "dimension drop" $\dim_H(\mathcal{K}^\lambda) < \min\{1, s_\lambda\}$ could be the presence of "exact overlaps", see the next subsection for a precise formulation. Surprisingly, we still do not know whether the sets $\mathcal{K}^\lambda$ have nonempty interior, even for a single parameter $\lambda \in (\frac{1}{3}, \frac{2}{5})$.

1.6.2. Exact overlap conjecture. It is still not fully understood what can cause the drop of the dimension, that is, under what circumstances we have $\dim_H(\Lambda_{\mathbf{r},\mathbf{t}}) < \min\{s(\mathbf{r}), 1\}$ for the IFS $\mathcal{F}_{\mathbf{r},\mathbf{t}} = (f_j(x) = r_j x + t_j)_{j=1}^m$ on $\mathbb{R}$, where $s(\mathbf{r})$ is the similarity dimension of $\mathcal{F}_{\mathbf{r},\mathbf{t}}$. It is easy to see that if $s(\mathbf{r}) < 1$, then exact overlap implies dimension drop. We say that $\mathcal{F}_{\mathbf{r},\mathbf{t}}$ has *exact overlap* if there exists $n \geqslant 1$ and finite words $\mathbf{i}', \mathbf{j}' \in \{1, \ldots, m\}^n$ such that $\mathbf{i}' \neq \mathbf{j}'$ but $f_{\mathbf{i}'} \equiv f_{\mathbf{j}'}$. In fact, the IFS $\mathcal{F}_{\mathbf{r},\mathbf{t}}$ and the IFS $\widetilde{\mathcal{F}}_{\mathbf{r},\mathbf{t}} := (f_{\mathbf{i}})_{\mathbf{i} \in \{1,\ldots,m\}^n \setminus \{\mathbf{j}'\}}$ have the same attractor $\Lambda_{\mathbf{r},\mathbf{t}}$, but

$$1 = \sum_j r_j^{s(\mathbf{r})} = \Big(\sum_j r_j^{s(\mathbf{r})}\Big)^n = \sum_{\mathbf{i} \in \{1,\ldots,m\}^n} r_{\mathbf{i}}^{s(\mathbf{r})} > \sum_{\substack{\mathbf{i} \in \{1,\ldots,m\}^n: \\ \mathbf{i} \neq \mathbf{j}'}} r_{\mathbf{i}}^{s(\mathbf{r})},$$

so the similarity dimension of $\widetilde{\mathcal{F}}_{\mathbf{r},\mathbf{t}}$, which is an upper bound for the Hausdorff dimension of $\Lambda_{\mathbf{r},\mathbf{t}}$, is strictly smaller than $s(\mathbf{r})$. It is an open problem whether the only possibility for the dimension drop is presence of exact overlap.

CONJECTURE 1.6.5 (Exact overlap conjecture). *If $\dim_H(\Lambda_{\mathbf{r},\mathbf{t}}) < \min\{s(\mathbf{r}), 1\}$, then $\mathcal{F}_{\mathbf{r},\mathbf{t}}$ has exact overlaps.*

It should be emphasized that the conjecture in this form applies only to self-similar IFS on $\mathbb{R}$; a straightforward generalization to higher dimensions is false, see Section 7.6.

Using methods of additive combinatorics, Hochman [**Hoc14b**] established the conjecture under the assumption that both contraction ratios and translation parameters are algebraic and in some other special cases; he also gave a sharp estimate on the dimension of the exceptional set of translation parameters for which the dimension drops.

THEOREM 1.6.6 (Hochman). *Let* $\mathcal{F}_{\mathbf{r},\mathbf{t}} = (f_j(x) = r_j x + t_j)_{j=1}^m$ *be a self-similar IFS on the line. Then*

$$\dim_H\big(\{\mathbf{t} \in \mathbb{R}^m : \dim_H(\Lambda_{\mathbf{r},\mathbf{t}}) < \min\{s(\mathbf{r}), 1\}\big) \leqslant m - 1.$$

It is easy to see that this bound is sharp in the case $s(\mathbf{r}) < 1$, since the set $\{\mathbf{t} \in \mathbb{R}^m : f_{\mathbf{i}} \equiv f_{\mathbf{j}}\}$ forms an $(m-1)$-dimensional subspace for all $\mathbf{i} \neq \mathbf{j} \in \{1, \ldots, m\}^n$. Indeed, $f_{\mathbf{i}} \equiv f_{\mathbf{j}}$ holds if and only if $r_{\mathbf{i}} = r_{\mathbf{j}}$ and

$$0 = \sum_{\ell=1}^{n} t_{i_\ell} r_{i_1 \ldots i_{\ell-1}} - \sum_{\ell=1}^{n} t_{j_\ell} r_{j_1 \ldots j_{\ell-1}} = \sum_{k=1}^{m} t_k \left(\sum_{\ell=1}^{n} \delta_{i_\ell,k} r_{i_1 \ldots i_{\ell-1}} - \delta_{j_\ell,k} r_{j_1 \ldots j_{\ell-1}} \right),$$

where $\delta_{i,j} = 1$ if $i = j$ and 0 otherwise. The set of such $\mathbf{t} = (t_1, \ldots, t_m)$ forms an $(m-1)$-dimensional subspace of $\mathbb{R}^m$.

Recently, Rapaport [**Rap22b**] verified the exact overlap conjecture in the case when the contracting ratios are algebraic.

THEOREM 1.6.7 (Rapaport). *Suppose that all* $r_1, \ldots, r_m$ *are algebraic numbers and that* $\mathcal{F}_{\mathbf{r},\mathbf{t}}$ *has no exact overlaps. Then* $\dim_H(\Lambda_{\mathbf{r},\mathbf{t}}) = \min\{s(\mathbf{r}), 1\}$.

Moreover, Rapaport and Varjú [**RV20**, Theorem A.1] showed that the conjecture holds also in the case of arbitrary homogeneous self-similar IFS with rational translations; in particular, this applies to the $\{0, 1, 3\}$-problem. More precisely:

THEOREM 1.6.8 (Rapaport and Varjú). *Let* $\mathcal{F}_{r,\mathbf{t}} = (f_j(x) = rx + t_j)_{j=1}^m$ *be a homogeneous self-similar IFS on the line, with* $t_j \in \mathbb{Q}$. *If* $\mathcal{F}_{r,\mathbf{t}}$ *has no exact overlaps, then* $\dim_H(\Lambda_{\mathbf{r},\mathbf{t}}) = \min\{s(\mathbf{r}), 1\}$.

For further details, see Section 6.8.

1.7. Self-similar graph directed IFS on the line*

Here we introduce the notion of Self-Similar Graph Directed IFS (self-similar GIFS) on $\mathbb{R}$. The theory in higher dimensions is analogous and can be found in [**MW88**] and [**Fal97**, Section 3]; then the closed intervals in the definition below are replaced by closed domains. This is part of the "enrichment" material, introduced for completeness, but not treated systematically in the book.

Let $\{I_i\}_{i=1}^N$ be closed intervals on the line with disjoint interiors. Suppose that we are given a directed graph $\mathcal{G} = (\mathcal{V}, \mathcal{E})$, where $\mathcal{V} = \{1, \ldots, N\}$ is the set of vertices and $\mathcal{E}$ is the set of edges, see Figure 1.14. The set of edges from i to j is denoted by $\mathcal{E}_{i,j}$. We allow multiple directed edges and loops. For every edge $e \in \mathcal{E}_{i,j}$, we are given a contracting similarity mapping

$$f_e : I_j \to I_i$$

with contraction ratio $r_e \in (0, 1)$. (This is a little bit counterintuitive since the edge e goes from i to j, but the function f_e maps I_j into I_i.) Our principal assumption is as follows:

(A1) *The graph* $\mathcal{G}$ *is strongly connected, that is, for every* $i, j \in \mathcal{V}$ *there is a directed path from* i *to* j *in the graph.*

It follows from the proof of [**MW88**, Theorem 1] that there is a unique tuple of nonempty compact sets $(\Lambda_1, \ldots, \Lambda_N)$ (we call them graph directed sets) such that

$$\Lambda_i = \bigcup_{j=1}^{N} \bigcup_{e \in \mathcal{E}_{i,j}} f_e(\Lambda_j), \quad i = 1, \ldots, N. \tag{1.82}$$

Moreover, we say that

$$\Lambda := \bigcup_{i=1}^{N} \Lambda_i$$

is the attractor of the self-similar GIFS $(f_e)_{e \in \mathcal{E}}$. For every $s \geqslant 0$ consider the $N \times N$ matrix $A^{(s)} := (a^{(s)}(i,j))_{i,j=1}^{N}$, where

$$a^{(s)}(i,j) = \begin{cases} 0, & \text{if } \mathcal{E}_{i,j} = \varnothing; \\ \sum_{e \in \mathcal{E}_{i,j}} r_e^s, & \text{otherwise.} \end{cases}$$

Denote by $\rho(A^{(s)})$ the spectral radius of the matrix $A^{(s)}$. It was proved by Mauldin and Williams in [**MW88**, Theorem 2] that the mapping $s \mapsto \rho(A^{(s)})$ is strictly decreasing, continuous, $\rho(A^{(0)}) \geqslant 1$, and $\rho(A^{(s)}) \to 0$ as $s \to \infty$. So, there exists a unique $s_0 \geqslant 0$ for which

$$\rho(A^{(s_0)}) = 1. \tag{1.83}$$

Similarly to Lemma 1.5.1 and Lemma 1.5.4, one can show that under our principal assumption (A1), but without assuming any separation conditions, the following holds:

$$\dim_{\mathrm{H}}(\Lambda) \leqslant \overline{\dim}_{\mathrm{B}}(\Lambda) \leqslant s_0. \tag{1.84}$$

However, to get the lower bound we need to assume that the cylinders are well-separated:

THEOREM 1.7.1 (Mauldin and Williams [**MW88**, Theorem 3]). *Assume that for every $i = 1, \ldots, N$, the family of sets $\{f_{i,j}(I_j) : (i,j) \in \mathcal{E}\}$ is nonoverlapping, that is, the interiors of the sets $f_{i,j}(I_j)$ are pairwise disjoint. Then for every $i = 1, \ldots, N$, we have*

$$\dim_{\mathrm{H}}(\Lambda_i) = \dim_{\mathrm{B}}(\Lambda_i) = s_0 \text{ and } 0 < \mathcal{H}^{s_0}(\Lambda_i) < \infty.$$

See [**Fal97**, page 50] for a proof of this theorem. We remark that if the directed graph consists of one vertex and m loops we are back in the case of the self-similar IFS considered above, and Theorem 1.5.3 is a special case of Theorem 1.7.1.

EXAMPLE 1.7.2. Let Λ be the attractor of the piecewise linear IFS $\mathcal{F} := (f_1, f_2)$, with $f_k : [0,1] \to [0,1]$, $k = 1, 2$, defined by the graphs of the functions of $\mathcal{F}$ in Figure 1.13.

Clearly, $\mathcal{F}$ is a "usual", and *not* a graph directed self-similar IFS on $\mathbb{R}$. However, its attractor Λ can be represented as an attractor of a graph directed IFS and then using Theorem 1.7.1 we can easily compute the dimension of Λ. Namely, let $I_k := f_k([0,1])$. Then $\Lambda \subset I_1 \cup I_2$. Let $f_{i,j}$ be the similarity mapping on $\mathbb{R}$ satisfying $f_{i,j}|_{I_j} := f_i|_{I_j}$. Let (Λ_1, Λ_2) be the unique family of nonempty compact sets defined in (1.82) for which

$$\Lambda_1 = f_{1,1}(\Lambda_1) \cup f_{1,2}(\Lambda_2), \quad \Lambda_2 = f_{2,1}(\Lambda_1) \cup f_{2,2}(\Lambda_2). \tag{1.85}$$

That is, $\Lambda = \Lambda_1 \cup \Lambda_2$ is the attractor of the self-similar GIFS $\{f_e\}_{e\in\{1,2\}^2}$. Then the s-powers of the contraction ratios of the mappings $f_{i,j}$ are the corresponding elements of the matrix

$$A^{(s)} := \begin{pmatrix} \left(\frac{1}{6}\right)^s & \left(\frac{1}{2}\right)^s \\ \left(\frac{7}{12}\right)^s & \left(\frac{1}{12}\right)^s \end{pmatrix}.$$

So, by Theorem 1.7.1 we obtain that $\dim_{\mathrm{H}}(\Lambda) = \dim_{\mathrm{B}}(\Lambda) = s_0$ and $0 < \mathcal{H}^{s_0}(\Lambda) < \infty$, where $\rho(A^{(s_0)}) = 1$. Using Wolfram Mathematica we obtain that $s_0 \approx 0.577295$.

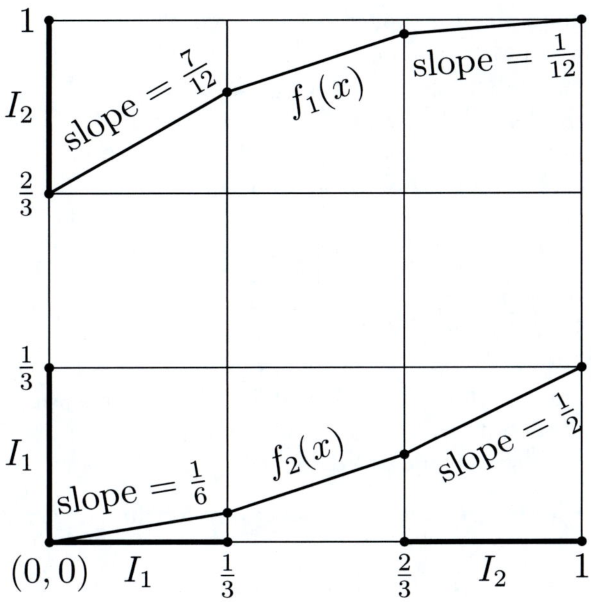

FIGURE 1.13. Piecewise linear IFS on $[0, 1]$

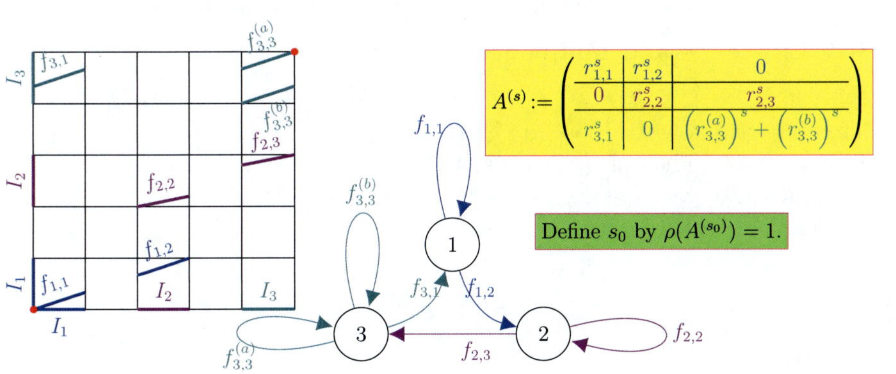

Always: $\dim_{\mathrm{H}} \Lambda \leq s_0$. If the interior of $\{f_{i,j}(I_j)\}_{(i,j)\in\mathcal{E}}$ are pairwise disjoint then $\dim_{\mathrm{H}} \Lambda = s_0$.

FIGURE 1.14. Graph Directed Self-similar IFS on $\mathbb{R}$. The contraction ratio of $f_{i,j}$ is $r_{i,j}$.

1.7.1. Self-similar graph directed IFS on $\mathbb{R}$ with overlaps. Consider a self-similar GIFS on the line. We neither assume any separation conditions, nor impose any restrictions on the contractions. Using [**KSS03**, Theorem 1], we can compute the dimension of the attractor for almost every translations (in the sense of Theorem 1.6.1).

Let $\mathcal{G}$ be a directed graph with the vertex set $\mathcal{V} = \{1, \ldots, N\}$ and the edge set $\mathcal{E}$. As in Section 1.7, we assume that $\mathcal{G}$ is strongly connected. For every $e \in \mathcal{E}$ we are given a contracting similarity $f_e : \mathbb{R} \to \mathbb{R}$:

$$f_e^{\mathbf{t}}(x) := r_e \cdot x + t_e, \quad 0 < r_e < 1.$$

We consider the contraction ratios r_e to be fixed and the vector of translations $\mathbf{t} = (t_e)_{e \in \mathcal{E}}$ as a parameter, similarly to Theorem 1.6.1. Let $\ell := \#\mathcal{E}$. By a fixed enumeration of $\mathcal{E}$ we assign to any vector of translations $\mathbf{t} \in \mathbb{R}^\ell$ a self-similar GIFS $\mathcal{F}^{\mathbf{t}} := (f_e^{\mathbf{t}})_{e \in \mathcal{E}}$. The family $\{\mathcal{F}^{\mathbf{t}}\}_{\mathbf{t} \in \mathbb{R}^\ell}$ is called a translation family of self-similar GIFS's. As stated in Section 1.7, for every $\mathbf{t} \in \mathbb{R}^\ell$ there exist unique nonempty compact sets $\Lambda_1^{\mathbf{t}}, \ldots, \Lambda_m^{\mathbf{t}}$ such that for every $1 \leqslant i \leqslant N$,

$$\Lambda_i^{\mathbf{t}} = \bigcup_{j=1}^{N} \bigcup_{e \in \mathcal{E}_{ij}} f_e^{\mathbf{t}}(\Lambda_j),$$

where $\mathcal{E}_{ij}$ is the set of directed edges in $\mathcal{G}$ connecting i to j. Then $(\Lambda_1^{\mathbf{t}}, \ldots, \Lambda_N^{\mathbf{t}})$ is a family of graph directed sets and $\Lambda^{\mathbf{t}} = \bigcup_{j=1}^N \Lambda_j^{\mathbf{t}}$ is the attractor of $\mathcal{F}^{\mathbf{t}}$. For every $s \geqslant 0$ we define the $N \times N$ matrix $A^{(s)}$ and the number s_0 according to (1.83). These are independent of $\mathbf{t}$, since they depend only on the contraction ratios of the translation family of the self-similar GIFS $\{\mathcal{F}^{\mathbf{t}}\}_{\mathbf{t} \in \mathbb{R}^\ell}$.

THEOREM 1.7.3 (Keane, Simon, Solomyak, [**KSS03**]). *For $\mathcal{L}^\ell$-almost every $\mathbf{t} \in \mathbb{R}^\ell$ holds*

(i) $\dim_{\mathrm{H}}(\Lambda^{\mathbf{t}}) = \dim_{\mathrm{B}}(\Lambda^{\mathbf{t}}) = \min\{1, s_0\}$,
(ii) *if* $s_0 > 1$ *then* $\mathcal{L}^1(\Lambda^{\mathbf{t}}) > 0$.

The proof uses a variant of the transversality method.

1.8. Stationary measures for iterated function systems

1.8.1. Existence and uniqueness.

DEFINITION 1.8.1. Let $\mathcal{F} = (f_1, \ldots, f_m)$ be an IFS on $\mathbb{R}^d$, where $f_i : \mathbb{R}^d \to \mathbb{R}^d$ are strict contractions, with Lipschitz constants $r_i < 1$, and let $\mathbf{p} = (p_1, \ldots, p_m)$ be a probability vector. The pair $(\mathcal{S}, \mathbf{p})$ is sometimes called a *probabilistic IFS*. A Borel probability measure ν is called the *stationary measure* for the IFS $\mathcal{F}$, corresponding to the probability vector $\mathbf{p}$, if

$$\nu = \sum_{j=1}^{m} p_j (f_j)_* \nu,$$

that is,

$$\nu(A) = \sum_{j=1}^{m} p_j \nu(f_j^{-1} A) \quad \text{for every Borel set } A \subset \mathbb{R}^d. \tag{1.86}$$

A stationary measure for an IFS of similitudes is called a *self-similar measure*, and if $\mathcal{F}$ is an IFS of affine maps, then it is called *self-affine*.

THEOREM 1.8.2 (Hutchinson [**Hut81**]). *For any strictly contracting IFS on* $\mathbb{R}^d$ *and a probability vector* $\mathbf{p}$*, there exists a unique stationary measure.*

We will give a brief sketch of the proof below (a more general result, due to Kravchenko [**Kra06**] will be proved in Section 14.1), but first we show the existence by a direct construction.

PROPOSITION 1.8.3. *Under the assumptions of Theorem* 1.8.2*, let* $\Pi : \Sigma \to \Lambda$ *be the natural projection, defined in* (1.11) *or* (1.42)*, and let* $\mu_{\mathbf{p}} = \mathbf{p}^{\mathbb{N}}$ *be the Bernoulli product measure, defined in Example* 1.3.4*. Then the measure* $\nu = \Pi_*(\mu_{\mathbf{p}})$ *satisfies* (1.86).

REMARK 1.8.4 (On terminology). There is no agreement in the literature on the meaning of the term "stationary measure for the IFS". The measure ν satisfying (1.86) is called "stationary" in [**BP17**, Section 2.1], however, it is often called the "invariant measure for the IFS", see [**Hut81**] and [**Fal97**, Section 2.2]. We would like to reserve the term "invariant measure for the IFS" for the push-forward of a σ-invariant measure under the natural projection. Such measures will be important later in the book. Similar terminology is used by Fraser [**Fra20**, Section 6.2], except that he refers to a measure ν satisfying (1.86) as a "Bernoulli invariant measure".

The following remark is immediate from the definitions.

REMARK 1.8.5 (On attractors and stationary measures for conjugate IFS's). Let $h : \mathbb{R}^d \to \mathbb{R}^d$ be a homeomorphism and let $\mathcal{F} = (f_i)_{i=1}^m$ and $\widetilde{\mathcal{F}} = (\widetilde{f}_i)_{i=1}^m$ be two contractive IFS's on $\mathbb{R}^d$, which are conjugate via h, that is, $\widetilde{f}_i = h \circ f_i \circ h^{-1}$ for all $i \leqslant m$, see Definition 1.3.8. If Λ is the attractor of $\mathcal{F}$, then $h(\Lambda)$ is the attractor of $\widetilde{\mathcal{F}}$. If $\nu_{\mathbf{p}}$ is the stationary measure for the IFS $\mathcal{F}$ with a probability vector $\mathbf{p}$, then $h_*(\nu_{\mathbf{p}}) = \nu_{\mathbf{p}} \circ h^{-1}$ is the corresponding stationary measure for the IFS $\widetilde{\mathcal{F}}$.

PROOF OF PROPOSITION 1.8.3. Let $A \subset \Sigma$ be a Borel set. We have

$$\nu(A) = \mu_{\mathbf{p}}(\Pi^{-1}A) = \sum_{j=1}^{m} \mu_{\mathbf{p}}(\Pi^{-1}A \cap [j]).$$

Observe that

$$\Pi^{-1}A \cap [j] = \{j\mathbf{i} : \Pi(j\mathbf{i}) \in A\} = \{j\mathbf{i} : \mathbf{i} \in \Pi^{-1}f_j^{-1}A\},$$

since $\Pi(j\mathbf{i}) = f_j\Pi(\mathbf{i})$. Thus $\Pi^{-1}A \cap [j]$ is the Cartesian product of $[j]$ and $\Pi^{-1}f_j^{-1}A$, and since $\mu_{\mathbf{p}}$ is a product measure, we obtain

$$\mu_{\mathbf{p}}(\Pi^{-1}A \cap [j]) = \mu([j]) \cdot \mu_{\mathbf{p}}(\Pi^{-1}f_j^{-1}A) = p_j\nu(f_j^{-1}A),$$

and (1.86) follows. □

PROOF SKETCH OF THEOREM 1.8.2. Let $M(\Lambda)$ be the set of Borel probability measures on the attractor Λ. One can show that $M(\Lambda)$ is compact with respect to the dual Lipschitz metric (or Kantorovich–Rubinstein metric) on $M(\Lambda)$ defined by

$$L(\mu, \mu') := \sup_{\mathrm{Lip}(g) \leqslant 1} \left| \int g d\mu - \int g d\mu' \right|, \tag{1.87}$$

where $\mathrm{Lip}(g)$ denotes the Lipschitz constant of g. (See, e.g., [**BP17**, Appendix A] for the proof.) Then one can prove that the function $\mathfrak{F}_{\mathcal{F},\mathbf{p}} : M(\Lambda) \to M(\Lambda)$, given by

$$\mathfrak{F}_{\mathcal{F},\mathbf{p}} : \nu \to \sum_{j=1}^{m} p_j \cdot (\nu \circ f_j^{-1}),$$

is a strict contraction (in fact, its Lipschitz constant is the maximum of Lipschitz constants of the functions of f_j). This is an easy argument which may be found in [**Hut81**] or in the book [**Fal97**, p. 37]. By the Banach Fixed Point Theorem, $\mathfrak{F}_{\mathcal{F},\mathbf{p}}$ has a unique fixed point, which must be the stationary measure. This concludes the proof of Theorem 1.8.2. □

REMARK 1.8.6. The above proof has a subtlety, which arises if one would like to work in an unbounded metric space. Let $M_{\rm loc}(X)$ be the set of Borel probability measures of bounded support on a complete metric space X. It was pointed out in [**Kra06**, Assertion 1.1] that $M_{\rm loc}(X)$ is not complete whenever X is unbounded. For example, the sequence of measures $\mu_n = 2^{-n}\delta_0 + \sum_{k=1}^n 2^{-k}\delta_k$ is a Cauchy sequence in $M_{\rm loc}(\mathbb{R})$, but has no limit. This does not affect the proof of Theorem 1.8.2, since it deals with $M(\Lambda)$ on the compact attractor. In order to resolve this issue for a general separable metric space (X, ϱ), Kravchenko [**Kra06**] defined $M(X)$ as the set of all Borel probability measures with $\int f d\mu < \infty$ for all Lipschitz function $f \in {\rm Lip}(X)$, equipped with the metric L defined in (1.87).

THEOREM 1.8.7 (Kravchenko [**Kra06**, Theorem 4.2]). *The metric space* $(M(X), L)$ *is complete if and only if the metric space* (X, ϱ) *is complete.*

This allowed for an extension of Theorem 1.8.2 to the setting of IFS with countable many maps on a metric space—see Section 14.1.

1.8.2. Self-similar measures. Here we consider self-similar measures, that is, stationary measures for an IFS of similarity maps. Let $\mathcal{S} = (S_1, \ldots, S_m)$ be a self-similar IFS on $\mathbb{R}^d$, with contraction ratios $0 < r_i < 1$, $\leqslant m$, and a probability vector $\mathbf{p} = (p_1, \ldots, p_m)$. If $p_i = 0$ for some i, then the map S_i can be removed from $\mathcal{S}$, without changing the self-similar measures. Thus we always assume that $p_i > 0$ for all $i \leqslant m$. By Theorem 1.8.2, there is a unique self-similar measure for the IFS $\mathcal{S}$, corresponding to the probability vector $\mathbf{p}$, which satisfies

$$\nu(A) = \sum_{j=1}^{m} p_j \nu(S_j^{-1} A) \quad \text{for every Borel set } A \subset \mathbb{R}^d. \tag{1.88}$$

By Proposition 1.8.3,

$$\nu = \Pi_*(\mu_{\mathbf{p}}), \quad \text{where } \mu_{\mathbf{p}} = (p_1, \ldots, p_m)^{\mathbb{N}}$$

is the infinite product measure on Σ. Since $p_i > 0$ for all $i \leqslant m$, the compact support of the self-similar measure ν is the attractor of the IFS Λ.

The simplest example of a self-similar measure is the restriction of the Lebesgue measure to the interval $[0, 1]$.

EXAMPLE 1.8.8. Let $p_1 = p_2 = 1/2$ and $\mathcal{S} := \left(S_i(x) = \frac{1}{2} \cdot x + \frac{i-1}{2}\right)_{i=1}^2$ on $\mathbb{R}$. Then $\Lambda = [0, 1]$ and ν is equal to $\mathcal{L}^1|_{[0,1]}$, since (1.88) obviously holds. Alternatively, $\nu := \Pi_*(\frac{1}{2}, \frac{1}{2})^{\mathbb{N}}$, so the ν-measure of every 2-adic interval satisfies

$$\nu\left(\left(\frac{k}{2^n}, \frac{k+1}{2^n}\right)\right) = \frac{1}{2^n}, \text{ for } k = 0, \ldots, 2^n - 1,$$

as it should.

The measure $\mathbf{p}^{\mathbb{N}}$ in the previous example corresponds to the coin tossing with a fair coin. If instead we consider the measure corresponding to a biased coin tossing, then we get a much more interesting example of a self-similar measure, which is singular with respect to the Lebesgue measure, but supported on the unit interval:

LEMMA 1.8.9. *Let $\mathcal{S}$ be as in Example 1.8.8 and $\nu_p := \Pi_* (p, 1-p)^{\mathbb{N}}$ for some $p \in (0,1)$, $p \neq \frac{1}{2}$. Then*

$$\nu_p \perp \mathcal{L}^1 \text{ and } \nu_p \perp \nu_q \text{ for } p \neq q. \tag{1.89}$$

To keep the presentation as elementary as possible, we refrain from using Birkhoff Ergodic Theorem in the introductory chapters. Instead, we apply the Strong Law of Large Numbers on appropriate probability spaces to prove this Lemma.

PROOF. We introduce the probability space $(\Sigma, \mathcal{B}, \mathbb{P}_p)$ for $p \in (0,1)$ as follows: The measurable space $(\Sigma, \mathcal{B})$ was introduced in Section 1.3.1. Put $\mu_p := (p, 1-p)^{\mathbb{N}}$ on $\mathcal{B}$. To complete the definition of our probability space we write $\mathbb{P}_p := \mu_p$. We consider a sequence of random variables $I_1, I_2, \dots$ taking values from the set $\{1, 2\}$ such that for all $j \in \mathbb{N}$, we have $I_j(\boldsymbol{\omega}) := \omega_j$ for an $\boldsymbol{\omega} = (\omega_1, \omega_2, \dots) \in \Sigma$. Using that μ_p is a product measure, we obtain that $I_1, I_2, \dots$ are independent random variables. Moreover, we obtain also from the definition of μ_p that the random variables $I_1, I_2, \dots$ are identically distributed satisfying

$$\mathbb{P}_p(I_j = 1) = p \text{ and } \mathbb{P}_p(I_j = 2) = 1 - p. \tag{1.90}$$

The expectation of I_j is

$$\mathbb{E}\left[I_j\right] = 1 \cdot p + 2 \cdot (1-p) = 2 - p. \tag{1.91}$$

By the Strong Law of Large Numbers (SLLN), for $\mathbb{P}_p$-a.e. $(\omega_1, \omega_2, \dots) \in \Sigma = \{1, 2\}^{\mathbb{N}}$

$$\lim_{n \to \infty} \frac{1}{n} \sum_{k=1}^{n} I_k = 2 - p,$$

which is different for different $p \in (0,1)$. This means that $\mathbb{P}_p \perp \mathbb{P}_q$ if $p \neq q$. This and Proposition 1.8.3 imply that $\nu_p \perp \nu_q$ if $p \neq q$. This completes the proof of the second part of (1.89). The proof of the first part follows from this and the comment in Example 1.8.8 that $\mathcal{L}^1|_{[0,1]} = \Pi_*(\mu_{1/2})$. □

EXAMPLE 1.8.10. Let $\mathcal{S}$ be a self-similar IFS having the Strong Separation Property, with contraction ratios $r_1, \dots, r_m$. This is the IFS for which we proved the Moran–Hutchinson Theorem in the previous section. In that proof we defined a measure ν, see (1.71). Observe that this ν is a self-similar measure for $\mathcal{S}$ with the probability vector

$$\mathbf{p} = (r_1^s, \dots, r_m^s), \tag{1.92}$$

where s is the similarity dimension.

DEFINITION 1.8.11. Let $\mathcal{S}$ be a self-similar IFS with contraction ratios $r_1, \dots$, r_m and similarity dimension s. Then the stationary measure of $\mathcal{S}$ with the probability vector $\mathbf{p} := (r_1^s, \dots, r_m^s)$ is called the *natural measure* for the IFS $\mathcal{S}$. Note that here we do not assume that $\mathcal{S}$ satisfies any separation properties.

PROPOSITION 1.8.12 (Hutchinson [**Hut81**]). *If the self-similar IFS $\mathcal{S}$ with similarity dimension s and attractor Λ satisfies the OSC, then there is $0 < c < \infty$ such that the natural measure for $\mathcal{S}$ is equal to $c \cdot \mathcal{H}^s|_\Lambda$.*

EXAMPLE 1.8.13 (Natural measure for the middle-α Cantor set). For $0 < \alpha < \frac{1}{2}$ consider the middle-α Cantor set Λ_α (Example 1.1.4). We recall that it is the attractor of the IFS

$$\mathcal{F}_\alpha := \left(f_1^\alpha(x) = \frac{1-\alpha}{2} \cdot x,\ f_2^\alpha(x) = \frac{1-\alpha}{2} \cdot x + \frac{1+\alpha}{2} \right).$$

In this case the natural measure ν_α is the push-forward measure of the evenly distributed measure $(\frac{1}{2}, \frac{1}{2})^{\mathbb{N}}$, that is, $\nu_\alpha := (\Pi_\alpha)_* \left(\frac{1}{2}, \frac{1}{2}\right)^{\mathbb{N}}$, where, as usual, the natural projection is defined by

$$\Pi_\alpha(\mathbf{i}) := \lim_{n \to \infty} f^\alpha_{i_1 \dots i_n}(0). \tag{1.93}$$

By Proposition 1.8.12, there is a constant $c_\alpha > 0$ such that $\nu_\alpha = c_\alpha \cdot \mathcal{H}^{s_\alpha}$, where $s_\alpha = \frac{\log 2}{-\log \frac{1-\alpha}{2}}$ is the similarity dimension.

EXAMPLE 1.8.14 (Bernoulli convolution). For $1/2 < \lambda < 1$ consider the IFS

$$\mathcal{S}^\lambda := \left(S_1^\lambda(x) := \lambda x - 1,\ S_2^\lambda(x) := \lambda x + 1 \right). \tag{1.94}$$

The fixed points of $S_1^\lambda(x)$ and $S_2^\lambda(x)$ are $-\frac{1}{1-\lambda}$ and $\frac{1}{1-\lambda}$ respectively. Clearly, for $I^\lambda := \left[-\frac{1}{1-\lambda}, \frac{1}{1-\lambda}\right]$ and $I_j^\lambda := S_j^\lambda(I^\lambda)$ we have $I_1^\lambda \cap I_2^\lambda \neq \emptyset$, so $I^\lambda = I_1^\lambda \cup I_2^\lambda$. Therefore, the attractor of $\mathcal{S}^\lambda$ is the interval I^λ itself, by Lemma 1.1.5. Observe that this homogeneous IFS is essentially the same as the one considered in Example 1.8.13, with the only (very important) difference that here the contraction ratio is greater than 1/2. If $\lambda < 1/2$, then the IFS $\mathcal{F}_\alpha$ and $\mathcal{S}^\lambda$, with appropriate α, are related by a linear change of variables. Indeed, for $\varphi : [0,1] \to \left[-\frac{1}{1-\lambda}, \frac{1}{1-\lambda}\right]$, $\varphi(t) = \frac{2t-1}{1-\lambda}$, and $\lambda = \frac{1-\alpha}{2}$, the following diagram commutes:

$$\begin{array}{ccc} [0,1] & \xrightarrow{\ f_i^\alpha\ } & [0,1] \\ \downarrow{\scriptstyle \varphi} & & \downarrow{\scriptstyle \varphi} \\ \left[-\frac{1}{1-\lambda}, \frac{1}{1-\lambda}\right] & \xrightarrow[\ S_i^\lambda\ ]{} & \left[-\frac{1}{1-\lambda}, \frac{1}{1-\lambda}\right] \end{array}$$

In Figure 1.15 we illustrate the IFS $\mathcal{S}^\lambda$ when $\lambda < \frac{1}{2}$ (the Strong Separation Property holds, the attractor is a Cantor set), $\lambda = \frac{1}{2}$ (Open Set Condition holds, the attractor is an interval), and $\lambda > \frac{1}{2}$ (overlapping case, the attractor is an interval).

Let μ be the evenly distributed measure $\mu := \left(\frac{1}{2}, \frac{1}{2}\right)^{\mathbb{N}}$ on the symbolic space $\Sigma := \{1,2\}^{\mathbb{N}}$. Then, by definition, the natural measure for the IFS is

$$\nu_\lambda := (\Pi_\lambda)_* \mu, \tag{1.95}$$

where Π_λ is the natural projection for this IFS. As we have assumed that $\lambda > 1/2$, the support of the natural measure, which is always the attractor, is the interval I^λ. So, the self-similar measure ν_λ may or may not be absolute continuous, see

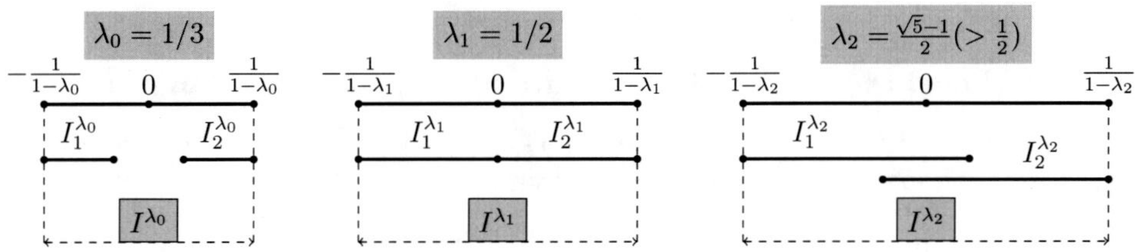

FIGURE 1.15. The first iterates of the IFS $\mathcal{S}^\lambda$ for different choices of the paramteter

Example 1.8.8 and Lemma 1.8.9. The following problem has attracted considerable attention since the 1930's:

PROBLEM 1.8.15. *For which* $\lambda \in (1/2, 1)$ *is the measure* ν_λ *absolutely continuous with respect to the Lebesgue measure?*

One of the motivations for this question comes from a link with random series and the work of P. Erdős from the late 1930's. Namely, assume that in the infinite random sum:

$$Y_\lambda := \sum_{n=0}^{\infty} \pm\lambda^n \tag{1.96}$$

the "+" and "−" signs are chosen independently with probabilities $(1/2, 1/2)$. Then it is easy to see that for all Borel sets A we have:

$$\nu_\lambda(A) = \mathbb{P}\{Y_\lambda \in A\}\,, \tag{1.97}$$

that is, ν_λ is the distribution of the random variable Y_λ. One can deduce from this that the self-similar measure ν_λ is the infinite convolution product of the measures $\frac{1}{2}(\delta_{\lambda^n}+\delta_{-\lambda^n})$. This is the reason that we call ν_λ the *(infinite) Bernoulli convolution.* Problem 1.8.15 is still not completely solved. It was proved by Erdős if $1/\lambda$ is a PV (or Pisot) numbers, then the measure ν_λ is singular [**Erd39**]. See Section D for the definition and basic properties of PV numbers; the golden ratio $\frac{1+\sqrt{5}}{2}$ is the best known example. The measure ν_λ for $\lambda = \frac{\sqrt{5}-1}{2}$ is sometimes called the "Erdős measure".

On the other hand, Erdős proved [**Erd40**] that there is $t < 1$ such that $\nu_\lambda \ll \mathcal{L}^1$ for almost all $\lambda \in (t, 1)$. (In fact, this t in Erdős' paper was very close to 1 and not estimated explicitly.) It was conjectured that $\nu_\lambda \ll \mathcal{L}$ holds for almost all $\lambda \in (1/2, 1)$, and this was proved by Solomyak [**Sol95**]. We give a complete proof of these results in Chapters 6 and 8. More recently, further dramatic progress was achieved, in particular, in the works of Hochman [**Hoc14b**], Shmerkin [**Shm14**, **Shm19**], and Varjú [**Var19a**]. These later results will be discussed, but mostly without proofs.

Another reason why the family of self-similar measures ν_λ attracted so much attention, is their connection with dynamical systems, discovered in the 1980's-1990's. We refer the reader to the original papers by Alexander and Yorke [**AY84**], Przytycki and Urbański [**PU89**], and Ledrappier [**Led92**].

1.9. Dimension of measures

1.9.1. Local dimension of a measure and exact dimensionality.

DEFINITION 1.9.1. Let μ be a finite Borel measure on a metric space and $x \in \operatorname{spt}(\mu)$. The local dimension of the measure μ at x is defined by

$$d_\mu(x) := \lim_{r\to 0} \frac{\log \mu(B(x,r))}{\log r}, \tag{1.98}$$

if the limit exists. Otherwise, we take lim inf and lim sup instead of lim and obtain the lower local dimension $\underline{d}_\mu(x)$ and the upper local dimension $\overline{d}_\mu(x)$ respectively.

DEFINITION 1.9.2. A finite Borel measure μ on a metric space is called *exact-dimensional* of dimension α if the local dimension $d_\mu(x)$ exists and is equal to α for μ-a.e. x.

It is a useful technicality that one can calculate the local dimension via exponential subsequences, just like the box dimension.

LEMMA 1.9.3. *For every finite Borel measure μ on a metric space and $x \in \operatorname{spt}(\mu)$, and for every strictly decreasing sequence r_n such that $r_n \to 0$ and $\frac{\log r_{n+1}}{\log r_n} \to 1$ as $n \to \infty$, we have*

$$\underline{d}_\mu(x) = \liminf_{n\to\infty} \frac{\log \mu(B(x,r_n))}{\log r_n} \quad \text{and} \quad \overline{d}_\mu(x) = \limsup_{n\to\infty} \frac{\log \mu(B(x,r_n))}{\log r_n}.$$

The proof is left to the reader.

1.9.2. Hausdorff and packing dimension of a measure.

DEFINITION 1.9.4. Let μ be a finite Borel measure μ on a metric space. The Hausdorff and the packing dimensions of μ are defined as follows:

$$\dim_{\mathrm{H}}(\mu) := \inf\{\dim_{\mathrm{H}}(E) : \mu(E^c) = 0\}. \tag{1.99}$$

and

$$\dim_{\mathrm{P}}(\mu) := \inf\{\dim_{\mathrm{P}}(E) : \mu(E^c) = 0\}. \tag{1.100}$$

We can compute both the Hausdorff and the packing dimensions of a measure in $\mathbb{R}^d$ in terms of its lower and upper local dimensions respectively:

THEOREM 1.9.5. *Let μ be a Borel probability measure in $\mathbb{R}^d$. Then*

(i) $\dim_{\mathrm{H}}(\mu) = \operatorname{ess\,sup}_{x\sim\mu} \underline{d}_\mu(x) = \inf\{\alpha : \underline{d}_\mu(x) \leqslant \alpha$ *for μ-almost all x*$\}$;
(ii) $\dim_{\mathrm{P}}(\mu) = \operatorname{ess\,sup}_{x\sim\mu} \overline{d}_\mu(x) = \inf\{\alpha : \overline{d}_\mu(x) \leqslant \alpha$ *for μ-almost all x*$\}$;
(iii) *if $\mu(E) = 1$ and $\underline{d}_\mu(x) = \alpha$ holds for all $x \in E$, then* $\dim_{\mathrm{H}}(E) = \alpha$;
(iv) *if $\mu(E) = 1$ and $\overline{d}_\mu(x) = \alpha$ holds for all $x \in E$, then* $\dim_{\mathrm{P}}(E) = \alpha$.

The proof follows from Theorem 1.4.20, see [**PU10**, Section 7.6].

The quantities $\dim_{\mathrm{H}}(\mu)$ and $\dim_{\mathrm{P}}(\mu)$ are sometimes called the *upper* Hausdorff and packing dimension of a measure respectively, to be distinguished from the *lower* Hausdorff and *lower* packing dimension of a measure, defined by

$$\underline{\dim}_{\mathrm{H}}(\mu) = \inf\{\dim_{\mathrm{H}}(A) : \mu(A) > 0\} \tag{1.101}$$

and

$$\underline{\dim}_{\mathrm{P}}(\mu) = \inf\{\dim_{\mathrm{P}}(A) : \mu(A) > 0\}.$$

One can show that

$$\underline{\dim}_{\mathrm{H}}(\mu) = \operatorname*{ess\,inf}_{x\sim\mu} \underline{d}_\mu(x) \tag{1.102}$$

and

$$\underline{\dim}_{\mathrm{P}}(\mu) = \operatorname*{ess\,inf}_{x\sim\mu} \overline{d}_\mu(x),$$

see [**PU10**, Theorem 7.6.5].

REMARK 1.9.6. If a measure μ is exact dimensional then

$$\underline{\dim}_{\mathrm{H}}(\mu) = \dim_{\mathrm{H}}(\mu) = \underline{\dim}_{\mathrm{P}}(\mu) = \dim_{\mathrm{P}}(\mu).$$

By (1.99) and (1.101), it is easy to see that if $\nu \ll \mu$ then

$$\dim_{\mathrm{H}}(\mu) \geqslant \dim_{\mathrm{H}}(\nu) \text{ and } \underline{\dim}_{\mathrm{H}}(\nu) \geqslant \underline{\dim}_{\mathrm{H}}(\mu).$$

Several other dimensions of measures are frequently used, among them the *correlation dimension* and *entropy dimension.*

1.9.3. Correlation dimension. Let μ be a Borel probability measure on a metric space (X, ϱ). For $\varepsilon > 0$ define

$$C_2(\mu, \varepsilon) = \int \mu(B(x, \varepsilon))\, d\mu(x).$$

DEFINITION 1.9.7. *The lower correlation dimension of μ is defined by*

$$D_2(\mu) = \liminf_{\varepsilon\to 0} \frac{\log C_2(\mu, \varepsilon)}{\log \varepsilon}.$$

Replacing lim inf *by* lim sup *we obtain the upper correlation dimension of μ, denoted $\overline{D}_2(\mu)$. If the limit exists, we say that the correlation dimension exists.*

The reason we emphasize the lower correlation dimension is an alternative, potential-theoretic definition. Recall the notion of s-energy of μ:

$$\mathcal{E}_s(\mu) = \int\int \frac{d\mu(x)\, d\mu(y)}{\varrho(x, y)^s}.$$

PROPOSITION 1.9.8 (Sauer and Yorke). *For a Borel probability measure μ on a metric space X, the lower correlation dimension is*

$$D_2(\mu) = \sup\{s \geqslant 0 : \ \mathcal{E}_s(\mu) < \infty\} = \inf\{s \geqslant 0 : \ \mathcal{E}_s(\mu) = \infty\}.$$

For the proof see [**SY97**, Proposition 2.3]. A frequently used method to estimate the Hausdorff dimension of a measure is based on the following lemma:

LEMMA 1.9.9. *Let μ be a finite Borel measure on $\mathbb{R}^d$. Then*

$$\dim_{\mathrm{H}}(\mu) \geqslant \sup\left\{s \geqslant 0 : \iint |x-y|^{-s}\, d\mu(x)\, d\mu(y) < \infty\right\} = D_2(\mu).$$

PROOF. This follows immediately from Lemma 1.4.15. □

The following table contains some of the important properties of the correlation, Hausdorff, and packing dimensions, based on the results of Mattila, Moran and Rey paper [**MMR00**]. All the measures ν, μ, μ_k in the table are nonnull, finite Borel measures on $\mathbb{R}^d$ and $g : \mathbb{R}^d \to \mathbb{R}^d$. The way to use the table is as follows: in the leftmost column some properties are listed. Each contains the word dim. If we

TABLE 2. Basic properties of different dimension concepts of measures

	$\underline{\dim}_{\mathrm{H}}$	$\dim_{\mathrm{H}}$	$\underline{\dim}_{\mathrm{P}}$	$\dim_{\mathrm{P}}$	$D_2(\cdot)$	$\overline{D}_2(\cdot)$
$\mu \ll \nu \Longrightarrow \dim(\mu) \geqslant \dim(\nu)$	Yes	No	Yes	No	No	No
$\mu \ll \nu \Longrightarrow \dim(\mu) \leqslant \dim(\nu)$	No	Yes	No	Yes	No*	No*
$(\mu \ll \nu \ \& \ \frac{d\mu}{d\nu} \in L^{\infty}) \Longrightarrow \dim(\mu) \geqslant \dim(\nu)$	Yes	Yes	Yes	Yes	Yes	Yes
g Lipschitz $\Longrightarrow \dim(g_*\mu) \leqslant \dim(\mu)$	Yes	Yes	Yes	Yes	Yes	Yes
$\dim(\mu) \leqslant d$	Yes	Yes	Yes	Yes	Yes	Yes
$\mu \ll \mathcal{L}^d \Longrightarrow \dim(\mu) = d$	Yes	Yes	Yes	Yes	No	No
$(\mu \ll \mathcal{L}^d \ \& \ d\mu/d\mathcal{L}^d \in L^{\infty}) \Longrightarrow \dim(\mu) = d$	Yes	Yes	Yes	Yes	Yes	Yes
μ is a dicrete measure $\Longrightarrow \dim(\mu) = 0$	Yes	Yes	Yes	Yes	Yes	Yes
$\dim(\nu + \mu) = \min\{\dim(\nu), \dim(\mu)\}$	Yes	No	Yes	No	Yes	No
$\dim(\nu + \mu) = \max\{\dim(\nu), \dim(\mu)\}$	No	Yes	No	Yes	No*	No*
$\dim(\sum_{k=1}^{\infty} \mu_k) = \inf_{k\in\mathbb{N}} \dim(\mu_k)$	Yes	No	Yes	No	No	No
$\dim(\sum_{k=1}^{\infty} \mu_k) = \sup_{k\in\mathbb{N}} \dim(\mu_k)$	No	Yes	No	Yes	No*	No*

replace dim with the dimension which corresponds to a given column, the table shows if the property holds for that dimension or not.

We note that the properties marked with the star were not given in [**MMR00**]; however, those are not hard to see, and for completeness, we present them here. Let $\mu := \mathcal{L}^1|_{[0,1]}$ and $\nu := \mathcal{L}^1|_{[0,1]} + \delta_2$, where δ_2 is the Dirac mass supported on 2. Clearly, $\mu \ll \nu$ and $\nu = \mu + \delta_2$. Simple calculations using Definition 1.9.7 show that $D_2(\mu) = \overline{D}_2(\mu) = 1$ and $D_2(\nu) = \overline{D}_2(\nu) = D_2(\delta_2) = 0$, hence the properties marked with the star fail.

1.9.4. Similarity dimension of a self-similar measure. Here we assume that $\mathcal{S}$ is a self-similar IFS satisfying the SSP (see Definition 1.5.2 (i)) given in the form (1.2) and μ is a Bernoulli measure, that is, $\mu_{\mathbf{p}}$ is the infinite product measure $\mu_{\mathbf{p}} = \mathbf{p}^{\mathbb{N}}$ for a probability vector $\mathbf{p} = (p_1, \ldots, p_m)$. We write ν for the corresponding self-similar (push-forward) measure, that is, $\nu = \Pi_*(\mu_{\mathbf{p}})$, where Π is the natural projection from the symbolic spave. Then, as it will be shown in Section 3.2, it follows from the Strong Law of Large Numbers that ν is exact-dimensional and the local dimension is given by the following formula:

$$d_\nu(x) = \frac{-\sum_{i=1}^{m} p_i \log p_i}{-\sum_{i=1}^{m} p_i \log r_i} \quad \text{for } \nu\text{-a.e. } x, \tag{1.103}$$

where $\mathbf{r} = (r_1, \ldots, r_m)$ is the vector of contraction ratios of the maps from the IFS. The expression

$$h_{\mathbf{p}} := -\sum_{i=1}^{m} p_i \log p_i \tag{1.104}$$

in the numerator is known as the *entropy* of the probability vector $\mathbf{p}$ (and of the measure $\mu_{\mathbf{p}}$). The sum in the denominator

$$\chi_{\mathbf{r}}^{\mathbf{p}} := -\sum_{i=1}^{m} p_i \log r_i \tag{1.105}$$

is called the *Lyapunov exponent*; it represents the average contraction rate of the probabilistic IFS. Observe that if ν is the natural measure for $\mathcal{S}$ then $d_\mu(x) \equiv s$, where s is the similarity dimension of $\mathcal{S}$. The ratio "entropy over the Lyapunov exponent" plays an important role even if $\mathcal{S}$ does not satisfies the SSP, and in much greater generality, e.g., for conformal IFS, see Chapter 14.

DEFINITION 1.9.10. Let ν be a self-similar measure corresponding to the self-similar IFS $\mathcal{S}$ with contraction ratios $\{r_i\}_{i=1}^m$ and probability vector $\mathbf{p}=(p_1,\ldots,p_m)$. The *similarity dimension of* ν is defined by

$$\dim_{\mathrm{S}}(\nu) := \frac{-\sum_{i=1}^{m} p_i \log p_i}{-\sum_{i=1}^{m} p_i \log r_i} = \frac{h_{\mathbf{p}}}{\chi_{\mathbf{r}}^{\mathbf{p}}}. \tag{1.106}$$

Here $h_{\mathbf{p}}$ is the *entropy* of the Bernoulli measure $\mathbf{p}^{\mathbb{N}}$ and $\chi_{\mathbf{r}}^{\mathbf{p}}$ is the *Lyapunov exponent* of the probabilistic IFS $(\mathcal{S}, \mathbf{p})$.

Using Theorem 1.9.5(i) and (1.103) we obtain:

COROLLARY 1.9.11. *Let $\mathcal{S}$ be a self-similar IFS on $\mathbb{R}^d$ satisfying the SSP, with contraction ratios given by the vector $\mathbf{r} = (r_1,\ldots,r_m)$. Let ν be the self-similar measure for $\mathcal{S}$ with a probability vector $\mathbf{p} = (p_1,\ldots,p_m)$. Then we have*

$$\underline{\dim}_{\mathrm{H}}(\nu) = \dim_{\mathrm{P}}(\nu) = \dim_{\mathrm{S}}(\nu) = \frac{h_{\mathbf{p}}}{\chi_{\mathbf{r}}^{\mathbf{p}}}. \tag{1.107}$$

The same is true if we assume that the OSC holds instead of the SSP, see for example [**Edg98**, Theorem 5.2.5]. We will prove this in Section 11.3, see Corollary 11.3.8. On the other hand, without assuming any separation conditions, the inequality $\dim_{\mathrm{H}}(\nu) \leqslant \dim_{\mathrm{S}}(\nu)$ holds, see Theorem 3.2.1 below. The analogous problem is much harder for self-affine measures, see Chapter 10.

1.9.5. Ergodic measures on the symbolic space. The notion of ergodicity is used rarely in the early chapters of this book. Nevertheless, here we present the definition (for measures on Σ) and two important properties of ergodic measures. Ergodicity will be treated in Chapter 9 in more detail.

DEFINITION 1.9.12. Let μ be a Borel probability measure on $\Sigma := \{1,\ldots,m\}^{\mathbb{N}}$ and let σ be the left shift (introduced in (1.14)).

(a) We say that μ is a *σ-invariant measure* if $\mu(\sigma^{-1}A) = \mu(A)$ for all Borel sets $A \subset \Sigma$.
(b) A Borel set $E \subset \Sigma$ is *σ-invariant* (or simply invariant) if $\sigma^{-1}E = E$. If an orbit $\{\sigma^n \mathbf{i}\}_{n=0}^{\infty}$, $\mathbf{i} \in \Sigma$, intersects an invariant set, then this orbit is entirely contained in this invariant set.
(c) The measure μ is *ergodic* if for every invariant set E holds $\mu(E) \in \{0, 1\}$.

We proved in Lemma 1.3.5 that for every probability vector $\mathbf{p} = (p_1, \ldots, p_m)$, the infinite product measure $\mathbf{p}^{\mathbb{N}}$ on Σ is σ-invariant. In fact, it is ergodic; see, e.g., [**Wal82**, Theorem 1.12] for a proof. Alternatively, this can be shown by Kolmogorov's zero-one law [**Dur96**, p. 62]. It follows from the Birkhoff Ergodic Theorem (Theorem 9.1.1) that for every ergodic σ-invariant measure μ on Σ and $f \in L^1(\mu)$ we have

$$\lim_{n\to\infty} \frac{1}{n} \sum_{k=0}^{n-1} f(\sigma^k \mathbf{i}) = \int f d\mu \quad \text{for } \mu\text{-a.e. } \mathbf{i}. \tag{1.108}$$

DEFINITION 1.9.13. Fix $m \geqslant 2$, $d \geqslant 1$, and an IFS $\mathcal{S} \in \mathscr{S}_m(d)$. Let μ be a σ-invariant ergodic Borel probability measure on Σ. The push-forward measure $\nu := \Pi_* \mu$ is called an *ergodic invariant measure for the IFS $\mathcal{S}$*.

Thus each self-similar measure for the IFS $\mathcal{S}$ is an ergodic invariant measure for $\mathcal{S}$. The following important result was obtained by Feng and Hu [**FH09**, Theorem 2.8].

THEOREM 1.9.14 (Feng and Hu). *Every ergodic invariant measure for a self-similar IFS is exact-dimensional.*

We will give a proof in the special case of self-similar homogeneous measures on the line in Section 11.3.

1.9.6. Entropy. The entropy $h_{\mathbf{p}} := -\sum_{i=1}^m p_i \log p_i$ appeared in (1.106), in the numerator of the formula for the similarity dimension of a self-similar measure. The notion of entropy, in different guises, plays an important role in the book, and so we present here some of its basic properties. For additional information, see Section 9.5 and any of the classical books on Ergodic Theory, including [**Wal82, Bow75, Pet83, PU10**].

We start with the *Shannon entropy* of a random variable X, with distinct outcomes $x_1, \ldots, x_m$, occuring with probabilities $\mathbb{P}(x_i)$ for $i = 1, \ldots, m$, given by

$$H(X) = -\sum_{i=1}^m \mathbb{P}(x_i) \log \mathbb{P}(x_i).$$

Notice that the Shannon entropy depends only on the probabilities of the distinct outcomes, and not on the outcomes themselves, so we have $H(X) = h_{\mathbf{p}}$, where $\mathbf{p} = (p_1, \ldots, p_m)$, with $p_i = \mathbb{P}(x_i)$. The intuitive meaning of the Shannon entropy is the amount of uncertainty in the outcome of the random variable. In measure-theoretic language we have, for a discrete probability measure $\mathfrak{m} = \sum_{i=1}^m p_i \delta_{x_i}$:

$$H(\mathfrak{m}) = -\sum_{i=1}^m \mathfrak{m}(\{x_i\}) \log \mathfrak{m}(\{x_i\}) = h_{\mathbf{p}}. \tag{1.109}$$

We will need yet another variant, called *entropy of a partition*.

DEFINITION 1.9.15. Let $(X, \mathcal{B}, \mu)$ be a probability measure space and $\xi = \{C_1, \ldots, C_m\}$ a measurable partition, that is, the sets $C_i \in \mathcal{B}$ are pairwise disjoint and $\bigcup_{i=1}^m C_i = X$. The entropy of the partition ξ is defined by

$$H(\mu, \xi) = -\sum_{i=1}^m \mu(C_i) \log \mu(C_i) = h_{\mathbf{p}}, \quad \text{where } \mathbf{p} = (\mu(C_1), \ldots, \mu(C_m)).$$

If the measure μ is fixed, we usually drop it from notation and simply write $H(\xi) = H(\mu, \xi)$.

Entropy provides a useful tool for estimating some kind of a dimension characteristic of a measure, by considering a sequence of finer and finer partitions and computing the (normalized) entropy of those. We will encounter many instances of this idea.

The proof of the following lemma is an exercise, or see [**Wal82**, Theorem 4.2].

LEMMA 1.9.16. *Consider the function* $\varphi : [0, \infty) \to \mathbb{R}$ *defined by*

$$\varphi(x) := \begin{cases} x \log \frac{1}{x}, & \text{if } x > 0; \\ 0, & \text{if } x = 0. \end{cases}$$

This function is strictly concave on $[0, 1]$, *and thus*

$$\varphi\Big(\sum_{i=1}^{n} a_i x_i\Big) \geqslant \sum_{i=1}^{n} a_i \varphi(x_i) \quad \text{whenever } a_i > 0, \ \sum_{i=1}^{n} a_i = 1, \quad \text{and } x_i \in [0, 1]. \tag{1.110}$$

Moreover, the equality in (1.110) *holds if and only if all* x_i *are equal to each other.*

Now the following is immediate.

COROLLARY 1.9.17. *Let* $(X, \mathcal{B}, \mu)$ *be a probability measure space and* $\xi = \{C_1, \ldots, C_m\}$ *a measurable partition. Then* $H(\mu, \xi) \leqslant \log m$, *with equality if and only if* $\mu(C_i) = \frac{1}{m}$ *for all* $i = 1, \ldots, m$.

Given two partitions ξ and η of the same space, we say that η is *finer* than ξ, and write $\xi \leqslant \eta$, if every element of ξ is a union of some elements of η. Further, we denote by $\xi \vee \eta$ the *join* of the partitions:

$$\xi \vee \eta = \{C_i \cap D_j\}_{i \leqslant m}^{j \leqslant k} \quad \text{where } \xi = \{C_1, \ldots, C_m\} \text{ and } \eta = \{D_1, \ldots, D_k\}.$$

In the proof of the following, very useful, property, we follow the exposition of [**Bow75**, Section 1.D].

LEMMA 1.9.18. *For any two finite measurable partitions* ξ *and* η,

$$H(\xi \vee \eta) \leqslant H(\xi) + H(\eta). \tag{1.111}$$

PROOF. Using that $\sum_j \mu(C_i \cap D_j) = \mu(C_i)$, since $\{D_j\}_{j \leqslant k}$ is a partition, we obtain

$$\begin{aligned} H(\xi \vee \eta) - H(\xi) &= -\sum_{i,j} \mu(C_i \cap D_j) \log \mu(C_i \cap D_j) + \sum_i \mu(C_i) \log \mu(C_i) \\ &= -\sum_{i,j} \mu(C_i \cap D_j) \log \frac{\mu(C_i \cap D_j)}{\mu(C_i)} \\ &= \sum_j \sum_i \mu(C_i) \left(\frac{-\mu(C_i \cap D_j)}{\mu(C_i)} \log \frac{\mu(C_i \cap D_j)}{\mu(C_i)} \right). \end{aligned} \tag{1.112}$$

Applying (1.110) to $a_i = \mu(C_i)$ and $x_i = \frac{\mu(C_i \cap D_j)}{\mu(C_i)}$, we get

$$\sum_i \mu(C_i) \varphi\left(\frac{\mu(C_i \cap D_j)}{\mu(C_i)}\right) \leqslant \varphi\left(\sum_i \mu(C_i \cap D_j)\right) = \varphi(\mu(D_j)).$$

Thus, $H(\xi \vee \eta) - H(\xi) \leqslant \sum_j \varphi(\mu(D_j)) = H(\eta)$, as desired. $\square$

REMARK 1.9.19.

(a) If the partitions ξ and η are *independent* with respect to the measure μ, that is, $\mu(C_i \cap D_j) = \mu(C_i)\mu(D_j)$, then (1.111) becomes equality, that is, $H(\xi \vee \eta) = H(\xi) + H(\eta)$. This follows from (1.112).

(b) From the elementary inequality $\varphi(\sum_{i=1}^k x_i) \leqslant \sum_{i=1}^k \varphi(x_i)$ for $k \geqslant 2$, $x_i > 0$, with $\sum_{i=1}^k x_i \leqslant 1$, it follows that

$$\xi \leqslant \eta \implies H(\xi) \leqslant H(\eta). \tag{1.113}$$

For the symbolic space $\Sigma = \{1, \ldots, m\}^{\mathbb{N}}$ there is a natural increasing sequence of partitions $\xi_n = \{[\mathbf{i}] : |\mathbf{i}| = n\}$. It is clear that $\xi_n \leqslant \xi_m$ for $m > n$.

PROPOSITION 1.9.20. *Let μ be a shift-invariant probability measure on Σ. Then the following limit exists:*

$$h_\mu := \lim_{n\to\infty} \frac{1}{n} H(\mu, \xi_n) = \lim_{n\to\infty} -\frac{1}{n} \sum_{|\mathbf{i}|=n} \mu([\mathbf{i}]) \log \mu([\mathbf{i}]). \tag{1.114}$$

Moreover, the sequence $n \mapsto \frac{1}{n} H(\mu, \xi_n)$ is nonincreasing.

The number h_μ is called the *entropy* of the σ-invariant measure μ. The fact that this definition agrees with a general definition of entropy for a measure-preserving system follows from the Kolmogorov-Sinai Theorem 9.5.1. We will show the existence of the limit in (1.114); for the proof of monotonicity see the comment at the beginning of Section 9.5.2 or [**Wal82**, Theorem 4.10].

PROOF OF (1.114). Since μ is fixed, we will write $H(\xi) = H(\mu, \xi)$ and show that the sequence $\{H(\xi_n)\}_{n \geqslant 1}$ is *subadditive*, that is,

$$H(\xi_{\ell+n}) \leqslant H(\xi_\ell) + H(\xi_n), \tag{1.115}$$

and then apply the following:

LEMMA 1.9.21 (Fekete's Lemma). *Let a_n be a subadditive sequence of real numbers, that is, for every $\ell, n \in \mathbb{N}$*

$$a_{\ell+n} \leqslant a_\ell + a_n.$$

Then the limit of the sequence a_n/n exists and

$$\lim_{n\to\infty} \frac{a_n}{n} = \inf_{n\in\mathbb{N}} \frac{a_n}{n}.$$

Note that the infinum on the right-hand side might be $-\infty$.

PROOF. It is enough to show that $\limsup_{n\to\infty} a_n/n \leqslant \inf_n a_n/n$. Let $q \in \mathbb{N}$ be arbitrary but fixed. Then for every $n \in \mathbb{N}$ there exists $k_n \in \mathbb{N}$ and $p_n \in \{0, \ldots, q-1\}$ such that $n = k_n q + p_n$. Hence, using the assumption on the sequence a_n,

$$\frac{a_n}{n} = \frac{a_{k_n q + p_n}}{k_n q + p_n} \leqslant \frac{k_n a_q + a_{p_n}}{k_n q + p_n} = \frac{a_q + a_{p_n}/k_n}{q + p_n/k_n}.$$

Since p_n is bounded and $k_n \to \infty$ as $n \to \infty$, we get that

$$\limsup_{n\to\infty} \frac{a_n}{n} \leqslant \frac{a_q}{q}.$$

Since $q \in \mathbb{N}$ was arbitrary, the claim follows. □

It remains to prove the subadditivity (1.115). Along with the partition ξ_n into cylinder sets $[\mathbf{i}]$ for $|\mathbf{i}| = n$, we consider the partition $\sigma^{-\ell}\xi_n$ for $\ell \geqslant 1$. Note that

$$\sigma^{-\ell}[\mathbf{i}] = \{\mathbf{j} \in \Sigma : j_{\ell+1} \dots j_{\ell+n} = i_1 \dots i_n\} \text{ for } \mathbf{i} \in \Sigma_n.$$

It follows that $\xi_{\ell+n} = \sigma^{-\ell}\xi_n \vee \xi_\ell$. Thus we have by Lemma 1.9.18,

$$H(\xi_{\ell+n}) = H(\sigma^{-\ell}\xi_n \vee \xi_\ell) \leqslant H(\sigma^{-\ell}\xi_n) + H(\xi_\ell).$$

However, $H(\sigma^{-\ell}\xi_n) = H(\xi_n)$ by the definition of entropy and σ-invariance of μ. The inequality (1.115) follows, and the proof is complete. □

EXAMPLE 1.9.22. Let $\mu = \mathbf{p}^{\mathbb{N}}$ be the infinite product (Bernoulli) measure of Σ for a probability vector $\mathbf{p} = (p_1, \dots, p_m)$. Then an easy computation shows that $H(\mu, \xi_n) = -n\sum_{i=1}^n p_i \log p_i = nh_{\mathbf{p}}$ for all $n \in \mathbb{N}$, and so $h_\mu = h_{\mathbf{p}}$, as was mentioned above. (In fact, $H(\mu, \xi_n) = nh_{\mathbf{p}}$ follows from Remark 1.9.19, since $\sigma^{-\ell}\xi_n$ and σ_ℓ are independent partitions with respect to the product measure μ.)

In the Euclidean space one can consider an increasing sequence of partitions using grid boxes. For example, let $\mathscr{D}_n$ be the partition of $\mathbb{R}^d$ into dyadic grid boxes:

$$\mathscr{D}_n := \prod_{i=1}^d [k_i 2^{-n}, (k_i + 1)2^{-n}), \quad \text{with } k_i \in \mathbb{Z}. \tag{1.116}$$

It is immediate that $\mathscr{D}_n \leqslant \mathscr{D}_\ell$ for $\ell > n$. Given a probability measure μ on $\mathbb{R}^d$, the *entropy dimension* of μ may be defined by

$$\dim_e(\mu) = \lim_{n\to\infty} \frac{-\sum_{D\in\mathscr{D}_n} \mu(D) \log \mu(D)}{n \log 2} = \lim_{n\to\infty} \frac{H(\mu, \mathscr{D}_n)}{n \log 2}, \tag{1.117}$$

if the limit exists; otherwise, taking lim inf and lim sup respectively, we obtain the lower and upper entropy dimensions $\underline{\dim}_e(\mu)$ and $\overline{\dim}_e(\mu)$. See Section 2.6 for more on entropy dimension. It turns out that the limit in (1.117) exists for any self-similar measure, without any separation assumptions; we will show this in Section 3.7.

1.10. Dimension of self-affine sets

Theorem 1.5.3 implies that for a self-similar IFS with contraction ratios $0 < r_1, \dots, r_m < 1$, the dimension (box or Hausdorff) of the attractor can be determined as the solution of the equation $r_1^s + \dots + r_m^s = 1$, if the cylinders are well separated. In particular, the relative positions of the cylinders cannot affect the dimension as long as the cylinders are disjoint. The following example shows that in the self-affine case the same does not hold.

EXAMPLE 1.10.1 (Falconer). Let $\mathcal{F}_1$ (respectively $\mathcal{F}_2$) be the self-affine IFS shown on the left (respectively right) side of Figure 1.16. More precisely, for some $0 < \tau < \frac{1}{2} < \lambda < 1$, let $A = \begin{bmatrix} \lambda & 0 \\ 0 & \tau \end{bmatrix}$, and let Λ_1 and Λ_2 be the attractors of the self-affine IFS

$$\mathcal{F}_1 := \left(A\mathbf{x},\ A\mathbf{x} + \begin{bmatrix} \varepsilon \\ 1-\tau \end{bmatrix}\right) \text{ and } \mathcal{F}_2 := \left(A\mathbf{x},\ A\mathbf{x} + \begin{bmatrix} 0 \\ 1-\tau \end{bmatrix}\right), \tag{1.118}$$

respectively.

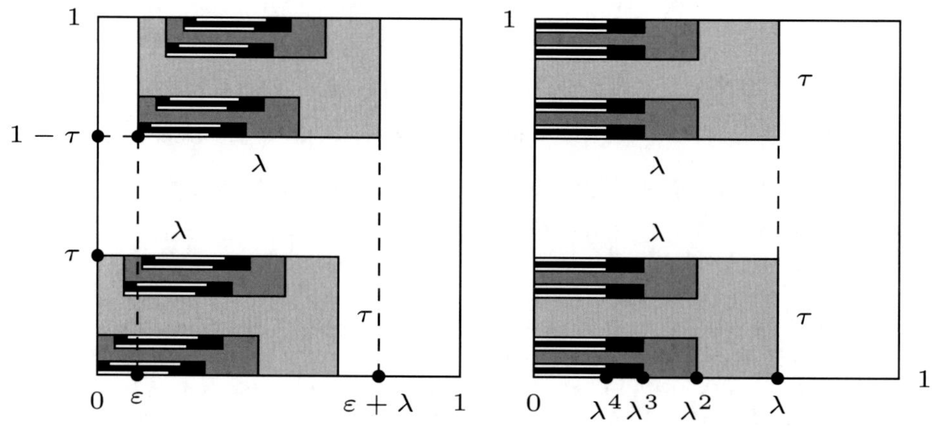

FIGURE 1.16. The attractors Λ_1 and Λ_2 on the left and right sides of the figure are constructed by maps contracting in the same way but, because of the difference between the relative positions of the cylinders we have $\dim_{\mathrm{H}}(\Lambda_2) = \frac{\log 2}{-\log \tau} < 1 \leqslant \dim_{\mathrm{H}}(\Lambda_1)$.

It is clear from Figure 1.16 that Λ_2 is a copy of the middle-$(1-2\tau)$ Cantor set on the y-axis. Therefore, $\dim_{\mathrm{H}}(\Lambda_2) = \frac{\log 2}{-\log \tau} < 1$. On the other hand, $\dim_{\mathrm{H}}(\Lambda_1) \geqslant 1$. Indeed, the projection π_1 to the x-axis is a Lipschitz map and $\pi_1(\Lambda_1) = [0, \frac{\varepsilon}{1-\lambda}]$. (The latter follows from the observation that $[0, \frac{\varepsilon}{1-\lambda}]$ is the attractor of the IFS $(\lambda x,\ \lambda x + \varepsilon)$ on the line.) Therefore, by Lemma 1.4.7 (**EP4**) we have that $\dim_{\mathrm{H}}(\Lambda_1) \geqslant 1$.

So we see that the relative positions of cylinders may affect the dimension in the self-affine case even if the cylinders are disjoint. This makes the dimension theory of self-affine IFS more difficult. For the rest of the section we use the notation of Section 1.2 and write Λ for the attractor of the self-affine IFS

$$\mathcal{F} := \left(f_i(\mathbf{x}) = A_i\mathbf{x} + \mathbf{t}_i\right)_{i=1}^{m}, \qquad \mathbf{x} \in \mathbb{R}^d. \tag{1.119}$$

We always assume that the maps $f_i : \mathbb{R}^d \to \mathbb{R}^d$, for $i \leqslant m$, are contractions, that is, the Euclidean operator norm of the nonsingular matrices $\|A_i\|$ is less than one for all i. In this section, first we introduce the notion of *affinity dimension*, which is the most natural guess for the dimension of a self-affine set. The affinity dimension is equal to the similarity dimension in the special case when the self-affine IFS is self-similar (that is, all of the matrices are of the form $A_i = r_i \cdot Q_i$ where $0 < r_i < 1$ and the matrix Q_i is orthogonal). Further, analogously to the similarity dimension for self-similar IFS, the affinity dimension is the natural upper bound on the Hausdorff dimension of the attractor Λ. The affinity dimension is defined in terms of singular values of some linear transformations.

Let $\overline{B}$ be the closed ball defined in (1.35). Then $f_i(\overline{B}) \subset \overline{B}$ and it follows from (1.36) that for every n the family of sets

$$\mathcal{E}_n := \left\{E_{i_1\dots i_n} = f_{i_1\dots i_n}(\overline{B}) : \ (i_1 \dots i_n) \in \{1 \dots m\}^n\right\}$$

is a cover of Λ by ellipsoids (ellipses in $\mathbb{R}^2$). In Figure 1.9, illustrating Example 1.2.1, we see $\mathcal{E}_1, \mathcal{E}_2$, and $\mathcal{E}_3$ as the collection of light gray, gray, and black ellipses respectively. For simplicity, assume that B is a unit ball (this can always be achieved by scaling, see Section 1.3.3). Then for an arbitrary $\mathbf{i} = (i_1, \ldots, i_n)$ the lengths of the mutually perpendicular half-axes of the ellipsoid $E_{\mathbf{i}}$ are the singular values

$$\alpha_1(A_{\mathbf{i}}) \geqslant \cdots \geqslant \alpha_d(A_{\mathbf{i}})$$

(see Appendix A.1). Thus the ellipsoid $E_{\mathbf{i}}$ can be covered by a congruent copy $R_{\mathbf{i}}$ of the d-dimensional box (parallelepiped) $[0, 2\alpha_1(\mathbf{i})] \times \cdots \times [0, 2\alpha_d(\mathbf{i})]$, in such a way that the edges of $R_{\mathbf{i}} \supset E_{\mathbf{i}}$ are parallel to the corresponding axes of $E_{\mathbf{i}}$. Since $|E_{\mathbf{i}}| < |R_{\mathbf{i}}| \leqslant \sqrt{d} \cdot |E_{\mathbf{i}}|$, the covering

$$\mathcal{R}_n := \{R_{\mathbf{i}} : \mathbf{i} = (i_1 \ldots i_n) \in \{1 \ldots m\}^n\} \tag{1.120}$$

is equivalent to $\mathcal{E}_n$ in regard to dimension calculation. Clearly, $|R_{\mathbf{i}}| < \delta_n := \sqrt{d} \cdot \max_i \|A_i\|^n$. To estimate the Hausdorff dimension of the attractor Λ we choose a small $\delta > 0$ (which is actually δ_n for a large n), and for $t > 0$ we want to estimate the highlighted sum in the definition of Hausdorff measure (1.51), for a cover which appears to be the most economical in the sense that it is close to the infimum. The most immediate idea could be to cover Λ by the d-dimensional boxes $\mathcal{R}_n$. However, in general, the upper estimate on the dimension obtained by substituting $\mathcal{R}_n$ for the cover in (1.51) would only give a crude upper bound. Therefore, we need to make one more step. For $1 \leqslant k \leqslant d$ we construct $\mathcal{C}_n^k$, a cover of Λ as follows:

Step 1: Cover Λ with $\mathcal{R}_n$.

Step 2: Then for all $|\mathbf{i}| = n$, partition $R_{\mathbf{i}} \in \mathcal{R}_n$ into the d-dimensional cubes $C_{\mathbf{i}}^1, \ldots, C_{\mathbf{i}}^{n_k(\mathbf{i})}$ (some of the cubes may reach out of $R_{\mathbf{i}}$) of side length $2\alpha_k(A_{\mathbf{i}})$, where

$$n_k(\mathbf{i}) := \prod_{\ell=1}^{k} \left\lceil \frac{\alpha_\ell(A_{\mathbf{i}})}{\alpha_k(A_{\mathbf{i}})} \right\rceil. \tag{1.121}$$

See Figure 1.17. If we substitute the cover $\{A_i\}$ in the definition of the Hausdorff dimension (1.51) by $\mathcal{C}_n^k = \{C_{\mathbf{i}}^r\}_{r=1}^{n_k(\mathbf{i})}$, then the sum in (1.51) is

$$\begin{aligned} S^t(\mathcal{C}_n^k) &:= \sum_{|\mathbf{i}|=n} \sum_{r=1}^{n_k(\mathbf{i})} |C_{\mathbf{i}}^r|^t \\ &= \sum_{|\mathbf{i}|=n} \underbrace{\prod_{\ell=1}^{k} \left\lceil \frac{\alpha_\ell(A_{\mathbf{i}})}{\alpha_k(A_{\mathbf{i}})} \right\rceil}_{n_k(\mathbf{i})} \cdot \Big(\underbrace{|\overline{B}| \cdot \alpha_k(A_{\mathbf{i}})}_{|C_{\mathbf{i}}^r|}\Big)^t \\ &\asymp \sum_{|\mathbf{i}|=n} \alpha_1(A_{\mathbf{i}}) \cdots \alpha_{k-1}(A_{\mathbf{i}}) \cdot \alpha_k(A_{\mathbf{i}})^{t-(k-1)}, \end{aligned} \tag{1.122}$$

where for $k = 1$, $\alpha_0(A_{\mathbf{i}})$ is assumed to be equal to 1. Thus we need to minimize the above expression over $1 \leqslant k \leqslant d$. An immediate calculation shows that for $0 < t \leqslant d$,

$$\begin{aligned} \varphi^t(A_{\mathbf{i}}) &:= \min_{1 \leqslant k \leqslant d} \alpha_1(A_{\mathbf{i}}) \cdots \alpha_{k-1}(A_{\mathbf{i}}) \cdot \alpha_k(A_{\mathbf{i}})^{t-(k-1)} \\ &= \alpha_1(A_{\mathbf{i}}) \cdots \alpha_{\lfloor t \rfloor}(A_{\mathbf{i}}) \alpha_{\lfloor t \rfloor+1}(A_{\mathbf{i}})^{t-\lfloor t \rfloor}. \end{aligned} \tag{1.123}$$

Thus, $\varphi^t(A_{\mathbf{i}})$ (called the *singular value function*) is the contribution of $R_{\mathbf{i}}$ to the sum in (1.51) when covering Λ with cubes corresponding to the (optimal) $\lceil t \rceil$-th largest side of $R_{\mathbf{i}}$. The definition of the singular value function is extended for $t > d$ continuously by

$$\varphi^t(A_{\mathbf{i}}) := (\alpha_1(A_{\mathbf{i}}) \cdots \alpha_d(A_{\mathbf{i}}))^{t/d} = |\det(A_{\mathbf{i}})|^{t/d}. \tag{1.124}$$

From (1.122) we obtain that if the cover in the definition of the Hausdorff dimension

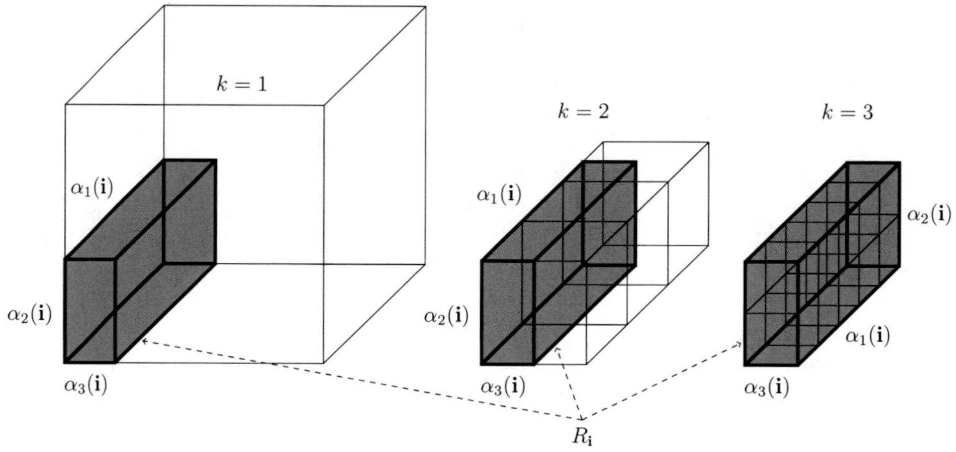

FIGURE 1.17. $R_{\mathbf{i}}$ and the cubes $C_{\mathbf{i}}^1, \dots, C_{\mathbf{i}}^{n_k(\mathbf{i})}$ for $k = 1,2,3$ respectively. In this case $n_1(\mathbf{i}) = 1, n_2(\mathbf{i}) = 3$, and $n_3(\mathbf{i}) = 10$.

(1.51) is $\mathcal{C}_n^{[t]}$, then the sum is

$$S^t(\mathcal{C}_n^{[t]}) \asymp \sum_{|\mathbf{i}|=n} \varphi^t(A_{\mathbf{i}}). \tag{1.125}$$

In the definition of the Hausdorff measure (1.51), $\delta \to 0$ corresponds to $n \to \infty$ in our case. So, according to (1.52), if $t_0 > 0$ satisfies

$$\forall t < t_0 : \lim_{n\to\infty} S^t(\mathcal{C}_n^{[t]}) = \infty \text{ and } \forall t > t_0, \lim_{n\to\infty} S^t(\mathcal{C}_n^{[t]}) = 0, \tag{1.126}$$

then this value of t_0 is a reasonable guess for the Hausdorff dimension of Λ.

DEFINITION 1.10.2. The *affinity dimension* of the self-affine IFS $\mathcal{F} = (A_i\mathbf{x} + \mathbf{t}_i)_{i=1}^m$ is defined by

$$s(A_1, \dots, A_m) := \dim_{\text{aff}}(\mathcal{F}) := \inf\Big\{t > 0 : \sum_{n=0}^{\infty} \sum_{|\mathbf{i}|=n} \varphi^t(A_{\mathbf{i}}) < \infty\Big\}. \tag{1.127}$$

We either denote it by $s(A_1, \dots, A_m)$ when we want to emphasize that the affinity dimension of $\mathcal{F}$ depends only on the linear parts of the mappings of $\mathcal{F}$, or by $\dim_{\text{aff}}(\mathcal{F})$ when we want a more compact expression for it. Note that the affinity dimension is sometimes called the *singularity dimension.*

Now the following is almost immediate from the above calculations:

PROPOSITION 1.10.3. *Let $\mathcal{F} = (A_i\mathbf{x} + \mathbf{t}_i)_{i=1}^m$ be a self-affine IFS in $\mathbb{R}^d$ with attractor Λ. Then*

$$\dim_{\text{H}}(\Lambda) \leqslant \min\{d, s(A_1, \dots, A_m)\}. \tag{1.128}$$

In fact, even $\overline{\dim}_B(\Lambda) \leqslant \min\{d, s(A_1, \ldots, A_m)\}$, and we will prove this in Proposition 10.2.6. A natural guess is that the inequality in (1.128) is actually an equality under the Strong Separation Property, but this cannot always hold. For example, for the two IFS shown in Figure 1.16 the attractors have different Hausdorff dimension, as shown above. However, their affinity dimensions are the same, since the size of their cylinders is the same, and the affinity dimension depends only on the size of the cylinders. Nevertheless, we can expect an "almost all"-type result here.

THEOREM 1.10.4. *For $m \geqslant 2$ we are given nonsingular $d \times d$ matrices $\{A_1, \ldots, A_m\}$ such that their Euclidean norm satisfies*

$$\|A_i\| < \frac{1}{2}, \qquad i = 1, \ldots, m. \tag{1.129}$$

For $\mathbf{t} := (\mathbf{t}_1, \ldots, \mathbf{t}_m) \in \mathbb{R}^d \times \cdots \times \mathbb{R}^d$ consider the m-parameter family of self-affine IFS acting on $\mathbb{R}^d$:

$$\mathcal{F}^{\mathbf{t}} := (A_i \mathbf{x} + \mathbf{t}_i)_{i=1}^m . \tag{1.130}$$

Let $\Lambda^{\mathbf{t}}$ be the attractor of $\mathcal{F}^{\mathbf{t}}$. Then for $\mathcal{L}^{md}$-almost all $\mathbf{t}$ we have

$$\dim_H(\Lambda^{\mathbf{t}}) = \dim_B(\Lambda^{\mathbf{t}}) = \min\{d, s(A_1, \ldots, A_m)\} . \tag{1.131}$$

This theorem was proved by Falconer [**Fal88**] with the upper bound on the norms equal to 1/3 instead of 1/2; it was shown in [**Sol98**] that a modification of the original Falconer's proof yields the result as stated. Edgar [**Edg92**] observed that the bound in (1.129) is sharp, in the sense that there exist self-affine IFS with norms of the linear maps less than $1/2 + \varepsilon$, for any $\varepsilon > 0$, such that for almost all $\mathbf{t}$ (in fact, for all except in a countable union of proper linear subspaces), the Hausdorff dimension of the attractor is strictly less than the box dimension. Note that Theorem 1.10.4 (as opposed to the Moran-Hutchinson Theorem about the dimension of self-similar sets) does not require any separation conditions. For self-similar IFS one may want to apply this theorem when the OSC does not hold (or we cannot check it) and when we are in $\mathbb{R}^d$ for $d > 1$, since on the line a stronger result is available, see Theorem 6.1.1. It was proved in [**SS02**] that the bound 1/2 imposed on the contraction ratios in Theorem 1.10.4 cannot be improved even in the self-similar case. To be precise, for all self-similar sets the Hausdorff dimension and the box dimension coincide, but it may happen that this value is less than the right-hand side of (1.131) for a.e. $\mathbf{t}$.

1.11. Self-affine sets with aligned cylinders

Self-affine IFS with aligned cylinders form an important and much studied class. One assumption is that all the linear maps A_i in (1.119) are diagonal (we call such IFS "diagonally self-affine"), to which other assumptions are added, usually indicating a certain alignment of cylinders. The best known among these is the class of *Bedford-McMullen carpets*, studied by Bedford [**Bed84**] and McMullen [**McM84**] independently. A typical example of such a carpet is shown in the leftmost picture in the top row of Figure 1.18. Here the linear maps $A_i = \begin{bmatrix} 1/m & 0 \\ 0 & 1/n \end{bmatrix}$ are all equal to each other. The IFS is genuinely self-affine if $n \neq m$; we assume that $n > m$. The shaded rectangles correspond to level-one cylinders (images of the unit square under the maps of the IFS). A detailed analysis of these carpets will be given

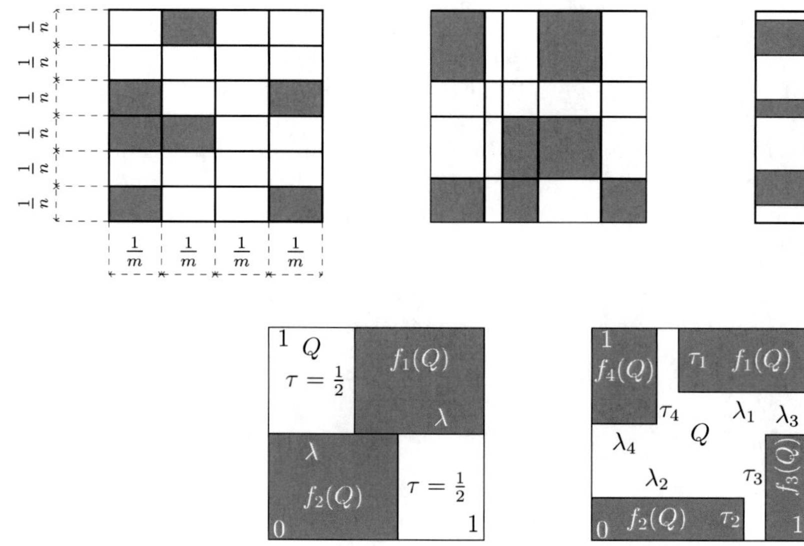

FIGURE 1.18. From left to right: In the first line, Bedford-McMullen, Barański, Gatzouras-Lalley carpets, and in the second line, Przytycki-Urbanski set and generalized four-corner set

in Section 11.5.2. Here we just mention an interesting fact (proved by Bedford and McMullen) that the Hausdorff dimension of the carpet is strictly less than the box dimension, whenever the number of shaded rectangles in "nonempty" columns is nonuniform, i.e., not constant. On the other hand, the box dimension is equal to the affinity dimension if and only if: either every column contains a chosen (shaded) rectangle, or alternatively, if every column contains at most one chose rectangle. Thus, these carpets fall into the exceptional set from Theorem 1.10.4. For further geometric properties of the Bedford-McMullen carpets, which will not discussed in Section 11.5.2, we direct the reader to the nice survey paper of Fraser [**Fra21**].

Other well-known self-affine carpets are Gatzouras-Lalley [**LG92**] and Barański [**Bar07**] carpets, shown (in special cases) in the middle and rightmost pictures in the top row of Figure 1.18. The dimension results for them are included in Section 11.5.3 and Section 11.5.4 without proof. Higher-dimensional generalizations of the carpets have also been considered, known as "self-affine sponges".

Another important class of diagonally self-affine sets is formed by *Przytycki-Urbański (PU)-sets* [**PU89**], shown in the leftmost picture in the bottom row of Figure 1.18. There, the IFS consists of only two maps, with the linear parts equal to $A_i = \begin{bmatrix} \lambda & 0 \\ 0 & 1/2 \end{bmatrix}$, $i = 1, 2$. As discovered in [**PU89**], there is a strong connection between this IFS and the Bernoulli convolution measure ν_λ, defined in Example 1.8.14. In fact, if $\dim_{\mathrm{H}}(\nu_\lambda) = 1$, which we now know holds for all transcendental $\lambda \in (\frac{1}{2}, 1)$ ([**Var19b**]), then the Hausdorff dimension of the PU-attractor equals its box dimension (which always equals the affinity dimension). On the other hand, if λ is a reciprocal of a PV-number (see Appendix D for the definition) then ν_λ is singular, and even has Hausdorff dimension strictly less than one [**Lal97**], and Przytycki and Urbański proved that the Hausdorff dimension of the corresponding self-affine attractor is strictly less than its box dimension. Since the set of reciprocals of PV-numbers has 1/2 as a limit point, these examples provided the source of the remark by Edgar [**Edg92**], mentioned in the previous section, to the effect that the norm bound of 1/2 in Theorem 1.10.4 cannot be impoved. The dimension theory of Przytycki-Urbański sets will be discussed in detail in Section 11.5.1. For the proof of their results, Przytycki and Urbański essentially used the Ledrappier-Young formula, which they had to adapt accordingly. This formula plays an important role in dimension theory of self-affine sets, see Chapter 11.

Another example of diagonally self-affine attractors, of a kind similar to PU-sets, but with three level one cylinders, is discussed in the next section.

1.11.1. Connection between box-like self-affine sets and self-similar attractors with overlaps. The main focus of this book is on self-similar sets and measures of overlapping constructions on one hand, and on self-affine sets and measures on the other hand. Here we point out an important connection between these two areas in a special case. A more general version of this argument is presented in Section 11.2.

EXAMPLE 1.11.1. Consider a one-parameter (the parameter is denoted by t) family of self-affine IFS $\mathcal{F}_t$ and the one-parameter family of self-similar IFS $\mathcal{S}_t$ generated by the projections of $I_i := \mathrm{proj}_x(f_i([0,1]^2))$ to the x-axis. More precisely, fix

$$0 < b < a < 1, \quad b < \frac{1}{3}, \quad b < h < h + b < 1 - b. \tag{1.132}$$

Using the diagonal matrix

$$A := \begin{bmatrix} a & 0 \\ 0 & b \end{bmatrix}, \tag{1.133}$$

we define a one-parameter family of self-affine IFS $\mathcal{F}_t$ with the attractor Λ_t on $\mathbb{R}^2$ as follows: $\mathcal{F}_t := (f_i)_{i=1}^3$, with $0 < t < a$, where

$$f_1(\mathbf{x}) := A\mathbf{x}, \quad f_2(\mathbf{x}) := A\mathbf{x} + \begin{bmatrix} t \\ h \end{bmatrix}, \quad \text{and} \quad f_3(\mathbf{x}) := A\mathbf{x} + \begin{bmatrix} 1-a \\ 1-b \end{bmatrix},$$

see Figure 1.19. This is a *box-like* self-affine IFS since its cylinders are axes-parallel boxes. Since $0 < b < a$, we say that direction x *dominates*. A simple calculation shows that (for every t) the affinity dimension is equal to

$$s = \begin{cases} \dfrac{\log 3}{-\log a}, & \text{if } 0 < a \leqslant \frac{1}{3}; \\ 1 + \dfrac{\log 3 + \log a}{-\log b}, & \text{if } \frac{1}{3} < a < 1. \end{cases} \tag{1.134}$$

The projection of the cylinders $\{f_i([0,1]^2)\}_{i=1}^3$ to the x-axis generates a one-parameter family of self-similar IFS on the line:

$$\mathcal{S}_t := \left(S_1(x) := ax,\ S_2(x) := ax + t,\ S_3(x) := ax + 1 - a\right).$$

Since $t < a$, the cylinder intervals $I_i := S_i[0,1]$, $i = 1, 2$, overlap. The natural measure on the attractor of $\mathcal{S}_t$ is the self-similar measure ν_t for the IFS $\mathcal{S}_t$ with the probabilities $\mathbf{p} = \left(\frac{1}{3}, \frac{1}{3}, \frac{1}{3}\right)$. Its similarity dimension (for every t) is

$$\dim_{\mathrm{S}}(\nu_t) = \min\left\{\frac{\log 3}{-\log a}, 1\right\}. \tag{1.135}$$

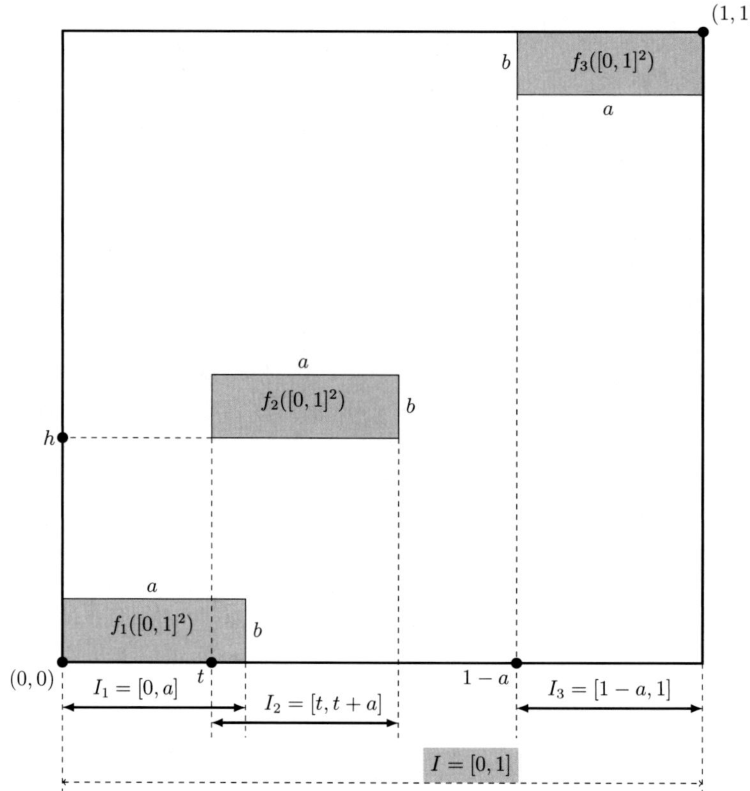

FIGURE 1.19. The level one cylinders of $\mathcal{F}$ of Example 1.11.1

A combination of the Feng–Hu Theorem [**FH09**] (see Theorem 11.2.1 below) and Hochman's Theorem [**Hoc14b**] (see Theorem 6.8.11 below, which is a variant of Theorem 1.6.6) yields that

$$\dim_{\mathrm{H}}\big(\{0 < t < a : \dim_{\mathrm{H}}(\Lambda_t) \neq s\}\big) = 0. \tag{1.136}$$

More precisely, we obtain from Feng-Hu Theorem that

$$\dim_{\mathrm{H}}(\Lambda_t) = s$$

whenever

$$\dim_{\mathrm{H}}(\nu_t) = \min\left\{\frac{\log 3}{-\log a}, 1\right\}. \tag{1.137}$$

On the other hand, Hochman's Theorem implies that the set of those $t \in (0, a)$ for which (1.137) does not hold has packing dimension 0.

1.12. Self-affine sets without alignment

As we have seen, self-affine sets with aligned cylinders might have strictly smaller Hausdorff (and box counting) dimension than the affinity dimension even without overlaps between cylinders. Essentially, this phenomenon occurs since the projection along the commonly preserved directions can have very heavy overlapping structure, which causes drop of the dimension in the projection, and this is inherited by the original system. This strong connection between the projection and original system is provided by the Feng-Hu Theorem 11.2.1 for aligned systems, that is, when the matrices are diagonal. A natural question is: can similar phenomena occur when the cylinders are not aligned?

One of the first results in this direction was due to Hueter and Lalley [**HL95**]. They studied planar self-affine IFS of the form $\mathcal{F} = (f_i(\mathbf{x}) = A_i\mathbf{x} + \mathbf{t}_i)_{i=1}^m$, where $A_i \in GL(2, \mathbb{R})$, $\mathbf{t}_i \in \mathbb{R}^2$ and $\|A_i\| < 1$ with the following properties:

HL-1 Each A_i satisfies the inequality $\alpha_1(A_i)^2 < \alpha_2(A_i)$.

HL-2 Let $\mathcal{Q}_2$ be the closed second quadrant of $\mathbb{R}^2$, then the sets $A_i^{-1}\mathcal{Q}_2$ are pairwise disjoint subsets of $\mathrm{int}(\mathcal{Q}_2)$.

HL-3 Each $\det(A_i)$ is strictly positive.

HL-4 The IFS $\mathcal{F}$ has the Strong Separation Property.

THEOREM 1.12.1 (Hueter and Lalley [**HL95**]). *Let $\mathcal{F} = (A_i\mathbf{x} + \mathbf{t}_i)_{i=1}^m$ be a self-affine IFS in $\mathbb{R}^2$ with attractor Λ. Suppose that $\mathcal{F}$ satisfies the conditions* **HL-1**–**HL-4**. *Then*

$$\dim_{\mathrm{H}}(\Lambda) = \dim_{\mathrm{B}}(\Lambda) = s(A_1, \ldots, A_m) < 1,$$

where $s(A_1, \ldots, A_m)$ is the affinity dimension defined in (1.127). *Moreover, $\mathcal{H}^s(\Lambda) < \infty$.*

Moreover, Hueter and Lalley showed that there exists an ergodic invariant measure for the IFS (see Definition 1.9.13), whose dimension attains the Hausdorff dimension of the set. The condition **HL-4** is quite natural in order to avoid overlaps and possible drop of the dimension; however, the other conditions seem very technical and ad hoc. In fact, **HL-1** and **HL-2** guarantee that the dimension of the projections of self-affine measures does not drop in those directions, which are invariant with respect to the matrices. For details, see Section 12.3.

In order to get a deeper understanding of the dimension theory of self-affine sets and measures, first we have to analyze what is this relation between the dimension of a self-affine measure and the dimension of its projections. Bárány and Käenmäki [**BK17**] obtained the *Ledrappier–Young formula* for planar self-affine measures generated by the IFS $\mathcal{F} = (f_i(\mathbf{x}) = A_i\mathbf{x} + \mathbf{t}_i)_{i=1}^m$. Along the way they showed that every self-affine measure on the plane is exact dimensional. Moreover, if $\mathcal{F}$ satisfies the Strong Open Set Condition (SOSC), then the dimension of the self-affine measure can be expressed in terms of entropy, Lyapunov exponents, and the dimension of the projection of the self-affine measure along "typical" strong stable directions. Very recently, this result was extended by Feng [**Fen23**] and shown to hold in complete generality. For the details in the planar case, see Chapter 12.

In particular, if a planar self-affine IFS $\mathcal{F}$ satisfies the SOSC, then the dimension of a self-affine measure drops compared to the *Lyapunov dimension* (see Chapter 10 for the definition) if and only if there exists a direction for which the dimension of the projection of the measure drops strictly below the minimum of 1 and the dimension of the self-affine measure (which is the dimension of the projection in a typical direction, see Theorem 2.3.3 below). Naively one can claim: if we can exclude the heavy overlaps in the projection, then the dimension of the measure does not drop.

DEFINITION 1.12.2. Let $\mathcal{A} = (A_1, \ldots, A_m)$ be a tuple of nonsingular $d \times d$ matrices. We say that $\mathcal{A}$ is *irreducible* if there is no proper subspace of $\mathbb{R}^d$, which is preserved by all the matrices in $\mathcal{A}$. We say that $\mathcal{A}$ is *strongly irreducible* if there is no finite collection $V_1, \dots, V_k$ of proper subspaces of $\mathbb{R}^d$, whose union is preserved by all the matrices in $\mathcal{A}$. That is, there is no finite collection $V_1, \dots, V_k$ of proper subspaces such that $A_i(\bigcup_{j=1}^k V_j) = \bigcup_{j=1}^k V_j$ for every $i = 1, \dots, m$.

The strong irreducibility of the tuple of matrices $(A_1, \dots, A_m)$ guarantees that there are no alignments between the cylinder sets of the attractor of the IFS $\mathcal{F} = (f_i(\mathbf{x}) = A_i\mathbf{x} + \mathbf{t}_i)_{i=1}^m$. Bárány, Hochman and Rapaport [**BHR19**] showed that this condition is already sufficient to prevent heavy overlaps in the projections.

THEOREM 1.12.3 (Bárány, Hochman and Rapaport [**BHR19**]). *Let $\mathcal{F} = (f_i(\mathbf{x}) = A_i\mathbf{x} + \mathbf{t}_i)_{i=1}^m$ be a planar self-affine IFS such that $\mathcal{F}$ satisfies the strong open set condition and the tuple of matrices $\mathcal{A} = (A_1, \dots, A_m)$ is strongly irreducible. Then*

$$\dim_{\mathrm{H}}(\Lambda) = \dim_{\mathrm{B}}(\Lambda) = s(A_1, \dots, A_m),$$

where Λ is the attractor of $\mathcal{F}$ and $s(A_1, \dots, A_m)$ is the affinity dimension defined in (1.127).

Hochman and Rapaport [**HR22**] extended this result, weakening the SOSC to exponential separation, which allows overlaps between the cylinder sets in the plane.

CHAPTER 2

Elements of Geometric Measure Theory

This chapter contains some results from Geometric Measure Theory. It is by no means comprehensive: we mainly focus on the topics that come up later in the book. Many results are stated without proof. We include proofs if they are short and instructive, or hard to find in the literature, or if they have elements which will be used later. We also include a section on dimension of measures, which discusses different notions of dimension and relations between them.

2.1. Densities for Radon measures

Let μ be a finite Borel measure on $\mathbb{R}^d$ and let $s \geqslant 0$. The upper and lower s-densities of the measure μ at a point $x \in \mathbb{R}^d$ are

$$\Theta^{*s}(\mu, x) := \limsup_{r\downarrow 0} \frac{\mu(B(x,r))}{(2r)^s}, \quad \Theta_*^s(\mu, x) := \liminf_{r\downarrow 0} \frac{\mu(B(x,r))}{(2r)^s}. \tag{2.1}$$

If they agree, then their common value is the s-density of μ at x and we denote it by $\Theta^s(\mu, x)$. According to Theorem B.4.2 (see [**Mat95**, Corollary 2.14] for the proof), for every Borel set $A \subset \mathbb{R}^d$ we have

$$\lim_{r\downarrow 0} \frac{\mu(A \cap B(x,r))}{\mu(B(x,r))} = \begin{cases} 1, & \text{if } x \in A; \\ 0, & \text{if } x \notin A, \end{cases} \quad \text{for } \mu\text{-almost all } x \in \mathbb{R}^d. \tag{2.2}$$

Hence for a Borel set $A \subset \mathbb{R}^d$ with $\mu(A) > 0$, for μ-a.e. $x \in A$,

$$\Theta^{*s}(\mu, x) = \Theta^{*s}(\mu|_A, x) \text{ and } \Theta_*^s(\mu, x) = \Theta_*^s(\mu|_A, x). \tag{2.3}$$

Using ideas similar to the ones in [**Mat95**, Remark 2.10], one can easily see that these densities are Borel functions of x. First we consider the special case when μ is the restriction of the Hausdorff or packing measure to a Borel set A of finite Hausdorff or packing measure respectively.

THEOREM 2.1.1. *Let $s \geqslant 0$ and $A, B \subset \mathbb{R}^d$ be Borel sets, with $\mathcal{H}^s(A) < \infty$ and $\mathcal{P}^s(B) < \infty$. Then for*

$$\Theta^{*s}(\mathcal{H}^s|_A, x) = \limsup_{r\downarrow 0} \mathcal{H}^s(A \cap B(x,r))/(2r)^s$$

and

$$\Theta_*^s(\mathcal{P}^s|_B, x) = \liminf_{r\downarrow 0} \mathcal{P}^s(B \cap B(x,r))/(2r)^s$$

we have

(i) $2^{-s} \leqslant \Theta^{*s}(\mathcal{H}^s|_A, x) \leqslant 1$ *for $\mathcal{H}^s$ almost all $x \in A$,*
(ii) $\Theta^{*s}(\mathcal{H}^s|_A, x) = 0$ *for $\mathcal{H}^s$-almost all $x \in \mathbb{R}^d \setminus A$,*
(iii) $\Theta_*^s(\mathcal{P}^s|_B, x) = 1$ *for $\mathcal{P}^s$-almost all $x \in B$.*

For the proof see [**Mat95**, Chapter 6]. The following density assertions are very important in fractal geometry:

THEOREM 2.1.2. *Let μ be a finite Borel measure on $\mathbb{R}^d$, $s \geqslant 0$, $0 < t < \infty$, and let $E \subset \mathbb{R}^d$ be a Borel set. Then the following implications hold:*

(i) $\Theta^{*s}(\mu, x) \leqslant t,\ \forall x \in E \Longrightarrow \mathcal{H}^s(E) \geqslant \frac{\mu(E)}{2^s t}$;
(ii) $\Theta^{*s}(\mu, x) \geqslant t,\ \forall x \in E \Longrightarrow \mathcal{H}^s(E) \leqslant 4^s \cdot \frac{\mu(E)}{t}$;
(iii) $\Theta^s_*(\mu, x) \leqslant t,\ \forall x \in E \Longrightarrow \mathcal{P}^s(E) \geqslant \frac{\mu(E)}{4^s \cdot t}$;
(iv) $\Theta^s_*(\mu, x) \geqslant t,\ \forall x \in E \Longrightarrow \mathcal{P}^s(E) \leqslant \frac{\mu(E)}{t}$.

Before we turn to the (partial) proof of this theorem we mention another interesting result of similar flavor without proof.

PROPOSITION 2.1.3. *Let μ be a Radon measure on $\mathbb{R}^d$. Then the following equivalences hold:*

(i) $\mu \ll \mathcal{H}^s \Longleftrightarrow \Theta^{*s}(\mu, x) < \infty$ *for μ-almost every $x \in \mathbb{R}^d$.*
(ii) $\mu \ll \mathcal{P}^s \Longleftrightarrow \Theta^s_*(\mu, x) < \infty$ *for μ-almost every $x \in \mathbb{R}^d$.*

These assertions appear as exercises in [**Mat95**, Chapter 6]. In order to prove Theorem 2.1.2 we need to use a covering lemma from [**Fal03**, Covering lemma 4.8], which we include as Lemma B.3.1 in Appendix B.3. We use the following notation: for a ball B in $\mathbb{R}^d$ we write $\widetilde{B}$ for the ball which has the same center as B, but with the radius four time larger.

PROOF OF THEOREM 2.1.2. The first two claims are proved in [**Fal03**, Proposition 4.9]. In particular, Theorem 2.1.2(i) may be derived from the Mass Distribution Principle Lemma 1.4.11. So we give only the proof of the last two.

Proof of part (iii). It is enough to show that for every $0 < t < \infty$,

$$\Theta^s_*(\mu, x) < t,\ \forall x \in E \Longrightarrow \mathcal{P}^s(E) \geqslant \frac{\mu(E)}{4^s \cdot t}. \tag{2.4}$$

For $\delta > 0$ set

$$\mathcal{C} = \mathcal{C}_\delta := \left\{\overline{B}(x, r) : x \in E,\ 0 < r \leqslant \delta,\ \mu\left(\overline{B}(x, 4r)\right) < t \cdot (8r)^s\right\}.$$

Note that $\mathcal{C}$ is a covering of E for every sufficiently small $\delta > 0$. Now we choose a countable subcollection $\{\overline{B}_i\}$ of $\mathcal{C}$ as in the 4-times covering lemma. Then $\{\overline{B}_i\}$ is a δ-packing and $\left\{\widetilde{B}_i\right\}$ is a covering of E. Moreover, by definition, for every i we have

$$\mu(\widetilde{B}_i) < t \cdot |\widetilde{B}_i|^s. \tag{2.5}$$

Hence

$$\mathcal{P}^s_\delta(E) \geqslant \sum_i |B_i|^s = \frac{1}{4^s} \sum_i |\widetilde{B}_i|^s > \frac{1}{4^s \cdot t} \sum_i \mu(\widetilde{B}_i) \geqslant \frac{1}{4^s \cdot t} \mu(E), \tag{2.6}$$

and therefore,

$$\widetilde{\mathcal{P}}^s(E) \geqslant \frac{1}{4^s \cdot t} \mu(E). \tag{2.7}$$

By definition, $\mathcal{P}^s(E) = \inf\left\{\sum_{i=1}^\infty \widetilde{\mathcal{P}}^s(E_i) :\ E \subset \bigcup_{i=1}^\infty E_i\right\}$. Let $\{E_i\}$ be a cover of E. We can apply the argument above for each $E \cap E_i$. This and the monotonicity of $\widetilde{\mathcal{P}}^s(\cdot)$ yield that $\widetilde{\mathcal{P}}^s(E_i) \geqslant \widetilde{\mathcal{P}}^s(E \cap E_i) \geqslant \frac{1}{4^s \cdot t}\mu(E \cap E_i)$. So, whenever $\{E_i\}$ is a covering of E we have

$$\sum_{i=1}^{\infty} \widetilde{\mathcal{P}}^s(E_i) \geqslant \frac{1}{4^s \cdot t} \sum_{i=1}^{\infty} \mu(E_i \cap E) \geqslant \frac{1}{4^s \cdot t}\mu(E).$$

That is, $\mathcal{P}^s(E) \geqslant \frac{1}{4^s \cdot t}\mu(E)$, as claimed.

Proof of part (iv). It is enough to show that for every $0 < t < \infty$ we have

$$\Theta_*^s(\mu, x) > t, \ \forall x \in E \Longrightarrow \mathcal{P}^s(E) \leqslant \frac{\mu(E)}{t}. \tag{2.8}$$

Fix an arbitrary $t' \in (0, t)$. Let $V \supset E$ be an open set. Put

$$E(V, \delta) := \left\{x \in E : \mu(B(x, r)) > t'(2r)^s, \forall r \in (0, \delta), \ B(x, \delta) \subset V\right\}.$$

Clearly,

$$\delta_1 < \delta_2 \Longrightarrow E(V, \delta_1) \supset E(V, \delta_2), \text{ and } \bigcup_{\delta > 0} E(V, \delta) = E. \tag{2.9}$$

For every $\rho \in (0, \delta)$ let $\mathcal{C}_{\rho,\delta}$ be the collection of ρ-packings of $E(V, \delta)$. Then for all $\{\overline{B}_i\} \in \mathcal{C}_{\rho,\delta}$ we have

$$\mu(\overline{B}_i) > t' \cdot |B_i|^s \text{ for all } i, \quad \bigcup_i \overline{B}_i \subset V \quad \text{and} \quad \{\overline{B}_i\} \text{ are disjoint.} \tag{2.10}$$

Hence

$$\mu(V) \geqslant \sum_i \mu\left(\overline{B}_i\right) > t' \sum_i |B_i|^s \tag{2.11}$$

holds for every ρ-packing $\{\overline{B}_i\}$ of $E(V, \delta)$, and so $\mu(V) \geqslant t' \cdot \mathcal{P}_\rho^s(E(V, \delta))$. Using that $\widetilde{\mathcal{P}}^s(E(V, \delta)) = \inf_{\rho > 0} \mathcal{P}_\rho^s(E(V, \delta))$ and $\widetilde{\mathcal{P}}^s(E(V, \delta)) \geqslant \mathcal{P}^s(E(V, \delta))$ we obtain that

$$\mu(V) \geqslant t' \cdot \widetilde{\mathcal{P}}^s(E(V, \delta)) \geqslant t' \cdot \mathcal{P}^s(E(V, \delta)). \tag{2.12}$$

Now observe that $E(V, \delta)$ are Borel sets. This easily follows from the simple fact that for every $r > 0$ the function $x \mapsto \mu(B(x, r))$ is lower semicontinuous, hence a Borel function. In view of (2.9) and since $\mathcal{P}^s$ is a Borel measure, we obtain that $\lim_{\delta \downarrow 0} \mathcal{P}^s\left(E(V, \delta)\right) = \mathcal{P}^s(E)$ and so

$$\mu(V) \geqslant t'\mathcal{P}^s(E).$$

It follows from the regularity of μ (see Appendix B.1) that $\inf_{E \subset V \text{open}} \mu(V) = \mu(E)$. From this and from the fact that $t' < t$ was arbitrary, we obtain (2.8). □

Recently considerable attention has been paid to the study of sets F such that the appropriate dimensional Hausdorff measure restricted to F has lower and upper densities uniformly (at every point) separated from both zero and infinity.

DEFINITION 2.1.4. Let $F \subset \mathbb{R}^d$ be a bounded set. We say that F is s-Ahlfors regular if there exists a constant $C \geqslant 1$ such that

$$C^{-1} \leqslant \frac{\mathcal{H}^s(F \cap B(x, r))}{r^s} \leqslant C, \quad \forall r \in (0, |F|) \text{ and } \forall x \in F. \tag{2.13}$$

In this case clearly $s = \dim_{\mathrm{H}}(F)$.

Some authors use the following equivalent definition:

PROPOSITION 2.1.5. *A set $F \subset \mathbb{R}^d$ is s-Ahlfors regular if and only if there exists a nontrivial Radon measure μ supported on F and a constant $c \geqslant 1$ such that*

$$c^{-1} \leqslant \frac{\mu(B(x,r))}{r^s} \leqslant c, \quad \forall r \in (0,|F|) \text{ and } \forall x \in F. \tag{2.14}$$

PROOF. We present the proof in the special case when $F \subset \mathbb{R}^d$ is a Borel set. If F is an s-Ahlfors regular set, then we choose $\mu := \mathcal{H}^s|_F$. Clearly, (2.14) holds with the choice $c = C$. On the other hand, assume that (2.14) holds for a nontrivial Radon measure μ and constant $c \geqslant 1$. The first two parts of Theorem 2.1.2 imply that

$$\frac{1}{c} \leqslant \frac{\mathcal{H}^s(F \cap B(x,r))}{\mu(B(x,r))} \leqslant 8^s c. \tag{2.15}$$

This and (2.14) imply that (2.13) holds for a suitable $C > 1$. □

2.2. Product theorems

A naive guess is that the dimension of a Cartesian product is the sum of the dimensions, but this is not always the case. The most basic result is due to Marstrand.

THEOREM 2.2.1 (Marstrand's product theorem [**Mar54**]). *Let $E \subset \mathbb{R}^n$ and $F \subset \mathbb{R}^m$ be Borel sets. Then*

$$\dim_{\mathrm{H}}(E) + \dim_{\mathrm{H}}(F) \leqslant \dim_{\mathrm{H}}(E \times F) \leqslant \dim_{\mathrm{H}}(E) + \overline{\dim}_{\mathrm{B}}(F). \tag{2.16}$$

COROLLARY 2.2.2. *If either $\dim_{\mathrm{H}}(E) = \overline{\dim}_{\mathrm{B}}(E)$ or $\dim_{\mathrm{H}}(F) = \overline{\dim}_{\mathrm{B}}(F)$, then*

$$\dim_{\mathrm{H}}(E \times F) = \dim_{\mathrm{H}}(E) + \dim_{\mathrm{H}}(F).$$

For a proof of Theorem 2.2.1 see [**Fal03**] or [**BP17**].

REMARK 2.2.3.

(a) The left inequality in (2.16) can be deduced from the following useful fact: given Borel sets $E \subset \mathbb{R}^n$ and $F \subset \mathbb{R}^m$,

$$\mathcal{H}^s(E) < \infty,\ \mathcal{H}^t(F) < \infty \implies \mathcal{H}^{s+t}(E \times F) \geqslant \mathcal{H}^s(E) \cdot \mathcal{H}^t(F).$$

(b) The left inequality in (2.16) may be strict: there are examples of $E, F \subset \mathbb{R}$ with $\dim_{\mathrm{H}}(E) = \dim_{\mathrm{H}}(F) = 0$, but $\dim_{\mathrm{H}}(E \times F) = 1$, see e.g., [**Fal03**, Ex. 7.8].

(c) The right inequality in (2.16) may be strengthened to

$$\dim_{\mathrm{H}}(E \times F) \leqslant \dim_{\mathrm{H}}(E) + \dim_{\mathrm{P}}(F),$$

see [**Tri82**].

For the upper box dimension we have the following (see [**Fal03**, (7.9)]):

PROPOSITION 2.2.4. *Given Borel sets $E \subset \mathbb{R}^n$ and $F \subset \mathbb{R}^m$,*

$$\overline{\dim}_{\mathrm{B}}(E \times F) \leqslant \overline{\dim}_{\mathrm{B}}(E) + \overline{\dim}_{\mathrm{B}}(F).$$

Theorem 2.2.1 has been extended to packing dimension, namely, for every Borel sets $E \subset \mathbb{R}^n$ and $F \subset \mathbb{R}^m$,

$$\dim_{\mathrm{P}}(E) + \dim_{\mathrm{H}}(F) \leqslant \dim_{\mathrm{P}}(E \times F) \leqslant \dim_{\mathrm{P}}(E) + \dim_{\mathrm{P}}(F). \tag{2.17}$$

The second inequality follows easily by Proposition 1.4.18 and Proposition 2.2.4, while the first inequality was shown by Tricot [**Tri82**].

2.3. Projection theorems

2.3.1. Projections onto lines. Denote by $\text{proj}_{\mathbf{v}}$ the orthogonal projection from $\mathbb{R}^d$ to the line through the origin determined by the vector $\mathbf{v}$, namely $\{t\mathbf{v} : t \in \mathbb{R}\}$. It is convenient to identify all such lines with $\mathbb{R}$, then we have

$$\text{proj}_{\mathbf{v}}(\mathbf{x}) = \langle \mathbf{x}, \mathbf{v} \rangle \quad \text{for } \|\mathbf{v}\| = 1. \tag{2.18}$$

We denote the unit sphere in $\mathbb{R}^d$ by $\mathbb{S}^{d-1}$. The $(d-1)$-dimensional surface measure σ_{d-1} can be defined as the normalized $(d-1)$-dimensional Hausdorff measure on $\mathbb{S}^{d-1}$. Alternatively, we can define

$$\sigma_{d-1}(A) = \alpha(d)^{-1}\mathcal{L}^d(\{tx : x \in A, 0 \leqslant t \leqslant 1\}), \quad A \subset \mathbb{S}^{d-1}, \tag{2.19}$$

where

$$\alpha(d) := \mathcal{L}^d(B(0,1)). \tag{2.20}$$

THEOREM 2.3.1 (Marstrand Projection Theorem). *For any Borel set $E \subset \mathbb{R}^d$,*
(i) $\dim_{\mathrm{H}}(\text{proj}_{\mathbf{v}}(E)) = \min\{\dim_{\mathrm{H}}(E), 1\}$ *for σ_{d-1}-a.e. $\mathbf{v} \in \mathbb{S}^{d-1}$;*
(ii) *if* $\dim_{\mathrm{H}}(E) > 1$, *then* $\mathcal{L}^1(\text{proj}_{\mathbf{v}}(E)) > 0$ *for σ_{d-1}-a.e. $\mathbf{v} \in \mathbb{S}^{d-1}$.*

This theorem for $d = 2$ was proved by Marstrand [**Mar54**]. There is an analogous theorem about projections to planes and higher-dimensional subspaces, due to Mattila [**Mat75**], which we state in at the end of Section 2.3.2. It follows from results of M. Järvenpää [**Jär94**] and Falconer, Howroyd [**FH96**] (see also [**Mat95**, 134]) that part (i) of Theorem 2.3.1 does not remain valid if we replace Hausdorff dimension with any of the dimensions $\underline{\dim}_{\mathrm{B}}, \overline{\dim}_{\mathrm{B}}, \underline{\dim}_{\mathrm{P}}, \overline{\dim}_{\mathrm{P}}$. However, although the packing dimension may decrease under projections, it is almost surely independent of the subspace onto which we project, see Falconer and Howroyd [**FH97**].

We sketch the proof of Theorem 2.3.1, because it introduces a method, due to Kaufman [**Kau68**], which we will use repeatedly. Broadly speaking, the idea is to integrate a certain quantity over the parameter space and show that it is finite. Then the quantity is finite for a.e. parameter value. In part (i) this quantity is the α-dimensional energy of the projection (push-forward) measure of a Frostman measure, and its finiteness yields the lower bound for the Hausdorff dimension. In part (ii) we follow Mattila and use the quantity, which involves the lower derivative of the same measure, and whose finiteness implies absolute continuity (an alternative method, due to Kaufman [**Kau68**], is to use the Fourier transform). These methods are very powerful in establishing "almost sure" results. The drawback is that we have no idea for which parameters the equality for the dimension and/or absolute continuity actually takes place.

PROOF. **(i)** The upper bound for the dimension of the projection trivially holds for all $\mathbf{v} \in \mathbb{S}^{d-1}$ (since the projection is a Lipschitz map), so we only need to establish the lower bound. Fix an arbitrary $0 < \alpha < \min\{\dim_{\mathrm{H}}(E), 1\}$. By the potential-theoretic characterization of the Hausdorff dimension (Theorem 1.4.16(i)), there exists a probability measure μ supported on E such that

$$\mathcal{E}_\alpha(\mu) = \iint |\mathbf{x} - \mathbf{y}|^{-\alpha}\, d\mu(\mathbf{x})\, d\mu(\mathbf{y}) < \infty. \tag{2.21}$$

Denote by $\mu_{\mathbf{v}}$ the push-forward measure onto the line $\{t\mathbf{v} : t \in \mathbb{R}\}$:

$$\mu_{\mathbf{v}} = (\mathrm{proj}_{\mathbf{v}})_*(\mu) = \mu \circ \mathrm{proj}_{\mathbf{v}}^{-1}.$$

We want to prove that

$$\mathcal{J}_1 := \int_{\mathbb{S}^{d-1}} \iint |\xi - \zeta|^{-\alpha}\, d\mu_{\mathbf{v}}(\xi)\, d\mu_{\mathbf{v}}(\zeta)\, d\sigma_{d-1}(\mathbf{v}) < \infty. \tag{2.22}$$

This would imply that $\mu_{\mathbf{v}}$, a probability measure supported on $\mathrm{proj}_{\mathbf{v}}(E)$, satisfies $\mathcal{E}_\alpha(\mu_{\mathbf{v}}) < \infty$ for a.e. $\mathbf{v}$, hence $\dim_{\mathrm{H}}(\mathrm{proj}_{\mathbf{v}}(E)) \geqslant \alpha$ for a.e. $\mathbf{v}$ by Lemma 1.4.15, which yields the desired claim.

In order to check (2.22), we write $\xi = \mathrm{proj}_{\mathbf{v}}(\mathbf{x}) = \langle \mathbf{x}, \mathbf{v}\rangle$, $\zeta = \mathrm{proj}_{\mathbf{v}}(\mathbf{y}) = \langle \mathbf{y}, \mathbf{v}\rangle$, make a change of variable (or rather, use the definition of the push-forward measure), and exchange the order of integration to obtain

$$\mathcal{J}_1 = \iint_{E\times E} \int_{\mathbb{S}^{d-1}} |\langle \mathbf{x} - \mathbf{y}, \mathbf{v}\rangle|^{-\alpha}\, d\sigma_{d-1}(\mathbf{v})\, d\mu(\mathbf{x})\, d\mu(\mathbf{y}).$$

For $\mathbf{x} \neq \mathbf{y}$ we have $\mathbf{x} - \mathbf{y} = |\mathbf{x} - \mathbf{y}| \cdot \mathbf{w}$ for some $\mathbf{w} \in \mathbb{S}^{d-1}$, so we can rewrite the inner integral as

$$|\mathbf{x} - \mathbf{y}|^{-\alpha} \int_{\mathbb{S}^{d-1}} |\langle \mathbf{w}, \mathbf{v}\rangle|^{-\alpha}\, d\sigma_{d-1}(\mathbf{v}) =: A_{\mathbf{w}} |\mathbf{x} - \mathbf{y}|^{-\alpha},$$

where $A_{\mathbf{w}}$ actually does not depend on $\mathbf{w}$ by spherical symmetry. Note that it is legitimate to exclude the diagonal from the domain of integration since $\mu \times \mu\left(\{(\mathbf{x}, \mathbf{y}) : \mathbf{x} = \mathbf{y}\}\right) = 0$ by (2.21).

The following simple geometric fact is left as an exercise:

There exists $C > 0$ such that for any $r > 0$ and any $\mathbf{w} \in \mathbb{S}^{d-1}$,

$$\sigma_{d-1}\left(\{\mathbf{v} \in \mathbb{S}^{d-1} : |\langle \mathbf{w}, \mathbf{v}\rangle| \leqslant r\}\right) \leqslant Cr. \tag{2.23}$$

This clearly implies

$$\begin{aligned}
A := A_{\mathbf{w}} &= \int_{\mathbb{S}^{d-1}} |\langle \mathbf{w}, \mathbf{v}\rangle|^{-\alpha}\, d\sigma_{d-1}(\mathbf{v}) \\
&= \int_0^\infty \sigma_{d-1}(\{\mathbf{v} \in \mathbb{S}^{d-1} : |\langle \mathbf{w}, \mathbf{v}\rangle|^{-\alpha} > r\})dr \\
&\leqslant \sigma_{d-1}(\mathbb{S}^{d-1}) + \int_1^\infty \sigma_{d-1}(\{\mathbf{v} \in \mathbb{S}^{d-1} : |\langle \mathbf{w}, \mathbf{v}\rangle|^{-\alpha} > r\})dr \\
&\leqslant \sigma_{d-1}(\mathbb{S}^{d-1}) + \int_1^\infty \sigma_{d-1}(\{\mathbf{v} \in \mathbb{S}^{d-1} : |\langle \mathbf{w}, \mathbf{v}\rangle| < r^{-1/\alpha}\})dr \\
&\leqslant \sigma_{d-1}(\mathbb{S}^{d-1}) + \int_1^\infty Cr^{-1/\alpha} dr < \infty
\end{aligned}$$

since $\alpha < 1$, and so $\mathcal{J}_1 = A\mathcal{E}_\alpha(\mu) < \infty$ by assumption (2.21), proving (2.22).

(ii) Since $\dim_{\mathrm{H}}(E) > 1$, Theorem 1.4.16(i) yields a probability μ supported on E, with $\mathcal{E}_1(\mu) < \infty$. Let $\mu_{\mathbf{v}}$ be the push-forward measure, as above, and consider the lower derivative of $\mu_{\mathbf{v}}$ with respect to the Lebesgue measure $\mathcal{L}^1$:

$$\underline{D}(\mu_{\mathbf{v}}, \zeta) := \liminf_{r\to 0} \frac{\mu_{\mathbf{v}}(B(\zeta, r))}{2r}.$$

We would like to show that

$$\mathcal{J}_2 := \int_{\mathbb{S}^{d-1}} \int \underline{D}(\mu_{\mathbf{v}}, \zeta)\, d\mu_{\mathbf{v}}(\zeta)\, d\sigma_{d-1}(\mathbf{v}) < \infty.$$

This would imply that for σ_{d-1}-a.e. $\mathbf{v}$ we have $\underline{D}(\mu_{\mathbf{v}}, \zeta) < \infty$ for $\mu_{\mathbf{v}}$-a.e. ζ, and then by [**Mat95**, p. 36] we would obtain that $\mu_{\mathbf{v}}$ is absolutely continuous with respect to the Lebesgue measure for a.e. $\mathbf{v}$. Since $\mu_{\mathbf{v}}$ is a probability measure on $\mathrm{proj}_{\mathbf{v}}(E)$, this would mean that $\mathcal{L}^1(\mathrm{proj}_{\mathbf{v}}(E)) > 0$ for a.e. $\mathbf{v}$, as desired.

In order to estimate $\mathcal{J}_2$, we first use Fatou's Lemma to get

$$\mathcal{J}_2 \leqslant \liminf_{r\to 0}\, (2r)^{-1} \int_{\mathbb{S}^{d-1}} \int \mu_{\mathbf{v}}(B(\zeta, r))\, d\mu_{\mathbf{v}}(\zeta)\, d\sigma_{d-1}(\mathbf{v}).$$

Further,

$$\begin{aligned} \int \mu_{\mathbf{v}}(B(\zeta, r))\, d\mu_{\mathbf{v}}(\zeta) &= \iint \mathbb{1}_{B(\zeta,r)}(\xi)\, d\mu_{\mathbf{v}}(\xi)\, d\mu_{\mathbf{v}}(\zeta) \\ &= \iint \mathbb{1}_{B(\mathrm{proj}_{\mathbf{v}}(\mathbf{y}),r)}(\mathrm{proj}_{\mathbf{v}}(\mathbf{x}))\, d\mu(\mathbf{x})\, d\mu(\mathbf{y}). \end{aligned}$$

Thus, exchanging the order of integration and integrating the characteristic function with respect to σ_{d-1}, we obtain

$$\mathcal{J}_2 \leqslant \liminf_{r\to 0}\, (2r)^{-1} \iint \sigma_{d-1}\left(\{\mathbf{v} :\ |\langle \mathbf{x} - \mathbf{y}, \mathbf{v}\rangle| \leqslant r\}\right) d\mu(\mathbf{x})\, d\mu(\mathbf{y}).$$

By the geometric fact (2.23) we have

$$\sigma_{d-1}\left(\{\mathbf{v} :\ |\langle \mathbf{x} - \mathbf{y}, \mathbf{v}\rangle| \leqslant r\}\right) \leqslant Cr|\mathbf{x} - \mathbf{y}|^{-1}, \tag{2.24}$$

hence $\mathcal{J}_2 \leqslant C\mathcal{E}_1(\mu) < \infty$, completing the proof. □

Observe that the proof of Theorem 2.3.1(ii) actually gives the following:

THEOREM 2.3.2. *Let μ be a finite Borel measure on $\mathbb{R}^d$ such that $\mathcal{E}_1(\mu) < \infty$. Then $\mu_{\mathbf{v}} \ll \mathcal{L}^1$ for σ_{d-1}-a.e. $\mathbf{v} \in \mathbb{S}^{d-1}$.*

We will also need a projection theorem for measures. Recall that for a Borel measure μ,

$$\dim_{\mathrm{H}}(\mu) = \inf\{\dim_{\mathrm{H}}(A) :\ \mu(A^c) = 0\}$$

and

$$\underline{\dim}_{\mathrm{H}}(\mu) = \inf\{\dim_{\mathrm{H}}(A) :\ \mu(A) > 0\},$$

see Section 1.9. Let μ be a finite Borel measure on $\mathbb{R}^d$. As before, denote by $\mu_{\mathbf{v}}$ the push-forward measure onto the line $\{t\mathbf{v} :\ t \in \mathbb{R}\}$:

$$\mu_{\mathbf{v}} = (\mathrm{proj}_{\mathbf{v}})_*(\mu) = \mu \circ \mathrm{proj}_{\mathbf{v}}^{-1}.$$

THEOREM 2.3.3 (X. Hu and S. J. Taylor [**HT94**, Section 6]). *For any finite Borel measure μ on $\mathbb{R}^d$,*

(i) $\dim_{\mathrm{H}}(\mu_{\mathbf{v}}) = \min\{\dim_{\mathrm{H}}(\mu), 1\}$ *for σ_{d-1}-a.e. $\mathbf{v} \in \mathbb{S}^{d-1}$;*
(ii) *if $\underline{\dim}_{\mathrm{H}}(\mu) > 1$, then $\mu_{\mathbf{v}} \ll \mathcal{L}^1$ for σ_{d-1}-a.e. $\mathbf{v} \in \mathbb{S}^{d-1}$.*

IDEA OF THE PROOF. (i) It is immediate from the definitions that the upper bound holds for all $\mathbf{v} \in \mathbb{S}^{d-1}$. For the lower bound apply Theorem 1.9.5(i), Egorov's Theorem, and Lemma 1.9.9 together with the monotonicity of the Hausdorff dimension of measures with respect to absolute continuity.

(ii) This follows from (1.102) and Theorem 2.3.2. □

2.3.2. Projections onto higher dimensional subspaces. Marstrand's projection theorem (Theorem 2.3.1) was extended by Mattila [**Mat75, Mat90**] to the case when the projection is onto a higher dimensional subspace of $\mathbb{R}^d$. Roughly speaking, Mattila proved that for a Borel set $A \subset \mathbb{R}^d$ with $\dim_{\mathrm{H}}(A) \leqslant m < d$, for almost all m-dimensional subspace V of $\mathbb{R}^d$ the following holds: the Hausdorff dimension of A is equal to the Hausdorff dimension of the orthogonal projection of A to V. To state this theorem precisely we need some further notation. First recall that a group G endowed with a topology is called a *topological group* if G is a Hausdorff space (any two distinct points can be separated by disjoint neighbourhoods) and the group operations ($(g_1, g_2) \mapsto g_1 \cdot g_2$ and $g \mapsto g^{-1}$) are continuous.

DEFINITION 2.3.4. Let G be a topological group. We say that a nonzero regular Borel measure μ is a left (respectively right) Haar measure on G if $\mu(gA) = \mu(A)$ (respectively $\mu(Ag) = \mu(A)$) holds for all $g \in G$ and for all Borel sets $A \subset G$.

The following theorem is well known. For the proof see [**Hal13**] or [**EW11**, Section 8.3].

THEOREM 2.3.5. *For a compact topological group there exists a unique left (respectively right) probability Haar measure.*

We use this theorem for the *orthogonal group*

$$O(d, \mathbb{R}) := \left\{ g : \mathbb{R}^d \to \mathbb{R}^d : \text{ is a linear isometry.} \right\}.$$

Equivalently, $O(d, \mathbb{R})$ is the group of orthogonal real $d \times d$ matrices. We define the metric on this group by $\operatorname{dist}(g_1, g_2) := \sup\limits_{\|x\|=1} \|g_1(x) - g_2(x)\|$, $g_1, g_2 \in O(d, \mathbb{R})$. Then $O(d, \mathbb{R})$ is a compact topological group with the composition $g_1 \circ g_2$ as the group operation. Denote by θ_n the left Haar measure on $O(d, \mathbb{R})$.

Now we introduce the Grassmannian manifold $G(d, m)$ for $1 \leqslant m < d$ as the set of all m-dimensional linear subspaces of $\mathbb{R}^d$. For instance, in the case $m = 1$ the manifold $G(d, 1)$ can be identified with the quotient of the sphere $\mathbb{S}^{d-1}/\sim$ under the equivalence relation $\mathbf{v} \sim -\mathbf{v}$, and then $\gamma_{d,1}$ is the push-forward of the surface measure σ_{d-1}. Also, $G(d, 1)$ corresponds to the projective space $\mathbb{PR}^{d-1}$. Denote by P_V the orthogonal projection $\mathbb{R}^d \to V$. Fix $1 \leqslant m < d$ and define a metric on $G(d, m)$ by

$$\operatorname{dist}(V, W) := \|P_V - P_W\|,$$

where $\|\cdot\|$ is the usual operator norm. The space $G(d, m)$ with this metric is a compact metric space on which $O(d, \mathbb{R})$ acts transitively, that is, for every $V, W \in G(d, m)$ there is $g \in O(d, \mathbb{R})$ such that $W = gV$. Fix an arbitrary $V \in G(d, m)$ and define $f_V : O(d, \mathbb{R}) \to G(d, m)$ by

$$f_V(g) := gV, \quad g \in O(d, \mathbb{R}). \tag{2.25}$$

Then we define the Radon probability measure $\gamma_{d,m}$ on $G(d, m)$ by $\gamma_{d,m} = (f_V)_* \theta_n$, that is,

$$\gamma_{d,m}(\mathcal{A}) := \theta_d(\{g : gV \in \mathcal{A}\}), \quad \mathcal{A} \subset G(d, m), \quad \mathcal{A} \text{ is Borel.} \tag{2.26}$$

It follows from the definition of the left Haar measure that

$$\gamma_{d,m}(h\mathcal{A}) = \gamma_{d,m}(\mathcal{A}), \quad h \in O(d, \mathbb{R}), \ \mathcal{A} \subset G(d, m), \ \mathcal{A} \text{ is Borel.}$$

It follows from the uniqueness of the Haar measure that the definition of $\gamma_{d,m}$ is independent of the choice of $V \in G(d, m)$. Alternatively, we can define $\gamma_{d,m}$ in a more direct way which is similar to the definition of σ_{d-1} in (2.19) (see [**Mat95**, p.53 Exercise 6]). Namely, denote by $L(v_1, \ldots, v_m)$ the linear subspace spanned by the vectors $v_1, \ldots, v_m \in \mathbb{R}^d$. Then for any Borel set $\mathcal{A} \subset G(n,m)$,

$$\gamma_{n,m}(\mathcal{A}) = \alpha(d)^{-m}\mathcal{L}^{m\cdot d}(\{(v_1, \ldots, v_m) : v_i \in \mathbb{R}^d,\ |v_i| \leqslant 1,\ L(v_1, \ldots, v_m) \in \mathcal{A}\}),$$

where we remind the reader that $\alpha(d)$ was defined as the volume of the unit ball in $\mathbb{R}^d$. The following theorem was proved by Mattila [**Mat75, Mat90**].

THEOREM 2.3.6 (Mattila). *Let $A \subset \mathbf{R}^d$ be a Borel set.*

(i) *If $\dim_{\mathrm{H}}(A) \leqslant m$, then $\dim_{\mathrm{H}}(P_V(A)) = \dim_{\mathrm{H}}(A)$ for $\gamma_{d,m}$ almost all $V \in G(d,m)$.*
(ii) *If $\dim_{\mathrm{H}}(A) > m$, then $\mathcal{L}^m(P_V(A)) > 0$ for $\gamma_{d,m}$ almost all $V \in G(d,m)$.*

2.4. Slicing theorems

As a "dual" question to the one on the dimension of orthogonal projections, it is interesting to know how much dimension "remains" on the slices. Recall that

$$\mathrm{proj}_{\mathbf{v}}(\mathbf{x}) = \langle \mathbf{v}, \mathbf{x}\rangle \text{ for } \|\mathbf{v}\| = 1.$$

The partition defined by the inverse branches of $\mathrm{proj}_{\mathbf{v}}$ is clearly measurable, see Proposition 9.4.11. Denote the slice of a Borel set $E \subset \mathbb{R}^d$ by $S_{\mathbf{v}}(x) = E \cap \mathrm{proj}_{\mathbf{v}}^{-1}(x)$. The next result says that almost every slice cannot have too large a dimension. The theorem is due to Marstrand [**Mar54**] for planar sets intersected with lines and it was generalized by Mattila for higher dimensional sets intersected with hyperplanes, see [**Mat95**, Ch. 10]. We state here a special case.

THEOREM 2.4.1 (Marstrand's Slicing Theorem). *Let $E \subset \mathbb{R}^d$ be a Borel set with $\dim_{\mathrm{H}}(E) > 1$. Then for every $\mathbf{v} \in \mathbb{S}^{d-1}$ and $\mathcal{L}^1$-almost every $x \in \mathrm{proj}_{\mathbf{v}}(E)$,*

$$\dim_{\mathrm{H}}(S_{\mathbf{v}}(x)) \leqslant \dim_{\mathrm{H}}(E) - 1.$$

This theorem follows from a more general result, which can be found in [**BP17**, Theorem 3.3.1]. They stated and proved only the planar case $d = 2$, but the proof there can be modified in a straightforward way.

PROPOSITION 2.4.2 (Generalized Marstrand's Slicing Theorem). *For every Borel set $E \subset \mathbb{R}^d$, every $\mathbf{v} \in \mathbb{S}^{d-1}$ and every probability measure μ on $\mathbb{R}$, with $\mu(I) \leqslant C|I|^s$ for every interval $I \subset \mathbb{R}$ and a uniform constant $C > 0$,*

$$\dim_{\mathrm{H}}(S_{\mathbf{v}}(x)) \leqslant \max\{\dim_{\mathrm{H}}(E) - s, 0\} \text{ for } \mu\text{-almost every } x.$$

PROOF. Applying an orthogonal transformation to E, we may assume that $\mathrm{proj}_{\mathbf{v}}$ is the orthogonal projection onto the 1st coordinate axis. Further, we may assume that $s \leqslant \dim_{\mathrm{H}}(E)$. Indeed, if $\dim_{\mathrm{H}}(E) < s$, then $\dim_{\mathrm{H}}(\mathrm{proj}(E)) < s$, and by Lemma 1.4.11 we have $\mu(\mathrm{proj}_{\mathbf{v}}(E)) = 0$, whence $S_{\mathbf{v}}(x) = \varnothing$ for μ-a.e. x. Denote $\alpha = \dim_{\mathrm{H}}(E)$. We have $\mathcal{H}^{\alpha+1/n}(E) = 0$, and hence by the definition of Hausdorff measure, for every $\varepsilon > 0$ and $n \geqslant 1$, there exists a countable collection of cubes $\{I_i\}$ in $\mathbb{R}^{d-1}$ and $\{J_i\}$ in $\mathbb{R}$ such that $E \subseteq \bigcup_i J_i \times I_i$, $|I_i| = |J_i| < \varepsilon$, and $\sum_i |J_i|^{\alpha+1/n} < \varepsilon$. Let

$$f(x, \mathbf{y}) = \sum_i \mathbb{1}_{J_i \times I_i}(x, \mathbf{y}) \cdot |J_i|^{\alpha - s - (d-1) + 1/n}.$$

Then

$$\iint f(x, \mathbf{y})\, d\mathbf{y}\, d\mu(x) = \sum_i |J_i|^{\alpha-s-(d-1)+1/n} \mathcal{L}^{d-1}(I_i)\mu(J_i)$$
$$\leqslant C \sum_i |J_i|^{\alpha+1/n} < C\varepsilon.$$

On the other hand,

$$\iint f(x, \mathbf{y})\, d\mathbf{y}\, d\mu(x) = \iint \mathbb{1}_{\mathrm{proj}_{\mathbf{v}}^{-1}(x)}(x, \mathbf{y}) f(x, \mathbf{y})\, d\mathbf{y}\, d\mu(x)$$
$$= \iint \sum_i \mathbb{1}_{(J_i \times I_i) \cap \mathrm{proj}_{\mathbf{v}}^{-1}(x)}(x, \mathbf{y}) \cdot |J_i|^{\alpha-s-(d-1)+1/n}\, d\mathbf{y}\, d\mu(x).$$

Since $J_i \times I_i \cap \mathrm{proj}_{\mathbf{v}}^{-1}(x) = I_i$ for $x \in J_i$ and $\{I_i : x \in J_i\}$ is an ε-cover of $S_{\mathbf{v}}(x)$, by Fubini's Theorem we have

$$\iint f(x, \mathbf{y})\, d\mathbf{y}\, d\mu(x) = \int \sum_i \mathbb{1}_{J_i}(x) \cdot |J_i|^{\alpha-s+1/n}\, d\mu(x)$$
$$\geqslant \int \mathcal{H}_\varepsilon^{\alpha-s+1/n}(S_{\mathbf{v}}(x))\, d\mu(x).$$

Letting $\varepsilon \to 0$ and using Fatou's Lemma, we get

$$\int \mathcal{H}^{\alpha-s+1/n}(S_{\mathbf{v}}(x)) d\mu(x) = 0 \text{ for every } n \geqslant 1.$$

Thus, $\mu(\{x : \dim_{\mathrm{H}}(S_{\mathbf{v}}(x)) \leqslant \alpha - s\}) = 0$, as desired. □

Marstrand [**Mar54**] and later Mattila [**Mat95**] showed that for a typical direction, there is a set of positive Lebesgue measure in the projection, for which the Hausdorff dimension of the slice is maximal possible.

THEOREM 2.4.3. *Let $t > 1$ and let $E \subseteq \mathbb{R}^d$ be a Borel set with $0 < \mathcal{H}^t(E) < \infty$. Then for σ_{d-1}-a.e. $\mathbf{v} \in \mathbb{S}^{d-1}$,*

$$\mathcal{L}^1(\{x : \dim_{\mathrm{H}}(S_{\mathbf{v}}(x)) = t - 1\}) > 0.$$

We prove this theorem below in Section 12.2; it uses the decomposition of a measure into an integral of conditional measures, which will be discussed in Section 9.4.

2.5. Besicovitch projection theorem and generalizations

One of the key concepts in Geometric Measure Theory is *rectifiability*. Although this is not a topic of our book, we would be amiss not to mention it altogether, and moreover, it will make its appearance a few times. We follow Mattila's book [**Mat95**] in our discussion; see also Falconer's book [**Fal85**] for a detailed treatment of the 1-dimensional case. We start with the most classical case of 1-dimensional sets in the plane, studied by Besicovitch, and then discuss higher-dimensional generalizations. Recall the classical notion of *curve length*.

DEFINITION 2.5.1. Let Γ be a curve with a parametrization $\psi : [a, b] \to \mathbb{R}^d$. The length of Γ is defined by $\ell(\Gamma) = \sup_P \sum_{j=1}^n |\psi(a_j) - \psi(a_{j-1})|$, where $P = \{a = a_0 < a_1 < \ldots < a_n = b\}$ is a finite partition of $[a, b]$ and the supremum is taken over all partitions. A curve Γ is *rectifiable* if its length is finite: $\ell(\Gamma) < \infty$.

One can show that the length measure on a curve coincides with the 1-dimensional Hausdorff measure $\mathcal{H}^1$ restricted to it.

DEFINITION 2.5.2. A set $E \subset \mathbb{R}^d$ is called 1-rectifiable if there are rectifiable curves Γ_i in $\mathbb{R}^d$, $i = 1, 2, \ldots$, such that

$$\mathcal{H}^1\Big(E \backslash \bigcup_i \Gamma_i\Big) = 0.$$

A set F is called purely 1-unrectifiable if $\mathcal{H}^1(F \cap \Gamma) = 0$ for any rectifiable curve Γ.

The following fundamental theorem was proved by Besicovitch [**Bes38**]. Recall that $\text{proj}_{\mathbf{v}}$ denotes the orthogonal projection from the plane $\mathbb{R}^2$ onto the line determined by the vector $\mathbf{v} \in \mathbb{S}^1$.

THEOREM 2.5.3 (Besicovitch). *Suppose that $A \subset \mathbb{R}^2$ is an $\mathcal{H}^1$-measurable set with $\mathcal{H}^1(A) < \infty$. Then the following are equivalent:*

(i) *A is purely 1-unrectifiable;*
(ii) *$\mathcal{L}^1(\text{proj}_{\mathbf{v}} A) = 0$ for a.e. $\mathbf{v} \in \mathbb{S}^1$;*
(iii) *$\mathcal{L}^1(\text{proj}_{\mathbf{v}} A) = 0$ for two nonparallel $\mathbf{v} \in \mathbb{S}^1$.*

See [**Fal85**, Theorem 6.13, Corollary 6.14] for the proof. Of course, (ii) $\Rightarrow$ (iii) is obvious and (iii) $\Rightarrow$ (i) is rather simple (in fact, a rectifiable curve has all projections of positive length, unless it is a straight line segment, in which case only one orthogonal projections has zero length). The implication (i) $\Rightarrow$ (ii) is a deep and difficult result, see [**Fal85**, Theorem 6.13].

EXAMPLE 2.5.4. Consider $A = \Lambda_{\frac{1}{2}} \times \Lambda_{\frac{1}{2}}$, where $\Lambda_{\frac{1}{2}}$ is the middle-half Cantor set. It is easy to see that A is an attractor of a self-similar IFS satisfying the SSP, with similarity dimension 1. Thus, $\dim_{\text{H}}(A) = 1$ and $0 < \mathcal{H}^1(A) < \infty$. But the vertical and horizontal projections of A, namely, $\text{proj}_0(A)$ and $\text{proj}_{\pi/2}(A)$, have zero length, being middle-half Cantor sets of dimension $1/2$. Thus A is purely 1-unrectifiable and hence its almost every projection has zero length. A simpler proof of this result, without the use of Besicovitch Theorem was given in [**PSS03**], see also [**BP17**].

Next we turn to higher-dimensional generalizations.

DEFINITION 2.5.5. A set $E \subset \mathbb{R}^d$ is called m-rectifiable if there exist subsets $A_i \subset \mathbb{R}^m$ and Lipschitz maps $f_i : A_i \to \mathbb{R}^d$, $i = 1, 2, \ldots$, such that

$$\mathcal{H}^m\Big(E \backslash \bigcup_i f_i(A_i)\Big) = 0,$$

where $\mathcal{H}^m$ is the m-dimensional Hausdorff measure. A set $F \subset \mathbb{R}^d$ is called purely m-unrectifiable if $\mathcal{H}^m(E \cap F) = 0$ for every m-rectifiable set E.

One can show that for $m = 1$ this definition agrees with the one above, which used the classical notion of rectifiability (this can be deduced from the observation that a curve Γ with a parametrization $\psi = (\psi_1, \ldots, \psi_d)$ is rectifiable if and only if every component ψ_j is a function of bounded variation).

It is an easy consequence of the definition that any set $A \subset \mathbb{R}^d$ with $\mathcal{H}^m(A) < \infty$ has a decomposition into a union of an m-rectifiable and purely m-unrectifiable sets, which is unique up to sets of $\mathcal{H}^m$ measure zero, see [**Mat95**, Theorem 15.6].

Now we state Federer's projection theorem [**Fed69**], which is a generalization of Theorem 2.5.3. One can show that if $\mathcal{H}^m(E) > 0$ and E is m-rectifiable, then $\mathcal{H}^m(P_V) > 0$ for $\gamma_{d,m}$-a.e. V.

THEOREM 2.5.6 (Federer). *Suppose that A is an $\mathcal{H}^m$-measurable set, with $\mathcal{H}^m(A) < \infty$. If A is purely m-unrectifiable, then $\mathcal{H}^m(P_V A) = 0$ for $\gamma_{d,m}$-a.e. V.*

The Federer–Besicovitch projection theorem has been extended to the case of a general *transversal family of maps* by Hovila, Järvenpäa, Järvenpäa, and Ledrappier [**HJJL12**]. We refer the reader to [**HJJL12**, Definition 2.4] for the definition of a transversal family. This notion is somewhat analogous to that of transversal families appearing in Section 6.6.

THEOREM 2.5.7 ([**HJJL12**]). *Suppose that $A \subset \mathbb{R}^d$ is an $\mathcal{H}^m$-measurable set, with $\mathcal{H}^m(A) < \infty$. Assume that $U \in \mathbb{R}^k$ is open and $\{P_{\mathbf{u}} : \mathbb{R}^d \to \mathbb{R}^m\}_{\mathbf{u}\in U}$ is a transversal family of maps. If A is purely m-unrectifiable then $\mathcal{H}^m(P_{\mathbf{u}} A) = 0$ for $\mathcal{L}^k$-a.e. $\mathbf{u} \in U$.*

This theorem had a number of nontrivial applications, in particular, to geodesic flows on Riemann surfaces, see [**HJJL12**].

2.6. L^q dimensions and entropy dimension

Here we continue the discussion of various notions of dimension for measures, started in Section 1.9. There is a useful generalization of correlation dimension, introduced in (1.9.7), which gives a whole "spectrum" of dimensions, parametrized by q; the correlation dimension corresponds to $q = 2$. There is also the "limiting case" $q = 1$, when the appropriate notion is that of *entropy*, or *information* dimension, see (1.117). Entropy dimension was introduced by Rényi [**Rén70**], and he was the first to consider L^q-dimensions as well.

DEFINITION 2.6.1. *Let μ be a Borel probability measure on a metric space (X, ρ). For $\varepsilon > 0$ let*

$$C_q(\mu, \varepsilon) = \int \big(\mu(B(x, \varepsilon))\big)^{q-1}\, d\mu(x).$$

For $q \geqslant 0$, $q \neq 1$, the lower L^q-dimension of μ is defined by

$$D_q(\mu) = \liminf_{\varepsilon\to 0} \frac{\log C_q(\mu, \varepsilon)}{(q-1)\log \varepsilon}.$$

Replacing lim inf *by* lim sup *we obtain the upper L^q-dimension of μ, denoted $\overline{D}_q(\mu)$.*

The potential-theoretic characterization for the lower L^q-dimension, analogous to Proposition 1.9.8, holds for $q > 1$. Similarly to the s-energy introduced in (1.63), we define the (s, q)-energy of μ by

$$\mathcal{E}_{s,q}(\mu) = \int \left(\int \frac{d\mu(y)}{\rho(x, y)^s}\right)^{q-1} d\mu(x).$$

The following statement was proven by Hunt and Kaloshin [**HK97**, Proposition 2.1].

PROPOSITION 2.6.2 (Hunt–Kaloshin). *Let $q > 1$. If μ is a Borel probability measure, then*

$$D_q(\mu) = \sup\{s \geqslant 0 :\ \mathcal{E}_{s,q}(\mu) < \infty\} = \inf\{s \geqslant 0 :\ \mathcal{E}_{s,q}(\mu) = \infty\}.$$

For $q = 1$ the appropriate notion is the *entropy* or *information* dimension, which measures the "average" local dimension.

DEFINITION 2.6.3. Given a probability measure μ on a metric space X, the r-scale entropy of μ is

$$H_r(\mu) = -\int \log(\mu(B(x,r))\, d\mu(x).$$

The entropy dimension of μ is

$$\dim_e(\mu) = \lim_{r\to 0} \frac{H_r(\mu)}{-\log r}, \tag{2.27}$$

if the limit exists; otherwise one considers the upper and lower entropy dimensions $\overline{\dim}_e(\mu)$ and $\underline{\dim}_e(\mu)$.

L.-S. Young [**You82**, Prop. 4.3] gave a sufficient condition for the existence of entropy dimension.

THEOREM 2.6.4 (Young). *Let μ be a compactly supported, exact-dimensional measure of dimension α. Then the entropy dimension $\dim_e(\mu)$ exists and equals α.*

It is immediate from the definition and Hölder's inequality that $q \mapsto D_q(\mu)$ is nonincreasing, both for $q < 1$ and for $q > 1$. Fan, Lau, and Rao [**FLR02**, Theorem 1.4] proved the following:

THEOREM 2.6.5 (Fan, Lau, and Rao). *For $0 < p < 1 < q$, we have*

$$D_q(\mu) \leqslant \underline{\dim}_{\mathrm{H}}(\mu) \leqslant \underline{\dim}_e(\mu) \leqslant \overline{\dim}_e(\mu) \leqslant \dim_{\mathrm{P}}(\mu) \leqslant \overline{D}_p(\mu).$$

There are several alternative (equivalent) definitions of the L^q-dimension and the entropy dimension. In the rest of the section we restrict ourselves to Borel probability measures in $\mathbb{R}^d$ of compact support. Let $\mathscr{D}_n$ be the partition of $\mathbb{R}^d$ into dyadic grid boxes $\prod_{i=1}^d [k_i 2^{-n}, (k_i+1)2^{-n})$, with $k_i \in \mathbb{Z}$. For $q \neq 0$ denote

$$\tau_n^{(q)}(\mu) = \sum_{Q\in\mathscr{D}_n} (\mu(Q))^q.$$

The L^q spectrum of μ is defined by

$$\tau(q) = \tau(\mu, q) = \liminf_{n\to\infty} \frac{\log \tau_n^{(q)}(\mu)}{-n\log 2} \quad \text{for } q > 0,\ q \neq 1, \tag{2.28}$$

One can also consider the upper limit, denoted $\overline{\tau}(q)$.

LEMMA 2.6.6. *We have*

$$D_q(\mu) = \frac{\tau(\mu,q)}{q-1} \quad \textit{and} \quad \overline{D}_q(\mu) = \frac{\overline{\tau}(\mu,q)}{q-1}.$$

The proof follows from the inequality

$$C_{d,q}^{-1}\tau_{n(\varepsilon)}^{(q)} \leqslant \int \left(\mu(B(x,\varepsilon))\right)^{q-1} d\mu(x) \leqslant C_{d,q}\tau_{n(\varepsilon)}^{(q)}, \tag{2.29}$$

for a constant $C_{d,q} > 1$ depending only on d and q, where $n = n(\varepsilon)$ is defined by $2^{-n-1} \leqslant \varepsilon < 2^{-n}$. This inequality is a straightforward exercise for $q > 1$; for $q < 1$ it is slightly more involved—a proof is given in [**PS00b**, Lemma 4.1].

The definition of the L^q spectrum (2.28) may be extended to $q \leqslant 0$ by using the convention that $0^q = 0$ for all $q \in \mathbb{R}$. This extension is an essential ingredient in the study of multifractal phenomena, see [**Rie95**]. In this book, we do not discuss the general theory of multifractal analysis, but instead provide a "window" into it, by analyzing the most basic case of self-similar measures with strong separation in Chapter 5.

We should note though that for $q < 0$ the situation is more subtle; in particular, Lemma 2.6.6 does not hold in general. For further details, see Section 5.1.6. The following definition is more flexible; it extends to all $q \in \mathbb{R}$:

DEFINITION 2.6.7. Let ν be a finite Borel measure on $\mathbb{R}^d$ with compact support. Then for every $q \in \mathbb{R}$, the L^q spectrum of μ is

$$\tau(q) = \tau(\mu, q) = \liminf_{\delta \downarrow 0} \frac{\log\left(\sup \sum_i \mu(B(x,\delta))^q\right)}{\log \delta}, \tag{2.30}$$

where the supremum is taken over all families of disjoint balls centered in the support of μ.

2.6.1. L^q-dimension and L^q-density. We say that an absolutely continuous measure $\mu \ll \mathcal{L}^d$ has a density in L^q if its Radon–Nikodym derivative $\frac{d\mu}{dx}$ belongs to $L^q(\mathbb{R}^d)$.

PROPOSITION 2.6.8. *If μ is a finite compactly supported Borel probability measure on $\mathbb{R}^d$ and $D_q(\mu) < d$ for $q > 1$, then μ (assuming it is absolutely continuous) does not have an L^q density.*

The proof uses the following alternative definition of the L^q-dimension for $q > 1$, which is analogous to that of entropy dimension in Definition 2.6.3.

LEMMA 2.6.9 (see [**SS16**, Lemma 4.1]). *For any finite compactly supported Borel probability measure μ on $\mathbb{R}^d$ and $q > 1$,*

$$D_q(\mu) = \liminf_{r \to 0} \frac{\log\left(r^{-d} \int \mu(B(x,r))^q \, dx\right)}{(q-1)\log r}.$$

The proof of Proposition 2.6.8 is immediate from Lemma 2.6.9 and the following classical result (see e.g. [**HL28**]). Let $c_d > 0$ be such that $\mathcal{L}^d(B(0,r)) = c_d r^d$.

LEMMA 2.6.10. *Suppose that $h \in L^q(\mathbb{R}^d)$ for $q > 1$. Then*

$$\lim_{r \to 0} \int_{\mathbb{R}^d} \left(\frac{1}{c_d r^d} \int_{B(x,r)} |h(t)| \, dt\right)^q dx = \int_{\mathbb{R}^d} |h(x)|^q \, dx.$$

2.6.2. More on entropy dimension. The entropy dimension in $\mathbb{R}^d$ has the following equivalent definition, which was, in fact, the one given in (1.117). This assertion appeared in Peres and Solomyak [**PS00b**].

PROPOSITION 2.6.11. *Let μ be any compactly supported measure in $\mathbb{R}^d$. Then*

$$\begin{aligned}\underline{\dim}_e(\mu) &= \liminf_{n \to \infty} \frac{-\sum_{D \in \mathcal{D}_n} \mu(D) \log \mu(D)}{n \log 2} = \liminf_{n \to \infty} \frac{H(\mu, \mathcal{D}_n)}{n \log 2}; \\ \overline{\dim}_e(\mu) &= \limsup_{n \to \infty} \frac{-\sum_{D \in \mathcal{D}_n} \mu(D) \log \mu(D)}{n \log 2} = \limsup_{n \to \infty} \frac{H(\mu, \mathcal{D}_n)}{n \log 2}.\end{aligned} \tag{2.31}$$

Proposition 2.6.11 immediately follows from Lemma 2.6.12, since it is easy to see that, like in Lemma 1.9.3, the limits can be taken along exponential sequences.

LEMMA 2.6.12 (see [**PS00b**, Lemma 2.3]). *Let μ be any compactly supported measure in $\mathbb{R}^d$. Then*

(i)

$$H(\mu,\mathscr{D}_n) \leqslant H(\mu,\mathscr{D}_{n+k}) \leqslant H(\mu,\mathscr{D}_n) + k\cdot 2^d \quad \text{for all } n,k\in\mathbb{N}; \tag{2.32}$$

(ii) *Let $C_1(d) = \max\{2^d\cdot\lceil\frac{1}{2}\log_2 d\rceil, 3^d\}$. Then we have for all $n\in\mathbb{N}$:*

$$|H_{2^{-n}}(\mu) - H(\mu,\mathscr{D}_n)| = \left|\int \log\frac{1}{\mu(B(x,2^{-n}))}\,d\mu(x) - H(\mu,\mathscr{D}_n)\right| \leqslant C_1(d). \tag{2.33}$$

Part (i) of the lemma will be derived from a more general fact.

FACT 2.6.13. *Let $(X,\mu,\mathcal{B})$ be a probability measure space, and let $\mathcal{F},\mathcal{G}$ be two finite measurable partitions.*

(i) *If $\mathcal{F}\leqslant\mathcal{G}$, then $H(\mu,\mathcal{F})\leqslant H(\mu,\mathcal{G})$.*
(ii) *Suppose that for some $K\in\mathbb{N}$ it holds that each element of $\mathcal{G}$ intersects at most K elements of $\mathcal{F}$. Then*

$$H(\mu,\mathcal{F}) - H(\mu,\mathcal{G}) \leqslant K.$$

We provide a proof, since its ingredients will be used in the sequel.

PROOF. Let us write $\mathcal{F}(x)$, $\mathcal{G}(x)$ for the elements of $\mathcal{F},\mathcal{G}$ containing x respectively. Furthermore, if $F\in\mathcal{F}$, let $\mathcal{G}(F)$ be the union of all elements of $\mathcal{G}$ intersecting F. Note that

$$H(\mu,\mathcal{F}) = \int\log\frac{1}{\mu(\mathcal{F}(x))}\,d\mu(x). \tag{2.34}$$

(i) If $\mathcal{F}\leqslant\mathcal{G}$, then $\mathcal{F}(x)\supset\mathcal{G}(x)$ for all x and hence $H(\mu,\mathcal{F})\leqslant H(\mu,\mathcal{G})$.

(ii) We can estimate

$$\begin{aligned} H(\mu,\mathcal{F}) - H(\mu,\mathcal{G}) &= \int\log\Big(\frac{\mu(\mathcal{G}(x))}{\mu(\mathcal{F}(x))}\Big)d\mu(x) \leqslant \int\frac{\mu(\mathcal{G}(x))}{\mu(\mathcal{F}(x))}d\mu(x)\\ &\leqslant \int\frac{\mu(\mathcal{G}(\mathcal{F}(x)))}{\mu(\mathcal{F}(x))}d\mu(x) = \sum_{F\in\mathcal{F}}\mu(\mathcal{G}(F))\leqslant K, \end{aligned}$$

since by assumption the measure of each element of $\mathcal{G}$ is counted at most K times in the last sum. Above we used that $\log(x)\leqslant x$ for every $x>0$. □

PROOF OF LEMMA 2.6.12. **(i)** We have $H(\mu,\mathscr{D}_n)\leqslant H(\mu,\mathscr{D}_{n+k})$ by the first assertion of the Fact. On the other hand, every set in $D'\in\mathscr{D}_n$ intersects exactly 2^d sets $D\in\mathscr{D}_{n+1}$, and the right inequality in (2.32) for $k=1$ is a special case of Fact 2.6.13(ii). The case of an arbitrary $k\in\mathbb{N}$ follows by induction.

(ii) Since $\mathscr{D}_n(x)\subset B(x,2^{-n}\sqrt{d})$, it follows from (2.34) that

$$H(\mu,\mathscr{D}_n)\geqslant\int\log\frac{1}{\mu(B(x,2^{-n}\sqrt{d}))}\,d\mu(x)\quad\text{for all } n\in\mathbb{N}.$$

Thus, if $2^{-k}\sqrt{d}\leqslant 1$, that is, $k\geqslant\lceil\frac{1}{2}\log_2 d\rceil$, then by (2.32),

$$H(\mu,\mathscr{D}_n)\geqslant H(\mu,\mathscr{D}_{n+k}) - k\cdot 2^d\geqslant\int\log\frac{1}{\mu(B(x,2^{-n}))}\,d\mu(x) - k\cdot 2^d.$$

On the other hand, denote by $\widetilde{D}$, for $D \in \mathscr{D}_n(x)$, the union of D with the neighboring boxes in $\mathscr{D}_n(x)$, that is, the ones whose closure intersects the closure of D. Using that $B(x, 2^{-n}) \subset \widetilde{\mathscr{D}}_n(x)$ for any $x \in \mathbb{R}^d$, we have

$$\begin{aligned} H_{2^{-n}}(\mu) \quad - \quad & \int \log \frac{1}{\mu(B(x,2^{-n}))}\, d\mu(x) = \int \log \frac{\mu(B(x,2^{-n}))}{\mu(\mathscr{D}_n(x))}\, d\mu(x) \\ \leqslant \quad & \int \frac{\mu(B(x,2^{-n}))}{\mu(\mathscr{D}_n(x))}\, d\mu(x) \leqslant \int \frac{\mu(\widetilde{\mathscr{D}}_n(x))}{\mu(\mathscr{D}_n(x))}\, d\mu(x) \leqslant \sum_{D \in \mathscr{D}_n} \mu(\widetilde{D}) \leqslant 3^d. \end{aligned}$$

The proof of the lemma, and hence of Proposition 2.6.11, is complete. □

2.7. Generalized Hausdorff and packing measures

Here is a quote from Mattila [**Mat95**, 4.9] (an excerpt): *"Although the Hausdorff dimension measures the metric size of any set, the values of the Hausdorff measures often do not give much extra information. This is so since there may be no value s for which the set has positive and finite $\mathcal{H}^s$ measure. But often replacing $|A| \mapsto |A|^s$ by some other function of the diameter, one can find measures measuring the given set in a more delicate manner".*

Let $\phi : [0, \infty) \to [0, \infty)$ be an increasing function with $\phi(0) = 0$. By analogy with (1.50), define for a set $E \subset \mathbb{R}^d$:

$$\mathcal{H}^\phi_\delta(E) = \inf \left\{ \sum_{i=1}^\infty \phi(|A_i|) : \ E \subset \bigcup_{i=1}^\infty A_i, \ |A_i| \leqslant \delta \right\}, \tag{2.35}$$

and then $\mathcal{H}^\phi(E) = \lim_{\delta \to 0} \mathcal{H}^\phi_\delta(E)$. This is an outer measure, which in the usual way, restricting to the σ-algebra of measurable sets, gives the *Hausdorff measure* $\mathcal{H}^\phi$, *corresponding to the gauge function* ϕ. Taking $\phi(t) = t^s$ results in the Hausdorff measure $\mathcal{H}^s$.

For example, sets of Hausdorff dimension s can be distinguished in a more delicate manner using the gauges

$$\phi_\alpha(t) = t^s (\log t^{-1})^\alpha, \qquad \phi_{\alpha,\beta}(t) = t^s (\log t^{-1})^\alpha (\log\log t^{-1})^\beta,$$

etc. Such gauges are especially useful for random fractals: e.g., the trajectories of the Brownian motion in $\mathbb{R}^d$ have positive and σ-finite $\mathcal{H}^\phi$ measure almost surely, with (for small t):

$$\phi(t) = t^2 \log\log t^{-1} \ \ (d \geqslant 3), \ \ \phi(t) = t^2 \log t^{-1} \log\log\log t^{-1} \ \ (d = 2),$$

see [**Mat95**, Section 4.9] and references there.

In a similar way, one can define generalized packing measures with respect to a gauge function. Given a gauge function ϕ, let

$$\mathcal{P}^\phi_\delta(E) = \sup \left\{ \sum_{i=1}^\infty \phi(|A_i|) : \ \{A_i\} \text{ is a } \delta\text{-packing of } E \right\},$$

and then continue, as in Section 1.4.6.

CHAPTER 3

General properties of self-similar sets and measures

In this chapter we start to develop the subject in depth. We begin with some general consequences of the stationarity, like Laws of Pure Type, the Frostman property and weak*-continuity. Furthermore, we show that self-similar measures are either supported on a hyperplane or any hyperplane has zero measure. Most of these properties hold for the more general self-affine case and so we state the results in that way.

Next, we show that the Hausdorff dimension of a self-similar measure is always bounded above by the ratio of entropy and the Lyapunov exponent, and if the Strong Separation Property (SSP) holds, then the inequality becomes equality and the measure is exact dimensional, thus proving completely Corollary 1.9.11. We also introduce the notion of Garsia entropy, which plays an important role in the case when there are *exact overlaps*, see Section 1.6.2. In the sections that follow we prove a number of results which hold always, independent of any separation conditions. They are: equality of the box and Hausdorff dimensions and finiteness of the Hausdorff measure at this dimension, lower semicontinuity of dimensions as a function of parameters (for self-similar sets), and then the existence of correlation and L^q-dimensions and lower semicontinuity of the entropy dimension (for self-similar measures).

Throughout this chapter, except in the first section, we always assume that Λ is the attractor of the self-similar IFS $\mathcal{S} = (S_1, \ldots, S_m)$ of m contracting similarity transformations which act on $\mathbb{R}^d$. The contraction ratios of S_i are $0 < r_i < 1$, $i = 1 \ldots m$. The corresponding natural measure (see Definition 1.8.11) is denoted by ν.

3.1. Consequences of stationarity

In this section, we assume that the IFS $(S_1, \ldots, S_m)$ contains contractive invertible affine maps on $\mathbb{R}^d$ of the form $S_i(\mathbf{x}) = A_i\mathbf{x} + \mathbf{t}_i$, with the attractor Λ defined in (1.36). Let $\mathbf{p} = (p_1, \ldots, p_m)$ be a probability vector and let ν be the unique measure, called self-affine measure, satisfying

$$\nu(A) = \sum_{i=1}^{m} p_i \nu(S_i^{-1}A) \text{ for every Borel set } A \subset \mathbb{R}^d. \tag{3.1}$$

Further, let μ be the Bernoulli measure on $\Sigma = \{1, \ldots, m\}^{\mathbb{N}}$ defined in Example 1.3.4, and let $\Pi\colon \Sigma \to \Lambda$ be the natural projection defined in (1.11) or (1.42). By Proposition 1.8.3, $\nu = \Pi_*\mu$.

3.1.1. Geometric properties. The equality (3.1) has some nice geometric consequences. Namely, a self-affine measure is either supported in a hyperplane or no hyperplane has positive measure. We adopt here the convention that a 0-dimensional hyperplane is a point and a set of dimension strictly smaller than 0 is an empty set.

PROPOSITION 3.1.1. *Let $\mathcal{S} = (S_i)_{i=1}^m$ be a self-affine IFS on $\mathbb{R}^d$, let $\mathbf{p} = (p_1, \dots, p_m)$ be a probability vector, and let $\nu = \Pi_* \mathbf{p}^{\mathbb{N}}$ be the corresponding self-affine measure satisfying* (3.1). *Then exactly one of the following holds:*

(i) *there exists at most one hyperplane D of $\mathbb{R}^d$ of dimension at most $d-1$, such that $\nu(D) = 1$;*
(ii) *for any hyperplane D of dimension at most $d-1$, holds $\nu(D) = 0$.*

Note that if (i) holds then $\Lambda \subset D$, where Λ is the attractor of the IFS $\mathcal{S}$.

PROOF. We argue by contradiction, and suppose that there exists a hyperplane D of dimension at most $d-1$, such that $0 < \nu(D) < 1$. We may suppose that D is the smallest in the sense of dimension, that is, $\dim D = t$ and for any hyperplane D' of dimension $k \leqslant t-1$, holds $\nu(D') = 0$. Since for any two t-dimensional hyperplanes $D \neq D'$ we have $\dim(D \cap D') < t$, there are at most countably many t-dimensional hyperplanes with positive measure. Let us choose the hyperplane D to be the one with the maximal measure.

Since $\nu(D) < 1$, there exists a point $\mathbf{x} \in \Lambda \backslash D$. Let $r > 0$ be such that $B(\mathbf{x}, r) \cap D = \varnothing$. By the definition of natural projection, there exists $\mathbf{i} \in \Sigma^*$ such that $S_{\mathbf{i}}(\Lambda) \subset B(\mathbf{x}, r)$, so that $S_{\mathbf{i}}(\Lambda) \cap D = \varnothing$. Let $n := |\mathbf{i}|$ and let $\Omega = \{\mathbf{j} \in \Sigma_n : S_{\mathbf{j}}(\Lambda) \cap D \neq \varnothing\}$. It is clear that $\mathbf{i} \in \Sigma_n \backslash \Omega$ and for any $\mathbf{j} \in \Sigma_n \backslash \Omega$,

$$\nu(S_{\mathbf{j}}^{-1}(D)) = \nu(S_{\mathbf{j}}^{-1}(D) \cap \Lambda) \leqslant \nu\Big(S_{\mathbf{j}}^{-1}(D \cap S_{\mathbf{j}}(\Lambda))\Big) = 0.$$

In view of (1.88), which can be extended by iteration to the level n (essentially, by Lemma 1.3.3 applied to the minimal cut-set Σ_n),

$$\begin{aligned} 0 < \nu(D) &= \sum_{\mathbf{j} \in \Sigma_n} p_{\mathbf{j}} \nu(S_{\mathbf{j}}^{-1}(D)) = \sum_{\mathbf{j} \in \Omega} p_{\mathbf{j}} \nu(S_{\mathbf{j}}^{-1}(D)) \\ &\leqslant \sum_{\mathbf{j} \in \Omega} p_{\mathbf{j}} \nu(D) \leqslant (1 - p_{\mathbf{i}}) \nu(D), \end{aligned}$$

where in the second line we used that $S_{\mathbf{j}}^{-1}(D)$ is a t-dimensional hyperplane and D has maximal measure among such. This is a contradiction. □

A simple corollary of the previous proposition is that every self-affine measure is either continuous (in other words, atom-free or nonatomic) or supported on a singleton.

3.1.2. Frostman property. Using similar methods to the proof of Proposition 3.1.1, one can show that any self-affine measure, which is not supported on a singleton, is a Frostman measure. We will follow the proof of Feng and Lau [**FL09**, Proposition 2.2].

PROPOSITION 3.1.2. *Let $\mathcal{S} = (S_i)_{i=1}^m$ be a self-affine IFS on $\mathbb{R}^d$, let $\mathbf{p} = (p_1, \dots, p_m)$ be a probability vector, and let $\nu = \Pi_* \mathbf{p}^{\mathbb{N}}$ be the corresponding self-affine measure satisfying* (3.1). *Suppose that ν is not supported on a singleton. Then there exist constants $C > 0$ and $s > 0$ such that for every $\mathbf{x} \in \mathbb{R}^d$ and $r > 0$*

$$\nu(B(\mathbf{x}, r)) \leqslant C r^s.$$

PROOF. Let Λ be the attractor of $\mathcal{S}$. Furthermore, let $\rho := \min_i \|A_i^{-1}\|^{-1}$. Note that $\rho \in (0,1)$. Then for every $i = 1, \ldots, m$,

$$\|S_i(\mathbf{x}) - S_i(\mathbf{y})\| = \|A_i(\mathbf{x}-\mathbf{y})\| \geqslant \|A_i^{-1}\|^{-1}\|\mathbf{x}-\mathbf{y}\| \geqslant \rho\|\mathbf{x}-\mathbf{y}\|$$

for every $\mathbf{x}, \mathbf{y} \in \mathbb{R}^d$. For every $r > 0$, let $\phi(r) = \sup_{\mathbf{x}\in\mathbb{R}^d} \nu(B(\mathbf{x}, r))$.

Since Λ is not a singleton, there exist $\mathbf{i}, \mathbf{j} \in \Sigma^*$ such that $|\mathbf{i}| = |\mathbf{j}|$, $S_{\mathbf{i}}(\Lambda) \cap S_{\mathbf{j}}(\Lambda) = \varnothing$ and $|S_{\mathbf{i}}(\Lambda)|, |S_{\mathbf{j}}(\Lambda)| < |\Lambda|/3$. Hence, for every $r < |\Lambda|/6$, the ball $B(x,r)$ can intersect at most one of $\{S_{\mathbf{i}}(\Lambda), S_{\mathbf{j}}(\Lambda)\}$. Thus, either $S_{\mathbf{i}}^{-1}(B(\mathbf{x},r)) \cap \Lambda = \varnothing$ or $S_{\mathbf{j}}^{-1}(B(\mathbf{x},r)) \cap \Lambda = \varnothing$. Let $q := \max\{1-p_{\mathbf{i}}, 1-p_{\mathbf{j}}\} < 1$ and $n := |\mathbf{i}| = |\mathbf{j}|$. Since $S_{\mathbf{k}}^{-1}(B(\mathbf{x},r)) \subset B(S_{\mathbf{k}}^{-1}(\mathbf{x}), r/\rho)$, we get for every $\mathbf{x} \in \mathbb{R}^d$ and $r < |\Lambda|/6$ that

$$\nu(B(\mathbf{x},r)) = \sum_{\mathbf{k}\in\Sigma_n} p_{\mathbf{k}}\nu(S_{\mathbf{k}}^{-1}(B(\mathbf{x},r))) \leqslant q\phi(r/\rho).$$

Hence, $\phi(r) \leqslant q\phi(r/\rho)$. Let $n(r) = \lceil \frac{\log(6r/|\Lambda|)}{\log\rho} \rceil$ and so for every $0 < r < |\Lambda|/6$

$$\phi(r) \leqslant q^{n(r)+1}\phi(r/\rho^{n(r)+1}) \leqslant \frac{6}{|\Lambda|} r^{\log q/\log\rho}.$$

For $r \geqslant |\Lambda|/6$, clearly $\phi(r) \leqslant 1 \leqslant \left(\frac{6}{|\Lambda|}\right)^{\log q/\log\rho} r^{\log q/\log\rho}$, which completes the proof. □

3.1.3. Weak* continuity of self-affine measures. Another important but quite easy corollary of Proposition 1.8.3 is continuous dependence of self-similar measures on the generating IFS and on the probabilities. Recall that a sequence of Borel probability measures μ_n on a metric space S converges in the weak* sense to a Borel probability measure μ if for every continuous and bounded function $f\colon S \to \mathbb{R}$,

$$\lim_{n\to\infty} \int f(x)\,d\mu_n(x) = \int f(x)\,d\mu(x). \tag{3.2}$$

In probability theory, the weak* convergence is often called simply the weak convergence of measures. By Alexandroff's Theorem [**Bog07**, Theorem 8.2.3] or Portmanteau Theorem [**Bil99**, Theorem 2.1], the following are equivalent:

- μ_n converges to μ in the weak* sense,
- $\liminf_{n\to\infty} \mu_n(U) \geqslant \mu(U)$ for every open set $U \subset S$,
- $\limsup_{n\to\infty} \mu_n(C) \leqslant \mu(C)$ for every closed set $C \subset S$,
- $\lim_{n\to\infty} \mu_n(A) = \mu(A)$ for every set $A \subset S$ such that $\mu(\partial A) = 0$, where $\partial A = \overline{A}\backslash A^o$ denotes the boundary of A.

This convergence induces the weak* topology, which is the topology generated by the open sets $\{\mu : |\int f d\mu - x| < \delta\}$, where f is continuous and bounded, $x \in \mathbb{R}$ and $\delta > 0$. Furthermore, if S is separable then the weak* topology is metrizable and separable, for example, by the Kantorovich–Rubinstein metric, see for example [**Bog07**, Theorem 8.2.18].

Now, let us consider the space $\mathscr{A}_m(d)$ of iterated function systems of m affinities on $\mathbb{R}^d$, see Notation 2. Furthermore, let $\mathcal{P}_m$ be the collection of probability vectors $\mathbf{p} = (p_1, \ldots, p_m)$ with $p_i > 0$ for all $i \in [m] := \{1, \ldots, m\}$. The topology on $\mathcal{P}_m$ is generated by the Euclidean distance and the topology of $\mathscr{A}_m(d)$ is generated by the product of topologies induced by the natural metric on the space of affine maps

$d(f,g) = \max\{\|f(0)-g(0)\|, \|Df(0)-Dg(0)\|\}$. Denote this induced metric on $\mathscr{A}_m(d)$ by $d_{\mathscr{A}}$.

PROPOSITION 3.1.3. *Let $\mathcal{F}_n = \left(f_i^{(n)}(\mathbf{x}) = A_i^{(n)}\mathbf{x} + \mathbf{t}_i^{(n)}\right)_{i=1}^m \in \mathscr{A}_m(d)$ and $\mathbf{p}_n \in \mathcal{P}_m$ be sequences such that $\mathcal{F}_n$ converges to $\mathcal{F} = (f_i(\mathbf{x}) = A_i\mathbf{x} + \mathbf{t}_i)_{i=1}^m \in \mathscr{A}_m(d)$ and $\mathbf{p}_n$ converges to $\mathbf{p} \in \mathcal{P}_m$. Furthermore, let ν_n and ν be the self-affine measures generated by the pairs $(\mathcal{F}_n, \mathbf{p}_n)$ and $(\mathcal{F}, \mathbf{p})$. Then ν_n converges to ν in the weak* sense.*

PROOF. Let $\mu_{\mathbf{p}_n}$ and $\mu_{\mathbf{p}}$ be the Bernoulli measures over the symbolic space $\Sigma = \{1,\dots,m\}^{\mathbb{N}}$. By Portmanteau Theorem, it is easy to see that $\mu_{\mathbf{p}_n}$ converges to $\mu_{\mathbf{p}}$ weak*.

Now, let Π_n and Π be the natural projections corresponding to the IFS's $\mathcal{F}_n$ and $\mathcal{F}$, furthermore let $\rho = \sup_{i,n}\|A_i^{(n)}\| < 1$ and $K = \sup_{i,n}\|\mathbf{t}_i\|$. It is easy to see that for every $\mathbf{i} \in \Sigma$,

$$\begin{aligned}
&|\Pi_n(\mathbf{i}) - \Pi(\mathbf{i})| \\
&= \left|\sum_{k=1}^{\infty} A_{i_1}^{(n)}\cdots A_{i_{k-1}}^{(n)}\mathbf{t}_{i_k}^{(n)} - \sum_{k=1}^{\infty} A_{i_1}\cdots A_{i_{k-1}}\mathbf{t}_{i_k}\right| \\
&\leqslant \left|\sum_{k=1}^{\infty} A_{i_1}\cdots A_{i_{k-1}}(\mathbf{t}_{i_k}^{(n)} - \mathbf{t}_{i_k})\right| + \left|\sum_{k=1}^{\infty}\left(A_{i_1}^{(n)}\cdots A_{i_{k-1}}^{(n)} - A_{i_1}\cdots A_{i_{k-1}}\right)\mathbf{t}_{i_k}^{(n)}\right| \\
&\leqslant \sum_{k=1}^{\infty}\rho^{k-1} d_{\mathscr{A}}(\mathcal{F}_n,\mathcal{F}) + \sum_{k=1}^{\infty}\sum_{\ell=0}^{k-2}\left|A_{i_1}^{(n)}\cdots A_{i_\ell}^{(n)}\left(A_{i_{\ell+1}} - A_{i_{\ell+1}}^{(n)}\right)A_{i_{\ell+2}}\cdots A_{i_{k-1}}\mathbf{t}_{i_k}^{(n)}\right| \\
&\leqslant \left(\sum_{k=1}^{\infty}\rho^{k-1} + \sum_{k=1}^{\infty}(k-1)\rho^{k-2}K\right) d_{\mathscr{A}}(\mathcal{F}_n,\mathcal{F}).
\end{aligned}$$

Thus, for every bounded and Lipshitz function $f\colon \mathbb{R}^d \to \mathbb{R}$, the sequence of bounded and continuous maps $g_n := f\circ\Pi_n\colon \Sigma \to \mathbb{R}$ converges uniformly to $g := f\circ\Pi\colon \Sigma \to \mathbb{R}$. Hence,

$$\left|\int f\,d\nu_n - \int f d\nu\right| = \left|\int g_n\,d\mu_{\mathbf{p}_n} - \int g\,d\mu_{\mathbf{p}}\right| \leqslant \|g_n - g\| + \left|\int g\,d\mu_{\mathbf{p}_n} - \int g\,d\mu_{\mathbf{p}}\right|,$$

where the left hand side converges to zero. □

3.1.4. Laws of pure type. Now, we turn to the case of self-similar measures. Jessen and Wintner [**JW35**] showed in 1935 that any convergent infinite convolution of discrete measures is of *pure type:* it is either singular or absolutely continuous with respect to Lebesgue measure. In particular, this applies to homogeneous self-similar measures. The purity of ν_λ can also be obtained from its self-similarity, as the following proposition demonstrates.

PROPOSITION 3.1.4. *Suppose that ν is a self-similar probability measure on $\mathbb{R}^d$ corresponding to the contracting similitudes $(S_1,\dots,S_m)$ and the (positive) probabilities $\{p_j : j = 1,\dots m\}$*, i.e.,

$$\nu = \sum_{j=1}^{m} p_j\left(\nu\circ S_j^{-1}\right). \tag{3.3}$$

Let Λ denote the closed support of ν, that is, Λ is the attractor of the IFS. If ν is not singular with respect to the Lebesgue measure $\mathcal{L}^d$ on $\mathbb{R}^d$, then

(i) *ν is absolutely continuous with respect to $\mathcal{L}^d$.*
(ii) *The restriction $\mathcal{L}^d|_\Lambda$ of $\mathcal{L}^d$ to Λ is absolutely continuous with respect to ν.*

Part **(i)** of this proposition is folklore; in fact, it holds in much greater generality: for any IFS of invertible contractions whose inverses preserve sets of zero Lebesgue measure. part **(ii)** was proved by Mauldin and Simon [**MS98**] for Bernoulli convolutions ν_λ, and we extend their proof to general self-similar measures below, following [**PSS00a**].

PROOF. **(i)** Apply the similitude S_j to the Lebesgue decomposition $\nu = \nu_{\mathrm{ac}} + \nu_{\mathrm{s}}$ of ν into an absolutely continuous part and a singular part. Averaging the result over j with weights p_j, we infer that ν_{ac} and ν_{s} also satisfy (3.3). Since ν is the only probability measure satisfying (3.3) (see Theorem 1.8.2), it follows that ν_{ac} and ν_{s} are proportional, hence one of them must vanish.

(ii) Since ν is not singular to $\mathcal{L}^d$, necessarily $\mathcal{L}^d(\Lambda) > 0$ and for some $\beta \in [0,1]$,

$$\sup\Big\{\mathcal{L}^d(A) : A \text{ Borel, } A \subset \Lambda,\ \nu(A) = 0\Big\} = \beta\mathcal{L}^d(\Lambda). \tag{3.4}$$

Observe that $\beta < 1$, since otherwise there exist Borel $A_n \subset \Lambda$ with $\nu(A_n) = 0$ and $\mathcal{L}^d(A_n) \geqslant (1-\frac{1}{n})\mathcal{L}^d(\Lambda)$, whereas $\widetilde{A} := \bigcup_{n=1}^\infty A_n$ has $\nu(\widetilde{A}) = 0$ and $\mathcal{L}^d(\widetilde{A}) = \mathcal{L}^d(\Lambda)$, and then ν is singular.

Let A_0 be a Borel subset of Λ such that $\nu(A_0) = 0$. Our strategy is to show that the set A_0 cannot have a Lebesgue density point and thus, necessarily has 0 Lebesgue measure.

Denote by $r_{\min}$ the minimal contraction ratio for the maps $\{S_j\}_{j=1}^m$. Fix $x \in \Lambda$ and $r \in (0, |\Lambda|)$. Since $\Lambda = \cup_{j=1}^m S_j(\Lambda)$, there exists a similitude $S_{\mathbf{i}} = S_{i_1} \circ S_{i_2} \circ \ldots \circ S_{i_\ell}$ such that $x \in \Lambda_{\mathbf{i}}$ and $\Lambda_{\mathbf{i}}$ is contained in the open ball $B(x,r)$. Choose such an $S_{\mathbf{i}}$ with ℓ minimal; clearly $|\Lambda_{\mathbf{i}}| = |\Lambda| \cdot r_{\mathbf{i}} \geqslant r \cdot r_{\min}$. We claim that $\mathcal{L}^d(\Lambda_{\mathbf{i}}) \geqslant \eta\mathcal{L}^d(B(x,r))$, where $\eta > 0$ does not depend on x and r. Indeed,

$$\frac{\mathcal{L}^d(\Lambda_{\mathbf{i}})}{\mathcal{L}^d(B(x,r))} = \frac{r_{\mathbf{i}}^d\mathcal{L}^d(\Lambda)}{r^d\mathcal{L}^d(B(0,1))} \geqslant \frac{r_{\min}^d\mathcal{L}^d(\Lambda)}{|\Lambda|^d \cdot \mathcal{L}^d(B(0,1))} =: \eta.$$

The preimage $S_{\mathbf{i}}^{-1}\big(A_0 \cap S_{\mathbf{i}}(\Lambda)\big)$ is a Borel subset of Λ, and ν assigns it measure zero, because

$$0 = \nu(A_0) = \sum_{|\mathbf{j}|=\ell} p_{\mathbf{j}}\nu(S_{\mathbf{j}}^{-1}A_0) \geqslant p_{\mathbf{i}}\nu(S_{\mathbf{i}}^{-1}A_0),$$

where we used self-similarity. By (3.4) and scaling,

$$\mathcal{L}^d\big(A_0 \cap \Lambda_{\mathbf{i}}\big) = r_{\mathbf{i}}^d\mathcal{L}^d\Big(S_{\mathbf{i}}^{-1}\big(A_0 \cap \Lambda_{\mathbf{i}}\big)\Big) \leqslant r_{\mathbf{i}}^d\beta\mathcal{L}^d(\Lambda) = \beta\mathcal{L}^d\big(\Lambda_{\mathbf{i}}\big),$$

and therefore,

$$\begin{aligned}\mathcal{L}^d\big(B(x,r)\backslash A_0\big) \geqslant \mathcal{L}^d\big(\Lambda_{\mathbf{i}}\backslash A_0\big) \ &\geqslant\ (1-\beta)\mathcal{L}^d\big(\Lambda_{\mathbf{i}}\big)\\ &\geqslant\ (1-\beta)\eta \cdot \mathcal{L}^d\big(B(x,r)\big)\,.\end{aligned}$$

Thus A_0 cannot have a Lebesgue density point, whence $\mathcal{L}^d(A_0) = 0$. □

3.2. On the dimension of self-similar measures.

The following result is essentially due to Billingsley [**Bil60, Bil61**].

THEOREM 3.2.1. *Let $\mathcal{S}$ be a self-similar IFS on $\mathbb{R}^d$ with contraction ratios given by the vector $\mathbf{r} = (r_1, \ldots, r_m)$, and let ν be the self-similar measure for $\mathcal{S}$ with the probability vector $\mathbf{p} = (p_1, \ldots, p_m)$. Then we have*

$$\dim_{\mathrm{H}}(\nu) \leqslant \dim_{\mathrm{P}}(\nu) \leqslant \frac{-\sum_{i=1}^m p_i \log p_i}{-\sum_{i=1}^m p_i \log r_i} = \frac{h_{\mathbf{p}}}{\chi_{\mathbf{r}}^{\mathbf{p}}} = \dim_{\mathrm{S}}(\nu). \tag{3.5}$$

Moreover, if $\mathcal{S}$ satisfies the SSP, then ν is exact dimensional and in particular, $\dim_{\mathrm{H}}(\nu) = \dim_{\mathrm{P}}(\nu) = \dim_{\mathrm{S}}(\nu)$.

Here $h_{\mathbf{p}}$ is the entropy (1.104) of the Bernoulli measure $\mu_{\mathbf{p}} = \mathbf{p}^{\mathbb{N}}$ on the symbolic space Σ and $\chi_{\mathbf{r}}^{\mathbf{p}}$ is the Lyapunov exponent (1.105) for the IFS. Recall that the ratio $h_{\mathbf{p}}/\chi_{\mathbf{r}}^{\mathbf{p}}$ is called the *similarity dimension of the measure*, see Definition 1.9.10. Theorem 3.2.1 asserts that the upper bound holds without any separation conditions, which is useful when studying self-similar IFS with overlaps, but also that (3.5) becomes equality if the SSP holds, proving Corollary 1.9.11. As we already mentioned, this last claim holds under the weaker assumption of the OSC, but the proof is more technical, see e.g., [**Edg98**, Theorem 5.2.5] and Corollary 11.3.8 in Section 11.3.

PROOF. Consider the sequence of independent identically distributed random variables $I_1, I_2, \ldots$, taking the values in $\{1, \ldots, m\}$, such that $\mathbb{P}(I_j = i) = p_i$ for $i = 1, \ldots, m$. Further, let us consider the sequence of random variables $f_k = -\log p_{I_k}$ and $g_k = -\log r_{I_k}$. By the Strong Law of Large Numbers,

$$\begin{aligned} \frac{1}{n}\sum_{k=1}^{n} f_k &\to \mathbb{E}(f_1) = -\sum_{i=1}^{m} p_i \log p_i = h_{\mathbf{p}} \text{ and} \\ \frac{1}{n}\sum_{k=1}^{n} g_k &\to \mathbb{E}(g_1) = -\sum_{i=1}^{m} p_i \log r_i = \chi_{\mathbf{r}}^{\mathbf{p}} \end{aligned} \tag{3.6}$$

almost surely. Observe that for any finite word $(i_1, \ldots, i_n) \in \Sigma^*$,

$$\mathbb{P}(I_1 = i_1, \ldots, I_n = i_n) = p_{i_1} \cdots p_{i_n} = \mu_{\mathbf{p}}([i_1, \ldots, i_n]),$$

where $[\mathbf{j}]$ is the cylinder set of sequences starting with $\mathbf{j} \in \Sigma^*$. Hence, the distribution of entire i.i.d. sequence is $\{I_j\}_{j \geqslant 1}$ is $\mu_{\mathbf{p}}$. Thus, (3.6) means in particular that

$$\mu_{\mathbf{p}}\left(\left\{\mathbf{i} = (i_1, i_2, \ldots) \in \Sigma : \frac{-\log(p_{i_1} \cdots p_{i_n})}{n} \to h_{\mathbf{p}}\right\}\right) = 1 \tag{3.7}$$

and

$$\mu_{\mathbf{p}}\left(\left\{\mathbf{i} = (i_1, i_2, \ldots) \in \Sigma : \frac{-\log(r_{i_1} \cdots r_{i_n})}{n} \to \chi_{\mathbf{r}}^{\mathbf{p}}\right\}\right) = 1. \tag{3.8}$$

Let $\Pi : \Sigma \to \mathbb{R}^d$ be the natural projection map associated with the IFS $\mathcal{S}$, so that the self-similar measure is $\nu = \Pi_*(\mu_{\mathbf{p}})$ by Proposition 1.8.3. To estimate the dimension of ν, we will use Theorem 1.9.5**(ii)** which says that $\dim_{\mathrm{P}}(\nu)$ is the essential supremum of the upper local dimensions $\overline{d}_\nu(x)$. Thus, we can take x from a set of full ν-measure and estimate $\overline{d}_\nu(x)$ from above. Since ν is the push-forward of $\mu_{\mathbf{p}}$, we can assume that $x = \Pi(\mathbf{i})$, where $\mathbf{i}$ is contained in the sets in (3.7) and (3.8). Let Λ be the attractor of the IFS of diameter $|\Lambda|$. We have

$$|\Pi([\mathbf{i}|_n])| = r_{i_1} \cdots r_{i_n}|\Lambda| =: r_n,$$

where we use the notation $\mathbf{i}|_n = i_1 \dots i_n$. Since $x = \Pi(\mathbf{i})$, we have $\Pi([\mathbf{i}|_n]) \subset B(\Pi(\mathbf{i}), r_n)$ and thus

$$\nu(B(x, r_n)) = \mu_{\mathbf{p}}\big(\{\mathbf{j} : \ |\Pi(\mathbf{j}) - \Pi(\mathbf{i})| \leqslant r_n\}\big) \geqslant \mu_{\mathbf{p}}([\mathbf{i}|_n]).$$

Therefore, keeping in mind that $\log(r_n)$ is negative for large n, we obtain by Lemma 1.9.3,

$$\begin{aligned}\overline{d}_\nu(x) &= \limsup_{r\to 0} \frac{\log \nu(B(x,r))}{\log r} = \limsup_{n\to\infty} \frac{\log \nu(B(x,r_n))}{\log r_n} \\ &\leqslant \limsup_{n\to\infty} \frac{\log \mu_{\mathbf{p}}([\mathbf{i}|_n])}{\log r_n} = \limsup_{n\to\infty} \frac{\log \mu_{\mathbf{p}}([\mathbf{i}|_n])}{n} \cdot \frac{n}{\log|\Lambda| + \log(r_{i_1}\cdots r_{i_n})} \\ &= \lim_{n\to\infty} \frac{\log \mu_{\mathbf{p}}([\mathbf{i}|_n])}{n} \cdot \lim_{n\to\infty} \frac{n}{\log(r_{i_1}\cdots r_{i_n})} = \frac{h_{\mathbf{p}}}{\chi_{\mathbf{r}}^{\mathbf{p}}},\end{aligned}$$

for $\nu_{\mathbf{p}}$-almost every x, as desired.

To finish the proof, suppose that the SSP holds. Then by compactness, there exists $\rho := \min_{i\neq j} \operatorname{dist}(S_i(\Lambda), S_j(\Lambda))/2 > 0$. Now, let us define R_n by

$$R_n = r_{i_1} \cdots r_{i_n}\rho.$$

Then we have $\Pi([\mathbf{i}|_n]) \supset B(\Pi(\mathbf{i}), R_n) \cap \Lambda$. Indeed, if $\mathbf{j} \in \{1, \dots, m\}^n$ is such that $\mathbf{i}|_n \neq \mathbf{j}$, then, denoting $\ell := \big|\mathbf{i}|_n \wedge \mathbf{j}\big|$, we have

$$\operatorname{dist}(S_{\mathbf{i}|_n}(\Lambda), S_{\mathbf{j}}(\Lambda)) = r_{\mathbf{i}|_n \wedge \mathbf{j}} \cdot \operatorname{dist}(S_{\sigma^\ell \mathbf{i}|_n}(\Lambda), S_{\sigma^\ell \mathbf{j}}(\Lambda)) > \rho r_{\mathbf{i}|_n \wedge \mathbf{j}} \geqslant R_n.$$

Hence, $\nu(B(x, R_n)) \geqslant \mu_{\mathbf{p}}([\mathbf{i}_n])$, and by Lemma 1.9.3, for ν-a.e. x,

$$\begin{aligned}\underline{d}_\nu(x) &= \liminf_{r\to 0} \frac{\log \nu(B(x,r))}{\log r} = \liminf_{n\to\infty} \frac{\log \nu(B(x,R_n)\cap\Lambda)}{\log R_n} \\ &\geqslant \liminf_{n\to\infty} \frac{\log \mu_{\mathbf{p}}([\mathbf{i}|_n])}{\log R_n} = \liminf_{n\to\infty} \frac{\log \mu_{\mathbf{p}}([\mathbf{i}|_n])}{n} \cdot \frac{n}{\log\rho + \log(r_{i_1}\cdots r_{i_n})} \\ &= \frac{h_{\mathbf{p}}}{\chi_{\mathbf{r}}^{\mathbf{p}}}.\end{aligned}$$

□

Theorem 3.2.1 extends to the projection of any *ergodic shift-invariant measure* μ on Σ (see Definition 1.9.12), and we state it for future use. We will need the notion of *entropy*, see Section 1.9.6 and, in particular, Proposition 1.9.20.

THEOREM 3.2.2. *Let $\mathcal{S}$ be a self-similar IFS on $\mathbb{R}^d$ with contraction ratios given by the vector $\mathbf{r} = (r_1, \dots, r_m)$, and let Π be the corresponding natural projection. Suppose that μ is an arbitrary ergodic σ-invariant measure on Σ with entropy*

$$h_\mu = \lim_{n\to\infty} \frac{-1}{n} \sum_{|\mathbf{i}|=n} \mu([\mathbf{i}]) \log \mu([\mathbf{i}]). \tag{3.9}$$

Consider the push-forward measure $\nu = \Pi_*(\mu) = \mu \circ \Pi^{-1}$. *Define the* Lyapunov exponent *of* μ *(with respect to the IFS) by*

$$\chi_\mu := \int_\Sigma f(\mathbf{i})\, d\mu(\mathbf{i}), \quad \textit{where } f(\mathbf{i}) = -\log r_{i_1}.$$

Then

$$\dim_{\mathrm{H}}(\nu) \leqslant \frac{h_\mu}{\chi_\mu},$$

and if the IFS satisfies the SSP, this becomes equality and ν *is exact dimensional.*

The proof is essentially the same as that of Theorem 3.2.1, except that (3.7) needs to be replaced by

$$-\frac{1}{n}\log \mu([\mathbf{i}|_n]) \to h_\mu \quad \text{for } \mu\text{-a.e. } \mathbf{i},$$

which is a consequence of the Shannon–McMillan–Breiman Theorem, see Corollary 9.5.4. Moreover, (3.8) should be replaced as well, by

$$\frac{1}{n}\sum_{k=1}^{n} g(\sigma^k \mathbf{i}) \to \chi_\mu \quad \text{for } \mu\text{-a.e. } \mathbf{i},$$

for $g(\mathbf{i}) = -\log r_{i_1}$, and this follows from Birkhoff's Ergodic Theorem, see (1.108) and Theorem 9.1.1.

In the special case when the IFS is homogeneous, that is, all the contraction ratios are equal to λ, we have $\chi_\mu = -\log\lambda$ for any μ. Thus we have

COROLLARY 3.2.3. *Let* $\mathcal{S}$ *be a homogeneous self-similar IFS on* $\mathbb{R}^d$ *with a common contraction ratio* λ, *and let* Π *be the corresponding natural projection. Suppose that* μ *is an arbitrary ergodic* σ*-invariant measure on* Σ *with entropy* h_μ. *Then*

$$\dim_{\mathrm{H}}(\Pi_*(\mu)) \leqslant \frac{h_\mu}{-\log\lambda}.$$

3.2.1. Random walk entropy. Let

$$\mathcal{S} = (S_1, \ldots, S_m)$$

be an IFS of contracting similitudes in $\mathbb{R}^d$, with the vector of contraction ratios $\mathbf{r} = (r_1, \ldots, r_m)$, and a probability vector $\mathbf{p} = (p_1, \ldots, p_m)$. There is a probabilistic interpretation of the attractor and the self-similar measure. It was developed by Diaconis–Shashahani [**DS86**] (see also Diaconis–Freedman [**DF99**] and references therein) and Barnsley–Demko [**BD85**] (see also Barnsley [**Bar93**]), called the *Chaos Game.* Pick any point $x_0 \in \mathbb{R}^d$ and apply one of the maps $S_i \in \mathcal{S}$ randomly and independently, with probability p_i. This way we get a random walk $(X_n)_{n\geqslant 1} = (S_{i_n} \circ \cdots \circ S_{i_1}(x_0))_{n\geqslant 1}$. The distribution of the random walk converges weak* to the self-similar measure $\nu = \Pi_*\mu_{\mathbf{p}}$, where $\Pi : \Sigma \to \Lambda$ is the natural projection and $\mu_{\mathbf{p}} = \mathbf{p}^{\mathbb{N}}$ is a Bernoulli measure on $\Sigma = \{1, \ldots, m\}^{\mathbb{N}}$. To be more precise, consider

$$\eta = \sum_{i=1}^{m} p_i \delta_{S_i}, \tag{3.10}$$

a discrete measure on the group of similarity transformations in $\mathbb{R}^d$, and its convolution powers

$$\eta^{*n} = \sum_{\mathbf{i}\in\Sigma_n} p_{\mathbf{i}}\delta_{S_{\mathbf{i}}}, \quad \text{where } p_{\mathbf{i}} = p_{i_1}\cdots p_{i_n}, \quad S_{\mathbf{i}} = S_{i_1}\cdots S_{i_n}. \tag{3.11}$$

Further, for $x_0 \in \mathbb{R}^d$ consider a sequence of discrete probability measures on $\mathbb{R}^d$ given by

$$\eta^{*n} \cdot x_0 = \sum_{\mathbf{i}\in\Sigma_n} p_{\mathbf{i}}\delta_{S_{\mathbf{i}}x_0}. \tag{3.12}$$

Then $\eta^{*n} \cdot x_0 \to \nu$ in the weak* topology, for any $x_0 \in \mathbb{R}^d$. The proof is left as an exercise. We do not assume familiarity with advanced probability theory and we will not use the Chaos Game in the book; it is mentioned only as a motivation. For a careful treatment of the Chaos Game, see, e.g., [**Edg98**, Chapter 5].

By definition (1.109), the *Shannon entropy* of the finitely-supported discrete measure η is

$$H(\eta) = -\int \log \eta(S)\, d\eta(S).$$

If $S_i \neq S_j$ for every $i \neq j$ then $H(\eta)$ coincides with $h_{\mathbf{p}}$, the entropy of the Bernoulli measure $\mathbf{p}^{\mathbb{N}}$ on the symbolic space, used above. Recall that by a *probabilistic IFS* we mean an IFS together with a probability vector.

LEMMA 3.2.4. *The sequence* $\{H(\eta^{*n})\}_{n\geqslant 1}$ *is subadditive.*

PROOF. We need to show that

$$H(\eta^{*(\ell+n)}) \leqslant H(\eta^{*\ell}) + H(\eta^{*n}) \quad \text{for any } \ell, n \in \mathbb{N}.$$

We consider a sequence of partitions α_n on the symbolic space Σ with the Bernoulli measure $\mathbf{p}^{\mathbb{N}}$. The partition α_n is obtained from the level-n cylinder partition ξ_n by combining together into one atom all those $[\mathbf{i}], [\mathbf{i}'] \in \xi_n$ for which $S_{\mathbf{i}} = S_{\mathbf{i}'}$. By the definition of entropy for a partition (Definition 1.9.15), holds $H(\eta^{*n}) = H(\alpha_n)$ for any $n \in \mathbb{N}$. We claim that

$$\alpha_{\ell+n} \leqslant \sigma^{-\ell}\alpha_n \vee \alpha_\ell. \tag{3.13}$$

Indeed, $\sigma^{-\ell}\alpha_n \vee \alpha_\ell$ is obtained from $\xi_{\ell+n}$ by combing together all those $[\mathbf{ij}]$ and $[\mathbf{i}'\mathbf{j}']$ in $\xi_{\ell+n}$, with $\mathbf{i}, \mathbf{i}' \in \Sigma_\ell$ and $\mathbf{j}, \mathbf{j}' \in \Sigma_n$, for which $S_{\mathbf{i}} = S_{\mathbf{i}'}$ and $S_{\mathbf{j}} = S_{\mathbf{j}'}$, whereas $\alpha_{\ell+n}$ must take account not only these, but all exact coincidences of length $\ell + n$. This implies (3.13), and therefore,

$$H(\alpha_{\ell+n}) \leqslant H(\sigma^{-\ell}\alpha_n \vee \alpha_\ell) \leqslant H(\sigma^{-\ell}\alpha_n) + H(\alpha_\ell) = H(\alpha_n) + H(\alpha_\ell),$$

concluding the proof. Here we used (1.113) and (1.111), as well as σ-invariance of the measure. (In fact, the second inequality above is equality, because the partitions $\sigma^{-\ell}\alpha_n$ and α_ℓ are independent with respect to the product measure $\mathbf{p}^{\mathbb{N}}$.) □

DEFINITION 3.2.5. The random walk entropy of the probabilistic IFS $(\mathcal{S}, \mathbf{p})$ is

$$h_{\mathrm{RW}}(\eta) = \lim_{n\to\infty} \frac{H(\eta^{*n})}{n}, \tag{3.14}$$

where η is given by (3.10).

The existence of the limit in (3.14) follows from Lemma 3.2.4 and Fekete's Lemma 1.9.21, which also implies that

$$h_{\mathrm{RW}}(\eta) = \inf_n \frac{H(\eta^{*n})}{n}.$$

In the context of Bernoulli convolutions the random walk entropy was defined by Garsia [**Gar63**], and so it is sometimes called the "Garsia entropy"; we will discuss it in Section 8.6. The random walk entropy takes into account exact coincidences in the IFS; the following lemma is immediate:

LEMMA 3.2.6. *We have*

$$h_{\mathrm{RW}}(\eta) = H(\eta) = H(\mathbf{p})$$

if and only if the semigroup generated by $\{S_1, \ldots, S_m\}$ *is free.*

Next comes the main statement of this subsection.

THEOREM 3.2.7. *Let* $\mathcal{S} = (S_1, \ldots, S_m)$ *be a self-similar IFS on* $\mathbb{R}^d$ *with contraction ratios given by the vector* $\mathbf{r} = (r_1, \ldots, r_m)$, *and let* ν *be the self-similar measure for* $\mathcal{S}$ *with the probability vector* $\mathbf{p} = (p_1, \ldots, p_m)$. *Consider* $\eta = \sum_{i=1}^m p_i \delta_{S_i}$. *Then we have*

$$\dim_{\mathrm{H}}(\nu) \leqslant \frac{h_{\mathrm{RW}}(\eta)}{-\sum_{i=1}^m p_i \log r_i} = \frac{h_{\mathrm{RW}}(\eta)}{\chi_{\mathbf{r}}^{\mathbf{p}}}. \tag{3.15}$$

PROOF. Fix $n \in \mathbb{N}$, and consider the associated order-n probabilistic IFS

$$(\mathcal{S}^{(n)}, \mathbf{p}^{(n)}) := \big((S_{\mathbf{i}})_{\mathbf{i}\in\Sigma_n}, (p_{\mathbf{i}})_{\mathbf{i}\in\Sigma_n}\big).$$

It is easy to see by induction that ν is also the self-similar measure for $(\mathcal{S}^{(n)}, \mathbf{p}^{(n)})$. Note that the "correct" alphabet for $\mathcal{S}^{(n)}$ is the support of the measure η^{*n}, which takes into account the coincidences at level n, and the probabilities are the corresponding weights of the atoms. We then obtain from Theorem 3.2.1:

$$\dim_{\mathrm{H}}(\nu) \leqslant \frac{H(\eta^{*n})}{\chi_{\mathbf{r}^{(n)}}^{\mathbf{p}^{(n)}}},$$

where $\mathbf{r}^{(n)} = (r_{\mathbf{i}})_{\mathbf{i}\in\Sigma^n}$ is the vector of contraction ratios for $\mathcal{S}^{(n)}$. Now, a standard computation yields

$$\chi_{\mathbf{r}^{(n)}}^{\mathbf{p}^{(n)}} = n\chi_{\mathbf{r}}^{\mathbf{p}},$$

and the desired inequality (3.15) follows from the definition (3.14). □

3.3. Equality of box and Hausdorff dimension

THEOREM 3.3.1 (Falconer). *Let* $\mathcal{S} = (S_1, \ldots, S_m)$ *be a self-similar IFS with attractor* Λ. *Then the box dimension* $\dim_{\mathrm{B}}(\Lambda)$ *exists and moreover,* $\dim_{\mathrm{B}}(\Lambda) = \dim_{\mathrm{H}}(\Lambda)$.

Note that we do not assume any separation conditions here, and the dimension(s) of Λ may be less than the similarity dimension. The result is from [**Fal89**], but that paper treats a more general class of sets (called "quasi self-similar") which makes it more technical.

PROOF. By introducing a proper conjugation, say $f(x) = |\Lambda|^{-1}x$, we may assume without loss of generality that $|\Lambda| = 1$. Recall the notion of "Moran cut-set" introduced in (1.73): for $0 < \delta \leqslant 1$,

$$\widetilde{\mathcal{M}}_\delta = \{\mathbf{i} \in \Sigma^* : \ r_{\mathbf{i}} \leqslant \delta < r_{\mathbf{i}^-}\},$$

where $\mathbf{i}^-$ denotes the finite word obtained from $\mathbf{i}$ by deleting the last symbol of $\mathbf{i}$, and $r_{\mathbf{i}} \in (0,1)$ is the contraction ratio of $S_{\mathbf{i}}$. Notice that

$$\delta r_{\min} < r_{\mathbf{i}} \leqslant \delta \text{ for } \ \mathbf{i} \in \widetilde{\mathcal{M}}_\delta,$$

where $r_{\min} = \min\{r_1, \dots, r_m\}$. By (1.7), we can find a *closed* ball $\overline{B} = \overline{B}(\mathbf{0}, R)$ such that

$$S_i(\overline{B}) \subset \overline{B} \text{ for all } i \leqslant m \text{ and } R > r_{\min}^{-1},$$

Denote $\overline{B}_{\mathbf{i}} = S_{\mathbf{i}}(\overline{B}) = \overline{B}(S_{\mathbf{i}}(\mathbf{0}), Rr_{\mathbf{i}})$ for $\mathbf{i} \in \Sigma^*$. Observe that

$$2\delta < 2R\delta r_{\min} \leqslant |\overline{B}_{\mathbf{i}}| = 2Rr_{\mathbf{i}} \leqslant 2R\delta \text{ for } \mathbf{i} \in \widetilde{\mathcal{M}}_\delta. \tag{3.16}$$

Note also that $S_i(\overline{B}) \subset \overline{B}$, for all $i \leqslant m$, implies that $\Lambda \subset \overline{B}$, see (1.36). Hence $\Lambda_{\mathbf{i}} \subset \overline{B}_{\mathbf{i}}$ for all $\mathbf{i} \in \Sigma^*$.

Now let $\widetilde{\mathcal{P}}_\delta$ be any maximal subset of $\widetilde{\mathcal{M}}_\delta$ with the property that $\{\overline{B}_{\mathbf{i}} : \mathbf{i} \in \widetilde{\mathcal{P}}_\delta\}$ is a disjoint family of balls. We claim that

$$\Lambda = \bigcup_{\mathbf{j} \in \widetilde{\mathcal{M}}_\delta} \Lambda_{\mathbf{j}} \subset \bigcup_{\mathbf{i} \in \widetilde{\mathcal{P}}_\delta} \overline{B}\big(S_{\mathbf{i}}(\mathbf{0}), (1+2R)Rr_{\mathbf{i}}\big). \tag{3.17}$$

Indeed, for any $\mathbf{j} \in \widetilde{\mathcal{M}}_\delta$ we can find $\mathbf{i} \in \widetilde{\mathcal{P}}_\delta$ such that $\overline{B}_{\mathbf{i}} \cap \overline{B}_{\mathbf{j}} \neq \varnothing$, by the maximality of $\widetilde{\mathcal{P}}_\delta$. Then the closed ball of radius $(r_{\mathbf{i}} + 2r_{\mathbf{j}})R$, concentric with $\overline{B}_{\mathbf{i}}$, covers the ball $\overline{B}_{\mathbf{j}}$. However,

$$r_{\mathbf{i}} + 2r_{\mathbf{j}} \leqslant (1+2R)r_{\mathbf{i}}$$

since $r_{\mathbf{j}} \leqslant \delta < r_{\min}^{-1} r_{\mathbf{i}} \leqslant Rr_{\mathbf{i}}$, hence $\Lambda_{\mathbf{j}} \subset \overline{B}_{\mathbf{j}} \subset \overline{B}\big(S_{\mathbf{i}}(\mathbf{0}), (1+2R)Rr_{\mathbf{i}}\big)$, and (3.17) is verified. Thus, we have, in view of (3.16), that

$$N_{2R(1+2R)\delta}(\Lambda) \leqslant \#\widetilde{\mathcal{P}}_\delta.$$

There is a constant C_d, depending only on the dimension d and R, such that any set of diameter $2R(1+2R)\delta$ can be covered by C_d balls of diameter δ, so that

$$N_\delta(\Lambda) \leqslant C_d \cdot \#\widetilde{\mathcal{P}}_\delta. \tag{3.18}$$

Next we consider a new iterated function system

$$\mathcal{S}_\delta := \big(\mathbf{x} \mapsto S_{\mathbf{i}}(\mathbf{x}) : \mathbf{i} \in \widetilde{\mathcal{P}}_\delta\big). \tag{3.19}$$

Let Λ_δ be its attractor. Observe that Λ is the attractor of the IFS $\big(\mathbf{x} \mapsto S_{\mathbf{i}}(\mathbf{x}) : \mathbf{i} \in \widetilde{\mathcal{M}}_\delta\big)$, since $\widetilde{\mathcal{M}}_\delta$ is a cut-set of Σ, and hence $\Lambda_\delta \subset \Lambda$. Thus,

$$S_{\mathbf{i}}(\Lambda_\delta) \subset S_{\mathbf{i}}(\Lambda) = \Lambda_{\mathbf{i}} \subset \overline{B}_{\mathbf{i}},$$

and hence the IFS $\mathcal{S}_\delta$ satisfies the Strong Separation Property (SSP) by the definition of $\widetilde{\mathcal{P}}_\delta$. Therefore,

$$\dim_{\mathrm{H}}(\Lambda) \geqslant \dim_{\mathrm{H}}(\Lambda_\delta) = s(\delta), \text{ where } \sum_{\mathbf{i} \in \widetilde{\mathcal{P}}_\delta} r_{\mathbf{i}}^{s(\delta)} = 1, \tag{3.20}$$

by Theorem 1.5.3. Using that $r_{\mathbf{i}} \geqslant \delta r_{\min}$ for $\mathbf{i} \in \widetilde{\mathcal{M}}_\delta$, we obtain from (3.20) that

$$\#\widetilde{\mathcal{P}}_\delta \cdot (\delta r_{\min})^{s(\delta)} \leqslant 1.$$

Combining this with (3.18) and taking logarithms yields

$$\log N_\delta(\Lambda) - \log C_d \leqslant \log(\#\widetilde{\mathcal{P}}_\delta) \leqslant s(\delta)[-\log\delta - \log r_{\min}].$$

Now, dividing by $-\log\delta$ (which is positive) and taking lim sup as $\delta \to 0$, we conclude that

$$\overline{\dim}_{\mathrm{B}}(\Lambda) \leqslant \limsup_{\delta \to 0} s(\delta) \leqslant \dim_{\mathrm{H}}(\Lambda) \leqslant \underline{\dim}_{\mathrm{B}}(\Lambda), \tag{3.21}$$

in view of Lemma 1.4.6, completing the proof. □

3.4. Hausdorff measure of self-similar sets

3.4.1. Finite Hausdorff measure. A slight modification of the proof of Theorem 3.3.1 yields the following:

THEOREM 3.4.1 (Falconer [**Fal97**, Cor. 3.3]). *Let $\mathcal{S} = (S_1, \dots, S_m)$ be a self-similar IFS with attractor Λ. Then $\mathcal{H}^{\dim_{\mathrm{H}}(\Lambda)}(\Lambda) < \infty$.*

As in the previous section, we assume that $|\Lambda| = 1$ without loss of generality. Before we prove Theorem 3.4.1, we need the following sharper bound on $\#\widetilde{\mathcal{P}}_\delta$ than in the proof of the previous theorem:

LEMMA 3.4.2. *For $0 < \delta \leqslant |\Lambda| = 1$, let $\widetilde{\mathcal{P}}_\delta$ be as in the proof of Theorem 3.3.1. Then $\#\widetilde{\mathcal{P}}_\delta \leqslant (r_{\min}\delta)^{-s}$, where $s = \dim_{\mathrm{H}}(\Lambda)$.*

PROOF. Let us argue by contradiction, namely, suppose that there exists $\delta \in (0, 1]$ such that $\#\widetilde{\mathcal{P}}_\delta > (r_{\min}\delta)^{-s}$. Let $\mathcal{S}_\delta$ be the IFS defined in (3.19), which satisfies the SSP by construction. Let $\widetilde{\nu}$ be the self-similar measure such that

$$\widetilde{\nu} = \frac{1}{\#\widetilde{\mathcal{P}}_\delta} \sum_{\mathbf{i} \in \widetilde{\mathcal{P}}_\delta} (S_{\mathbf{i}})_* \widetilde{\nu}.$$

By Theorem 3.2.1, we have

$$\dim_{\mathrm{H}}(\widetilde{\nu}) = \dim_{\mathrm{S}}(\widetilde{\nu}) = \frac{\#\widetilde{\mathcal{P}}_\delta \log \#\widetilde{\mathcal{P}}_\delta}{-\sum_{\mathbf{i} \in \widetilde{\mathcal{P}}_\delta} \log r_{\mathbf{i}}} \geqslant \frac{\log \#\widetilde{\mathcal{P}}_\delta}{-\log(r_{\min}\delta)} > s,$$

which is a contradiction, since $\dim_{\mathrm{H}}(\widetilde{\nu}) \leqslant \dim_{\mathrm{H}}(\Lambda) = s$. □

PROOF OF THEOREM 3.4.1. Let $\widetilde{\mathcal{P}}_\delta$ be as in the proof of Theorem 3.3.1. Then by (3.16) and (3.17), $\{\overline{B}(S_{\mathbf{i}}(\mathbf{0}), (1 + 2R)Rr_{\mathbf{i}})\}_{\mathbf{i} \in \widetilde{\mathcal{P}}_\delta}$ is a $2R(1 + 2R)\delta$-cover of Λ. Hence, for $s = \dim_{\mathrm{H}}(\Lambda)$,

$$\begin{aligned}\mathcal{H}^s_{2R(1+2R)\delta}(\Lambda) &\leqslant \sum_{\mathbf{i} \in \widetilde{\mathcal{P}}_\delta} |(1 + 2R)B_{\mathbf{i}}|^s \leqslant (2R)^s(1 + 2R)^s\delta^s \cdot \#\widetilde{\mathcal{P}}_\delta \\ &\leqslant (2R)^s(1 + 2R)^s r_{\min}^{-s},\end{aligned}$$

where in the last inequality we applied Lemma 3.4.2. Taking $\delta \to 0$ implies the desired statement. □

3.4.2. Ahlfors regularity and Hausdorff content. In this section, we show that self-similar sets with positive measure are $\dim_{\mathrm{H}}(\Lambda)$-Ahlfors regular (see Definition 2.1.4). This was first shown in the special case when $\dim_{\mathrm{H}}(\Lambda)$ is equal to the similarity dimension, by Bandt and Graf [**BG92**], and was extended by Farkas and Fraser [**FF15**].

THEOREM 3.4.3 (Farkas and Fraser). *Let $\mathcal{S} = (S_1, \dots, S_m)$ be a self-similar IFS with attractor Λ with Hausdorff dimension $\dim_{\mathrm{H}}(\Lambda) = s$. Then $\mathcal{H}^s(\Lambda) > 0$ if and only if Λ is s-Ahlfors regular.*

First we study the relation between the Hausdorff measure and the Hausdorff content (see (1.58) for the definition), which was also obtained by Farkas and Fraser [**FF15**] in the general setting. The same assertion was proven by Bandt and Graf [**BG92**] for the case when $\dim_{\mathrm{H}}(\Lambda)$ is equal to the similarity dimension.

THEOREM 3.4.4 (Farkas, Fraser). *Let $\Lambda \subset \mathbb{R}^d$ be a self-similar set with Hausdorff dimension* $\dim_{\mathrm{H}}(\Lambda) = s$. *Then $\mathcal{H}^s(E) = \mathcal{H}^s_\infty(E)$ for every $\mathcal{H}^s$-measurable set $E \subset \Lambda$, where $\mathcal{H}^s_\infty$ is the Hausdorff content.*

PROOF OF THEOREM 3.4.3 ASSUMING THEOREM 3.4.4. It is immediate from Definition 2.1.4 that if a set Λ is Ahlfors regular then it has positive Hausdorff measure, so it is enough to show the other direction.

Let Λ be a self-similar set for an IFS $\mathcal{S} = (S_i)_{i \leqslant m}$ with contraction ratios $r_i \in (0,1)$, $i \leqslant m$, and suppose that $\mathcal{H}^s(\Lambda) > 0$ for $s = \dim_{\mathrm{H}}(\Lambda)$. Again we can assume that $|\Lambda| = 1$ without loss of generality. Let $x \in \Lambda$ be arbitrary, and let $\mathbf{i} \in \Sigma = \{1, \ldots, m\}^{\mathbb{N}}$ be a natural coding of x, i.e. $\{x\} = \bigcap_{n=1}^{\infty} S_{\mathbf{i}|_n}(\Lambda)$. Next, for every $\rho > 0$ let $n \geqslant 1$ be the smallest integer such that $|S_{\mathbf{i}|_n}(\Lambda)| = r_{\mathbf{i}|_n} \leqslant \rho$, so that $r_{\mathbf{i}|_n} \geqslant r_{\min}\rho$, and then, using (1.57) we have

$$\mathcal{H}^s(\Lambda \cap B(x,\rho)) \geqslant \mathcal{H}^s(S_{\mathbf{i}|_n}(\Lambda)) = r_{\mathbf{i}|_n}^s \mathcal{H}^s(\Lambda) \geqslant r_{\min}^s \rho^s \mathcal{H}^s(\Lambda).$$

On the other hand, by Theorem 3.4.4,

$$\mathcal{H}^s(\Lambda \cap B(x,\rho)) = \mathcal{H}^s_\infty(\Lambda \cap B(x,\rho)) \leqslant 2^s \rho^s,$$

and s-Ahlfors regularity follows. □

The following lemma is a special case of Theorem 3.4.4 in which E is the self-similar set Λ. We will later see that the general case of Theorem 3.4.4 is easily reduced to this special case.

LEMMA 3.4.5. *Let $\Lambda \subset \mathbb{R}^d$ be a self-similar set with* $\dim_{\mathrm{H}}(\Lambda) = s$. *Then* $\mathcal{H}^s(\Lambda) = \mathcal{H}^s_\infty(\Lambda)$.

PROOF. Let $\mathcal{S} = \left(S_i(x) = r_i O_i x + t_i\right)_{i=1}^m$ be a self-similar IFS, with attractor Λ. If $\mathcal{H}^s(\Lambda) = 0$ then $\mathcal{H}^s_\infty(\Lambda) = 0$, see Section 1.4.3, so we may assume that $\mathcal{H}^s(\Lambda) > 0$ without loss of generality. The collection of the sets $\{S_{\mathbf{i}}(\Lambda)\}_{\mathbf{i} \in \Sigma^*}$ is a Vitali cover of Λ, thus, applying Vitali Covering Theorem B.3.4 we have that there exists a subset $\mathcal{I} \subset \Sigma^*$ such that $S_{\mathbf{i}}(\Lambda) \cap S_{\mathbf{j}}(\Lambda) = \varnothing$ for $\mathbf{i} \neq \mathbf{j} \in \mathcal{I}$ and

$$\text{either} \quad \mathcal{H}^s\Big(\Lambda \backslash \bigcup_{\mathbf{i} \in \mathcal{I}} S_{\mathbf{i}}(\Lambda)\Big) = 0 \quad \text{or} \quad \sum_{\mathbf{i} \in \mathcal{I}} |S_{\mathbf{i}}(\Lambda))|^s = |\Lambda|^s \sum_{\mathbf{i} \in \mathcal{I}} r_{\mathbf{i}}^s = \infty.$$

But by Theorem 3.4.1 and (1.57),

$$\mathcal{H}^s(\Lambda) \sum_{\mathbf{i} \in \mathcal{I}} r_{\mathbf{i}}^s = \sum_{\mathbf{i} \in \mathcal{I}} \mathcal{H}^s(S_{\mathbf{i}}(\Lambda)) = \mathcal{H}^s\Big(\bigcup_{\mathbf{i} \in \mathcal{I}} S_{\mathbf{i}}(\Lambda)\Big) \leqslant \mathcal{H}^s(\Lambda) < \infty.$$

Hence $\mathcal{H}^s\left(\Lambda \backslash \bigcup_{\mathbf{i} \in \mathcal{I}} S_{\mathbf{i}}(\Lambda)\right) = 0$, and then clearly,

$$\sum_{\mathbf{i} \in \mathcal{I}} r_{\mathbf{i}}^s = 1.$$

By induction,

$$\mathcal{H}^s(\Lambda) = \mathcal{H}^s\Big(\bigcup_{\mathbf{i}_1, \ldots, \mathbf{i}_n \in \mathcal{I}} S_{\mathbf{i}_1} \circ \cdots \circ S_{\mathbf{i}_n}(\Lambda)\Big)$$

for every $n \geqslant 1$ and hence by continuity of the measure,

$$\mathcal{H}^s(\Lambda) = \mathcal{H}^s(\widetilde{\Lambda}), \quad \text{where } \widetilde{\Lambda} := \bigcap_{n=1}^{\infty} \bigcup_{\mathbf{i}_1, \ldots, \mathbf{i}_n \in \mathcal{I}} S_{\mathbf{i}_1} \circ \cdots \circ S_{\mathbf{i}_n}(\Lambda).$$

Now, let $\{U_i\}_i$ be a cover of Λ such that $\sum_i |U_i|^s \leqslant \mathcal{H}^s_\infty(\Lambda) + \varepsilon$, where $\varepsilon > 0$ is arbitrary but fixed. For any $\delta > 0$ let $n \geqslant 1$ be such that

$$(\max_{\mathbf{i}\in\mathcal{I}} r_{\mathbf{i}})^n \max_i |U_i| < \delta.$$

Then $\{S_{\mathbf{i}_1} \circ \cdots \circ S_{\mathbf{i}_n}(U_i)\}_{\mathbf{i}_1,\ldots,\mathbf{i}_n\in\mathcal{I},\, i\leqslant m}$ is a δ-cover of $\widetilde{\Lambda}$ and therefore,

$$\mathcal{H}^s_\delta(\widetilde{\Lambda}) \leqslant \sum_{\mathbf{i}_1,\ldots,\mathbf{i}_n\in\mathcal{I},\, i\leqslant m} |S_{\mathbf{i}_1} \circ \cdots \circ S_{\mathbf{i}_n}(U_i)|^s = \Big(\sum_{\mathbf{i}\in\mathcal{I}} r^s_{\mathbf{i}}\Big)^n \sum_i |U_i|^s \leqslant \mathcal{H}^s_\infty(\Lambda) + \varepsilon.$$

Taking $\delta \to 0$ we have $\mathcal{H}^s(\widetilde{\Lambda}) \leqslant \mathcal{H}^s_\infty(\Lambda) + \varepsilon$. Since $\varepsilon > 0$ was arbitrary and $\mathcal{H}^s(\Lambda) = \mathcal{H}^s(\widetilde{\Lambda})$, the statement follows. □

PROOF OF THEOREM 3.4.4. Suppose that $E \subset \Lambda$ is $\mathcal{H}^s$-measurable. Then by Lemma 3.4.5,

$$\mathcal{H}^s_\infty(E) \leqslant \mathcal{H}^s(E) = \mathcal{H}^s(\Lambda) - \mathcal{H}^s(\Lambda\backslash E) \leqslant \mathcal{H}^s_\infty(\Lambda) - \mathcal{H}^s_\infty(\Lambda\backslash E) \leqslant \mathcal{H}^s_\infty(E),$$

where we used that $\mathcal{H}^s_\infty$ is σ-subadditive, and the claim follows. □

3.5. Lower semicontinuity of the Hausdorff dimension of self-similar sets.

Essentially the same idea as the one used in the proofs of Theorem 3.3.1 and Theorem 3.4.1, yields that the dimension is lower semicontinuous on the set of self-similar IFS. We should point out that it is an **open question** whether the Hausdorff (or upper/lower box) dimension is lower semicontinuous on the larger family of self-affine IFS. First, we need to define the topology on the space of IFS. Recall (see Notation 1) that $\mathscr{S}_m(d)$ denotes the collection of self-similar IFS of the form $\mathcal{S} := (S_1, \ldots, S_m)$ on $\mathbb{R}^d$, such that the attractor is not a singleton.

DEFINITION 3.5.1. The topology on $\mathscr{S}_m(d)$ is inherited from the topological group of similarity transformations $\mathcal{G}$ on $\mathbb{R}^d$. The natural topology on $\mathcal{G}$ is generated by a neighborhood base of the identity map Id in $\mathcal{G}$: For every $\varepsilon > 0$ and arbitrary $d+1$ points of $\mathbb{R}^d$ of general position $x_0, \ldots, x_d \in \mathbb{R}^d$, let

$$U_\varepsilon(x_0, \ldots, x_d) := \{g \in \mathcal{G} : |g(x_i) - x_i| < \varepsilon, i = 0, \ldots, d\}. \tag{3.22}$$

Since $g \in \mathcal{G}$ is determined by its values on $x_0, \ldots, x_d$, this topology coincides with the topology of uniform convergence on bounded sets and the topology of pointwise convergence on an arbitrary subset of $\mathbb{R}^d$ with at least $d+1$-many elements in general position. Alternatively, we can write a similarity transformation as $S(\mathbf{x}) = rA\mathbf{x} + \mathbf{t}$, where $r > 0$, A is an orthogonal matrix, and $\mathbf{t} \in \mathbb{R}^d$. The topology on $\mathcal{G}$ coincides with the one induced from $(0, \infty) \times O(d, \mathbb{R}) \times \mathbb{R}^d$, where $O(d, \mathbb{R})$ is the orthogonal group.

Below we denote by $\Lambda_{\mathcal{S}}$ the attractor of a self-similar IFS $\mathcal{S} = (S_1, \ldots, S_m)$.

THEOREM 3.5.2 (Jonker and Veerman). *The function $\mathcal{S} \mapsto \dim_{\mathrm{H}}(\Lambda_{\mathcal{S}})$ is lower semicontinuous on the set $\mathscr{S}_m(d)$ of self-similar IFS in $\mathbb{R}^d$. That is, by definition, for any $\mathcal{S} \in \mathscr{S}_m(d)$ and any $\varepsilon > 0$ there exists a neighborhood U of $\mathcal{S}$ in $\mathcal{G}^m$ such that*

$$\mathcal{S}' \in U \implies \dim_{\mathrm{H}}(\Lambda_{\mathcal{S}'}) \geqslant \dim_{\mathrm{H}}(\Lambda_{\mathcal{S}}) - \varepsilon. \tag{3.23}$$

PROOF. This proof is from [**JV02**], but our special case is easier, since [**JV02**] treats a more general case of "semi-conformal" IFS. It can also be deduced from [**Fal89**].

Since the attractor of $\mathcal{S} \in \mathscr{S}_m(d)$ is not a singleton, by applying an appropriate conjugation, it is enough to show lower semicontinuity of the Hausdorff dimension at the points $\mathcal{S} \in \mathscr{S}_m(d)$ for which the attractor $\Lambda_{\mathcal{S}}$ has diameter equal to 1. We repeat the construction from the proof of Theorem 3.3.1 and, given $0 < \delta \leqslant 1 = |\Lambda|$, consider the iterated function system (3.19):

$$\mathcal{S}_\delta = \left(\mathbf{x} \mapsto S_{\mathbf{i}} : \mathbf{i} \in \widetilde{\mathcal{P}}_\delta\right),$$

which satisfies the SSP and whose attractor Λ_δ is contained in $\Lambda_{\mathcal{S}}$, see (3.19). In view of (3.21), we can choose $\delta > 0$ in such a way that

$$s(\delta) > \dim_{\mathrm{H}}(\Lambda_{\mathcal{S}}) - \varepsilon/2,$$

where $s(\delta)$ is the similarity dimension of $\mathcal{S}_\delta$. Now, for any other IFS $\mathcal{S}' \in \mathscr{S}_m(d)$ (not necessarily satisfying $|\Lambda_{\mathcal{S}'}| = 1$), we can consider the IFS

$$\mathcal{S}_\delta' := \left(\mathbf{x} \mapsto S_{\mathbf{i}}' : \mathbf{i} \in \widetilde{\mathcal{P}}_\delta\right),$$

with attractor Λ_δ'. We claim that $\mathcal{S}_\delta'$ also satisfies the SSP whenever $\mathcal{S}'$ is sufficiently close to $\mathcal{S}$. Indeed, for any $\mathbf{i} \in I^*$, the function $\mathcal{S}' \mapsto S_{\mathbf{i}}'$ from $\mathcal{G}^m$ to $\mathcal{G}$ is continuous. Moreover, the mapping $\mathcal{S}' \mapsto \Lambda_{\mathcal{S}'}$ is continuous in the Hausdorff metric, see Proposition 1.3.7. Thus, since the distance between the cylinders $S_{\mathbf{i}}(\Lambda_{\mathcal{S}})$ and $S_{\mathbf{j}}(\Lambda_{\mathcal{S}})$, with $\mathbf{i} \neq \mathbf{j}$ in $\widetilde{\mathcal{P}}_\delta$, is positive, it will remain positive for all pairs of distinct cylinders $S_{\mathbf{i}}(\Lambda_{\mathcal{S}'})$ and $S_{\mathbf{j}}(\Lambda_{\mathcal{S}'})$, whenever $\mathcal{S}'$ is sufficiently close to $\mathcal{S}$.

The SSP is verified, and hence $\dim_{\mathrm{H}}(\Lambda_\delta') = s'(\delta)$, the similarity dimension of the IFS $\mathcal{S}_\delta'$. Clearly, the similarity dimension depends continuously on the self-similar IFS. Thus, for $\mathcal{S}'$ in a sufficiently small neighborhood of $\mathcal{S}$ we have

$$\dim_{\mathrm{H}}(\Lambda_{\mathcal{S}'}) \geqslant \dim_{\mathrm{H}}(\Lambda_\delta') = s'(\delta) \geqslant s(\delta) - \varepsilon/2 > \dim_{\mathrm{H}}(\Lambda_{\mathcal{S}}) - \varepsilon,$$

as desired. □

3.6. Existence of L^q dimensions

It was shown in [**PS00b**] that the correlation dimension, and more generally, L^q dimension, exists for any self-similar (and even self-conformal) measure in $\mathbb{R}^d$, without assuming any separation condition. Here we present the proof for $q > 1$ (the argument for $q < 1$ is similar, but slightly more involved, see [**PS00b**]). We refer the reader to Section 2.6 for the definitions and basic properties of the q-spectrum and L^q dimensions. Recall that $\mathscr{D}_n$ is the dyadic partition of $\mathbb{R}^d$ into grid boxes of side 2^{-n}.

THEOREM 3.6.1. *Let ν be a self-similar measure in $\mathbb{R}^d$. Then the following limit exists, for $q > 0$, $q \neq 1$:*

$$D_q(\nu) = \lim_{n\to\infty} \frac{\log \tau_n^{(q)}(\nu)}{n(1-q)\log 2}, \quad \textit{where } \tau_n^{(q)}(\nu) = \sum_{Q\in\mathscr{D}_n} (\nu(Q))^q.$$

The plan of the proof is to show that there exists $C \geqslant 1$ such that

$$\tau_{\ell+n}^{(q)}(\nu) \leqslant C\tau_\ell^{(q)}(\nu)\tau_n^{(q)}(\nu) \text{ for all } \ell, n \in \mathbb{N}. \tag{3.24}$$

Then the sequence $\left\{\log\left(C\tau_n^{(q)}(\nu)\right)\right\}_{n\geqslant 1}$ is subadditive, hence the limit

$$\tau(\nu,q) = \lim_{n\to\infty} \frac{\log \tau_n^{(q)}(\nu)}{-n\log 2}$$

exists by Fekete's Lemma 1.9.21, as desired.

In order to implement this plan, we need some preliminaries.

DEFINITION 3.6.2. Let K be a compact set in $\mathbb{R}^d$. Fix M and $n, N \in \mathbb{N}$. Say that a covering $\{G_i\}_{i=1}^k$ of K by Borel sets is $(M, 2^{-n}, N)$-*good* if $|G_i| \leqslant M \cdot 2^{-n}$ for all $i \leqslant k$ and any cube $Q \in \mathscr{D}_n$ intersects at most N elements of the covering.

LEMMA 3.6.3. *Let $M > 0$, $N, d \in \mathbb{N}$, $q > 1$. There exists $C_1 = C_1(M,N,d,q) > 1$ such that for any $n \in \mathbb{N}$ and any compactly supported measure ν on R^d, for any $(M, 2^{-n}, N)$-good covering $\{G_i\}_{i=1}^k$ of* spt(ν)*, we have*

$$C_1^{-1}\tau_n^{(q)}(\nu) \leqslant \sum_{i=1}^k (\nu(G_i))^q \leqslant C_1\tau_n^{(q)}(\nu). \tag{3.25}$$

PROOF. Observe that any set of diameter $\leqslant M2^{-n}$ intersects at most $(M+1)^d$ grid boxes in $\mathscr{D}_n$. Therefore, we have for all $i \leqslant k$:

$$\left(\nu(G_i)\right)^q = \Big(\sum_{Q\in\mathscr{D}_n,\, Q\cap G_i\neq\varnothing} \nu(Q)\Big)^q \leqslant (M+1)^{d(q-1)} \sum_{Q\in\mathscr{D}_n,\, Q\cap G_i\neq\varnothing} \nu(Q)^q,$$

by Hölder's inequality. Since every $Q \in \mathscr{D}_n$ intersects at most N of the sets G_i, we obtain the right-hand side inequality of (3.25), with $C_1 = N\cdot(M+1)^{d(q-1)}$. The left-hand side inequality is proved similarly, starting with $\nu(Q) \leqslant \sum_{G_i:\, Q\cap G_i\neq\varnothing} \nu(G_i)$. □

Now suppose that $\mathcal{S} = (S_1, \ldots, S_m)$ is a self-similar IFS in $\mathbb{R}^d$, with contraction ratios $r_i \in (0,1)$. Let $\mathbf{p} = (p_1, \ldots, p_m)$ be a probability vector and ν the corresponding self-similar measure. Denote the attractor of $\mathcal{S}$ by $\Lambda = \Lambda(\mathcal{S})$ and recall that spt$(\nu) = \Lambda$. Recall that for $\mathbf{i} = i_1 \ldots i_k$ we write $S_{\mathbf{i}} = S_{i_1} \circ \cdots S_{i_k}$, $r_{\mathbf{i}} = r_{i_1} \cdots r_{i_k}$, $p_{\mathbf{i}} = p_{i_1} \cdots p_{i_k}$ and $\Lambda_{\mathbf{i}} = S_{\mathbf{i}}(\Lambda)$.

Consider the minimal cut-set

$$W_n := \{\mathbf{i} \in \Sigma^* : \ |\Lambda_{\mathbf{i}}| \leqslant 2^{-n},\ |\Lambda_{\mathbf{i}^-}| > 2^{-n}\}. \tag{3.26}$$

It is a version of the Moran cut-set, see (1.73); we have $W_n := \widetilde{\mathcal{M}}_{2^{-n}}$ when $|\Lambda| = 1$, which we do not assume here. As we have seen many times already,

$$\nu = \sum_{\mathbf{i}\in W_n} p_{\mathbf{i}}(\nu \circ S_{\mathbf{i}}^{-1}) \quad \text{for all } n \in \mathbb{N}. \tag{3.27}$$

We write $r_{\min} = r_{\min}^{\mathcal{S}} = \min_{i\leqslant m} r_i$ and note that

$$2^{-n} r_{\min} \leqslant |\Lambda_{\mathbf{i}}| = r_{\mathbf{i}}|\Lambda| \leqslant 2^{-n} \quad \text{for all } \mathbf{i} \in W_n.$$

LEMMA 3.6.4. *There exist $M > 0$ and $N \in \mathbb{N}$ such that for any $\ell, n \in \mathbb{N}$ and any $\mathbf{i} \in W_n$, the collection*

$$\{S_{\mathbf{i}}^{-1}Q :\ Q \in \mathscr{D}_{\ell+n},\ Q \cap \Lambda_{\mathbf{i}} \neq \varnothing\}$$

is an $(M, 2^{-\ell}, N)$-good covering of Λ. In fact, one can choose

$$M = M(\mathcal{S}) = \left(r_{\min}^{\mathcal{S}}\right)^{-1} \cdot |\Lambda(\mathcal{S})| \cdot \sqrt{d} \ \textit{and}\ N = N(\mathcal{S}) = \left\lceil |\Lambda(\mathcal{S})|^{-1} \cdot \sqrt{d} + 1 \right\rceil^d \tag{3.28}$$

REMARK. The formulas (3.28), which show dependence on the IFS $\mathcal{S}$ will only be needed in the next section. In this section we will write $\Lambda = \Lambda(\mathcal{S})$ and $r_{\min} = \min_i r_i$.

PROOF. This is a covering of Λ: indeed, we took a covering of $\Lambda_{\mathbf{i}}$ by dyadic boxes from $\mathscr{D}_{\ell+n}$ and then applied $S_{\mathbf{i}}^{-1}$ to it. We have

$$|S_{\mathbf{i}}^{-1}Q| = r_{\mathbf{i}}^{-1}\sqrt{d}\cdot 2^{-\ell-n} \leqslant r_{\min}^{-1}\cdot|\Lambda|\cdot\sqrt{d}\cdot 2^{-\ell} \quad \text{for all } \mathbf{i}\in W_n \text{ and all } Q\in\mathscr{D}_{\ell+n},$$

by the definition of (3.26). So we can choose $M = r_{\min}^{-1}\cdot|\Lambda|\cdot\sqrt{d}$. On the other hand, given a box $P\in\mathscr{D}_\ell$, the number of sets $S_{\mathbf{i}}^{-1}(Q)$, with $Q\in\mathscr{D}_{\ell+n}$, intersecting P, equals the number of $Q\in\mathscr{D}_{\ell+n}$ intersecting $S_{\mathbf{i}}(P)$. Since

$$|S_{\mathbf{i}}(P)| = r_{\mathbf{i}}\sqrt{d}\cdot 2^{-\ell} \leqslant |\Lambda|^{-1}\sqrt{d}\cdot 2^{-\ell-n}$$

by (3.26), this number is at most $N = \left\lceil |\Lambda|^{-1}\sqrt{d}+1\right\rceil^d$, and the lemma is proved. □

PROOF OF THEOREM 3.6.1 FOR $q > 1$. As already mentioned, it suffices to prove the inequality (3.24). Combining Lemmas 3.6.3 and 3.6.4 we obtain that there exists $C_1 > 1$ independent of $\ell, n\in\mathbb{N}$ such that for any $\mathbf{i}\in W_n$,

$$C_1^{-1}\tau_\ell^{(q)} \leqslant \sum_{Q\in\mathscr{D}_{\ell+n}} \nu(S_{\mathbf{i}}^{-1}(Q))^q \leqslant C_1\tau_\ell^{(q)}. \tag{3.29}$$

For a set $A\subset\mathbb{R}^d$, define

$$W_n(A) := \{\mathbf{i}\in W_n : \Lambda_{\mathbf{i}}\cap A \neq \varnothing\}. \tag{3.30}$$

For $Q\in\mathscr{D}_{\ell+n}$ let $\widetilde{Q}$ be the unique box in $\mathscr{D}_n$ such that $Q\subset\widetilde{Q}$. Using (3.27), we have for any $Q\in\mathscr{D}_{\ell+n}$:

$$\nu(Q) = \sum_{\mathbf{i}\in W_n} p_{\mathbf{i}}\nu(S_{\mathbf{i}}^{-1}(Q)) = \sum_{\mathbf{i}\in W_n(\widetilde{Q})} p_{\mathbf{i}}\nu(S_{\mathbf{i}}^{-1}(Q)). \tag{3.31}$$

Denote

$$p_+(\widetilde{Q}) = \sum_{\mathbf{i}\in W_n(\widetilde{Q})} p_{\mathbf{i}}. \tag{3.32}$$

Using convexity of x^q for $q > 1$, we obtain from (3.31):

$$\begin{aligned}(\nu(Q))^q &= p_+(\widetilde{Q})^q\Big(\sum_{\mathbf{i}\in W_n(\widetilde{Q})} \frac{p_{\mathbf{i}}}{p_+(\widetilde{Q})}\,\nu(S_{\mathbf{i}}^{-1}(Q))\Big)^q \\ &\leqslant p_+(\widetilde{Q})^{q-1}\sum_{\mathbf{i}\in W_n(\widetilde{Q})} p_{\mathbf{i}}\,\nu(S_{\mathbf{i}}^{-1}(Q))^q.\end{aligned}$$

Summing over $Q\subset\widetilde{Q}$, $Q\in\mathscr{D}_{\ell+n}$, we obtain

$$\begin{aligned}\sum_{Q\subset\widetilde{Q}}(\nu(Q))^q &\leqslant p_+(\widetilde{Q})^{q-1}\sum_{\mathbf{i}\in W_n(\widetilde{Q})} p_{\mathbf{i}}\sum_{Q\subset\widetilde{Q}}\nu(S_{\mathbf{i}}^{-1}(Q))^q \\ &\leqslant p_+(\widetilde{Q})^{q-1}\sum_{\mathbf{i}\in W_n(\widetilde{Q})} p_{\mathbf{i}}\cdot C_1\tau_\ell^{(q)}(\nu) \\ &= C_1 p_+(\widetilde{Q})^q\cdot\tau_\ell^{(q)}(\nu),\end{aligned}$$

where in the second displayed line we used the right inequality in (3.29). Now, summing over $\widetilde{Q} \in \mathscr{D}_n$ yields

$$\tau^{(q)}_{\ell+n}(\nu) \leqslant C_1 \tau^{(q)}_{\ell}(\nu) \sum_{\widetilde{Q} \in \mathscr{D}_n} p_+(\widetilde{Q})^q. \tag{3.33}$$

Write $\widetilde{B} \sim \widetilde{Q}$ if $\widetilde{B} \in \mathscr{D}_n$ and the closures of $\widetilde{B}$ and $\widetilde{Q}$ intersect. There are 3^d such boxes. Recall that $|\Lambda_{\mathbf{i}}| \leqslant 2^{-n}$ for $\mathbf{i} \in W_n$, hence $\Lambda_{\mathbf{i}} \cap \widetilde{Q} \neq \varnothing$ implies $\Lambda_{\mathbf{i}} \subset \bigcup_{\widetilde{B} \sim \widetilde{Q}} \widetilde{B}$. Recall also that $\nu = \Pi_* \mu = \mu \circ \Pi^{-1}$, where $\mu = \mathbf{p}^{\mathbb{N}}$. Since $p_{\mathbf{i}} = \mu([\mathbf{i}])$ and $\Lambda_{\mathbf{i}} = \Pi([\mathbf{i}])$, we have by (3.32),

$$p_+(\widetilde{Q}) = \sum_{\mathbf{i} \in W_n(\widetilde{Q})} p_{\mathbf{i}} \leqslant \nu\Big(\bigcup_{\mathbf{i} \in W_n(\widetilde{Q})} \Lambda_{\mathbf{i}} \Big) \leqslant \sum_{\widetilde{B} \sim \widetilde{Q}} \nu(\widetilde{B}).$$

By Hölder's inequality,

$$p_+(\widetilde{Q})^q \leqslant 3^{d(q-1)} \sum_{\widetilde{B} \sim \widetilde{Q}} (\nu(\widetilde{B}))^q.$$

Therefore,

$$\sum_{\widetilde{Q} \in \mathscr{D}_n} p_+(\widetilde{Q})^q \leqslant 3^d 3^{d(q-1)} \sum_{\widetilde{B} \in \mathscr{D}_n} (\nu(\widetilde{B}))^q = 3^{dq} \tau^{(q)}_n(\nu).$$

Combining this with (3.33) yields (3.24), and the proof is finished. □

3.7. Existence of entropy dimension and lower semicontinuity of the entropy and Hausdorff dimensions of self-similar measures

Hochman and Shmerkin proved in [**HS12**, Theorem 1.8] that the Hausdorff dimension of the Bernoulli convolution is a lower semicontinuous function of the parameters. Although the authors of [**HS12**] use a direct method to prove this theorem, they also hinted at an alternative proof. In fact, the idea comes from [**PS00b**], where the existence of entropy dimension for self-similar measures was shown. In this section we prove this latter assertion, and using analogous tools obtain lower semicontinuity of the entropy dimension for self-similar measures. Then lower semicontinuity of the Hausdorff dimension of self-similar measures follows by Young's Theorem 2.6.4 and the theorem of Feng and Hu on exact dimensionality of self-similar measures, which will be shown below in Theorem 11.3.1, in the case of self-similar IFS on $\mathbb{R}$.

First we introduce a natural topology on the space of *probabilistic* self-similar IFS and recall the notation.

Fix $d \geqslant 1$ and $m \geqslant 2$. Let $\mathcal{P}_m$ be the collection of probability vectors $\mathbf{p} = (p_1, \ldots, p_m)$ with $p_i > 0$ for all $i \in [m] := \{1, \ldots, m\}$. The topology on $\mathcal{P}_m$ is generated by the Euclidean distance.

Recall that $\mathscr{S}_m(d)$ denotes the collection of self-similar IFS of the form $\mathcal{S} := (S_1, \ldots, S_m)$ on $\mathbb{R}^d$, such that the attractor is not a singleton. The topology on $\mathscr{S}_m(d)$ was given in Definition 3.5.1, and the topology on the space of probabilistic self-similar IFS in $\mathbb{R}^d$ consisting of m maps is the topology of $\mathscr{S}_m(d) \times \mathcal{P}_m$.

We denote by $\nu_{\mathcal{S},\mathbf{p}}$ the self-similar measure corresponding to $\mathcal{S}$ and $\mathbf{p}$. Recall that the entropy dimension of a measure was introduced in (1.117), see also (2.31).

THEOREM 3.7.1. *For any self-similar IFS $\mathcal{S} \in \mathscr{S}_m(d)$ and a probability vector $\mathbf{p} \in \mathcal{P}_m$, the entropy dimension $\dim_e(\nu_{\mathcal{S},\mathbf{p}})$ exists.*

THEOREM 3.7.2. *For every $d \geqslant 1$ and $m \geqslant 2$ the following mapping is lower semicontinuous:*

$$(\mathcal{S}, \mathbf{p}) \mapsto \dim_{\mathrm{H}}(\nu_{\mathcal{S},\mathbf{p}}), \qquad \mathcal{S} \in \mathscr{S}_m(d) \text{ and } \mathbf{p} \in \mathcal{P}_m, \tag{3.34}$$

where $\nu_{\mathcal{S},\mathbf{p}}$ is the self-similar measure corresponding to $\mathcal{S}$ and $\mathbf{p}$.

The proof of both theorems depends on a lemma, stated below, which is a combination of results from [**PS00b**]. Part of this lemma will be used later in this book. The similarity ratio of S_i is denoted by r_i and $r_{\min}^{\mathcal{S}} = \min_{i \leqslant m} r_i$.

Recall that $\mathscr{D}_n$ denotes the partition of $\mathbb{R}^d$ into the dyadic cubes $\prod_{i=1}^d \left[\frac{k_i}{2^n}, \frac{k_i+1}{2^n}\right)$, $k_i \in \mathbb{Z}$. For an arbitrary compactly supported probability Borel measure $\mathfrak{m}$ on $\mathbb{R}^d$ the entropy of the partition $\mathscr{D}_n$ is

$$h_n(\mathfrak{m}) := H(\mathfrak{m}, \mathscr{D}_n) = -\sum_{Q \in \mathscr{D}_n} \mathfrak{m}(Q) \log \mathfrak{m}(Q), \tag{3.35}$$

see Definition 1.9.15.

LEMMA 3.7.3 ([**PS00b**]). *There exists a mapping*

$$C : \mathscr{S}_m(d) \to \mathbb{R}_+, \qquad C(\mathcal{S}) = K(d) + 2^d \log_2\left((r_{\min}^{\mathcal{S}})^{-1} \cdot |\Lambda(\mathcal{S})| \cdot \sqrt{d} \right), \tag{3.36}$$

where $K(d)$ depends only on d, such that for any $\mathbf{p} \in \mathcal{P}_m$ and any $n, \ell \in \mathbb{N}$,

$$h_{n+\ell}(\nu_{\mathcal{S},\mathbf{p}}) \geqslant h_n(\nu_{\mathcal{S},\mathbf{p}}) + h_\ell(\nu_{\mathcal{S},\mathbf{p}}) - C(\mathcal{S}),$$

that is, the sequence $\{h_n(\nu_{\mathcal{S},\mathbf{p}}) - C(\mathcal{S})\}_{n=1}^\infty$ is super-additive.

We prove this lemma after the derivation of Theorems 3.7.1 and 3.7.2.

PROOF OF THEOREM 3.7.1 ASSUMING LEMMA 3.7.3. By super-additivity, the following limit exists:

$$h_{\nu_{\mathcal{S},\mathbf{p}}} := \lim_{n\to\infty} \frac{h_n(\nu_{\mathcal{S},\mathbf{p}})}{n \log 2} = \lim_{n\to\infty} \frac{h_n(\nu_{\mathcal{S},\mathbf{p}}) - C(\mathcal{S})}{n \log 2} = \sup_n \frac{h_n(\nu_{\mathcal{S},\mathbf{p}}) - C(\mathcal{S})}{n \log 2}. \tag{3.37}$$

Observe that the quantity $h_{\nu_{\mathcal{S},\mathbf{p}}}$ equals the entropy dimension $\dim_e(\nu_{\mathcal{S},\mathbf{p}})$ by (2.31), and the proof is complete. □

PROOF OF THEOREM 3.7.2 ASSUMING LEMMA 3.7.3. Observe first that the mapping $C : \mathscr{S}_m(d) \to \mathbb{R}_+$, with $\mathcal{S} \mapsto C(\mathcal{S})$ defined in Lemma 3.7.3, is continuous.

Let $\partial\mathcal{D}_n = \bigcup_{Q \in \mathcal{D}_n} \partial Q$, where ∂Q denotes the boundary of Q.

CLAIM. *For every n, the mapping $(\mathcal{S}, \mathbf{p}) \mapsto h_n(\nu_{\mathcal{S},\mathbf{p}})$ has a continuity point at $(\mathcal{S}, \mathbf{p})$ if $\nu_{\mathcal{S},\mathbf{p}}(\partial\mathcal{D}_n) = 0$.*

We leave the standard proof of this claim as an exercise for the reader. The combination of the last two observations yields that for every $n \geqslant 1$ the mapping

$$(\mathcal{S}, \mathbf{p}) \mapsto \frac{h_n(\nu_{\mathcal{S},\mathbf{p}}) - C(\mathcal{S})}{n \log 2} \tag{3.38}$$

is also continuous on the set $E = \{(\mathcal{S}, \mathbf{p}) : \nu_{\mathcal{S},\mathbf{p}}(\bigcup_{n\geqslant 1} \partial\mathscr{D}_n) = 0\}$. It follows from (3.37) that $h_{\nu_{\mathcal{S},\mathbf{p}}}$ is represented as the supremum of a sequence of continuous functions, hence $h_{\nu_{\mathcal{S},\mathbf{p}}}$ is lower semicontinuous at every point in E.

By Proposition 3.1.1, $(\mathcal{S}, \mathbf{p}) \notin E$ if and only if there exists a $(d-1)$-dimensional hyperplane P, perpendicular to one of the main axes with a dyadic coordinate, such that $\nu_{\mathcal{S},\mathbf{p}}(P) = 1$. To show that $(\mathcal{S}, \mathbf{p}) \mapsto h_{\nu_{\mathcal{S},\mathbf{p}}}$ is lower semicontinuous at every pair $(\mathcal{S},\mathbf{p})$, let us fix d vectors $\mathbf{t}_k = (t_1^k,\dots,t_d^k)$ with irrational coordinates, such that $t_i^k - t_i^\ell \notin \mathbb{Q}$ for all $1 \leqslant i,k,\ell \leqslant d$. Define d other dyadic-like partitions $\mathscr{D}_n^{(k)}$ formed by cubes $\prod_{i=1}^d [\frac{k_i}{2^n} + t_i^k, \frac{k_i+1}{2^n} + t_i^k)$, using the convention that $\mathscr{D}_n^{(0)} = \mathscr{D}_n$. It is clear that $\bigcap_{k=0}^d \partial\mathscr{D}_n^{(k)} = \varnothing$. It follows that for any $(\mathcal{S},\mathbf{p})$ there exists a k such that

$$(\mathcal{S},\mathbf{p}) \in E_k = \Big\{(\mathcal{S},\mathbf{p}) : \nu_{\mathcal{S},\mathbf{p}}\Big(\bigcup_{n\geqslant 1}\partial\mathscr{D}_n^{(k)}\Big) = 0\Big\}.$$

Since the claim of Lemma 3.7.3 remains valid for every partition $\mathscr{D}_n^{(k)}$ and the function $h_{\nu_{\mathcal{S},\mathbf{p}}}$ is independent of the choice of the partitions, it follows that $(\mathcal{S},\mathbf{p}) \mapsto h_{\nu_{\mathcal{S},\mathbf{p}}}$ is lower semicontinuous.

As observed above, the quantity $h_{\nu_{\mathcal{S},\mathbf{p}}}$ equals the entropy dimension $\dim_e(\nu_{\mathcal{S},\mathbf{p}})$ by (2.31). The equality of the Hausdorff dimension to the entropy dimension follows by Young's Theorem 2.6.4 and Feng and Hu's theorem regarding the exact dimensionality of the self-similar measures, which will be proven later in Section 11.3, Theorem 11.3.1, in the special case $d = 1$. □

In order to prove Lemma 3.7.3 we need some facts related to the entropy of the partition $\mathscr{D}_n$. The first one is simply Lemma 2.6.12 in the current notation.

FACT 3.7.4. *For an arbitrary compactly supported probability Borel measure* $\mathfrak{m}$ *on* $\mathbb{R}^d$ *we have*

(a) $h_n(\mathfrak{m}) \leqslant h_{n+k}(\mathfrak{m}) \leqslant h_n(\mathfrak{m}) + 2^{kd}$ *for all* k,n.

(b) *Let* $C_1(d) = \max\{2^d \cdot (1 + \lceil \frac{1}{2}\log_2 d\rceil), 3^d\}$. *Then we have*

$$\left|\int \log \frac{1}{\mathfrak{m}(B(x,2^{-n}))} d\mathfrak{m}(x) - h_n(\mathfrak{m})\right| \leqslant C_1(d), \quad \textit{for all } n \in \mathbb{N}. \tag{3.39}$$

The next fact is Lemma 3.6.4, which we restate for convenience. Recall the notation for the cut-set:

$$W_n := \{\mathbf{i} \in \Sigma^* : |\Lambda_{\mathbf{i}}| \leqslant 2^{-n} \text{ and } 2^{-n} < |\Lambda_{\mathbf{i}^-}|\}, \tag{3.40}$$

where $\mathbf{i}^- = (i_1,\dots,i_{|\mathbf{i}|-1})$.

FACT 3.7.5. *Let* $\mathcal{S} = (S_1,\dots,S_m) \in \mathscr{S}_m(d)$ *and let*

$$M = M(\mathcal{S}) = (r_{\min}^{\mathcal{S}})^{-1} \cdot |\Lambda(\mathcal{S})| \cdot \sqrt{d} \quad \textit{and} \quad N = N(\mathcal{S}) = \left\lceil |\Lambda(\mathcal{S})|^{-1} \cdot \sqrt{d} + 1\right\rceil^d.$$

Then for every $n,\ell \in \mathbb{N}$ *and* $\mathbf{i} \in W_n$, *the collection*

$$\{S_{\mathbf{i}}^{-1}Q : Q \in \mathcal{D}_{\ell+n},\ Q \cap \Lambda_{\mathbf{i}} \neq \varnothing\} \tag{3.41}$$

is an $(M, 2^{-\ell}, N)$*-good covering of* Λ. *This means, according to Definition* 3.6.2, *that for every* $\mathbf{i} \in W_n$:

- $|S_{\mathbf{i}}^{-1}Q| \leqslant M \cdot 2^{-\ell}$ *and*
- *any cube in* $\mathscr{D}_\ell$ *intersects at most* N *sets from the collection* (3.41).

PROOF OF LEMMA 3.7.3. In this proof we use the notation of Theorem 3.7.2. Let

$$\varphi(x) := \begin{cases} x\log\frac{1}{x}, & \text{if } x > 0; \\ 0, & \text{if } x = 0. \end{cases}$$

Then one can easily see that

$$\varphi(xy) = y\cdot\varphi(x) + x\cdot\varphi(y), \quad \text{if } x,y > 0. \tag{3.42}$$

Fix an $\mathcal{S} \in \mathscr{S}_m(d)$ and $\mathbf{p} \in \mathcal{P}_m$. We will simply write ν for the corresponding self-similar measure $\nu_{\mathcal{S},\mathbf{p}}$. By self-similarity, for every k, n and $Q \in \mathscr{D}_k$ we have

$$\nu(Q) = \sum_{\mathbf{i} \in W_n(Q)} p_{\mathbf{i}} \cdot \nu\left(S_{\mathbf{i}}^{-1} Q\right) \tag{3.43}$$

where $W_n(Q) = \{\mathbf{i} \in W_n : \Lambda_{\mathbf{i}} \cap Q \neq \varnothing\}$ has the same meaning as in (3.30), and W_n was defined in (3.40). Further, for every k, the entropy of the partition $\mathscr{D}_k$ with respect to the measure ν is denoted by h_k, that is,

$$h_k = \sum_{Q \in \mathscr{D}_k} \varphi(\nu(Q)). \tag{3.44}$$

In the rest of the proof we verify that

$$h_{\ell+n} \geqslant h_n + h_\ell - C, \tag{3.45}$$

where $C = C(\mathcal{S})$ can be given in the form (3.36). Fix an arbitrary ℓ and n. For all $Q \in \mathscr{D}_{\ell+n}$ we introduce

$$p(Q) := \sum_{\mathbf{i} \in W_n(Q)} p_{\mathbf{i}} \leqslant \nu\Big(\bigcup_{\mathbf{i} \in W_n(Q)} \Lambda_{\mathbf{i}} \Big).$$

Observe that

$$\nu(B(x, 2^{-n})) \leqslant p(Q) \leqslant \nu(B(x, 2^{-n}(\sqrt{d}+1))), \quad \text{for } x \in Q \in \mathcal{D}_{\ell+n}. \tag{3.46}$$

Here we used that $|\Lambda_{\mathbf{i}}| \leqslant 2^{-n}$ for $\mathbf{i} \in W_n$ and

$$\Lambda_{\mathbf{i}} \cap Q \neq \varnothing \implies \Lambda_{\mathbf{i}} \subset \overline{B}(x, 2^{-n} + \sqrt{d} \cdot 2^{-\ell-n}) \subset B(x, 2^{-n}(\sqrt{d}+1)).$$

As above, in the rest of the proof whenever we write Q we always mean that $Q \in \mathcal{D}_{\ell+n}$. First we apply the identity (3.42) with $x = p(Q)$ and $y = \nu(Q)/p(Q)$. This yields

$$\varphi(\nu(Q)) = A_1(Q) + A_2(Q), \tag{3.47}$$

where

$$A_1(Q) = \frac{\nu(Q)}{p(Q)} \cdot \varphi\left(p(Q)\right) = \nu(Q) \cdot \log \frac{1}{p(Q)} \tag{3.48}$$

and

$$A_2(Q) = p(Q) \cdot \varphi\left(\nu(Q)/p(Q)\right) = p(Q) \cdot \varphi \left(\sum_{\mathbf{i} \in W_n(Q)} \frac{p_{\mathbf{i}}}{p(Q)} \nu\left(S_{\mathbf{i}}^{-1} Q\right) \right),$$

where we used (3.43) in the second equality. Hence, by the concavity of φ,

$$\begin{aligned} A_2(Q) &\geqslant p(Q) \cdot \sum_{\mathbf{i} \in W_n(Q)} \frac{p_{\mathbf{i}}}{p(Q)} \cdot \varphi\left(\nu\left(S_{\mathbf{i}}^{-1} Q\right)\right) \\ &= \sum_{\mathbf{i} \in W_n(Q)} p_{\mathbf{i}} \cdot \varphi\left(\nu\left(S_{\mathbf{i}}^{-1} Q\right)\right). \end{aligned} \tag{3.49}$$

So, by (3.44) and (3.47) we have

$$h_{\ell+n} = \sum_{Q \in \mathcal{D}_{\ell+n}} \left[A_1(Q) + A_2(Q)\right]. \tag{3.50}$$

Let k be the smallest integer for which $\sqrt{d}+1 \leqslant 2^k$. Combining (3.48) and (3.46) with Fact 3.7.4, we obtain that

$$\begin{aligned}\sum_{Q\in\mathcal{D}_{\ell+n}} A_1(Q) &\geqslant \int \log \frac{1}{\nu(B(x,2^{-n}(\sqrt{d}+1)))} d\nu(x) \\ &\geqslant h_{n-k} - C_1 \geqslant h_n - C_1 - 2^{kd},\end{aligned} \tag{3.51}$$

where $C_1 = C_1(d)$ was introduced in part (b) of Fact 3.7.4. Observe that k also depends only on d. So we obtained that

$$A_1(Q) \geqslant h_n - C_2(d). \tag{3.52}$$

On the other hand, (3.49) yields

$$\sum_{Q\in\mathcal{D}_{\ell+n}} A_2(Q) \geqslant \sum_{\mathbf{i}\in W_n} p_{\mathbf{i}} \sum_{Q:\ Q\cap\Lambda_{\mathbf{i}}\neq\varnothing} \varphi\left(\nu\left(S_{\mathbf{i}}^{-1}Q\right)\right). \tag{3.53}$$

By the Fact 3.7.5, the collection $\left\{S_{\mathbf{i}}^{-1}Q :\ Q\in\mathcal{D}_{\ell+n},\ Q\cap\Lambda_{\mathbf{i}}\neq\varnothing\right\}$ provides an $(M, 2^{-\ell}, N)$-good covering of Λ for every ℓ, n and $\mathbf{i}\in W_n$, where $M = M(\mathcal{S})$ and $N = N(\mathcal{S})$ were defined in (3.28). Therefore, $S_{\mathbf{i}}^{-1}Q \subset B(x, M\cdot 2^{-\ell})$ for any $\mathbf{i}\in W_n$ and $x\in S_{\mathbf{i}}^{-1}Q$. Using Fact 3.7.4 we get

$$\begin{aligned}\sum_{Q:\ Q\cap\Lambda_{\mathbf{i}}\neq\varnothing} \varphi\left(\nu\left(S_{\mathbf{i}}^{-1}Q\right)\right) &= \sum_{Q:\ Q\cap\Lambda_{\mathbf{i}}\neq\varnothing} \nu\left(S_{\mathbf{i}}^{-1}Q\right)\log\frac{1}{\nu\left(S_{\mathbf{i}}^{-1}Q\right)} \\ &\geqslant \int \log\frac{1}{\nu\left(B(x, M\cdot 2^{-\ell})\right)} d\nu(x) \\ &\geqslant h_\ell - C_1 - 2^{\widehat{k}d},\end{aligned}$$

where $\widehat{k} = \widehat{k}(\mathcal{S})$ is the smallest integer for which $M(\mathcal{S}) \leqslant 2^{\widehat{k}}$. Note that $\sum_{\mathbf{i}\in W_n} p_{\mathbf{i}} = 1$. Hence,

$$\sum_{\mathbf{i}\in W_n} p_{\mathbf{i}} \sum_{Q:\ Q\cap\Lambda_{\mathbf{i}}\neq\varnothing} \varphi\left(\nu\left(S_{\mathbf{i}}^{-1}Q\right)\right) \geqslant h_\ell - C_1 - 2^{\widehat{k}d}.$$

Combining this with (3.53), (3.52) and (3.50), we obtain that (3.45) holds, which completes the proof of the Lemma. □

CHAPTER 4

Separation properties for self-similar IFS

In this chapter we first focus on criteria that are equivalent to the Open Set Condition (OSC), presenting important work of Bandt, Graf [**BG92**], and Schief [**Sch94**]. This is followed by an in-depth study of what became known as the Weak Separation Property (WSP) and the Finite Type Condition, including a large number of representative examples. In Section 4.4.3 we show that Finite Type implies WSP.

4.1. Conditions that are equivalent to the OSC

Recall that we consider a self-similar IFS $\mathcal{S} = (S_1, \ldots, S_m)$ of m contracting similarity transformations on $\mathbb{R}^d$. The contraction ratios of S_i are $0 < r_i < 1$, $i = 1, \ldots, m$. Theorem 1.5.3 asserts that whenever the IFS $\mathcal{S}$ satisfies the OSC (Definition 1.5.2), the similarity dimension $s = s(\mathcal{S})$ is equal to the Hausdorff dimension of the attractor Λ and the s-dimensional Hausdorff measure of Λ is positive and finite:

$$0 < \mathcal{H}^s(\Lambda) < \infty. \tag{4.1}$$

In fact, it was shown by Bandt and Graf [**BG92**], and Schief [**Sch94**] that the OSC and the positivity of the s-dimensional Hausdorff measure of the attractor are equivalent. We will prove this, following [**BP17**, Section 9.6] in part, but also using Zerner's result [**Zer96**] on the Weak Separation Property (see the next Section 4.2), and some developments from the work of Farkas and Fraser [**FF15**].

4.1.1. Bandt–Graf conditions. In [**BG92**] Bandt and Graf introduced some useful conditions equivalent to the OSC, and we are going to present them now. First we need to fix some notation.

As usual, denote the set of finite words over the alphabet $\{1, \ldots, m\}$ by $\Sigma^* := \{1, \ldots, m\}^*$ and the elements of Σ^* by $\mathbf{i}, \mathbf{j}, \mathbf{k}$. We say that $\mathbf{i} = (i_1, \ldots, i_n)$ is an initial word of $\mathbf{j} = (j_1, \ldots, j_m)$ if $m \geqslant n$ and $i_k = j_k$ for all $1 \leqslant k \leqslant n$. The words $\mathbf{i}$ and $\mathbf{j}$ are incomparable if neither of them is an initial word of the other. Geometrically this means that neither of $\Lambda_{\mathbf{i}}$ and $\Lambda_{\mathbf{j}}$ is a sub-cylinder of the other, that is, we cannot get one of them as the image of the other by a map $S_{\mathbf{k}}$ for some $\mathbf{k} \in \Sigma^*$. If the SSP (the Strong Separation Property, see Definition 1.5.2) holds, then the distance between such $\Lambda_{\mathbf{i}}$ and $\Lambda_{\mathbf{j}}$ with incomparable words is positive: $d(\Lambda_{\mathbf{i}}, \Lambda_{\mathbf{j}}) > r_{|\mathbf{i} \wedge \mathbf{j}|} \cdot \text{const}$. The same is not true if only the OSC is assumed, but the conditions below are meant to express that for incomparable $\mathbf{i}$ and $\mathbf{j}$ the corresponding cylinder sets are sufficiently far away from each other (in an appropriate sense) relative to their sizes.

Let $\mathcal{G}$ be the topological group of similarity transformations on $\mathbb{R}^d$ with the natural topology defined by the neighborhood base of the identity map Id as in (3.22).

Condition BG1. Consider the set of *neighbor maps*:

$$\mathcal{N} := \left\{ S_{\mathbf{i}}^{-1} \circ S_{\mathbf{j}} : \mathbf{i} \neq \mathbf{j} \in \Sigma^* \right\}. \tag{4.2}$$

Thus $\mathcal{N} \subset \mathcal{G}$. We say that Condition BG1 holds if

$$\mathrm{Id} \notin \mathrm{clos}(\mathcal{N}). \tag{4.3}$$

Condition BG2. For $\varepsilon > 0$ and $\mathbf{i}, \mathbf{j} \in \Sigma^*$ the maps $S_{\mathbf{i}}$ and $S_{\mathbf{j}}$ are said to be *ε-relatively close* if

$$|S_{\mathbf{i}}(x) - S_{\mathbf{j}}(x)| < \varepsilon \min\{|\Lambda_{\mathbf{i}}|, |\Lambda_{\mathbf{j}}|\} \text{ for all } x \in \Lambda. \tag{4.4}$$

We say that Condition BG2 holds if there exists $\varepsilon > 0$ such that for all distinct $\mathbf{i}, \mathbf{j} \in \Sigma^*$ the maps $S_{\mathbf{i}}, S_{\mathbf{j}}$ are not ε-relatively close. In this form this condition was stated in [**PRSS01**].

Condition BG3. Fix $\varepsilon \in (0,1)$ satisfying $(1+2\varepsilon)r_{\max} < 1$. Recall that the open ε-neighborhood of Λ is denoted by $B(\Lambda,\varepsilon)$. For $\mathbf{i} \in \Sigma^*$, consider the set of cylinders $\Lambda_{\mathbf{j}}$ of size approximately equal to that of $\Lambda_{\mathbf{i}}$, which intersect $S_{\mathbf{i}}B(\Lambda,\varepsilon) = B(\Lambda_{\mathbf{i}}, \varepsilon r_{\mathbf{i}})$. More precisely, let

$$\widetilde{\Gamma}_\varepsilon(\mathbf{i}) := \left\{ \mathbf{j} \in \Sigma^* : |\Lambda_{\mathbf{j}}| \leqslant |S_{\mathbf{i}}B(\Lambda,\varepsilon)| < |\Lambda_{\mathbf{j}^-}|,\ \Lambda_{\mathbf{j}} \cap S_{\mathbf{i}}B(\Lambda,\varepsilon) \neq \varnothing \right\} \tag{4.5}$$

and

$$\widetilde{\gamma}_\varepsilon := \sup_{\mathbf{i} \in \Sigma^*} \#\widetilde{\Gamma}_\varepsilon(\mathbf{i}). \tag{4.6}$$

We say that Condition BG3 holds if $\widetilde{\gamma}_\varepsilon < \infty$ for some $0 < \varepsilon < \min\{1, (r_{\max}^{-1} - 1)/2\}$.

Note that the first condition on $\mathbf{j}$ in (4.5), namely, the restriction on $|\Lambda_{\mathbf{j}}|$, means than $\mathbf{j}$ belongs to a minimal cut-set, the Moran cut-set $\widetilde{\mathcal{M}}_{|S_{\mathbf{i}}B(\Lambda,\varepsilon)|/|\Lambda|}$, in fact. Observe that

$$\mathbf{j} \in \widetilde{\Gamma}_\varepsilon(\mathbf{i}) \implies |\Lambda_{\mathbf{j}}| \leqslant |S_{\mathbf{i}}B(\Lambda,\varepsilon)| < r_{\min}^{-1} \cdot |\Lambda_{\mathbf{j}}| = r_{\min}^{-1} r_{\mathbf{j}} \cdot |\Lambda|. \tag{4.7}$$

REMARK 4.1.1. It is easy to see that although the value of $\widetilde{\gamma}_\varepsilon$ may depend on ε, as long as $(1+2\varepsilon)r_{\max} < 1$ is satisfied, the resulting $\widetilde{\gamma}_\varepsilon$ is going to be finite or infinite independently of the actual value of ε.

Now we state the main theorem of this section. Recall that we introduced the Strong Open Set Condition (SOSC) in Definition 1.5.2.

THEOREM 4.1.2 (Bandt, Graf and Schief). *Let $\mathcal{S}$ be a self-similar IFS and let s be the similarity dimension. Then the following are equivalent:*

- (i) *Condition BG1;*
- (ii) *Condition BG2;*
- (iii) *Condition BG3;*
- (iv) *SOSC;*
- (v) *OSC;*
- (vi) $0 < \mathcal{H}^s(\Lambda)$;
- (vii) *Λ is s-Ahlfors regular.*

COROLLARY 4.1.3 (Schief). *Suppose that a self-similar IFS $\mathcal{S}$ in $\mathbb{R}^d$ satisfies the OSC and has similarity dimension equal to the dimension of the space: $s = d$. Then the attractor Λ has nonempty interior.*

PROOF OF THE COROLLARY. We present a short derivation of the corollary from Theorem 4.1.2, following [**BP17**]. By the theorem, OSC implies $\mathcal{H}^d(\Lambda) > 0$, hence $\mathcal{L}^d(\Lambda) > 0$. Let $V \neq \varnothing$ be an open and bounded set such that the sets $S_i(V) \subset V$ are pairwise disjoint for $i \leqslant m$. Let r_i be the contraction ratio of the similarity map S_i. We have

$$\mathcal{L}^d\Big(\bigcup_{i=1}^m S_i(V)\Big) = \sum_{i=1}^m \mathcal{L}^d(S_i(V)) = \sum_{i=1}^m r_i^d \mathcal{L}^d(V) = \mathcal{L}^d(V),$$

since $\sum_{i=1}^m r_i^d = 1$. It follows that the open set

$$V \setminus \bigcup_{i=1}^m S_i(\overline{V}) \subset V \setminus \bigcup_{i=1}^m S_i(V)$$

has zero Lebesgue measure, hence it is empty. It follows that $\overline{V} = \bigcup_{i=1}^m S_i(\overline{V})$, whence $\overline{V} = \Lambda$ by the uniqueness of the attractor, and the proof is complete. □

REMARK 4.1.4. It follows from the proof that under the conditions of the corollary, the OSC holds with respect to the open set $V = \mathrm{int}(\Lambda)$.

The scheme of the proof of Theorem 4.1.2 is as follows. First, we show the following chain of implications for the conditions (iii)–(vii) in Theorem 4.1.2:

$$\text{(iii)} \Rightarrow \text{(iv)} \Rightarrow \text{(v)} \Rightarrow \text{(vi)} \Rightarrow \text{(vii)} \Rightarrow \text{(iii)}.$$

In the next Section 4.2, we prove the equivalence of conditions BG1, BG2 and BG3 = (iii).

4.1.2. Condition BG3. First, let us show that Ahlfors regularity = (vii) implies (iii) = BG3.

LEMMA 4.1.5. *Let $\mathcal{S} = \big(S_i(x) = r_i O_i x + t_i\big)_{i=1}^m$ be a self-similar IFS on $\mathbb{R}^d$ and let s be the similarity dimension. Suppose that the attractor Λ of $\mathcal{S}$ is s-Ahlfors regular. Then $\mathcal{S}$ satisfies the condition BG3.*

PROOF. Since Λ is s-Ahlfors regular, there exists $C > 0$ such that

$$C^{-1} \leqslant \frac{\mathcal{H}^s(\Lambda \cap B(x,r))}{r^s} \leqslant C, \quad \forall r \in (0, |\Lambda|) \text{ and } \forall x \in \Lambda.$$

Ahlfors regularity clearly implies that $\mathcal{H}^s(\Lambda) > 0$, and therefore,

$$0 < \mathcal{H}^s(\Lambda) = \mathcal{H}^s\Big(\bigcup_{i=1}^m S_i\Lambda\Big) \leqslant \sum_{i=1}^m \mathcal{H}^s(S_i\Lambda) = \sum_{i=1}^m r_i^s \mathcal{H}^s(\Lambda) = \mathcal{H}^s(\Lambda) < \infty.$$

Thus, $\mathcal{H}^s\big(S_i(\Lambda) \cap S_j(\Lambda)\big) = 0$ for every $i \neq j$, and similarly,

$$\mathcal{H}^s\big(S_{\mathbf{j}}(\Lambda) \cap S_{\mathbf{j}'}(\Lambda)\big) = 0 \text{ for every incomparable } \mathbf{j}, \mathbf{j}' \in \Sigma^*. \tag{4.8}$$

Fix any $\mathbf{i} \in \Sigma^*$, and let $x \in \Lambda_{\mathbf{i}}$ be an arbitrary point. For any $\mathbf{j} \in \widetilde{\Gamma}_\varepsilon(\mathbf{i})$, we have $|\Lambda_{\mathbf{j}}| \leqslant |B(\Lambda_{\mathbf{i}}, r_{\mathbf{i}}\varepsilon)|$ and $\Lambda_{\mathbf{j}} \cap B(\Lambda_{\mathbf{i}}, r_{\mathbf{i}}\varepsilon) \neq \varnothing$, whence

$$\Lambda_{\mathbf{j}} \subset B(x, 2|B(\Lambda_{\mathbf{i}}, r_{\mathbf{i}}\varepsilon)|).$$

Thus, by s-Ahlfors regularity and (4.8),

$$\begin{aligned} C2^s|B(\Lambda_{\mathbf{i}}, r_{\mathbf{i}}\varepsilon)|^s &\geqslant \mathcal{H}^s\big(\Lambda \cap B(x, 2|B(\Lambda_{\mathbf{i}}, r_{\mathbf{i}}\varepsilon)|)\big) \\ &\geqslant \mathcal{H}^s\Big(\bigcup_{\mathbf{j}\in\widetilde{\Gamma}_\varepsilon(\mathbf{i})} \Lambda_{\mathbf{j}}\Big) = \sum_{\mathbf{j}\in\widetilde{\Gamma}_\varepsilon(\mathbf{i})} r_{\mathbf{j}}^s \mathcal{H}^s(\Lambda) \\ &\geqslant r_{\min}^s |S_{\mathbf{i}} B(\Lambda, \varepsilon)|^s \cdot |\Lambda|^{-s} \cdot \mathcal{H}^s(\Lambda) \cdot \#\widetilde{\Gamma}_\varepsilon(\mathbf{i}). \end{aligned}$$

Above we used that any $\mathbf{j}, \mathbf{j}' \in \widetilde{\Gamma}_\varepsilon(\mathbf{i})$ are incomparable, and then the inequality (4.7). Since $\mathbf{i}$ was arbitrary, we get $\sup_{\mathbf{i}\in\Sigma^*} \#\widetilde{\Gamma}_\varepsilon(\mathbf{i}) \leqslant \frac{C2^s|\Lambda|^s}{r_{\min}^s \mathcal{H}^s(\Lambda)}$, and BG3 follows. □

LEMMA 4.1.6. *Let $\mathcal{S} = \big(S_i(x) = r_i O_i x + t_i\big)_{i=1}^m$ be a self-similar IFS on $\mathbb{R}^d$ and let s be the similarity dimension. If $\mathcal{S}$ satisfies the condition BG3, then it satisfies the SOSC.*

PROOF. Let $\varepsilon > 0$ be arbitrary but fixed, such that $(1 + 2\varepsilon)r_{\max} < 1$ and $\sup_{\mathbf{k}\in\Sigma^*} \#\widetilde{\Gamma}_\varepsilon(\mathbf{k}) < \infty$, where $\widetilde{\Gamma}_\varepsilon$ was defined in (4.5). Now let $\boldsymbol{\ell} \in \Sigma^*$ be such that

$$\#\widetilde{\Gamma}_\varepsilon(\boldsymbol{\ell}) = \max_{\mathbf{j}\in\Sigma^*} \#\widetilde{\Gamma}_\varepsilon(\mathbf{j}).$$

Note that for every $\mathbf{k} \in \Sigma^*$ we have

$$\widetilde{\Gamma}_\varepsilon(\mathbf{jk}) \supseteq \left\{\mathbf{ji} : \mathbf{i} \in \widetilde{\Gamma}_\varepsilon(\mathbf{k})\right\}.$$

So, by the maximality property of $\boldsymbol{\ell}$ we have

$$\widetilde{\Gamma}_\varepsilon(\mathbf{j}\boldsymbol{\ell}) = \left\{\mathbf{ji} : \mathbf{i} \in \widetilde{\Gamma}_\varepsilon(\boldsymbol{\ell})\right\} \text{ for every } \mathbf{j} \in \Sigma^*. \tag{4.9}$$

Therefore, there is no word starting with $a \neq j_1$ such that the cylinder Λ_a would intersect $S_{\mathbf{j}\boldsymbol{\ell}}(B(\Lambda, \varepsilon))$, hence for every $\mathbf{j} \in \Sigma^*$ and $a \in \{1, \ldots, m\}$ with $a \neq j_1$, we have

$$\operatorname{dist}(\Lambda_a, \Lambda_{\mathbf{j}\boldsymbol{\ell}}) \geqslant \varepsilon r_{\mathbf{j}\boldsymbol{\ell}}. \tag{4.10}$$

Now we can construct the open set V for which the SOSC holds:

$$V := \bigcup_{\mathbf{j}\in\Sigma^*} S_{\mathbf{j}\boldsymbol{\ell}} B(\Lambda, \varepsilon/2). \tag{4.11}$$

By definition, $V \cap \Lambda \neq \varnothing$ and $S_a(V) \subset V$ for all $a \in \{1, \ldots, m\}$. On the other hand, the fact that $S_a(V) \cap S_b(V) = \varnothing$ for every distinct $a, b \in \{1, \ldots, m\}$ easily follows from (4.10). □

Now we are ready to prove a part of Theorem 4.1.2.

PROPOSITION 4.1.7. *The conditions* (iii)–(vii) *in Theorem* 4.1.2 *are equivalent.*

PROOF. The direction (iii)⇒(iv) was proven in Lemma 4.1.6. The implication (iv)⇒(v) is immediate. We have already demonstrated (v)⇒(vi) in Theorem 1.5.3. The direction (vi)⇒(vii) follows from Theorem 3.4.3. Finally, the implication (vii)⇒(iii) was shown in Lemma 4.1.5. □

Theorem 4.1.2 will follow from Proposition 4.1.7 and from the equivalence of (i)–(iii), which will be proven in the next Section 4.2 in Proposition 4.2.12.

4.1.3. Further remarks. Intuitively, it seems that if the cylinders of a self-similar set Λ_i do not intersect "too much", then the OSC should hold. In particular, a natural assumption is that

$$\#\Big(\bigcup_{i\neq j}(\Lambda_i \cap \Lambda_j)\Big) < \infty, \tag{4.12}$$

which may be called the "Finite Intersection Property", or FIP. Bandt and Rao [**BR07**] proved that for an IFS in $\mathbb{R}^2$, if the attractor Λ is connected, then FIP implies OSC, and they conjectured that this is false in higher dimensions. This was confirmed by Tetenov, Kamalutdinov, and Vaulin [**TKV15**], who constructed a self-similar Jordan arc in $\mathbb{R}^3$, which has one-point intersection property, but does not satisfy the OSC. More recently, Kamalutdinov and Tetenov [**KT20**] proved that FIP does not imply OSC even in $\mathbb{R}$.

THEOREM 4.1.8. *There is a self-similar IFS on $\mathbb{R}$, whose attractor is not a singleton, with one-point intersection property, i.e., $\#\left(\bigcup_{i\neq j}(\Lambda_i \cap \Lambda_j)\right) = 1$, but without the OSC.*

In Section 7.3.2 we sketch a proof of this theorem.

Another interesting example of self-similar IFS on the line was constructed by Farkas [**Far**]. It shows that even on the line it can happen that the OSC holds only with open sets having infinitely many connected components.

EXAMPLE 4.1.9 (Farkas). Let Λ be the attractor of $\mathcal{S} := (S_i(x))_{i=1}^5$, where

$$S_1(x) = \frac{1}{3}x,\ S_2(x) = \frac{1}{9}x + \frac{4}{9},\ S_3(x) = \frac{1}{3}x + \frac{2}{3},\ S_4(x) = \frac{1}{9}x + \frac{1}{3},\ S_5(x) = -\frac{1}{9}x + \frac{4}{9}.$$

(i) The IFS $\mathcal{S}$ satisfies the OSC. The similarity dimension is clearly $s = 1$, hence $O = \operatorname{int}(\Lambda) \neq \varnothing$ by Corollary 4.1.3 and by Remark 4.1.4, the OSC holds with respect to O.

(ii) The open set O has the property that between any two of its connected components there are infinitely many other connected components of O.

Examples with similar properties can be obtained from projections of the "1-dimensional Sierpiński gasket" to lines with rational slopes which form "self-similar tiles" having infinitely many connected components, see [**Ken97**] and Section 7.7.1. However, the property (ii) was not shown there explicitly.

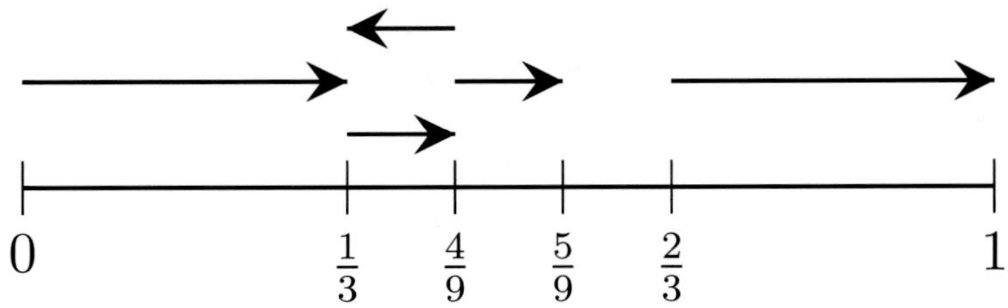

FIGURE 4.1. The images of the unit interval $[0, 1]$ by the maps of the IFS $\mathcal{S}$ in Example 4.1.9, with orientation shown.

PROOF. Simple algebra shows that for every $y \in [0,1]$,

$$S_1(y) = 1 - S_3(1-y), \; S_2(y) = 1 - S_2(1-y), \tag{4.13}$$

furthermore,

$$S_5^{-1}(x) = 1 - S_4^{-1}(x) \in [0,1] \text{ for every } x \in [1/3, 4/9]. \tag{4.14}$$

(i) Let $O := (1/3, 4/9) \cup \bigcup_{\mathbf{i}\in\{1,2,3\}^*} S_{\mathbf{i}}\big((1/3,4/9)\big) \subset (0,1)$. We claim that $\mathcal{S}$ satisfies the OSC with respect to O (see Definition 1.5.2). It is straightforward that $S_i(O) \subseteq O$ for $i = 1,2,3$ and $S_j(O) \subset S_j((0,1)) \subset (1/3,4/9) \subset O$ for $j = 4,5$. Moreover, $S_1(O), S_2(O), S_3(O)$ and $S_j(O)$ are mutually disjoint for $j = 4,5$, hence it remains to show that $S_4(O) \cap S_5(O) = \varnothing$. We argue by contradiction and suppose that $S_4(O) \cap S_5(O) \neq \varnothing$. Then there exist $\mathbf{i},\mathbf{j} \in \{1,2,3\}^*$ such that $x \in S_4 \circ S_{\mathbf{i}}\big((1/3,4/9)\big) \cap S_5 \circ S_{\mathbf{j}}\big((1/3,4/9)\big)$. By (4.14), we get $y = S_5^{-1}(x) \in S_{\mathbf{j}}\big((1/3,4/9)\big)$ and $1 - y = S_4^{-1}(x) \in S_{\mathbf{i}}\big((1/3,4/9)\big)$, thus by (4.13), $y \in S_{\mathbf{i}'}\big((5/9,2/3)\big)$, where $\mathbf{i}' = (4-i_1, \ldots, 4-i_n)$ for $\mathbf{i} = (i_1,\ldots,i_n)$. However, it is again clear from Figure 4.1 that $S_{\mathbf{i}'}\big((5/9,2/3)\big) \cap S_{\mathbf{j}}\big((1/3,4/9)\big) = \varnothing$ for any $\mathbf{i}',\mathbf{j} \in \{1,2,3\}^*$.

(ii) It remains to show that between any two connected components of O there exist infinitely many connected component of O. Indeed, for any $\mathbf{i} \neq \mathbf{j} \in \{1,2,3\}^*$ with $S_{\mathbf{i}}(4/9) < S_{\mathbf{j}}(1/3)$, the components $S_{\mathbf{i}\wedge\mathbf{j}i_{|\mathbf{i}\wedge\mathbf{j}|+1}\underbrace{3,\ldots,3}_{k}}\big((1/3,4/9)\big)$ and $S_{\mathbf{i}\wedge\mathbf{j}j_{|\mathbf{i}\wedge\mathbf{j}|+1}\underbrace{1,\ldots,1}_{k}}\big((1/3,4/9)\big)$ are between $S_{\mathbf{i}}\big((1/3,4/9)\big)$ and $S_{\mathbf{j}}\big((1/3,4/9)\big)$ for every sufficiently large k. □

4.2. Weak Separation Property

Consider the following two examples of IFS on the line.

EXAMPLE 4.2.1. Let $\mathcal{S} = \big(S_i(x)\big)_{i=0,1,3}$ be a homogeneous IFS, where

$$S_0(x) := \frac{1}{3}x, \; S_1(x) := \frac{1}{3}x + \frac{2}{9}, \text{ and } S_3(x) := \frac{1}{3}x + \frac{2}{3}, \quad x \in \mathbb{R}. \tag{4.15}$$

Note that $S_{03} \equiv S_{10}$.

EXAMPLE 4.2.2. For $0 < \lambda < 1$, define the IFS (S_1, S_2) on the line by

$$S_1(x) := \lambda x \quad \text{and} \quad S_2(x) := \lambda x + (1-\lambda),$$

and let $\lambda = \frac{\sqrt{5}-1}{2}$. Here the attractor is the entire unit interval, but the invariant measure with equal weights has been the subject of intensive research. In fact, this is, up to an affine change of variable, the classical Bernoulli convolution measure, corresponding to the golden ratio parameter, see Example 1.8.14. Note that $S_{100} \equiv S_{011}$.

The OSC does not hold in either of these examples. In both cases, the size of all level n-cylinders is λ^n. Let $n \in \mathbb{N}$, and let us fix two distinct $\mathbf{i},\mathbf{j} \in \Sigma_n$. The interesting common feature of these examples is as follows: either $\Lambda_{\mathbf{i}} = \Lambda_{\mathbf{j}}$ or $\Lambda_{\mathbf{i}}$ is shifted away from $\Lambda_{\mathbf{j}}$ by a distance which is greater than a positive constant (independent of n) times λ^n. That is, if $\Lambda_{\mathbf{i}}$ and $\Lambda_{\mathbf{j}}$ do not coincide, then they are sufficiently separated from each other relative to their size. More precisely,

$$\text{either } \Lambda_{\mathbf{i}} = \Lambda_{\mathbf{j}} \text{ or } \frac{|S_{\mathbf{i}}(0) - S_{\mathbf{j}}(0)|}{|\Lambda_{\mathbf{i}}|} > c. \tag{4.16}$$

Checking (4.16) in Example 4.2.1 is an easy exercise, and in Example 4.2.2 it follows from a result of Garsia, see Corollary D.0.11.

The Weak Separation Property (WSP) was introduced by Lau and Ngai [**LN99**], and independently by Zerner [**Zer96**], to capture this phenomenon in greater generality. In the homogeneous case on the line (that is, when all the contraction ratios are identical) the WSP holds if and only if (4.16) is satisfied.

4.2.1. Equivalent conditions for WSP. Let $\mathcal{S} = (S_i)_{i=1}^m$ be a self-similar IFS. Recall the set of neighbor maps $\mathcal{N} = \{S_{\mathbf{i}}^{-1} \circ S_{\mathbf{j}} : \mathbf{i} \neq \mathbf{j}\}_{\mathbf{i},\mathbf{j}\in\Sigma^*}$ defined in (4.2).

DEFINITION 4.2.3. We say that the **Weak Separation Property** (WSP) holds for the self-similar IFS $\mathcal{S}$ if

$$\mathrm{Id} \notin \mathrm{clos}(\mathcal{N}\backslash\{\mathrm{Id}\}). \tag{4.17}$$

In the next theorem we give several equivalent conditions for the WSP. In order to state it, we first recall the notation introduced in (4.5) and (4.6), using any $R > 0$ instead of $\varepsilon \in (0,1)$ and $\mathbf{k} \in \Sigma^*$ playing the role of $\mathbf{i}$ there:

$$\widetilde{\Gamma}_R(\mathbf{k}) := \{\mathbf{i} \in \Sigma^* : \ |\Lambda_{\mathbf{i}}| \leqslant |S_{\mathbf{k}}B(\Lambda, R)| < |\Lambda_{\mathbf{i}^-}|, \ \Lambda_{\mathbf{i}} \cap S_{\mathbf{k}}B(\Lambda, R) \neq \varnothing\}$$

and

$$\widetilde{\gamma}_R = \sup_{\mathbf{k}\in\Sigma^*} \#\widetilde{\Gamma}_R(\mathbf{k}).$$

Now we define

$$\Gamma_R(\mathbf{k}) := \left\{S_{\mathbf{i}} : \mathbf{i} \in \widetilde{\Gamma}_R(\mathbf{k})\right\} \quad \text{and} \quad \gamma_R := \sup_{\mathbf{k}\in\Sigma^*} \#\Gamma_R(\mathbf{k}). \tag{4.18}$$

Note that $\gamma_R \leqslant \widetilde{\gamma}_R$ always holds. However, $\gamma_R < \widetilde{\gamma}_R = \infty$ in each of the last two examples.

In the theorem below it will be convenient to have more flexibility, and we define for any $c > 0$, $R > 0$, and $\mathbf{k} \in \Sigma^*$:

$$\Gamma_{c,R}(\mathbf{k}) := \{S_{\mathbf{i}} : \ |\Lambda_{\mathbf{i}}| \leqslant c r_{\mathbf{k}} < |\Lambda_{\mathbf{i}^-}|, \ \Lambda_{\mathbf{i}} \cap S_{\mathbf{k}}B(\Lambda, R) \neq \varnothing\} \tag{4.19}$$

and

$$\gamma_{c,R} := \sup_{\mathbf{k}\in\Sigma^*} \#\Gamma_{c,R}(\mathbf{k}).$$

Note that $\Gamma_R(\mathbf{k}) = \Gamma_{c,R}(\mathbf{k})$ for $c = |B(\Lambda, R)|$. Below we write r_h for the contraction/dilation coefficient of a similarity map h.

THEOREM 4.2.4 (Zerner [**Zer96**]). *Assume that Λ is not contained in a hyperplane. Then the following conditions are equivalent:*

(i) *For any $\rho > 1$ there exist $x \in \mathbb{R}^d$ and $\varepsilon > 0$ such that*

$$\forall h \in \mathcal{N} \text{ with } r_h \in [\rho^{-1}, \rho], \ h(x) \neq x \Longrightarrow |h(x) - x| > \varepsilon. \tag{4.20}$$

(ii) *There exist $x_0, x_1, \ldots, x_d \in \mathbb{R}^d$ in general position and there exists $\varepsilon > 0$ such that*

$$\forall h \in \mathcal{N}\backslash\{\mathrm{Id}\}, \ \exists j \in \{0, \ldots, d\} \text{ with } |h(x_j) - x_j| \geqslant \varepsilon. \tag{4.21}$$

(iii) *The identity is an isolated point of $\mathcal{N}$ (in the topology defined in (3.22)).*

(iv-a) *For all $c > 0$ and $R > 0$, we have $\gamma_{c,R} < \infty$.*

(iv-b) *There exist $c > 0$ and $R > 0$ such that $\gamma_{c,R} < \infty$.*

(v-a) *There exist* $x \in \mathbb{R}^d$ *and* $\ell \in \mathbb{N}$ *such that for all* $\mathbf{i} \in \Sigma^*$ *and for any ball of radius* b *(denoted* B_b*) we have*

$$\#\left(B_b \cap \left\{S_{\mathbf{ji}}(x) : \mathbf{j} \in \widetilde{\mathcal{M}}_b\right\}\right) \leqslant \ell, \tag{4.22}$$

where we recall that $\widetilde{\mathcal{M}}_b$ *is defined in* (1.73).

(v-b) *There exist* $c \geqslant 1$, $\rho \in (0,1)$, $x \in \mathbb{R}^d$ *and* $\widetilde{\ell} \in \mathbb{N}$ *such that for all* $\mathbf{i} \in \Sigma^*$ *and* $k \in \mathbb{N}$ *holds*

$$\#\left(B_{c\rho^k} \cap \left\{S_{\mathbf{ji}}(x) : \mathbf{j} \in \widetilde{\mathcal{M}}_{c\rho^k}\right\}\right) \leqslant \widetilde{\ell}.$$

The WSP was originally introduced by Lau and Ngai [**LN99**] as condition (v-b), with $c = 1$. We note that Zerner's formulation is slightly different from ours. Recall that $\widetilde{\mathcal{M}}_b$ denotes the Moran cut-set, see (1.73), i.e., $\widetilde{\mathcal{M}}_b = \{\mathbf{k} : |\Lambda_{\mathbf{k}}| \leqslant b < |\Lambda_{\mathbf{k}^-}|\}$.

We start by showing

PROOF OF **(v-a)** $\iff$ **(v-b)**. It is obvious that **(v-a)** $\implies$ **(v-b)** with any $c \geqslant 1$, and $\rho \in (0,1)$. Conversely, suppose that **(v-b)** holds for some $c \geqslant 1$ and $\rho \in (0,1)$ with a bound $\widetilde{\ell}$. Let $b > 0$ be arbitrary. We may assume that $b \leqslant c$ since $\#\widetilde{\mathcal{M}}_b$ is uniformly bounded for every $b > c$. Then there is a unique $k \geqslant 0$ such that $c\rho^{k+1} < b \leqslant c\rho^k$, so for every $\mathbf{j} \in \widetilde{\mathcal{M}}_b$ and for the prefix $\mathbf{j}' \in \widetilde{\mathcal{M}}_{c\rho^k}$ of $\mathbf{j}$ we have $|\mathbf{j}| - |\mathbf{j}'| \leqslant \lceil \frac{\log \rho r_{\min}}{\log r_{\max}} \rceil$. Hence for every $b > 0$ and $y \in \mathbb{R}^d$,

$$\#\left(B(y,b) \cap \{S_{\mathbf{ji}}(0) : \mathbf{j} \in \widetilde{\mathcal{M}}_b\}\right) \leqslant m^{\lceil \frac{\log \rho r_{\min}}{\log r_{\max}} \rceil} \widetilde{\ell}.$$

□

The following technical lemma will be useful in the proof of Theorem 4.2.4.

LEMMA 4.2.5. *Let* $0 < c_1 < c_2$ *and* $R > 0$. *There exists an increasing continuous function* $\phi(t) > 1$ *on* $[1, \infty)$*, depending only on the IFS, such that for all* $\mathbf{k} \in \Sigma^*$,

$$\frac{1}{\phi(c_2/c_1)} \cdot \#\Gamma_{c_2,R}(\mathbf{k}) \leqslant \#\Gamma_{c_1,R}(\mathbf{k}) \leqslant \phi(c_2/c_1) \cdot \#\Gamma_{c_2,R}(\mathbf{k}). \tag{4.23}$$

Therefore,

$$\frac{1}{\phi(c_2/c_1)} \cdot \gamma_{c_2,R} \leqslant \gamma_{c_1,R} \leqslant \phi(c_2/c_1) \cdot \gamma_{c_2,R}. \tag{4.24}$$

PROOF. We start with the left inequality in (4.23). Suppose that $f \in \Gamma_{c_2,R}(\mathbf{k})$. Then there exists $\mathbf{i} \in \Sigma^*$ such that $|\Lambda_{\mathbf{i}}| \leqslant c_2 r_{\mathbf{k}} < |\Lambda_{\mathbf{i}^-}|$, $f = S_{\mathbf{i}}$ and $\Lambda_{\mathbf{i}} \cap S_{\mathbf{k}} B(\Lambda, R) \neq \varnothing$. The word $\mathbf{i}$ may be nonunique; we fix one of them, for instance, take the lexicographically smallest word representing f. The cylinder set $[\mathbf{i}]$ is partitioned into cylinder sets corresponding to words in $\widetilde{\mathcal{M}}_{c_1 r_{\mathbf{k}}/|\Lambda|}$, all of which start with $\mathbf{i}$. For at least one of these words $\mathbf{ii}'$ we have

$$\Lambda_{\mathbf{ii}'} \cap S_{\mathbf{k}} B(\Lambda, R) \neq \varnothing,$$

and then $S_{\mathbf{ii}'} \in \Gamma_{c_1,R}(\mathbf{k})$. Fix such a word $\mathbf{i}'$. Observe that $r_{\mathbf{i}}|\Lambda| \leqslant c_2 r_{\mathbf{k}}$, hence

$$c_1 r_{min} r_{\mathbf{k}} < r_{\mathbf{ii}'}|\Lambda| \leqslant c_2 r_{\max}^{|\mathbf{i}'|} r_{\mathbf{k}},$$

and therefore,

$$|\mathbf{i}'| \leqslant \left\lceil \frac{\log(r_{\min}^{-1} c_2/c_1)}{-\log r_{\max}} \right\rceil =: \psi(c_2/c_1) = p.$$

Let

$$\phi(t) = \psi(t) m^{\psi(t)}.$$

We claim that the map $f \mapsto S_{\mathbf{i}\mathbf{i}'}$ is at most $\phi(c_2/c_1) = pm^p$-to-1. Indeed, if $S_{\mathbf{i}} \neq S_{\mathbf{j}}$ in $\Gamma_{c_2,R}(\mathbf{k})$, it may happen that $S_{\mathbf{i}\mathbf{i}'} = S_{\mathbf{j}\mathbf{j}'} \in \Gamma_{c_1,R}$, but there are no more than pm^p possibilities for $\mathbf{i}'$ or $\mathbf{j}'$, and the claim follows. This concludes the proof of the left inequality in (4.23).

Now we prove the right inequality in (4.23). Suppose that $f \in \Gamma_{c_1,R}(\mathbf{k})$. Then there exists $\mathbf{i} \in \Sigma^*$ such that $|\Lambda_{\mathbf{i}}| \leqslant c_1 r_{\mathbf{k}} < |\Lambda_{\mathbf{i}^-}|$, $f = S_{\mathbf{i}}$ and $\Lambda_{\mathbf{i}} \cap S_{\mathbf{k}} B(\Lambda, R) \neq \varnothing$. As before, we fix one specific $\mathbf{i}$ with these properties. There is a unique prefix $\mathbf{i}' \in \widetilde{\mathcal{M}}_{c_2 r_{\mathbf{k}}/|\Lambda|}$ of $\mathbf{i}$, and of course, $\Lambda_{\mathbf{i}'} \cap S_{\mathbf{k}}(\Lambda, R) \neq \varnothing$, so $S_{\mathbf{i}'} \in \Gamma_{c_2,R}$. Now $\mathbf{i} = \mathbf{i}'\mathbf{i}''$ and, similarly to the above,

$$|\mathbf{i}''| \leqslant \psi(c_2/c_1).$$

Now, if $S_{\mathbf{i}} \neq S_{\mathbf{j}}$ but $S_{\mathbf{i}'} = S_{\mathbf{j}'}$ in this procedure, then the suffixes $\mathbf{i}''$ and $\mathbf{j}''$ must be distinct, and we conclude that that the map $f \mapsto S_{\mathbf{i}'}$ is at most $\phi(c_2/c_1)$-to-1, as above. □

For $0 < b < 1$ define

$$\mathcal{S}_b := \{S_{\mathbf{i}} : \mathbf{i} \in \widetilde{\mathcal{M}}_{b/|\Lambda|}\} = \{S_{\mathbf{i}} : \ \mathbf{i} \in \Sigma^*, \ |\Lambda_{\mathbf{i}}| \leqslant \rho < |\Lambda_{\mathbf{i}^-}|\}. \tag{4.25}$$

We will need another lemma, which is a consequence of **(v-a)**.

LEMMA 4.2.6. *If Λ is not contained in a hyperplane, then* **(v-a)** *implies that*

$$C^* := \sup_{y\in\mathbb{R}^d} \sup_{b>0} \# \{f \in \mathcal{S}_b : f(x) = y\} < \infty, \tag{4.26}$$

where x is from condition **(v-a)**.

PROOF OF LEMMA 4.2.6. Let x and ℓ be as in **(v-a)**. Since $\{S_{\mathbf{i}}(x)\}_{\mathbf{i}\in\Sigma^*}$ is dense in Λ, which is not contained in a hyperplane, we can choose $\mathbf{i}^{(0)}, \ldots, \mathbf{i}^{(d)} \in \Sigma^*$ such that the $d+1$ points $x_k := S_{\mathbf{i}^{(k)}}(x)$, $k = 0, \ldots, d$, are in general position. For $y \in \mathbb{R}^d$ and $b > 0$ let

$$\mathcal{S}_b^{-1}(y) := \{f \in \mathcal{S}_b : f(x) = y\}.$$

As mentioned above, a similarity on $\mathbb{R}^d$ is determined by its values taken at $d+1$ points of general position. So,

$$\#\mathcal{S}_b^{-1}(y) = \# \{(f(x_0), \ldots, f(x_d)) : f \in \mathcal{S}_b^{-1}(y)\}.$$

Therefore,

$$\begin{aligned}
\sup_{y,b} \#\mathcal{S}_b^{-1}(y) &= \sup_{y,b} \# \{(f(x_0), \ldots, f(x_d)) : \ f \in \mathcal{S}_b^{-1}(y)\} \\
&\leqslant \prod_{k=0}^{d} \sup_{y,b} \# \{f(x_k) : \ f \in \mathcal{S}_b^{-1}(y)\} \\
&\leqslant \prod_{k=0}^{d} \sup_{y,b} \# \{\{f(S_{\mathbf{i}^{(k)}}(x)) : f \in \mathcal{S}_b\} \cap \overline{B}(y, b\rho)\},
\end{aligned}$$

where $\rho := \max_k |x_k - x|$. In the last step we used that

$$y = f(x) \Rightarrow |f(x_k) - y| = r_f |x_k - x| \leqslant b\rho.$$

Using this and **(v-a)** and denoting by c_ρ the number of unit balls needed to cover a ball of radius ρ, we obtain

$$\sup_{y,b} \# \mathcal{S}_b^{-1}(y) \leqslant (c_\rho \ell)^{d+1} < \infty.$$

□

PROOF OF THEOREM 4.2.4. Without loss of generality we may assume that $|\Lambda| = 1$, and then $\widetilde{\mathcal{M}}_{b/|\Lambda|} = \widetilde{\mathcal{M}}_b$.

(i) $\Longrightarrow$ (ii) Fix any $\rho > 1$ and then chose x and ε for which (4.20) holds. As in the proof of Lemma 4.2.6, since $\{S_{\mathbf{i}}(x)\}_{\mathbf{i} \in \Sigma^*}$ is dense in Λ, which is not contained in a hyperplane, we can choose $\mathbf{i}^{(0)}, \ldots, \mathbf{i}^{(d)} \in \Sigma^*$ such that the $d+1$ points $x_k := S_{\mathbf{i}^{(k)}}(x)$, $k = 0, \ldots, d$, are in general position.

Chose an arbitrary $h \in \mathcal{N} \setminus \{\mathrm{Id}\}$. If $r_h \notin [\rho^{-1}, \rho]$, then

$$|h(x_0) - x_0| + |h(x_1) - x_1| \geqslant \big||h(x_0) - h(x_1)| - |x_0 - x_1|\big| > (1 - \rho^{-1})|x_0 - x_1|,$$

and (4.21) follows for $\varepsilon = \frac{1}{2}(1 - \rho^{-1})|x_0 - x_1|$.

If $r_h \in [\rho^{-1}, \rho]$, then chose x_k such that $h(x_k) \neq x_k$, which exists by the assumption $h \neq \mathrm{Id}$. We have

$$|h(x_k) - x_k| = r_{\mathbf{i}^{(k)}} |S_{\mathbf{i}^{(k)}}^{-1} h S_{\mathbf{i}^{(k)}}(x) - x| \geqslant \min_k r_{\mathbf{i}^{(k)}} \varepsilon,$$

where we used that the contraction/dilation ratio of $S_{\mathbf{i}^{(k)}}^{-1} h S_{\mathbf{i}^{(k)}} \in \mathcal{N}$ is equal to $r_h \in [\rho^{-1}, \rho]$, so (4.20) applies.

(ii) $\Longrightarrow$ (iii) and (iii) $\Longrightarrow$ (ii) These are immediate from the definition of the topology.

(ii) $\Longrightarrow$ (iv-a) Choose $\varepsilon > 0$ which satisfies (4.21) and fix any $c > 0$ and $R > 0$. Fix $\mathbf{k} \in \Sigma^*$ and two distinct $f, g \in \Gamma_{c,R}(\mathbf{k})$. Note that

$$c r_{\mathbf{k}} r_{\min} < r_f \leqslant c r_{\mathbf{k}}, \tag{4.27}$$

since $f = S_{\mathbf{i}}$, with $\mathbf{i} \in \widetilde{\mathcal{M}}_{c r_{\mathbf{k}}}$. Further $f^{-1} g \in \mathcal{N} \setminus \{\mathrm{Id}\}$, so it follows from (4.21) that there is $\ell = \ell(f, g) \in \{0, \ldots, d\}$ such that

$$|f^{-1} g(x_\ell) - x_\ell| \geqslant \varepsilon \Longrightarrow |g(x_\ell) - f(x_\ell)| \geqslant \varepsilon r_f > \varepsilon c r_{\mathbf{k}} r_{\min}.$$

Thus,

$$B(f(x_\ell), \varepsilon \delta r_{\mathbf{k}}) \cap B(g(x_\ell), \varepsilon \delta r_{\mathbf{k}}) = \varnothing, \tag{4.28}$$

where $\delta = c r_{\min}/3$. For any $f \in \Gamma_{c,R}(\mathbf{k})$ we have $f(\Lambda) \cap S_{\mathbf{k}} B(\Lambda, R) \neq \varnothing$, whence

$$f(\Lambda) \subset B(S_{\mathbf{k}} B(\Lambda, R), r_f).$$

Therefore, using that $\mathrm{dist}(f(x_\ell), f(\Lambda)) = r_f \cdot \mathrm{dist}(x_\ell, \Lambda)$ and keeping (4.27) in mind, we obtain for all $f \in \Gamma_{c,R}(\mathbf{k})$ and $\ell \in \{0, \ldots, d\}$:

$$\begin{aligned} B(f(x_\ell), \varepsilon \delta r_{\mathbf{k}}) &\subset B(S_{\mathbf{k}} B(\Lambda, R), r_f + r_f \,\mathrm{dist}(x_\ell, \Lambda) + \varepsilon \delta r_{\mathbf{k}}) \\ &\subset B\big(\Lambda_{\mathbf{k}}, c r_{\mathbf{k}}(1 + \mathrm{dist}(x_\ell, \Lambda) + \varepsilon \delta) + r_{\mathbf{k}} R\big) \\ &\subset B(\Lambda_{\mathbf{k}}, C_1 r_{\mathbf{k}}), \end{aligned} \tag{4.29}$$

where $C_1 = c\big(1 + \max_\ell \operatorname{dist}(x_\ell, \Lambda) + \varepsilon r_{\min}/3\big) + R$. Using Lebesgue measure and comparing (4.28) with (4.29), we obtain that there is a constant $L = L(c, R)$ (independent of $\mathbf{k}$) such that

$$\max_{p \in \{0,\dots,d\}} \{\#A : \ A \subset \Gamma_{c,R}(\mathbf{k}),\ \ell(f,g) = p \ \text{ for all } \ f \neq g \text{ in } A\} \leqslant L.$$

Consider the complete labeled graph with vertices $\Gamma_{c,R}(\mathbf{k})$ and edges labeled by $\ell(f,g)$. A straightforward application of Ramsey's theorem [**Ber73**, Theorem 19.2.3] yields that there exists L^* such that

$$\forall\, \mathbf{k},\ \#\Gamma_{c,R}(\mathbf{k}) \leqslant L^*,$$

as desired. (Instead of Ramsey's theorem, we could use the Blaschke selection principle and separation of the sets $f(\Lambda)$ for $f \in \Gamma_{c,R}(\mathbf{k})$ in the Hausdorff metric, as in [**BP17**, Section 9.6].)

(iv-a) $\Longrightarrow$ (v-a) Actually we prove that (iv-a) implies a stronger assertion (replacing in the condition $\exists\, x$ with $\forall\, x$):

(v*): $\forall x \in \mathbb{R}^d$, $\exists\, \ell \in \mathbb{N}$ such that for all $\mathbf{i}$ and $b > 0$

$$\#\left\{B_b \cap \left\{S_{\mathbf{ji}}(x), \mathbf{j} \in \widetilde{\mathcal{M}}_b\right\}\right\} \leqslant \ell \tag{4.30}$$

(recall that $|\Lambda| = 1$ by assumption).

In fact, let B_b be any ball of radius b, as in (v-a). Fix any $x \in \mathbb{R}^d$ and let $R > 0$ be such that $x \in B(\Lambda, R)$. Since every $S_\mathbf{i}$ is a contraction, we have $S_\mathbf{i}(x) \in B(\Lambda, R)$ for all $\mathbf{i} \in \Sigma^*$. It follows that

$$\begin{aligned}
\mathcal{J} \ &:= \ \#\left\{B_b \cap \left\{S_{\mathbf{ji}}(x),\ \mathbf{j} \in \widetilde{\mathcal{M}}_b\right\}\right\} \\
&\leqslant \ \#\left\{S_\mathbf{j} :\ \mathbf{j} \in \widetilde{\mathcal{M}}_b,\ S_\mathbf{j}(S_\mathbf{i}(x)) \in B_b\right\} \\
&\leqslant \ \#\left\{S_\mathbf{j} :\ \mathbf{j} \in \widetilde{\mathcal{M}}_b,\ S_\mathbf{j}(B(\Lambda, R)) \cap B_b \neq \varnothing\right\} \\
&\leqslant \ \#\left\{S_\mathbf{j} :\ \mathbf{j} \in \widetilde{\mathcal{M}}_b,\ B_{b+r_\mathbf{j} R} \cap \Lambda_\mathbf{j} \neq \varnothing\right\} \\
&\leqslant \ \#\left\{S_\mathbf{j} :\ \mathbf{j} \in \widetilde{\mathcal{M}}_b :\ B_{b(1+R)} \cap \Lambda_\mathbf{j} \neq \varnothing\right\},
\end{aligned}$$

where all the balls are concentric with B_b. If $B_{b(1+R)} \cap \Lambda = \varnothing$, we have $\mathcal{J} = 0$; otherwise, we can chose $\mathbf{k} \in \widetilde{\mathcal{M}}_b$ such that $B_{b(1+R)} \cap \Lambda_\mathbf{k} \neq \varnothing$. Then

$$B_{b(1+R)} \subset B(\Lambda_\mathbf{k}, 2b(1+R)) \subset S_\mathbf{k} B(\Lambda, \widetilde{R}), \quad \text{where} \quad \widetilde{R} = 2(1+R) r_{\min}^{-1},$$

since $r_\mathbf{k} \geqslant b r_{\min}$. Thus,

$$\mathcal{J} \leqslant \#\left\{S_\mathbf{j} :\ \mathbf{j} \in \widetilde{\mathcal{M}}_b,\ \Lambda_\mathbf{j} \cap S_\mathbf{k} B(\Lambda, \widetilde{R}) \neq \varnothing\right\} = \Gamma_{b/r_\mathbf{k}, \widetilde{R}}(\mathbf{k}).$$

Since $b/r_\mathbf{k} \in [1, r_{\min}^{-1}]$, the claim (4.30) follows from (iv-a) and Lemma 4.2.5.

(v-a) $\Longrightarrow$ (iv-b) Here we use Lemma 4.2.6. Fix $x \in \mathbb{R}^d$ such that (4.22) holds. Let $R > 0$ be such that $x \in B(\Lambda, R)$. We are going to show that $\gamma_{1,R} < \infty$. Fix any $\mathbf{k} \in \Sigma^*$, and let $b = r_\mathbf{k} = |\Lambda_\mathbf{k}|$. Recall that, (in view of $|\Lambda| = 1$),

$$\Gamma_{1,R}(\mathbf{k}) = \left\{S_\mathbf{j} :\ \mathbf{j} \in \widetilde{\mathcal{M}}_b : \Lambda_\mathbf{j} \cap B(\Lambda_\mathbf{k}, R r_\mathbf{k}) \neq \varnothing\right\}.$$

Let B_b be a ball of radius b which contains $\Lambda_\mathbf{k}$. Then $S_\mathbf{j}(x) \in B(\Lambda_\mathbf{j}, R r_\mathbf{j})$, hence

$$\left(\mathbf{j} \in \widetilde{\mathcal{M}}_b,\ \Lambda_\mathbf{j} \cap B(\Lambda_\mathbf{k}, R r_\mathbf{k}) \neq \varnothing\right) \ \Longrightarrow \ S_\mathbf{j}(x) \in B_{(2R+1)b}. \tag{4.31}$$

Here $B_{(2R+1)b}$ is a ball concentric with B_b, and we use that $r_{\mathbf{j}} \leqslant r_{\mathbf{k}} = b$.

Let L be the minimal number of unit balls needed to cover a ball of radius $(2R+1)$ in $\mathbb{R}^d$. Then we can choose balls $B'_1, \dots, B'_L$ of radius b which cover $B_{(2R+1)b}$. It follows from (4.31) that

$$\#\Gamma_{1,R}(\mathbf{k}) \leqslant \sum_{p=1}^{L} \#\left\{S_{\mathbf{j}} : \mathbf{j} \in \widetilde{\mathcal{M}}_b,\ S_{\mathbf{j}}(x) \in B'_p\right\}. \tag{4.32}$$

To estimate the cardinalities in the right-hand side, note that by (v-a), for every p, there are at most ℓ points $y_1, \dots, y_\ell \in B'_p$ for which

$$\exists \mathbf{j} \in \widetilde{\mathcal{M}}_b, \quad \text{such that} \quad S_{\mathbf{j}}(x) = y_q. \tag{4.33}$$

On the other hand, by Lemma 4.2.6, for every $q = 1, \dots, \ell$, the maximum number of $\mathbf{j}$'s that satisfy (4.33) is less than or equal to C^*. Thus the cardinalities on the right-hand side of (4.32) are all less than or equal to $C^*\ell$ and hence

$$\#\Gamma_{1,R}(\mathbf{k}) \leqslant L \cdot C^*\ell,$$

which is independent of $\mathbf{k}$. This yields $\gamma_{1,R} < \infty$.

(iv-b) $\Longrightarrow$ (i) This step resembles the proof of Lemma 4.1.6. Let $c > 0$ and $R > 0$ be such that $\gamma_{c,R} < \infty$. In view of Lemma 4.2.5, we have $\gamma_{1,R} < \infty$. Fix an arbitrary $\rho > 1$. We need to find $x \in \mathbb{R}^d$ such that (4.20) holds for all $h \in \mathcal{N}\backslash\{\mathrm{Id}\}$, with $r_h \in [\rho^{-1}, \rho]$. Choose $\boldsymbol{\ell} \in \Sigma^*$ such that

$$\#\Gamma_{1,R}(\boldsymbol{\ell}) = \max_{\mathbf{k}\in\Sigma^*} \#\Gamma_{1,R}(\mathbf{k}).$$

Observe that for any $\mathbf{v} \in \Sigma^*$ we have

$$\Gamma_{1,R}(\mathbf{v}\boldsymbol{\ell}) = \{S_{\mathbf{v}}S_{\mathbf{j}} :\ S_{\mathbf{j}} \in \Gamma_{1,R}(\boldsymbol{\ell})\}. \tag{4.34}$$

Indeed, the inclusion $\supseteq$ holds by definition, and then equality follows by the maximality property of $\boldsymbol{\ell}$.

Let $b := r_{\boldsymbol{\ell}}$. Then by definition,

$$\Gamma_{1,R}(\boldsymbol{\ell}) = \left\{S_{\mathbf{i}} : \mathbf{i} \in \widetilde{\mathcal{M}}_b,\ \Lambda_{\mathbf{i}} \cap S_{\boldsymbol{\ell}}B(\Lambda, R) \neq \varnothing\right\}.$$

Pick an arbitrary $y \in \Lambda$ and let $x := S_{\boldsymbol{\ell}}(y)$. Clearly $B := B(x, Rr_{\boldsymbol{\ell}}) \subset S_{\boldsymbol{\ell}}B(\Lambda, R)$. Fix an integer $q > 0$ such that $r_{\max}^q < \rho^{-1}$. Without loss of generality, in view of (4.34), we may assume that $|\boldsymbol{\ell}| > q$. Let

$$\varepsilon_1 := \min\left\{|x - gS_{\mathbf{k}}(y)| : g \in \Gamma_{1,R}(\boldsymbol{\ell}),\ 0 \leqslant |\mathbf{k}| \leqslant q \text{ and } gS_{\mathbf{k}}(y) \neq x\right\}, \tag{4.35}$$

where $S_{\mathbf{k}} = \mathrm{Id}$ when $|\mathbf{k}| = 0$, and let

$$\varepsilon := \min\left\{Rr_{\boldsymbol{\ell}}, \varepsilon_1\right\}.$$

For $h \in \mathcal{N}\backslash\{\mathrm{Id}\}$ we have $h = S_{\mathbf{v}}^{-1}S_{\mathbf{w}}$. We can assume that $r_{\mathbf{w}} \leqslant r_{\mathbf{v}}$, since otherwise $|h(x)-x| \geqslant |x-h^{-1}(x)|$ and the following works for h^{-1} instead of h. By assumption on the contraction ratio, $\rho^{-1} \leqslant r_h \leqslant 1$. Then $r_{\mathbf{w}\boldsymbol{\ell}} \leqslant r_{\mathbf{v}\boldsymbol{\ell}}$, and we can choose a prefix $\boldsymbol{\ell}'$ of $\boldsymbol{\ell}$ such that

$$\mathbf{w}\boldsymbol{\ell}' \in \widetilde{\mathcal{M}}_{r_{\mathbf{v}\boldsymbol{\ell}}}.$$

Immediate calculation yields that if $\boldsymbol{\ell} = \boldsymbol{\ell}'\boldsymbol{\ell}''$, then

$$0 \leqslant |\boldsymbol{\ell}''| \leqslant q. \tag{4.36}$$

We distinguish between two possibilities:

(a) $S_{\mathbf{w}\boldsymbol{\ell}'} \in \Gamma_{1,R}(\mathbf{v}\boldsymbol{\ell})$. Then by (4.34) we have $S_{\mathbf{w}\boldsymbol{\ell}'} = S_{\mathbf{vi}}$ for some $S_{\mathbf{i}} \in \Gamma_{1,R}(\boldsymbol{\ell})$. So,

$$\begin{aligned} |h(x) - x| &= \left|S_{\mathbf{v}}^{-1} S_{\mathbf{w}} S_{\boldsymbol{\ell}'} S_{\boldsymbol{\ell}''} y - x\right| = \left|S_{\mathbf{v}}^{-1} S_{\mathbf{v}} S_{\mathbf{i}} S_{\boldsymbol{\ell}''} y - x\right| \\ &= \left|S_{\mathbf{i}} S_{\boldsymbol{\ell}''} y - x\right|, \end{aligned}$$

which is either 0 or at least $\varepsilon_1 \geqslant \varepsilon$ by (4.35) and (4.36).

(b) $S_{\mathbf{w}\boldsymbol{\ell}'} \notin \Gamma_{1,R}(\mathbf{v}\boldsymbol{\ell})$. Using that $\Lambda_{\mathbf{w}\boldsymbol{\ell}} \subset \Lambda_{\mathbf{w}\boldsymbol{\ell}'}$, by the definition of $\Gamma_{1,R}(\mathbf{v}\boldsymbol{\ell})$ we obtain that

$$\operatorname{dist}(\Lambda_{\mathbf{w}\boldsymbol{\ell}}, \Lambda_{\mathbf{v}\boldsymbol{\ell}}) \geqslant r_{\mathbf{v}} r_{\boldsymbol{\ell}} R,$$

which implies, since $x \in \Lambda_{\ell}$, that $|S_{\mathbf{w}}x - S_{\mathbf{v}}x| \geqslant r_{\mathbf{v}} r_{\boldsymbol{\ell}} R$, and hence

$$|h(x) - x| = |S_{\mathbf{v}}^{-1} S_{\mathbf{w}}(x) - x| = r_{\mathbf{v}}^{-1} \left|S_{\mathbf{w}}(x) - S_{\mathbf{v}}(x)\right| \geqslant R r_{\boldsymbol{\ell}} \geqslant \varepsilon.$$

□

REMARK 4.2.7.

(a) When we proved **(iv-a)** $\Longrightarrow$ **(v-a)**, we have actually shown that **(iv)** implies even the stronger assertion **(v*)**. Hence, any of the conditions of Theorem 4.2.4 implies that (4.26) holds for every x (assuming that Λ is not contained in a hyperplane).

(b) In the proof of **(iv-b)** $\Longrightarrow$ **(i)** we only assumed about x that $x \in S_{\ell}(\Lambda)$. So, each of the conditions of Theorem 4.2.4 implies that **(i)** holds for uncountably many x. In particular, for any $\rho > 1$,

$$\exists\, x \in \Lambda,\ \exists\, \varepsilon > 0,\ \forall h \in \mathcal{N} \setminus \{\mathrm{Id}\}, \quad r_h \in [\rho^{-1}, \rho] \implies |h(x) - x| > \varepsilon,$$

since $\mathcal{N}$ is countable.

Using Remark 4.2.7(a) and Lemma 4.2.6 we obtain the following:

COROLLARY 4.2.8. *Suppose that a self-similar IFS $\mathcal{S}$ has the WSP, and let $\mathcal{S}_b$ be defined by* (4.25). *Further, denote*

$$\mathcal{S}_b(x) := \{g(x) : g \in \mathcal{S}_b\}. \tag{4.37}$$

If the attractor Λ is not contained in a hyperplane, then for every $x \in \mathbb{R}^d$ there exists $C > 0$ such that for every $b > 0$,

$$\#\mathcal{S}_b(x) \leqslant \#\mathcal{S}_b \leqslant C \cdot \#\mathcal{S}_b(x). \tag{4.38}$$

A further equivalent characterization of WSP was obtained in [**LW04**, Proposition 2.1].

PROPOSITION 4.2.9 (Lau and Wang). *The WSP holds for the IFS $(S_i)_{i=1}^m$ if and only if there exists $\gamma \in \mathbb{N}$ and a compact set $D \subset \mathbb{R}^d$ with a nonempty interior satisfying $\bigcup_{i=1}^m S_i(D) \subset D$, such that*

$$\forall x \in \mathbb{R}^d,\ \forall b \in (0,1), \quad \#\{S \in \mathcal{S}_b :\ x \in S(D)\} \leqslant \gamma.$$

Recall that the original definition of OSC (Definition 1.5.2) requires the existence of a nonempty bounded open set V such that (a) $\bigcup\limits_{i=1}^{m} S_i(V) \subset V$ and (b) every point of the space is covered by at most one of the sets $\{S_i(V)\}_{i=1}^m$. If we take the interior $V := D^\circ$ of the set D in Proposition 4.2.9 then V satisfies condition (a). On the other hand, instead of condition (b) we require in Proposition 4.2.9 only that every point of the space is covered by at most γ of the sets $\{S(\overline{V})\}_{S\in\mathcal{S}_b}$ for any $b > 0$.

Finally, we present one more equivalent characterization of the WSP:

LEMMA 4.2.10. *The following are equivalent:*

(i) *The WSP holds for the self-similar IFS* $\mathcal{S} = (S_i)_{i=1}^m$.

(ii) *There exists* $\gamma < \infty$ *such that for every ball* B_b *of radius* b,

$$\#\left\{S_{\mathbf{j}} : \mathbf{j} \in \widetilde{\mathcal{M}}_{b/|\Lambda|} \text{ and } B_b \cap \Lambda_{\mathbf{j}} \neq \varnothing\right\} \leqslant \gamma. \tag{4.39}$$

(iii) *For any* $C_1 < 1 < C_2$ *there exists* $L = L(C_1, C_2)$ *such that for every ball* B *we have*

$$\#\left\{S_{\mathbf{j}} : C_1 \leqslant \frac{|\Lambda_{\mathbf{j}}|}{|B|} \leqslant C_2,\ \Lambda_{\mathbf{j}} \cap B \neq \varnothing\right\} \leqslant L. \tag{4.40}$$

PROOF OF LEMMA 4.2.10. We only prove that WSP=(i) implies (ii). The implication **(ii)** $\Longrightarrow$ **(iii)** is left to the reader as an easy exercise, as well as the proof of the implication **(iii)** $\Longrightarrow$ **(iv-a)** of Theorem 4.2.4.

As before, assume that $|\Lambda| = 1$ without loss of generality. Fix an arbitrary ball B_b of radius b. We can assume that $b \leqslant 1$ and $\Lambda \cap B_b \neq \varnothing$; otherwise, the claim is obvious. Then there exists $\mathbf{k} \in \Sigma^*$ such that $B_b \cap \Lambda_{\mathbf{k}} \neq \varnothing$ and

$$br_{\min} < r_{\mathbf{k}} \leqslant b.$$

Then $B_b \subset B(\Lambda_{\mathbf{k}}, 2b) \subset B(\Lambda_{\mathbf{k}}, 2r_{\mathbf{k}}r_{\min}^{-1}) = S_{\mathbf{k}}B(\Lambda, 2r_{\min}^{-1})$, and so

$$\mathbf{j} \in \widetilde{\mathcal{M}}_b,\ \Lambda_{\mathbf{j}} \cap B_b \neq \varnothing \quad\Longrightarrow\quad \Lambda_{\mathbf{j}} \cap S_{\mathbf{k}}B(\Lambda, 2r_{\min}^{-1}) \neq \varnothing.$$

We fix $\mathbf{k}$ and let $c = br_{\mathbf{k}}^{-1}$, so that $\widetilde{\mathcal{M}}_b = \widetilde{\mathcal{M}}_{cr_{\mathbf{k}}}$. It follows that the set in (4.39) is contained in $\Gamma_{c,2r_{\min}^{-1}}(\mathbf{k})$, see the definition (4.19). By the condition **(iv-a)** of Theorem 4.2.4, we have that

$$\gamma_{1,2r_{\min}^{-1}} = \sup_{\mathbf{i}\in\Sigma^*} \#\Gamma_{1,2r_{min}^{-1}}(\mathbf{i}) < \infty.$$

It remains to note that $c \in [1, r_{\min}^{-1}]$, and so (4.39) is satisfied with $\gamma = \gamma_{1,2r_{\min}^{-1}} \cdot \phi(r_{\min}^{-1})$, where ϕ is from Lemma 4.2.5. □

4.2.2. Relation between the WSP and the OSC. Recall that we say that the self-similar IFS $\mathcal{S} = (S_i)_{i=1}^m$ has an *exact overlap* if there are distinct $\mathbf{i}, \mathbf{j} \in \{1, \ldots, m\}^*$ such that $S_{\mathbf{i}} \equiv S_{\mathbf{j}}$. Combining Proposition 4.1.7 and Theorem 4.2.4 we obtain the following:

COROLLARY 4.2.11. *The OSC holds if and only if the WSP holds and there are no exact overlaps, that is,* $S_{\mathbf{i}} \neq S_{\mathbf{j}}$ *whenever* $\mathbf{i} \neq \mathbf{j} \in \Sigma^*$.

PROOF. Observe that if there are no exact overlaps, then

$$\#\widetilde{\Gamma}_\varepsilon(\mathbf{i}) = \#\Gamma_\varepsilon(\mathbf{i}),$$

where these quantities are defined in (4.5) and (4.18). Thus, by Proposition 4.1.7, the OSC holds.

On the other hand, the OSC clearly implies that there are no exact overlaps and thus, by the previous observation and Theorem 4.2.4, the WSP holds. □

PROPOSITION 4.2.12. *The conditions* (i)–(iii) *in Theorem* 4.1.2 *are equivalent.*

PROOF. Under the assumption that there are no exact overlaps, the condition BG1 is equivalent to **(iii)** in Theorem 4.2.4; BG2 is equivalent to (i), and BG3 is equivalent to **(iv-b)** in the same theorem. Hence, the claim follows. □

We would like to draw the attention of the reader to a common misunderstanding caused by a misinterpretation of Corollary 4.2.11. Namely, let $\mathcal{S}$ be a self-similar IFS with an exact overlap such that the WSP holds. Then in general, we cannot get rid of the overlapping feature of $\mathcal{S}$ by passing to a higher iterate of the system $\mathcal{S}^n := (S_{\mathbf{i}})_{|\mathbf{i}|=n}$ for a suitable n, and removing some of those mappings $S_{\mathbf{j}}$ for which there is an $\mathbf{i}$ such that $S_{\mathbf{i}} = S_{\mathbf{j}}$. Now we develop this with more details.

The easiest way to construct an IFS with an exact overlap for which the WSP holds is as follows:

EXAMPLE 4.2.13. Let Λ be the attractor of the self-similar IFS $\mathcal{S} = (S_k)_{k=1}^m$ on $\mathbb{R}^d$ which satisfies the OSC. Let $\mathbf{i} \in \{1, \ldots, m\}^*$ be an arbitrary finite word and we define $S_{m+1} := S_{\mathbf{i}}$. Finally let $\widetilde{\Lambda}$ be the attractor of the IFS $\widetilde{\mathcal{S}} = (S_k)_{k=1}^{m+1}$. Then clearly $\Lambda = \widetilde{\Lambda}$, and $\widetilde{\mathcal{S}}$ has an exact overlap and satisfies the WSP.

We say that a self-similar IFS has a *simple exact overlap* if it is analogous to the one in Example 4.2.13. More precisely:

DEFINITION 4.2.14. Let $\mathcal{S} = (S_k)_{k=1}^M$ be a self-similar IFS with an exact overlap. We say that $\mathcal{S}$ has a *simple exact overlap* if there exists a $\Gamma \subset \{1, \ldots, M\}^*$ such that

(i) the attractor of $\widetilde{\mathcal{S}} = (S_{\mathbf{i}})_{\mathbf{i} \in \Gamma}$ and the attractor of $\mathcal{S}$ is the same set Λ, and
(ii) $\widetilde{\mathcal{S}}$ satisfies the OSC.

In Example 4.2.13 $M = m + 1$ and $\Gamma = \{1, \ldots, m\}$. However, in general, self-similar IFS's with an exact overlap, satisfying the WSP, have much more complicated structure.

Farkas [**Far16**, Example 8.6] gave an example of a self-similar IFS $\mathcal{S}$ with an exact overlap on the line, which satisfies the WSP, such that its attractor Λ does not coincide with the attractor of any self-similar IFS satisfying the OSC (in particular, it is not one with a simple exact overlap).

EXAMPLE 4.2.15. Let $r, g \in (0, 1)$ be such that $3r + 2g = 1$. We start with a self-similar IFS on the line $\widehat{\mathcal{S}} := (\widehat{S}_1, \widehat{S}_2, \widehat{S}_3)$ such that $\widehat{S}_i(x) = rx + (i-1)(r+g)$. The attractor $\widehat{\Lambda}$ has the convex hull $\widehat{I} = [0, 1]$. The level-one cylinder intervals are $\widehat{I}_i = \widehat{S}_i(\widehat{I})$, $i = 1, 2, 3$. The distance (gap) between two consecutive level-one cylinders is $g > 0$, so the SSP holds. The right endpoint of $\widehat{I}_2 = \widehat{S}_2(\widehat{I})$ is $2r + g$. Let

$$\Lambda := \widehat{\Lambda} \cap [0, 2r + g].$$

Farkas proved in [**Far16**, Example 8.6] the following:

(i) Λ is the attractor of the IFS $\mathcal{S} = (S_i(x) = rx + t_i)_{i=1}^4$, where

$$t_1 = 0,\ t_2 = r(r+g),\ t_3 = r + g,\ t_4 = r + g + r(r+g).$$

(see Figure 4.2), which satisfies the WSP.
(ii) No self-similar IFS with attractor Λ satisfies the OSC.

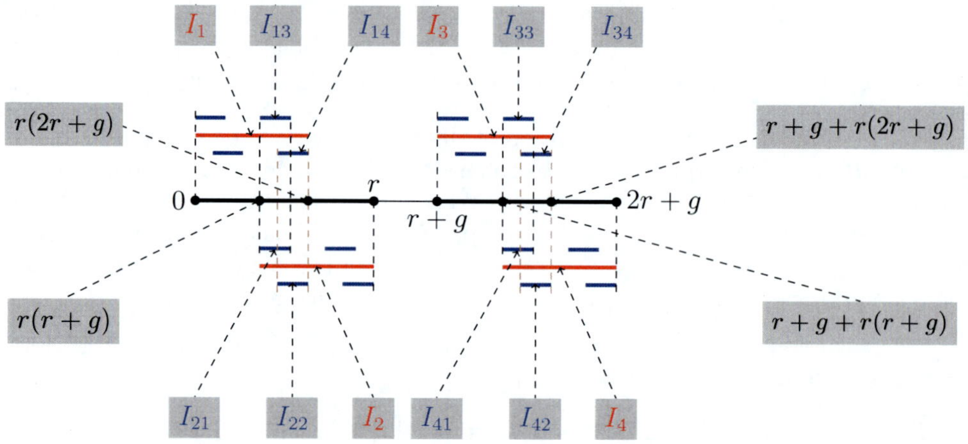

FIGURE 4.2. The first and the second level cylinder-intervals of $\mathcal{S}$ from Example 4.2.15

4.2.3. Dimension of the attractor of an IFS satisfying the WSP. We will show that the most natural estimate of the dimension gives a correct answer. Namely, given a self-similar IFS $\mathcal{S} = (S_i)_{i=1}^m$, for $0 < b < 1$ consider the IFS $\mathcal{S}_b$ formed from the higher iterates of $\mathcal{S}$ as defined in (4.25). Let s_b be the similarity dimension of $\mathcal{S}_b$, that is,

$$\sum_{f \in \mathcal{S}_b} r_f^{s_b} = 1, \tag{4.41}$$

where r_f is the contraction ratio of the similarity map f. Clearly,

$$\frac{\log \#\mathcal{S}_b}{-\log b r_{\min}} \leqslant s_b \leqslant \frac{\log \#\mathcal{S}_b}{-\log b}. \tag{4.42}$$

THEOREM 4.2.16 (Zerner [**Zer96**]). *Let Λ be the attractor of $\mathcal{S}$ and $\beta := \lim_{b\to 0} s_b$. We have*

(i)

$$\dim_{\mathrm{H}}(\Lambda) = \dim_{\mathrm{B}}(\Lambda) \leqslant \beta. \tag{4.43}$$

(ii) *If the IFS $\mathcal{S}$ satisfies the WSP and Λ is not contained in a hyperplane, then*

(a) *the inequality above becomes equality:*

$$\dim_{\mathrm{H}}(\Lambda) = \dim_{\mathrm{B}}(\Lambda) = \beta; \tag{4.44}$$

(b) *the corresponding Hausdorff measure is positive and finite:*

$$0 < \mathcal{H}^{\beta}(\Lambda) < \infty; \tag{4.45}$$

(c) *If $\dim_{\mathrm{H}}(\Lambda) = d$, then Λ contains interior points.*

REMARK 4.2.17. Note that the limit $\beta = \lim_{b\to 0} s_b$ exists. In fact, for any finite partition of the symbolic space into cylinder sets $\mathcal{P}$ define $\mathcal{S}_{\mathcal{P}} = \{S_{\mathbf{i}} : [\mathbf{i}] \in \mathcal{P}\}$ and $s_{\mathcal{P}} > 0$ such that

$$\sum_{S_{\mathbf{i}} \in \mathcal{S}_{\mathcal{P}}} r_{\mathbf{i}}^{s_{\mathcal{P}}} = 1.$$

The family $\mathcal{S}_b$ and the number s_b are obtained in the special case when the partition corresponds to the symbolic Moran cover $\widetilde{\mathcal{M}}_b$. It is easy to see that if $\mathcal{P}' \leqslant \mathcal{P}$ are two partitions, that is, $\mathcal{P}'$ is a refinement of $\mathcal{P}$, then $s_{\mathcal{P}'} \leqslant s_{\mathcal{P}}$. It follows that the function $b \mapsto s_b$ is monotone decreasing, so the limit exists. Moreover, we can obtain β as a limit of any monotone sequence of partitions whose join is the partition into points; for example, partitions into cylinders of length n.

PROOF. As before, we assume that $|\Lambda| = 1$.

(i) Fix an arbitrary $x \in \Lambda$. Then $N_b(\Lambda)$, the minimal number of balls of diameter b required to cover Λ, is at most $\#\mathcal{S}_b(x)$. This and (4.42) together yield (4.43), in view of Theorem 3.3.1.

(ii)(a) To verify (4.44) assume that the WSP holds and Λ is not contained in a hyperplane. Then it follows from Remark 4.2.7(b) that there exists $x \in \Lambda$ and $\varepsilon > 0$ such that for every $b > 0$ and $f, g \in \mathcal{S}_b$ distinct, holds

$$|f(x) - g(x)| \geqslant \varepsilon b r_{\min}. \tag{4.46}$$

Let $\mathcal{C} := \{B(f(x), br_{\min}\varepsilon/2)\}_{f \in \mathcal{S}_b}$. Then $\mathcal{C}$ consists of $\#\mathcal{S}_b$ pairwise disjoint balls centered at Λ of radius const$\cdot b$. By Lemma 1.4.2(e), this implies that $\underline{\dim}_{\mathrm{B}}(\Lambda) \geqslant \beta$, and (4.44) follows.

The proof of (ii)(b) follows [**Fal89**, Theorem 2]. By Lemma 4.2.10(ii),

$$\#\left\{S_{\mathbf{j}} : \mathbf{j} \in \widetilde{\mathcal{M}}_b \text{ and } B_b \cap S_{\mathbf{j}}(\Lambda) \neq \varnothing\right\} \leqslant \gamma < \infty.$$

We want to show that

$$0 < \gamma^{-1} \leqslant \mathcal{H}^{\beta}(\Lambda). \tag{4.47}$$

Let us argue by contradiction, that is, suppose that $\mathcal{H}^{\beta}(\Lambda) < \gamma^{-1}$. Given $\delta \in (0,1)$, there exists a cover $\Lambda \subset \bigcup_i U_i$ by open balls U_i such that $\sum_i |U_i|^{\beta} < \gamma^{-1}$ and $|U_i| < \delta$. By continuity, we can find $t < \beta$ such that

$$\sum_i |U_i|^t < \gamma^{-1}.$$

For every U_i, let $f_{i,1}, \ldots, f_{i,j} \in \mathcal{S}_{|U_i|}$ be all the functions such that $U_i \cap f_{i,k}(\Lambda) \neq \varnothing$ for $k = 1, \ldots, j$. Thus, $j \leqslant \gamma$ and $r_{\min}|U_i| < r_{f_{i,k}} \leqslant |U_i|$.

We claim that $\{f_{i,k}(U_q)\}_{i,k,q}$ is a cover of Λ of diameter at most δ^2. Indeed, $|f_{i,k}(U_q)| = r_{f_{i,k}}|U_q| \leqslant |U_i||U_q| \leqslant \delta^2$, and if $x \in \Lambda$ is arbitrary then $x \in U_i$ for some i and there exists $f \in \mathcal{S}_{|U_i|}$ such that $x \in f(\Lambda)$. Then necessarily $f = f_{i,k}$ for some $k \leqslant j$. Hence $x \in f_{i,k}(\Lambda)$, and since $\{U_q\}_q$ is a cover of Λ, the claim follows. Moreover,

$$\sum_{i,k,q} |f_{i,k}(U_q)|^t = \sum_{i,q} \sum_{k=1}^{j} r^t_{f_{i,k}} |(U_q)|^t < \gamma \sum_{i,q} |U_i|^t |U_q|^t < \gamma^{-1}.$$

Repeating this argument, if we have already defined a cover $\{U_i\}$ of Λ by balls of diameter at most $\delta^n > 0$ with $\sum_i |U_i|^t < \gamma^{-1}$, then $\{f_{i,k}(U_q)\}_{i,k,q}$ is a cover of diameter at most δ^{n+1} and with $\sum_{i,q,k} |f_{(i,k)}(U_q)|^t < \gamma^{-1}$, where $\{f_{i,k}\}_k$ are the maps in $\mathcal{S}_{|U_i|}$ such that $U_i \cap f_{i,k}(\Lambda) \neq \varnothing$. Thus, $\mathcal{H}^t(\Lambda) < \gamma^{-1}$, which is impossible since $t < \beta = \dim_{\mathrm{H}} \Lambda$.

The proof of (ii)(c) follows [**Zer96**, Theorem 3]. Since $\beta = d$ we have $\mathcal{L}^d(\Lambda) > 0$ by (ii)(b), and the goal is to show that Λ contains interior points. Fix

$\rho = r_{\min}^{-1}$, let $x \in \Lambda, \varepsilon$ be as in Remark 4.2.7(b), and write $U := B(x, \frac{\varepsilon}{2})$. Denote $\mathcal{F} := \{S_{\mathbf{i}} : \mathbf{i} \in \Sigma^*\}$. For $g \in \mathcal{F}$ let

$$I_U(g) := \left\{h \in S_{r_g} : g(U) \cap h(\Lambda) \neq \varnothing\right\}.$$

Then $\#I_U(g) \leqslant \gamma_{1,\varepsilon/2} < \infty$ by Theorem 4.2.4(iv-a), hence there exists $f \in \mathcal{F}$ such that

$$\#I_U(f) = \max_{g \in \mathcal{F}} \#I_U(g). \tag{4.48}$$

Consider the concentric balls

$$f(U) = B(f(x), r_f\varepsilon/2) \text{ and } B_\varrho := B(f(x), \varrho),$$

where

$$\varrho := \frac{1}{3}\left(r_f\varepsilon/2 - \max\left\{\operatorname{dist}(f(x), h(\Lambda)) : h \in I_U(f)\right\}\right).$$

Since $h(\Lambda)$ intersects $f(U) = B(f(x), r_f\varepsilon/2)$, we have $\varrho > 0$. (Note that ϱ is relatively small compared to $r_f\varepsilon/2$.) Let $\widehat{r} := r_f\varepsilon/2 - \varrho$. Our goal is to show that there exists $y \in B_\varrho$ such that $B(y, \widehat{r}) \subset \Lambda$.

Observe that, by construction, for every $y \in B_\varrho$, we have

$$I_U(f) = I_{U_y}(f), \tag{4.49}$$

where $U_y := f^{-1}B(y, \widehat{r}) \subset U$. Indeed,

$$B(f(y), \widehat{r}) \supset B(f(x), r_f\varepsilon/2 - 2\varrho),$$

whereas $h(\Lambda)$ intersects $B(f(x), r_f\varepsilon/2 - 3\rho)$ for any $h \in I_U(f)$ by the definition of ϱ. Thus,

$$h(\Lambda) \cap f(U) \neq \varnothing \iff h(\Lambda) \cap f(U_y) \neq \varnothing$$

confirming (4.49).

It follows that f has a maximality property similar to (4.48) for U_y for every $y \in B_\varrho$, and hence

$$I_{U_y}(gf) = gI_{U_y}(f), \quad \text{for all } g \in \mathcal{F},$$

which is proved as in (4.34). Therefore,

$$B(g(y), r_g\widehat{r}) \cap \Lambda = g\left[B(y, \widehat{r}) \cap \Lambda\right], \quad \text{for all } g \in \mathcal{F}.$$

We obtain that the *approximate density*, defined by

$$b(a, r) := \frac{\mathcal{L}^d\left(B(a, r) \cap \Lambda\right)}{\mathcal{L}^d(B(a, r))},$$

satisfies

$$\forall y \in B_\varrho \cap \Lambda, \forall g \in \mathcal{F}, \quad b\big(g(y), r_g\widehat{r}\big) = b(y, \widehat{r}). \tag{4.50}$$

We want to show that

$$\exists\, y \in B_\varrho \cap \Lambda : \; b(y, \widehat{r}) = 1, \tag{4.51}$$

which would imply $B(y, \widehat{r}) \subset \Lambda$, as desired.

Suppose that (4.51) is false. Then there exists $\delta > 0$ such

$$\mathcal{L}^d(E_\delta) > 0, \quad \text{where } E_\delta := \{y \in B_\varrho \cap \Lambda : \; b(y, \widehat{r}) \leqslant 1 - \delta\},$$

since $\mathcal{L}^d(B_\varrho \cap \Lambda) > 0$.

CLAIM. *There exists $c > 0$ such that*

$$\mathcal{L}^d\Big(\bigcup_{g\in\mathcal{S}_b} g(E_\delta)\Big) \geqslant c, \quad \textit{for any } 0 < b < 1. \tag{4.52}$$

PROOF OF THE CLAIM. Fix $b \in (0,1)$. Notice that for any $g_1, g_2 \in \mathcal{S}_b$, the balls $g_1(B_\varrho)$ and $g_2(B_\varrho)$ are either concentric, if $g_1(f(x)) = g_2(f(x))$, or disjoint otherwise. It is here that we use the property of $x \in \Lambda$ and ε from Remark 4.2.7(b):

$$g_1(f(x)) \neq g_2(f(x)) \implies |g_1(f(x)) - g_2(f(x))| \geqslant \max\{r_{g_1}, r_{g_2}\} r_f \cdot \varepsilon$$

(note also that $\varrho < r_f\varepsilon/2$). Let $\mathcal{S}_b' \subset \mathcal{S}_b$ be such that $g(f(x))$ are all distinct for $g \in \mathcal{S}_b'$, so that $\#\mathcal{S}_b' = \#\mathcal{S}_b(f(x))$. Then we have, using that $E_\delta \subset B_\varrho$,

$$\begin{aligned}
\mathcal{L}^d\Big(\bigcup_{g\in\mathcal{S}_b} g(E_\delta)\Big) &\geqslant \sum_{g\in\mathcal{S}_b'} \mathcal{L}^d(g(E_\delta)) = \mathcal{L}^d(E_\delta) \sum_{g\in\mathcal{S}_b'} r_g^d \\
&\geqslant \mathcal{L}^d(E_\delta)\cdot(br_{\min})^d \cdot \#\mathcal{S}_b/C \\
&\geqslant \mathcal{L}^d(E_\delta)\cdot r_{\min}^d/C =: c > 0.
\end{aligned}$$

We applied Corollary 4.2.8 in the second displayed line, and the last inequality follows from

$$0 < \mathcal{L}^d(\Lambda) = \mathcal{L}^d\Big(\bigcup_{g\in\mathcal{S}_b} g(\Lambda)\Big) \leqslant \sum_{g\in\mathcal{S}_b} \mathcal{L}^d(g(\Lambda)) \leqslant \mathcal{L}^d(\Lambda)\cdot \#\mathcal{S}_b \cdot b^d.$$

□

Now we conclude the proof of part (ii)(c) of the theorem. By the Lebesgue density theorem and Egorov's theorem, we can find $r_0 > 0$ and a subset $A \subset \Lambda$ such that $\mathcal{L}^d(\Lambda\backslash A) < c$ and

$$b(a,r) > 1-\delta, \quad \text{for all } a \in A \text{ and all } r \in (0, r_0).$$

However, by the Claim there exists $y \in E_\delta$ and $g \in \mathcal{S}_{r_0}$, with $g(y) \in A$, and so $b(g(y)), r_g\widehat{r}) \leqslant 1-\delta$ by (4.50), which is a contradiction. □

EXAMPLE 4.2.18 ([**LN07**]). Let F be the attractor of the following self-similar IFS in the plane:

$$\begin{aligned}
\mathcal{S} := (S_1(\mathbf{x}) = \rho x, S_2(\mathbf{x}) &= r\mathbf{x} + (\rho - \rho r, 0), \\
S_3(\mathbf{x}) = rx + (1-r, 0), S_4(\mathbf{x}) &= r\mathbf{x} + (0, 1-r))
\end{aligned} \tag{4.53}$$

Since $S_1S_3 = S_2S_1$, this IFS does not satisfy the OSC. However, it was shown by Lau and Ngai [**LN07**] that $\mathcal{S}$ satisfies the WSP. They calculated

$$\dim_{\mathrm{H}}(F) = \dim_{\mathrm{B}}(F) = \alpha,$$

where α is the solution of the equation

$$\rho^\alpha + 3r^\alpha - (r\rho)^\alpha = 1.$$

4.2.4. Singularity and absolute continuity of self-similar measures. Interesting results on this theme were obtained by Lau and Wang. For the proofs we refer to their paper [**LW04**].

THEOREM 4.2.19 (Lau and Wang [**LW04**]). *Assume that the self-similar IFS $\mathcal{S}$ satisfies the WSP. Let μ be the self-similar measure defined by the probability vector $\mathbf{p} = (p_1, \ldots, p_m)$. Then μ is singular with respect to $\mathcal{H}^\beta|_\Lambda$, where β was defined in Theorem* 4.2.16, *if and only if there exists a finite word $\mathbf{i} \in \Sigma^*$ such that*

$$r_{\mathbf{i}}^\beta < \sum\{p_{\mathbf{j}} : S_{\mathbf{j}} = S_{\mathbf{i}}\},$$

where $r_{\mathbf{i}}$ is the contraction ratio of $S_{\mathbf{i}}$.

THEOREM 4.2.20 (Lau and Wang [**LW04**]). *Under the assumptions of Theorem* 4.2.19, *if the self-similar measure is absolutely continuous with respect to $\mathcal{H}^\beta|_\Lambda$, then the Radon–Nikodym derivative of μ is bounded.*

EXAMPLE 4.2.21. Let $\mathcal{S} := (S_i)_{i=1}^3$ be an IFS on $\mathbb{R}$ defined by

$$S_1(x) = \rho x,\ S_2(x) = rx + \rho(1-r),\ S_3(x) = rx + (1-r), \tag{4.54}$$

where $0 < \rho < 1$, $0 < r < 1$, and $\rho + 2r - \rho r \leqslant 1$.

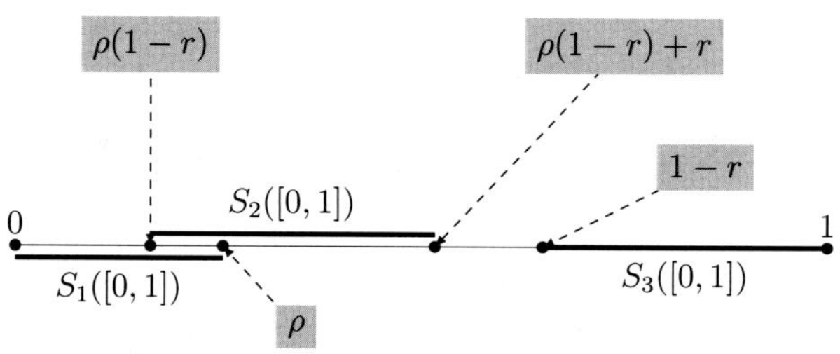

FIGURE 4.3. See [**LW04**, Fig. 4.1]

Note that $S_{13^k} = S_{2^k1}$ for every k, hence OSC does not hold. Lau and Wang in [**LW04**, Proposition 4.4] proved that $\mathcal{S}$ satisfies the WSP. Note that in the special case when $\rho = r = \frac{1}{3}$ we get Example 4.3.1 once again.

PROPOSITION 4.2.22 (Lau and Wang, [**LW04**]). *Let $\mathcal{S}$ be as in Example* 4.2.21 *and let α be the solution of*

$$\rho^\alpha + 2r^\alpha - (r\rho)^\alpha = 1. \tag{4.55}$$

Then the Hausdorff dimension of the attractor Λ is $\dim_{\mathrm{H}}(\Lambda) = \alpha$. Furthermore, every self-similar measure is singular with respect to $\mathcal{H}^\alpha|_\Lambda$.

4.3. Example: The $\{0, 1, 3\}$-IFS with contraction ratio $\lambda = 1/3$

In Section 1.6.1, in relation with the so-called $\{0, 1, 3\}$ problem, we studied the one-parameter family of self-similar IFS on the line $\mathcal{S}^\lambda = (S_i(x) = \lambda x + i)_{i \in \{0,1,3\}}$. Here we consider the special case when $\lambda = 1/3$. We first show that the IFS $\mathcal{S}^{1/3}$ satisfies the WSP and has an exact overlap, but not a *simple* exact overlap in the sense of Definition 4.2.14. We then compute the dimension of the attractor and construct an ergodic shift-invariant measure in the symbolic space whose push-forward measure has the same dimension as the attractor. We call such a measure a *dimension maximizing measure.*

Throughout this subsection we write $\Sigma^* := \bigcup_{n=0}^{\infty} \{0,1,3\}^n$.

EXAMPLE 4.3.1. Let Λ be the attractor of the self-similar IFS on the line

$$\mathcal{S} = \left(S_i(x) = \frac{1}{3}x + i\right)_{i\in\{0,1,3\}}. \tag{4.56}$$

See Figure 4.4. Observe that $\mathcal{S}$ is conjugate to the IFS in Example 4.2.1.

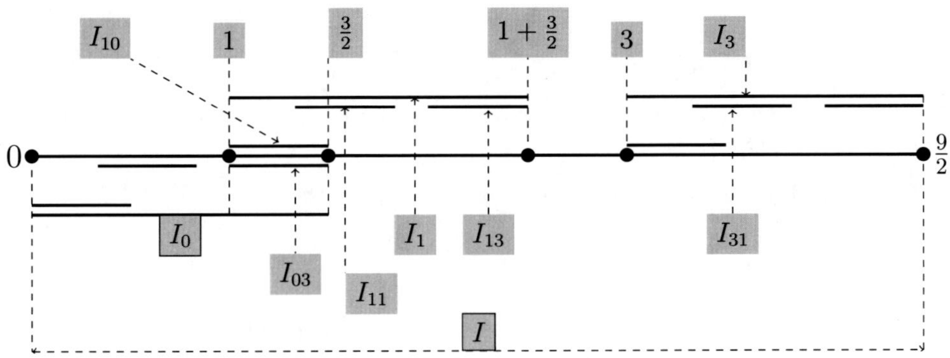

FIGURE 4.4. The first and second level cylinder intervals of $\mathcal{S}$ from Example 4.3.1

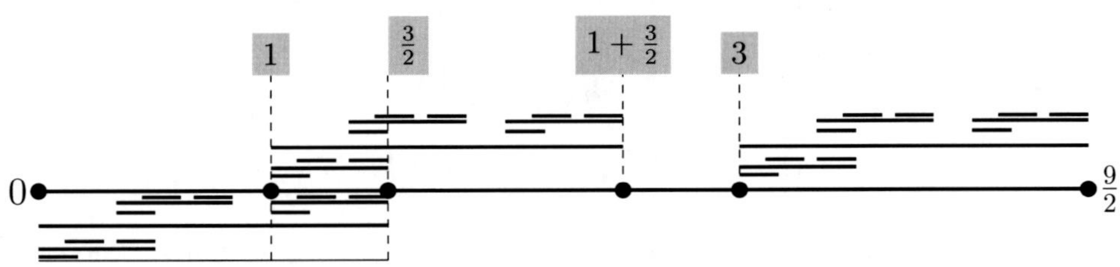

FIGURE 4.5. The first, second and third level cylinder intervals of $\mathcal{S}$ from Example 4.3.1

FACT 4.3.2. *The WSP holds for the IFS $\mathcal{S}$.*

PROOF. For $\mathbf{i} = (i_1, \ldots, i_n) \in \Sigma^*$ we have

$$S_{\mathbf{i}}(x) = 3^{-n} \cdot x + \sum_{k=1}^{n} i_k 3^{-k+1}. \tag{4.57}$$

Note that (4.57) implies that for every $\mathbf{i} = (i_1, \ldots, i_n)$, $\mathbf{j} = (j_1, \ldots, j_m) \in \Sigma^*$ we have

$$S_{\mathbf{i}}^{-1} \circ S_{\mathbf{j}}(x) - x = (3^{n-m} - 1)x + \sum_{k=1}^{m} j_k 3^{n+1-k} - \sum_{\ell=1}^{n} i_\ell 3^{n+1-\ell}.$$

Hence the neighbor maps are of the form

$$\mathcal{N} = \Big\{S_{\mathbf{i}}^{-1} \circ S_{\mathbf{j}}(x) = 3^{n-m}x + \sum_{k=1}^{m} j_k 3^{n+1-k} - \sum_{\ell=1}^{n} i_\ell 3^{n+1-\ell} : \mathbf{i}, \mathbf{j} \in \Sigma^*, |\mathbf{i}| = n, |\mathbf{j}| = m\Big\}. \tag{4.58}$$

Using the condition (i) of Theorem 4.2.4, we get that the WSP holds for $\mathcal{S}$. Indeed, we can take $x = 0$ and note that if $r_h \in [\rho^{-1}, \rho]$ for $h \in \mathcal{N}$, then $|n - m| \leqslant \log_3 \rho$, hence condition (i) holds with $\varepsilon = \rho^{-1}$. □

The IFS $\mathcal{S}$ has an exact overlap. Namely, we obtain from (4.57) that

$$S_{03} \equiv S_{10}. \tag{4.59}$$

However, we have the following.

LEMMA 4.3.3. *The IFS $\mathcal{S}$ in Example* 4.3.1 *does not have a simple exact overlap in the sense of Definition* 4.2.14.

We will adopt the following notation:

$$y_0.y_1 \dots y_n = \sum_{j=0}^{n} y_j 3^{-j}.$$

For $n \geqslant 1$ and $a \in \{0,1,3\}$ we will write $a^n := (\underbrace{a, \dots, a}_{n})$ and $\overline{a} = a^\infty = aaa\dots$ for $a \in \{0,1,3\}$. Further, let $I := \left[0, \frac{9}{2}\right]$ be the convex hull of Λ, in view of $9/2 = \sum_{k=1}^{\infty} 3 \cdot 3^{-k+1}$. For $\mathbf{k} \in \{0,1,3\}^*$ denote by $I_{\mathbf{k}} := S_{\mathbf{k}}(I)$ the corresponding cylinder interval. To prove Lemma 4.3.3 we need the following simple fact.

FACT 4.3.4. *For every $n \geqslant 1$, if*

$$\frac{3}{2} \in I_{\mathbf{k}}, \quad \textit{with} \quad |\mathbf{k}| = n,$$

then either

(i) $I_{\mathbf{k}} = I_{03^{n-1}} = I_{1^{n-1}0}$, *and $\frac{3}{2}$ is the right endpoint of this cylinder interval; or*

(ii) $I_{\mathbf{k}} = I_{1^n}$, $\frac{3}{2} \in I_{1^n} \cap \Lambda$, *and $\frac{3}{2}$ is not an endpoint of this cylinder interval.*

More generally, for $n \geqslant \ell \geqslant 1$, if

$$3^{1-\ell} \cdot \frac{3}{2} \in I_{\mathbf{k}}, \quad \textit{with} \quad |\mathbf{k}| = n,$$

then either

(i′) $I_{\mathbf{k}} = I_{0^\ell 3^{n-\ell}} = I_{0^{\ell-1}1^{n-\ell}0}$, *and $3^{1-\ell} \cdot \frac{3}{2}$ is the right endpoint of this cylinder interval; or*

(ii′) $I_{\mathbf{k}} = I_{0^\ell 1^{n-\ell}}$, $3^{1-\ell} \cdot \frac{3}{2} \in I_{1^n} \cap \Lambda$, *and $3^{1-\ell} \cdot \frac{3}{2}$ is not an endpoint of this cylinder interval.*

An easy verification of this fact is left to the reader; Figure 4.5 is a good illustration of it. As a hint for the first claim, note that, using the notation introduced above, we have

$$\frac{3}{2} = 0.3^{n-1}\overline{3} = 1.1^{n-2}0\overline{3} = 1.\overline{1}.$$

The second claim follows from the first one by self-similarity (see Figure 4.5).

PROOF OF LEMMA 4.3.3. Let Λ be the attractor of $\mathcal{S}$. Assume, for contradiction, that $\mathcal{F} = (f_i)_{i=1}^M$ is a self-similar IFS such that

- For all $k \in \{1, \ldots, M\}$ there exists an $\mathbf{i} \in \{0,1,3\}^*$ such that $f_k = S_{\mathbf{i}}$,
- Λ is the attractor of $\mathcal{F}$ is Λ, that is,

$$\Lambda = \bigcup_{k=1}^{M} f_k(\Lambda), \quad \text{and} \tag{4.60}$$

- the OSC holds for $\mathcal{F}$.

Observe first that there exists $k \leqslant M$ such that $f_k = S_{0^\ell}$ for some $\ell \geqslant 1$, since $0 = \min \Lambda$ is the left endpoint of I and has a unique representation $0 = 0.\overline{0}$ in the alphabet $\{0,1,3\}$ (similarly, there exists $k' \leqslant M$ such that $f_{k'} = S_{3^{\ell'}}$ for some $\ell' \geqslant 1$, but we will not need this). The right endpoint of $f_k(I)$ is

$$\gamma := 0.0^{\ell-1}\overline{3} = 0.0^{\ell-2}\overline{1} = 3^{1-\ell} \cdot \frac{3}{2}.$$

Note that the points b_n defined by

$$b_n = 0.0^{\ell-2}1^n 3 \in \Lambda \backslash f_k(\Lambda), \ n \geqslant 1,$$

satisfy $b_n \downarrow \gamma$. Since $\mathcal{F}$ is a finite IFS, infinitely many of the b_n's must belong to some cylinder of the attractor $f_i(\Lambda)$, where $f_i = S_{\mathbf{j}}$. Then $f_i(\Lambda) \ni \gamma$, hence $\gamma \in I_{\mathbf{j}}$ and γ is not an endpoint of $I_{\mathbf{j}}$. By Fact 4.3.4, $\mathbf{j} = 0^{\ell-1}1^m$ for some $m \geqslant 1$. Note that

$$f_i(\Lambda) \cap f_k(\Lambda) = \Lambda_{0^{\ell-1}1^m} \cap \Lambda_{0^\ell} \supset \Lambda_{0^{\ell-1}1^m 0} = \Lambda_{0^\ell 3^m}.$$

Let s be the similarity dimension of $\mathcal{F}$. The OSC assumption implies that

$$\mathcal{H}^s(\Lambda_{0^\ell 3^m}) = 3^{-s(\ell+m)} \cdot \mathcal{H}^s(\Lambda) > 0,$$

and so we obtain that

$$\mathcal{H}^s(\Lambda_{0^{\ell-1}1^m} \cap \Lambda_{0^\ell}) > 0.$$

However, this contradicts (4.8), completing the proof. $\square$

4.3.1. Dimension of the attractor. Repeated application of (4.59) yields that if $\mathbf{j} = (j_1, \ldots, j_n)$ is obtained from $\mathbf{i} := (i_1, \ldots, i_n) \in \{0,1,3\}^n$ in such a way that we substitute all blocks of the form $0\underbrace{3\ldots3}_{k}$ in $\mathbf{i}$ by the block $\underbrace{1\ldots1}_{k}0$, where $k \geqslant 1$, then $S_{\mathbf{i}} = S_{\mathbf{j}}$. Hence, for

$$\mathcal{T}_n := \left\{(i_1, \ldots, i_n) \in \{0,1,3\}^n : (i_k, i_{k+1}) \neq (0,3), \forall k = 1, \ldots, n-1\right\}.$$

we have

$$\left\{S_{\mathbf{j}} : \mathbf{j} \in \mathcal{T}_n\right\} = \left\{S_{\mathbf{i}} : \mathbf{i} \in \{0,1,3\}^n\right\}. \tag{4.61}$$

Moreover, it is immediate from Fact 4.3.5 below that

$$S_{\mathbf{i}} \neq S_{\mathbf{j}} \text{ for distinct } \mathbf{i}, \mathbf{j} \in \mathcal{T}_n. \tag{4.62}$$

FACT 4.3.5. *For distinct $\mathbf{i} = (i_1, \ldots, i_n)$, $\mathbf{j} = (j_1, \ldots, j_n) \in \mathcal{T}_n$ if $I_{\mathbf{i}} \bigcap I_{\mathbf{j}} \neq \emptyset$, then $|\mathbf{i} \wedge \mathbf{j}| = n-1$ and $\{i_n, j_n\} = \{0,1\}$.*

PROOF. For $n = 1$ it is obvious that the assertion holds. We assume that it is true up to $n-1$ for some $n \geqslant 2$. Fix distinct $\mathbf{i} = (i_1, \ldots, i_n)$, $\mathbf{j} = (j_1, \ldots, j_n) \in \mathcal{T}_n$ with $I_{\mathbf{i}} \bigcap I_{\mathbf{j}} \neq \varnothing$, and let $k-1 := |\mathbf{i} \wedge \mathbf{j}|$. To get a contradiction we assume that $k < n$. Then $i_k, j_k = \{0,1\}$ by the induction hypothesis. We may assume that $i_k = 0$ and $j_k = 1$. Let $\boldsymbol{\omega} := \mathbf{i} \wedge \mathbf{j}$. Then $I_{\boldsymbol{\omega} 0 i_{k+1}} \bigcap I_{\boldsymbol{\omega}1} \neq \varnothing$ since we assumed that $I_{\mathbf{i}} \bigcap I_{\mathbf{j}} \neq \varnothing$. Hence, $S_{\boldsymbol{\omega}}^{-1}\left(I_{\boldsymbol{\omega}0i_{k+1}} \bigcap I_{\boldsymbol{\omega}1}\right) = I_{0i_{k+1}} \bigcap I_1 \neq \varnothing$. This is possible only if $i_{k+1} = 3$ since $(I_{00} \bigcup I_{01}) \bigcap I_1 = \varnothing$. This is a contradiction. Namely, in this case $(i_k, i_{k+1}) = (0,3)$ which is impossible since $\mathbf{i} \in \mathcal{T}_n$. That is, $I_{\mathbf{i}} \bigcap I_{\mathbf{j}} \neq \varnothing$ implies that $|\mathbf{i} \wedge \mathbf{j}| = n-1$. Then $\{i_n, j_n\} = \{0,1\}$ is immediate from the construction. □

In order to compute the dimension of the attractor Λ with the help of Theorem 4.2.16, consider the 3×3 matrix

$$A := \begin{pmatrix} 1 & 1 & 0 \\ 1 & 1 & 1 \\ 1 & 1 & 1 \end{pmatrix}. \tag{4.63}$$

It can be verified by induction that the (k,ℓ)-th entry of A^n satisfies $A^n(k,\ell) = \#\{(i_1, \ldots, i_n) \in \mathcal{T}_n :\ i_1 = k, \text{ and } i_n = \ell\}$. Therefore,

$$\#\mathcal{T}_n = \|A^n\|_1. \tag{4.64}$$

Observe that the dominant eigenvalue of A is $\lambda := \frac{3+\sqrt{5}}{2}$. Hence,

$$\lim_{n\to\infty} \frac{1}{n} \log \#\mathcal{T}_n = \log \lambda. \tag{4.65}$$

FACT 4.3.6.

$$\dim_{\mathrm{H}}(\Lambda) = \dim_{\mathrm{B}}(\Lambda) = \frac{\log \lambda}{\log 3} = \frac{\log(3+\sqrt{5}) - \log 2}{\log 3} \approx 0.876036. \tag{4.66}$$

PROOF. According to Remark 4.2.17, it suffices to count the number of distinct cylinders of level n, corresponding to the functions $S_{\mathbf{i}}$, with $|\mathbf{i}| = n$: In view of (4.61),

$$\mathcal{S}_{3^{-n}d} = \left\{S_{\mathbf{i}} : \mathbf{i} \in \{0,1,3\}^n\right\} = \left\{S_{\mathbf{i}} : \mathbf{i} \in \mathcal{T}_n\right\}, \tag{4.67}$$

with $d = 9/2$, since $[0, 9/2]$ is the convex hull of the attractor. (Recall that $\mathcal{S}_b$ was defined in (4.25).) Combining (4.67) and (4.62) we get

$$\#\mathcal{S}_{3^{-n}d} = \#\mathcal{T}_n. \tag{4.68}$$

Putting together this, (4.65), Fact 4.3.2 and part (ii) of Theorem 4.2.16 we obtain the desired claim. □

We remark that one can explicitly compute $\#\mathcal{T}_n$ using the recurrence relation:

$$\#\mathcal{T}_n = \frac{5+3\sqrt{5}}{10} \cdot \left(\frac{3+\sqrt{5}}{2}\right)^n + \frac{5-3\sqrt{5}}{10} \cdot \left(\frac{3-\sqrt{5}}{2}\right)^n.$$

We leave this computation as an exercise for the reader.

4.3.2. Dimension maximizing measure*. Here our goal is to describe the dimension maximizing measure for $\mathcal{S}$. This subsection requires some concepts and results from Ergodic Theory; we refer the reader to Chapter 9 and to the classical sources.

The set of all infinite sequences over the alphabet $\{0,1,3\}$ that do not contain the "forbidden word" $(0,3)$ will be denoted by Σ_A, where A is the 3×3 matrix introduced in (4.63). That is,

$$\Sigma_A := \{\mathbf{j} \in \Sigma : (j_k, j_{k+1}) \neq (0,3) \text{ for all } k \geqslant 1\}.$$

Then (Σ_A, σ), where the left shift σ was defined in (1.14), is a *subshift of finite type*. In particular, $\Sigma_A \subset \Sigma$ is compact and $\sigma(\Sigma_A) = \Sigma_A$. Let Π be the natural projection from Σ to the attractor Λ and denote $\Pi_A := \Pi|_{\Sigma_A}$. Then by (4.61) we have

$$\Pi_A(\Sigma_A) = \Pi(\Sigma) = \Lambda. \tag{4.69}$$

The nonnegative primitive matrix A generates a stochastic matrix $P=(p_{i,j})_{i,j\in\{0,1,3\}}$. The Markov measure on Σ_A determined by P is called the *Parry measure*; it is known to be the measure of maximal entropy for (Σ_A, σ), see [**Wal82**, Theorem 8.9].

Now we explain this in detail. First we define the stochastic matrix $P = (p_{i,j})_{i,j\in\{0,1,3\}}$ by

$$p_{i,j} := \frac{a_{i,j} \cdot v_j}{\lambda v_i}, \tag{4.70}$$

where $\mathbf{v} = (v_0, v_1, v_3) = \left(\frac{\sqrt{5}-1}{2}, 1, 1\right)$ is a right eigenvector of A corresponding to the largest eigenvalue $\lambda = \frac{3+\sqrt{5}}{2}$. So,

$$P := \frac{1}{\sqrt{5}+3} \begin{pmatrix} 2 & \frac{4}{(\sqrt{5}-1)} & 0 \\ \sqrt{5}-1 & 2 & 2 \\ \sqrt{5}-1 & 2 & 2 \end{pmatrix}.$$

The stationary measure for the corresponding Markov chain (see [**Dur96**, p. 300]) is given by the probability vector $\mathbf{p}$ satisfying $\mathbf{p}^* \cdot P = \mathbf{p}^*$. After a standard computation we get that $\mathbf{p}^* = (p_0, p_1, p_3) = \left(\frac{1}{10}\left(5-\sqrt{5}\right), \frac{1}{\sqrt{5}}, \frac{1}{10}\left(5-\sqrt{5}\right)\right)$.

The Parry measure μ on Σ_A is defined as follows: for $(i_1, \dots, i_n) \in \mathcal{T}_n$ let

$$\mu([i_1, \dots i_n]) := p_{i_1} \cdot \prod_{k=1}^{n-1} p_{i_k, i_{k+1}}, \tag{4.71}$$

where $[i_1, \dots i_n] := \{\mathbf{j} : j_k = i_k, \text{ for all } k \leqslant n\}$ is the usual symbolic cylinder set. By the Kolmogorov's extension principle we extend μ to the Borel subsets of Σ_A. It follows from [**Wal82**, Theorem 1.13] that the measure μ is invariant and ergodic (see Section 1.9.5 for definitions). Observe that putting together (4.71) and (4.70) we obtain

$$\mu([i_1, \dots i_n]) = \frac{p_{i_1} v_{i_n}}{v_{i_1}} \cdot \lambda^{-(n-1)}, \text{ for all } (i_1, \dots, i_n) \in \mathcal{T}_n. \tag{4.72}$$

This implies that

$$\lim_{n\to\infty} -\frac{1}{n} \log \mu([i_1, \dots, i_n]) = \log \lambda, \text{ for all } \mathbf{i} = (i_1, i_2, \dots) \in \Sigma_A. \tag{4.73}$$

Define the measure

$$\nu := (\Pi_A)_* \mu.$$

Now we can show that ν is a dimension maximizing measure, that is, the following holds:

FACT 4.3.7. $\dim_{\mathrm{H}}(\nu) = \dim_{\mathrm{H}}(\Lambda)$.

PROOF. As we have already mentioned, the convex hull of the attractor Λ is $I := \left[0, \frac{9}{2}\right]$. We write $I_{i_1 \ldots i_n} := S_{i_1 \ldots i_n}(I)$ for $(i_1, \ldots, i_n) \in \{0, 1, 3\}^n$. Using that $\mathrm{dist}(I_1, I_3) = \frac{1}{2}$ and Fact 4.3.5 we obtain that for $x \in I_{\mathbf{i}}$, $\mathbf{i} = (i_1, \ldots, i_n) \in \mathcal{T}_n$, and $\mathbf{j} = (j_1, \ldots, j_n) \in \mathcal{T}_n$, holds

$$B(x, \rho_n) \cap I_{\mathbf{j}} \neq \varnothing \Longrightarrow I_{\mathbf{j}} \subset I_{i_1 \ldots i_{n-1}}, \tag{4.74}$$

where $\rho_n = 2^{-1} 3^{-(n-1)}$ and $B(x, \rho_n) = (x - \rho_n, x + \rho_n)$. This implies that

$$\nu(B(x, \rho_n)) \leqslant \mu([i_1, \ldots, i_{n-1}]). \tag{4.75}$$

Combining this with (4.73) yields that for ν-a.e. $x \in \Lambda$ we have

$$\liminf_{n \to \infty} \frac{\log \nu(B(x, \rho_n))}{\log \rho_n} \geqslant \frac{\log \lambda}{\log 3}. \tag{4.76}$$

Now by Theorem 1.9.5(i) we obtain $\dim_{\mathrm{H}}(\nu) \geqslant \frac{\log \lambda}{\log 3}$. The opposite inequality follows from Fact 4.3.6 and the observation that $\mathrm{spt}(\nu) \subset \Lambda$. Thus,

$$\dim_{\mathrm{H}}(\nu) = \dim_{\mathrm{H}}(\Lambda) = \frac{\log \lambda}{\log 3}. \tag{4.77}$$

□

We have confirmed that ν is a dimension maximizing measure. It follows from [**Wal82**, 8.10] that μ is the *unique* invariant measure on (Σ_A, σ) of maximal entropy. This can be used to show that ν is the unique dimension maximizing invariant measure.

4.4. Finite type condition

Lalley [**Lal97**] computed the Hausdorff dimension of the attractor of an IFS which consists of finitely many contractions on the line of the form:

$$S_i(x) = \theta^{-1} x + b_i, \quad b_j \in D \subset \mathbb{Z}[\theta], \quad i = 1, \ldots, m, \tag{4.78}$$

where θ is a PV (Pisot–Vijayaraghavan) number (see Definition D.0.2) and $\mathbb{Z}[\theta]$ is the smallest ring containing $\mathbb{Z}$ and θ. Lalley constructed a finite directed graph from the IFS in (4.78) and expressed the dimension of the attractor in terms of the dominant eigenvalue of its adjacency matrix. The idea of representing the attractor of an overlapping IFS as the attractor of a nonoverlapping graph-directed system, was generalized by Ngai and Wang [**NW01**]. They introduced a separation condition, which they called the *Finite Type Condition* (FTC). We will see that the FTC implies the WSP, but the converse is not known in general, see Section 4.4.3. The FTC holds for the family in (4.78) and for many other interesting examples. Under the finite type condition the dimension theory of the attractor has been developed, and in many cases the multifractal spectra of invariant measures have been described (at least, in part, see Section 5.2 below).

Throughout this section we assume that

$$\mathcal{S} := (S_i(x) = r_i Q_i x + b_i)_{i=1}^m, \tag{4.79}$$

where $0 < r_i < 1$ and Q_i is an orthogonal matrix, for $i = 1, \dots, m$. As usual, we write Π for the natural projection from the symbolic space $\Sigma = \{1, \dots, m\}^{\mathbb{N}}$ to the attractor Λ. Here we follow [**JY05**] presenting the definition. First we introduce the necessary notation. Let

$$V_k := \{S_{\mathbf{i}} : \mathbf{i} \in \Sigma^*, \ r_{\mathbf{i}} \leqslant r_{\min}^k < r_{\mathbf{i}^-}\}. \tag{4.80}$$

Further, define

- $V := \bigcup_k V_k$,

so in fact, $V = \{S_{\mathbf{i}} : \mathbf{i} \in \Sigma^*\}$. Let $\Omega \neq \varnothing$ be an open set satisfying $\cup_{j=1}^m S_j(\Omega) \subset \Omega$. Such Ω is said to be an *invariant set* for the IFS.

- Two mappings $f_1, f_2 \in V_k$ are called *neighbors* (with respect to Ω) if $f_1(\Omega) \cap f_2(\Omega) \neq \varnothing$.

DEFINITION 4.4.1 (Finite type condition). For $f \in V_k$ the collection of neighbors

$$\Omega(f) := \{g \in V_k : g \text{ is a neighbor of } f\}$$

is called the *neighborhood of f* (w.r.t. Ω). We say that $f_1 \in V_k$ and $f_2 \in V_{k'}$ have the same *neighborhood type* and write $f_1 \overset{\tau}{\sim} f_2$ if there is a similarity τ such that

- the scaling ratio of τ is $r_{\min}^{k-k'}$,
- $\Omega(f_1) = \tau\Omega(f_2)$,
- $f_1 = \tau \circ f_2$.

This defines an equivalence relation on the set of functions $f \in V = \cup_k V_k$. If there are only finitely many classes for this equivalence relation (that is, finitely many neighborhood types), then we say that the IFS $\mathcal{S}$ from (4.79) is of finite type with respect to Ω, and Ω is a finite type condition set. We denote by $[\Omega(f)]$ the equivalence class of $\Omega(f)$.

Finally, we say that $\mathcal{S}$ *is of finite type* if there exists a nonempty $\mathcal{S}$-invariant open set Ω with respect to which the FTC is satisfied.

REMARK 4.4.2.

(a) The definition of finite type is rather "rigid", in the sense that equivalence between neighborhood types is necessarily realized by a similarity map with a contraction ratio equal to an integer power of the minimal contraction ratio $r_{\min}$ of the IFS. As a consequence, the logarithms of the contraction ratios of a finite type IFS are necessarily commensurable, see Theorem 4.4.3 below.

(b) If the IFS $\mathcal{S}$ satisfies the Open Set Condition with an open set Ω and the contraction ratios have commensurable logarithms, then the FTC holds with respect to Ω, see [**NW01**, Theorem 2.6]. In this case every neighborhood $\Omega(f)$ consists of a single vertex.

(c) A related, but different definition of "finite type IFS" was suggested by Bandt, see [**BM09**] and references therein. His notion of neighborhood type is more algebraic and is based on the "neighbor maps" $S_{\mathbf{i}}^{-1} \circ S_{\mathbf{j}}$ for intersecting cylinders of the attractor of roughly equal sizes. It does not pose any restrictions on the similarity maps realizing the equivalence. Similar generalized finite type conditions were defined by Lau and Ngai [**LN07**] and by Hare, Hare and Rutar [**HHR21**].

THEOREM 4.4.3 (Jin and Yau). *If an IFS of the form* (4.79) *is of finite type, then* $\log r_1, \dots, \log r_m$ *are commensurable, i.e.,* $\log r_i / \log r_j \in \mathbb{Q}$ *for all* $i, j \leqslant m$. *Equivalently, there exist* $0 < r_0 < 1$ *and a sequence of natural numbers* $\{n_i\}_{i=1}^m$ *such that the contractions ratios satisfy* $r_i = r_0^{n_i}$, $i \leqslant m$.

4.4.1. Examples of IFS of Finite Type. We first consider the special case of homogeneous IFS on $\mathbb{R}$ with a positive contraction ratio:

$$(4.81)\quad \mathcal{S} = (S_j)_{j\leqslant m}, \quad \text{where } S_j(x) = \rho x + b_j, \quad 0 < \rho < 1, \quad b_j \in \mathbb{R}, \quad j = 1, \dots, m.$$

The following observation is due to Feng [**Fen03**, p. 103].

LEMMA 4.4.4. *The IFS $\mathcal{S}$ from (4.81) satisfies the finite type condition if and only if there exists a finite set Γ such that for each $n \in \mathbb{N}$ and any two $\mathbf{i} = (i_1, \dots, i_n)$ and $\mathbf{j} = (j_1, \dots, j_n)$ in Σ^*,*

$$(4.82)\qquad \textit{either}\quad \rho^{-n}|S_{\mathbf{i}}(0) - S_{\mathbf{j}}(0)| \geqslant |\Lambda| \quad \textit{or} \quad \rho^{-n}|S_{\mathbf{i}}(0) - S_{\mathbf{j}}(0)| \in \Gamma,$$

where Λ is the attractor of $\mathcal{S}$.

Note that the diameter of the attractor is equal to $|\Lambda| = (1-\rho)^{-1}\cdot|b_{\max} - b_{\min}|$, where $b_{\min} = \min_j b_j$ and $b_{\max} = \max_j b_j$ (here it is important that the contraction coefficient ρ is positive). Since the IFS on $\mathbb{R}$ is homogeneous, one can replace the point 0 in (4.82) by any other $x \in \mathbb{R}$; the current form is most convenient when $b_{\min} = 0$.

PROOF SKETCH. First we check that (4.82) implies the FTC. As an open set in the definition we can take the open line segment

$$\Omega := \big((1-\rho)^{-1}\cdot b_{\min}, (1-\rho)^{-1}\cdot b_{\max}\big),$$

which is the interior of the convex hull of the attractor. It is clear that Ω is invariant under the IFS. Furthermore, as for any homogeneous IFS, the set V_n is simply

$$V_n = \{S_{\mathbf{i}} : \ |\mathbf{i}| = n\}.$$

Denote $\Omega_{\mathbf{i}} := S_{\mathbf{i}}(\Omega)$. Then $|\Omega_{\mathbf{i}}| = \rho^{-n}|\Lambda|$, so $S_{\mathbf{i}}$ and $S_{\mathbf{j}}$ are neighbors if and only $\rho^{-n}|S_{\mathbf{i}}(0) - S_{\mathbf{j}}(0)| < |\Lambda|$. Thus, each neighborhood in the sense of Definition 4.4.1 corresponds to a finite subset of $\Gamma \cup (-\Gamma)$. Moreover, if two neighborhoods correspond to the same finite subset, then they have the same type, so finiteness of Γ implies the FTC. Conversely, if Γ is infinite, then there are infinitely many neighborhood types if we use the open set Ω. However, Ω, being the interior of the convex hull of Λ, is clearly contained in any $\mathcal{S}$-invariant open set, so the number of neighborhood types will be infinite for any other $\mathcal{S}$-invariant open set as well. □

Lemma 4.4.4 implies, in particular, that the PV-type IFS (4.78), considered by Lalley, is of finite type, in view of Corollary D.0.11.

Next we present examples of IFS of finite type in higher dimensions. Denote the set of $d \times d$ matrices with entries in $\mathbb{R}$ and $\mathbb{Z}$ by $M_d(\mathbb{R})$ and $M_d(\mathbb{Z})$ respectively.

EXAMPLE 4.4.5 (Generalized integer IFS). Let $\mathcal{S}$ be an IFS on $\mathbb{R}^d$ of the form

$$(4.83)\qquad \mathcal{S} = \left(S_i(x) = r^{n_i}Q^{n_i}x + b_i\right)_{i=1}^{m},$$

where $n_i \in \mathbb{N}$, $0 < r < 1$, and Q is a $d \times d$ orthogonal matrix. We assume that

- $r^{-1}\cdot Q^{-1} \in M_d(\mathbb{Z})$
- $b_i \in \mathbb{Z}^d$ for all $1 \leqslant i \leqslant m$.

Ngai and Wang [**NW01**, Theorem 2.7] showed that the IFS $\mathcal{S}$ is of finite type with respect to any invariant open set Ω.

EXAMPLE 4.4.6 (Generalized Pisot–Vijayaraghavan IFS). Let $\mathcal{S}$ be an IFS on $\mathbb{R}^d$ of the form

$$\mathcal{S} = \left(S_i(x) = \theta^{-n_i} Q_i x + b_i\right)_{i=1}^{m}, \tag{4.84}$$

where θ is a PV number,

- $Q_1, \ldots, Q_m$ are orthogonal matrices generating a finite matrix group G;
- there exist $r_1, \ldots r_d \in \mathbb{R}$ such that

$$G \cdot \{b_j : 1 \leqslant j \leqslant m\} \subset r_1\mathbb{Z}[\theta] \times \cdots \times r_d\mathbb{Z}[\theta] \subset \mathbb{R}^d.$$

Ngai and Wang [**NW01**, Theorem 2.9] proved that $\mathcal{S}$ is an IFS of finite type with respect to any bounded invariant open set Ω.

4.4.2. Dimension of the attractor for finite type IFS. Here we describe an algorithm for computing the dimension of the attractor, following Ngai and Wang [**NW01**]. It is based on a construction of a directed graph whose incidence matrix encodes geometric information about the attractor. This algorithm generalizes earlier work, which developed a similar approach in special cases, especially that of Lalley [**Lal97**].

First we define a directed graph $\mathcal{G}_R$ corresponding to a finite type IFS. This will be done in two steps. The set of vertices of $\mathcal{G}_R$ is V.

Step 1: Draw a directed edge from $f \in V_k$ to $g \in V_{k+1}$ corresponding to each $\boldsymbol{\ell} \in \Sigma^*$ satisfying

$$S_{\boldsymbol{\ell}} = f^{-1} \circ g. \tag{4.85}$$

We say that f is a "parent" of g and g is an "$\boldsymbol{\ell}$-child" of f. This way we obtain a directed graph, which we denote by $\mathcal{G}$. We call $g \in V_k$ an offspring of $f \in V_{k'}$ for $k' < k$ if there exists a sequence $g_0 := f, g_1, \ldots, g_{k-k'} := g$ such that g_n is a $\boldsymbol{\ell}_n$-child of g_n with some $\boldsymbol{\ell}_n \in \Sigma^*$ for all $n = 1, \ldots, k - k'$.

Step 2: It may happen that for some $g \in V_{k+1}$ there is more than one word $\boldsymbol{\ell} \in \Sigma^*$ for which there exists $f \in V_k$ satisfying (4.85). Arrange these words $\boldsymbol{\ell}$ in an increasing lexicographical order to get $\boldsymbol{\ell}_1, \ldots, \boldsymbol{\ell}_p$; the corresponding parents are denoted by $f_1, \ldots f_p \in V_k$. Among these p edges keep the first one, (f_1, g), corresponding to $\boldsymbol{\ell}_1$, and delete all the other $p-1$ edges. After that, delete all the vertices in the remaining part of the graph which have only finitely many offsprings. This way we obtain the *reduced directed graph* $\mathcal{G}_R$.

Observe that if g is an $\boldsymbol{\ell}$-child of $f_2 \in V$ and $f_1 \overset{\tau}{\sim} f_2$, that is, $f_1 = \tau \circ f_2$, then $\tau \circ g$ is an $\boldsymbol{\ell}$-child of f_1. In fact,

$$f_1^{-1} \circ \tau \circ g = f_2^{-1} \circ \tau^{-1} \circ \tau \circ g = f_2^{-1} \circ g = S_{\boldsymbol{\ell}}.$$

So, if there is a directed edge corresponding to $\boldsymbol{\ell}$ between f_1 and g, then there is a directed edge, also corresponding to $\boldsymbol{\ell}$, between any f_2 having the same neighborhood type as f_1 to some element having the same neighborhood type as g. This generates an adjacency matrix M on the set of neighborhood types.

More precisely, assume that there are m neighborhood types. Then M is the $m \times m$ matrix such that for two neighborhood types α, β the entry $M_{\alpha,\beta}$ is defined as follows: if $f \in V$ has the neighborhood of type α, that is, $[\Omega(f) = \alpha]$, then

$$M_{\alpha,\beta} := \#\{g : g \text{ is a child of } f \text{ in } \mathcal{G}_R, \ \ [\Omega(g)] = \beta\} \tag{4.86}$$

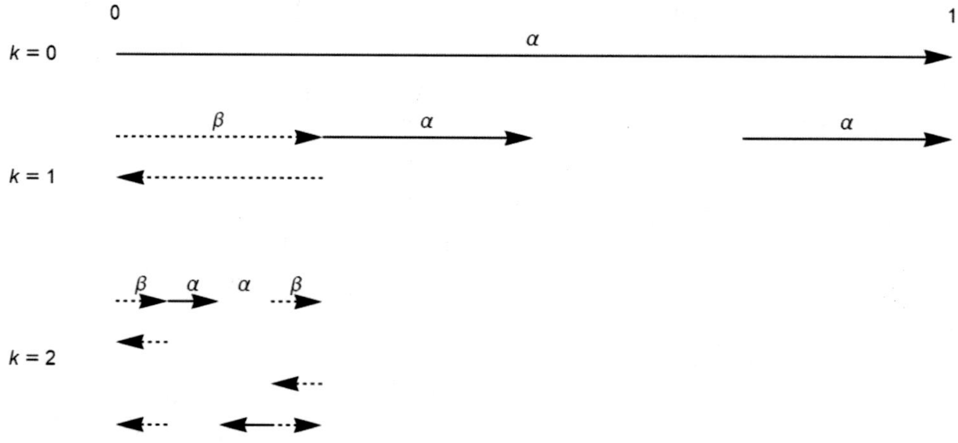

FIGURE 4.6. The iterations and neighborhood types for the IFS in Example 4.4.8.

This does not depend on the choice of f.

THEOREM 4.4.7 (Ngai, Wang, [**NW01**]). *Let $\mathcal{S}$ be an IFS of the form* (4.79) *satisfying the Finite Type Condition. Then*

$$\dim_{\mathrm{H}}(\Lambda) = \frac{\log \lambda}{-\log r_{\min}} \quad \textit{and} \quad 0 < \mathcal{H}^{\frac{\log \lambda}{-\log r_{\min}}}(\Lambda) < \infty, \tag{4.87}$$

where λ is the dominant eigenvalue of M.

We give a proof of Theorem 4.4.7 below, in Section 4.4.3. Here we illustrate the theorem on several examples, calculating the adjacency matrix in some special cases.

For the details of the following calculation see [**NW01**, Example 5.1].

EXAMPLE 4.4.8. Consider the IFS $\mathcal{S} = (S_i)_{i\leqslant 4}$, where

$$S_1(x) = \tfrac{1}{4}x, \quad S_2(x) = \tfrac{1}{4}(-x+1), \quad S_3(x) = \tfrac{1}{4}(x+1), \quad S_4(x) = \tfrac{1}{4}(x+3). \tag{4.88}$$

It is clear that $V_1 = \{S_1, S_2, S_3, S_4\}$. From Figure 4.6, one can see that there are only two neighborhood types with respect to $\Omega = (0,1)$; we denote them by α and β. Let α be the neighborhood type of the root vertex. Then S_1, S_2 are of type β and S_3, S_4 are of type α. Thus a type-α has two type-α children and one type-β child. One can show that every type-β has two type-β children and two type-α children, hence the adjacency matrix is

$$M = \begin{bmatrix} 2 & 1 \\ 2 & 2 \end{bmatrix}.$$

It has spectral radius $\lambda = 2+\sqrt{2}$, so, by Theorem 4.4.7 the Hausdorff dimension of the attractor is $\frac{\log(2+\sqrt{2})}{\log 4} = 0.88577\ldots$, whereas the similarity dimension is equal to 1.

EXAMPLE 4.4.9. Let $\rho := \frac{\sqrt{5}-1}{2}$ (the reciprocal of the golden ration $\frac{\sqrt{5}+1}{2}$, which is a PV number). Define a self-similar IFS on the plane (see Figure 4.7):

$$\mathcal{S} := \big(S_1(x,y) = (\rho x, \rho y), \quad S_2(x,y) = (\rho x + \rho^2, \rho y), \quad S_3(x,y) = (\rho^2 x, \rho^2 y + \rho)\big).$$

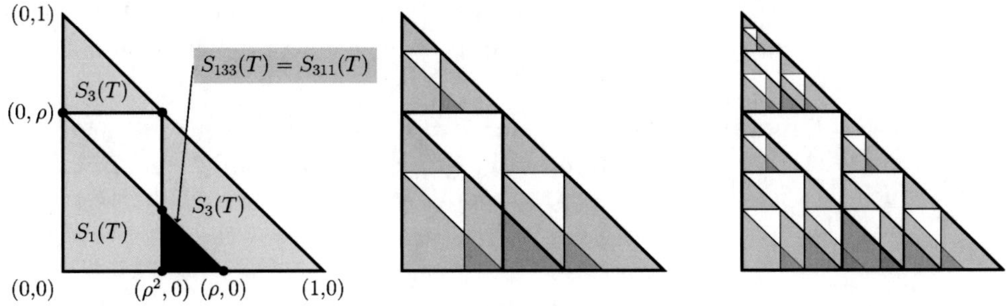

FIGURE 4.7. T is the triangle $(0,0),(0,1),(1,0)$

This is an IFS of finite type since the translations satisfy $\{(0,0),(\rho^2,0),(0,\rho)\} \subset \mathbb{Z}[\rho^{-1}] \times \mathbb{Z}[\rho^{-1}]$. The open triangle with the vertices $(0,0),(1,0),(0,1)$ can be chosen as Ω in Definition 4.4.1. A careful analysis (see [**NW01**, Example 5.4]) shows that there are seven neighborhood types and the adjacency matrix is

$$\begin{bmatrix} 1 & 2 & 1 & 1 & 1 & 1 & 0 \\ 0 & 1 & 1 & 0 & 0 & 1 & 0 \\ 1 & 2 & 1 & 1 & 1 & 0 & 0 \\ 1 & 1 & 0 & 1 & 0 & 0 & 1 \\ 1 & 2 & 0 & 1 & 2 & 0 & 0 \\ 1 & 2 & 0 & 1 & 1 & 1 & 1 \\ 1 & 2 & 0 & 1 & 1 & 0 & 1 \end{bmatrix}$$

Computing the spectral radius and applying Theorem 4.4.7 yields that the Hausdorff dimension of the attractor Λ is

$$\dim_{\mathrm{H}}(\Lambda) = 1.682391\ldots < 1.83157\ldots = \dim_{\mathrm{S}}(\mathcal{S}).$$

4.4.3. Relation between the Finite Type Condition and the Weak Separation Property; proof of Theorem 4.4.7. Nguyen [**Ngu02**] proved that the FTC implies the WSP. In general, it is not known whether these conditions are equivalent. Feng [**Fen16**] showed that for a self-similar homogeneous IFS on the line, which has attractor $[0,1]$ and a positive contraction ratio, the weak separation property implies finite type. This was recently extended by Hare, Hare and Rutar [**HHR21**] who removed the hypotheses of homogeneity and positive contraction ratio, but still assumed that the IFS is on the line with attractor $[0,1]$. For further details, see [**HHR21**].

THEOREM 4.4.10 (Nguyen[**Ngu02**]). *If an IFS of the form* (4.79) *is of finite type, see Definition* 4.4.1, *then it has the Weak Separation Property, see Definition* 4.2.3.

REMARK 4.4.11. The proof is rather simple in the special case of a homogeneous IFS on the line with a positive contraction ratio ρ, see (4.81), in which case there is a simple criterion for FTC, Lemma 4.4.4. One can show using Theorem 4.2.4 (v-b) that in such a case the WSP is equivalent to $\sup_n \#\Gamma_n < \infty$, where $\Gamma_n := \{\rho^n|S_{\mathbf{i}}(0) - S_{\mathbf{j}}(0)| : \mathbf{i},\mathbf{j} \in \Sigma^n\} \cap [-|\Lambda|,|\Lambda|]$. This clearly shows the difference between the two conditions: whereas the WSP (in the homogeneous case) requires only that the number of nearby cylinder sets of the same level be uniformly bounded (for all levels), the FTC requires also the number of possible configurations to be finite, up to scaling.

Proof of Theorem 4.4.10. Without loss of generality, we may assume that $|\Lambda| = 1$ and $0 \in \Lambda$. We will prove that an IFS $(S_i(x) = r_i O_i x + b_i)_{i=1}^m$ of similarities on $\mathbb{R}^d$, which satisfies the FTC, fulfils the condition (v-b) of Theorem 4.2.4. In fact, we will show the following: there exists $\ell \geqslant 1$ such that for every $\mathbf{i} \in \Sigma^*$, $k \geqslant 0$, and $y \in \mathbb{R}^d$,

$$\#\Big(B(y, r_{\min}^k) \cap \{S_{\mathbf{ji}}(0) : \mathbf{j} \in \widetilde{\mathcal{M}}_{r_{\min}^k}\}\Big) \leqslant \ell, \tag{4.89}$$

where $r_{\min} = \min\{r_1, \ldots, r_m\}$.

Let Ω be the open set with respect to which the Finite Type Condition holds in Definition 4.4.1. Since there are only finitely many neighborhood types,

$$C := \sup\{\#\Omega(f) : k \geqslant 1,\ f \in V_k\} < \infty, \tag{4.90}$$

where we recall that $V_k = \{S_{\mathbf{j}} : \mathbf{j} \in \widetilde{\mathcal{M}}_{r_{\min}^k}\}$ and $\Omega(f) = \{g \in V_k : g(\Omega) \cap f(\Omega) \neq \varnothing\}$. Let $y \in \mathbb{R}^d$ and $\mathbf{i} \in \Sigma^*$ be arbitrary but fixed, and denote

$$F_k := \{S_{\mathbf{j}} \in V_k : S_{\mathbf{ji}}(0) \in B(y, r_{\min}^k)\}.$$

Since $\#F_k$ is clearly an upper bound for the left-hand side of (4.89), it is enough to show that $\#F_k$ is uniformly bounded.

First, we choose a "maximal disjoint subset" G_k of F_k. Namely, take an arbitrary $f_1 \in F_k =: F_k^{(0)}$ and let $F_k^{(1)} := F_k^{(0)} \backslash \Omega(f_1)$. If $F_k^{(1)}$ is empty then let $G_k = \{f_1\}$, otherwise let $f_2 \in F_k^{(1)}$ be arbitrary and let $F_k^{(2)} = F_k^{(1)} \backslash \Omega(f_2)$. Continue the procedure inductively until $F_k^{(j)}$ is empty, and let $G_k = \{f_1, \ldots, f_j\}$. By construction, $f_i(\Omega) \cap f_{i'}(\Omega) = \varnothing$ for $i \neq i' \in \{1, \ldots, j\}$ and $F_k \subset \bigcup_{i=1}^j \Omega(f_i)$, hence

$$\#F_k \leqslant C \cdot \#G_k,$$

where C is from (4.90).

Since Ω is invariant, we have $\Lambda \subset \overline{\Omega}$. Hence, if $S_{\mathbf{ji}}(0) \in B(y, r_{\min}^k)$ then $S_{\mathbf{j}}(\overline{\Omega}) \cap B(y, r_{\min}^k) \neq \varnothing$, in view of the assumption $0 \in \Lambda$, and so $S_{\mathbf{j}}(\Omega) \subset B(y, r_{\min}^k(|\Omega|+1))$. Now, since $\{f(\Omega)\}_{f \in G_k}$ are pairwise disjoint and contained in $B(y, r_{\min}^k(|\Omega|+1))$, we have

$$\begin{aligned} r_{\min}^{kd}(|\Omega|+1)^d \mathcal{L}^d(B(0,1)) &= \mathcal{L}^d(B(y, r_{\min}^k(|\Omega|+1))) \\ &\geqslant \mathcal{L}^d\left(\bigcup_{f \in G_k} f(\Omega)\right) \\ &= \sum_{f \in G_k} \mathcal{L}^d\left(f(\Omega)\right) \\ &\geqslant \#G_k \cdot r_{\min}^{(k+1)d} \mathcal{L}^d\left(\Omega\right). \end{aligned} \tag{4.91}$$

Therefore,

$$\begin{aligned} \#\left(B(y, r_{\min}^k) \cap \{S_{\mathbf{ji}}(0) : \mathbf{j} \in \widetilde{\mathcal{M}}_{r_{\min}^k}\}\right) &\leqslant \#F_k \leqslant C \cdot \#G_k \\ &\leqslant C \cdot \frac{(|\Omega|+1)^d \mathcal{L}^d(B(0,1))}{r_{\min}^d \mathcal{L}^d(\Omega)} =: \widetilde{C}. \end{aligned}$$

□

Now, using Theorem 4.4.10, we prove Theorem 4.4.7. The proof will rely on the following lemma.

LEMMA 4.4.12. *Let $(S_i(x) = r_i O_i x + b_i)_{i=1}^m$ be an IFS of similarities satisfying the Finite Type Condition. Denote the equivalence classes from Definition 4.4.1 by $\alpha_1, \ldots, \alpha_k$, and let M be the $k \times k$ matrix defined in (4.86). Let $e \in \mathbb{R}^k$ be the vector corresponding to the classes of the first level, that is,*

$$e = \begin{pmatrix} \#\{f \in \mathcal{S}_{r_{\min}} : f \in \alpha_1\} \\ \cdots\cdots\cdots\cdots\cdots\cdots \\ \#\{f \in \mathcal{S}_{r_{\min}} : f \in \alpha_k\} \end{pmatrix}.$$

Then for every $n \geqslant 1$,

$$\#\{f \in \mathcal{S}_{r_{\min}^n} : f \in \alpha_j\} = \left((M^*)^{n-1} e\right)_j,$$

where M^* is the transpose of the matrix M defined in (4.86).

PROOF. We prove this by induction. For $n = 1$, the statement obviously holds. Suppose that it holds up to n. Let $f \in \mathcal{S}_{r_{\min}^n}$ and suppose that f is in class α_j. Then by the finite type condition and the definition of M, for every α_i class, there are exactly M_{α_i,α_j}-many functions in $\mathcal{S}_{r_{\min}^{n+1}}$ such that their parent is f. So the number of α_j-type maps in $\mathcal{S}_{r_{\min}^{n+1}}$ is

$$\sum_{i=1}^k M_{\alpha_i,\alpha_j} \left((M^*)^{n-1} e\right)_i = \left((M^*)^n e\right)_j.$$

□

PROOF OF THEOREM 4.4.7. By Theorem 4.4.10, the self-similar set satisfies the WSP, and so by Theorem 4.2.16,

$$\dim_{\mathrm{H}}(\Lambda) = \dim_{\mathrm{B}}(\Lambda) = \beta = \lim_{b \to 0} \frac{\log \#\mathcal{S}_b}{-\log b},$$

and $0 < \mathcal{H}^\beta(\Lambda) < \infty$. Thus, it is enough to show that $\beta = \frac{\log \lambda}{-\log r_{\min}}$.

It is easy to see that it is enough to take the limit along the subsequence $r_{\min}^n$. On the other hand, by Lemma 4.4.12,

$$\#\mathcal{S}_{r_{\min}^n} = \| (M^*)^{n-1} e\|_1,$$

where $\|.\|_1$ denotes the usual ℓ^1 norm on $\mathbb{R}^k$. It is easy to see, since every class is nonempty by definition, that there exists $n_0 > 0$ such that for every $n \geqslant n_0$, the vector $(M^*)^{n-1} e$ has strictly positive entries. Hence, there exists $C > 0$ such that

$$\| (M^*)^{n-n_0} \|_1 \leqslant \| (M^*)^{n-1} e\|_1 \leqslant C\| (M^*)^{n-n_0} \|_1, \tag{4.92}$$

where $\|.\|_1$ denotes the induced ℓ^1 norm. By the Perron–Frobenius theorem, since M^* has nonnegative entries, we get that the spectral radius is the maximal eigenvalue λ of M. Thus, combining (4.92) with Lemma 4.4.12 we get

$$\lim_{n \to \infty} \frac{\#\mathcal{S}_{r_{\min}^n}}{n} = \lim_{n \to \infty} \frac{\| (M^*)^{n-n_0} \|_1}{n} = \log \lambda,$$

which had to be proven. □

For more recent results related to the connection between some generalizations of the finite type condition and the WSP see [**DWX21**].

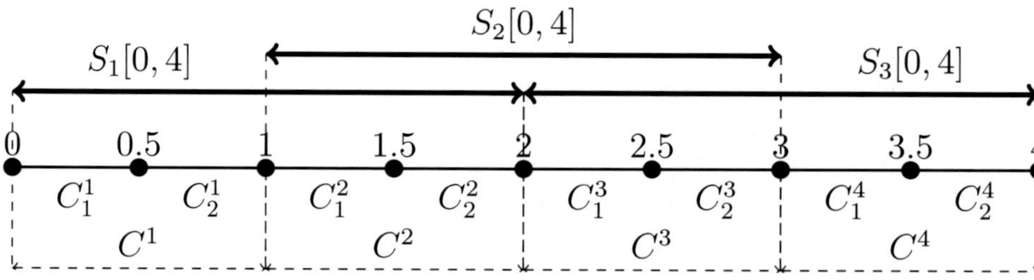

FIGURE 4.8. Definitions in the case of the IFS of Example 4.5.1

4.5. Special family of finite type*

A special family satisfying the FTC is the lattice centered homogeneous IFS on $\mathbb{R}$ having the reciprocal of a natural number as contraction ratio. In this section we study the dimension and absolute continuity of measures that are self-similar with respect to such an IFS. This section is marked with an asterisk since it uses some elements of Ergodic Theory. Where necessary, we refer the reader to Chapter 9 and to the classical book [**Wal82**].

Throughout this section we always assume that $\mathcal{N}$ is a lattice on $\mathbb{R}$, that is, $\mathcal{N} = \{\alpha + k\beta : k \in \mathbb{Z}\}$ for some $\alpha, \beta \in \mathbb{R}$. We say that $\mathcal{F}$ is a *self-similar integer IFS* on $\mathbb{R}$ if

$$\mathcal{F} = \left(S_i(x) := \frac{1}{L} \cdot x + t_i\right)_{i=1}^{m}, \quad t_i \in \mathcal{N}, \quad L,\ m \in \mathbb{N}, \quad L, m \geqslant 2. \tag{4.93}$$

Note that one can define self-similar integer IFS on $\mathbb{R}^d$ for any $d \geqslant 1$ in an analogous way. The first two assertions, Lemma 4.5.2 and Theorem 4.5.3, remain valid for self-similar integer IFS on $\mathbb{R}^d$ (see [**Rui09**, Proposition 10]). In Section 4.3 we studied the example of $\mathcal{S} = \left(S_i(x) = \frac{1}{3}x + i\right)_{i\in\{0,1,3\}}$, which is a self-similar integer IFS. The following example of a self-similar integer IFS is naturally obtained by taking the orthogonal projection of the Sierpiński gasket to the x-axis.

EXAMPLE 4.5.1. Let

$$\mathcal{F} = \left(S_1(x) = x/2,\ S_2(x) = x/2 + 1,\ S_3(x) = x/2 + 2\right),$$

with $\mathbf{p} := \left(\frac{1}{3}, \frac{1}{3}, \frac{1}{3}\right)$, and $\mu := \mathbf{p}^{\mathbb{N}}$, $\nu := \Pi_*\mu$. The attractor is the interval $[0,4]$. See Figure 4.8.

LEMMA 4.5.2. *All self-similar integer IFS on $\mathbb{R}$ satisfy the Finite Type Condition.*

The proof is given in Fact 4.5.5.

As usual, we denote the natural projection from $\Sigma := \{1,\ldots,m\}^{\mathbb{N}}$ to the attractor Λ by Π. For every probability vector $\mathbf{p} = (p_1,\ldots,p_m)$ the associated self-similar measure $\nu_{\mathbf{p}}$ is the push forward measure of $\mathbf{p}^{\mathbb{N}}$ by Π, that is, $\nu_{\mathbf{p}} := \mathbf{p}^{\mathbb{N}} \circ \Pi^{-1}$,

THEOREM 4.5.3 (Lau and Wang, Ruiz). *Let $\mathcal{F}$ be a self-similar integer IFS on $\mathbb{R}$ and $\mathbf{p}$ be a probability vector. Then the following are equivalent:*

(i) $\dim_{\mathrm{H}}(\nu_{\mathbf{p}}) = 1$;
(ii) $\nu_{\mathbf{p}} \ll \mathcal{L}^1$;
(iii) $\nu_{\mathbf{p}} \ll \mathcal{L}^1$ *with a bounded density.*

In Section 4.5.2 we state and prove Proposition 4.5.12 which is a more detailed version of this theorem. The equivalence of (i) and (ii) was proved by Ruiz [**Rui09**]. The equivalence of (ii) and (iii) was verified by Lau and Wang [**LW04**]; it is a special case of Theorem 4.2.20.

Next we give a condition which guarantees that $\dim_{\mathrm{H}}(\nu_{\mathbf{p}}) < 1$ for a self-similar integer IFS on $\mathbb{R}$. Recall that "gcd" and "lcm" stand for "greatest common divisor" and "least common multiple" respectively. Moreover, we write $L \nmid R$ if $L, R \in \mathbb{N}$ and R is not divisible by L.

THEOREM 4.5.4. *Let $\mathcal{F} = (S_i)_{i=1}^m$ be a self-similar integer IFS on $\mathbb{R}$ with a common contraction ratio $1/L$, as in (4.93). Suppose we are given a probability vector $\mathbf{p} = (p_1, \ldots, p_m)$ whose elements are rational numbers*

$$p_i = \frac{k_i}{\ell_i}, \ k_i, \ell_i \in \mathbb{Z}\backslash \{0\}, \ \gcd(k_i, \ell_i) = 1, \quad i \in \{1, \ldots, m\}.$$

Denote $R = \operatorname{lcm}(\ell_1, \ldots, \ell_m)$. Let ν be the corresponding self-similar measure $\nu = \sum_{i=1}^m p_i(\nu \circ S_i^{-1})$. Then

$$L \nmid R \implies \dim_{\mathrm{H}}(\nu) < 1. \tag{4.94}$$

Theorem 4.5.4 was stated in [**Tor**, Theorem 44] as an immediate extension of [**BR14**, Lemma 3.5]. The proof will be given in Section 4.5.3.

Recall that if the OSC holds, then $\dim_{\mathrm{H}}(\nu_{\mathbf{p}}) = \frac{-\sum_{i=1}^{m} p_i \log p_i}{\log L}$ (see Corollary 1.9.11 and later Corollary 11.3.8). Any IFS of the from (4.93) is conjugate by Remark 1.8.5 to one with the lattice $\mathcal{N} = (1 - L^{-1}) \cdot \mathbb{Z}$. So, without loss of generality in the rest of this section we always assume that $\mathcal{F}$ is of the form

$$\mathcal{F} = \left(S_i(x) := \frac{1}{L} \cdot x + t_i \right)_{i=1}^{m}, \ t_i \in (1 - L^{-1}) \cdot \mathbb{N}, \ 0 = t_1 < t_2 < \cdots < t_m, \tag{4.95}$$

so that the fixed point of S_i is $t_i \in \mathbb{Z}$. The convex hull of the attractor Λ is the interval having integer endpoints:

$$C := \left[0, \frac{t_m}{1 - L^{-1}}\right]. \tag{4.96}$$

FACT 4.5.5. *$\mathcal{F}$ is an IFS of finite type.*

PROOF. We use the observation due to Feng, see Lemma 4.4.4. Our IFS $\mathcal{F}$ is of the form of (4.95). To apply Lemma 4.4.4, observe that

$$L^n(S_{\mathbf{j}}(0) - S_{\mathbf{i}}(0)) = \sum_{k=1}^{m} (j_k - i_k) L^{n+1-k} \in \mathbb{Z}. \tag{4.97}$$

Using that there are only finitely many integers in the interval $(-|\Lambda|, |\Lambda|)$, we obtain that any integer IFS satisfies the finite type condition. □

For the proof of Theorem 4.5.3 the self-similar measure ν was represented by Ruiz [**Rui09**] using a matrix expression which he interpreted as a "hidden Markov chain". Ruiz [**Rui09**, p. 355] pointed out that his matrix expression for the self-similar measure is similar to the one obtained by the method of "second-order identities" developed in [**LN00**]. We do not use the terminology of hidden Markov chains here, but develop everything from scratch. First we need to introduce a family of integer matrices, which will be denoted $Z_1, \ldots, Z_L$.

4.5.1. Matrices $Z_1, \ldots, Z_L$ and the measures ν^* and η. We are going to use the notation $[K] = \{1, \ldots, K\}$ for $K \in \mathbb{N}$. Fix a probability vector $\mathbf{p} = (p_1, \cdots, p_m)$ and write $\nu := \nu_{\mathbf{p}}$ for the corresponding self-similar measure. For all $k \geqslant 0$ let $\mathcal{D}_k$ be the standard grid of half-open intervals of side $1/L^k$:

$$\mathcal{D}_k := \left\{ \left[\frac{q}{L^k}, \frac{q+1}{L^k} \right) : q \in \mathbb{Z} \right\}.$$

It follows from Proposition 3.1.1 that

$$\nu\left(\left\{ \frac{q}{L^k} : k \geqslant 0,\ q \in \mathbb{Z} \right\} \right) = 0. \tag{4.98}$$

Since $\operatorname{spt}(\nu)$ is compact, there are only finitely many, say N, intervals from $\mathcal{D}_0$ having positive ν-measure. We call these intervals level-0 cylinders and denote them $C^1, \ldots, C^N$ in increasing order. It follows from (4.95) that $C^1 = [0,1]$. For every k, and $u = 1, \ldots, N$, the level-0 cylinder C^u is partitioned into L^k L-adic intervals: $C^u_{\mathbf{i}} \in \mathcal{D}_k$, with $\mathbf{i} \in [L]^k$, where we label these intervals according to the lexicographic order starting from the left endpoint of C^u. We refer to these L-adic intervals as level-k cylinders. Now we list the obvious properties of this partition:

FACT 4.5.6.

(a) *For every $k \in [m]$, $u \in [N]$ we have $S_k(C^u) = C^v_j$ for some $v \in [N]$ and $j \in [1, L]$.*

(b) $S_k(C^u) = C^v_j$ *if and only if* $S_k(C^u_{\mathbf{i}}) = C^v_{j\mathbf{i}}$ *holds for all* $\mathbf{i} \in [L]^*$.

Define the $N \times N$ matrices $Z_1, \ldots Z_L$ and Z by

$$Z_j(u,v) = \begin{cases} p_k, & \text{if } S_k(C^v) = C^u_j; \\ 0, & \text{otherwise.} \end{cases} \quad \text{and } Z := \sum_{j=1}^{N} Z_j. \tag{4.99}$$

In the case of the IFS in Example 4.5.1, we have

$$Z_1 = \frac{1}{3} \begin{bmatrix} 1 & 0 & 0 & 0 \\ 1 & 0 & 1 & 0 \\ 1 & 0 & 1 & 0 \\ 0 & 0 & 1 & 0 \end{bmatrix}, \quad Z_2 = \frac{1}{3} \begin{bmatrix} 0 & 1 & 0 & 0 \\ 0 & 1 & 0 & 1 \\ 0 & 1 & 0 & 1 \\ 0 & 0 & 0 & 1 \end{bmatrix}.$$

It follows from the definition that

$$Z^n(u,v) > 0 \text{ if and only if } \exists \mathbf{i} \in \Sigma^*, |\mathbf{i}| = n, \text{ such that } S_{\mathbf{i}}(C^v) \subset C^u. \tag{4.100}$$

FACT 4.5.7. *The matrix Z^T is stochastic and primitive.*

PROOF. It is immediate from the definition that Z^T is a stochastic matrix. To see that Z is primitive, for every $u \in [N]$ pick an arbitrary point $x_u \in \operatorname{int}(C^u) \cap \Lambda$. Let $\rho := \min_{1 \leqslant u \leqslant N} \{\operatorname{dist}(x_u, \mathbb{N})\}$. Fix $K \in \mathbb{N}$ such that $L^{-K}|\Lambda| < \rho$ and for every u pick $\mathbf{i}_u \in [m]^K$ satisfying $x_u \in \Lambda_{\mathbf{i}_u}$. Then $|\Lambda_{\mathbf{i}_u}| < \rho$, hence $\Lambda_{\mathbf{i}_u} \subset \operatorname{int}(C_u)$ and we obtain that $S_{\mathbf{i}_u}(C^v) \subset C^u$ for all $v \in [N]$ since $\operatorname{int}(C^v) \cap \Lambda \neq \varnothing$ by construction. In view of part (a) of Fact 4.5.6 and (4.100), this implies that all elements of the matrix Z^K are positive, hence Z^T is a primitive matrix. □

Combining (4.99) with Fact 4.5.6 yields that

$$Z_j(u,v) = p_k \iff S_k^{-1}(C^u_{j\mathbf{i}}) = C^v_{\mathbf{i}}, \forall k \in [m],\ j \in [L],\ u, v \in [N],\ \mathbf{i} \in [L]^*. \tag{4.101}$$

Hence,

$$\nu\left(C^u_{j\mathbf{i}}\right) = \sum_{k=1}^{m} p_k \nu\left(S_k^{-1} C^u_{j\mathbf{i}}\right) = \sum_{v=1}^{N} Z_j(u,v)\cdot \nu(C^v_{\mathbf{i}}),\ v \in [N],\ j \in [L],\ \mathbf{i} \in [L]^*.$$

Using this, by induction one can see that

$$\mathbf{\Psi}_{\mathbf{i}} = Z_{\mathbf{i}} \cdot \mathbf{\Psi}, \quad \mathbf{i} \in [L]^*, \tag{4.102}$$

where $Z_{\mathbf{i}} := Z_{i_1} \cdots Z_{i_n}$ for $\mathbf{i} = (i_1, \ldots, i_n) \in [m]^n$ and

$$\mathbf{\Psi}_{\mathbf{i}} := \begin{bmatrix} \nu(C^1_{\mathbf{i}}) \\ \vdots \\ \nu(C^N_{\mathbf{i}}) \end{bmatrix} \text{ and } \mathbf{\Psi} := \begin{bmatrix} \nu(C^1) \\ \vdots \\ \nu(C^N) \end{bmatrix}.$$

From this, by the definition of the matrix Z we obtain the equality:

$$\mathbf{\Psi} = Z \cdot \mathbf{\Psi}. \tag{4.103}$$

Thus $\lambda = 1$ is an eigenvalue of the matrix Z and the corresponding eigenvectors are $\mathbf{\Psi}$ and $\mathbb{1}$ (where $\mathbb{1} \in \mathbb{R}^N$ is the vector with all components equal to 1) have strictly positive entries. Using that the matrix Z is primitive, we obtain from the Perron–Frobenius Theorem (see e.g. [**Wal82**, Theorem 0.16]) that 1 is the leading eigenvalue of Z. We have shown

FACT 4.5.8. *The leading eigenvalue of the matrix Z is equal to* 1. *The corresponding left and right eigenvectors are $\mathbb{1}$ and $\mathbf{\Psi}$ respectively.*

If we multiply both sides of (4.102) by $\mathbf{e}_u^T$, the u-th coordinate unit vector, then we get that

$$\nu(C^u_{\mathbf{i}}) = \mathbf{e}_u^T Z_{\mathbf{i}} \mathbf{\Psi}, \quad u \in [N],\ \mathbf{i} \in [L]^*. \tag{4.104}$$

Now we construct the auxiliary measure ν^* on the interval $[0,1]$ by adding up the ν measure of every C^u.

More precisely, let ξ_u be defined by $C^u = \xi_u + [0,1]$. Then for a Borel set $H \subset [0,1]$ let

$$\nu^*(H) = \sum_{u=1}^{N} \nu\left(\xi_u + H\right). \tag{4.105}$$

Thus ν^* is a probability Borel measure. Now recall that $C^1 = [0,1]$ since we assumed in (4.95) that $t_1 = 0$. We write $C_{\mathbf{i}} := C^1_{\mathbf{i}}$. Adding up both sides of (4.104) for all $u \in [N]$ we get

$$\nu^*(C_{\mathbf{i}}) = \mathbb{1}^T Z_{\mathbf{i}} \mathbf{\Psi}. \tag{4.106}$$

We can represent ν^* as a projection of a shift-invariant measure on a symbolic space. The relevant symbolic space is $\Sigma := \{1, \ldots, L\}^{\mathbb{N}}$ endowed with the metric $\rho(\mathbf{i}, \mathbf{j}) := L^{-|\mathbf{i} \wedge \mathbf{j}|}$ for $\mathbf{i}, \mathbf{j} \in \Sigma$. The natural projection $\Gamma : \Sigma \to [0,1]$, given by

$$\Gamma(i_1, i_2, \ldots) = \sum_{n=1}^{\infty} \frac{i_n - 1}{L^n}, \tag{4.107}$$

is 1-to-1 except at the points of $\Gamma^{-1}A$, where

$$A := \left\{ \frac{k}{L^\ell} \in (0,1) : k, \ell \in \mathbb{N} \right\}. \tag{4.108}$$

Hence, $\nu^* = \Gamma_* \eta$ where the measure η is defined on Σ by

$$\eta([\mathbf{i}]) := \mathbb{1}^T Z_{\mathbf{i}} \mathbf{\Psi}, \quad \mathbf{i} \in [L]^*. \tag{4.109}$$

This is because the measure ν, and consequently ν^*, are nonatomic (see Proposition 3.1.1).

LEMMA 4.5.9. *The measure η is σ-invariant on Σ and mixing and hence ergodic.*

Recall (1.34) that σ-invariance of the measure η means that

$$\eta(\sigma^{-1}A) = \eta(A) \text{ for all Borel set } A \subset \Sigma. \tag{4.110}$$

Recall also (see Section 1.9.5) that η is ergodic if $\sigma^{-1}A = A$ implies $\eta(A) \in \{0,1\}$ for $A \subset \Sigma$ Borel. The *mixing property* of η means that

$$\lim_{n\to\infty} \eta\left(\sigma^{-n}A \cap B\right) = \eta(A)\eta(B) \text{ for all Borel sets } A, B \subset \Sigma. \tag{4.111}$$

Taking $B = A$ in (4.111) we immediately see that mixing implies ergodicity.

PROOF. It is enough to check (4.110) for cylinder sets, see Theorem B.1.7. Let $\mathbf{i} \in [L]^*$

$$\eta(\sigma^{-1}[\mathbf{i}]) = \sum_{k=1}^{L} \eta([k,\mathbf{i}]) = \mathbb{1}^T \sum_{k=1}^{L} Z_k Z_{\mathbf{i}} \mathbf{\Psi} = \mathbb{1}^T Z Z_{\mathbf{i}} \mathbf{\Psi} = \mathbb{1}^T Z_{\mathbf{i}} \mathbf{\Psi} = \eta([\mathbf{i}]). \tag{4.112}$$

The proof of mixing mimics the standard proof of mixing for Markov measures on mixing shifts of finite type. We include the argument for the reader's convenience. In order to check that the measure is mixing, it is enough to check the property (4.111) for cylinder sets A and B, see e.g., [**Wal82**, Theorem 1.17], as was the case when checking σ-invariance. That is, the measure η is mixing if for every $\mathbf{i} = (i_1, \ldots, i_k)$ and $\mathbf{j} = (j_1, \ldots, j_\ell) \in [L]^*$ we have

$$\lim_{n\to\infty} \eta\left([\mathbf{i}] \cap \sigma^{-n}[\mathbf{j}]\right) = \eta([\mathbf{i}]) \cdot \eta([\mathbf{j}]). \tag{4.113}$$

Observe that $[\mathbf{i}] \cap \sigma^{-n}[\mathbf{j}] = \bigcup_{\boldsymbol{\xi} \in [L]^{n-k}} [\mathbf{i}\boldsymbol{\xi}\mathbf{j}]$ for $n > k$. Hence, by (4.109), for any $n > k$,

$$\begin{aligned} \eta\left([\mathbf{i}] \cap \sigma^{-n}[\mathbf{j}]\right) &= \sum_{\boldsymbol{\xi} \in [L]^{n-k}} \mathbb{1}^T Z_{\mathbf{i}} Z_{\boldsymbol{\xi}} Z_{\mathbf{j}} \mathbf{\Psi} \\ &= \mathbb{1}^T Z_{\mathbf{i}} \left(\sum_{\boldsymbol{\xi} \in [L]^{n-k}} Z_{\boldsymbol{\xi}} \right) Z_{\mathbf{j}} \mathbf{\Psi} = \mathbb{1}^T Z_{\mathbf{i}} Z^{n-k} Z_{\mathbf{j}} \mathbf{\Psi}. \end{aligned} \tag{4.114}$$

Recall that Z is a primitive matrix whose leading eigenvalue is 1 and the corresponding right and left eigenvectors are $\mathbf{\Psi}$ and $\mathbb{1}$ respectively. Thus, by a corollary of the Perron–Frobenius Theorem (see e.g., [**Wal82**, Theorem 0.17]), we have $\lim_{n\to\infty} Z^{n-k} = \mathbf{\Psi} \cdot \mathbb{1}^T$. Hence, by (4.114),

$$\lim_{n\to\infty} \eta\left([\mathbf{i}] \cap \sigma^{-n}[\mathbf{j}]\right) = \left(\mathbb{1}^T Z_{\mathbf{i}} \mathbf{\Psi}\right) \cdot \left(\mathbb{1}^T Z_{\mathbf{j}} \mathbf{\Psi}\right) = \eta([\mathbf{i}]) \cdot \eta([\mathbf{j}]),$$

as desired. □

Since η is ergodic, by the Shannon–McMillan–Breiman Theorem, see Theorem 9.5.3, we obtain

$$-\frac{1}{n}\log\eta([\mathbf{i}|_n]) \to h_\eta \qquad \text{for } \eta\text{-a.e. } \mathbf{i}\in\Sigma=\{1,\dots,L\}^{\mathbb{N}}, \tag{4.115}$$

where $h_\eta = h_\eta(\sigma)$ is the entropy of the shift-invariant measure η. Using that $\nu^* = \Gamma_*\eta$ and $\Gamma : \Sigma \to [0,1]$ is 1-to-1 except at the points of $\Gamma^{-1}A$, where A is defined in (4.108), and $\nu^*(A) = 0$, we obtain that

$$\lim_{n\to\infty} -\frac{1}{n}\log\nu^*(D_k(x)) = h_\eta \qquad \text{for a.e. } x\in[0,1], \tag{4.116}$$

where $D_k(x)$ is the element of the partition $\mathcal{D}_k$ that contains x. The following technical lemma allows us to find the almost sure local dimension applying (4.116).

LEMMA 4.5.10. *For every $\alpha > 0$,*

$$\nu^*\Big(\{x\in[0,1] : \operatorname{dist}(x,\partial D_k(x)) < L^{-k(1+\alpha)} \textit{ for infinitely many } k\text{'s}\}\Big) = 0.$$

PROOF. Using the definition of Γ (4.107), we obtain that

$$\Gamma^{-1}(\{x\in[0,1] : \operatorname{dist}(x,\partial D_k(x)) < L^{-k(1+\alpha)}\}) \\ \subset \{\mathbf{i}=(i_1,\dots)\in\Sigma : i_{k+1} = \dots = i_{k+\lfloor k\alpha\rfloor} \in \{1, L\}\}.$$

Denote $\mathbf{1}_k = (\overbrace{1,\dots,1}^{k\text{-times}})$ and $\mathbf{L}_k = (\overbrace{L,\dots,L}^{k\text{-times}})$ for $k\in\mathbb{N}$. Then by the σ-invariance (4.110), we have

$$\sum_{k=1}^{\infty}\nu^*(\{x\in[0,1] : \operatorname{dist}(x,\partial D_k(x)) < L^{-k(1+\alpha)}\}) \leqslant \sum_{k=1}^{\infty}\eta([\mathbf{1}_{\lfloor\alpha k\rfloor}]) + \sum_{k=1}^{\infty}\eta([\mathbf{L}_{\lfloor\alpha k\rfloor}]) \\ = \sum_{k=1}^{\infty}\mathbb{1}^T Z_1^{\lfloor\alpha k\rfloor}\mathbf{\Psi} + \sum_{k=1}^{\infty}\mathbb{1}^T Z_L^{\lfloor\alpha k\rfloor}\mathbf{\Psi}.$$

A simple consequence of (4.99) and Fact 4.5.8 is that there exist $0 < \lambda < 1$ and $C > 0$ such that $\mathbb{1}^T Z_i^k\mathbf{\Psi} \leqslant C\lambda^k$ for every $k\in\mathbb{N}$, and the claim of the lemma follows by Borel–Cantelli. □

Note that for all x in the set defined in Lemma 4.5.10, we have for every $\alpha > 0$ and k sufficiently large,

$$B(x, L^{-k(1+\alpha)}/2) \subseteq D_k(x) \subseteq B(x, L^{-k}).$$

Thus, combining (4.116) with Lemma 4.5.10 we obtain that the local dimension $d_{\nu^*}(x)$ of the measure ν^* is equal to

$$d_{\nu^*}(x) = \frac{h_\eta}{\log L} \qquad \text{for } \nu^*\text{-a.e. } x, \tag{4.117}$$

hence $\dim_{\mathrm{H}}(\nu^*) = \frac{h_\eta}{\log L}$. Using this, we prove the following:

LEMMA 4.5.11. *The local dimension $d_\nu(x)$ of the measure ν is equal to*

$$d_\nu(x) = \frac{h_\eta}{\log L} \qquad \textit{for } \nu\textit{-a.e. } x, \tag{4.118}$$

hence $\dim_{\mathrm{H}}(\nu) = \frac{h_\eta}{\log L}$.

PROOF. Let $\nu_u := \nu|_{C^u}$ and recall that $C^u = \xi_u + [0,1]$ for $u \in [N]$. We write $g_u(x) := x + \xi_u$, $u \in [N]$. Now we define the measure ν_u^* on $[0,1]$ by $\nu_u^*(H) := \nu_u(\xi_u + H)$, that is, $\nu_u^* = (g_u^{-1})_* \nu_u$. Recall that ν^* is a measure of exact dimension $\frac{h_\eta}{\log L}$. By (4.105) we have $\nu_u^* \ll \nu^*$. Using these facts and Remark 1.9.6, we obtain that ν_u^* is also a measure of exact dimension $\frac{h_\eta}{\log L}$. Since g_u is bi-Lipschitz, we get that ν_u is also a measure of exact dimension $\frac{h_\eta}{\log L}$ for all $u \in [N]$. By definition, $\nu = \sum_{u=1}^{N} \nu_u$, and hence ν is also a measure of exact dimension $\frac{h_\eta}{\log L}$. □

4.5.2. Absolute continuity of self-similar measures corresponding to $\mathcal{F}$. Theorem 4.5.3 follows from the following proposition which is due to Ruiz [**Rui09**, Proposition 10].

PROPOSITION 4.5.12. *The following are equivalent:*

(i) $\nu \ll \mathcal{L}^1$;
(ii) $\dim_{\mathrm{H}}(\nu) = 1$;
(iii) $\mathbb{1}^T Z_{\mathbf{i}} \mathbf{\Psi} = L^{-k}$ *for all* k *and* $\mathbf{i} \in [L]^k$;
(iv) ν^* *is the Lebesgue measure on* $[0,1]$;
(v) ν *is not singular.*

PROOF. The implications **(i)** $\Longrightarrow$ **(ii)** and **(iv)** $\Longrightarrow$ **(v)** are immediate, **(v)** $\Longleftrightarrow$ **(i)** follows from Proposition 3.1.4(i) (the law of pure type). The equivalence **(iii)** $\Longleftrightarrow$ **(iv)** follows from (4.106), (4.109) , and the fact that the Lebesgue measure on $[0,1]$ is the unique probability measure giving equal mass L^{-k} to every L-adic subinterval of $[0,1]$ of length L^{-k}. It remains to show **(ii)** $\Longleftrightarrow$ **(iii)**.

By Lemma 4.5.11, $\dim_{\mathrm{H}}(\nu) = 1$ is equivalent to $h_\eta = \log L$. Assume that this is the case, and consider

$$a_k := \frac{1}{k} \cdot H(\eta, \xi_k) = \frac{-1}{k} \sum_{\boldsymbol{\omega} \in [L]^k} \eta[\boldsymbol{\omega}] \log \eta[\boldsymbol{\omega}], \ k \in \mathbb{N},$$

where $H(\eta, \xi_k)$ is the entropy of the partition ξ_k of Σ into cylinders of length k, see Definition 1.9.15. By Proposition 1.9.20, $a_k \searrow h_\eta = \log L$. In particular, $a_k \geqslant \log L$ for all k. On the other hand, it follows from Corollary 1.9.17 that $a_k \leqslant \log L$, since ξ_k is a partition with L^k elements. It follows that $a_k = \log L$ for all k, and moreover, every element of the partition has equal mass. This means that $\eta[\boldsymbol{\omega}] = L^{-k}$ for all $\boldsymbol{\omega} \in [L]^k$, hence **(iii)** holds by (4.109). Conversely, if **(iv)** holds, then $\eta[\boldsymbol{\omega}] = L^{-k}$ for all $\boldsymbol{\omega} \in [L]^k$, hence $h_\nu = \log L$, completing the proof. □

4.5.3. The dimension of self-similar measures with rational weights. We follow the proof of Bárány and Rams in [**BR14**, Lemma 3.5]

PROOF OF THEOREM 4.5.4. We argue by contradiction. Assume that $\dim_{\mathrm{H}}(\nu) = 1$. Then by Proposition 4.5.12 we have

$$\mathbb{1}^T Z_{\boldsymbol{\omega}} \mathbf{\Psi} = L^{-n}, \text{ for all } n \in \mathbb{N}, \ \boldsymbol{\omega} = (\omega_1, \dots, \omega_n) \in [L]^n. \tag{4.119}$$

Recall that we defined the matrix Z in (4.99) and we proved in Fact 4.5.7 that Z is a primitive matrix. Hence we can choose $K_0 > 0$ such that for every $K > K_0$ all entries of the matrix Z^K are positive. Fix $K > K_0$ satisfying $L^K \geqslant N^2 + 1$.

We claim that one can choose a word $\boldsymbol{\tau} = (\tau_1, \dots, \tau_K) \in [L]^K$ such that for $\mathcal{A} := [L]^K \setminus \{\boldsymbol{\tau}\}$ the $N \times N$ matrix

$$Y := \sum_{\boldsymbol{\xi} \in \mathcal{A}} Z_{\boldsymbol{\xi}} = Z^K - Z_{\boldsymbol{\tau}} \tag{4.120}$$

has all its entries positive. Indeed, for every $\boldsymbol{\omega} \in [L]^K$ we have $Z_{\boldsymbol{\omega}} \leqslant Z^K$, where the inequality is understood entrywise. If the strict inequality does not hold, then there is an entry in $Z_{\boldsymbol{\omega}}$ and Z^K which is the same. But note that there are N^2 entries altogether and the number of words in $[L]^K$ is greater. Thus if the strict inequality $Z_{\boldsymbol{\omega}} < Z$ holds for every $\boldsymbol{\omega} \in [L]^K$, then there exist $\boldsymbol{\xi} \neq \boldsymbol{\omega}$ in $[L]^K$ such that $Z_{\boldsymbol{\omega}}$ and $Z_{\boldsymbol{\xi}}$ have the same entry as Z. However, $Z \geqslant Z_{\boldsymbol{\omega}} + Z_{\boldsymbol{\xi}}$, which is a contradiction, and the claim is proved.

Let $\rho > 0$ be the leading eigenvalue of the primitive matrix Y and let $\mathbf{u}$ and $\mathbf{v}$ be the corresponding left and right eigenvectors respectively. That is $\mathbf{u}^T \cdot Y = \rho \mathbf{u}^T$ and $Y \cdot \mathbf{v} = \rho \mathbf{v}$. Then by a corollary of the Perron–Frobenius Theorem (see e.g., [**Wal82**, Theorem 0.17]) we have $\lim\limits_{n\to\infty} \rho^{-n} Y^n = \mathbf{v}\mathbf{u}^T$ and

$$\lim_{n\to\infty} \frac{1}{n} \log\left(\mathbb{1}^T Y^n \boldsymbol{\Psi}\right) = \log \rho. \tag{4.121}$$

Observe that for every n, in view of (4.119),

$$\mathbb{1}^T Y^n \boldsymbol{\Psi} = \mathbb{1}^T \left(\sum_{\boldsymbol{\xi} \in \mathcal{A}} Z_{\boldsymbol{\xi}} \right)^n \boldsymbol{\Psi} \sum_{\boldsymbol{\xi}_1 \dots \boldsymbol{\xi}_n \in \mathcal{A}} \underbrace{\mathbb{1}^T Z_{\boldsymbol{\xi}_1 \dots \boldsymbol{\xi}_n} \boldsymbol{\Psi}}_{L^{-nK}} = \frac{(L^K - 1)^n}{L^{nK}}. \tag{4.122}$$

Putting together (4.121) and (4.122) we get that $\rho = \frac{L^K - 1}{L^K}$. The matrix Y is the sum of matrices $Z_{\boldsymbol{\omega}}$ for $\boldsymbol{\omega} \in \mathcal{A} \subset [L]^K$ and all entries of $Z_{\boldsymbol{\omega}}$ are sums of rational numbers of the form $p_{i_1} \cdots p_{i_K} = \frac{k_{i_1} \cdots k_{i_K}}{\ell_{i_1} \cdots \ell_{i_K}}$. By assumption, R is the least common multiple of ℓ_r for $r \in [m]$. Notice that $R^K \rho$ is an eigenvalue of the $N \times N$ integer matrix $R^K \cdot Y$. However, this is impossible since we have also assumed that $L \nmid R$. Hence $R^K \rho = \left(\frac{R}{L}\right)^K (L^K - 1) \in \mathbb{Q} \setminus \mathbb{Z}$, but no element of $\mathbb{Q} \setminus \mathbb{Z}$ can be the root of the characteristic polynomial of an integer matrix. This is because the characteristic polynomial of an integer matrix is a monic polynomial (a polynomial whose leading coefficient is 1) with integer coefficients. The roots of such polynomials are algebraic integers, and the only rational numbers which are algebraic integers are integers, see Proposition D.0.1. $\square$

CHAPTER 5

Multifractal Analysis for self-similar measures

In this book, we do not discuss the general theory of multifractal analysis, which aims, quoting Falconer [**Fal97**, p. 185], *"to quantify the singularity structure of measures and provide a model for phenomena in which scaling occurs with a range of different power laws"*. Instead, we focus on the simplest nontrivial case of self-similar measures with strong separation, in order to introduce the basic concepts and ideas. After proving the basic results on multifractal spectrum and multifractal formalism in this case, we discuss some extensions and modifications without proof or with partial proofs.

5.1. Multifractal spectrum and multifractal formalism

In this section we consider a self-similar measure ν and study the size (Hausdorff dimension) of the set K_α where the local dimension $d_\nu(x) = \lim_{r\to 0} \frac{\log \nu(B(x,r))}{\log r}$ (see Section 1.9) of the measure ν exists and is equal to a given number α, that is,

$$K_\alpha := \{x \in \Lambda : d_\nu(x) = \alpha\}. \tag{5.1}$$

Roughly speaking, the function

$$\mathcal{D} : \alpha \mapsto \dim_{\mathrm{H}}(K_\alpha) \tag{5.2}$$

shows the size of the set of x for which $\nu(B(x,\rho))$ is approximately ρ^α for a small ρ. It is usually called the *multifractal spectrum* of ν.

To be more precise, consider a self-similar IFS in $\mathbb{R}^d$

$$\mathcal{S} = (S_1, \ldots, S_m), \tag{5.3}$$

where

$$\forall i \leqslant m,\ \forall \mathbf{x}, \mathbf{y} \in \mathbb{R}^d,\ \|S_i(\mathbf{x}) - S_i(\mathbf{y})\| = r_i\|\mathbf{x} - \mathbf{y}\|,\ 0 < r_i < 1.$$

As usual, we write Λ for the attractor of $\mathcal{S}$ and Π for the natural projection (defined in (1.11)) from the symbolic space $\Sigma := \{1, \ldots, m\}^{\mathbb{N}}$ to Λ. For a probability vector $\mathbf{p} := (p_1, \ldots, p_m)$, let ν be the corresponding self-similar measure ν, given by

$$\nu = \Pi_*(\mu), \quad \text{where} \quad \mu := \mathbf{p}^{\mathbb{N}}.$$

Throughout this section we make the following

Principal assumptions:

(A1) $\mathcal{S}$ satisfies the SSP, that is,

$$\delta_* := \min_{i \neq j} \operatorname{dist}\{S_i(\Lambda), S_j(\Lambda)\} > 0; \tag{5.4}$$

(A2) $\mathbf{p} \neq (r_1^s, \ldots, r_m^s)$,

where s is the similarity dimension. We require (A2) to avoid a trivial situation. Namely, if $\mathbf{p} = (r_1^s, \dots, r_m^s)$, then ν is the natural measure and by the definition (1.98) of the local dimension we obtain that

$$\forall x \in \Lambda,\ d_\mu(x) \equiv s \implies K_s = \Lambda \text{ and } K_\alpha = \varnothing \text{ for } \alpha \neq s. \tag{5.5}$$

In what follows the function $T : \mathbb{R} \to \mathbb{R}$ will play a central role:

$$\boxed{T(q) \text{ is the unique solution of the equation } \sum_{i=1}^{m} p_i^q r_i^{T(q)} = 1,\ q \in \mathbb{R}.} \tag{5.6}$$

THEOREM 5.1.1. *Assume that* (A1) *and* (A2) *hold for the IFS* $\mathcal{S}$. *Let*

$$\alpha_{\min} := \min_{1\leqslant i\leqslant m} \frac{\log p_i}{\log r_i}, \qquad \alpha_{\max} := \max_{1\leqslant i\leqslant m} \frac{\log p_i}{\log r_i}. \tag{5.7}$$

Then the multifractal spectrum of the self-similar measure $\nu = \Pi_* \left(\mathbf{p}^{\mathbb{N}}\right)$ *for every* $\alpha \in [\alpha_{\min}, \alpha_{\max}]$ *is given by*

$$\mathcal{D}(\alpha) = \dim_{\mathrm{H}} \{x : d_\nu(x) = \alpha\} = T^*(\alpha),$$

where T^* *is the Legendre transform of the convex function* T, *that is,*

$$T^*(\alpha) := \inf_q \left(T(q) + \alpha \cdot q\right). \tag{5.8}$$

Moreover, for $\alpha \notin [\alpha_{\min}, \alpha_{\max}]$ *we have* $K_\alpha = \varnothing$.

The multifractal spectrum for conformal systems was first determined by Rand [**Ran89**]. Theorem 5.1.1 is due to Cawley and Mauldin [**CM92**], which was later extended by Arbeiter and Patzschke [**AP96**] to systems satisfying the Strong Open Set Condition. For simplicity, we restrict ourselves to the SSP (A1).

5.1.1. The first two derivatives of the function $T(q)$. We start with a few simple observations. We differentiate (5.6) to obtain

$$T'(q) = -\frac{\sum_{i=1}^{m} p_i^q r_i^{T(q)} \log p_i}{\sum_{i=1}^{m} p_i^q r_i^{T(q)} \log r_i}. \tag{5.9}$$

Hence,

$$T'(q) < 0 \qquad \text{for all } q \in \mathbb{R}. \tag{5.10}$$

Now we differentiate (5.6) one more time to get

$$\sum_{i=1}^{m} p_i^q r_i^{T(q)} \left(T''(q) \cdot \log r_i + \left(\log p_i + T'(q) \log r_i\right)^2\right) = 0, \tag{5.11}$$

FACT 5.1.2. *For every* $q \in \mathbb{R}$ *we have*

$$T''(q) > 0. \tag{5.12}$$

PROOF. Fix an arbitrary $q \in \mathbb{R}$. First assume that

$$T''(q) \cdot \log r_i + \left(\log p_i + T'(q) \log r_i\right)^2 = 0, \quad \forall\ 1 \leqslant i \leqslant m. \tag{5.13}$$

It follows from (A2) that $\alpha_{\min} < \alpha_{\max}$. So, there exists at least one index i such that $\log p_i + T'(q) \log r_i \neq 0$. In view of (5.13), for this value of i we have $T''(q) \cdot \log r_i < 0$, which implies that $T''(q) > 0$.

Now we assume that (5.13) does not hold. Then there exists an $i \in \{1, \ldots, m\}$ such that $T''(q) \cdot \log r_i + (\log p_i + T'(q) \log r_i)^2 < 0$. Hence, $T''(q) \cdot \log r_i < 0$ which implies that $T''(q) > 0$. □

It follows from (5.6), (5.10) and (5.12) that the graph of the function $T(q)$ is similar to the one in Figure 5.1.

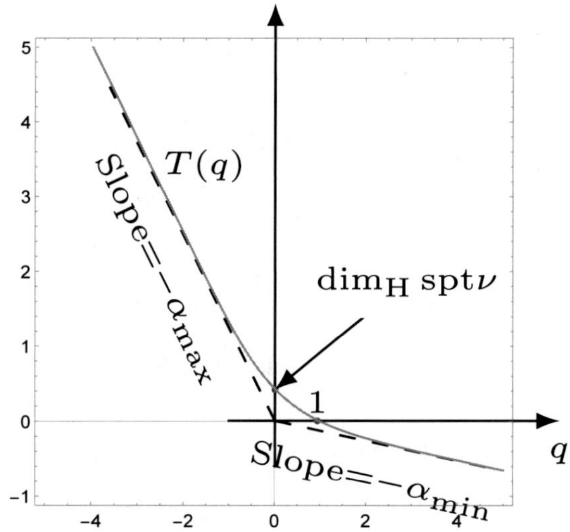

FIGURE 5.1. The $T(q)$ function.

5.1.2. Legendre transform. The Legendre transform $T^*(\alpha)$ of $T(q)$ is defined by

$$T^*(\alpha) := \inf_{q \in \mathbb{R}} (T(q) + \alpha q) \tag{5.14}$$

To get a better understanding of $T^*(\alpha)$ one should follow our argument in Figure 5.2. Let $\alpha \in (\alpha_{\min}, \alpha_{\max})$ and $q \in \mathbb{R}$ be arbitrary. Let ℓ be the line through the point $(q, T(q))$ with slope $-\alpha$. One can easily see that $T(q) + \alpha q$ is the y-intercept of the line ℓ. Now we translate the line ℓ down until its translated copy (call it $\ell(\alpha)$) will be a tangent line to the graph of the function $T(q)$. Then by elementary geometry, the y-intercept of $\ell(\alpha)$ is equal to $T^*(\alpha) = \min_{q \in \mathbb{R}} (T(q) + \alpha q)$. We write $q(\alpha)$ for the abscissa of the point of intersection of the graph of $T(q)$ and its tangent line $\ell(\alpha)$. Then $\ell(\alpha)$ is the line through the point $(q(\alpha), T(q(\alpha)))$ with slope $-\alpha$. By elementary geometry, the y-intercept of $\ell(\alpha)$ (which is $T^*(\alpha)$ as we saw above) is equal to $T(q(\alpha)) + \alpha q(\alpha)$. Thus, we obtained that

$$T^*(\alpha) = T(q(\alpha)) + \alpha q(\alpha). \tag{5.15}$$

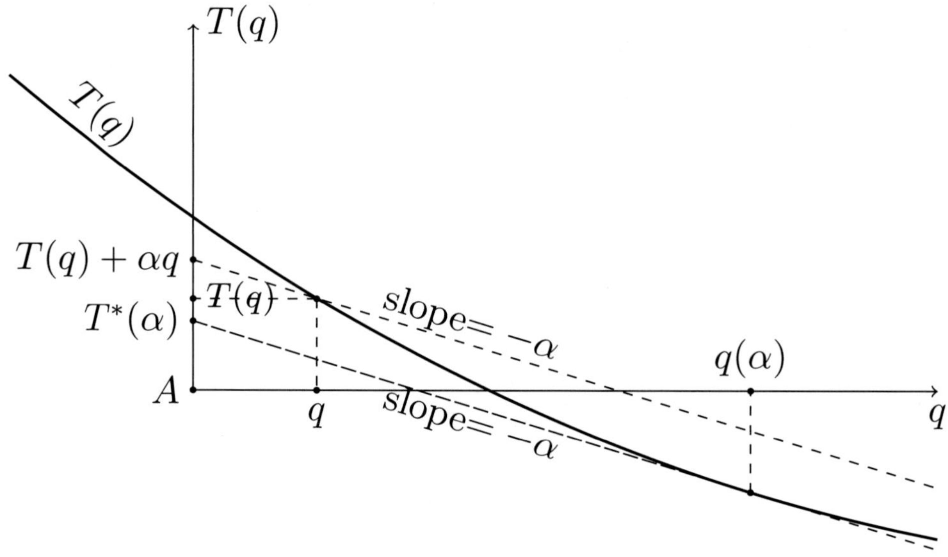

FIGURE 5.2. Legendre transformation

Denote the inverse function of $q(\alpha)$ by $\alpha(q)$; this is possible, in view of Fact 5.1.2. That is,

$$T'(q(\alpha)) = -\alpha, \quad \alpha(q) := -T'(q) = \frac{\sum_{i=1}^{m} p_i^q r_i^{T(q)} \log p_i}{\sum_{i=1}^{m} p_i^q r_i^{T(q)} \log r_i}, \tag{5.16}$$

where the second equality follows from (5.9). Moreover,

$$\alpha_0 = \alpha(q(\alpha_0)), \qquad \forall \alpha_0 \in [\alpha_{\min}, \alpha_{\max}]. \tag{5.17}$$

5.1.3. An auxiliary measure and its local dimension. To prove Theorem 5.1.1 we define the probability measures

$$\mu_q := \left\{ p_1^q \cdot r_1^{T(q)}, \ldots, p_m^q \cdot r_m^{T(q)} \right\}^{\mathbb{N}} \text{ and } \nu_q := \Pi_*(\mu_q).$$

Before we state the next lemma, recall the formula (1.107), which says that

$$\text{for } \nu\text{-a.e. } x, \quad d_\nu(x) = \frac{h_{\mathbf{p}}}{\chi_{\mathbf{r}}^{\mathbf{p}}} = \frac{\sum_{i=1}^{m} p_i \log p_i}{\sum_{i=1}^{m} p_i \log r_i}. \tag{5.18}$$

LEMMA 5.1.3.

(i) *For every* $q \in \mathbb{R}$ *and every* $x \in \Lambda$,

$$d_{\nu_q}(x) = T(q) + q \cdot d_\nu(x). \tag{5.19}$$

More precisely, if any of the two local dimensions in formula (5.19) *exists, then the other one exists as well and the equality in* (5.19) *holds.*

(ii) *Moreover, for ν_q-a.e. x*

$$d_\nu(x) = \alpha(q) \text{ and } d_{\nu_q}(x) = T(q) + q \cdot \alpha(q). \tag{5.20}$$

PROOF.

(i) Let $x \in \Lambda$ be arbitrary and let $\mathbf{i} \in \Sigma$ be the unique coding of x, that is, $\Pi(\mathbf{i}) = x$. As in the proof of Theorem 3.2.1, it is easy to see that

$$d_\nu(x) = \lim_{n\to\infty} \frac{\log \mu([i_1 \dots i_n])}{\log r_{i_1\dots i_n}} \text{ and } d_{\nu_q}(x) = \lim_{n\to\infty} \frac{\log \mu_q([i_1, \dots, i_n])}{\log r_{i_1\dots i_n}}. \tag{5.21}$$

Thus,

$$\begin{aligned} d_{\nu_q}(x) &= \lim_{n\to\infty} \frac{\log(p_{i_1}^q r_{i_1}^{T(q)} \cdots p_{i_n}^q r_{i_n}^{T(q)})}{\log r_{i_1\dots i_n}} \\ &= q \lim_{n\to\infty} \frac{\log(p_{i_1} \cdots p_{i_n})}{\log r_{i_1\dots i_n}} + T(q) = q d_\nu(x) + T(q), \end{aligned}$$

which shows (5.19).

(ii) By (5.21), the equation (5.20) is an easy consequence of the Law of Large Numbers, or Birkhoff Ergodic Theorem for the Bernoulli system (Σ, σ, μ_q). Define the functions $\varphi, \psi : \Sigma \to \mathbb{R}$ by $\varphi(\mathbf{j}) := \log p_{j_1}$, $\psi(\mathbf{j}) := \log r_{j_1}$. Then by the Birkhoff Ergodic Theorem (1.108) (see Section 9.1 for more details), for μ_q-a.e. $\mathbf{j} \in \Sigma$ we have

$$\lim_{n\to\infty} \frac{1}{n} \sum_{k=0}^{n-1} \log p_{j_k} = \lim_{n\to\infty} \frac{1}{n} \sum_{k=0}^{n-1} \varphi(\sigma^k \mathbf{j}) = \int \varphi \, d\mu_q = \sum_{k=1}^{m} p_k^q r_k^{T(q)} \log p_k.$$

Similarly, for μ_q-a.e. $\mathbf{j} \in \Sigma$ we obtain

$$\lim_{n\to\infty} \frac{1}{n} \sum_{k=0}^{n-1} \log r_{j_k} = \lim_{n\to\infty} \frac{1}{n} \sum_{k=0}^{n-1} \psi(\sigma^k \mathbf{j}) = \int \psi \, d\mu_q = \sum_{k=1}^{m} p_k^q r_k^{T(q)} \log r_k.$$

Using the last two equations, we obtain that for ν_q-typical $x = \Pi(\mathbf{i})$,

$$\begin{aligned} d_\nu(x) &= \lim_{n\to\infty} \frac{\log \mu([i_1 \dots i_n])}{\log r_{i_1\dots i_n}} \\ &= \lim_{n\to\infty} \frac{\log p_{i_1\dots i_n}}{\log r_{i_1\dots i_n}} \\ &= \frac{\lim_{n\to\infty} \frac{1}{n} \sum_{k=1}^{n} \log p_{i_k}}{\lim_{n\to\infty} \frac{1}{n} \sum_{k=1}^{n} \log r_{i_k}} \\ &= \frac{\sum_{k=1}^{m} p_k^q r_k^{T(q)} \log p_k}{\sum_{k=1}^{m} p_k^q r_k^{T(q)} \log r_k} = \alpha(q). \end{aligned} \tag{5.22}$$

Finally, applying Theorem 3.2.1, we get that for ν_q-almost every x,

$$d_{\nu_q}(x) = \frac{\sum_{k=1}^{m} p_k^q r_k^{T(q)} \log(p_k^q r_k^{T(q)})}{\sum_{k=1}^{m} p_k^q r_k^{T(q)} \log r_k} = T(q) + q\alpha(q). \tag{5.23}$$

□

PROOF OF THEOREM 5.1.1. For $\alpha \in (\alpha_{\min}, \alpha_{\max})$ define

$$f(\alpha) := T(q(\alpha)) + \alpha \cdot q(\alpha) \quad \text{and} \quad E_\alpha := \left\{x \in \Lambda : d_{\nu_{q(\alpha)}}(x) = f(\alpha)\right\}.$$

Observe that (5.19) implies $E_\alpha = \{x : d_\nu(x) = \alpha\}$. The conditions of Theorem 1.9.5(iii) hold for the set E_α and for the measure $\nu_{q(\alpha)}$. Indeed, we know that $\nu_{q(\alpha)}(E_\alpha) = 1$ and by definition, for all $x \in E_\alpha$ we have $d_{\nu_{q(\alpha)}}(x) = f(\alpha)$. Therefore, by Theorem 1.9.5(iii) we obtain that

$$f(\alpha) = \dim_{\mathrm{H}}(E_\alpha) = \dim_{\mathrm{H}} \{x : d_\nu(x) = \alpha\}.$$

It remains to check that

$$f(\alpha) = T^*(\alpha). \tag{5.24}$$

To see this, observe by differentiation that the function $q \mapsto T(q) + \alpha \cdot q$ attains its minimum at the point q where $\alpha = -T'(q)$, which is equal to $q(\alpha)$ by definition. Thus, by (5.8) we have

$$T^*(\alpha) = T(q(\alpha)) + \alpha \cdot (q(\alpha)) = f(\alpha). \tag{5.25}$$

We also point out that $K_\alpha = \varnothing$ for $\alpha \notin [\alpha_{\min}, \alpha_{\max}]$ by the first equation in the formula (5.22), which holds *for all* x, such that the limit exists. □

5.1.4. The shape of the graph of the function $\mathcal{D}(\alpha)$. See Figure 5.3 for a typical graph of the function $\mathcal{D}(\alpha) = \dim_{\mathrm{H}}(K_\alpha)$. Above we have proved that

$$\mathcal{D}(\alpha) = T(q(\alpha)) + \alpha \cdot q(\alpha). \tag{5.26}$$

By the chain rule and (5.16), we have

$$\mathcal{D}'(\alpha) = q(\alpha). \tag{5.27}$$

Here we list some features of the graph $\mathcal{D}(\alpha)$; they are shown in Figure 5.3.

- *$\mathcal{D}(\alpha)$ is concave and it attains its maximum, equal to s, the similiarity dimension of the IFS $\mathcal{S}$.*

Indeed, since $q(\alpha)$ (the inverse of $\alpha(q)$) is strictly monotone decreasing, we obtain that the function $\mathcal{D}(\alpha)$ is strictly concave on the interval $[\alpha_{\min}, \alpha_{\max}]$ and it attains its maximum at a point $\alpha_2 \in (\alpha_{\min}, \alpha_{\max})$ where $q(\alpha_2) = 0$. Substituting this into (5.6) we get that $T(q(\alpha_2)) = s$. Thus, $\mathcal{D}(\alpha_2) = s$. Using (5.16) again, we obtain that

$$\alpha_2 = \alpha(q(\alpha_2)) = \frac{\sum_{i=1}^{m} r_i^s \log p_i}{\sum_{i=1}^{m} r_i^s \log r_i}. \tag{5.28}$$

- *The bisector of the first quadrant is a tangent line to $\mathcal{D}(\alpha)$ at the point (α_1, α_1), with $\alpha_1 = \dim_{\mathrm{H}}(\nu)$.*

Indeed, let $\alpha_1 := \alpha(1)$, that is, α_1 is the α that corresponds to $q = 1$. Then (5.6) implies that $T(1) = 0$. By (5.16), we obtain that

$$\alpha_1 = \alpha(1) = \frac{\sum_{i=1}^{m} p_i \log p_i}{\sum_{i=1}^{m} p_i \log r_i} = \dim_{\mathrm{H}}(\nu).$$

Further, using that $q(\alpha_1) = 1$ and $T(q(\alpha_1)) = 0$, we obtain from (5.26) that $\mathcal{D}(\alpha_1) = \alpha_1$. It follows from (5.27) that $\mathcal{D}'(\alpha_1) = q(\alpha_1) = 1$. The last two observations together imply that the tangent line of $\mathcal{D}(\alpha)$ at $\alpha = \alpha_1$ is the line through the origin with a slope $\pi/4$.

- *The behavior of the graph of $\mathcal{D}(\alpha)$ around the endpoints: If there are unique i and j such that $\alpha_{\min} = \frac{\log p_i}{\log r_i}$ and $\alpha_{\max} = \frac{\log p_j}{\log r_j}$ then we have:*

$$\lim_{\alpha \to \alpha_{\min}^+} \mathcal{D}(\alpha) = \lim_{\alpha \to \alpha_{\max}^-} \mathcal{D}(\alpha) = \mathcal{D}(\alpha_{\min}) = \mathcal{D}(\alpha_{\max}) = 0. \tag{5.29}$$

The fact that $\mathcal{D}(\alpha_{\min}) = \mathcal{D}(\alpha_{\max}) = 0$ is left as an exercise to the reader.

EXAMPLE 5.1.4. Let $m = 2$ and $\mathbf{p} = \left(\frac{3}{4}, \frac{1}{4}\right)$, $\mathbf{r} = \left(\frac{1}{9}, \frac{1}{3}\right)$; that is, we consider the IFS

$$\mathcal{S} = \left(\frac{1}{9} \cdot x,\ \frac{1}{3} \cdot x + \frac{2}{3}\right),$$

and let ν be the self-similar measure with probabilities $\mathbf{p}$. See Figure 5.3 for the graph of $\mathcal{D}(\alpha)$.

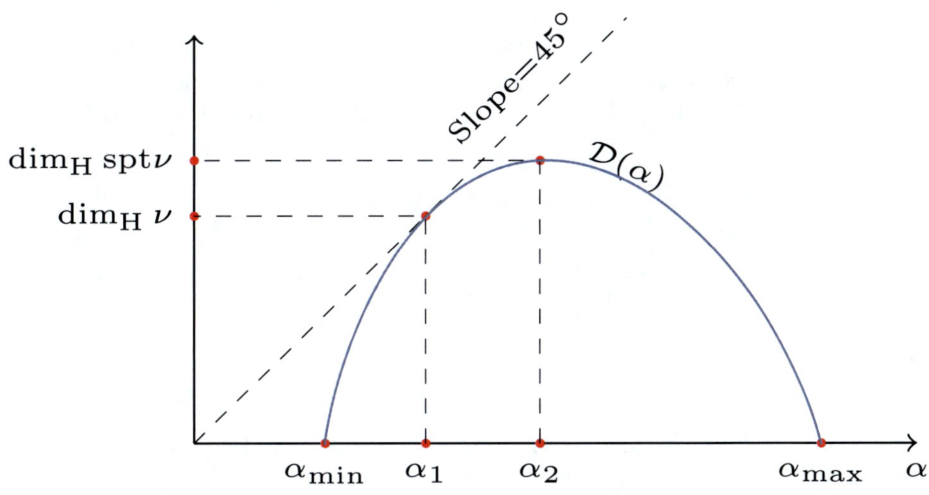

FIGURE 5.3. Dimension spectrum $\mathcal{D}(\alpha)$.

We will next show how to find the dimension spectrum from a system of equations, following Riedi [**Rie95**, Corollary 17], in the setting of Theorem 5.1.1.

LEMMA 5.1.5. *For $\alpha \in (\alpha_{\min}, \alpha_{\max})$ consider the following system of equations, where the unknowns are γ and q:*

$$\sum_{i=1}^{m} \left(\frac{p_i}{r_i^{\alpha}}\right)^q \cdot r_i^{\gamma} = 1 \tag{5.30}$$

$$\sum_{i=1}^{m} r_i^{\gamma} \left(\frac{p_i}{r_i^{\alpha}}\right)^q \cdot \log\left(\frac{p_i}{r_i^{\alpha}}\right) = 0. \tag{5.31}$$

The system has a unique solution, given by

$$\gamma = f(\alpha) = \mathcal{D}(\alpha) \quad \textit{and} \quad q = f'(\alpha). \tag{5.32}$$

PROOF. We may assume that $\mathbf{p} \neq (r_1^s, \ldots, r_m^s)$. We have seen that the range of $-T'$ is $[\alpha_{\min}, \alpha_{\max}]$. Now we show that there is a unique solution for all α from this range. Assume that for $\alpha \in (\alpha_{\min}, \alpha_{\max})$ the pair (γ_0, q_0) is a solution of the system. Then

$$\sum_{i=1}^{m} p_i^{q_0} r_i^{\gamma_0 - \alpha q_0} = 1, \tag{5.33}$$

$$\sum_{i=1}^{m} p_i^{q_0} r_i^{\gamma_0 - \alpha q_0} \log p_i = \alpha \sum_{i=1}^{m} p_i^{q_0} r_i^{\gamma_0 - \alpha q_0} \log r_i. \tag{5.34}$$

The first equation yields that $T(q_0) = \gamma_0 - \alpha q_0$ by (5.6). Substituting this into the second equation yields

$$\alpha = \frac{\sum_{i=1}^{m} p_i^{q_0} r_i^{T(q_0)} \log p_i}{\sum_{i=1}^{m} p_i^{q_0} r_i^{T(q_0)} \log r_i}.$$

However, by (5.16) this means that $\alpha = -T'(q_0)$. On the other hand, one can check that whenever $\alpha = -T'(q_0)$ holds, $(T(q_0) + \alpha q_0, q_0)$ is a solution of the system. The uniqueness of the solution follows from $T''(q) < 0$. Note that for $\alpha = -T'(q_0)$ we have $f(\alpha) = T(q_0) + \alpha q_0 = \gamma_0$ by (5.15). Finally, differentiating (5.15) yields $f'(\alpha) = q_0$. □

For $m = 2$ one can solve this system of equations explicitly by introducing the new variables $x_i = p_i^q r_i^{\gamma - q\alpha}$, to obtain

$$f(\alpha) = \frac{c_2 \log(-c_2) + (c_1 - c_2)\log(c_1 - c_2) - c_1 \log c_1}{\log r_1 \log p_2 - \log r_2 \log p_1}, \tag{5.35}$$

where

$$c_i = \log p_i - \alpha \log r_i$$

5.1.5. L^q spectrum and multifractal formalism.

In Section 2.6 the L^q *spectrum* $\tau(q) = \tau(q, \nu)$ was defined for all $q \in \mathbb{R}$ by the formula

$$\tau(q) = \tau(\nu, q) = \liminf_{\delta \downarrow 0} \frac{\log\left(\sup \sum_i \nu(B(x,\delta))^q\right)}{\log \delta}, \tag{5.36}$$

where the supremum is taken over all families of disjoint balls centered in the support of ν.

DEFINITION 5.1.6. We say that the *multifractal formalism* holds for $q > 0$ for ν if the Legendre transform

$$\tau^*(\alpha) = \inf\{\alpha t - \tau(t) : t \in \mathbb{R}\}$$

is equal to the Hausdorff dimension of

$$K_\alpha := \left\{ x \in \operatorname{spt}(\nu) : \lim_{\delta \to 0} \frac{\log \nu(B(x,\delta))}{\log \delta} = \alpha \right\}.$$

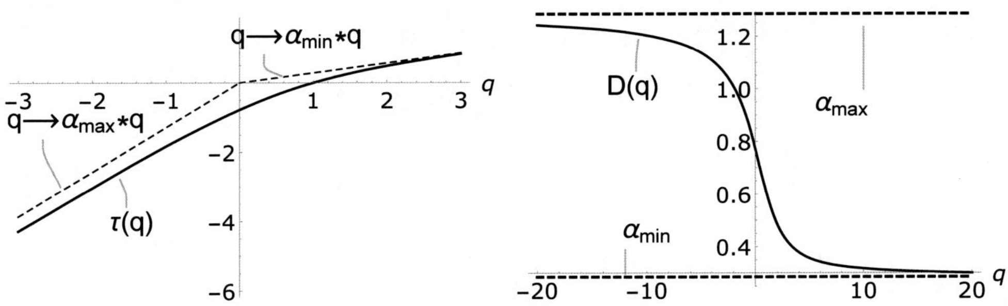

FIGURE 5.4. $m = 4$, $r_1 = \cdots = r_4 = \frac{1}{6}$, $\mathbf{p} = (0.1, 0.1, 0.2, 0.6)$

That is, for all $q > 0$ we have

(5.37) $$\dim_{\mathrm{H}}(K_\alpha) = \inf\{\alpha t - \tau(t) : t \in \mathbb{R}\} = \alpha q - \tau(q), \quad \text{where } \alpha = \tau'(q).$$

If (5.37) is satisfied for all $q \in J$ for some set $J \subset \mathbb{R}$, then we say that the multifractal formalism holds on J. Extending earlier results of Cawley and Mauldin [**CM92**] and Riedi [**Rie95**] for IFS satisfying the SSP, Lau and Ngai [**LN99**] proved

THEOREM 5.1.7. *If $\mathcal{S}$ is a self-similar IFS with contraction ratios $r_1, \ldots r_m$, satisfying the OSC, and ν is the self-similar measure corresponding to the probabilities $p_1, \ldots, p_m$, then*

$$\sum_{i=1}^{m} p_i^q r_i^{-\tau(q)} = 1,$$

that is, $T(q) = -\tau(q)$, and the multifractal formalism holds for $q \geqslant 0$.

In Figure 5.4, for a specific example of self-similar measure, we show the graphs of the q-spectrum and the L^q dimension, as functions of q.

PROOF SKETCH IN THE CASE WHEN THE SSP HOLDS, FOR $q>0$. We will use the definition of L^q spectrum from Section 2.6 for $q > 0$:

$$\tau(q) = \tau(\nu, q) = \liminf_{\delta \to 0} \frac{\log C_q(\nu, \delta)}{\log \delta},$$

where

$$C_q(\nu, \delta) = \int \left(\nu(B(x, \delta))\right)^{q-1} d\nu(x).$$

As usual, denote the attractor by Λ and its cylinders by $\Lambda_{\mathbf{i}}$ for $\mathbf{i} \in \Sigma^*$. For $\mathbf{i} = i_1 \ldots i_n \in \Sigma^*$ we write

$$r_{\mathbf{i}} = r_{i_1} \cdot \ldots \cdot r_{i_n} \quad \text{and} \quad p_{\mathbf{i}} = p_{i_1} \cdot \ldots \cdot p_{i_n}.$$

Recall that by the Strong Separation Property (5.4), we have

$$\delta_* = \min_{i \neq j} \mathrm{dist}(\Lambda_i, \Lambda_j) > 0.$$

Note also that

$$\nu(\Lambda_{\mathbf{i}}) = p_{\mathbf{i}},$$

by the SSP and the definition of self-similar measure. Assume that $|\Lambda| = 1$ without loss of generality. Fix $\delta > 0$ and consider the Moran cut-set $\widetilde{\mathcal{M}}_\delta$:

$$\widetilde{\mathcal{M}}_\delta = \{\mathbf{i} \in \Sigma^* : r_{\mathbf{i}} \leqslant \delta < r_{\mathbf{i}^-}\},$$

where $\mathbf{i}^-$ is obtained from $\mathbf{i}$ by deleting the last symbol. Observe that

$$\forall\, \mathbf{i} \neq \mathbf{j} \text{ in } \widetilde{\mathcal{M}}_\delta, \quad \mathrm{dist}(\Lambda_{\mathbf{i}}, \Lambda_{\mathbf{j}}) > \delta_* \delta,$$

since $r_{\mathbf{i}\wedge\mathbf{j}} > \delta$ by the definition of the Moran cut-set. Thus we have for any $\mathbf{i} \in \widetilde{\mathcal{M}}_\delta$ and and any $x \in \Lambda_{\mathbf{i}}$:

$$\nu(B(x,\delta)) \geqslant \nu(\Lambda_{\mathbf{i}}) = p_{\mathbf{i}} \ \text{ and } \ \nu(B(x,\delta_*\delta)) \leqslant \nu(\Lambda_{\mathbf{i}}) = p_{\mathbf{i}},$$

hence

$$C_q(\nu,\delta) = \int \nu(B(x,\delta))^{q-1} d\nu(x) \geqslant \sum_{\mathbf{i}\in\widetilde{\mathcal{M}}_\delta} p_{\mathbf{i}}^{q-1}\nu(\Lambda_{\mathbf{i}}) = \sum_{\mathbf{i}\in\widetilde{\mathcal{M}}_\delta} p_{\mathbf{i}}^q \tag{5.38}$$

and

$$C_q(\nu,\delta_*\delta) \leqslant \sum_{\mathbf{i}\in\widetilde{\mathcal{M}}_\delta} p_{\mathbf{i}}^q. \tag{5.39}$$

Recall that $\sum_{i=1}^m p_i^q r_i^{T(q)} = 1$ by the definition (5.6), and therefore,

$$\sum_{\mathbf{i}\in\widetilde{\mathcal{M}}_\delta} p_{\mathbf{i}}^q r_{\mathbf{i}}^{T(q)} = 1,$$

as for any other minimal cut-set of the symbolic space, see Lemma 1.3.3. It follows that

$$\sum_{\mathbf{i}\in\widetilde{\mathcal{M}}_\delta} p_{\mathbf{i}}^q \in \left[\delta^{-T(q)},\ (r_{\min}\delta)^{-T(q)}\right] \quad \text{or} \quad \sum_{\mathbf{i}\in\widetilde{\mathcal{M}}_\delta} p_{\mathbf{i}}^q \in \left[(r_{\min}\delta)^{-T(q)},\ \delta^{-T(q)}\right], \tag{5.40}$$

depending on whether $q > 1$ or $q < 1$. In any case, combining (5.38), (5.39), and (5.40) immediately yields

$$\tau(q) = -T(q).$$

The second claim, that the multifractal formalism holds in this case, is (5.8), which was proved in Theorem 5.1.1. □

We note the following proposition that will be needed in Chapter 6:

PROPOSITION 5.1.8. *Consider the symbolic space $\Sigma = \{1,\dots,m\}^{\mathbb{N}}$, with the metric $\varrho(\omega,\tau) = m^{-|\omega\wedge\tau|}$, equipped with the product measure $\mu_{\mathbf{p}} = \mathbf{p}^{\mathbb{N}}$ for some probability vector $\mathbf{p} = (p_1 \dots, p_m)$. Then the L^q-dimension of $\mu_{\mathbf{p}}$ is given by*

$$D_q(\mu_{\mathbf{p}}) = \frac{\tau(\mu_{\mathbf{p}})}{q-1} = \frac{\log \sum_{i=1}^m p_i^q}{(q-1)\log m}.$$

The proof is the same as for the self-similar measure of an IFS satisfying the SSP; we just work in the symbolic space. In fact, it is even easier, because all symbolic cylinders of order n have the same diameter m^{-n}, so this is more like the case of a homogeneous self-similar IFS, when $T(q)$ is computed explicitly.

5.1.6. An improved multifractal formalism. In Section 2.6 a definition, equivalent to the following, was given for the L^q spectrum for $q > 0$: Let $\mathcal{B}_\delta$ be the set of δ-mesh boxes:

$$\mathcal{B}_\delta := \left\{ \prod_{k=1}^{d} [j_k \cdot \delta, (j_k + 1) \cdot \delta) : j_1, \dots, j_d \in \mathbb{Z} \right\}. \tag{5.41}$$

The L^q spectrum of the measure μ for $q > 0$ is given by

$$\tau(q) = \liminf_{\delta \downarrow 0} \frac{\log S_\delta(q, \mu)}{\log \delta}, \quad q > 0, \tag{5.42}$$

where

$$S_\delta(q, \mu) := \sum_{B \in \mathcal{B}_\delta,\ \mu(B) \neq 0} \mu(B)^q$$

is sometimes called the *q-variation* of the measure μ.

Actually, with a clever modification a similar method can be applied also for $q < 0$. It was introduced by R. Riedi [**Rie95**]. The problem with the definition (5.42) for $q < 0$ is that the right-hand side may "blow up" if a box "just touches" the support of the measure, and it may depend on the choice of the coordinate system. The idea of Riedi is to consider the sum of the measures of the boxes dilated by a factor of three, but only those for which the "middle sub-box" has positive measure.

Let μ be a Borel measure on $\mathbb{R}^d$ of bounded support. For a box of the form

$$B = \prod_{k=1}^{d} [n_k \cdot \delta, (n_k + 1) \cdot \delta]$$

we denote

$$\widehat{B} := \prod_{k=1}^{d} [(n_k - 1) \cdot \delta, (n_k + 2) \cdot \delta].$$

The *modified q-variation* is given by

$$\widehat{S}_\delta(q, \mu) := \sum_{B \in \mathcal{B}_\delta,\ \mu(B) > 0} \mu(\widehat{B})^q. \tag{5.43}$$

It was proved by Riedi [**Rie95**] (see also [**Lau95**]) that if we let

$$\widehat{\tau}(q) = \liminf_{\delta \to 0} \frac{\log \widehat{S}_\delta(q, \mu)}{\log \delta}, \quad \forall q \in \mathbb{R}, \tag{5.44}$$

then

$$\tau(q) = \widehat{\tau}(q) \ \text{ for } q \in \mathbb{R}, \tag{5.45}$$

where the left-hand side is from (5.36).

5.2. Multifractal analysis for IFS satisfying the weak separation or finite type condition

This is a vast area of research; we mention a few results without proof, far from being comprehensive.

Assume that ν is a self-similar measure corresponding to an IFS in $\mathbb{R}^d$, which satisfies the Weak Separation Property (WSP). Feng and Lau [**FL09**] proved that there is an open ball U_0 such that

- the multifractal formalism is valid for $\nu|_{U_0}$ in the whole range of dimension spectrum,
- the dimension spectra for ν and $\nu|_{U_0}$ coincide for $q > 0$,
- the distribution of $\nu|_{U_0}$ can be described by the product of (finitely many) nonnegative matrices.

Recently, Rutar [**Rut21b, Rut21a**] considered IFS of finite type on $\mathbb{R}$ and introduced conditions on the graph induced by the finite type condition, under which the multifractal formalism is valid for the whole ν, i.e. when $U_0 = \Lambda$, for all $q \in \mathbb{R}$. He gave a large family of examples for which these conditions hold.

To avoid technical difficulties in stating the results, we next restrict ourselves to some special families of IFS satisfying the finite type condition, namely, to self-similar IFS on $\mathbb{R}$ of the form:

$$\mathcal{S} = (S_i(x) = \rho x + b_i)_{i=1}^{m}. \tag{5.46}$$

An explicit formula for the L^q-spectrum of the Erdős measure (the Bernoulli convolution measure for $\rho = \frac{\sqrt{5}-1}{2}$, see Example 1.8.14) was given and validity of the multifractal formalism was shown by Lau and Ngai [**LN98**] for $q > 0$. Then the same authors [**LN99**] studied the L^q-spectrum for self-similar IFS satisfying the Weak Separation Property; in fact, the WSP was introduced in this paper. Their results were used by Feng [**Fen03**] in his proof of the following theorem:

THEOREM 5.2.1 (Feng). *Let $\mathcal{S}$ be an IFS on the line of the form* (5.46). *Assume that $\mathcal{S}$ satisfies the finite type condition, and let ν be a self-similar measure. Then*

(i) *The function $q \mapsto \tau(q)$ is differentiable for $q > 0$;*
(ii) *multifractal formalism holds for ν for $q > 0$;*
(iii) *the measure ν is exact dimensional with* $\dim_{\mathrm{H}}(\nu) = \dim_{\mathrm{e}}(\nu) = \tau'(1)$.

We remark that (ii) follows from (i) and [**LN99**, Theorem B] and (iii) follows from (i) and [**Nga97**, Theorem 1.1].

The L^q spectrum for $q < 0$ has been found only in special cases. Feng [**Fen05**] considered the so-called *"simple", or "multinacci" PV numbers.* Namely let $\lambda_k > 0$ be the solution of the polynomial equation

$$x^k + x^{k-1} + \cdots + x = 1.$$

Then λ_k is the reciprocal of a PV (Pisot) number, see [**Bra51**]. Feng [**Fen05**] proved

THEOREM 5.2.2 (Feng). *Let ν be the Bernoulli convolution measure for $\rho = \lambda_k$.*

(i) *If $k = 2$ (the case of the Erdős measure), then there exists q_0 such that for any $q \neq q_0$, the multifractal formalism holds, i.e.,*

$$\dim_{\mathrm{H}} K(\alpha(q)) = \alpha(q)q - \tau(q), \quad \alpha(q) = \tau'(q).$$

However, $\tau(q)$ is not differentiable at q_0.

(ii) *If $k \geqslant 3$ then* (5.37) *holds for every $q \in \mathbb{R}$.*

5.3. Divergence points of self-similar measures

Throughout the previous sections, we were studying the structure of points where the local dimension exists. It is natural to ask how large is the set of points for which the behavior of the local dimension is irregular, i.e., the local dimension does not exist. The dimension of this set was determined by Barreira and Schmeling [**BS00**] and Chen and Xiong [**CX99**]. They showed that the set of such points has full dimension.

THEOREM 5.3.1 (Barreira–Schmeling; Chen–Xiong). *Suppose that the assumptions* (A1) *and* (A2) *hold for the IFS $\mathcal{S}$ in* (5.3). *Denote the attractor of $\mathcal{S}$ by Λ and let $\nu = \Pi_* \left(\mathbf{p}^{\mathbb{N}}\right)$. Then*

$$\dim_{\mathrm{H}}\left(\{x \in \Lambda : \underline{d}_\nu(x) < \overline{d}_\nu(x)\}\right) = \dim_{\mathrm{H}}(\Lambda).$$

Later, Olsen and Winter [**OW03**] studied the local dimension in a more detailed way. Namely, they studied the dimension of the set on which the local dimension has a given set of accumulation points. For $x \in \Lambda$ let

$$A_\nu(x) = \left\{\alpha \in \mathbb{R} : \exists\, r_n \to 0 \text{ such that } \lim_{n\to\infty} \frac{\log \nu(B(x, r_n))}{\log r_n} = \alpha\right\}.$$

THEOREM 5.3.2 (Olsen, Winter). *Suppose that the assumptions* (A1) *and* (A2) *hold for the IFS $\mathcal{S}$ in* (5.3). *Let Λ be the attractor of $\mathcal{S}$ and $\nu = \Pi_* \left(\mathbf{p}^{\mathbb{N}}\right)$. Denote*

$$\alpha_{\min} := \min_{1\leqslant i\leqslant m} \frac{\log p_i}{\log r_i} \quad \text{and} \quad \alpha_{\max} := \max_{1\leqslant i\leqslant m} \frac{\log p_i}{\log r_i},$$

as in (5.7). *Let T be the convex function defined in* (5.6) *and let T^* be the Legendre transform of T defined in* (5.8).

(i) *If $I \subseteq \mathbb{R}$ is* **not** *a closed subinterval of $[\alpha_{\min}, \alpha_{\max}]$, then*

$$\{x \in \Lambda : A_\nu(x) = I\} = \varnothing.$$

(ii) *If $I \subseteq [\alpha_{\min}, \alpha_{\max}]$ is a closed subinterval, then*

$$\dim_{\mathrm{H}}\left(\{x \in \Lambda : A_\nu(x) = I\}\right) = \inf_{\alpha\in I} T^*(\alpha).$$

(iii) *If $C \subseteq \mathbb{R}$ is an arbitrary subset, then*

$$\dim_{\mathrm{H}}\left(\{x \in \Lambda : A_\nu(x) \subseteq C\}\right) = \sup_{\alpha\in C} T^*(\alpha).$$

In the remaining part of the section we give a simplified version of the proof of Theorem 5.3.1 by Chen and Xiong [**CX99**]. The strategy is as follows: we construct a measure on Σ, which is not σ-invariant, but at certain scales the lower and upper local dimension is arbitrarily close to given values. Then we show that the Mass Distribution Principle may be applied for such measures.

Let us fix a fast increasing sequence n_k such that $\lim_{k\to\infty} \frac{1}{n_k}\sum_{\ell=1}^{k-1} n_\ell = 0$. For example, $n_k = 2^{k^2}$ will do. Further, for every $q > 0$ and $i \in \mathbb{N}$, let $\mathbf{z}^{q,i}$ be a probability vector such that

$$\mathbf{z}^{q,i} = (z_1^{q,i}, \ldots, z_m^{q,i}), \text{ where } z_j^{q,i} = \begin{cases} p_j^q r_j^{T(q)}, & \text{if } n_{2k} < i \leqslant n_{2k+1} \\ p_j^{-q} r_j^{T(-q)}, & \text{if } n_{2k+1} < i \leqslant n_{2k+2}, \end{cases}$$

for some $k \in \mathbb{N}$. Here $T(q)$ is the convex function from (5.6), that is, $\sum_{j=1}^m p_j^q r_j^{T(q)} = 1$ for all $q \in \mathbb{R}$. Now, define for $q > 0$ a measure $\widetilde{\mu}_q$ on Σ by $\widetilde{\mu}_q = \prod_{i=1}^\infty \mathbf{z}^{q,i}$. That is, for every finite word $\mathbf{i} = (i_1, \ldots, i_n)$,

$$\widetilde{\mu}_q([\mathbf{i}]) = z_{i_1}^{q,1} \cdots z_{i_n}^{q,n}.$$

Note that $\widetilde{\mu}_q$ is well-defined by Lemma 1.3.2. The measure $\widetilde{\mu}_q$ is a probability measure on Σ, such that the random variables $I_n(\mathbf{i}) := i_n$ are independent and identically distributed at certain scales.

LEMMA 5.3.3. *Let $(a_1, \ldots, a_m)$ be an arbitrary vector. Then for $\widetilde{\mu}_q$-almost every $\mathbf{i} = (i_1 i_2 \ldots)$,*

$$\begin{aligned}\lim_{k\to\infty} \frac{1}{n_{2k}} \sum_{\ell=1}^{n_{2k}} a_{i_\ell} &= \sum_{j=1}^{m} p_j^{-q} r_j^{T(-q)} a_j, \\ \lim_{k\to\infty} \frac{1}{n_{2k+1}} \sum_{\ell=1}^{n_{2k+1}} a_{i_\ell} &= \sum_{j=1}^{m} p_j^{q} r_j^{T(q)} a_j.\end{aligned}$$

PROOF. Since a_{i_ℓ} are identically distributed for $n_{2k-1} < \ell \leqslant n_{2k}$, $k \in \mathbb{N}$, we get by the Strong Law of Large Numbers that

$$\lim_{k\to\infty} \frac{1}{\sum_{m=1}^{k} (n_{2m} - n_{2m-1})} \sum_{m=1}^{k} \sum_{\ell=n_{2m-1}+1}^{n_{2m}} a_{i_\ell} = \sum_{j=1}^{m} p_j^{-q} r_j^{T(-q)} a_j.$$

Our assumption on n_k implies that

$$\lim_{k\to\infty} \frac{n_{2k}}{\sum_{m=1}^{k} (n_{2m} - n_{2m-1})} = 1, \tag{5.47}$$

and similarly,

$$\lim_{k\to\infty} \frac{\sum_{m=1}^{k} (n_{2m-1} - n_{2m-2})}{n_{2k}} = 0. \tag{5.48}$$

By (5.48)

$$\left| \frac{1}{n_{2k}} \sum_{m=1}^{k} \sum_{\ell=n_{2m-2}+1}^{n_{2m-1}} a_{i_\ell} \right| \leqslant \frac{\sum_{m=1}^{k} (n_{2m-1} - n_{2m-2})}{n_{2k}} \max_i |a_i| \to 0 \text{ as } k \to \infty.$$

Thus, by decomposing

$$\begin{aligned}\frac{1}{n_{2k}} \sum_{\ell=1}^{n_{2k}} a_{i_\ell} &= \frac{\sum_{m=1}^{k} (n_{2m} - n_{2m-1})}{n_{2k} \sum_{m=1}^{k} (n_{2m} - n_{2m-1})} \sum_{m=1}^{k} \sum_{\ell=n_{2m-1}+1}^{n_{2m}} a_{i_\ell} \\ &+ \frac{1}{n_{2k}} \sum_{m=1}^{k} \sum_{\ell=n_{2m-2}+1}^{n_{2m-1}} a_{i_\ell},\end{aligned}$$

we obtain the first claim of the lemma. The proof of the second equality is identical, so we omit it. □

PROOF OF THEOREM 5.3.1. Let $\widetilde{\nu}_q = \Pi_*(\widetilde{\mu}_q)$. We prove the theorem in two steps. First, we show that

$$\widetilde{\nu}_q(\{x \in \Lambda : \underline{d}_\nu(x) < \overline{d}_\nu(x)\}) = 1, \tag{5.49}$$

then we show that $\widetilde{\nu}_q$ is a *Frostman measure.* That is, for every $q > 0$ there exists $C > 0$ such that for every $x \in \Lambda$ and every $r > 0$,

$$\widetilde{\nu}_q(B(x, r)) \leqslant C r^{d_q}, \tag{5.50}$$

where $d_q = \min\{\alpha_{\min} q + T(q), -q\alpha_{\max} + T(-q)\}$ and $\alpha_{\min}$ and $\alpha_{\max}$ were defined in (5.7). By Lemma 1.4.11,

$$\dim_{\mathrm{H}}\{x \in \Lambda : \underline{d}_\nu(x) < \overline{d}_\nu(x)\} \geqslant d_q$$

for every $q > 0$, and using that $\lim_{q\to 0} T(q) = s$, where s is the similarity dimension of $\mathcal{S}$, we obtain the claim of the theorem, since $\dim_{\mathrm{H}}(\Lambda) = s$ when the SSP holds.

First, we show (5.49). Similarly to the proof of Lemma 5.1.3, for $\widetilde{\nu}_q$-almost every x,

$$\begin{aligned}\overline{d}_\nu(x) &= \limsup_{n\to\infty} \frac{\log \mu([i_1 \dots i_n])}{\log r_{i_1\dots i_n}} \geqslant \limsup_{k\to\infty} \frac{\log p_{i_1\dots i_{n_{2k}}}}{\log r_{i_1\dots i_{n_{2k}}}} \\ &= \limsup_{k\to\infty} \frac{\frac{1}{n_{2k}}\sum_{j=1}^{n_{2k}} \log p_{i_j}}{\frac{1}{n_{2k}}\sum_{j=1}^{n_{2k}} \log r_{i_j}} = \frac{\sum_{j=1}^m p_j^{-q} r_j^{T(-q)} \log p_j}{\sum_{j=1}^m p_j^{-q} r_j^{T(-q)} \log r_j},\end{aligned}$$

where in the last equation we used Lemma 5.3.3. Similarly, for $\widetilde{\nu}_q$-almost every x,

$$\underline{d}_\nu(x) \leqslant \liminf_{k\to\infty} \frac{\log p_{i_1\dots i_{n_{2k+1}}}}{\log r_{i_1\dots i_{n_{2k+1}}}} = \frac{\sum_{j=1}^m p_j^{q} r_j^{T(q)} \log p_j}{\sum_{j=1}^m p_j^{q} r_j^{T(q)} \log r_j}.$$

By Fact 5.1.2, the map $-T'(q) = \frac{\sum_{j=1}^m p_j^q r_j^{T(q)} \log p_j}{\sum_{j=1}^m p_j^q r_j^{T(q)} \log r_j}$ is strictly monotone decreasing, and so $\overline{d}_\nu(x) > \underline{d}_\nu(x)$, for $\widetilde{\nu}_q$-almost every x. This proves (5.49).

Now we turn to the proof of (5.50). Since the SSP holds, there exists $\delta_* := \min\{\operatorname{dist}(\Lambda_i, \Lambda_j) : i \neq j\} > 0$. Fix $x \in \Lambda$ and $0 < r < \delta_*$. Let $\mathbf{i} = (i_1, i_2, \dots)$ be the symbolic code of the point x, that is, $x = \Pi(\mathbf{i})$. Fix $n \geqslant 1$ so that

$$r_{i_1\dots i_n}\delta_* \leqslant r < r_{i_1\dots i_{n-1}}\delta_*.$$

Then $B(x,r) \cap \Lambda = B(x,r) \cap \Lambda_{i_1\dots i_n}$, and so

$$\widetilde{\nu}_q(B(x,r)) \leqslant \widetilde{\nu}_q(\Lambda_{i_1\dots i_n}) = \prod_{j=1}^n z_{i_j}^{q,j}. \tag{5.51}$$

Simple calculations show that for $n_{2k} < j \leqslant n_{2k+1}$,

$$z_i^{q,j} = p_i^q r_i^{T(q)} = r_i^{\frac{\log p_i}{\log r_i} q + T(q)} \leqslant r_i^{\alpha_{\min} q + T(q)} \leqslant r_i^{d_q},$$

and for $n_{2k+1} < j \leqslant n_{2k+2}$, $z_i^{q,j} \leqslant r_i^{-\alpha_{\max} q + T(-q)} \leqslant r_i^{d_q}$. Hence, combining this with (5.51), we obtain

$$\prod_{j=1}^n z_{i_j}^{q,j} \leqslant r_{i_1,\dots,i_n}^{d_q} \leqslant \delta_*^{-d_q} r^{d_q},$$

which proves (5.50). □

CHAPTER 6

Transversality techniques for self-similar IFS

As we have seen in Chapter 1, if the IFS has "heavy overlaps", that is, the Open Set Condition does not hold, or maybe we do not know if it holds, the dimension calculation for self-similar sets becomes much more difficult. Sometimes the Weak Separation Property is satisfied, but this is usually an exception. Thus one may be interested in obtaining *almost sure* results, for "typical" (in the measure-theoretic sense) or "generic" (in the topological sense) systems. In order to make this precise, our IFS should depend on one or more parameters from a certain interval or region. The *transversality method* was developed to handle such situations. The precursor was the work of Falconer [**Fal87**, **Fal88**] where the dependence on parameters was *linear*. The setting of [**Fal87**] and Theorem 6.1.1, which considers a family of self-similar IFS on the line, is particularly transparent. It can be handled by *lifting the IFS* to a higher-dimensional space and projecting it to lines. Then the almost sure results follow by Marstrand–Mattila Projection Theorems 2.3.1. The "transversality" is then manifested by the simple geometric fact (2.23), which in the planar case reduces to the statement that in the plane distinct lines through the origin intersect at a nonzero angle. The first section of this chapter is devoted to Theorem 6.1.1, with a complete proof. The second section deals with *self-similar measures* on the line, still with the translations being the parameters. The questions we address are the calculation of Hausdorff dimension and absolute continuity vs. singularity. Recall that the Hausdorff dimension of a measure is only a lower bound of the dimension of its support, and the measure may be singular even if its support is an interval. In the simple situation of IFS in the first two sections, we can use projection results for measures to obtain almost sure results on absolute continuity. After that we move to the "core" of the transversality method, which was first introduced by Pollicott and Simon [**PS95**] (and in some ways goes back to [**Sim93**]) to investigate also a class of self-similar IFS on the line, but with an important distinction that the *contraction ratio* is used as a parameter. This is more difficult, since the dependence of the IFS on the parameter is now nonlinear. However, it turns out that one can still view the attractor as a "projection", in fact, the natural projection from the symbolic space, and the methods of Kaufman and Mattila from the proofs of classical projection theorems can still be applied, appropriately modified. By now there are many variants of "projection schemes", the parameter space may be multi-dimensional, and the dependence on the parameter may be rather general. For the purposes of exposition, we consider the basic case where there is one parameter—the contraction ratio—but with an arbitrary finite number of maps. This includes the two much-studied classes: expansions with deleted digits and Bernoulli convolutions. The *transversality condition* then becomes, roughly speaking, the condition that the graphs of two functions

$$\lambda \mapsto \Pi_\lambda(\mathbf{i}) \quad \text{and} \quad \lambda \mapsto \Pi_\lambda(\mathbf{j}), \quad \text{for } \mathbf{i} \neq \mathbf{j} \in \Sigma,$$

can only intersect at a nonzero angle (transversally). Here Π_λ is the natural projection corresponding to the parameter λ and $\mathbf{i}, \mathbf{j}$ are elements of the symbolic space. Precise definition is given in Section 6.3. We then use it to prove almost sure results on the dimension/Lebesgue measure of the attractor (self-similar set), as well as those on absolute continuity (with a density in L^2) for self-similar measure on an interval of transversality. In the following Section 6.4 we show how transversality may be verified, with the help of the "$(*)$-function method" which originated in the work of Solomyak [**Sol95**]). In Section 6.5 this is used to give a proof of the Pollicott–Simon Theorem from [**PS95**] on sets with deleted digits and the proof of absolute continuity of Bernoulli convolutions for a.e. parameter in $(1/2, 1)$, which requires some additional "tricks". Section 6.6 contains a "Generalized Pojection Scheme", which is based on an extended version of [**Sol98**]. This scheme "distills" some of the key features of the transversality method, combining [**Sol98**] with [**PS98**]. Then, in Section 6.7 we indicate how this Generalized Pojection Scheme may be used to obtain results on L^q densities for homogeneous self-similar measures (including Bernoulli convolutions), as well as extensions to the nonhomogeneous case, following in part [**Neu01, NW05**], but leaving the details to the reader. In Section 6.7.3 we present a more recent application, with a partial proof: computing the dimension of graphs of Weierstrass-type functions for typical parameters, following [**BBR14**]. The last Section 6.8 of the chapter is devoted to important further developments. These include the work of Peres and Schlag [**PS00a**] on Sobolev smoothness and dimension of exceptions for families of self-similar sets and measures with overlaps, and the more recent breakthrough by Hochman [**Hoc14b, Hoc15**], Shmerkin [**Shm14, Shm19**], and Varjú [**Var19a, Var19b**]. The proofs of these results are beyond the scope of the book, but we describe them in some detail.

6.1. Self-similar sets on the line with overlaps (linear dependence)

Consider the self-similar IFS $\mathcal{F}_{\mathbf{r},\mathbf{t}} = (f_j)_{j=1}^m$ on $\mathbb{R}$, where

$$f_j(x) = r_j x + t_j, \quad 0 < |r_j| < 1 \text{ and } t_j \in \mathbb{R}, \ j = 1, \dots, m, \tag{6.1}$$

and we write $\mathbf{r} = (r_1, \dots, r_m)$, $\mathbf{t} = (t_1, \dots, t_m)$. Denote its attractor by $\Lambda_{\mathbf{r},\mathbf{t}}$, and let $s(\mathbf{r})$ be the similarity dimension.

THEOREM 6.1.1.

(i) $\dim_{\mathrm{H}}(\Lambda_{\mathbf{r},\mathbf{t}}) = \dim_{\mathrm{B}}(\Lambda_{\mathbf{r},\mathbf{t}}) = \min\{1, s(\mathbf{r})\}$ *for* $\mathcal{L}^m$*-a.e.* $\mathbf{t} \in \mathbb{R}^m$*;*
(ii) *if* $s(\mathbf{r}) > 1$*, then* $\mathcal{L}^1(\Lambda_{\mathbf{r},\mathbf{t}}) > 0$ *for* $\mathcal{L}^m$*-a.e.* $\mathbf{t} \in \mathbb{R}^m$.

REMARK 6.1.2.

(a) We do not know if $s(\mathbf{r}) > 1$ implies that $\Lambda_{\mathbf{r},\mathbf{t}}$ has nonempty interior for a.e. $\mathbf{t}$. This is trivially true for $m = 2$, but remains a difficult open problem for $m \geqslant 3$. See Section 7.5 for more on this topic.

(b) As Falconer [**Fal87**] points out, "the m degrees of freedom are slightly illusory". First, observe that by making a change of variable $x = x' - (t_1/r_1)$ we can assume without loss of generality that $t_1 = 0$. Thus, "for $\mathcal{L}^m$-almost every $\mathbf{t} \in \mathbb{R}^m$" in the theorem can be replaced by "for a.e. $\mathbf{t} \in \{0\} \times \mathbb{R}^{m-1}$". Using that

$$\Lambda_{\mathbf{r},c\mathbf{t}} = c\Lambda_{\mathbf{r},\mathbf{t}} \quad \text{for } c \neq 0, \tag{6.2}$$

one can make a further reduction, for instance, to the case when $t_2 = 1$, or choose another normalization.

The theorem is essentially due to Falconer [**Fal87**], with an addition from [**SS02**]. We present a slightly different proof here.

Without loss of generality assume that $|r_1| \geqslant |r_2| \geqslant \cdots \geqslant |r_m|$. If $|r_1|+|r_2| \geqslant 1$, then $\Lambda_{\mathbf{r},\mathbf{t}}$ contains an interval (as long as the fixed points of f_1 and f_2 are distinct, so for all $\mathbf{t}$ outside a hyperplane). If $|r_1| + |r_2| < 1$, then we obtain $\Lambda_{\mathbf{r},\mathbf{t}}$ as a projection of a self-similar set in $\mathbb{R}^d$ satisfying the Open Set Condition, and can use projection theorems. Now for the details.

Observe that $\Lambda_{\mathbf{r},\mathbf{t}}$ contains the attractor of the IFS $(r_j x + t_j)_{j=1}^2$ (in general, if an IFS $\mathcal{F}_2$ is obtained by removing some of the maps from $\mathcal{F}_1$, then their attractors satisfy $\Lambda_1 \supset \Lambda_2$).

LEMMA 6.1.3. *Let $1 > |r_1| \geqslant |r_2| > 0$ and $|r_1| + |r_2| \geqslant 1$, and suppose that $\frac{t_1}{1-r_1} \neq \frac{t_2}{1-r_2}$; in other words, the functions f_1 and f_2 do not have the same fixed point. Then the attractor of (f_1, f_2) is a closed interval of positive length.*

PROOF. By applying an appropriate conjugation, see Remark 1.8.5, we may assume without loss of generality that $t_1 = 0$ and $t_2 = 1$. Let

$$I := \begin{cases} [0, \frac{1}{1-r_2}], & \text{if } r_1 > 0,\ r_2 > 0 \\ [0, 1], & \text{if } r_1 > 0,\ r_2 < 0 \\ [\frac{r_1}{1-r_2}, \frac{1}{1-r_2}], & \text{if } r_1 < 0,\ r_2 > 0 \\ [\frac{r_1}{1-r_1 r_2}, \frac{1}{1-r_1 r_2}], & \text{if } r_1 < 0,\ r_2 < 0. \end{cases}$$

It is an easy exercise to check directly that $I = f_1(I) \cup f_2(I)$, hence I is the attractor. $\square$

PROOF OF THEOREM 6.1.1. In view of Lemma 6.1.3, in the case $|r_1| + |r_2| \geqslant 1$ the theorem is proved, since $\Lambda_{\mathbf{r},\mathbf{t}}$ contains an interval for a.e. $\mathbf{t}$ (observe that $s(\mathbf{r}) \geqslant 1$ in this case). Assuming $|r_1| + |r_2| \leqslant 1$, we will "lift" the IFS to the space $\mathbb{R}^{m-1}$ (although we already know what happens when $|r_1| + |r_2| = 1$, it is convenient to include this case for later use). For simplicity, assume that $r_j > 0$ for all $j = 1, \dots, m$; the general case is left as an exercise.

Consider the IFS $\widetilde{\mathcal{F}}_{\mathbf{r}} = (\widetilde{f}_j)_{j=1}^m$ in $\mathbb{R}^{m-1}$, with $\mathbf{r} = (r_1, \dots, r_m)$, defined by

$$\widetilde{f}_1(\mathbf{x}) = r_1 \mathbf{x},\ \widetilde{f}_j(\mathbf{x}) = r_j \mathbf{x} + (1 - r_j)\mathbf{e}_{j-1},\ j = 2, \dots, m, \tag{6.3}$$

where $\mathbf{e}_j$ is the j-th unit vector in $\mathbb{R}^{m-1}$.

Recall that $\text{proj}_{\mathbf{v}}$ denotes the orthogonal projection from $\mathbb{R}^{m-1}$ to the line through the origin determined by the vector $\mathbf{v}$, that is, $\text{proj}_{\mathbf{v}}(\mathbf{x}) = \langle \mathbf{x}, \mathbf{v} \rangle$ for $\|\mathbf{v}\| = 1$. Clearly, we have the following commutative diagram

$$\begin{array}{ccc} \mathbb{R}^{m-1} & \xrightarrow{\ \widetilde{f}_j\ } & \mathbb{R}^{m-1} \\ \text{proj}_{\mathbf{v}} \big\downarrow & & \big\downarrow \text{proj}_{\mathbf{v}} \\ \mathbb{R} & \xrightarrow{\ f_j\ } & \mathbb{R} \end{array}$$

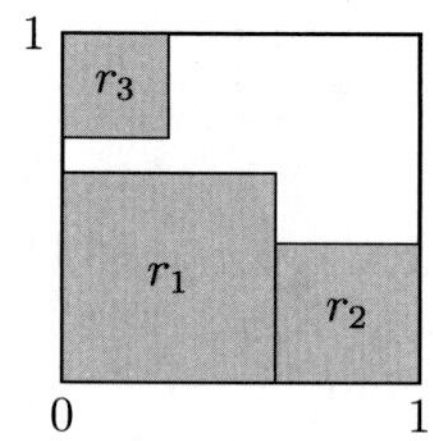

FIGURE 6.1. Images of $V = (0,1)^2$ under the IFS $\widetilde{\mathcal{F}}_r$ for $m = 3$.

whenever $f_j(x) = r_j x + t_j$ and

$$t_1 = 0,\ t_j = (1 - r_j)v_{j-1},\ j = 2, \ldots, m. \tag{6.4}$$

It follows that

$$\Lambda_{\mathbf{r},\mathbf{t}} = \mathrm{proj}_{\mathbf{v}}(\widetilde{\Lambda}_{\mathbf{r}}) \quad \text{for } \mathbf{t} = (t_j)_{j=1}^m \text{ from (6.4)}, \tag{6.5}$$

where $\widetilde{\Lambda}_{\mathbf{r}}$ is the attractor of the IFS $\widetilde{\mathcal{F}}_{\mathbf{r}}$.

LEMMA 6.1.4. *The IFS $\widetilde{\mathcal{F}}_{\mathbf{r}}$ defined in* (6.3) *satisfies the Open Set Condition with the open set $V = (0,1)^{m-1}$ (the open unit cube).*

PROOF. To verify that $\widetilde{f}_j(V)$ and $\widetilde{f}_k(V)$ are disjoint for $j < k$ we note that

$$\widetilde{f}_j(V) \subset \{\mathbf{x} \in \mathbb{R}^{m-1} : (\mathbf{x}, \mathbf{e}_k) < r_j\},\ \widetilde{f}_k(V) \subset \{\mathbf{x} \in \mathbb{R}^{m-1} : (\mathbf{x}, \mathbf{e}_k) > 1 - r_k\},$$

and observe that $r_j \leqslant 1 - r_k$ by the assumptions $r_1 + r_2 \leqslant 1$ and $r_1 \geqslant r_2 \geqslant \cdots \geqslant r_m \geqslant 1$. □

Now we can conclude the proof of Theorem 6.1.1. By Theorem 1.5.3, $\dim_H(\widetilde{\Lambda}_{\mathbf{r}}) = s(\mathbf{r})$. By Projection Theorem 2.3.1, we have

$$\dim_H(\mathrm{proj}_{\mathbf{v}}(\widetilde{\Lambda}_{\mathbf{r}})) = \min\{1, s(\mathbf{r})\} \quad \text{for a.e. } \mathbf{v} \in \mathbb{S}^{m-2},$$

and if $s(\mathbf{r}) > 1$, then $\mathcal{L}^1(\mathrm{proj}_{\mathbf{v}}(\widetilde{\Lambda}_{\mathbf{r}})) > 0$ for a.e. $\mathbf{v} \in \mathbb{S}^{m-2}$. By (6.5), we obtain the claim of Theorem 6.1.1 for a.e. $\mathbf{t}$ of the form (6.4), and in view of Remark 6.1.2(b), we obtain the claim for a.e. $\mathbf{t} \in \mathbb{R}^m$. □

6.2. Self-similar measures on the line with overlaps (linear dependence)

Next we turn to self-similar measures for the same family of IFS, given by (6.1). Let $\nu^{\mathbf{p}}_{\mathbf{r},\mathbf{t}}$ be the stationary measure for the IFS $\mathcal{F}_{\mathbf{r},\mathbf{t}}$ with the probability vector $\mathbf{p} = (p_1, \ldots, p_m)$. Let

$$h_{\mathbf{p}} = -\sum_{j=1}^m p_j \log p_j$$

be the entropy of the Bernoulli measure $\mu_{\mathbf{p}} = \mathbf{p}^{\mathbb{N}}$, and let

$$\chi^{\mathbf{p}}_{\mathbf{r}} := -\sum_{j=1}^m p_j \log r_j$$

be the Lyapunov exponent of the IFS.

THEOREM 6.2.1. *Suppose that* $|r_1| + |r_2| \leqslant 1$. *Then*
(i)

$$\dim_H(\nu^{\mathbf{p}}_{\mathbf{r},\mathbf{t}}) = \min\left\{1, \frac{h_{\mathbf{p}}}{\chi^{\mathbf{p}}_{\mathbf{r}}}\right\} \quad \text{for a.e. } \mathbf{t} \in \mathbb{R}^m. \tag{6.6}$$

(ii) *If* $\frac{h_{\mathbf{p}}}{\chi^{\mathbf{p}}_{\mathbf{r}}} > 1$ *then* $\nu^{\mathbf{p}}_{\mathbf{r},\mathbf{t}} \ll \mathcal{L}^1$ *for a.e.* $\mathbf{t} \in \mathbb{R}^m$.

Proof sketch. We use the same argument as in the proof of Theorem 6.1.1, lifting the IFS to the IFS $\widetilde{\mathcal{F}}_{\mathbf{r}}$ in $\mathbb{R}^{m-1}$. Under the Open Set Condition, the self-similar measure for the lifted IFS, with the probability vector $\mathbf{p}$, has dimension $\frac{h_{\mathbf{p}}}{\chi^{\mathbf{p}}_{\mathbf{r}}}$. Then we apply the Projection Theorem 2.3.3. □

The case $|r_1| + |r_2| > 1$ is more difficult for measures than for sets, since having an interval in the support of the measure does not imply that the measure is absolutely continuous. Here we consider only the homogeneous case $r_j = r$, $j = 1, \dots, m$, and after some preliminary remarks also assume the "natural" choice of weights $p_j = m^{-1}$, $j = 1, \dots, m$. For a stronger statement (with an exceptional set of zero Hausdorff dimension) in the case of general weights, see [**SS16**, Theorem B]. It should be pointed out that in the homogeneous case the stationary measure for the IFS is an **infinite convolution measure**, and this property is exploited repeatedly. In the nonhomogeneous case, a strong result was recently obtained by Käenmäki and Orponen [**KO20**], see the discussion in Section 6.8.2.2.

Let $\nu^{\mathbf{p}}_{r,\mathbf{t}}$ be the stationary measure for the IFS $(rx + t_j)_{j=1}^m$, with the probability vector $\mathbf{p} = (p_1, \dots, p_m)$, that is,

$$\nu^{\mathbf{p}}_{r,\mathbf{t}} = (\Pi_{r,\mathbf{t}})_*\mu_{\mathbf{p}},$$

where $\mu_{\mathbf{p}} = \mathbf{p}^{\mathbb{N}}$ is the product (Bernoulli) measure on $\Sigma = \{1, \dots, m\}^{\mathbb{N}}$ and $\Pi_{r,\mathbf{t}} : \Sigma \to \Lambda_{r,\mathbf{t}}$ is the natural projection, given by

$$\Pi_{r,\mathbf{t}}(\mathbf{i}) = \sum_{n=1}^{\infty} t_{i_n} r^{n-1}.$$

For $A \subset \mathbb{N}$ consider $\Sigma_A := \{1, \dots, m\}^A$ and the projection

$$\Pi_{r,\mathbf{t},A} : \Sigma_A \to \mathbb{R}, \quad \Pi_{r,\mathbf{t},A}(\mathbf{i}) = \sum_{n \in A} t_{i_n} r^{n-1}.$$

Further, let

$$\nu^{\mathbf{p}}_{r,\mathbf{t},A} = (\Pi_{r,\mathbf{t},A})_*\mathbf{p}^A.$$

Let $A^c = \mathbb{N} \backslash A$, then $\Sigma = \Sigma_A \times \Sigma_{A^c}$, $\mu = \mathbf{p}^{\mathbb{N}} = \mathbf{p}^A \times \mathbf{p}^{A^c}$, and

$$\Pi_{r,\mathbf{t}}(\mathbf{i}) = \Pi_{r,\mathbf{t},A}(\mathbf{i}|_A) + \Pi_{r,\mathbf{t},A^c}(\mathbf{i}|_{A^c}),$$

where $\mathbf{i}|_A$ denotes the subsequence of $\mathbf{i} = (i_n)_{n \geqslant 1}$ corresponding to $n \in A$.

LEMMA 6.2.2. *Let* (X_1, μ_1) *and* (X_2, μ_2) *be two finite measure spaces, and let* $f_j : X_j \to \mathbb{R}$, *for* $j = 1, 2$, *be bounded measurable functions. Let* $\nu_j = \mu_j \circ f_j^{-1}$ *be the corresponding push-forward measures on the line. Further, let* $F : X_1 \times X_2 \to \mathbb{R}$ *be given by* $F(x_1, x_2) = f_1(x_1) + f_2(x_2)$. *Then*

$$(\mu_1 \times \mu_2) \circ F^{-1} = \nu_1 * \nu_2.$$

PROOF. This is an easy computation using Fubini's Theorem. In the language of Probability, this says that the distribution of the sum of two independent random variables is the convolution of their distributions. □

COROLLARY 6.2.3. *For any $r \in (0,1), \mathbf{t}, \mathbf{p}$, and $A \subset \mathbb{N}$ we have*

$$\nu^{\mathbf{p}}_{r,\mathbf{t}} = \nu^{\mathbf{p}}_{r,\mathbf{t},A} * \nu^{\mathbf{p}}_{r,\mathbf{t},A^c}.$$

Denote $\nu_{r,\mathbf{t}} = \nu^{\mathbf{p}}_{r,\mathbf{t}}$, where $\mathbf{p} = (\frac{1}{m}, \ldots, \frac{1}{m})$ is the uniform probability vector (which gives the "natural" self-similar measure in the homogenous case). Then $h_{\mathbf{p}} = \log m$ and $\chi^{\mathbf{p}}_{\mathbf{r}} = \log(1/r)$, so the threshold for absolute continuity of $\nu_{r,\mathbf{t}}$ should be at $r = 1/m$.

THEOREM 6.2.4. *Let $m \geqslant 3$, and suppose that $|r| > \frac{1}{m}$. Then*

$$\nu_{r,\mathbf{t}} \ll \mathcal{L}^1 \quad \textit{for a.e. } \mathbf{t} \in \mathbb{R}^m.$$

We note that the statement of the theorem is false for $m = 2$; this will be discussed later.

PROOF. By Theorem 6.2.1, we already know the result for $|r| \leqslant 1/2$. Since our IFS is homogeneous, we can reduce the case of $-r$ to r (this is an easy exercise), so we assume without loss of generality that $r > 0$.

First suppose that $m \geqslant 4$; this case is easier to handle. We have

$$\nu_{r,\mathbf{t}} = \nu_{r,\mathbf{t},2\mathbb{N}} * \nu_{r,\mathbf{t},2\mathbb{N}-1}.$$

Moreover, it is immediate from the definitions that

$$\nu_{r,\mathbf{t},2\mathbb{N}} = \nu_{r^2,\mathbf{t}}.$$

If $r \in (\frac{1}{2}, 2^{-1/2}]$, then $r^2 \in (\frac{1}{4}, \frac{1}{2}]$, so $r^2 > 1/m$, and by Theorem 6.2.1, $\nu_{r^2,\mathbf{t}} \ll \mathcal{L}^1$ for a.e. $\mathbf{t}$. Now we use the following standard

LEMMA 6.2.5. *Let ν_1, ν_2 be finite Borel measures on $\mathbb{R}$. If $\nu_1 \ll \mathcal{L}^1$ then $\nu_1 * \nu_2 \ll \mathcal{L}^1$.*

Thus, we obtain that $\nu_{r,\mathbf{t}} \ll \mathcal{L}^1$ for a.e. $\mathbf{t}$, when $r \in (\frac{1}{2}, 2^{-1/2}]$. More generally, for $k \geqslant 2$, we have

$$\nu_{r,\mathbf{t}} = \nu_{r,\mathbf{t},k\mathbb{N}} * \nu_{r,\mathbf{t},(k\mathbb{N})^c}, \quad \nu_{r,\mathbf{t},k\mathbb{N}} = \nu_{r^k,\mathbf{t}}.$$

If $r \in (2^{-1/(k-1)}, 2^{-1/k}]$, then $r^k \in (2^{-k/(k-1)}, \frac{1}{2}] \subset (m^{-1}, \frac{1}{2}]$, so by Theorem 6.2.1, $\nu_{r^k,\mathbf{t}} \ll \mathcal{L}^1$ for a.e. $\mathbf{t}$, and hence $\nu_{r,\mathbf{t}} \ll \mathcal{L}^1$ for a.e. $\mathbf{t}$ by Lemma 6.2.5. Since

$$\bigcup_{k \geqslant 2} (2^{-1/(k-1)}, 2^{-1/k}] = (2^{-1}, 1),$$

the proof of the theorem for $m \geqslant 4$ is complete.

The remaining case $m = 3$ is proved similarly, but is a bit more subtle. We write

$$\nu_{r,\mathbf{t}} = \nu_{r,\mathbf{t},3\mathbb{N}} * \nu_{r,\mathbf{t},(3\mathbb{N})^c}$$

and are going to analyze when $\nu_{r,\mathbf{t},(3\mathbb{N})^c}$ is absolutely continuous.

We can "lift" the IFS $\mathcal{F}$ to $\mathbb{R}^2$ similarly to the proof of Theorem 6.2.1 and consider the IFS $\widetilde{\mathcal{F}} = \left(\widetilde{f}_j\right)_{j=1}^3$ given by

$$\begin{aligned}
\widetilde{f}_1(\mathbf{x}) &= r\mathbf{x}, \\
\widetilde{f}_2(\mathbf{x}) &= r\mathbf{x} + \mathbf{e}_1, \\
\widetilde{f}_3(\mathbf{x}) &= r\mathbf{x} + \mathbf{e}_2.
\end{aligned}$$

For $r > 1/2$, this IFS will have overlaps, but for certain r we can get rid of them by throwing away every third term of the sequence. To this end, consider the following IFS $\widetilde{\mathcal{G}}_r = (\widetilde{g}_{ij})_{i,j=1}^3$:

$$\begin{aligned}
\widetilde{g}_{11}(\mathbf{x}) &= r^3\mathbf{x},\\
\widetilde{g}_{12}(\mathbf{x}) &= r^3\mathbf{x} + r\mathbf{e}_1,\\
\widetilde{g}_{13}(\mathbf{x}) &= r^3\mathbf{x} + r\mathbf{e}_2,\\
\widetilde{g}_{21}(\mathbf{x}) &= r^3\mathbf{x} + \mathbf{e}_1,\\
\widetilde{g}_{22}(\mathbf{x}) &= r^3\mathbf{x} + \mathbf{e}_1 + r\mathbf{e}_1,\\
\widetilde{g}_{23}(\mathbf{x}) &= r^3\mathbf{x} + \mathbf{e}_1 + r\mathbf{e}_2,\\
\widetilde{g}_{31}(\mathbf{x}) &= r^3\mathbf{x} + \mathbf{e}_2,\\
\widetilde{g}_{32}(\mathbf{x}) &= r^3\mathbf{x} + \mathbf{e}_2 + r\mathbf{e}_1,\\
\widetilde{g}_{33}(\mathbf{x}) &= r^3\mathbf{x} + \mathbf{e}_2 + r\mathbf{e}_2.
\end{aligned}$$

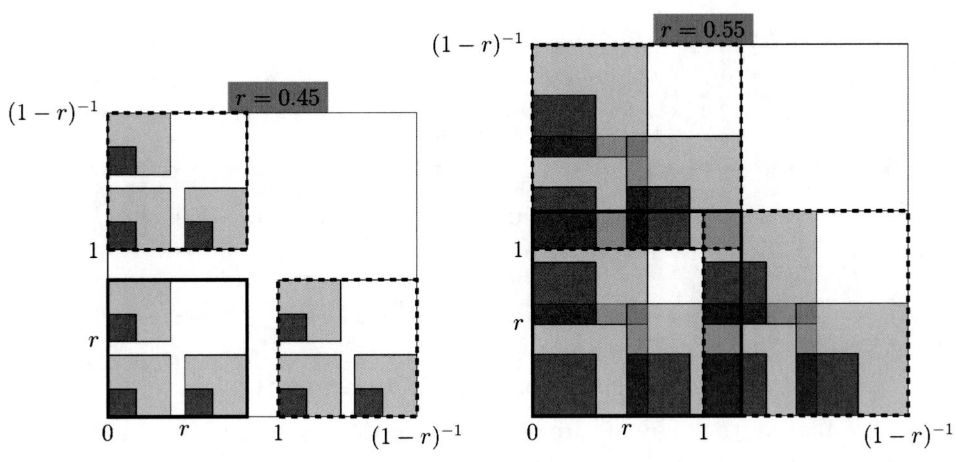

FIGURE 6.2. The darker gray squares are the images of $Q = \left(0, \frac{1}{1-r}\right)^2$ under the IFS $\widetilde{\mathcal{G}}_r$

LEMMA 6.2.6. *The IFS $\widetilde{\mathcal{G}}$ has the Strong Separation Property whenever*

$$1 < r + \frac{r^3(1+r)}{1-r^3}, \tag{6.7}$$

which holds for $r < 0.589$.

Proof of the lemma. Let $V = [0, \frac{1+r}{1-r^3}]^2$ (the closed square). It is easy to see that $\widetilde{g}_{ij}(V) \subset V$. We have

$$\widetilde{g}_{12}(V) \subset \{\mathbf{x} \in \mathbb{R}^2 : \langle \mathbf{x}, \mathbf{e}_1 \rangle \leqslant r^3(1+r)(1-r^3)^{-1} + r\},$$
$$\widetilde{g}_{21}(V) \subset \{\mathbf{x} \in \mathbb{R}^2 : \langle \mathbf{x}, \mathbf{e}_1 \rangle \geqslant 1\},$$

hence $\widetilde{g}_{12}(V) \cap \widetilde{g}_{21}(V) = \varnothing$ whenever (6.7) holds. The disjointness of other images follows by self-similarity. $\square$

Let $\widetilde{\eta}_r$ be the stationary measure for the IFS $\widetilde{\mathcal{G}}_r$ with the probability vector $(\frac{1}{9}, \ldots, \frac{1}{9})$. It follows from Lemma 6.2.6 and Corollary 1.9.11 that

$$\dim_{\mathrm{H}}(\widetilde{\eta}_r) = \frac{\log 9}{\log(1/r^3)},$$

which is greater than one for $r > 3^{-2/3} > 0.48$. It is clear that

$$\nu_{r,\mathbf{t},(3\mathbb{N})^c} = (\mathrm{proj}_{\mathbf{v}})_*\widetilde{\eta}_r \quad \text{for } \mathbf{v} \in \mathbb{S}^1,\ \mathbf{t} = (0, v_1, v_2).$$

Therefore, by Theorem 2.3.3, Lemma 6.2.5, and Remark 6.1.2,

$$r \in [0.5, 0.589) \implies \nu_{r,\mathbf{t}} \ll \nu_{r,\mathbf{t},(3\mathbb{N})^c} \ll \mathcal{L}^1 \quad \text{for a.e. } \mathbf{t} \in \mathbb{R}^3.$$

Next we repeat the construction used above for $m \geqslant 4$ and write

$$\nu_{r,\mathbf{t}} = \nu_{r,\mathbf{t},2\mathbb{N}} * \nu_{r,\mathbf{t},2\mathbb{N}-1}. \tag{6.8}$$

If $r \in (3^{-1/2}, 2^{-1/2})$, then $r^2 > 1/3 = 1/m$ and $r^2 < 1/2$, so we can apply Theorem 6.2.1 to obtain

$$r \in (3^{-1/2}, 2^{-1/2}) \implies \nu_{r,\mathbf{t}} \ll \nu_{r,\mathbf{t},2\mathbb{N}} = \nu_{r^2,\mathbf{t}} \ll \mathcal{L}^1 \quad \text{for a.e. } \mathbf{t} \in \mathbb{R}^3.$$

Since $3^{-1/2} \approx 0.577 < 0.589$, we have obtained the desired result for all $r \in [\frac{1}{2}, 2^{-1})$. Now, repeating the argument based on (6.8) inductively, we obtain it for all $r \in \bigcup_{k=1}^{\infty} [2^{-1/k}, 2^{-1/2k}] = [\frac{1}{2}, 1)$ completing the proof. □

6.3. Self-similar sets and measures on the line (nonlinear dependence)

Consider the family of homogeneous self-similar IFS on the line

$$\mathcal{F}_{\lambda,D} := (S_1^\lambda(x) = \lambda x + d_1, \ldots, S_m^\lambda(x) = \lambda x + d_m) \tag{6.9}$$

for $m \geqslant 2$ and $\lambda \in (0,1)$. We emphasize that $D = \{d_1, \ldots, d_m\} \subset \mathbb{R}$, called the "digit set", is fixed, and

the contraction ratio λ is a parameter.

We assume that all the digits d_j are distinct; they do not have to be integers. The attractor $\mathcal{C}_\lambda$ can be explicitly written as

$$\mathcal{C}_\lambda = \Big\{ \sum_{n=0}^{\infty} a_n \lambda^n : a_n \in D \Big\}, \tag{6.10}$$

the set of all expansions in "base" λ with the digits in the set D. This follows from the uniqueness of the attractor and the observation that $\mathcal{C}_\lambda = \bigcup_{j=1}^m S_j^\lambda(\mathcal{C}_\lambda)$, which is immediate from (6.10).

The natural projection from the symbolic space $\Sigma = \{1, \ldots, m\}^{\mathbb{N}}$ to $\mathcal{C}_\lambda$ will be denoted by Π_λ to emphasize dependence on parameter. For our IFS it takes a particularly simple form:

$$\Pi_\lambda(\mathbf{i}) = \sum_{j=1}^{\infty} d_{i_j} \lambda^{j-1} \quad \text{for } \mathbf{i} = (i_1, i_2, \ldots) \in \Sigma. \tag{6.11}$$

Recall that the IFS (6.9) has the Strong Separation Property (SSP) when $(d_i + \lambda\mathcal{C}_\lambda) \cap (d_j + \lambda\mathcal{C}_\lambda) = \varnothing$ for all $i \neq j$. It follows the SSP holds for λ in an interval $J \subset (0,1)$ if and only if

the function $\lambda \mapsto \Pi_\lambda(\mathbf{i}) - \Pi_\lambda(\mathbf{j})$ has no zeros in J for $\mathbf{i} \neq \mathbf{j}$ in Σ.

Now we define the **transversality condition** introduced in [**PS95**], and in some ways going back to [**Sim93**].

DEFINITION 6.3.1. We say that the IFS family (6.9) satisfies the *transversality condition* on an interval $J \subset [0,1)$ if the function

$$\lambda \mapsto \Pi_\lambda(\mathbf{i}) - \Pi_\lambda(\mathbf{j}) \text{ has no double zeros in } J \text{ for } \mathbf{i} \neq \mathbf{j} \text{ in } \Sigma.$$

We will also say that J is a transversality interval for the IFS family (6.9).

Here a "double zero" for f means λ_0 such that $f(\lambda_0) = f'(\lambda_0) = 0$, so it includes zeros of higher order as well. An equivalent way to state the transversality condition is to say that the graphs of the functions $\lambda \mapsto \Pi_\lambda(\mathbf{i})$ and $\lambda \mapsto \Pi_\lambda(\mathbf{j})$ defined on J intersect at nonzero angles (transversally), if at all.

Now we can state the result about the dimension and Lebesgue measure of the attractor; it should be compared to Theorem 6.1.1.

THEOREM 6.3.2. *Suppose that the IFS family* (6.9) *satisfies the transversality condition on an interval* $J \subset [0,1)$. *Let* $s(\lambda) = \frac{\log m}{-\log \lambda}$ *be the similarity dimension. Then*

(i) $\dim_H(\mathcal{C}_\lambda) = \min\{1, s(\lambda)\}$ *for* $\mathcal{L}^1$*-a.e.* $\lambda \in J$;
(ii) $\mathcal{L}^1(\mathcal{C}_\lambda) > 0$ *for* $\mathcal{L}^1$*-a.e.* $\lambda \in J \cap (1/m, 1)$.

Next we state the result about absolute continuity of a self-similar measure. In fact, part (ii) of Theorem 6.3.2 will follow from it.

A self-similar measure for the homogeneous IFS $\mathcal{F}_{\lambda,D}$, with the probability vector $\mathbf{p} = (p_1, \ldots, p_m)$, is given by

$$\nu_{\lambda,D}^{\mathbf{p}} := \mu_{\mathbf{p}} \circ \Pi_\lambda^{-1}$$

where $\mu_{\mathbf{p}} = \mathbf{p}^{\mathbb{N}}$. If a measure ν on $\mathbb{R}$ is absolutely continuous with respect to the Lebesgue measure $\mathcal{L}^1$, we will denote by $\frac{d\nu}{dx}$ its Radon-Nikodym derivative, and will sometimes call it the "density of ν".

THEOREM 6.3.3. *Suppose that the IFS family* (6.9) *satisfies the transversality condition on an interval* $J \subset [0,1)$. *Then* $\nu_{\lambda,D}^{\mathbf{p}} \ll \mathcal{L}^1$ *for* $\mathcal{L}^1$*-a.e.* $\lambda \in J \cap (\sum_{i=1}^m p_i^2, 1)$. *Moreover, the Radon-Nikodym derivative* $\frac{d\nu_{\lambda,D}^{\mathbf{p}}}{dx}$ *belongs to* $L^2(\mathbb{R})$ *for* $\mathcal{L}^1$*-a.e.* $\lambda \in J \cap (\sum_{i=1}^m p_i^2, 1)$.

The "natural measure" on the attractor of the IFS is obtained when all the probabilities are equal $p_i = 1/m$. We denote it $\nu_{\lambda,D}$, omitting the probability vector; thus

$$\nu_{\lambda,D} = \mu \circ \Pi_\lambda^{-1}, \quad \text{where } \mu = (1/m, \ldots, 1/m)^{\mathbb{N}}. \tag{6.12}$$

Then we have $\sum_{i=1}^m p_i^2 = 1/m$, and so $\nu_{\lambda,D} \ll \mathcal{L}^1$ for a.e. $\lambda \in J \cap (1/m, 1)$, whence $\mathcal{L}^1(\mathcal{C}_\lambda) > 0$ for such λ, implying Theorem 6.3.2(ii). However, for future applications it is important to have the flexibility of choosing an arbitrary probability vector.

Before proving the theorems, we need to derive some consequences of the transversality condition. Let $D' := D - D$, and

$$\mathcal{B}_{D'} := \Big\{ \sum_{n=0}^\infty a_n x^n : \ a_n \in D', \ a_0 \neq 0 \Big\}. \tag{6.13}$$

The reason we consider this class of power series is that $\lambda \mapsto \Pi_\lambda(\mathbf{i}) - \Pi_\lambda(\mathbf{j})$ belongs to $\mathcal{B}_{D'}$, assuming $i_1 \neq j_1$. More generally, denote by $\mathbf{i} \wedge \mathbf{j}$ the common initial segment of the sequences $\mathbf{i}$ and $\mathbf{j}$, so that

$$|\mathbf{i} \wedge \mathbf{j}| := \min\{k \geqslant 1 : \ i_k \neq j_k\} - 1.$$

Then

$$\Pi_\lambda(\mathbf{i}) - \Pi_\lambda(\mathbf{j}) = \lambda^{|\mathbf{i}\wedge\mathbf{j}|} f(\lambda), \quad \text{where } f \in \mathcal{B}_{D'}.$$

LEMMA 6.3.4. *A closed interval $J \subset [0,1)$ is a transversality interval for the IFS family* (6.9) *if and only if there exists $\delta > 0$ such that for all $f \in \mathcal{B}_{D'}$ and all $\lambda \in J$,*

$$|f(\lambda)| \leqslant \delta \implies |f'(\lambda)| \geqslant \delta. \tag{6.14}$$

This was the form in which transversality appeared in [**PS95**, Lemma 1].

PROOF. One direction, the sufficiency of (6.14) is clear: if the transversality fails, then there exists $f \in \mathcal{B}_{D'}$ with a double zero in J, contradicting (6.14).

Conversely, suppose that the condition in the corollary fails. Then we can find a sequence $\delta_n \to 0$ and a sequence $f_n \in \mathcal{B}_{D'}$ such that $|f(\lambda_n)| \leqslant \delta_n$, $|f'(\lambda_n)| \leqslant \delta_n$, for some $\lambda_n \in J$. Since J is closed, we can assume that $\lambda_n \to \lambda \in J$ without loss of generality. Further, we can find a subsequence f_{n_k} which converges coefficient-wise to some power series f, and the definition of $\mathcal{B}_{D'}$ implies that $f \in \mathcal{B}_{D'}$ as well. All the power series in $\mathcal{B}_{D'}$ converge absolutely in $(0,1)$ with all derivatives, which easily implies that $f(\lambda) = f'(\lambda) = 0$. This contradicts the transversality condition, and the proof is complete. □

LEMMA 6.3.5. *Suppose that the IFS family* (6.9) *satisfies the transversality condition on a closed interval $[\lambda_0, \lambda_1] \subset [0,1)$. Then there exists a constant $C_1 \geqslant 1$ such that*

$$\mathcal{L}^1\{\lambda \in [\lambda_0,\lambda_1] : \ |\Pi_\lambda(\mathbf{i}) - \Pi_\lambda(\mathbf{j})| \leqslant r\} \leqslant C_1 r \lambda_0^{-|\mathbf{i}\wedge\mathbf{j}|} \quad \textit{for all } r > 0. \tag{6.15}$$

PROOF. Let $k = |\mathbf{i} \wedge \mathbf{j}|$. We have $\Pi_\lambda(\mathbf{i}) - \Pi_\lambda(\mathbf{j}) = \lambda^k f(\lambda)$ for some $f \in \mathcal{B}_{D'}$. If we show that

$$\mathcal{L}^1\{\lambda \in [\lambda_0,\lambda_1] : \ |f(\lambda)| \leqslant \rho\} \leqslant C_1 \rho \ \text{ for all } \rho > 0 \tag{6.16}$$

and the constant C_1 independent of f, then the desired estimate will follow by taking $\rho = \lambda_0^{-k} r$. Let $\delta > 0$ be from Lemma 6.3.4. If $\rho \geqslant \delta$, then (6.16) holds with $C_1 = 1/\delta$. Otherwise,

$$\{\lambda \in [\lambda_0,\lambda_1] : \ |f(\lambda)| \leqslant \rho\} \subset \Psi_\delta := \{\lambda \in [\lambda_0,\lambda_1] : \ |f(\lambda)| \leqslant \delta\}.$$

The property (6.14) implies that Ψ_δ is a union of intervals of monotonicity for f, and on these intervals the graph of f has a slope at least δ in absolute value. Thus, in each of these intervals, the portion where $|f(\lambda)| < \rho$ has Lebesgue measure at most $2\rho/\delta$. On the other hand, $|f'(\lambda)|$ has a uniform upper bound on $[\lambda_0, \lambda_1]$, equal to $C' := \max_{i,j} |d_i - d_j| \cdot (1-\lambda_1)^{-2}$, hence each of these intervals of monotonicity (except maybe the first and the last one) has length at least $2\delta/C'$. Thus, the number of these intervals is not greater than $C'/(2\delta) + 2$, and we obtain

$$\mathcal{L}^1\{\lambda \in [\lambda_0,\lambda_1] : \ |f(\lambda)| \leqslant \rho\} \leqslant (C'/(2\delta) + 2)(2\rho/\delta).$$

This proves (6.16). □

PROOF OF THEOREM 6.3.2(i). The upper bound for $\dim_H(\mathcal{C}_\lambda)$ holds for all λ by Lemma 1.5.4, so we can focus on the lower bound. Moreover, we can assume that $s(\lambda) < 1$, that is, $\lambda < 1/m$, since otherwise the dimension formula will follow from part (ii).

Clearly, it suffices to show that, for an arbitrary $\varepsilon > 0$, we have $\dim_H(\mathcal{C}_\lambda) \geqslant s(\lambda) - \varepsilon$ for $\mathcal{L}^1$-a.e. $\lambda \in J \cap (0, 1/m)$, and this will follow if we prove that for every $\lambda' \in \mathrm{Int}(J) \cap (0, 1/m)$ there exists a neighborhood $U = (\lambda_0, \lambda_1) \ni \lambda'$ such that $\dim_H(\mathcal{C}_\lambda) \geqslant s(\lambda') - \varepsilon$ for $\mathcal{L}^1$-a.e. $\lambda \in U$. To this end, fix $\varepsilon > 0$ and $\lambda' \in \mathrm{Int}(J) \cap (0, 1/m)$, and choose a neighborhood $U = (\lambda_0, \lambda_1) \subset \mathrm{Int}(J) \cap (0, 1/m)$ containing λ', such that

$$s(\lambda_0) > s' := s(\lambda') - \varepsilon. \tag{6.17}$$

Now we follow the scheme reminiscent of Kaufman's proof of Theorem 2.3.1(i) and consider the integral

$$\mathcal{J}_1 := \int_U \iint_{\mathbb{R}^2} |x - y|^{-s'} \, d\nu_{\lambda,D}(x) \, d\nu_{\lambda,D}(y) \, d\lambda = \int_U \mathcal{E}_{s'}(\nu_{\lambda,D}) \, d\lambda, \tag{6.18}$$

where $\nu_{\lambda,D}$ is the natural measure on the attractor $\mathcal{C}_\lambda$, see (6.12), and $\mathcal{E}_{s'}(\nu_{\lambda,D})$ is its s'-energy. If we show that $\mathcal{J}_1 < \infty$, then for a.e. $\lambda \in U$, the energy $\mathcal{E}_{s'}(\nu_{\lambda,D})$ is finite, hence $\dim_H(\mathcal{C}_\lambda) \geqslant s'$ by the potential-theoretic characterization of Hausdorff dimension (Theorem 1.4.16), as desired. We have

$$\mathcal{E}_{s'}(\nu_{\lambda,D}) = \iint_{\Sigma^2} |\Pi_\lambda(\mathbf{i}) - \Pi_\lambda(\mathbf{j})|^{-s'} \, d\mu(\mathbf{i}) \, d\mu(\mathbf{j}).$$

by the definition of the natural measure $\nu_{\lambda,D}$. Thus,

$$\mathcal{J}_1 = \iint_{\Sigma^2} \Big(\int_{\lambda_0}^{\lambda_1} |\Pi_\lambda(\mathbf{i}) - \Pi_\lambda(\mathbf{j})|^{-s'} \, d\lambda \Big) d\mu(\mathbf{i}) \, d\mu(\mathbf{j}),$$

exchanging the order of integration. Next, in view of Theorem B.5.1,

$$\begin{aligned}
&\int_{\lambda_0}^{\lambda_1} |\Pi_\lambda(\mathbf{i}) - \Pi_\lambda(\mathbf{j})|^{-s'} \, d\lambda \\
&= \int_0^\infty \mathcal{L}^1\{\lambda \in [\lambda_0, \lambda_1] : |\Pi_\lambda(\mathbf{i}) - \Pi_\lambda(\mathbf{j})|^{-s'} \geqslant t\} \, dt \\
&\leqslant C_1 \int_0^\infty \min\{1, \lambda_0^{-|\mathbf{i} \wedge \mathbf{j}|} t^{-1/s'}\} \, dt \\
&= C_1 \Big(\int_0^{\lambda_0^{-|\mathbf{i} \wedge \mathbf{j}| s'}} 1 \, dt + \lambda_0^{-|\mathbf{i} \wedge \mathbf{j}|} \int_{\lambda_0^{-|\mathbf{i} \wedge \mathbf{j}| s'}}^\infty t^{-1/s'} \, dt \Big) \\
&= C_1 (1 - s')^{-1} \lambda_0^{-|\mathbf{i} \wedge \mathbf{j}| s'},
\end{aligned}$$

where we used Lemma 6.3.5 (i.e. transversality) and the fact that $C_1 \geqslant 1$. Therefore,

$$\begin{aligned}
\mathcal{J}_1 &\leqslant C_1 (1 - s')^{-1} \iint_{\Sigma^2} \lambda_0^{-|\mathbf{i} \wedge \mathbf{j}| s'} \, d\mu(\mathbf{i}) \, d\mu(\mathbf{j}) \\
&= C_1 (1 - s')^{-1} \sum_{n=0}^\infty \lambda_0^{-ns'} (\mu \times \mu)\{(\mathbf{i}, \mathbf{j}) : |\mathbf{i} \wedge \mathbf{j}| = n\} \\
&\leqslant C_1 (1 - s')^{-1} \sum_{n=0}^\infty \lambda_0^{-ns'} m^{-n} < \infty,
\end{aligned}$$

since $\lambda_0^{-s'} < m$ by (6.17). The proof is complete. $\square$

Proof of Theorem 6.3.3, *which implies Theorem* 6.3.2*(ii)*. Here we use the scheme of Mattila [**Mat90, Mat95**]. We are going to prove that for any $U = [\lambda_0, \lambda_1] \subset J \cap (\sum_{i=1}^m p_i^2, 1)$,

$$\mathcal{J}_2 := \int_U \int_{\mathbb{R}} \underline{D}(\nu_{\lambda,D}^{\mathbf{p}}, x)\, d\nu_{\lambda,D}^{\mathbf{p}}(x)\, d\lambda < \infty, \tag{6.19}$$

where

$$\underline{D}(\nu_{\lambda,D}^{\mathbf{p}}, x) := \liminf_{r \to 0} \frac{\nu_{\lambda,D}^{\mathbf{p}}[x - r, x + r]}{2r} \tag{6.20}$$

is the lower derivative of the self-similar measure $\nu_{\lambda,D}^{\mathbf{p}}$ with respect to the Lebesgue measure at x, see Definition B.2.4 in Appendix B. (Note that D has two different meanings in the formula above, but this should not cause a confusion.) Once (6.19) is established, we can conclude that for $\mathcal{L}^1$-a.e. $\lambda \in U$ we have $\underline{D}(\nu_{\lambda,D}^{\mathbf{p}}, x) < \infty$ for $\nu_{\lambda,D}^{\mathbf{p}}$-a.e. x, and then by [**Mat95**, p.36] (see also Theorem B.2.5 in the Appendix B) for such λ the measure $\nu_{\lambda,D}^{\mathbf{p}}$ is absolutely continuous with respect to Lebesgue $\mathcal{L}^1$. Moreover, we then have for such λ that

$$\underline{D}(\nu_{\lambda,D}^{\mathbf{p}}, x) = \frac{d\nu_{\lambda,D}^{\mathbf{p}}}{dx} \quad \text{and} \quad d\nu_{\lambda,D}^{\mathbf{p}} = \frac{d\nu_{\lambda,D}^{\mathbf{p}}}{dx}\, dx \ \text{ for } \mathcal{L}^1\text{-a.e. } x,$$

hence $\mathcal{J}_2 = \int_U \|\frac{d\nu_{\lambda,D}^{\mathbf{p}}}{dx}\|^2\, d\lambda$, and its finiteness implies that $\frac{d\nu_{\lambda,D}^{\mathbf{p}}}{dx} \in L^2(\mathbb{R})$ for a.e. $\lambda \in U$. Thus it remains to show (6.19).

We first use Fatou's Lemma to obtain

$$\mathcal{J}_2 \leqslant \liminf_{r \to 0} (2r)^{-1} \int_U \int_{\mathbb{R}} \nu_{\lambda,D}^{\mathbf{p}}([x - r, x + r])\, d\nu_{\lambda,D}^{\mathbf{p}}(x)\, d\lambda.$$

Further,

$$\begin{aligned} \int_{\mathbb{R}} \nu_{\lambda,D}^{\mathbf{p}}[x - r, x + r]\, d\nu_{\lambda,D}^{\mathbf{p}}(x) &= \iint_{\mathbb{R}^2} \mathbb{1}_{[x-r,x+r]}(y)\, d\nu_{\lambda,D}^{\mathbf{p}}(y)\, d\nu_{\lambda,D}^{\mathbf{p}}(x) \\ &= \iint_{\Sigma^2} \mathbb{1}_{[\Pi_\lambda(\mathbf{i})-r,\Pi_\lambda(\mathbf{i})+r]}(\Pi_\lambda(\mathbf{j}))\, d\mu_{\mathbf{p}}^2(\mathbf{i}, \mathbf{j}), \end{aligned}$$

by the definition of $\nu_{\lambda,D}^{\mathbf{p}}$. We can exchange the order of integration and integrate the characteristic function to obtain

$$\mathcal{J}_2 \leqslant \liminf_{r \to 0} (2r)^{-1} \iint_{\Sigma^2} \mathcal{L}^1\{\lambda \in U : \ |\Pi_\lambda(\mathbf{i}) - \Pi_\lambda(\mathbf{j})| \leqslant r\}\, d\mu_{\mathbf{p}}(\mathbf{i})\, d\mu_{\mathbf{p}}(\mathbf{j}).$$

Now we apply transversality in the form (6.15) to get rid of lim inf and obtain

$$\begin{aligned} \mathcal{J}_2 &\leqslant (C_1/2) \iint_{\Sigma^2} \lambda_0^{-|\mathbf{i} \wedge \mathbf{j}|}\, d\mu_{\mathbf{p}}(\mathbf{i})\, d\mu_{\mathbf{p}}(\mathbf{j}) \\ &= (C_1/2) \sum_{n=0}^{\infty} \lambda_0^{-n} (\mu_{\mathbf{p}} \times \mu_{\mathbf{p}})\{(\mathbf{i}, \mathbf{j}) : \ |\mathbf{i} \wedge \mathbf{j}| = n\} \\ &\leqslant (C_1/2) \sum_{n=0}^{\infty} \lambda_0^{-n} \Big(\sum_{i=1}^m p_i^2\Big)^n < \infty, \end{aligned} \tag{6.21}$$

since $\lambda_0 > \sum_{i=1}^m p_i^2$ by assumption. Above we made a simple computation: denoting by $[u]$ the cylinder set of sequences in Σ starting with u, we have

$$\begin{aligned} & (\mu_{\mathbf{p}} \times \mu_{\mathbf{p}})\{(\mathbf{i},\mathbf{j}) : \ |\mathbf{i} \wedge \mathbf{j}| = n\} \\ \leqslant \ & \sum_{\mathbf{k}:\ |\mathbf{k}|=n} (\mu_{\mathbf{p}} \times \mu_{\mathbf{p}})\{(\mathbf{i},\mathbf{j}) : \ \mathbf{i} \in [\mathbf{k}],\ \mathbf{j} \in [\mathbf{k}]\} \\ = \ & \sum_{\mathbf{k}:\ |\mathbf{k}|=n} (\mu[\mathbf{k}])^2 \\ = \ & \sum_{i_1 \dots i_n} (p_{i_1} \cdots p_{i_n})^2 = \Big(\sum_{i=1}^m p_i^2\Big)^n. \end{aligned}$$

The proof is complete. □

6.3.1. Dimension of exceptions. The results of Theorem 6.3.2 and 6.3.3 are "almost-sure" type results, since they make claims for almost every parameter value. It is natural to inquire about the size of the exceptional set. The following result has essentially appeared in [**PS95**]; it provides estimates for the Hausdorff dimension of exceptions for the dimension formula in Theorem 6.3.2. Results on dimension of exceptions for absolute continuity are more difficult; the first ones are due to Peres and Schlag [**PS00a**]. Even stronger results on dimension and absolute continuity were recently obtained by Hochman, Shmerkin, and others; we will return to all these later in Section 6.8, but without proof.

THEOREM 6.3.6. *Suppose that the IFS family*

$$\mathcal{F}_{\lambda,D} := (S_1^\lambda(x) = \lambda x + d_1, \dots, S_m^\lambda(x) = \lambda x + d_m)$$

satisfies the transversality condition on an interval $J \subset [0,1)$. Let $s(\lambda) = \frac{\log m}{-\log \lambda}$ be the similarity dimension and $b < 1/m$. Consider the following "exceptional set":

$$E_b := \{\lambda \in (0,b) \cap J : \ \dim_{\mathrm{H}}(\mathcal{C}_\lambda) < s(\lambda)\}.$$

Then $\dim_{\mathrm{H}}(E_b) \leqslant s(b) < 1$.

PROOF SKETCH. It is along the lines of [**PS95**, Section 4]. It suffices to prove the claim for the set

$$E_b^\varepsilon := \{\lambda \in (0,b) \cap J : \ \dim_{\mathrm{H}}(\mathcal{C}_\lambda) < s(\lambda) - \varepsilon\}$$

for any $\varepsilon > 0$. Fix $\varepsilon > 0$ and suppose that $\dim_{\mathrm{H}}(E_b^\varepsilon) > s(b)$. Choose $\alpha \in \big(s(b), \dim_{\mathrm{H}}(E_b^\varepsilon)\big)$. Then $\mathcal{H}^\alpha(E_b^\varepsilon) = +\infty$, and by Theorem 1.4.13 there exists a Borel subset $F \subset E_b^\varepsilon$ such that $0 < \mathcal{H}^\alpha(F) < \infty$ and

$$\mathcal{H}^\alpha(F \cap B(x,r)) \leqslant \mathrm{const} \cdot r^\alpha, \quad \text{for all } x \in J \text{ and } r > 0. \tag{6.22}$$

Let $\eta = \mathcal{H}^\alpha|_F$ be the restriction of the α-Hausdorff measure to the set F (it is a Frostman measure, see Definition 1.4.12). We want to repeat the proof of Theorem 6.3.2(i) with the measure η replacing the Lebesgue measure. Namely, instead of (6.18) we consider

$$\widetilde{\mathcal{J}}_1 := \int_{\lambda_0}^{\lambda_1} \mathcal{E}_{s'}(\nu_{\lambda,D})\, d\eta(\lambda),$$

where $s' = s(\lambda') - \varepsilon/2 > s(\lambda) - \varepsilon$ for $\lambda \in (\lambda_0, \lambda_1)$. If we show that $\widetilde{\mathcal{J}}_1 < \infty$, then

$$\dim_{\mathrm{H}}(\mathcal{C}_\lambda) \geqslant s' > s(\lambda) - \varepsilon \ \text{ for } \eta\text{-a.e. } \lambda \in F \subset E_b^\varepsilon,$$

by Lemma 1.5.3, which is a contradiction. We then essentially repeat the calculation of Theorem 6.3.2(i), but instead of Lemma 6.3.5 we use the estimate

$$\eta(\{\lambda \in [\lambda_0, \lambda_1] : |\Pi_\lambda(\mathbf{i}) - \Pi_\lambda(\mathbf{j})| \leqslant r\}) \leqslant \widetilde{C}_1 r^\alpha \lambda_0^{-|\mathbf{i}\wedge\mathbf{j}|\alpha}, \quad r > 0,$$

which follows from the Frostman condition (6.22) and the proof of Lemma 6.3.5. The convergence of the corresponding integral, namely,

$$\int_{\lambda_0^{-|\mathbf{i}\wedge\mathbf{j}|s'}}^{\infty} t^{-\alpha/s'}\,dt < \infty,$$

follows from $\alpha > s(b) \geqslant s'$. The rest of the proof is unchanged. □

6.4. How to check transversality?

Recall that transversality on an interval J for the IFS (6.9) was introduced in Definition 6.3.1. It can be stated as absence of double zeros for functions in $\mathcal{B}_{D'}$ from (6.13). How can we find such intervals? For practical purposes, in many situations the following "method of $(*)$-functions" from [**Sol95**], is sufficient. It usually does not give the best possible interval, but one which is often "good enough". The first idea is that it should be easier to deal with power series with "interval" bounds on the coefficients, rather than those with finitely many possible coefficients. To this end, we define, for $\gamma > 0$,

$$\mathcal{B}_\gamma = \Big\{g(x) = 1 + \sum_{n=1}^{\infty} a_n x^n : |a_n| \leqslant \gamma,\ n \geqslant 1\Big\}. \tag{6.23}$$

We will say that J is a transversality interval for $\mathcal{B}_\gamma$ if a function from $\mathcal{B}_\gamma$ cannot have a double zero in J. Note that

$$\mathcal{B}_{D'} \subset \mathcal{B}_{\gamma(D)},$$

where

$$\gamma(D) = \max\Big\{\frac{|d_i - d_j|}{|d_k - d_\ell|} : 1 \leqslant i,j,k,\ell \leqslant m,\ k \neq \ell\Big\}. \tag{6.24}$$

Thus we obtain

LEMMA 6.4.1. *Any transversality interval for $\mathcal{B}_{\gamma(D)}$ is also a transversality interval for the IFS $(d_i + \lambda x)_{i=1}^m$.*

The second idea is that the function in $\mathcal{B}_\gamma$ with a smallest possible double zero (which exists by compactness) must have exactly two coefficient sign changes and at most one coefficient in the interior of the allowed interval $[-\gamma, \gamma]$.

DEFINITION 6.4.2. Let $\gamma > 0$. A power series $h(x)$ is called a $(*)$-function for $\mathcal{B}_\gamma$ if for some $k \geqslant 1$ and $a_k \in \mathbb{R}$,

$$h(x) = 1 - \gamma \sum_{i=1}^{k-1} x^i + a_k x^k + \gamma \sum_{i=k+1}^{\infty} x^i. \tag{6.25}$$

Notice that h is an element of $\mathcal{B}_\gamma$ if and only if $a_k \in [-\gamma, \gamma]$, but this is not required in the definition.

It is sometimes useful to strengthen the transversality condition a bit further, following [**PS96**].

DEFINITION 6.4.3. Let J be a closed subinterval of $[0, 1)$ and let $\gamma, \delta > 0$. We say that $\mathcal{B}_\gamma$ satisfies the **δ-transversality condition** on J if

$$\forall\, g \in \mathcal{B}_\gamma,\ (\lambda \in J,\ g(\lambda) < \delta) \implies g'(\lambda) < -\delta.$$

Observe that δ-transversality means that the graph of each $g \in \mathcal{B}_\gamma$ crosses transversally, with slope at most $-\delta$, all horizontal lines below height δ that it meets. It is immediate that δ-transversality implies transversality.

LEMMA 6.4.4. *Suppose that h is a $(*)$-function for $\mathcal{B}_\gamma$ such that*

$$h(x_0) > \delta \quad \text{and} \quad h'(x_0) < -\delta \tag{6.26}$$

for some $x_0 \in (0, 1)$ and $\delta \in (0, \gamma)$. Then $\mathcal{B}_\gamma$ satisfies the δ-transversality condition on $[0, x_0]$.

PROOF. First we claim that

$$h(x) > \delta \quad \text{and} \quad h'(x) < -\delta \quad \text{for all } x \in [0, x_0]. \tag{6.27}$$

We have to consider the cases $k = 1$ and $k \geqslant 2$ separately. If $k = 1$, then $h(x) = 1 + a_1 x + \gamma \sum_{i=2}^\infty x^i$, with $a_1 < 0$ by (6.26). Obviously, h' is increasing on $[0, 1)$, hence $h'(x) \leqslant h'(x_0) < -\delta$, and $h(x) \geqslant h(x_0) > \delta$, since h is decreasing on $[0, x_0]$. Now, if $k \geqslant 2$ then $h'(0) = -\gamma < -\delta$. If $h'(x) \geqslant -\delta$ for any $x \in (0, x_0)$, then h' has a local maximum in $(0, x_0)$ and a local minimum in $(x_0, 1)$, due to the fact that $\lim_{x\uparrow 1} h'(x) = +\infty$. Then h'' has at least two zeros in $(0, 1)$, which is impossible for a power series with one coefficient sign change by the Decartes' rule of signs. Thus, $h'(x) < -\delta$ for all $x \in [0, x_0]$, hence h is decreasing on $[0, x_0]$, and (6.27) follows.

Now let $g \in \mathcal{B}_\gamma$. Consider $f(x) = g(x) - h(x)$. Then (6.25) and (6.23) imply that $f(x) = \sum_{i=1}^\ell c_i x^i - \sum_{i=\ell+1}^\infty c_i x^i$, where $c_i \geqslant 0$ and $\ell = k-1$ or $\ell = k$, depending on the sign of a_k. We have for any $x \in [0, x_0]$, by (6.27),

$$g(x) < \delta \implies f(x) < 0 \implies f'(x) < 0 \implies g'(x) < -\delta,$$

proving δ-transversality, see Definition 6.3.1. The middle implication is a consequence of one coefficient sign change:

$$f(x) < 0 \implies \sum_{i=1}^{\ell} c_i x^i < \sum_{i=\ell+1}^{\infty} c_i x^i \implies \sum_{i=1}^{\ell} c_i i x^{i-1} < \sum_{i=\ell+1}^{\infty} c_i i x^{i-1} \implies f'(x) < 0.$$

□

COROLLARY 6.4.5.

(i) *$\mathcal{B}_1$ satisfies the transversality condition on $[0, 0.649]$;*
(ii) *$\mathcal{B}_2$ satisfies the transversality condition on $[0, 0.5)$, and this is sharp (i.e. fails at 0.5);*
(iii) *$\mathcal{B}_3$ satisfies the transversality condition on $[0, 0.415]$;*
(iv) *$\mathcal{B}_\gamma$ satisfies the transversality condition on $[0, (1 + \sqrt{\gamma})^{-1})$ for all $\gamma \geqslant 1$, and this is sharp for $\gamma \geqslant 3 + \sqrt{8}$.*

REMARK 6.4.6.

(a) As already mentioned, finding the "maximal" transversality interval for a family of IFS is harder. In the paper [**SS06**] a rigorous computer-assisted argument shows that for the family of IFS $(d_1 + \lambda x, d_2 + \lambda x)$, with $d_1 \neq d_2$, equivalently, for the class of power series $\mathcal{B}_{\{-1,0,1\}}$, the maximal transversality interval is approximately $[0, 0.6684]$.

(b) It is known (see [**Sol98**]) that every $\lambda \in (2^{-1/2}, 1)$ is a double zero of a $\{-1,0,1\}$-power series, hence transversality fails there. Also, $0.682328\ldots$, the positive root of $x^3+x+1=0$, is a double zero of such a series. In [**SS06**] it is shown (with some help from *Mathematica*) that there is such a double zero, which is less that 0.6685, and this is close to being optimal. Moreover, as shown in [**SS06**], the set of double zeros has a complicated structure near its minimum; in particular, it is disconnected.

PROOF OF COROLLARY 6.4.5.

(i) Let

$$h_1(x) = 1 - x - x^2 - x^3 + 0.08753x^4 + \sum_{n=5}^{\infty} x^n.$$

This is a $(*)$-function for $\mathcal{B}_1$, and one can check that $h_1(0.649) > 0$, $h_1'(0.649) < 0$. The δ-transversality condition on $[0, 0.649]$ follows by Lemma 6.4.4. (The function h_1 was found with the help of *Mathematica*. Actually, for the proof of absolute continuity of the classical Bernoulli convolution ν_λ it suffices to know transversality on $[0, 0.64]$, and then this can be checked even by hand, using the $(*)$-function $1 - x - x^2 - x^3 + 0.1x^4 + \sum_{n=5}^{\infty} x^n$.)

(ii) Let

$$\begin{aligned} h_2(x) &= 1 - 2x - 2x^2 + 2\sum_{n=3}^{\infty} x^n = 1 - 2x - 2x^2 + 2x^3(1-x)^{-1} \\ &= (4x^3 - 3x + 1)(1-x)^{-1} = (2x-1)^2(x+1)(1-x)^{-1}. \end{aligned}$$

We see that h_2 is a $(*)$-function for $\mathcal{B}_2$ and $h_2(x_0) > 0$, $h_2'(x_0) < 0$ for all $x_0 \in [0, \frac{1}{2})$. It remains to apply Lemma 6.4.4.

(iii) The function $h_3(x) = 1 - 3x - 0.7x^2 + \sum_{n=3}^{\infty} 3x^n$ is a $(*)$-function for $\mathcal{B}_3$, and it satisfies $h_3(0.415) > 0$, $h_3'(0.415) < 0$, and it remains to apply Lemma 6.4.4.

(iv) Let $h_\gamma(x) = 1 - (1 + 2\sqrt{\gamma})x + \sum_{n=2}^{\infty} \gamma x^n$. It is easy to see that $h_\gamma(x) > 0$ and $h_\gamma'(x) < 0$ for all $x \in (0, (1+\sqrt{\gamma})^{-1})$, so Lemma 6.4.4 implies the desired estimate. Furthermore, $(1+\sqrt{\gamma})^{-1}$ is a double zero for h_γ, hence this interval is best possible provided $1 + 2\sqrt{\gamma} \leqslant \gamma$ (so that all coefficients of h_γ are in $[-\gamma, \gamma]$), and this is equivalent to $\gamma \geqslant 3 + \sqrt{8}$. □

6.5. First applications

6.5.1. Expansions with deleted digits: The $\{0, 1, 3\}$ problem.

PROOF OF THEOREMS 1.6.3(I) AND 1.6.4. We recall that the "$\{0, 1, 3\}$ problem" is concerned with the structure of the sets $\mathcal{K}^\lambda$, defined as attractors of the IFS $(d_i + \lambda x)_{i=1}^3$ on the line, where the set of "digits" is $D = \{0, 1, 3\}$. We are going to apply Theorem 6.3.2. The relevant class of power series is $\mathcal{B}_3$, see (6.23), since the functions $\Pi_\lambda(\mathbf{i}) - \Pi_\lambda(\mathbf{j})$ in this case are power series with coefficients in $D - D = \{0, \pm 1, \pm 2, \pm 3\}$. By Corollary 6.4.5, $\mathcal{B}_3$ satisfies the transversality condition on $[0, 0.415]$. Thus, $\dim_{\mathrm{H}}(\mathcal{K}^\lambda) = \frac{\log 3}{-\log \lambda}$ for a.e. $\lambda \in (\frac{1}{4}, \frac{1}{3})$, and $\mathcal{L}^1(\mathcal{K}^\lambda) > 0$ for a.e. $\lambda \in (\frac{1}{3}, 0.4)$. The nontrivial parameter range in this example is $(\frac{1}{4}, \frac{2}{5})$ by Lemma 1.6.2, since for $\lambda \leqslant \frac{1}{4}$ the Open Set Condition holds, whereas for $\lambda \in (\frac{2}{5}, 1)$ the attractor $\mathcal{K}^\lambda$ is an in interval. □

Part (ii) of Theorem 1.6.3 (that is, dimension drop for a dense set of parameters) will be proved in Section 7.3, where we will also show that $\mathcal{H}^{s(\lambda)}(\mathcal{K}_\lambda) = 0$ for all $\lambda \in (\frac{1}{4}, \frac{1}{3})$.

It should be mentioned that "almost every λ" in the Theorems 1.6.3(i) and 1.6.4 can be replaced by "for all λ outside a set of Hausdorff dimension zero", in view of the recent progress; see Theorem 6.8.11 and Theorem 8.4.3 discussed below.

6.5.2. Bernoulli convolutions. Recall (1.97) that the (infinite) Bernoulli convolution measure ν_λ is defined as the distribution of the random series $Y_\lambda = \sum_{n=0}^{\infty} a_n \lambda^n$, where a_n are i.i.d. Bernoulli random variables, with the values $\{-1, 1\}$ taken with probabilities $(\frac{1}{2}, \frac{1}{2})$. Equivalently, ν_λ is the natural self-similar measure for the IFS $(-1 + \lambda x, 1 + \lambda x)$. Since Y_λ is the sum of independent random variables, its distribution ν_λ is an infinite convolution product of discrete measures $\frac{1}{2}(\delta_{-\lambda^n} + \delta_{\lambda^n})$, for $n = 0, 1, 2, \ldots$ Taking the Fourier transform, we get

$$\widehat{\nu}_\lambda(\xi) = \int_{\mathbb{R}} e^{it\xi}\, d\nu_\lambda(t) = \prod_{n=0}^{\infty} \tfrac{1}{2}\widehat{(\delta_{\lambda^n} + \delta_{-\lambda^n})}(\xi) = \prod_{n=0}^{\infty} \cos(\lambda^n \xi). \tag{6.28}$$

Notice that (6.28) implies

$$\widehat{\nu}_\lambda(\xi) = \widehat{\nu}_{\lambda^m}(\xi)\widehat{\nu}_{\lambda^m}(\lambda\xi) \cdot \ldots \cdot \widehat{\nu}_{\lambda^m}(\lambda^{m-1}\xi) \quad \text{for } m \geqslant 2. \tag{6.29}$$

We postpone the discussion of early results on Bernoulli convolutions to Chapter 8, and apply here the results of the two previous sections to obtain the definitive "almost every parameter" statement on absolute continuity.

THEOREM 6.5.1.

(i) $\nu_\lambda \ll \mathcal{L}^1$ *with the Radon–Nikodym derivative in* $L^2(\mathbb{R})$ *for a.e.* $\lambda \in (1/2, 2^{-1/2})$;

(ii) $\nu_\lambda \ll \mathcal{L}^1$ *with a continuous Radon–Nikodym derivative for a.e.* $\lambda \in (2^{-1/2}, 1)$.

PROOF. (Compare with the proof of Theorem 6.2.4.) (i) By Corollary 6.4.5 and Lemma 6.4.1, the IFS $(-1 + \lambda x, 1 + \lambda x)$ satisfies the transversality condition on $[0, 0.649]$, hence $\nu_\lambda \ll \mathcal{L}^1$ with $\frac{d\nu^{\mathbf{p}}_{\lambda,D}}{dx} \in L^2(\mathbb{R})$ for a.e. $\lambda \in (1/2, 0.649)$ by Theorem 6.3.3. We need to establish the analogous statement for the interval $(0.649, 2^{-1/2})$. Consider

$$\eta_\lambda = \nu_\lambda * \nu_\lambda.$$

This is the distribution of the sum of two independent realizations of the random series $\sum_{n=0} \pm\lambda^n$. It follows that η_λ is the distribution of $Z_\lambda = \sum_{n=0}^{\infty} c_n \lambda^n$ where $c_n \in \{-2, 0, 2\}$ with probabilities $\mathbf{p} = (\frac{1}{4}, \frac{1}{2}, \frac{1}{4})$. Thus, $\eta_\lambda = \nu^{\mathbf{p}}_{\lambda,D}$, where $D = \{-2, 0, 2\}$, so $\gamma(D) = 2$. We compute $p_1^2 + p_2^2 + p_3^2 = 3/8$. By Theorem 6.3.3 and Corollary 6.4.5(ii), $\eta_\lambda \ll \mathcal{L}^1$ with $\frac{d\eta_\lambda}{dx} \in L^2(\mathbb{R})$ for a.e. $\lambda \in (3/8, 1/2)$. By Plancherel's theorem C.0.3 and Lemma C.0.7, a function is in $L^2(\mathbb{R})$ if and only if its Fourier transform is in $L^2(\mathbb{R})$. Thus, $\widehat{\eta}_\lambda \in L^2(\mathbb{R})$ for a.e. $\lambda \in (3/8, 1/2)$. But $\widehat{\eta}_\lambda = (\widehat{\nu}_\lambda)^2$, so $\widehat{\nu}_\lambda \in L^4(\mathbb{R})$ for a.e. $\lambda \in (3/8, 1/2)$. Recall (6.29), which for $m = 2$ becomes

$$\widehat{\nu}_\lambda(\xi) = \widehat{\nu}_{\lambda^2}(\xi)\widehat{\nu}_{\lambda^2}(\lambda\xi).$$

It follows that $\widehat{\nu}_\lambda \in L^2(\mathbb{R})$ for a.e. $\lambda \in ((3/8)^{1/2}, 2^{-1/2})$ and hence $\frac{d\nu_\lambda}{dx} \in L^2(\mathbb{R})$ by Lemma C.0.7. Since $(3/8)^{1/2} \approx 0.61237 < 0.649$, the proof of (i) is complete.

By part (i) and Plancherel's theorem (see Theorem C.0.3 and Lemma C.0.7), $\widehat{\nu}_\lambda \in L^2(\mathbb{R})$ for a.e. $\lambda \in (1/2, 2^{-1/2})$. In view of (6.29), $\widehat{\nu}_\lambda \in L^{2/m}(\mathbb{R})$ for a.e. $\lambda \in (2^{-1/m}, 2^{-1/(2m)})$ where $m = 2, 3, \ldots$ This implies that $\widehat{\nu}_\lambda \in L^1(\mathbb{R})$ for a.e. $\lambda \in (2^{-1/2}, 1)$ (here we use that $|\widehat{\nu}_\lambda|$ is always bounded by one, being the Fourier transform of a probability measure). For such λ the density of ν_λ is continuous, being the inverse Fourier transform of an L^1 function. The proof is finished. $\square$

6.6. Generalized projection scheme

Here we present a generalization of the method used by Kaufman [**Kau68**] and Mattila [**Mat90**] to prove projection theorems. It should be mentioned that different projection schemes have also been used, notably the one by Peres and Schlag [**PS00a**]. We follow [**Sol98**] in part, but modify the scheme slightly to make it more flexible and add more, in particular, results on projections of measures.

Let $(\mathcal{X}, \rho)$ be a compact metric space (the one that we will "project"), and let U be a locally separable compact metric space (this will be the set of parameters), equipped with a Borel measure η, finite on compact subsets. We further suppose that there is a family of maps

$$\Phi_\lambda : \mathcal{X} \to \mathbb{R}^n, \quad \lambda \in U,$$

satisfying the following properties: the *Hölder condition:*

$$\|\Phi_\lambda(x) - \Phi_\lambda(y)\| \leqslant C(\lambda) \cdot \rho(x, y)^{\alpha(\lambda)} \quad \text{for } x, y \in \mathcal{X} \text{ and } \lambda \in U, \tag{6.30}$$

where it is assumed that the functions $C(\lambda)$ and $\alpha(\lambda)$ are continuous on U, and the

Generalized Transversality Condition: for any $\lambda_0 \in U$ and $\varepsilon > 0$ there exists a neighborhood $B_\varepsilon(\lambda_0)$ such that for all $r > 0$,

$$\eta(\{\lambda \in B_\varepsilon(\lambda_0) : \|\Phi_\lambda(x) - \Phi_\lambda(y)\| \leqslant r\}) \leqslant \frac{C_{\lambda_0,\varepsilon} \cdot r^n}{\rho(x, y)^{n(\alpha(\lambda_0)+\varepsilon)}}. \tag{6.31}$$

THEOREM 6.6.1. *Assuming that the Hölder condition* (6.30) *and the Generalized transversality condition* (6.31) *hold, we have*

(i) $\dim_{\mathrm{H}}(\Phi_\lambda(\mathcal{X})) = \min\left\{n, \frac{\dim_{\mathrm{H}}(\mathcal{X})}{\alpha(\lambda)}\right\}$ *for η-a.e. $\lambda \in U$;*

(ii) $\mathcal{L}^n(\Phi_\lambda(\mathcal{X})) > 0$ *for η-a.e. $\lambda \in U$ such that* $\frac{\dim_{\mathrm{H}}(\mathcal{X})}{\alpha(\lambda)} > n$.

In Marstrand's projection theorem 2.3.1, assuming the projected set is compact (the general case may be derived by approximation), we have $\mathcal{X} = E \subset \mathbb{R}^d$, with the induced metric, the parameter λ is a unit vector in $\mathbb{R}^d$, and Φ_λ is the orthogonal projection onto the line through λ, considered as a map into $\mathbb{R}$. Then (6.30) holds with $\alpha(\lambda) \equiv 1$, because projections are Lipschitz, and (6.31) reduces to (2.24). More generally, in Mattila's projection theorems [**Mat90**] from $\mathbb{R}^d$ to $\mathbb{R}^n$, the parameter space U is the Grassmanian $G(d, n)$. For homogeneous self-similar IFS with the contraction ratio being the parameter, see (6.9), the space $\mathcal{X} = \Sigma$ is the symbolic space, and $\Phi_\lambda = \Pi_\lambda$ is the natural projection. If we use the metric $\rho(\mathbf{i}, \mathbf{j}) = m^{-|\mathbf{i} \wedge \mathbf{j}|}$ on Σ, then Π_λ is Hölder, with exponent $\alpha(\lambda) = -\frac{\log \lambda}{\log m}$. Generalized transversality follows from Lemma 6.3.5. Although literally the case of hyperbolic nonlinear IFS with overlaps is not covered, a minor modification of the scheme applied there as well, and this will be presented in Chapter 14.

Theorem 6.6.1 will be deduced from a generalized projection theorem for measures, which we now state. We will need the *(lower) correlation dimension* $D_2(\mu)$ of a measure, see Definition 1.9.7 and Proposition 1.9.8. Recall also the *lower Hausdorff dimension* $\underline{\dim}_{\rm H}(\mu)$ of a measure, see (1.101).

THEOREM 6.6.2. *In the general setting above, let μ be a finite Borel measure on $\mathcal{X}$ and let $\nu_\lambda = \mu \circ \Phi_\lambda^{-1}$ be the push-forward measure on $\mathbb{R}^n$ depending on the parameter λ. Then, assuming the Hölder condition* (6.30) *and the Generalized transversality condition* (6.31) *hold, we have*

(i) $\dim_{\rm H}(\nu_\lambda) = \min\left\{n, \frac{\dim_{\rm H}(\mu)}{\alpha(\lambda)}\right\}$ *for η-a.e. $\lambda \in U$;*
(ii) $\nu_\lambda \ll \mathcal{L}^n$ *for η-a.e. $\lambda \in U$ such that* $\underline{\dim}_{\rm H}(\mu)/\alpha(\lambda) > n$.
(iii) $D_2(\nu_\lambda) = \min\left\{n, \frac{D_2(\mu)}{\alpha(\lambda)}\right\}$ *for η-a.e. $\lambda \in U$;*
(iv) $\nu_\lambda \ll \mathcal{L}^n$ *with a density in $L^2(\mathbb{R}^n)$ for η-a.e. $\lambda \in U$ such that* $\frac{D_2(\mu)}{\alpha(\lambda)} > n$.

In many details, the proof is quite similar to that of Theorems 6.3.2 and 6.3.3 and may be skipped on first reading.

PROOF OF THEOREM 6.6.2. (iii) The upper bound is easy and holds for all λ. Indeed, by (6.30), $\mathcal{E}_t(\nu_\lambda) \geqslant C(\lambda)^{-t}\mathcal{E}_{t\alpha(\lambda)}(\mu)$ for all $\lambda \in U$ and $t \geqslant 0$. Thus $\mathcal{E}_s(\mu) = \infty$ implies $\mathcal{E}_{s/\alpha(\lambda)}(\nu_\lambda) = \infty$, and hence $D_2(\nu_\lambda) \leqslant D_2(\mu)/\alpha(\lambda)$ by Proposition 1.9.8.

For the lower bound, the case $\frac{D_2(\mu)}{\alpha(\lambda)} > n$ will follow from part (iv), so assume $\frac{D_2(\mu)}{\alpha(\lambda)} \leqslant n$. It is enough to show the following

CLAIM: *for any $\lambda_0 \in U$ and any $s < \frac{D_2(\mu)}{\alpha(\lambda_0)} \leqslant n$, we have $D_2(\nu_\lambda) \geqslant s$ for η-a.e. λ in some neighborhood of λ_0.*

We will next explain why this is sufficient (the same argument will apply in several other places). If the desired claim does not hold, then there exists $\varepsilon > 0$ such that

$$\eta(F) > 0, \quad \text{where } F := \{\lambda \in U : D_2(\nu_\lambda) < \frac{D_2(\mu)}{\alpha(\lambda)} - \varepsilon < n\}. \tag{6.32}$$

Since U is a locally compact separable metric space, there exists a compact subset $F' \subset F$ such that $\eta(F') > 0$. Clearly, we can then find $\lambda_0 \in F'$ such that every open neighborhood V of λ_0 satisfies

$$\eta(F' \cap V) > 0.$$

Fix such a λ_0. According to the Claim, we will have $D_2(\nu_\lambda) \geqslant \frac{D_2(\mu)}{\alpha(\lambda_0)} - \frac{\varepsilon}{2}$ for η-a.e. λ in some neighborhood G of λ_0. By continuity of $\alpha(\lambda)$, there exists an open neighborhood $U_\varepsilon \subset G$ of λ_0 such that $\frac{D_2(\mu)}{\alpha(\lambda)} - \varepsilon < \frac{D_2(\mu)}{\alpha(\lambda_0)} - \frac{\varepsilon}{2}$ for all $\lambda \in U_\varepsilon$. This will be a contradiction, since by the choice of λ_0 and (6.32) we would have

$$D_2(\nu_\lambda) < \frac{D_2(\mu)}{\alpha(\lambda)} - \varepsilon < \frac{D_2(\mu)}{\alpha(\lambda_0)} - \frac{\varepsilon}{2}$$

for $\lambda \in F' \cap U_\varepsilon$, which has positive η-measure.

We will now prove the Claim. Let $\varepsilon > 0$ be such that $s < \frac{D_2(\mu)}{\alpha(\lambda_0)+\varepsilon}$, and choose a neighborhood B_ε of λ_0 for which (6.31) holds. Denote $\alpha_0 = \alpha(\lambda_0)$. Since $s(\alpha_0+\varepsilon) <$

$D_2(\mu)$, by the basic property of the correlation dimension (Proposition 1.9.8), we have

$$\int_{\mathcal{X}}\int_{\mathcal{X}} \rho(x,y)^{-s(\alpha_0+\varepsilon)}\,d\mu(x)\,d\mu(y) < \infty. \tag{6.33}$$

We would like to show that

$$\mathcal{S} := \int_{B_\varepsilon}\int_{\mathbb{R}^n}\int_{\mathbb{R}^n} \|u-v\|^{-s}\,d\nu_\lambda(u)\,d\nu_\lambda(v)\,d\eta(\lambda) < \infty.$$

Then for η-a.e. $\lambda \in B_\varepsilon$ we will have $D_2(\nu_\lambda) \geqslant s$, as desired. Recalling that $\nu_\lambda = \mu \circ \Phi_\lambda^{-1}$, making a change of variable and exchanging the order of integration yields

$$\mathcal{S} = \int_{\mathcal{X}}\int_{\mathcal{X}}\int_{B_\varepsilon} \|\Phi_\lambda(x) - \Phi_\lambda(y)\|^{-s}\,d\eta(\lambda)\,d\mu(x)\,d\mu(y). \tag{6.34}$$

From (6.31) we have, for all $r > 0$:

$$\eta(\{\lambda \in B_\varepsilon : \|\Phi_\lambda(x) - \Phi_\lambda(y)\| \leqslant r\}) \leqslant C_{\lambda_0,\varepsilon} \min\Big\{1, \frac{r^n}{\rho(x,y)^{n(\alpha_0+\varepsilon)}}\Big\}.$$

Then

$$\begin{aligned}\int_{B_\varepsilon} \frac{d\eta(\lambda)}{\|\Phi_\lambda(x) - \Phi_\lambda(y)\|^s} &= \int_0^\infty \eta(\{\lambda \in G : \|\Phi_\lambda(x) - \Phi_\lambda(y)\|^{-s} \geqslant t\})\,dt \\ &\leqslant C_{\lambda_0,\varepsilon} \int_0^\infty \min\{1, \rho(x,y)^{-n(\alpha_0+\varepsilon)} t^{-n/s}\}\,dt.\end{aligned}$$

The last integral is equal to

$$\rho(x,y)^{-s(\alpha_0+\varepsilon)} + \rho(x,y)^{-n(\alpha_0+\varepsilon)} \int_{\rho(x,y)^{-s(\alpha_0+\varepsilon)}}^\infty t^{-n/s}\,dt = C' \cdot \rho(x,y)^{-s(\alpha_0+\varepsilon)},$$

in view of $s < n$. Substituting this into (6.34) and using (6.33), we obtain $\mathcal{S} < \infty$, as desired. □

PROOF OF THEOREM 6.6.1. (i) The case of $\frac{\dim_H(\mathcal{X})}{\alpha(\lambda)} > n$ will follow from part (ii) of that theorem, so we assume $\frac{\dim_H(\mathcal{X})}{\alpha(\lambda)} \leqslant n$. The inequality $\dim_H(\Phi_\lambda(\mathcal{X})) \leqslant \frac{\dim_H(\mathcal{X})}{\alpha(\lambda)}$ holds for all λ; in fact, this is (1.56). To establish the lower bound, it is enough to show that for any $\lambda_0 \in U$, such that $\frac{\dim_H(\mathcal{X})}{\alpha(\lambda_0)} \leqslant n$, and any $s < \frac{\dim_H(\mathcal{X})}{\alpha(\lambda_0)}$ we have $\dim_H(\Phi_\lambda(\mathcal{X})) \geqslant s$ for η-a.e. λ in some neighborhood of λ_0. (That this is sufficient is verified similarly to the proof above, reducing it to the Claim.) Denote $\alpha_0 = \alpha(\lambda_0)$. Let $\varepsilon > 0$ be such that $s < \frac{\dim_H(\mathcal{X})}{\alpha_0+\varepsilon}$. By Frostman's Lemma in a compact metric space (Theorem 1.4.13(ii)), together with Lemma 1.4.14, which holds in a compact metric space as well, since $s(\alpha_0+\varepsilon) < \dim_H(\mathcal{X})$, there exists a Borel probability measure μ on $\mathcal{X}$ such that $\mathcal{E}_{s(\alpha_0+\varepsilon)}(\mu) < \infty$. Then $D_2(\mu) \geqslant s(\alpha_0+\varepsilon)$. Consider the push-forward measure $\nu_\lambda = \mu \circ \Phi_\lambda^{-1}$. By Theorem 6.6.2(iii), which we just proved, $D_2(\nu_\lambda) \geqslant \frac{s(\alpha_0+\varepsilon)}{\alpha(\lambda)}$ for η-a.e. $\lambda \in U$. By continuity of $\alpha(\lambda)$ we then have $D_2(\nu_\lambda) \geqslant s$ for η-a.e. λ in a neighborhood of λ_0. Finally,

$$\dim_H(\Phi_\lambda(\mathcal{X})) \geqslant \dim_H(\nu_\lambda) \geqslant D_2(\nu_\lambda)$$

for all λ, and the proof is complete. □

PROOF OF THEOREM 6.6.2. (iv) Let $\lambda_0 \in U$ be such that $\frac{D_2(\mu)}{\alpha(\lambda_0)} > n$. Denote $\alpha_0 = \alpha(\lambda_0)$. Choose $\beta < D_2(\mu)$ so that $\frac{\beta}{\alpha_0} > n$. Then $\mathcal{E}_\beta(\mu) < \infty$. It is enough to show that ν_λ is absolutely continuous with respect to $\mathcal{L}^n$, with a density in L^2, for η-a.e. λ in some neighborhood of λ_0. Fix $\varepsilon > 0$ such that $\beta > n(\alpha_0 + \varepsilon)$ and then B_ε, a neighborhood of λ_0 in which (6.31) holds. Let

$$\underline{D}(\nu_\lambda, u) = \liminf_{r\to 0} \frac{\nu_\lambda\left(B(u,r)\right)}{\mathcal{L}^n\left(B(u,r)\right)}$$

be the lower derivative of ν_λ, see Definition B.2.4. If we show that

$$\mathcal{J} := \int_{B_\varepsilon} \int_{\mathbb{R}^n} \underline{D}(\nu_\lambda, u)\, d\nu_\lambda(u)\, d\eta(\lambda) < \infty, \tag{6.35}$$

then for η-a.e. $\lambda \in B_\varepsilon$,

$$\mathcal{I}(\lambda) := \int_{\mathbb{R}^n} \underline{D}(\nu_\lambda, u)\, d\nu_\lambda(u) < \infty.$$

For such λ we have $\underline{D}(\nu_\lambda, u) < \infty$ for ν_λ-a.e. $u \in \mathbb{R}^n$, and hence ν_λ is absolutely continuous, see Mattila [**Mat95**, Lemma 2.12] and Theorem B.2.5 in the Appendix. Moreover, we can then write

$$\underline{D}(\nu_\lambda, u) = \frac{d\nu_\lambda}{du}$$

(the Radon–Nikodym derivative), and

$$\mathcal{I}(\lambda) = \int_{\mathbb{R}^n} \left(\frac{d\nu_\lambda}{du}\right)^2 du < \infty,$$

that is, $\frac{d\nu_\lambda}{du} \in L^2(\mathbb{R}^n)$, as desired. In order to establish (6.35), we first apply Fatou's Lemma:

$$\mathcal{J} \leqslant C_n \cdot \liminf_{r\to 0}\ r^{-n} \int_{B_\varepsilon} \int_{\mathbb{R}^n} \nu_\lambda(B(u,r))\, d\nu_\lambda(u)\, d\eta(\lambda).$$

Next,

$$\begin{aligned} \int_{\mathbb{R}^n} \nu_\lambda(B(u,r))\, d\nu_\lambda(u) &= \int_{\mathbb{R}^n} \int_{\mathbb{R}^n} \mathbb{1}_{B(u,r)}\, d\nu_\lambda(v)\, d\nu_\lambda(u) \\ &= \int_{\mathcal{X}} \int_{\mathcal{X}} \mathbb{1}_{\{y:\ \|\Phi_\lambda(x) - \Phi_\lambda(y)\| \leqslant r\}}\, d\mu(y)\, d\mu(x). \end{aligned}$$

By Fubini, integrating with respect to λ, we obtain

$$\mathcal{J} \leqslant C_n \cdot \liminf_{r\to 0}\ r^{-n} \iint\limits_{\mathcal{X}\times\mathcal{X}} \eta\big(\{\lambda \in B_\varepsilon : \|\Phi_\lambda(x) - \Phi_\lambda(y)\| \leqslant r\}\big)\, d\mu(x)\, d\mu(y).$$

By (6.31),

$$\mathcal{J} \leqslant C_n \cdot \iint\limits_{\mathcal{X}\times\mathcal{X}} \rho(x,y)^{-n(\alpha_0+\varepsilon)}\, d\mu(x)\, d\mu(y),$$

which is finite since $n(\alpha_0 + \varepsilon) < \beta$. □

PROOF OF THEOREM 6.6.1. (ii). Let $\lambda_0 \in U$ be such that $\frac{\dim_{\mathrm{H}}(\mathcal{X})}{\alpha(\lambda_0)} > n$. Choose $\beta < \dim_{\mathrm{H}}(\mathcal{X})$ so that $\frac{\beta}{\alpha(\lambda_0)} > n$. There exists a probability measure μ on $\mathcal{X}$ satisfying $\int_{\mathcal{X}} \int_{\mathcal{X}} \rho(x,y)^{-\beta} d\mu(x)\, d\mu(y) < \infty$. Let $\nu_\lambda = \mu \circ \Phi_\lambda^{-1}$. It is enough to show that ν_λ is absolutely continuous with respect to $\mathcal{L}^n$ for η-a.e. λ in some neighborhood of λ_0. But this follows from Theorem 6.6.2(iv), which we just proved. □

PROOF OF THEOREM 6.6.2. (ii) We will also reduce this claim to part (iv), with the help of a useful trick based on Egorov's Theorem. Let $\lambda_0 \in U$ be such that $\underline{\dim}_{\mathrm{H}}(\mu)/\alpha(\lambda_0) > n$. We can find a neighborhood G of λ_0 and $\gamma > n$ such that

$$\underline{\dim}_{\mathrm{H}}(\mu)/\alpha_G > \gamma,$$

where $\alpha_G := \max\{\alpha(\lambda) : \lambda \in G\}$. Recall that $\underline{\dim}_{\mathrm{H}}(\mu) = \operatorname{ess\,inf} \underline{d}_\mu(x)$, so we have $\underline{d}_\mu(x) > \gamma\alpha_G > n\alpha_G$ for μ-a.e. x. Recall also that t he lower local dimension of μ at x is

$$\underline{d}_\mu(x) = \lim_{s\to 0}\left(\inf_{r\leqslant s}\frac{\log\mu(B(x,r))}{\log r}\right). \tag{6.36}$$

By Egorov's Theorem, for every $\varepsilon > 0$ there exists a Borel set A_ε with $\mu(A_\varepsilon) > 1-\varepsilon$, such that the convergence in (6.36) is uniform. Thus there exists $s_0 > 0$ such that

$$\forall\, x \in A_\varepsilon,\ \inf_{r\leqslant s_0}\frac{\log\mu(B(x,r))}{\log r} \geqslant \gamma\alpha_G.$$

Therefore, for all $x \in A_\varepsilon$, for all $r \leqslant s_0$,

$$\log\mu(B(x,r)) \leqslant \gamma\alpha_G \log r,$$

or $\mu(B(x,r)) \leqslant r^{\gamma\alpha_G}$. Define $\mu_\varepsilon := \mu|_{A_\varepsilon}$. The last inequality implies that for all $r \leqslant s_0$ we have

$$C_2(\mu_\varepsilon) = \int \mu_\varepsilon(B(x,r))\,d\mu_\varepsilon(x) \leqslant r^{\gamma\alpha_G},$$

hence

$$D_2(\mu) = \liminf_{r\to 0}\frac{\log C_2(\mu,r)}{\log r} \geqslant \gamma\alpha_G > n\alpha_G.$$

Then $\mu_\varepsilon \circ \Phi_\lambda^{-1} \ll \mathcal{L}^n$ for η-a.e. $\lambda \in G$ by the part (iv) of the theorem. (Even with a density in L^2, but now this is irrelevant, since this property will be lost when we pass to the limit $\varepsilon \to 0$.) Letting $\varepsilon \to 0$ along a countable set, we obtain a set of λ of full measure in G for which $\mu_\varepsilon \circ \Phi_\lambda^{-1} \ll \mathcal{L}^n$. Then we also have $\nu_\lambda = \mu \circ \Phi_\lambda^{-1} \ll \mathcal{L}^n$ for such λ, by the definition of μ_ε. Indeed, for any Borel set $E \subset \mathbb{R}^n$ we have $\mu_\varepsilon \circ \Phi_\lambda^{-1}(E) \to \mu \circ \Phi_\lambda^{-1}(E)$, as $\varepsilon \to 0$. This completes the proof. □

There is a useful extension of Theorem 6.6.2(iii),(iv).

THEOREM 6.6.3. *In the setting of Theorem 6.6.2, the following hold for any* $q \in (1,2)$*:*

(i) $D_q(\nu_\lambda) = \min\{n, \frac{D_q(\mu)}{\alpha(\lambda)}\}$ *for* η*-a.e.* $\lambda \in U$*;*

(ii) $\nu_\lambda \ll \mathcal{L}^n$ *with a density in* $L^q(\mathbb{R}^n)$ *for* η*-a.e.* $\lambda \in U$ *such that* $\frac{D_q(\mu)}{\alpha(\lambda)} > n$*. On the other hand,* ν_λ *cannot have a density in* L^q *if* $\frac{D_q(\mu)}{\alpha(\lambda)} < n$*.*

Part (i) is due to Hunt and Kaloshin [**HK97**]. They also pointed out that the statement breaks down for $q > 2$. A counter-example is obtained from the uniform (Haar) measure μ on a circle in $\mathbb{R}^2$ and its projections onto lines. Then $D_q(\mu) = 1$ but $D_q(\mathrm{proj}_\theta\,\mu) < 1$ for any $q > 2$ and any θ (this is an easy exercise). A counter-example for part (ii), in the case $q > 2$, may be obtained from the uniform measure on a "Cantor target" $T_a = \{(r,\theta) : r \in C_a,\ \theta \in [0,2\pi)\}$, where C_a is a self-similar Cantor set contained in $[0,1]$.

PROOF. The proof is similar to that of Theorem 6.6.2(iv)–(vi), with some minor modifications. It uses (s,q)-energy instead of s-energy, and Proposition 2.6.2 of Hunt and Kaloshin. The condition $q \in (1,2)$ is crucial in one place, where the Hölder inequality is applied. We sketch the proof of part (ii), and leave part (i) as an exercise.

(ii) Recall from Section 2.6 that $D_q(\mu) = \sup\{s : \mathcal{E}_{s,q}(\mu) < \infty\}$, where

$$\mathcal{E}_{s,q}(\mu) = \int \left(\int \frac{d\mu(y)}{\rho(x,y)^s} \right)^{q-1} d\mu(x).$$

Let $\lambda_0 \in U$ be such that $\frac{D_q(\mu)}{\alpha(\lambda_0)} > n$. It suffices to show that $\nu_\lambda \ll \mathcal{L}^n$, with a density in L^q, for η-a.e. λ in a neighborhood of λ_0. Denote $\alpha_0 = \alpha(\lambda_0)$ and fix $\beta < D_q(\mu)$ satisfying $\frac{\beta}{\alpha_0} > n$. Choose $\varepsilon > 0$ so that $\beta > n(\alpha_0 + \varepsilon)$, and then find $B_\varepsilon \ni \lambda_0$ in which (6.31) holds.

Consider the lower derivative of ν_λ with respect to $\mathcal{L}^n$:

$$\underline{D}(\nu_\lambda, u) = \liminf_{r\to 0} \frac{\nu_\lambda(B(u,r))}{\mathcal{L}^n(B(u,r))}.$$

It is enough to prove that

$$\mathcal{J}_q := \int_{B_\varepsilon} \int_{\mathbb{R}^n} (\underline{D}(\nu_\lambda, u))^{q-1} d\nu_\lambda(u)\, d\eta(\lambda) < \infty. \tag{6.37}$$

By Fatou's Lemma and a change of variable,

$$\begin{aligned} \mathcal{J}_q &\leqslant C \cdot \liminf_{r\to 0}\, r^{-n(q-1)} \int_{B_\varepsilon} \int_{\mathbb{R}^n} (\nu_\lambda(B(u,r))^{q-1} d\nu_\lambda(u)\, d\eta(\lambda) \\ &= C \cdot \liminf_{r\to 0}\, r^{-n(q-1)} \int_{B_\varepsilon} \int_{\mathcal{X}} (\mu(B(\Phi_\lambda(x),r))^{q-1} d\mu(x)\, d\eta(\lambda). \end{aligned}$$

We next reverse the order of integration and apply Hölder's inequality $\int_{B_\varepsilon} f^{q-1}\, d\eta \leqslant \eta(B_\varepsilon)^{2-q}(\int_{B_\varepsilon} f)^{q-1}$ (note that $q \in (1,2)$) to obtain

$$\mathcal{J}_q \leqslant C \cdot \liminf_{r\to 0}\, r^{-n(q-1)} \int_{\mathcal{X}} \left(\int_{B_\varepsilon} \nu_\lambda(B(\Phi_\lambda(x),r))\, d\eta(\lambda) \right)^{q-1} d\mu(x).$$

The rest of the proof proceeds almost exactly as in Theorem 6.6.2(iv):

$$\int_{B_\varepsilon} \nu_\lambda(B(\Phi_\lambda(x),r))\, d\eta(\lambda) = \int_{\mathcal{X}} \eta(\{\lambda \in B_\varepsilon : \|\Phi_\lambda(x) - \Phi_\lambda(y)\| \leqslant r\})\, d\mu(y),$$

hence, in view of (6.31),

$$\mathcal{J}_q \leqslant C \cdot \int_{\mathcal{X}} \left(\int_{\mathcal{X}} \rho(x,y)^{-n(\alpha_0+\varepsilon)}\, d\mu(y) \right)^{q-1} d\mu(x),$$

which is finite, because $n(\alpha_0 + \varepsilon) < \beta < D_q(\mu)$. This completes the proof of the main claim. It remains to note that if $\frac{D_q(\mu)}{\gamma(\lambda)} < n$, then $D_q(\nu_\lambda) < n$ by part (i) (left as an exercise), and hence ν_λ cannot have a density in L^q by Proposition 2.6.8. □

6.7. Applications of the Generalized Projection Scheme

6.7.1. L^q densities for Bernoulli convolutions. Here we return to the IFS

$$\mathcal{F}_{\lambda,D} := (S_1^\lambda(x) = \lambda x + d_1, \ldots, S_m^\lambda(x) = \lambda x + d_m),$$

see (6.9), where $m \geqslant 2$ and $\lambda \in (0,1)$.

THEOREM 6.7.1. *Let $\Pi_\lambda : \Sigma \to \mathbb{R}$ be the natural projection (6.11) associated with the IFS $\mathcal{F}_{\lambda,D}$. Let μ be **any** Borel probability measure on Σ and let J be a closed subinterval of $[0,1)$, which is a transversality interval for this IFS. Consider the push-forward measure $\nu_\lambda := \mu \circ \Pi_\lambda^{-1}$.*

(i) *Let $q \in (1,2]$. Then for Lebesgue-a.e. $\lambda \in J \cap (m^{-D_q(\mu)}, 1)$ we have $\nu_\lambda \ll \mathcal{L}^1$, with $\frac{d\nu_\lambda}{dx} \in L^q(\mathbb{R})$.*

(ii) *For any $q > 1$ and all $\lambda \in (0,1)$, we have $D_q(\nu_\lambda) \leqslant \min\left\{1, \frac{D_q(\mu)\log m}{-\log\lambda}\right\}$. As a consequence, if ν_λ is absolutely continuous, with $\frac{d\nu_\lambda}{dx} \in L^q(\mathbb{R})$, then $\lambda \geqslant m^{-D_q(\mu)}$.*

(iii) *Suppose that μ is a shift-invariant and ergodic measure of entropy h_μ. Then ν_λ is absolutely continuous for Lebesgue-a.e. $\lambda \in J \cap (e^{-h_\mu}, 1)$ and singular for all $\lambda < e^{-h_\mu}$.*

The theorem follows by an application of the GPS (Theorem 6.6.2(ii) and Theorem 6.6.3(ii)); details are left as an exercise. We only recall that the natural projection $\Pi : \Sigma \to \mathbb{R}$ is Hölder-continuous with exponent $\alpha = \frac{-\log\lambda}{\log m}$.

Next we specialize Theorem 6.7.1 to the case of self-similar measures. Then $\nu_\lambda = \nu_{\lambda,D}^{\mathbf{p}}$, where $\mathbf{p}$ is a probability vector. The L^q-dimension of $\mu_{\mathbf{p}} = \mathbf{p}^{\mathbb{N}}$ was computed in Proposition 5.1.8:

$$D_q(\mu_{\mathbf{p}}) = \frac{\tau(\mu_{\mathbf{p}})}{q-1} = \frac{\log\sum_{i=1}^m p_i^q}{(q-1)\log m},$$

and we obtain the following:

COROLLARY 6.7.2. *Suppose that the IFS family $\mathcal{F}_{\lambda,D}$ satisfies the transversality condition on an interval $J \subset [0,1)$. Let $\mathbf{p}$ be a probability vector and $h_{\mathbf{p}} = -\sum_{i=1}^m p_i \log p_i$. Then*

(i) *the self-similar measure $\nu_{\lambda,D}^{\mathbf{p}}$ is absolutely continuous for Lebesgue-a.e. $\lambda \in J \cap (e^{-h_{\mathbf{p}}}, 1)$, and is singular for all $\lambda < e^{-h_{\mathbf{p}}}$;*

(ii) *given $q \in (1,2]$, the measure $\nu_{\lambda,D}^{\mathbf{p}}$ is absolutely continuous with $\frac{d\nu_{\lambda,D}^{\mathbf{p}}}{dx} \in L^q(\mathbb{R})$ for Lebesgue-a.e. $\lambda \in J \cap \left((\sum_{i=1}^m p_i^q)^{1/(q-1)}, 1\right)$, and $\frac{d\nu_{\lambda,D}^{\mathbf{p}}}{dx}$, if it exists, cannot belong to $L^q(\mathbb{R})$ for any $\lambda < (\sum_{i=1}^m p_i^q)^{1/(q-1)}$. This last claim also holds for $q > 2$.*

REMARK 6.7.3.

(a) Observe that the function $q \mapsto (\sum_{i=1}^m p_i^q)^{1/(q-1)}$ is increasing in q for $q > 1$ and converges to $\prod_{i=1}^m p_i^{p_i} = e^{-h_{\mathbf{p}}}$ as $q \to 1$. This allows one to deduce the a.e. absolute continuity statement in part (i) from the one in part (ii) by letting $q \to 1$. This function is strictly increasing if and only if $\mathbf{p} \neq (\frac{1}{m}, \ldots, \frac{1}{m})$. Thus, for such measures the thresholds for almost sure existence of densities in L^q change with q. As a result, we get a gradual change in the typical properties of the density of $\nu_{\lambda,D}^{\mathbf{p}}$ as λ increases.

(b) What happens in the "borderline" case? It was proved in [**NW05**] that for $\lambda = e^{-h_{\mathbf{p}}}$ the self-similar measure $\nu^{\mathbf{p}}_{\lambda,D}$ is necessarily singular when $\mathbf{p} \neq (\frac{1}{m}, \dots, \frac{1}{m})$.

6.7.2. Self-similar measures with nonuniform contraction ratios. We start with IFS consisting of two linear maps: $\mathcal{F}_{\lambda,\gamma,D} := (d_1 + \lambda x, d_2 + \gamma x)$, where $D = \{d_1, d_2\}$. We assume that the fixed points are distinct (otherwise, the IFS is trivial), that is,

$$\frac{d_1}{1-\lambda} \neq \frac{d_2}{1-\gamma}.$$

One can often apply a global coordinate change and get a conjugate IFS of a simpler form, see Definition 1.3.8 and Remark 1.8.5. The next lemma as an easy exercise.

LEMMA 6.7.4. *Given the contraction ratios λ, γ, any two linear IFS $\mathcal{F}_{\lambda,\gamma,D} := (d_1 + \lambda x, d_2 + \gamma x)$ and $\mathcal{F}_{\lambda,\gamma,\widetilde{D}} := (\widetilde{d}_1 + \lambda x, \widetilde{d}_2 + \gamma x)$ are conjugate via an affine-linear map $h(x) = ax + b$, as long as both are nontrivial, that is, $\frac{d_1}{1-\lambda} \neq \frac{d_2}{1-\gamma}$ and $\frac{\widetilde{d}_1}{1-\lambda} \neq \frac{\widetilde{d}_2}{1-\gamma}$.*

In view of the lemma, we can assume, without loss of generality, that

$$\mathcal{F}_{\lambda,\gamma,D} = \mathcal{F}_{\lambda,\gamma} = (\lambda x, 1 + \gamma x), \quad \text{with } 0 < |\gamma| \leqslant |\lambda|.$$

We will also assume that γ and λ are positive. Since we focus on the nonuniform case here, we have $0 < \gamma < \lambda < 1$. Denote by $\nu^p_{\lambda,\gamma}$ the self-similar measure for the IFS $\mathcal{F}_{\lambda,\gamma}$, with the probability vector $(p, 1-p)$.

THEOREM 6.7.5 (Neunhäuserer). *Let $p \in (0,1)$ and $q \in (1,2]$. For $\mathcal{L}^2$-a.e. $(\lambda, \gamma) \in (0, 0.649)^2$, we have*

(i)

$$\left(\frac{p}{\lambda}\right)^p \left(\frac{1-p}{\gamma}\right)^{1-p} < 1 \implies \nu^p_{\lambda,\gamma} \ll \mathcal{L}^1;$$

(ii)

$$\frac{p^q}{\lambda^{q-1}} + \frac{(1-p)^q}{\gamma^{q-1}} < 1 \implies \frac{d\nu^p_{\lambda,\gamma}}{dx} \in L^q(\mathbb{R}).$$

REMARK 6.7.6.

(a) In the proof, $(0, 0.649)$ can be replaced by any transversality interval for $\mathcal{B}_1$. Note that the condition in the first claim can be rewritten as

$$\dim_S(\nu^p_{\lambda,\gamma}) = \frac{-p \log p - (1-p)\log(1-p)}{-p \log \lambda - (1-p)\log \gamma} > 1.$$

More recently, Saglietti, Shmerkin, and Solomyak [**SSS18**] obtained the first claim for a.e. $(\lambda, \gamma) \in (0,1)^2$, by a different method; however, no information on density is obtained in that paper.

(b) The proof of the "almost sure" results gives a stronger statement: absolute continuity and the density properties hold for $\nu^p_{\lambda,c\lambda}$ for $\mathcal{L}^1$-a.e. $\lambda \in (0, 0.649)$, for any fixed $c \in (0,1]$.

PROOF SKETCH. Let $(S_i(x) = \lambda_i x + d_i)_{i=1}^m$ be an IFS on $\mathbb{R}$ and let $\Pi_{\boldsymbol{\lambda}} : \Sigma \to \mathbb{R}$ be the corresponding natural projection map, where $\boldsymbol{\lambda} = (\lambda_1, \lambda_2)$. We can write it explicitly as follows:

$$\Pi_{\boldsymbol{\lambda}}(\mathbf{i}) = d_{i_1} + \lambda_{i_1} d_{i_2} + \lambda_{i_1 i_2} d_{i_3} + \cdots \tag{6.38}$$

We have, by assumption, the IFS $\mathcal{F}_{\lambda,\gamma}$, with $d_1 = 0, d_2 = 1$, $\lambda_1 = \gamma < \lambda = \lambda_2$, so we can write $\gamma = c\lambda$ for some $c \in (0,1)$. We will show that for any $c \in (0,1)$, the 1-parameter family of IFS $\mathcal{F}_{\lambda,c\lambda}$ satisfies the Generalized transversality condition (6.31) for $\lambda \in [0, 0.649]$, and then the claim follows by Theorem 6.6.2(ii).

To verify transversality, observe that for our IFS $\Pi_{\boldsymbol{\lambda}}$ is a power series in λ, with the kth coefficient in the set $\{0, 1, c, c^2, \dots, c^k\}$, in view of (6.38). Since $c \in (0,1)$, we obtain that for $\mathbf{i}, \mathbf{j} \in \Sigma$ with $i_1 = 2, j_1 = 1$,

$$\Pi_{\boldsymbol{\lambda}}(\mathbf{i}) - \Pi_{\boldsymbol{\lambda}}(\mathbf{j}) \in \mathcal{B}_1,$$

where $\mathcal{B}_1$ was defined in (6.23). Now absence of double zeros on $[0, 0.649]$ follows from Corollary 6.4.5(i), and it is an easy exercise to show that this implies (6.14). □

Theorem 6.7.5 is a special case of a more general result on self-similar measures with nonuniform contraction ratios by Ngai and Wang (2005). In fact, they obtained the result independently: [**Neu01**] appeared in 2001, whereas the [**NW05**] was already circulated in 2001 as a preprint. It appears that the authors of these papers were not aware of each other's work at the time of publication.

In [**NW05**] the authors consider familes of IFS on $\mathbb{R}$ of the form

$$(6.39)\qquad \left(S_i^{\boldsymbol{\lambda}}(x) = \rho_i(\boldsymbol{\lambda})x + d_i(\boldsymbol{\lambda})\right)_{i=1}^m, \quad \boldsymbol{\lambda} \in \Omega \subset \mathbb{R}^n,$$

where $\rho_i(\boldsymbol{\lambda}) = \rho_i(\lambda_1, \dots, \lambda_n) \in (0,1)$ and $d_i(\boldsymbol{\lambda}) = d_i(\lambda_1, \dots, \lambda_n) \in \mathbb{R}$ for $i = 1, \dots, m$, depend on n real parameters. It is assumed that $\rho_i, d_i \in C^2(\Omega)$, with Ω open. Let $\Pi_{\boldsymbol{\lambda}} : \Sigma \to \mathbb{R}$ be the natural projection associated with this IFS. Say that an IFS family satisfies the *NW-transversality condition* on Ω if for all pairs $\mathbf{i} \neq \mathbf{j}$ in Σ,

$$(6.40)\qquad \|\nabla(\Pi_{\boldsymbol{\lambda}}(\mathbf{i}) - \Pi_{\boldsymbol{\lambda}}(\mathbf{j}))\| > 0 \quad \text{whenever} \quad \Pi_{\boldsymbol{\lambda}}(\mathbf{i}) = \Pi_{\boldsymbol{\lambda}}(\mathbf{j}),$$

where ∇ is the gradient with respect to $\boldsymbol{\lambda}$.

THEOREM 6.7.7 (Ngai and Wang). *Let* (6.39) *be a family of IFS on* $\mathbb{R}$, *with* C^2*-smooth dependence on parameters, satisfying the transversality condition* (6.40) *on an open set* $\Omega \subset \mathbb{R}^n$. *For a fixed probability vector* $\mathbf{p} = (p_1, \dots, p_m)$ *let* $\nu_{\boldsymbol{\lambda}}^{\mathbf{p}}$ *be the corresponding self-similar measure. Then for* $\mathcal{L}^n$*-a.e.* $\lambda \in \Omega$,

$$\prod_{i=1}^m \left(\frac{p_i}{\rho_i(\boldsymbol{\lambda})}\right)^{p_i} < 1 \implies \nu_{\boldsymbol{\lambda}}^{\mathbf{p}} \ll \mathcal{L}^1,$$

$$\forall\, q \in (1,2], \quad \sum_{i=1}^m p_i \left(\frac{p_i}{\rho_i(\boldsymbol{\lambda})}\right)^{q-1} < 1 \implies \frac{d\nu_{\boldsymbol{\lambda}}^{\mathbf{p}}}{dx} \in L^q(\mathbb{R}).$$

This theorem also can be deduced from the Generalized Projection Scheme.

REMARK 6.7.8.

(a) Let $(S_i(x) = \rho_i x + d_i)_{i=1}^m$ be an IFS on $\mathbb{R}$. By Theorem 3.2.1 it follows that if $\prod_{i=1}^m (\frac{p_i}{\rho_i})^{p_i} > 1$, then the corresponding self-similar measure is singular. Ngai and Wang [**NW05**] proved that in the case of equality $\prod_{i=1}^m (\frac{p_i}{\rho_i})^{p_i} = 1$, the self-similar measure is singular, provided $p_i \neq \rho_i$ for some i; and, moreover, if $p_i = \rho_i$ for all i and the equality to 1 holds, then the self-similar measure is absolutely continuous if and only if the IFS satisfies the OSC. Moreover, they obtained an analogous result for IFS in $\mathbb{R}^d$.

(b) It is not so easy to verify the multi-dimensional NW-transversality condition. The proof of [**SSS18**] does not use transversality.

6.7.3. Distributions of dynamically defined infinite sums and Weierstrass functions. Bernoulli convolutions are the distributions of random variables of the form $\sum_{k=1}^{\infty} \xi_n \lambda^n$, where ξ_n are independent and identically distributed random variables. One can consider a generalization by assuming that ξ_n might depend on $\{\xi_1, \ldots, \xi_{n-1}\}$.

Let $b \geqslant 2$ be an integer and let $\phi\colon \mathbb{R} \to \mathbb{R}$ be a 1-periodic, piecewise continuous function with continuity intervals $\left(\frac{i}{b}, \frac{i+1}{b}\right)$, $i \in \mathbb{Z}$ such that $\phi|_{\left(\frac{i}{b}, \frac{i+1}{b}\right)}$ can be continuously extended to $\left[\frac{i}{b}, \frac{i+1}{b}\right]$. Consider the symbolic space

$$\Sigma = \{0, \ldots, b-1\}^{\mathbb{N}}$$

and a Bernoulli measure μ on Σ with probabilities $\{p_0, \ldots, p_{b-1}\}$. Then for every $0 < \lambda < 1$ and $x \in \mathbb{R}$, let

$$\Pi^{\phi}_{x,\lambda}(\mathbf{i}) = \sum_{n=1}^{\infty} \lambda^n \phi\left(\frac{x}{b^n} + \frac{i_1}{b^n} + \cdots + \frac{i_n}{b}\right). \tag{6.41}$$

The measures $\nu^{\mathbf{p}}_{\lambda,x} := (\Pi^{\phi}_{\lambda,x})_* \mu_{\mathbf{p}}$ can be considered as a natural generalization of Bernoulli convolutions.

Distributions such as $\nu^{\mathbf{p}}_{\lambda,x}$ play an important role in the dimension theory of Weierstrass-type functions. Namely, let $\psi\colon \mathbb{R} \to \mathbb{R}$ be a piecewise C^1, 1-periodic function with continuity intervals $\left(\frac{i}{b}, \frac{i+1}{b}\right)$ such that $\psi|_{\left(\frac{i}{b}, \frac{i+1}{b}\right)}$ can be continuously extended to $\left[\frac{i}{b}, \frac{i+1}{b}\right]$. Then for every $0 < \lambda < 1$ the function

$$\Phi(x) = \sum_{k=0}^{\infty} \gamma^n \psi(b^n x) \tag{6.42}$$

is well defined. There is an interesting dichotomy if $\lambda b > 1$, namely, either Φ is piecewise C^1, or it is not differentiable at every point, but Hölder continuous with exponent $\frac{-\log \gamma}{\log b}$, see Hu and Lau [**HL93**]. The first example of a function of the form (6.42) was given by Karl Weierstrass in the 19th century, with $\psi(x) = \cos(2\pi x)$, as an example of a nowhere differentiable function on the line.

It is an interesting question to determine what is the Hausdorff dimension of the graph of Φ, i.e., the Hausdorff dimension of the set $\mathrm{graph}(\Phi) = \{(x, \Phi(x)) : x \in [0,1]\}$. For the classical Weierstrass function, Mandelbrot conjectured that $\dim_{\mathrm{H}}(\mathrm{graph}(\Phi)) = 2 + \frac{\log \gamma}{\log b}$, however, this conjecture has not been verified until recently.

Ledrappier [**Led92**] gave a sufficient condition in order to verify Mandelbrot's conjecture using random sums of the form (6.41).

PROPOSITION 6.7.9 (Ledrappier [**Led92**]). *Let $\phi = \psi'$, $\lambda = 1/(\gamma b)$ and let $\mathbf{p}$ be the probability vector with uniform probabilities b^{-1}. If $\dim_{\mathrm{H}}(\nu^{\mathbf{p}}_{\lambda,x}) = 1$ for Lebesgue-almost every x then $\dim_{\mathrm{H}}(\mathrm{graph}(\Phi)) = 2 + \log \gamma / \log b$.*

Using this condition, Barański, Bárány and Romanowska [**BBR14**] studied the dimension of a "typical" Weierstrass function. Mandelbrot's conjecture was finally verified by Shen [**She18**]. Shen showed that $\dim_{\mathrm{H}}(\mathrm{graph}(\Phi)) = 2 + \frac{\log \gamma}{\log b}$ for every $b \geqslant 2$ integer and $\gamma \in (b^{-1}, 1)$ using higher order transversality methods of Tsujii [**Tsu01**]. More recently, Ren and Shen [**RS21**] used methods of additive combinatorics to prove a dichotomy for the graph of a general Weierstrass-type function (6.42) with an analytic map ψ: either $\Phi(x)$ is real analytic, or its graph has Hausdorff dimension $2 + \log \gamma / \log b$.

Next we will show how Ledrappier's condition can be checked by using the generalized projection scheme. We consider the special case when $b \geqslant 3$ and $\phi(x) = \sin(2\pi x)$. Let

$$h_b(\lambda) = \frac{\lambda^2}{(1-\lambda)^2} + \frac{\lambda^2}{(b-\lambda)^2} - \sin^2 \frac{\pi}{b}.$$

It is easy to see that for $b \geqslant 3$, the function h_b is strictly monotone increasing on $(0,1)$, $\lim_{\lambda\to 1-} h_b(\lambda) = \infty$ and $h_b(b^{-1}) < 0$. Thus there exists a unique $\lambda_b > b^{-1}$ such that $h_b(\lambda_b) = 0$.

For short, denote the projection $\Pi^{\phi}_{x,\lambda}$ defined in (6.41) by

$$\Pi_x(\mathbf{i}) = \sum_{n=1}^{\infty} \lambda^n \sin\left(2\pi\Big(\frac{x}{b^n} + \frac{i_1}{b^n} + \cdots + \frac{i_n}{b}\Big)\right). \tag{6.43}$$

LEMMA 6.7.10. *If $\lambda \in (b^{-1}, \lambda_b)$, then there exists $\delta > 0$ such that for every $\mathbf{i} = (i_1, i_2, \ldots)$, $\mathbf{j} = (j_1, j_2, \ldots) \in \Sigma$ with $i_1 \neq j_1$ and every $x \in [0,1]$,*

$$|\Pi_x(\mathbf{i}) - \Pi_x(\mathbf{j})| > \delta \quad \textit{or} \quad \left|\frac{d}{dx}\Pi_x(\mathbf{i}) - \frac{d}{dx}\Pi_x(\mathbf{j})\right| > \delta.$$

PROOF. Fix $\lambda \in (b^{-1}, \lambda_b)$. Suppose the assertion does not hold. Then for every $\delta > 0$ there exist $\mathbf{i} = (i_1, i_2, \ldots)$, $\mathbf{j} = (j_1, j_2, \ldots) \in \Sigma$ with $i_1 \neq j_1$ and $x \in [0,1]$, such that

$$|\Pi_x(\mathbf{i}) - \Pi_x(\mathbf{j})| \leqslant \delta \quad \text{and} \quad \left|\frac{d}{dx}\Pi_x(\mathbf{i}) - \frac{d}{dx}\Pi_x(\mathbf{j})\right| \leqslant \delta. \tag{6.44}$$

By the definition of Π_x,

$$\begin{aligned} |\Pi_x(\mathbf{i}) - \Pi_x(\mathbf{j})| &\geqslant \lambda \left|\sin\Big(2\pi\frac{x+i_1}{b}\Big) - \sin\Big(2\pi\frac{x+j_1}{b}\Big)\right| - 2\sum_{n=2}^{\infty}\lambda^n \\ &= 2\lambda \sin\Big(2\pi\frac{|i_1-j_1|}{2b}\Big)\left|\cos\Big(2\pi\frac{2x+i_1+j_1}{2b}\Big)\right| - \frac{2\lambda^2}{1-\lambda} \\ &\geqslant 2\lambda\sin\frac{\pi}{b}\left|\cos\Big(2\pi\frac{2x+i_1+j_1}{2b}\Big)\right| - \frac{2\lambda^2}{1-\lambda}, \end{aligned} \tag{6.45}$$

as $1 \leqslant |i_1 - j_1| \leqslant b-1$. Similarly, since

$$\frac{d}{dx}\Pi_x(\mathbf{i}) = 2\pi\sum_{n=1}^{\infty}\Big(\frac{\lambda}{b}\Big)^n \cos\left(2\pi\Big(\frac{x}{b^n} + \frac{i_1}{b^n} + \cdots + \frac{i_n}{b}\Big)\right),$$

we obtain

$$\begin{aligned}
&\left|\frac{d}{dx}\Pi_x(\mathbf{i}) - \frac{d}{dx}\Pi_x(\mathbf{j})\right| \\
&\geqslant \frac{2\pi\lambda}{b}\left|\cos\left(2\pi\frac{x+i_1}{b}\right) - \cos\left(2\pi\frac{x+j_1}{b}\right)\right| - 4\pi\sum_{n=2}^{\infty}\left(\frac{\lambda}{b}\right)^n \\
&= \frac{4\pi\lambda}{b}\sin\left(2\pi\frac{|i_1-j_1|}{2b}\right)\left|\sin\left(2\pi\frac{2x+i_1+j_1}{2b}\right)\right| - \frac{4\pi\lambda^2}{b(b-\lambda)} \\
&\geqslant \frac{4\pi\lambda}{b}\sin\frac{\pi}{b}\left|\sin\left(2\pi\frac{2x+i_1+j_1}{2b}\right)\right| - \frac{4\pi\lambda^2}{b(b-\lambda)}.
\end{aligned} \tag{6.46}$$

By (6.44), (6.45) and (6.46),

$$\sin\frac{\pi}{b}\left|\cos\left(2\pi\frac{2x+i_1+j_1}{2b}\right)\right| \leqslant \frac{\lambda}{1-\lambda} + \frac{\delta}{2\lambda},$$

$$\sin\frac{\pi}{b}\left|\sin\left(2\pi\frac{2x+i_1+j_1}{2b}\right)\right| \leqslant \frac{\lambda}{b-\lambda} + \frac{\delta b}{4\pi\lambda}.$$

Taking the sum of the squares of the two inequalities, we get

$$\sin^2\frac{\pi}{b} \leqslant \left(\frac{\lambda}{1-\lambda} + \frac{\delta}{2\lambda}\right)^2 + \left(\frac{\lambda}{b-\lambda} + \frac{\delta b}{4\pi\lambda}\right)^2.$$

Since δ is arbitrarily small, in fact this implies

$$0 \leqslant \frac{\lambda^2}{(1-\lambda)^2} + \frac{\lambda^2}{(b-\lambda)^2} - \sin^2\frac{\pi}{b},$$

which is impossible for $\lambda \in (b^{-1}, \lambda_b)$. □

Using Lemma 6.7.10, we can check the generalized transversality condition (6.31). An immediate consequence of Lemma 6.7.10 is that for every $0 < r < \delta$ and $\mathbf{i}, \mathbf{j} \in \Sigma$ with $i_1 \neq j_1$ and interval $I \subset [0,1]$

$$\mathcal{L}^1\big(\{x \in I : |\Pi_x(\mathbf{i}) - \Pi_x(\mathbf{j})| < r\}\big) \leqslant \mathcal{L}^1(I)\cdot\delta^{-1}r. \tag{6.47}$$

Observe if $\mathbf{i}\wedge\mathbf{j} = (i_1, \ldots, i_n)$ then

$$|\Pi_x(\mathbf{i}) - \Pi_x(\mathbf{j})| = \lambda^n\left|\Pi_{\frac{x}{b^n}+\frac{i_1}{b^n}+\cdots+\frac{i_n}{b}}(\sigma^n\mathbf{i}) - \Pi_{\frac{x}{b^n}+\frac{i_1}{b^n}+\cdots+\frac{i_n}{b}}(\sigma^n\mathbf{j})\right| \leqslant \frac{\lambda}{1-\lambda}\lambda^n.$$

Thus, for every $\mathbf{i} \neq \mathbf{j}$ with $\mathbf{i}\wedge\mathbf{j} = (i_1, \ldots, i_n)$

$$\begin{aligned}
&\mathcal{L}^1\big(\{x \in [0,1] : |\Pi_x(\mathbf{i}) - \Pi_x(\mathbf{j})| < r\}\big) \\
&= b^n\mathcal{L}^1\Big(\Big\{y \in \Big[\frac{i_1}{b^n} + \cdots + \frac{i_n}{b}, \frac{i_1+1}{b^n} + \cdots + \frac{i_n}{b}\Big] : \\
&\qquad\qquad |\Pi_y(\sigma^n\mathbf{i}) - \Pi_y(\sigma^n\mathbf{j})| < \lambda^{-n}r\Big\}\Big) \\
&\leqslant b^n b^{-n}\delta^{-1}r\lambda^{-n} = \delta^{-1}\lambda^{-n}r.
\end{aligned}$$

Hence, (6.31) holds with respect to the Lebesgue measure and metric $\rho(\mathbf{i}, \mathbf{j}) = \lambda^{|\mathbf{i}\wedge\mathbf{j}|}$. Applying Theorem 6.6.2 we obtain the following proposition:

PROPOSITION 6.7.11. *Let $b \geqslant 3$ be an integer, $\lambda \in (b^{-1}, \lambda_b)$ where λ_b is the unique root of h_b in $(0,1)$, and let $\Pi_x : \Sigma \to \mathbb{R}$ be defined as in (6.43). Then for $\mathcal{L}^1$-a.e. $x \in [0,1]$, $(\Pi_x)_*\mu_b \ll \mathcal{L}^1$, where μ_b is the Bernoulli measure with uniform probabilities on Σ.*

Now Proposition 6.7.9 implies that $\dim_H(\mathrm{graph}(\Phi)) = 2 + \frac{\log\gamma}{\log b}$ for every $(b\lambda_b)^{-1} < \gamma < 1$.

6.8. A survey of more recent results

In this section we give a brief survey of some of the advances achieved after those discussed in the earlier sections of this chapter. The proofs are omitted, although we try to indicate connections between various results.

6.8.1. Sobolev smoothness and dimension of exceptions. In [**PS00a**], Peres and Schlag developed the transversality method further. The original motivation was to improve the "almost every λ" statement in Theorem 6.5.1 on the typical absolute continuity of Bernoulli convolutions. Can we say more about the exceptional set of $\lambda \in (1/2, 1)$ for which ν_λ is singular? Specifically, is there a non-trivial upper bound for its Hausdorff dimension? Again, results about orthogonal projections serve as a guide. In particular, Falconer [**Fal82**] proved the following (Falconer's result concerns projections from $\mathbb{R}^n$ to k-dimensional subspaces; we state the planar special case for simplicity):

THEOREM 6.8.1. *Let $E \subset \mathbb{R}^2$ be a set with* $\dim_H(E) > 1$. *Then*

$$\dim_H\left(\{\mathbf{v} \in \mathbb{S}^1 : \mathcal{L}^1(\mathrm{proj}_{\mathbf{v}}(E)) = 0\}\right) \leqslant 2 - \dim_H(E).$$

Falconer's proofs uses Fourier-analytic methods. (See a comprehensive survey on projections by Falconer, Fraser, and Jin [**FFJ15**].) Based on the intuition that the family of Bernoulli convolutions $\lambda \mapsto \nu_\lambda$ can be viewed as a kind of "nonlinear projection", one expects an analogous result for them as well.

Peres and Schlag established this in their highly influential paper [**PS00a**]. An important tool used there is the notion *Sobolev dimension*, which they introduced. The homogeneous Sobolev γ-norm for a finite measure ν on $\mathbb{R}^n$ is given by

$$\|\nu\|_{2,\gamma}^2 = \int_{\mathbb{R}^n} |\widehat{\nu}(\xi)|^2 |\xi|^{2\gamma}\, d\xi.$$

Finiteness of $\|\nu\|_{2,\gamma}^2$ for $\gamma > 0$ means that ν has a fractional derivative of order γ in $L^2(\mathbb{R}^n)$ (by the definition of the fractional derivative). We refer the reader to [**AF03**] for a comprehensive monograph on this topic and to [**DNPV12**] for an article "Hitchhiker's guide to fractional Sobolev spaces". The Sobolev dimension is defined by

$$\dim_s(\nu) = \sup\left\{\alpha > 0 : \int_{\mathbb{R}^n} |\widehat{\nu}(\xi)|^2 |(1 + |\xi|)^{\alpha-n}\, d\xi < \infty\right\}. \tag{6.48}$$

When the Sobolev dimension of a finite measure ν in $\mathbb{R}^n$ is less than n, it coincides with the *correlation dimension.* If $\sigma = \dim_s(\nu) > n$, then ν is absolutely continuous and the density $\frac{d\nu}{dx}$ has $(\sigma - n)/2$ fractional derivatives.

In fact, Peres and Schlag proved [**PS00a**, Theorem 5.4]:

THEOREM 6.8.2 (Peres and Schlag). *Suppose that $J = [\lambda_0, \lambda_1] \subset (\frac{1}{2}, 1)$ is an interval of δ-transversality for $\{-1, 0, 1\}$ power series. Then*

(i)

$$\dim_s(\nu_\lambda) \geqslant \frac{\log 2}{-\log\lambda} \quad \textit{for a.e. } \lambda \in J;$$

(ii)

$$\dim_H\{\lambda \in J : \ \nu_\lambda \textit{ does not have } L^2\textit{-density }\} \leqslant 2 - \frac{\log 2}{\log \lambda_0}.$$

At the first stage of the proof (i) is established, and then it is proved, roughly speaking, that Sobolev smoothness a.e. can be "traded" for absolute continuity with an L^2-density on a set outside of Hausdorff dimension less than one. This result was obtained as a special case of a "generalized projection theorem" (different from our Generalized Projection Scheme), which has many other applications, in particular, to biased Bernoulli convolutions, to families of self-similar sets and measures, as well as to "pinned distance sets". We do not include a detailed treatment of this theory, since the book of Mattila [**Mat15**, Ch. 18] covers this topic thoroughly.

Theorem 6.8.2, together with a "thinning and convolving argument" in the spirit of [**Sol95**] and the proof of Theorem 6.5.1, yields

COROLLARY 6.8.3 ([**PS00a**]). *For every $\lambda_0 > \frac{1}{2}$ there exists $\varepsilon(\lambda_0) > 0$ such that*

$$\dim_H\{\lambda \in (\lambda_0, 1) : \ \nu_\lambda \textit{ does not have } L^2\textit{-density }\} \leqslant 1 - \varepsilon(\lambda_0).$$

Lower bounds for the almost sure value of Sobolev dimension are obtained as well.

These techniques were further developed by Lindenstrauss, Peres, and Schlag in [**LPS02**]. In particular, they introduced the idea of "analytically extendable measures" and used it to show that there exists an exceptional set for absolute continuity of biased Bernoulli convolutions of zero Lebesque measure that is independent of p (in the appropriate parameter region), see [**LPS02**, Prop. 4.2].

6.8.2. Dimension of self-similar measures via additive combinatorics. Consider a self-similar IFS on the line

$$\mathcal{S} = (S_i(x) = r_i x + d_i)_{i=1}^m$$

with attractor Λ. Recall that

$$\dim_H(\Lambda) \leqslant \min\{s, 1\}, \tag{6.49}$$

where s is the similarity dimension, with equality when the IFS satisfies the OSC. There is a folklore conjecture that "dimension drop", that is, strict inequality in (6.49), implies that the IFS has "exact overlaps", i.e. the semigroup generated by $(S_i)_{i=1}^m$ is not free, or in other words, there exist distinct words $\mathbf{i}, \mathbf{j}$ such that $S_\mathbf{i} \equiv S_\mathbf{j}$. Hochman [**Hoc14b**] has made a breakthrough in this direction, proving that dimension drop implies *super-exponential concentration of cylinders*, which can be viewed as "almost" exact overlaps. We now state the result precisely. Denote $\mathcal{A} = \{1, \ldots, m\}$ and define the distance between the cylinders associated to $\mathbf{i}, \mathbf{j} \in \mathcal{A}^n$ by

$$d(\mathbf{i}, \mathbf{j}) = \begin{cases} \infty, & r_\mathbf{i} \neq r_\mathbf{j}, \\ |S_\mathbf{i}(0) - S_\mathbf{j}(0)|, & r_\mathbf{i} = r_\mathbf{j}. \end{cases}$$

In the definition 0 may be replaced by any other point; this results in an equivalent condition. For $n \in \mathbb{N}$, let

$$\Delta_n = \min\{d(\mathbf{i}, \mathbf{j}) : \ \mathbf{i}, \mathbf{j} \in \mathcal{A}^n, \ \mathbf{i} \neq \mathbf{j}\}.$$

Note that exact overlaps occur if and only if $\Delta_n = 0$ for some n.

DEFINITION 6.8.4. We say that the IFS $\mathcal{S}$ has *exponential separation*, or equivalently, does not have *super-exponential concentration*, if there exists $\varepsilon > 0$ such that

$$\Delta_n \geqslant \varepsilon^n \quad \text{for infinitely many } n. \tag{6.50}$$

Exponential separation is easier to state for a homogeneous IFS

$$\mathcal{F}_{\lambda,D} := (S_1^\lambda(x) = \lambda x + d_1, \ldots, S_m^\lambda(x) = \lambda x + d_m),$$

with $D = \{d_1, \ldots, d_m\}$ being the digit set. Recall the space of power series, which already appeared many times, e.g., in the transversality condition:

$$\mathcal{B}_{D'} := \Big\{ \sum_{n=0}^{\infty} a_n x^n : \ a_n \in D', \ a_0 \neq 0 \Big\}, \quad \text{where } D' = D - D.$$

Let $\mathcal{B}_{D',n}$ be the set of polynomials of degree n in $\mathcal{B}_{D'}$. Observe that exponential separation holds for $\mathcal{F}_{\lambda,D}$ if and only if

$$\exists\, \varepsilon > 0 : \min_{P \in \mathcal{B}_{D',n}} |P(\lambda)| \geqslant \varepsilon^n, \quad \text{for infinitely many } n. \tag{6.51}$$

It was a natural question whether super-exponential concentration can even happen without exact overlaps, i.e., $\frac{1}{n}\log \Delta_n \to -\infty$ but $S_{\mathbf{i}} \neq S_{\mathbf{j}}$ for any $\mathbf{i} \neq \mathbf{j} \in \mathcal{A}^*$. This question was answered positively by Baker [**Bak21**] and Bárány and Käenmäki [**BK21**] independently. In particular, Baker [**Bak21**] showed that this is possible even in the case when $r_1 = \cdots = r_m$ is rational.

In fact, Hochman proved the corresponding result for self-similar measures, which implies the result about their attractors (self-similar sets). Recall (1.106) that for a self-similar measure corresponding to the IFS $\mathcal{S}$ with a probability vector $\mathbf{p} = (p_1, \ldots, p_m)$, the similarity dimension is defined by

$$\dim_{\mathrm{S}}(\nu) = \frac{-\sum_{i=1}^{m} p_i \log p_i}{-\sum_{i=1}^{m} p_i \log r_i} = \frac{h_{\mathbf{p}}}{\chi_{\mathbf{r}}^{\mathbf{p}}}.$$

Recall that we always assume that $p_i > 0$ for all $i \leqslant m$ when considering a probabilistic IFS, even if this is not stated explicitly. If we do mention it, we write it as $\mathbf{p} > 0$.

THEOREM 6.8.5 (Hochman [**Hoc14b**]). *Let $\mathcal{S}$ be a linear contracting IFS on $\mathbb{R}$, which satisfies the exponential separation condition* (6.50) *for some $\varepsilon > 0$. Let ν be the self-similar measure on $\mathbb{R}$ corresponding to a probability vector $\mathbf{p}$. Then $\dim_{\mathrm{H}}(\nu) = \min\{\dim_{\mathrm{S}}(\nu), 1\}$ for any probability vector $\mathbf{p}$.*

Taking the "natural" weights $p_i = r_i^s$, we obtain

COROLLARY 6.8.6 (Hochman [**Hoc14b**]). *If* (6.50) *holds, then*

$$\dim_{\mathrm{H}}(\Lambda) = \min\{s, 1\}.$$

These results have many interesting applications and special cases. We only mention two of them.

COROLLARY 6.8.7 (Hochman [**Hoc14b**]). *If the IFS $\mathcal{S} = (r_i x + d_i)_{i=1}^m$ on $\mathbb{R}$ has algebraic parameters, i.e. r_i and d_i are algebraic numbers for $i = 1, \ldots, m$, then there is a dichotomy: either* (6.49) *holds, or there are exact overlaps.*

This follows by a standard argument; we show it for *rational* parameters. If r_i and d_i are all rationals and q is the lowest common multiple of all their denominators, then $|S_{\mathbf{i}}(0) - S_{\mathbf{j}}(0)|$ can be written as a rational number with a denominator q^n, when $|\mathbf{i}| = |\mathbf{j}| = n$. Thus, either $\Delta_n = 0$, or $\Delta_n \geqslant q^{-n}$, whence (6.50) holds. □

REMARK 6.8.8. For algebraic parameters, when there are exact overlaps, a minor modification of Theorem 6.8.5 shows that $\dim_{\mathrm{H}}(\nu_\lambda) = \min\{1, \frac{h_\lambda}{-\log \lambda}\}$, where h_λ is the random walk entropy, usually called in this context the *Garsia entropy*, see (3.14) and Definition 8.6.1. This is stated in Theorem 8.6.4, as part of a discussion of ν_λ with algebraic parameters.

More recently, Varjú [**Var19b**] obtained the following spectacular result.

THEOREM 6.8.9 (Varjú [**Var19b**]). *Let ν_λ be the Bernoulli convolution measure with a contraction ratio λ. If $\lambda \in (1/2, 1)$ is transcendental, then* $\dim_{\mathrm{H}} \nu_\lambda = 1$.

Thus the exact overlap conjecture has been verified for Bernoulli convolutions (since for algebraic parameters it follows from Corollary 6.8.6) and hence for homogeneous self-similar IFS with two maps. The proof builds in a clever way on many of the earlier recent results; most crucially, on those of Hochman [**Hoc14b**] and Breuillard and Varjú [**BV20, BV19**].

Extending the result to homogeneous self-similar IFS with more than two maps is still open, but highly nontrivial results were obtained by Rapaport and Varjú [**RV20**]. Extending the techniques of [**Hoc14b**] and [**Var19b**] they proved, in particular, that the exact overlaps conjecture holds for the IFS $(\lambda x, \lambda x + 1, \lambda x + \tau)$ when $\lambda \geqslant 2^{-2/3}$. They also verified the conjecture for arbitrary homogeneous self-similar IFS with rational translations in [**RV20**, Theorem A.1].

In his recent paper, Rapaport [**Rap22b**] verified the exact overlap conjecture for an arbitrary self-similar IFS when the contraction ratios are algebraic, for any number of maps.

THEOREM 6.8.10 (Rapaport [**Rap22b**]). *Let $\mathcal{S} = (r_i x + d_i)_{i=1}^m$ be a linear contracting IFS on $\mathbb{R}$. Suppose that $r_1, \dots, r_m$ are all algebraic numbers and that $\mathcal{S}$ has no exact overlaps. Let ν be the self-similar measure on $\mathbb{R}$ corresponding to a probability vector $\mathbf{p}$. Then* $\dim_{\mathrm{H}}(\nu) = \min\{\dim_{\mathrm{S}}(\nu), 1\}$ *for any probability vector* $\mathbf{p}$.

Combining this result with Baker [**Bak21**], we obtain that there exists a self-similar IFS on the line for which the cylinders are super-exponentially concentrated but the dimension does not drop for any self-similar measure.

The second application of Theorem 6.8.5 is for parametric families of IFS. We state here the special case of [**Hoc15**, Theorem 1.10] for self-similar IFS on the line.

THEOREM 6.8.11 (Hochman [**Hoc15**]). *Let $I \subset \mathbb{R}^d$ be a connected and compact domain, let $r_i : I \to (-1,1)\backslash\{0\}$ and $d_i : I \to \mathbb{R}$ real-analytic. Consider $(S_{i,t} = r_i(t)x + d_i(t))_{i=1}^m$, a parametric family of self-similar IFS, and let $\Pi_{\mathbf{t}} : \Sigma \to \Lambda_t$ be the associated natural projection. Suppose that*

$$\forall\, \mathbf{i}, \mathbf{j} \in \Sigma, \;\; \Pi_t(\mathbf{i}) \equiv \Pi_t(\mathbf{j}) \text{ on } I \iff \mathbf{i} = \mathbf{j}. \tag{6.52}$$

For a probability vector $\mathbf{p}$ let $\nu_t^{\mathbf{p}}$ be the corresponding self-similar measure. Then the set $\{t \in I : \exists\, \mathbf{p} > 0,\ \dim_{\mathrm{H}}(\nu_t^{\mathbf{p}}) < \min\{\dim_{\mathrm{S}}(\nu_t^{\mathbf{p}}), 1\}$ has Hausdorff and packing dimension at most $d - 1$.

As a special case, we consider the translations as parameters.

COROLLARY 6.8.12. *Let $\mathcal{S} = (r_i x + d_i)_{i=1}^m$ be a self-similar IFS on the line and let* **p** *be a probability vector and let $\nu_{\mathbf{d}}^{\mathbf{p}}$ be the corresponding self-similar measure, where $\mathbf{d} = (d_1, \ldots, d_m)$ is the vector of the translation parameters. Then*

$$\dim_{\mathrm{P}}\{\mathbf{d} \in \mathbb{R}^m : \exists \mathbf{p} > 0,\ Dh(\nu^{\mathbf{p}}) < \min\{\dim_{\mathrm{S}}(\nu), 1\} \leqslant m - 1.$$

This result clearly implies Theorem 1.6.6 by applying it for the natural measure.

Hochman extended his theory to the case of self-similar IFS in $\mathbb{R}^d$ for $d \geqslant 2$, see [**Hoc15**]. There are many additional complications, and we do not state these results here. The methods used by Hochman are based on entropy techniques and ideas of additive combinatorics. A crucial step is a powerful "inverse theorem for entropy", which provides structural information about measures, under the assumption that the entropy almost does not grow under convolution.

6.8.2.1. *Results on absolute continuity.* Hochman's results were a breakthrough in our knowledge of the dimension of self-similar measures with overlaps, but they didn't tell anything new about absolute continuity in the "super-critical" regime, when the similarity dimension is greater than one. Shmerkin [**Shm14**] realized that, combining Hochman's Theorem with an old one, due to Erdős and Kahane, it is possible to obtain sharp results for the exceptional set for absolute continuity, in parametrized families of measures having a convolution structure.

THEOREM 6.8.13 (Shmerkin [**Shm14**]). *Let $m \geqslant 2$ and $d_1, \ldots, d_m$ are distinct real numbers. Consider the homogeneous self-similar IFS $\mathcal{F}_{\lambda,D} = (\lambda x + d_i)_{i=1}^m$. For $\lambda \in (0,1)$ and $\mathbf{p} = (p_1, \ldots, p_m)$ a probability vector, denote by $\nu_\lambda^{\mathbf{p}}$ the self-similar measure for the IFS $\mathcal{F}_{\lambda,D}$, corresponding to* **p**. *There exists a set $\mathcal{E} \subset (0,1)$ of zero Hausdorff dimension such that if $\lambda \in (0,1)\backslash\mathcal{E}$ and* **p** *is such that the similarity dimension $\dim_{\mathrm{S}}(\nu_\lambda^{\mathbf{p}}) > 1$, then $\nu_\lambda^{\mathbf{p}}$ is absolutely continuous.*

In the special case of $m = 2$ we obtain that the exceptional set for absolute continuity of the classical Bernoulli convolution ν_λ^p has zero Hausdorff dimension in the entire interval $(p^p(1-p)^{1-p}, 1)$. We point out several features of Theorem 6.8.13.

(a) Unlike Hochman's result, this theorem doesn't say anything about specific parameters and doesn't say anything about the *packing* measure of exceptions;

(b) it crucially uses the convolution structure of the self-similar measure, which is present only in the case of uniform contraction ratio;

(c) on the other hand, compared with earlier transversality results, it doesn't "care" about the weights p_i, as long as the similarity dimension of the measure is greater than one.

We sketch the proof, assuming Hochman's Theorem, in Chapter 8, where Fourier-analytic methods are discussed. The main new ingredient is a proposition saying that a convolution of two finite measures on the line, one of which has Hausdorff dimension and the other one exhibits a power Fourier decay at infinity, is absolutely continuous

Shmerkin's theorem above doesn't say anything about the density of the absolutely continuous self-similar measure; however, in a later work [**SS16**] it was shown, in particular, that under the conditions of Theorem 6.8.13 the measure $\nu_\lambda^{\mathbf{p}}$ actually has an L^q density for some $q > 1$. Even more recently, Shmerkin obtained the following striking result:

THEOREM 6.8.14 (Shmerkin [**Shm19**]). *In the setting of Theorem 6.8.13, there exists a set $\mathcal{E}' \subset (0,1)$ of zero Hausdorff dimension such that if $\lambda \in (0,1)\backslash\mathcal{E}'$ and $\lambda^{q-1} > \sum_{i=1}^m p_i^q$ for $q > 1$, then $\nu_\lambda^{\mathbf{p}}$ is absolutely continuous with a density in L^q.*

In particular, for the classical Bernoulli convolution ν_λ, the density is in L^q for all $q \in (1,\infty)$ for all $\lambda > 1/2$ outside a set of Hausdorff dimension zero. As a corollary, we also obtain that ν_λ has continuous density for all $\lambda \in (2^{-1/2}, 1)$ outside a set of Hausdorff dimension zero.

In fact, Shmerkin [**Shm19**] obtained analogous results for parametrized families of homogeneous self-similar measures on the line with a more general dependence on parameters (e.g., using the translation parameters instead of the contraction ratio).

THEOREM 6.8.15 (Shmerkin [**Shm19**, Theorem 9.2]). *Let*

$$u \mapsto (\lambda(u), d_1(u), \ldots, d_m(u))$$

be a real-analytic map from an open domain $U \subset \mathbb{R}^\ell$ to $((-1,0) \cup (0,1)) \times \mathbb{R}^m$. Assume that the family of IFS is nondegenerate in the following sense: for all $\mathbf{i} \neq \mathbf{j} \in \{1,\ldots,m\}^{\mathbb{N}}$ there is $u \in U$ such that

$$\sum_{n=0}^{\infty} d_{i_n}(u)\lambda(u)^n \neq \sum_{n=0}^{\infty} d_{j_n}(u)\lambda(u)^n$$

(this is a generalization of (6.52)). Given a probability vector $\mathbf{p} = (p_1,\ldots,p_m)$, let $\nu_u(\lambda,\mathbf{p})$ be the self-similar measure for the IFS $(x \mapsto \lambda(u)x + d_j(u))_{j=1}^m$ on $\mathbb{R}$. Then there exists a set $\mathcal{E} \subset U$ of Hausdorff dimension at most $\ell - 1$ such that if $u \in U\backslash\mathcal{E}$ and $|\lambda(u)|^{q-1} > \sum_{j=1}^m p_j^q$ for $q > 1$, then $\nu_u(\lambda,\mathbf{p})$ is absolutely continuous with a density in L^q.

A simple but useful consequence of having a density in L^q for all $q > 1$ is the following fact, where the crucial point is that the inequality is uniform in x.

LEMMA 6.8.16. *Suppose that a finite positive measure ν on $\mathbb{R}$ has a density which belongs to L^q for all $q > 1$. Then for every $\varepsilon > 0$ there exists $C_\varepsilon > 0$ such that for all $x \in \mathbb{R}$ and $r > 0$,*

$$\nu(B(x,r)) \leqslant C_\varepsilon r^{1-\varepsilon}. \tag{6.53}$$

PROOF. Let $h(x)$ be the density of ν. We write for $E = B(x,r)$, using Hölder's inequality:

$$\nu(E) = \int_E h(x)\,dx \leqslant \|h\|_q \cdot \mathcal{L}(E)^{1-\frac{1}{q}},$$

and the claim follows. □

The overall strategy of the proof of Theorem 6.8.14 and 6.8.15 is inspired by the ideas of [**Hoc14b**]. The key difference is that, whereas Hochman's method is based on entropy, Shmerkin works with L^q norms. He proved an inverse theorem for the decay of L^q norms under convolutions, using tools from additive combinatorics. This inverse theorem is a crucial step in the proof of the following result, which parallels Hochman's Theorem 6.8.5. (We state only the homogenous case, although [**Shm19**] treats the nonhomogeneous IFS case as well.)

THEOREM 6.8.17 (Shmerkin [**Shm19**]). *Let $m \geqslant 2$ and $d_1, \ldots, d_m$ be distinct real numbers. Consider the homogeneous self-similar IFS $\mathcal{F}_{\lambda,D} = (\lambda x + d_i)_{i=1}^m$. For $\lambda \in (0,1)$ and $\mathbf{p} = (p_1, \ldots, p_m)$ a probability vector, denote by $\nu_\lambda^{\mathbf{p}}$ the self-similar measure for the IFS $\mathcal{S}_\lambda$, corresponding to $\mathbf{p}$. Suppose that the IFS satisfies the exponential separation property (Definition 6.8.4), or equivalently,* (6.51). *Then for all $q \in (1, +\infty)$,*

$$D_q(\nu_\lambda^{\mathbf{p}}) = \min\left\{ \frac{-\log \sum_{j=1}^m p_j^q}{(q-1)\log \lambda}, 1 \right\}.$$

Recall that $\frac{-\log \sum_{j=1}^m p_j^q}{(q-1)} = D_q(\mu)$, where $\mu = \mathbf{p}^{\mathbb{N}}$, so the upper bound for the L^q-dimension is standard. Once Theorem 6.8.17 is proved, in order to obtain Theorem 6.8.14, Shmerkin used the scheme analogous to that in Theorem 6.8.13: namely, the convolution structure and a result from [**SS16**], asserting that the convolution of a measure on the line having L^q-dimension equal to one with another measure having power Fourier decay at infinity, is a.c. with an L^q-density.

For the natural measure, namely, $p_i = \frac{1}{m}$ for all $i \leqslant m$ we obtain the following.

COROLLARY 6.8.18. *Suppose that ν is the natural self-similar measure for the IFS $\mathcal{F}_{\lambda,D} = (\lambda x + d_i)_{i=1}^m$, which satisfies the exponential separation property* (6.51). *Then for all $q \in (1, +\infty)$,*

$$D_q(\nu) = \min\left\{ \frac{\log m}{-\log \lambda}, 1 \right\}.$$

This is a strengthening of Hochman's Theorem 6.8.5 in this special case, because $\dim_S(\nu) = \frac{\log m}{-\log \lambda}$ and $D_q(\nu) \leqslant \dim_H(\nu)$ for $q > 1$.

P. Varjú [**Var19a**] developed the methods of additive combinatorics, combined with entropy techniques, even further. His work builds on that of Hochman, but it is even more delicate. Focusing on the case of (biased) Bernoulli convolutions ν_λ^p as the most studied and arguably important class of self-similar measures with overlaps, he managed to prove absolute continuity for algebraic parameters under certain Diophantine conditions, in particular, providing the first explicit examples of absolute continuity in the biased case. We present just one consequence of Varjú's results as an illustration.

THEOREM 6.8.19 (Varjú [**Var19a**]). *The following (biased) Bernoulli convolution measures are absolutely continuous: $\nu_\lambda^p = \nu_{1-a/b}^p$ where $p \in (0.25, 0.75)$ and*

$$a, b \in \mathbb{N}, \quad a < 10^{-37} \frac{b}{\log(b+1)(\log\log(b+2))^3}.$$

Other results of Varjú and Breuillard–Varjú, concerning the dimension of Bernoulli convolutions for algebraic parameters, will be discussed below, in Section 8.6.

In a recent preprint Kittle [**Kit21**] substantially improved the method of Varjú [**Var19a**], using a new ingredient, which he calls the "detail of a measure around a scale". We refer to the original paper for precise statements, but mention that the smallest parameter for which he succeeded to show that $\nu_\lambda^{\frac{1}{2}}$ is absolutely continuous is $\lambda \approx 0.78207$ with minimal polynomial $x^8 - 2x^7 - x + 1$.

6.8.2.2. *Nonhomogeneous self-similar measures.* In Section 6.7.2 we mentioned the result of Saglietti, Shmerkin, and Solomyak [**SSS18**] on typical absolute continuity for nonhomogeneous self-similar measures on the line in the super-critical case. Here we state it precisely.

THEOREM 6.8.20 ([**SSS18**]). *Let $\mathcal{S} = (S_i(x) = \lambda_i x + d_i)_{i=1}^m$ be a self-similar IFS on the line with distinct translations d_i. Denote $\boldsymbol{\lambda} = (\lambda_1, \ldots, \lambda_m) \in (0,1)^m$ and fix a probability vector $\mathbf{p} = (p_1, \ldots, p_m)$. Consider the corresponding self-similar measure, denoted by $\nu_{\boldsymbol{\lambda}}^{\mathbf{p}}$, with the similarity dimension $\dim_S(\nu_{\boldsymbol{\lambda}}^{\mathbf{p}}) = h_{\mathbf{p}}/\chi_{\boldsymbol{\lambda}}^{\mathbf{p}}$. There exists a set*

$$E_{\mathbf{p}} \subset \mathcal{R}_{\mathbf{p}} := \{\boldsymbol{\lambda} \in (0,1)^m : \dim_S(\nu_{\boldsymbol{\lambda}}^{\mathbf{p}}) > 1\}$$

such that $\mathcal{L}^m(E_{\mathbf{p}}) = 0$ and for any $\boldsymbol{\lambda} \in \mathcal{R}_{\mathbf{p}} \backslash E_{\mathbf{p}}$ we have $\nu_{\boldsymbol{\lambda}}^{\mathbf{p}} \ll \mathcal{L}^1$.

In the nonhomogeneous setting of this theorem, the self-similar measures no longer have an infinite convolution structure, and hence the methods of [**Shm14**] are not directly applicable. A way around this problem was found in [**GSSY16**]: Galicer, Saglietti, Shmerkin, and Yavicoli (see [**GSSY16**, Lemma 6.6]) were able to express nonhomogeneous self-similar measures as averages over measures with infinite convolution structure. However, these infinite convolutions are only "statistically self-similar" in some sense. In the paper [**SSS18**], additive combinatorics entropy techniques of Hochman and estimates of the Fourier decay, going back to Erdős and Kahane, were extended to this random setting. Then it was possible to use Lemma 8.4.1 of Shmerkin to reach the necessary conclusion.

Modifying the techniques of [**SSS18**], Käenmäki and Orponen [**KO20**] obtained a result analogous to Theorem 6.8.20, in which the contraction parameters are fixed, but the translations $d_i(u)$ depend real-analytically on a real parameter. Under some conditions of nondegeneracy and *transversality of order K* (a variant of a condition used by Hochman [**Hoc14b**]), see [**KO20**] for details, they showed that absolute continuity holds in the "super-critical" parameter region (i.e., where $\dim_S(\nu_{\boldsymbol{\lambda}}^{\mathbf{p}}) > 1$), everywhere outside of an exceptional set of Hausdorff dimension zero. One of the main motivations for Käenmäki and Orponen was to prove the following result on absolute continuity of projections of self-similar measures.

THEOREM 6.8.21 (Käenmäki and Orponen [**KO20**, Prop.1.7]). *Let $\Psi = (\lambda_i x_i + t_i)_{i=1}^m$ be a self-similar IFS in $\mathbb{R}^2$ with $\boldsymbol{\lambda} = (\lambda_1, \ldots, \lambda_m) \in (0,1)^m$ and $t_i \in \mathbb{R}^2$, such that the fixed points of the contractions do not lie on the same line, which satisfies the exponential separation condition in the plane, defined similarly to Definition 6.8.4. Consider the corresponding planar self-similar measure $\nu_{\boldsymbol{\lambda}}^{\mathbf{p}}$ for a probability vector $\mathbf{p}$ and suppose that $\dim_S(\nu_{\boldsymbol{\lambda}}^{\mathbf{p}}) > 1$. Then the orthogonal projection of the planar self-similar measure $\nu_{\boldsymbol{\lambda}}^{\mathbf{p}}$ to the line in the direction $\mathbf{v} \in \mathbb{S}^1$ is absolutely continuous except for a set of directions of Hausdorff dimension zero, that is, $(\mathrm{proj}_{\mathbf{v}})_* \nu_{\boldsymbol{\lambda}}^{\mathbf{p}} \ll \mathcal{L}^1$ for all $\mathbf{v} \in \mathbb{S}^1 \backslash E$, where $\dim_H(E) = 0$.*

For projections of homogeneous self-similar measures analogous result was obtained in [**SS16**], but with the strong separation condition (in the plane) instead of the much weaker, exponential separation condition in the last theorem.

6.8.2.3. *The case of general ergodic measures on the symbolic space.* It is often important to consider general push-forward measures under the natural projection, and not just self-similar measures. In this direction T. Jordan and A. Rapaport obtained the following interesting result.

THEOREM 6.8.22 (Jordan and Rapaport [**JR21**]). *Let $\mathcal{S}$ be a linear contracting IFS on $\mathbb{R}$, which satisfies the exponential separation condition* (6.50) *for some* $\varepsilon > 0$. *Let μ be any ergodic shift-invariant measure on the symbolic space Σ and let $\Pi_*\mu$ be the push-forward measure under the natural projection. Then*

$$\dim_{\mathrm{H}}(\Pi_*\mu) = \min\left\{1, \frac{h_\mu}{\chi_\mu}\right\},$$

where h_μ and χ_μ are the entropy and the Lyapunov exponent of μ respectively.

The proof relies on approximation of ergodic measures by self-similar measures; the implementation is quite nontrivial—it uses the result of Shmerkin [**Shm19**, Theorem 6.6] on L^q-dimensions of self-similar measures satisfying the exponential separation condition, which is a general version of Theorem 6.8.17.

CHAPTER 7

Further properties of self-similar IFS with overlaps

Here we continue to study families of self-similar IFS on $\mathbb{R}$ with overlaps. In the last chapter we proved that, under the assumption of transversality, for almost every parameter the Hausdorff (and hence packing) dimension of the attractor equals the minimum of 1 and the similarity dimension. A natural question arises about the Hausdorff and packing *measure* of the attractor, in the interval of parameters where the similarity dimension is $\leqslant 1$. It turns out that in many cases the Hausdorff measure (in the similarity dimension) is typically zero, whereas the packing measure is positive and finite, and as a consequence, the natural self-similar measure is typically (a.s.) equivalent to the packing measure. We will also see that the positivity of the Hausdorff measure is equivalent to the weak separation property in dimension one, but not in higher dimensions.

This chapter contains many examples. Several of them are especially important, since they provide counter-examples to what might be natural conjectures. In Section 7.3.2 we present an example of Kalamutdinov and Tetenov, demonstrating that a self-similar IFS may have the "one-point intersection property", while the Open Set Condition fails. In Section 7.4, we present an example of Simon and Vágó, showing that the analog of the exact overlaps conjecture does not hold for the absolute continuity of self-similar measures. In Section 7.5, an example of Jordan et al. demonstrates that in dimension 2 a self-similar set may have positive Lebesgue measure, but empty interior.

7.1. Hausdorff and packing measure for self-similar IFS: Introduction

Consider a one-parameter family of homogeneous self-similar IFS on $\mathbb{R}$ of the form

$$\mathcal{S}^t := \left(S_i^t(x) = \lambda(t)\cdot x + a_i(t)\right)_{i=1}^m, \quad t \in J, \tag{7.1}$$

where $J \subset \mathbb{R}$ is an interval. We assume that $t \mapsto \lambda(t)$ and $t \mapsto a_i(t)$ belong to $\mathcal{C}^1(J)$, and there exist q_1, q_2 such that

$$0 < q_1 < \lambda(t) < q_2 < \frac{1}{m} \qquad \text{for all } t \in J. \tag{7.2}$$

As usual, we denote the symbolic space by $\Sigma := \{1,\dots,m\}^{\mathbb{N}}$ and the set of finite words by $\Sigma^* := \bigcup_{n=1}^{\infty}\Sigma_n$, where $\Sigma_n = \{1,\dots,m\}^n$. We write Λ^t for the attractor of $\mathcal{S}^t$ and let $\Pi^t : \Sigma \to \Lambda^t$ be the natural projection. Observe that

$$\Pi^t(\mathbf{i}) = \sum_{n=0}^{\infty} a_{i_n}(t)\cdot\lambda(t)^{n-1}, \quad \text{where } \mathbf{i} = i_0 i_1 \ldots \in \Sigma. \tag{7.3}$$

It follows from (7.2) that the similarity dimension satisfies $s(t) = \frac{\log m}{-\log\lambda(t)} < 1$.

Next we state the transversality condition. It is very similar, but slightly more general, than the condition (6.3.1) we introduced in the beginning of Chapter 6, and is a very special case of the generalized transversality condition (6.31).

DEFINITION 7.1.1. For $\mathbf{i}, \mathbf{j} \in \Sigma$ let $f_{\mathbf{i},\mathbf{j}}(t) = \Pi^t(\mathbf{i}) - \Pi^t(\mathbf{j})$. We say that the family of IFS $\{\mathcal{S}^t\}_{t\in J}$ from (7.1) satisfies the transversality condition on J if there exists $\delta > 0$ such that for all $\mathbf{i}, \mathbf{j} \in \Sigma$, with $i_1 \neq j_1$,

$$t \in J,\ |f_{\mathbf{i},\mathbf{j}}(t)| \leqslant \delta \implies |f'_{\mathbf{i},\mathbf{j}}(t)| \geqslant \delta. \tag{7.4}$$

Similarly to Lemma 6.3.4, using compactness of Σ, one shows that this condition is equivalent to

$$\text{if there exists } t \in J \text{ such that } f_{\mathbf{i},\mathbf{j}}(t) = f'_{\mathbf{i},\mathbf{j}}(t) = 0, \quad \text{then } \mathbf{i} \equiv \mathbf{j}. \tag{7.5}$$

THEOREM 7.1.2. *Assume that the transversality condition* (7.4) *holds for a family of IFS* $\{\mathcal{S}^t\}_{t\in J}$ *of the form* (7.1), *satisfying* (7.2). *Then the packing measure of the attractor* Λ^t *is positive and finite for almost all parameters:*

$$0 < \mathcal{P}^{s(t)}(\Lambda^t) < \infty, \qquad \text{for Lebesgue-a.e. } t \in J. \tag{7.6}$$

Moreover, for such t *the measure* $\mathcal{P}^{s(t)}|_{\Lambda^t}$ *is a constant multiple of the natural self-similar measure* ν_t.

In fact, finiteness of the packing measure in the similarity dimension always holds; it is included for completeness.

The theorem was first proved in [**PSS00b**]. Rams [**Ram05**] significantly simplified the proof of this theorem, and we present his approach in the next section. In fact, he used the *intersection number method*, which originated in his earlier paper [**Ram02**] and which has other applications as well. Both [**PSS00b**] and [**Ram05**] also cover some cases of *nonhomogeneous* self-similar IFS. This will be discussed, but not with full details, also following Rams' approach.

Next we state the results on zero Hausdorff measure, but first we need a definition.

DEFINITION 7.1.3. Let $\{\mathcal{S}^t\}_{t\in J}$ be a family of IFS given by (7.1). We say that $t \in J$ is an *intersection parameter* if there exist $i \neq j$ such that the corresponding first level cylinders of the attractor intersect: $\Lambda^t_i \cap \Lambda^t_j \neq \varnothing$. The set of intersection parameters will be denoted by $\mathcal{IP}$.

Notice that $t \in J$ is an intersection parameter if and only if $\mathcal{S}^t$ does not have the Strong Separation Property.

THEOREM 7.1.4. *Let* $\{\mathcal{S}^t\}_{t\in J}$ *be a family of IFS given by* (7.1) *and satisfying* (7.2). *If the transversality condition* (7.4) *holds and* $\mathcal{IP}$ *has positive Lebesgue measure, then* $\mathcal{H}^{s(t)}(\Lambda^t) = 0$ *for Lebesgue-a.e.* $t \in \mathcal{IP}$.

In [**PSS00b**] this was proved in the more general case of nonhomogeneous one-parameter families of IFS.

The assumption (7.2) implies that $s(t) < \frac{\log m}{-\log q_2} < 1$; however, the proof given below applies also in the case when $\lambda(t) = m^{-1}$ for all $t \in J$, so that $s(t) = 1$. In that case we obtain under the transversality condition that $\mathcal{L}^1(\Lambda^t) = 0$ for a.e. $t \in \mathcal{IP}$.

In the next proposition we assume that $a_i(t) \equiv a_i$ are constants and the contraction rate $\lambda(t) = t$ is the parameter.

PROPOSITION 7.1.5. *Consider the IFS*

$$\mathcal{S}^t = \left(S_i^t(x) = tx + a_i\right)_{i=1}^m, \quad t \in J. \tag{7.7}$$

Suppose that the set of intersection parameters $\mathcal{IP}$ contains an interval $\widetilde{J}$. Then

(i) *there exists a dense countable set $\mathcal{E} \subset \widetilde{J}$ for which the attractor has a "dimension drop", that is, $\dim_H(\Lambda^t) < s(t)$ for $t \in \mathcal{E}$;*

(ii) $\mathcal{H}^{s(t)}(\Lambda^t) = 0$ *for t in a dense G_δ subset of $\widetilde{J}$.*

In this generality the proposition seems to be new; it was proved in [**PSS00b**] under the transversality assumption.

7.2. Positive packing measure and intersection numbers

In this section we follow Michał Rams' papers [**Ram02**] and [**Ram05**].

7.2.1. The homogeneous case.

PROOF OF THEOREM 7.1.2. Recall the set-up for the reader's convenience. We have the family of IFS

$$\mathcal{S}^t := \left(S_i^t(x) = \lambda(t) \cdot x + a_i(t)\right)_{i=1}^m, \quad t \in J,$$

where $0 < q_1 < \lambda(t) < q_2 < \frac{1}{m}$ for all $t \in J$. The attractor is denoted by Λ^t and its cylinder sets are $\Lambda_{\mathbf{i}}^t$ for $\mathbf{i} \in \Sigma_n := \{1, \ldots, m\}^n$.

Without loss of generality we may assume that the parameter interval J is so small that there exists an open set U and $0 < \varepsilon_1 < \varepsilon_2$ such that $B(\Lambda^t, \varepsilon_1) \subset U \subset B(\Lambda^t, \varepsilon_2)$ hods for every $t \in J$. Then for any $\mathbf{i} \in \Sigma_n$,

$$B(\Lambda_{\mathbf{i}}^t, \lambda(t)^n \varepsilon_1) \subset U^t(\mathbf{i}) \subset B(\Lambda_{\mathbf{i}}^t, \lambda(t)^n \varepsilon_2), \tag{7.8}$$

where $U^t(\mathbf{i}) := S_{\mathbf{i}}^t(U)$. In view of the transversality condition (7.4), we obtain from (7.2) that there exists Const > 0 such that for all $n \in \mathbb{N}$,

$$\mathcal{L}^1 \left\{t \in J : U^t(\mathbf{i}) \cap U^t(\mathbf{j}) \neq \varnothing\right\} < \text{Const} \cdot q_2^n, \quad \forall \mathbf{i}, \mathbf{j} \in \Sigma_n, \ i_1 \neq j_1. \tag{7.9}$$

Let ν_t be the natural measure on Λ^t, that is, $\nu_t := \Pi_*^t(\mu)$ is the push-forward measure of the uniform measure $\mu := \left(\frac{1}{m}, \ldots, \frac{1}{m}\right)^{\mathbb{N}}$. Clearly,

$$\begin{aligned} \frac{\#\left\{\mathbf{j} \in \Sigma_n : \Lambda_{\mathbf{j}}^t \subset B(x,r)\right\}}{m^n} &\leqslant \nu_t(B(x,r)) \\ &\leqslant \frac{\#\left\{\mathbf{j} \in \Sigma_n : \Lambda_{\mathbf{j}}^t \cap B(x,r) \neq \varnothing\right\}}{m^n}. \end{aligned} \tag{7.10}$$

Observe that setting $C_1 = \varepsilon_1, C_2 = 2|U|$, for every $\mathbf{j} \in \Sigma_n$, $t \in J$, and $x \in \Lambda_{\mathbf{i}}^t$, we have

$$\Lambda_{\mathbf{j}}^t \cap B(x, C_1 \lambda^n(t)) \neq \varnothing \Longrightarrow U^t(\mathbf{i}) \cap U^t(\mathbf{j}) \neq \varnothing \tag{7.11}$$

and

$$U^t(\mathbf{j}) \cap U^t(\mathbf{i}) \neq \varnothing \Longrightarrow \Lambda_{\mathbf{j}}^t \subset B(x, C_2 \lambda^n(t)). \tag{7.12}$$

Therefore, for every $t \in J$, $\mathbf{i} \in \Sigma_n$ and $x \in \Lambda_{\mathbf{i}}^t$ we have

$$\nu_t\left(B(x, C_1 \lambda^n(t))\right) \leqslant \frac{\widetilde{A}^t(\mathbf{i})}{m^n} \leqslant \nu_t\left(B(x, C_2 \lambda^n(t))\right), \tag{7.13}$$

where

$$\widetilde{A}^t(\mathbf{i}) := \#\left\{\mathbf{j} \in \Sigma_n : U^t(\mathbf{i}) \cap U^t(\mathbf{j}) \neq \varnothing\right\}. \tag{7.14}$$

The intersection number, for $n \in \mathbb{N}$, is defined as the number of pairs of words of length n for which the corresponding images of U intersect:

$$\widetilde{A}^t_n := \#\left\{(\mathbf{i},\mathbf{j}) \in \Sigma_n \times \Sigma_n : U^t(\mathbf{i}) \cap U^t(\mathbf{j}) \neq \varnothing\right\} = \sum_{\mathbf{i}\in\Sigma_n} \widetilde{A}^t(\mathbf{i}). \tag{7.15}$$

In order to estimate the intersection numbers, we will need

$$A^t_n := \left\{(\mathbf{i},\mathbf{j}) \in \Sigma_n \times \Sigma_n : U^t(\mathbf{i}) \cap U^t(\mathbf{j}) \neq \varnothing,\ i_1 \neq j_1\right\}, \tag{7.16}$$

where $\mathbf{i} = i_1 \dots i_n$ and $\mathbf{j} = j_1 \dots j_n$. It follows from the definitions that $\widetilde{A}^t_n = A^t_n + m \cdot \widetilde{A}^t_{n-1}$, hence

$$\frac{\widetilde{A}^t_n}{m^n} = 1 + \sum_{\ell=1}^{n} m^{-\ell} A^t_\ell. \tag{7.17}$$

LEMMA 7.2.1. *Assume that for $t \in J$ there exists a sequence of natural numbers $n_i(t)$ such that $\lim_{n\to\infty} n_i(t) \to \infty$ and a constant $M(t) > 0$ such that*

$$\widetilde{A}^t_{n_i(t)} < M(t) \cdot m^{n_i(t)}. \tag{7.18}$$

Then the $s(t)$-dimensional packing measure of the attractor Λ^t is positive:

$$\mathcal{P}^{s(t)}(\Lambda^t) > 0. \tag{7.19}$$

PROOF OF LEMMA 7.2.1. Fix $t \in J$ for which (7.18) holds. To simplify notation, we will drop the index t in the proof of this lemma, so that $\Lambda = \Lambda^t$, $\nu = \nu_t$, $\widetilde{A}_n = \widetilde{A}^t_n$, $M = M(t)$, etc. First observe that by (7.14) and (7.15) for any $n \in \mathbb{N}$ we have

$$\begin{aligned} \int \widetilde{A}_n(\mathbf{i}|_n)d\mu(\mathbf{i}) &= \sum_{\mathbf{i}\in\Sigma_n} \#\left\{\mathbf{j} \in \Sigma_n : U(\mathbf{j}) \cap U(\mathbf{i}) \neq \varnothing\right\} \cdot m^{-n} \\ &= \widetilde{A}_n \cdot m^{-n}. \end{aligned} \tag{7.20}$$

Recall that $s = \log m/(-\log\lambda)$ is the similarity dimension of the IFS. We consider the lower s-density of the measure ν at $x = \Pi(\mathbf{i})$ for $\mathbf{i} \in \Sigma$ (see (2.1)):

$$\Theta^s_*(\nu, x) := \liminf_{n\to\infty} \frac{\nu(B(x, C_1\lambda^n))}{(C_1\lambda^n)^s}, \tag{7.21}$$

and use (7.13) to relate it to $\widetilde{A}_n(\mathbf{i}|_n)$, as follows:

$$\Theta^s_*(\nu, x) \leqslant \liminf_{n\to\infty} \frac{\frac{\widetilde{A}_n(\mathbf{i}|_n)}{m^n}}{C_1^s \cdot m^{-n}} = C_1^{-s} \cdot \liminf_{n\to\infty} \widetilde{A}_n(\mathbf{i}|_n). \tag{7.22}$$

By the definition of ν as the push-forward of μ, Fatou's Lemma, and (7.20), we obtain

$$\begin{aligned} \int \Theta^s_*(\nu, x)\, d\nu(x) &\leqslant C_1^{-s} \int \liminf_{n\to\infty} \widetilde{A}_n(\mathbf{i}|_n)\, d\mu(\mathbf{i}) \\ &\leqslant C_1^{-s} \liminf_{n\to\infty} \int \widetilde{A}_n(\mathbf{i}|_n)\, d\mu(\mathbf{i}) \\ &= C_1^{-s} \liminf_{n\to\infty} \widetilde{A}_n \cdot m^{-n} \\ &\leqslant C_1^{-s} \cdot M. \end{aligned} \tag{7.23}$$

Then by Markov's inequality we obtain that for an arbitrary $N > 0$:

$$\nu\{x : \Theta^s_*(\nu, x) > N\} \leqslant \frac{C_1^s \cdot M}{N}, \tag{7.24}$$

whence

$$\nu\{x : \Theta^s_*(\nu, x) = \infty\} = 0.$$

This implies that $\mathcal{P}^{s(t)}(\Lambda^t) > 0$, see Theorem 2.1.2 in Section 2.1 or [**TT85**, Theorem 5.4]. □

To conclude the proof of the theorem we need one more lemma:

LEMMA 7.2.2. *Assuming the transversality condition* (7.4), *for almost all* $t \in J$ *there exist* $n_i(t) \to \infty$ *and* $M(t) > 0$ *such that* (7.18) *holds.*

PROOF OF LEMMA 7.2.2. First observe that by transversality (7.9), for an arbitrary n we have

$$\int_J A_n^t \, dt = \sum_{\substack{\mathbf{i},\mathbf{j} \in \Sigma_n \\ i_1 \neq j_1}} \mathcal{L}^1 \{t \in J : U(\mathbf{i}) \cap U(\mathbf{j}) \neq \varnothing\} < \text{const} \cdot m^{2n} q_2^n. \tag{7.25}$$

Combining this with (7.17) and then using (7.2), we obtain that there exists $0 < \text{Const}' = \text{Const}'(t) < \infty$ such that for all n we have

$$\int_J \frac{\widetilde{A}_n^t}{m^n} \, dt \leqslant |J| + \text{Const} \cdot \sum_{\ell=1}^n (m \cdot q_2)^\ell < \text{Const}'. \tag{7.26}$$

Then by Fatou's Lemma,

$$\int_J \liminf_{n\to\infty} \frac{\widetilde{A}_n^t}{m^n} \, dt \leqslant \liminf_{n\to\infty} \int_J \frac{\widetilde{A}_n^t}{m^n} \, dt < \text{Const}'. \tag{7.27}$$

It follows that for every $N \in \mathbb{N}$,

$$\mathcal{L}^1 \left\{ t \in J : \liminf_{n\to\infty} \frac{\widetilde{A}_n^t}{m^n} > N \right\} \leqslant \frac{\text{Const}'}{N}, \tag{7.28}$$

and the proof of the desired claim is finished. □

Lemmas 7.2.1 and 7.2.2 together immediately imply the assertion of Theorem 7.1.2.

□

7.2.2. Intersection numbers for nonhomogeneous IFS. We start with the definitions for a single self-similar IFS of the form

$$\mathcal{S} := (S_i(x) = \lambda_i x + a_i)_{i=1}^m, \quad x \in \mathbb{R}. \tag{7.29}$$

and later consider families of such IFS, depending on parameter. Intersection numbers are defined using a version of Moran covers. We have already introduced Moran cut-sets and the corresponding Moran covers in Lemma 1.5.4. When we consider parametrized families, it will be convenient to have the cut-sets independent of the parameter. Thus it is reasonable to modify the definition slightly, allowing more flexibility, following [**Ram02**, p. 128].

Let Λ be the attractor of $\mathcal{S}$ and $\Pi : \Sigma \to \Lambda$ the natural projection. Fix an open interval $U \subset \mathbb{R}$ such that $\Lambda \subset U$. For $\mathbf{i} \in \Sigma^*$ we write $U(\mathbf{i}) = S_{\mathbf{i}}(U)$; note that $|U(\mathbf{i})| = |U| \cdot \lambda_{\mathbf{i}}$.

DEFINITION 7.2.3. For $\delta > 0$ and $v > 1$, we say that $\{U(\mathbf{i}) : \mathbf{i} \in Z_\delta \subset \Sigma^*\}$ is a *Moran cover of* Λ *of size* δ *and variation* v if

- for all $\mathbf{i} \in Z_\delta$ we have $|U(\mathbf{i})| \in \left(v^{-1}\delta, v\delta\right)$ and
- $\left\{[\mathbf{i}] : \mathbf{i} \in Z_\delta\right\}$ is a partition of Σ, so that Z_δ is a cut-set.

A *Moran collection* Z is a choice of cut-sets $Z = \{Z_\delta\}_{\delta \leqslant |U|}$ such that $\{U(\mathbf{i}) : \mathbf{i} \in Z_\delta \subset \Sigma^*\}$ is a choice of Moran covers of Λ having the same variation v for all $\delta \leqslant |U|$.

Below we fix a Moran collection Z with a variation $v > 1$. Let s be the similarity dimension of the IFS $\mathcal{S}$, and consider the corresponding natural measure μ on Σ:

$$\mu = \{\lambda_1^s, \ldots, \lambda_m^s\}^{\mathbb{N}}$$

Observe that there exists $c > 1$ such that for all $\mathbf{i} \in \Sigma^*$ we have

$$\frac{|U(\mathbf{i})|^s}{\mu([\mathbf{i}])} \in (c^{-1}, c),$$

and hence

$$\mu([\mathbf{i}]) \in \left((cv)^{-s} \cdot \delta^s, (cv)^s \cdot \delta^s\right) \text{ for all } \mathbf{i} \in Z_\delta. \tag{7.30}$$

DEFINITION 7.2.4 (Intersection numbers). By analogy with (7.15) and (7.16), we define the intersection numbers in the nonhomogeneous case. Let $Z = \{Z_\delta\}_{\delta \leqslant |U|}$ be a Moran collection. For $0 < \delta \leqslant |U|$ let

$$\widetilde{A}_\delta := \#\{(\mathbf{i}, \mathbf{j}) \in Z_\delta \times Z_\delta : U(\mathbf{i}) \cap U(\mathbf{j}) \neq \varnothing\} \tag{7.31}$$

and

$$A_\delta := \#\{(\mathbf{i}, \mathbf{j}) \in Z_\delta \times Z_\delta : U(\mathbf{i}) \cap U(\mathbf{j}) \neq \varnothing,\ i_1 \neq j_1\}. \tag{7.32}$$

There is a relation between the intersection numbers $\widetilde{A}_\delta$ and A_δ, similar to (7.17).

PROPOSITION 7.2.5 ([**Ram05**, Proposition 4.3]). *There exist constants* $L_2, L_3 > 0$ *such that for every* $0 < \delta < |U|$ *we have*

$$L_2^{-1}\delta^{-s}\left(1 + \sum_{\delta_i \geqslant L_3^{-1}\delta} \delta_i^s A_{\delta_i}\right) \leqslant \widetilde{A}_\delta \leqslant L_2\delta^{-s}\left(1 + \sum_{\delta_i \geqslant L_3\delta} \delta_i^s A_{\delta_i}\right). \tag{7.33}$$

7.2.3. Connection with the correlation dimension. Before turning to families of IFS, we mention an interesting application of intersection numbers. The following theorem was proved by Rams in [**Ram02**, Proposition 3.2]:

THEOREM 7.2.6. *Let* $\mathcal{S}$ *be a self-similar IFS* (7.29) *of similarity dimension* s *and let* ν *be the natural self-similar measure, that is,* $\nu = \Pi_*(\mu)$, *where* $\mu = \{\lambda_1^s, \ldots, \lambda_m^s\}^{\mathbb{N}}$. *Let* $Z = \{Z_\delta\}_{\delta \leqslant |U|}$ *be a Moran collection, and consider the intersection numbers* $\widetilde{A}_\delta$ *defined by* (7.31) *for* $0 < \delta \leqslant |U|$. *Then the limit*

$$\widetilde{s} := \lim_{\delta \to 0} \frac{\log \widetilde{A}_\delta}{-\log \delta} \tag{7.34}$$

exists and $2s - \widetilde{s} = D_2(\nu)$, *the correlation dimension of* ν.

Recall that the correlation dimension always exists for a self-similar measure, see Theorem 3.6.1.

PROOF SKETCH. For $\mathbf{i} \in \Sigma^*$ and $\delta > 0$ let

$$\widetilde{A}_\delta(\mathbf{i}) = \#\{\mathbf{j} \in Z_\delta : \; U(\mathbf{i}) \cap U(\mathbf{j}) \neq \varnothing\}.$$

Similarly to (7.13), it is not hard to show that there exist positive constants C_1', C_2' such that for all $\mathbf{i} \in Z_\delta$ and $x \in \Lambda_\mathbf{i}$ holds

$$\nu\big(B(x, C_1'\delta)\big) \leqslant \widetilde{A}_\delta(\mathbf{i}) \cdot \delta^s \leqslant \nu\big(B(x, C_2'\delta)\big). \tag{7.35}$$

By the Definition 1.9.7 of the correlation measure,

$$D_2(\nu) = \lim_{r \to 0} \frac{\log \int \nu(B(x,r))\, d\nu(x)}{\log r}.$$

where the existence of the limit follows from Theorem 3.6.1 and (2.29). In view of (7.30) and (7.35), summing over all $\mathbf{i} \in Z_\delta$ we obtain

$$\int \nu\big(B(x, C_1'\delta)\big)\, d\nu(x) \leqslant \sum_{\mathbf{i} \in Z_\delta} \widetilde{A}_\delta(\mathbf{i}) \cdot \delta^s \cdot \mu([\mathbf{i}]) \leqslant \widetilde{A}_\delta \cdot (cv)^s \delta^{2s},$$

and similarly,

$$\int \nu\big(B(x, C_2'\delta)\big)\, d\nu(x) \geqslant \widetilde{A}_\delta \cdot (cv)^{-s} \delta^{2s}.$$

Now the desired claim is immediate. □

REMARK 7.2.7. A related result was obtained by Simon and Solomyak in [**SS98**, Theorem 1] under some additional assumptions. Namely, if the attractor Λ has Newhouse thickness greater than 1 (see [**PT93**] for the definition), or if $\mathcal{S}$ is a homogeneous IFS such that $\Lambda - \Lambda$ is interval, then

$$D_2(\nu) = 2s - \overline{\dim}_{\mathrm{B}}(\mathcal{Z}), \tag{7.36}$$

where s is the similarity dimension, $\mathcal{Z} = \{(\mathbf{i}, \mathbf{j}) \in \Sigma^2 : \Pi(\mathbf{i}) = \Pi(\mathbf{j})\}$, and the upper box dimension $\overline{\dim}_{\mathrm{B}}$ is considered with respect to the metric $\varrho(\mathbf{i}, \mathbf{j}) = |\Lambda_{\mathbf{i} \wedge \mathbf{j}}|$.

In order to explain the link with Theorem 7.2.6, assume that $\mathcal{S}$ is a homogeneous IFS with contraction ratio $\lambda \in (0,1)$. Then

$$\overline{\dim}_{\mathrm{B}}(\mathcal{Z}) = \limsup_{n \to \infty} \frac{\log \widetilde{T}_n}{-n \log \lambda},$$

where $\widetilde{T}_n = \#\big\{(\mathbf{i}, \mathbf{j}) \in \Sigma_n \times \Sigma_n : \; \Lambda_\mathbf{i} \cap \Lambda_\mathbf{j} \neq \varnothing\big\}$. The assumption that $\Lambda - \Lambda$ is an interval guarantees that

$$\Lambda_\mathbf{i} \cap \Lambda_\mathbf{j} \neq \varnothing \iff I_\mathbf{i} \cap I_\mathbf{j} \neq \varnothing,$$

where $I_\mathbf{i}, I_\mathbf{j}$ are the smallest closed intervals containing $\Lambda_\mathbf{i}, \Lambda_\mathbf{j}$ respectively, which shows a close connection of $\widetilde{T}_n$ to the intersection number $\widetilde{A}_n$ from (7.15) (with a parameter t fixed). For further details and the exact statement with the proof, see Corollary 12.4.6 below.

Next we mention a consequence of Theorem 7.2.6 in the homogeneous case. So now we again assume that our self-similar IFS $\mathcal{S}$ on $\mathbb{R}$ is of the form

$$\mathcal{S} := \left(S_i(x) = \lambda \cdot x + a_i\right)_{i=1}^m, \tag{7.37}$$

and ν is the push-forward measure of the uniformly distributed measure $\mu := \left(\frac{1}{m}, \ldots, \frac{1}{m}\right)^{\mathbb{N}}$ on the symbolic space Σ. The following was pointed out in [**Sim02**, Corollary 7].

COROLLARY 7.2.8. *Let I be the smallest interval that contains the attractor Λ and let $z \geqslant |I|$ be arbitrary. By analogy with* (7.16), *let*

$$A_n(z) := \#\left\{(\mathbf{i},\mathbf{j}) \in \Sigma_n \times \Sigma_n : i_1 \neq j_1, \left|\sum_{k=1}^{n} i_k\lambda^k - \sum_{k=1}^{n} j_k\lambda^k\right| \leqslant z\lambda^n\right\}$$

and consider the exponential growth rate of $A_n(z)$:

$$E_n(z) := \frac{\log A_n(z)}{n}. \tag{7.38}$$

If

$$\limsup_{n\to\infty} E_n(z) > \log m, \tag{7.39}$$

then $\lim_{n\to\infty} E_n(z)$ exists and equals $(2s - D_2(\nu))\log\lambda$. This means that the correlation dimension of ν drops, that is, $D_2(\nu) < s$. On the other hand, if $\limsup_{n\to\infty} E_n(z) \leqslant \log m$, then $D_2(\nu) = s$.

Comparing this proposition with a recent result of Shmerkin, see Corollary 6.8.18, we obtain

COROLLARY 7.2.9. *Let $z \geqslant |I|$. Whenever the IFS $\mathcal{S}$ of the form* (7.37) *satisfies the Exponential Separation Condition* (6.50), *we have*

$$\limsup_{n\to\infty} E_n(z) \leqslant \log m. \tag{7.40}$$

7.2.4. Intersection numbers for families of nonhomogeneous IFS. For a nontrivial parameter interval $J \subset \mathbb{R}$ we consider the parameter-dependent IFS

$$\mathcal{S}^t := \left(S_i^t(x) = \lambda_i(t)\cdot x + a_i(t)\right)_{i=1}^m, \qquad t \in J. \tag{7.41}$$

We restrict ourselves to the special case when there is a function $b(t)$ with the property that there exists $\widetilde{t} \in J$ such that for all $i \in \{1,\ldots,m\}$:

$$\lambda_i(t) = \lambda_i(\widetilde{t})^{b(t)}, \qquad t \in J. \tag{7.42}$$

The benefit of assumption (7.42) is that, although the contraction ratio depends on the parameter t, the quantity $\lambda(t)^{s(t)}$, where $s(t)$ is the similarity dimension, is independent of t. This provides a kind of homogeneity to the system. In particular, the natural measure is constructed with the same weights for all $t \in J$. Moreover, a Moran collection (see Definition 7.2.3) for one parameter is a Moran collection for all other parameters. More precisely, it follows from (7.42) that if Z is a Moran collection for $\widetilde{t} \in J$ with variation v, then there exists v' such that Z is also a Moran collection for any other $t \in J$ with variation v'.

NOTATION. Although the similarity dimension $s(t) = \frac{s(\widetilde{t})}{b(t)}$ may differ for different $t \in J$, the quantity $\lambda_i(t)^{s(t)} \equiv \lambda_i(\widetilde{t})^{s(\widetilde{t})}$ is independent of $t \in J$ and hence the natural weights $\lambda_1(t)^{s(t)},\ldots,\lambda_m(t)^{s(t)}$ are identical for all $t \in J$. So we can define

$$\mu := \left\{\lambda_1(t)^{s(t)},\ldots,\lambda_m(t)^{s(t)}\right\}^{\mathbb{N}},$$

and the natural measure is

$$\nu_t := \Pi_*^t(\mu), \tag{7.43}$$

where $\Pi^t : \Sigma \to \Lambda^t$ is the natural projection. For the rest of this section fix an open interval U such that

$$\Lambda^t \subset U, \qquad \forall t \in J. \tag{7.44}$$

As before, we write $U^t(\mathbf{i}) := S^t_{\mathbf{i}}(U)$ for $\mathbf{i} \in \Sigma^\star$. Then $|U^t(\mathbf{i})| = |U| \cdot \lambda_{\mathbf{i}}(t)$.

Intersection numbers. The intersection numbers in the nonhomogeneous case were already defined in (7.31) and (7.32). Here we repeat the definition for convenience emphasizing the dependence on parameter t. It is important, however, that $Z = \{Z_\delta\}_{\delta \leqslant |U|}$ is a Moran collection for all parameters $t \in J$. For $0 < \delta < |U|$ let

$$\widetilde{A}^t_\delta := \#\left\{(\mathbf{i}, \mathbf{j}) \in Z_\delta \times Z_\delta : U^t(\mathbf{i}) \cap U^t(\mathbf{j}) \neq \varnothing\right\} \tag{7.45}$$

and

$$A^t_\delta := \#\left\{(\mathbf{i}, \mathbf{j}) \in Z_\delta \times Z_\delta : U^t(\mathbf{i}) \cap U^t(\mathbf{j}) \neq \varnothing,\ i_1 \neq j_1\right\}. \tag{7.46}$$

Usually, we use the properties $\widetilde{A}^t_\delta$ to get some information about Λ^t, but we can apply the transversality condition directly only for A^t_δ. So, it is vital to establish a connection (similar to (7.17)) between them. When we use the transversality method, the following parameter interval plays an important role. For $\mathbf{i}, \mathbf{j} \in \Sigma^\star$ let

$$I(\mathbf{i}, \mathbf{j}) := \left\{t \in J : U^t(\mathbf{i}) \cap U^t(\mathbf{j}) \neq \varnothing.\right\} \tag{7.47}$$

Note that

$$A^t_\delta = \sum_{\substack{\mathbf{i},\mathbf{j} \in Z_\delta \\ i_1 \neq j_1}} \mathbb{1}_{I(\mathbf{i},\mathbf{j})}(t).$$

The following is a key lemma:

LEMMA 7.2.10 ([**Ram05**, Lemma 4.5]). *Assume that the one-parameter family of self-similar IFS of the form* (7.41) *satisfies the transversality condition and* (7.42) *holds. Then for almost all $t \in J$ there exists a positive $M = M(t)$ and a sequence $z_i = z_i(t)$, $\lim_{i\to\infty} z_i = 0$, such that*

(i) *if $s(t) < 1$, then $\widetilde{A}_{z_i(t)}(t) \leqslant M \cdot z_i(t)^{-s(t)}$,*
(ii) *if $s(t) > 1$, then $\widetilde{A}_{z_i(t)}(t) \leqslant M \cdot z_i(t)^{-2s(t)+1}$,*
(iii) *if $s(t) = 1$, then $\widetilde{A}_{z_i(t)}(t) \leqslant -M \cdot z_i(t)^{-s(t)} \log z_i(t)$.*

We do not sketch the proof, but only mention that part (i) is analogous to Lemma 7.2.2.

Applications of Lemma 7.2.10 lead to three interesting results which will be discussed below.

The case when $s(t) \leqslant 1$. Rams proved the following theorem [**Ram05**, Theorem 5.6] (the first part of which appeared earlier in [**PSS00b**] with a different proof).

THEOREM 7.2.11 (Rams). *Under the assumptions of Lemma* 7.2.10 *for almost all parameters $t \in J$, with $s(t) \leqslant 1$, we have*

(i) *if $s(t) < 1$, then the s-dimensional packing measure of Λ^t is positive and $\mathcal{P}^{s(t)}|_{\Lambda^t}$ is a (positive) constant multiple ν_t.*
(ii) *if $s(t) = 1$, then the packing measure with the gauge function $\phi(x) = x\log(1/x)$ of Λ^t is positive and ν_t is absolutely continuous with respect to this measure.*

For the definition of packing measure with respect to a gauge function, see Section 2.7. To appreciate part (ii) of the theorem, consider the four-corner Cantor set $F \subset [0,1]^2 = C_{1/2} \times C_{1/2}$, where $C_{1/2}$ is the middle-$\frac{1}{2}$ Cantor set. This is a set of Hausdorff dimension 1, so for the 1-parameter family of self-similar sets F_α, $\alpha \in [0,\pi)$, obtained as orthogonal projections of F to the line making an angle α with the horizontal, the similarity dimension is $s(\alpha) \equiv 1$. However, it was proved by Kenyon [**Ken97**] that for all α such that $\tan\alpha \notin \mathbb{Q}$, we have $\mathcal{H}^1(F_\alpha) = \mathcal{L}^1(F_\alpha) = 0$. (Actually, Kenyon proved this for the projections of the "1-dimensional Sierpiński gasket", see Section 7.7.1, but the proof is the same.) So, for all α with irrational $\tan\alpha$, the natural measure on F_α cannot be the (1-dimensional) Hausdorff measure, but for almost all α it is absolutely continuous with respect to the packing measure with the gauge function $x\log(1/x)$.

The case when $s(t) > 1$. In the case when $s(t) > 1$ we are interested in whether ν_t is absolute continuous with L^2 density. The following theorem is a generalization of Theorem 6.3.3, because it covers nonhomogeneous families of IFS satisfying the property (7.42). It can be proved with the methods of Section 6.3; a different proof was given by Rams in [**Ram05**, Theorem 5.4] using the intersection numbers, in particular, Lemma 7.2.10.

THEOREM 7.2.12. *Under the assumptions of Lemma* 7.2.10, *for almost all parameter values for which $s(t) > 1$, the natural measure ν_t is absolutely continuous with respect to the Lebesgue measure with L^2 density.*

This theorem follows from part (ii) of Lemma 7.2.10 and the following two facts. The first one is from [**Mat95**, p. 36 and p. 43] and was already used in the book (see (6.19) and the beginning of the proof of Theorem 6.3.3), and the second one is proved in [**Ram05**, Proposition 5.3]:

(a) The measure ν_t is absolutely continuous with respect to the Lebesgue measure with L^2 density if and only if $\int_{\Lambda^t} \underline{D}(\nu_t, x)\, d\nu_t(x) < \infty$;

(b) if $\int_{\Lambda^t} \underline{D}(\nu_t, x)\, d\nu_t(x) = \infty$ then $\lim_{\delta\to 0} \delta^{2s-1} \cdot \widetilde{A}_\delta = \infty$. We recall that $\underline{D}(\nu, x) := \liminf_{n\to\infty} \frac{\nu[x-r,x+r]}{2r}$.

These, together with Lemma 7.2.10, yield Theorem 7.2.12.

7.3. Zero Hausdorff measure and dimension drop

We are going to use the Bandt, Graf and Schief Theorem 4.1.2. More specifically, we will use that the Hausdorff measure of a self-similar set in the similarity dimension is zero if and only if for every $\varepsilon > 0$ there exist distinct ε-relatively close cylinders of the IFS, see condition BG2 (4.4). For convenience, we state this explicitly below, in the special case of a self-similar homogeneous IFS on the line.

PROPOSITION 7.3.1 (Bandt and Graf [**BG92**]). *Let*

$$(S_i(x) = \lambda x + a_i)_{i=1}^m, \quad \lambda \in (0,1),$$

be a self-similar IFS on the line, let Λ be the attractor, and let $s = \frac{\log m}{-\log\lambda}$ be the similarity dimension. Then $\mathcal{H}^s(\Lambda) = 0$ if and only if for every $\varepsilon > 0$ there exists $n \in \mathbb{N}$ and $\mathbf{i} \neq \mathbf{j}$ in Σ_n such that

$$|S_{\mathbf{i}}(0) - S_{\mathbf{j}}(0)| < \varepsilon\lambda^n |\Lambda|. \tag{7.48}$$

REMARK. It is obvious that the right-hand side of (7.48) may be replaced by $c\varepsilon\lambda^n$, for any $c > 0$ independent of n.

NOTATION. Consider a family of homogeneous IFS $\{\mathcal{S}^t\}_{t\in J}$ given by (7.1), having the contraction ratio $\lambda(t)$. Let $\mathcal{V}_\varepsilon$ be the set of $t \in J$ such that there exist $n \in \mathbb{N}$ and distinct $\mathbf{i}, \mathbf{j} \in \Sigma_n$, satisfying

$$|f_{\mathbf{i}\bar{1},\mathbf{j}\bar{1}}(t)| = |\Pi^t(\mathbf{i}\bar{1}) - \Pi^t(\mathbf{j}\bar{1})| < \varepsilon\lambda(t)^n. \tag{7.49}$$

Here $\bar{1} = 111\ldots$ Observe that, since the self-similar IFS is homogeneous, by (7.3) we have

$$f_{\mathbf{i}\bar{1},\mathbf{j}\bar{1}}(t) = S^t_{\mathbf{i}}(0) - S^t_{\mathbf{j}}(0). \tag{7.50}$$

PROOF OF THEOREM 7.1.4. By Proposition 7.3.1, $\mathcal{H}^{s(t)}(\Lambda^t) = 0$ if and only if $t \in \mathcal{V}_\varepsilon$ for all $\varepsilon > 0$. Thus it is enough to prove that $\mathcal{IP}\backslash\mathcal{V}_\varepsilon$ has zero Lebesgue measure for any $\varepsilon > 0$. To this end, we will show that $\mathcal{IP}\backslash\mathcal{V}_\varepsilon$ has no Lebesgue density points in the interior of J.

Fix $\varepsilon > 0$ and $t_0 \in \mathcal{IP}$ in the interior of J. Since $t_0 \in \mathcal{IP}$, there exist infinite sequences $\mathbf{i}, \mathbf{j}$ in Σ, with $i_1 \neq j_1$, such that $\Pi^{t_0}(\mathbf{i}) = \Pi^{t_0}(\mathbf{j})$, whence the corresponding cylinders of the attractor of level n intersect for all n:

$$\Lambda^{t_0}_{\mathbf{i}|n} \cap \Lambda^{t_0}_{\mathbf{j}|n} \neq \varnothing.$$

It follows from (7.3) and (7.2) that the diameter of Λ^t satisfies

$$|\Lambda^t| \leqslant \max_{j\leqslant m} \|a_j\|_\infty \cdot (1 - q_2)^{-1} =: C_1,$$

hence

$$|\Pi^{t_0}(\mathbf{i}|_n\bar{1}) - \Pi^{t_0}(\mathbf{j}|_n\bar{1})| \leqslant |\Lambda^{t_0}_{\mathbf{i}|n}| \leqslant C_1\lambda(t_0)^n, \quad n \in \mathbb{N}.$$

Consider the function

$$f(t) := f_{(\mathbf{i}|_n)\bar{1},(\mathbf{j}|_n)\bar{1}}(t) = \Pi^t(\mathbf{i}|_n\bar{1}) - \Pi^t(\mathbf{j}|_n\bar{1}). \tag{7.51}$$

For n sufficiently large we have

$$|f(t_0)| \leqslant C_1\lambda(t_0)^n < \delta, \tag{7.52}$$

where $\delta > 0$ is the constant from the definition of transversality (7.1.1). By transversality,

$$|f'(t_0)| \geqslant \delta. \tag{7.53}$$

Thus $f(t)$ is monotonic in a neighborhood of t_0. Without loss of generality, assume that $f(t)$ is increasing and positive at $t = t_0$. Then it follows from (7.53) and (7.52) that $f(t_1) = 0$ for some $t_1 \in I_{t_0} := [t_0 - \frac{C_1}{\delta}\lambda(t_0)^n, t_0]$. (Here we assume that n is sufficiently large, so that $I_{t_0} \subset J$.) In view of (7.3),

$$\|f'\|_{C(J)} \leqslant C_2 := 2\Big(\frac{\max_{j\leqslant m}\|a'_j\|_\infty}{1 - q_2} + \frac{\max_{j\leqslant m}\|a_j\|_\infty}{(1 - q_2)^2}\Big).$$

It follows that

$$|f(t)| \leqslant (\varepsilon/2)\lambda(t_0)^n, \quad \text{for all } t \in I'_{t_0} := [t_1, t_1 + C_2^{-1}(\varepsilon/2)\lambda(t_0)^n].$$

It remains to check that

$$\lambda(t)^n \leqslant 2\lambda(t_0)^n \quad \text{for all } t \in I'_{t_0},\ n \text{ sufficiently large.} \tag{7.54}$$

Then $|f(t)| \leqslant \varepsilon\lambda(t)^n$ for all $t \in I'_{t_0}$, and by construction, in view of (7.51) and (7.50), this means that $I'_{t_0} \subset \mathcal{V}_\varepsilon$. Since the ratio $\frac{|I'_{t_0}|}{|I_{t_0}|} = \frac{\varepsilon\delta}{C_1C_2}$ does not depend on n, it follows that t_0 is not a density point of $\mathcal{IP}\backslash\mathcal{V}_\varepsilon$. In order to verify (7.54), observe that

$$|\lambda(t) - \lambda(t_0)| \leqslant \|\lambda'\|_{C(J)} \cdot |t - t_0| \leqslant \text{const} \cdot \lambda(t_0)^n \leqslant C_3 \cdot q_2^n, \quad \text{for all } t \in I'_{t_0}.$$

But then

$$\left|1 - \frac{\lambda(t)}{\lambda(t_0)}\right| \leqslant \frac{C_3 \cdot q_2^n}{q_1},$$

whence

$$(\lambda(t)/\lambda(t_0))^n \leqslant (1 + (C_3/q_1)q_2^n)^n < 2,$$

for n sufficiently large, since $q_2 < 1$. The proof is complete. □

Next we turn to the proof of Proposition 7.1.5. Recall that the attractor of the IFS (7.7) can be written explicitly as the set of sums of power series in t:

$$\Lambda^t = \Big\{\sum_{n=0}^{\infty} b_n t^n : b_n \in D\Big\}, \quad D := \{a_1, \ldots, a_m\}.$$

It follows that $t \in \mathcal{IP}$ if and only if t is a zero of a power series from

$$\mathcal{B}_{D'} = \Big\{\sum_{n=0}^{\infty} b_n x^n : b_n \in D' = D - D\Big\},$$

see (6.13). First we note the following

LEMMA 7.3.2. *Under the assumptions of Proposition 7.1.5, if $t_0 \in \mathrm{Int}(\widetilde{J})$ is a zero of a power series $f \in \mathcal{B}_{D'}$, such that the graph of $x \mapsto f(x)$ changes the sign at $x = t_0$, then t_0 is a limit point of parameters for which $\dim_{\mathrm{H}}(\Lambda^t) < s(t)$.*

PROOF OF THE LEMMA. Observe that if t is a zero of a polynomial from $\mathcal{B}_{D'}$ of degree $n-1$ (note that $0 \in D'$ so the truncation of any power series from $\mathcal{B}_{D'}$ is a polynomial belonging to $\mathcal{B}_{D'}$), then there exist $\mathbf{i} \neq \mathbf{j}$, words of length n in the alphabet D, such that the corresponding cylinders of the attractor coincide: $\Lambda^t_{\mathbf{i}} = \Lambda^t_{\mathbf{j}}$. Or, to be more precise, two of the level-n iterates of the IFS: $S^t_{\mathbf{i}}$ and $S^t_{\mathbf{j}}$ are identical. Then, passing to the n-level iterates, we see that the attractor Λ^t is also the attractor of an IFS $\mathcal{S}'$ consisting of $m^n - 1$ maps, with a common contraction ratio t^n. Therefore,

$$\dim_{\mathrm{H}}(\Lambda^t) \leqslant s(\mathcal{S}') = \frac{\log(m^n - 1)}{-n\log t} < s(t).$$

Now, since $t_0 \in (0,1)$, it is clear from the Intermediate Value Theorem that changing the "tail" of the power series f starting from $n = N$ will result in a power series having a zero t_N near t_0, for N sufficiently large, and $t_N \to t_0$. (This is a consequence of the assumption that the graph of f changes the sign at t_0.) We can change the tail so that the resulting power series is a polynomial (if f is not a polynomial, we simply truncate it at level N; if f is a polynomial, we add a nonzero coefficient at level N). From the above, we see that $\dim_{\mathrm{H}}(\mathcal{C}^{t_N}) < s(t_N)$, hence we obtained a sequence $t_N \to t_0$ of parameters at which the attractor has dimension drop, as desired. □

PROOF OF PROPOSITION 7.1.5. (i) Let $d'_{\max} := \max(D')$; clearly, $\min(D') = -d'_{\max}$ by symmetry. Let

$$\widetilde{\mathcal{B}}_{D'}^{+} = \{f \in \mathcal{B}_{D'} : \exists\, N \in \mathbb{N},\ \forall\, n \geqslant N,\ b_n = d'_{\max}\},$$

$$\widetilde{\mathcal{B}}_{D'}^{-} = \{f \in \mathcal{B}_{D'} : \exists\, N \in \mathbb{N},\ \forall\, n \geqslant N,\ b_n = -d'_{\max}\},$$

and finally, $\widetilde{\mathcal{B}}_{D'} = \widetilde{\mathcal{B}}_{D'}^{+} \cup \widetilde{\mathcal{B}}_{D'}^{-}$. Note that $\widetilde{\mathcal{B}}_{D'}$ is a countable set of rational functions. Let $t_0 \in \mathrm{Int}(\widetilde{\mathcal{J}})\backslash\mathcal{Z}$, where $\mathcal{Z}$ is a countable set of zeros of all functions in $\widetilde{\mathcal{B}}_{D'}$. By construction, t_0 is a zero of a power series $f \in \mathcal{B}_{D'}\backslash\widetilde{\mathcal{B}}_{D'}$. If the graph of f changes sign at t_0, then Lemma 7.3.2 gives the needed claim. Otherwise, f is strictly positive or strictly negative in some punctured neighborhood $B(t_0, \varepsilon)\backslash\{t_0\}$. Assume it is strictly positive without loss of generality. Since $f \notin \widetilde{\mathcal{B}}_{D'}$, there is a sequence $n_k \to \infty$ such that $b_{n_k} \neq -d'_{\max}$. We can then consider a sequence of perturbed power series, replacing the coefficient b_{n_k} by $-d'_{\max}$. This “lowers” the graph a little, and so for n_k sufficiently large, we will obtain a zero γ_{n_k} satisfying the conditions of Lemma 7.3.2, with $\gamma_{n_k} \to t_0$. This allows us to conclude the proof by applying the lemma.

(ii) By Proposition 7.3.1,

$$\{t \in \widetilde{\mathcal{J}} : \mathcal{H}^{s(t)}(\Lambda^t) = 0\} = \bigcap_{n \in \mathbb{N}} (\mathcal{V}_{1/n} \cap \widetilde{\mathcal{J}}).$$

It follows from (7.49) and continuity of $t \mapsto \Pi^t(\mathbf{i})$ that $\mathcal{V}_\varepsilon$ is an open set, hence $\{t \in \widetilde{\mathcal{J}} : \mathcal{H}^{s(t)}(\Lambda^t) = 0\}$ is a G_δ set, and it is dense by part (i). □

7.3.1. Set of intersection parameters. For Theorem 7.1.4 and Proposition 7.1.5 to be meaningful, we need to know when the set of intersection parameters $\mathcal{IP}$ has positive Lebesgue measure and/or contains an interval. This is sometimes difficult to decide; here we provide some easy partial results.

Below we denote by $\mathsf{Conv}(A)$ the convex hull of a set A and by Λ_i^t, $i \in \mathcal{A}$, the cylinders of the attractor. We always assume that the family of IFS is nondegenerate, that is,

$$\Pi^t(\mathbf{i}) = \Pi^t(\mathbf{j}),\ \forall\, t \in J \ \Rightarrow\ \mathbf{i} \equiv \mathbf{j}. \tag{7.55}$$

PROPOSITION 7.3.3. *Let $\{\mathcal{S}^t\}_{t \in J}$ be a family of IFS given by* (7.1) *and satisfying* (7.2). *Suppose that for some $i \neq j \in \mathcal{A}$ there exists a subinterval $\widetilde{J} \subset J$ such that*

$$\mathsf{Conv}(\Lambda_i^t) \cap \mathsf{Conv}(\Lambda_j^t) \neq \varnothing \quad \textit{for all } t \in \widetilde{J} \tag{7.56}$$

and

$$\Lambda^t - \Lambda^t \quad \textit{is an interval for all } t \in \widetilde{J}. \tag{7.57}$$

Then $\widetilde{J} \subset \mathcal{IP}$.

PROOF. Let $\mathsf{Conv}(\Lambda^t) = [b_\ell(t), b_r(t)]$. Then from (7.57) it follows that for all $t \in \widetilde{J}$,

$$\Lambda^t - \Lambda^t = [b_\ell(t) - b_r(t), b_r(t) - b_\ell(t)] = \mathsf{Conv}(\Lambda^t) - \mathsf{Conv}(\Lambda^t).$$

For any set $A \subset \mathbb{R}$ we have $A - A = \{x \in \mathbb{R} : A \cap (A + x) \neq \varnothing\}$. Therefore, for all $t \in \widetilde{J}$,

$$\Lambda^t \cap (\Lambda^t + x) \neq \varnothing \iff \mathsf{Conv}(\Lambda^t) \cap [\mathsf{Conv}(\Lambda^t) + x] \neq \varnothing.$$

Since the IFS is homogeneous, Λ_i^t and Λ_j^t are both translates of $\lambda(t)\Lambda^t$; thus (7.56) implies $\Lambda_i^t \cap \Lambda_j^t \neq \varnothing$ for all $t \in \widetilde{J}$, and the claim follows. □

REMARK 7.3.4. Conditions (7.56) and (7.57) are easy to verify. Indeed, let Λ be the attractor of a homogeneous IFS $(\lambda x + a_i)_{i=1}^m$, with $\lambda \in (0,1)$. Assume, without loss of generality, that $a_1 = 0 < a_2 < \ldots < a_m$. Then $\Lambda - \Lambda$ is also a self-similar set:

$$\Lambda - \Lambda = \left\{ \sum_{n=0}^{\infty} c_n \lambda^n : c \in \Gamma \right\},$$

where $\Gamma = \{a_i - a_j\}_{i,j \leqslant m}$. Let $g := \min_{j \leqslant m}(a_j - a_{j-1})$ and let G be the maximal gap between two consecutive elements of Γ (in increasing order). Then it is an easy exercise to check that there exist $i \neq j$ for which $\mathsf{Conv}(\Lambda_i) \cap \mathsf{Conv}(\Lambda_j) \neq \varnothing$ if and only if

$$\frac{\lambda}{1-\lambda} \max \Gamma \geqslant g, \tag{7.58}$$

and $\Lambda - \Lambda$ is an interval if and only if

$$\frac{\lambda}{1-\lambda} \max \Gamma \geqslant G/2. \tag{7.59}$$

PROPOSITION 7.3.5. *Let $\{\mathcal{S}^t\}_{t \in J}$ be a family of IFS given by* (7.1) *and satisfying* (7.2). *If the transversality condition* (7.4) *holds, then* $\dim_{\mathrm{H}}(\mathcal{IP}) \leqslant 2s_{\max}$, *where* $s_{\max} := \max\{s(t) : t \in J\}$. *Thus, if* $s_{\max} \in (0, \frac{1}{2})$ *then* $\mathcal{L}^1(\mathcal{IP}) = 0$.

This was proved in [**PSS00b**, Prop. 2.4] in the more general case of not-necessarily homogenous IFS families.

PROOF SKETCH. Recall that $t_0 \in \mathcal{IP}$ means that there exist $\mathbf{i}, \mathbf{j} \in \Sigma$, with $i_1 \neq j_1$, such that $f_{\mathbf{i},\mathbf{j}}(t_0) = 0$. By the transversality condition (7.4), there exists $\delta > 0$ such that $|f_{\mathbf{i},\mathbf{j}}(t)| \leqslant \delta$ implies $|f'_{\mathbf{i},\mathbf{j}}(t)| \geqslant \delta$; in particular, $|f'_{\mathbf{i},\mathbf{j}}(t_0)| \geqslant \delta$. For $w \in \Sigma^*$ denote $I_w^t = \mathsf{Conv}(\Lambda_w^t)$. If $t_0 \in \mathcal{IP}$ and $\mathbf{i}, \mathbf{j} \in \Sigma$ are as above, then the intervals I_u^t and I_v^t move relative to each other with a speed bounded below by $\delta/2$ as t varies close to t_0, provided that u and v are prefixes of $\mathbf{i}, \mathbf{j}$ of length n, for n sufficiently large. Therefore, the set $\{t \in J : I_u^t \cap I_v^t \neq \varnothing\}$ is a union of at most $\frac{C}{\delta}$ intervals of length

$$\frac{C}{\delta} \cdot \max\{|I_u^t| + |I_v^t| : t \in J\} = \frac{C' \lambda_{\max}^n}{\delta}$$

each. This is a consequence of transversality. It follows that $\mathcal{IP}$ may be covered by $\frac{C}{\delta} m^{2n}$ intervals of length $\frac{C' \lambda_{\max}^n}{\delta}$, hence

$$\dim_{\mathrm{H}}(\mathcal{IP}) \leqslant \frac{2 \log m}{-\log \lambda_{\max}},$$

as desired. □

7.3.2. No OSC with one-point intersection. Here we sketch a proof of Theorem 4.1.8 by Kamalutdinov and Tetenov [**KT20**], with the help of a transversality lemma. We recall the result for convenience:

THEOREM 7.3.6 (Kamalutdinov and Tetenov [**KT20**]). *There is a self-similar IFS on $\mathbb{R}$ with one-point intersection property, but without the OSC.*

We start with a version of a transversality lemma, which is a generalization of Proposition 7.3.5 and [**PSS00b**, Prop.2.4]. It replaces a more technical, as we believe, General Positioning Theorem used in [**KT20**].

For a nontrivial parameter interval $J \subset \mathbb{R}$ consider the IFS family

$$\mathcal{S}^t := \left(S_i^t(x) = \lambda_i(t) \cdot x + a_i(t)\right)_{i=1}^m, \quad t \in J. \tag{7.60}$$

Assume that $\lambda_i(t) \in (-1,1)$ and $a_i(t) \in \mathbb{R}$ are $\mathcal{C}^1$-functions on J. Denote by Λ^t the attractor and by $s(t)$ the similarity dimension of $\mathcal{S}^t$. As usual, $\Sigma = \{1,\dots,m\}^{\mathbb{N}}$ is the symbolic space, $\Pi^t : \Sigma \to \Lambda^t$ is the natural projection, and $\Lambda_w^t = S_w^t(\Lambda)$ is the cylinder of the attractor corresponding to a finite word w.

LEMMA 7.3.7. *Fix two distinct finite words u, v, not necessarily of the same length. Suppose that the transversality condition on J holds for the pair of cylinders $[u], [v]$, namely, there exists $\delta > 0$ such that for all $\mathbf{i} \in [u]$, $\mathbf{j} \in [v]$,*

$$\forall\, t \in J,\ \left|\Pi^t(\mathbf{i}) - \Pi^t(\mathbf{j})\right| \leqslant \delta \implies \left|\frac{\partial}{\partial t}\left(\Pi^t(\mathbf{i}) - \Pi^t(\mathbf{j})\right)\right| \geqslant \delta. \tag{7.61}$$

Then

$$\dim_{\mathrm{H}}\left(\{t \in J : \ \Lambda_u^t \cap \Lambda_v^t \neq \varnothing\}\right) \leqslant 2 s_{\max} := \max_{t \in J} s(t).$$

Of course, this lemma has content only when $s_{\max} < \frac{1}{2}$, in which case we will say that this is a family of *small* IFS. The proof of the lemma is essentially contained in [**PSS00b**, Prop. 2.4], where transversality on all distinct cylinders is considered. It proceeds along the lines of the sketch of the proof of Proposition 7.3.5, except that we use a Moran cover for a specific $t_0 \in J$ and take a sufficiently small subinterval of the parameter interval, or alternatively, work with a Moran collection, see Definition 7.2.3, for a small parameter interval.

Now we turn to the example.

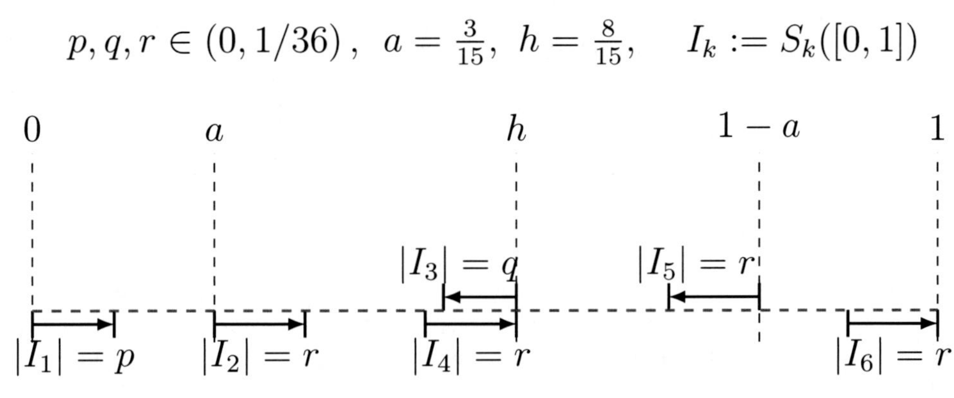

FIGURE 7.1. One point intersection property with no OSC

EXAMPLE 7.3.8 (Kalamutdinov and Tetenov). Let $h := \frac{8}{15}$, $a := \frac{3}{15}$, and choose $p, q, r \in \left(0, \frac{1}{36}\right)$. For the parameter $q \in \left(0, \frac{1}{36}\right)$ we define the one-parameter family $\mathcal{S}^q$ of small IFS on the interval $[0,1]$ as follows:

$$\begin{gathered} S_1^q(x) := px, \quad S_2^q(x) := a + rx, \quad S_3^q(x) := h - qx, \\ S_4^q(x) := h - r + rx, \quad S_5^q(x) := 1 - a - rx, \quad S_6^q(x) := 1 - r + rx. \end{gathered} \tag{7.62}$$

Observe that only S_3^q depends on the parameter q, but we put the superscript q for all functions of the IFS for convenience. This IFS is denoted by

$$\mathcal{S}_{p,r}^q := (S_1^q, \dots, S_6^q).$$

Note that in this example there is always an overlap in at least one point, since $S_3^q(0) = S_4^q(1) = h$. So in order to prove Theorem 7.3.6, we show that the IFS $\mathcal{S}_{p,r}^q$ does not satisfy the WSP for a specific choice of p, r and hence it does not satisfy OSC (see Corollary 4.2.11). Furthermore, we show that $\mathcal{S}_{p,r}^q$ satisfies the transversality assumption of Lemma 7.3.7 in parameter q, and so we prove that $S_3^q(0) = S_4^q(1)$ is the only intersection for Lebesgue almost every q.

LEMMA 7.3.9. *Assume that*

$$\frac{\log p}{\log r} \notin \mathbb{Q}. \tag{7.63}$$

Then the IFS $\mathcal{S}_{p,r}^q$ from Example 7.3.8 *does not satisfy the WSP.*

PROOF. In Example 7.3.8, for some fixed n and m we choose

$$\mathbf{i} := (3, \underbrace{1, \ldots, 1}_{m}, 5) \quad \text{and} \quad \mathbf{j} := (4, \underbrace{6, \ldots, 6}_{n}, 2),$$

and calculate:

$$H_m(x) := S_{\mathbf{i}}^q(x) = h - p^m q + p^m qa - p^m qr \cdot x,$$

and

$$G_n(x) := S_{\mathbf{j}}^q(x) = h - r^{n+1} + r^{n+1}a - r^{n+2} \cdot x.$$

In view of (7.63), for each q we can find $(m_k, n_k) \in \mathbb{N}^2$ such that

$$\lim_{k \to \infty} p^{-m_k} r^{n_k+1} = q. \tag{7.64}$$

In order to refute the WSP, consider the following sequence of functions

$$G_{n_k}^{-1} H_{m_k}(x) = \frac{\left(r^{n_k+1} - p^{m_k} q\right)(a-1)}{r^{n_k+2}} + \frac{p^{m_k} q}{r^{n_k+1}} x.$$

It follows from (7.64) that these linear functions tends to the identity Id : $x \to x$ as $k \to \infty$. This implies that (4.17) does not hold, and hence by Theorem 4.2.4 the WSP is not satisfied. □

We now fix p, r satisfying (7.63) and denote the attractor of $\mathcal{S}_{p,r}^q$ by Λ^q. In order to conclude the proof of the one-point intersection property, it remains to verify that for Lebesgue-a.e. $q \in (0, \frac{1}{36})$ we have $\Lambda_3^q \cap \Lambda_4^q = \{h\}$, and this is done in [**KT20**]. In order to simplify our task, we will not try to get precise bounds and instead assume that $p, q, r \in (0, \varepsilon)$ for $\varepsilon > 0$ sufficiently small. It is then enough to show that for any $m, n \geqslant 0$, the set

$$\Delta_{m,n}(p, r) = \bigcup_{i \neq 1, j \neq 6} \{q \in (0, \varepsilon) : \Lambda_{31^m i}^q \cap \Lambda_{46^n j}^q \neq \varnothing\} \tag{7.65}$$

has Lebesgue measure zero, provided $\varepsilon > 0$ is small enough. In fact, we will deduce from Lemma 7.3.7 that

$$\dim_{\mathrm{H}}\left(\Delta_{m,n}(p, r)\right) \leqslant 2s_{\max} < 1.$$

In order to justify this claim, we need to check the transversality condition (7.61). Observe that

$$S_{31^m i}^q(0) = h - qp^m a_i \quad \text{and} \quad S_{46^n j}^q(0) = h - r^{n+1} + r^{n+1} a_j,$$

where $a_i \geqslant \frac{1}{5}$ and $a_j \leqslant \frac{4}{5}$. Thus for $\Lambda^q_{31^m i}$ and $\Lambda^q_{46^n j}$ to have a chance to intersect, we must have

$$\begin{aligned} |r^{n+1}(a_j - 1) + qp^m a_i| &= |S^q_{31^m i}(0) - S^q_{46^n j}(0)| \\ &\leqslant |\Lambda^q_{31^m i}| + |\Lambda^q_{46^n j}| \\ &\leqslant \varepsilon(qp^m + r^{n+1}). \end{aligned} \tag{7.66}$$

On the other hand, for every $\mathbf{i} \in \Sigma$

$$\frac{\partial}{\partial q}\Pi^q(\mathbf{i}) = \delta_{i_1}\Pi^q(\sigma\mathbf{i}) + \lambda_{i_1}\frac{\partial}{\partial q}\Pi^q(\sigma\mathbf{i}),$$

where $\delta_i = 1$ if $i = 3$ and 0 otherwise. Hence $\left|\frac{\partial}{\partial q}\Pi^q(\mathbf{i})\right| \leqslant \frac{1}{1-\varepsilon}$. So after a little computation we obtain, for $\mathbf{i}', \mathbf{j}' \in \Sigma$:

$$\begin{aligned} &\frac{\partial}{\partial q}\left[\Pi^q(46^n j\mathbf{j}') - \Pi^q(31^m i\mathbf{i}')\right] \\ &= \frac{\partial}{\partial q}\left[qp^m a_i - \lambda_i p^m q\Pi^q(\mathbf{i}') - r^{n+1} + r^{n+1}a_j + \lambda_j r^{n+1}\Pi^q(\mathbf{j}')\right] \\ &= p^m a_i - \frac{\partial\lambda_i}{\partial q}p^m q\Pi^q(\mathbf{i}') - \lambda_i p^m \Pi^q(\mathbf{i}') - \lambda_i p^m q\frac{\partial}{\partial q}\Pi^q(\mathbf{i}') \\ &\quad + \frac{\partial\lambda_j}{\partial q}r^{n+1}\Pi^q(\mathbf{j}') + \lambda_j r^{n+1}\frac{\partial}{\partial q}\Pi^q(\mathbf{j}') \\ &\geqslant p^m a_i - p^m q - \lambda_i p^m - \frac{\lambda_i p^m q + \lambda_j r^{n+1}}{1-\varepsilon} \\ &\geqslant p^m\left(\frac{1}{5} - \frac{2\varepsilon}{1-\varepsilon}\right) - r^{n+1}\frac{\varepsilon}{1-\varepsilon}, \end{aligned} \tag{7.67}$$

keeping in mind that $a_i \geqslant \frac{1}{5}$. From (7.66) it follows that

$$p^m a_i \geqslant \frac{r^{n+1}(1 - a_j - \varepsilon)}{q} - \varepsilon p^m,$$

hence by $a_j \leqslant 4/5$

$$p^m \geqslant \frac{r^{n+1}(1 - a_j - \varepsilon)}{q(a_i + \varepsilon)} \geqslant \frac{r^{n+1}(\frac{1}{5} - \varepsilon)}{\varepsilon(\frac{1}{5} + \varepsilon)}.$$

It follows that

$$\begin{aligned} &p^m\left(\frac{1}{5} - \frac{2\varepsilon}{1-\varepsilon}\right) - r^{n+1}\frac{\varepsilon}{1-\varepsilon} \\ &\geqslant r^{n+1}\left(\frac{\frac{1}{5} - \varepsilon}{\varepsilon(\frac{1}{5} + \varepsilon)}\left(\frac{1}{5} - \frac{2\varepsilon}{1-\varepsilon}\right) - \frac{\varepsilon}{1-\varepsilon}\right) \geqslant r^{n+1} \end{aligned}$$

for $0 < \varepsilon \leqslant \frac{1}{20}$. Now the transversality condition (7.61) on cylinders $[46^n j], [31^m i]$ holds by (7.67) for $\varepsilon > 0$ sufficiently small. □

7.4. Singularity of self-similar measures without exact overlaps

By analogy with the exact overlap conjecture (Conjecture 1.6.5) one can ask the following question:

QUESTION 7.4.1. Is it true for a self-similar measure ν on the line of similarity dimension greater than 1, that ν can be singular only if there is an exact overlap?

This question was answered negatively by Simon and Vágó in [**SV19**, Theorem 14] by a Baire category type argument applied to the family of self-similar measures $\{\nu_\alpha\}$ introduced in the following example.

EXAMPLE 7.4.2. Let ℓ_α be the straight line on the plane through the origin that makes an angle α with the positive ray of x-axis. Let $\mathcal{S} := \left(S_i(x) = \frac{1}{3}x + t_i\right)_{i=1}^{8}$ be the self-similar IFS on the plane that generates the Sierpiński carpet. Moreover, we write $\mathcal{S}_\alpha$ for the self-similar IFS on the line obtained by the orthogonal projection of $\mathcal{S}$ to ℓ_α (see Figure 7.2). Let ν be the self-similar measure on the Sierpiński carpet with equal probabilities. The family of measures obtained by the orthogonal projection of ν to ℓ_α is denoted by $\{\nu_\alpha\}_{\alpha\in[0,\pi)}$.

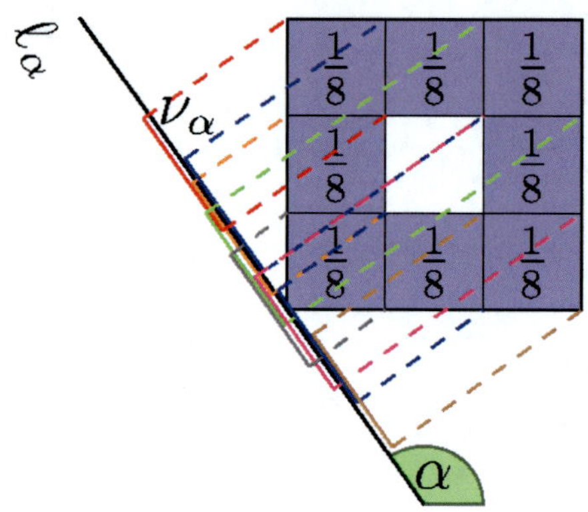

FIGURE 7.2. The projected system

REMARK 7.4.3.

(a) It is immediate that for all $\alpha \in [0,\pi)$ the measure ν_α is self-similar for the self-similar IFS $\mathcal{S}_\alpha$.

(b) Note that we have an exact overlap for the self-similar IFS $\mathcal{S}_\alpha$ if and only if the projection to ℓ_α of two different level n cylinder squares of the Sierpiński carpet coincide for some $n \in \mathbb{N}$. It is clear that $\tan\alpha \in \mathbb{Q}$ for such α, whence we have exact overlap only for countably many $\alpha \in [0,\pi)$. In fact, it is not hard to show (we leave this as an exercise) that exact overlap happens for $\mathcal{S}_\alpha$ for *every* α with a rational slope.

(c) It follows from a result of Shmerkin [**Shm19**, Theorem 9.2] (see Theorem 6.8.15) that $\nu_\alpha \ll \mathcal{L}^1$ with L^q density for all $q > 1$ for all but a set of zero Hausdorff dimension $\alpha \in [0,\pi)$.

The following theorem shows that the exceptional set in Remark 7.4.3 part (c) is large in topological sense.

THEOREM 7.4.4 ([**SV19**]). *There is a dense G_δ set of $\alpha \in [0,\pi)$ such that ν_α is singular to $\mathcal{L}^1$.*

To answer Question 7.4.1 negatively, combine this theorem with Remark 7.4.3(b) to obtain that there is $\alpha \in [0,\pi)$ such that there are no exact overlaps in $\mathcal{S}_\alpha$ but the measure ν_α is singular and the similarity dimension of ν_α is greater than 1.

Theorem 7.4.4 is a corollary of two results: on the one hand, Manning and Simon [**MS13**, Corollary 10] proved that for any α with a rational slope, ν_α is singular. On the other hand, the set of α for which ν_α is singular is a G_δ set.

REMARK 7.4.5.

(a) In fact, if ν_α is singular, then the stronger property holds: $\dim_{\rm H}(\nu_\alpha) < 1$. This was not stated in [**MS13**] explicitly, but it follows from a result of Ruiz [**Rui09**], see Theorem 4.5.3, since S_α is an integer IFS in the sense of (4.93).

(b) Alternatively, the fact that $\dim_{\rm H}(\nu_\alpha) < 1$ for α with a rational slope, follows from Theorem 4.5.4, since S_α is an integer IFS with rational weights and condition (4.94) can be easily verified. We leave the details as an exercise for the reader.

The claim about the set of singular parameters being G_δ is much more general. We sketch the proof of the following proposition, which is a minor generalization of [**PSS00a**, Prop. 8.1].

PROPOSITION 7.4.6. *Let $\mathcal{S}^{\boldsymbol{\lambda}} = \big(S_i^{\boldsymbol{\lambda}}(x) = \rho_i(\boldsymbol{\lambda})x + d_i(\boldsymbol{\lambda})\big)_{i\leqslant m}$ be a family of self-similar IFS on the line, for $\boldsymbol{\lambda} \in \Omega \subset \mathbb{R}^n$, where Ω is a compact set and $\rho_i : \Omega \to (-1,1)$, $d_i(\Omega) \to \mathbb{R}$ are continuous functions, such that the attractor of $\mathcal{S}^{\boldsymbol{\lambda}}$ is not a single point for every $\boldsymbol{\lambda} \in \Omega$. Let $\mu = \mathbf{p}^{\mathbb{N}}$ be a Bernoulli measure on $\Sigma = \{1,\dots,m\}^{\mathbb{N}}$ and $\nu_{\boldsymbol{\lambda}}$ the self-similar measure corresponding to $\mathbf{p}$. Then the set of "singular" parameters*

$$\mathcal{E}_\perp := \{\boldsymbol{\lambda} \in \Omega : \ \nu_\lambda \perp \mathcal{L}^1\} \quad \textit{is a } G_\delta \textit{ set.}$$

LEMMA 7.4.7. *For any open interval $(a,b) \subset \mathbb{R}$ the function $\boldsymbol{\lambda} \mapsto \nu_{\boldsymbol{\lambda}}(a,b)$ is continuous.*

The lemma follows from two facts: (a) the function $\boldsymbol{\lambda} \mapsto \nu_{\boldsymbol{\lambda}}$ is weak* continuous, see Proposition 3.1.3, and (b) the self-similar measure $\nu_{\boldsymbol{\lambda}}$ is continuous (has no atoms) by Proposition 3.1.1. We leave the details as an exercise for the reader.

PROOF SKETCH. Let $\mathcal{G}$ be the set of all finite unions of open intervals in $\mathbb{R}$. We claim that

$$\mathcal{E}_\perp = \bigcap_{n\geqslant 1} \ \bigcup_{\substack{G\in\mathcal{G}:\\ \mathcal{L}^1(G)<n^{-2}}} \{\boldsymbol{\lambda} \in \Omega : \ \nu_{\boldsymbol{\lambda}}(G) > 1/2\}. \tag{7.68}$$

Then we will be done, since the right-hand side is a G_δ set by the lemma.

Suppose first that $\boldsymbol{\lambda}$ is in the right-hand side of (7.68). Then for any $n \geqslant 1$ there exists $G_n \in \mathcal{G}$ such that $\mathcal{L}^1(G_n) < n^{-2}$ but $\nu_{\boldsymbol{\lambda}}(G_n) > 1/2$. We obtain that $G_\infty := \limsup G_n = \bigcap_{N\geqslant 1}\bigcup_{n\geqslant N} G_n$ satisfies $\mathcal{L}^1(G_\infty) = 0$, whereas $\nu_{\boldsymbol{\lambda}}(G_\infty) \geqslant 1/2$. This implies that $\nu_{\boldsymbol{\lambda}}$ is not absolutely continuous, but then by the Law of Pure Types, see Proposition 3.1.4(i), it is singular.

Now let $\boldsymbol{\lambda} \in \mathcal{E}_\perp$. Then there exists a measurable set H such that $\mathcal{L}^1(H) = 0$ but $\nu_{\boldsymbol{\lambda}}(H) = 1$. For any $n \in \mathbb{N}$ there is an open set $V \supset H$ such that $\mathcal{L}^1(V) < n^{-2}$. Clearly, $\nu_{\boldsymbol{\lambda}}(V) = 1$. We can write $V = \bigcup_{k\geqslant 1} V_k$, where V_k are (at most countably many) disjoint open intervals. Then $\nu_{\boldsymbol{\lambda}}(V) = \sum_{k\geqslant 1} \nu_{\boldsymbol{\lambda}}(V_k)$, and so we can find G, a finite union of open intervals V_k such that $\nu_{\boldsymbol{\lambda}}(G) > 0.5$. This shows that $\boldsymbol{\lambda}$ is in the right-hand side of (7.68), and the proof is complete. □

A much more general result was obtained by Simon and Vágó [**SV19**, Theorem 10].

THEOREM 7.4.8. *Consider a one-parameter families of measures ν_α on $\mathbb{R}^d$ for some $d \geqslant 1$, which are constructed as follows: The parameter space U is a nonempty compact metric space. We are given a continuous mapping $\Pi : U \times \Omega \to R \subset \mathbb{R}^d$, where R is an open ball in $\mathbb{R}^d$ and Ω is a compact metric space. Furthermore, let μ be a probability Radon measure on Ω. For every $\alpha \in U$ define $\nu_\alpha := (\Pi_\alpha)_*\mu$. Finally let λ be a Radon (reference) measure whose support is also contained in R. Then the set of parameters of singularity $\{\alpha \in U : \nu_\alpha \perp \lambda\}$ is a G_δ set.*

The following example is another interesting application of this theorem.

EXAMPLE 7.4.9. Let $\mathcal{S} := (S_i(x))_{i=1}^m$ be a self-similar IFS on the plane. Let ν be an arbitrary self-similar measure, and we write ν_α for the orthogonal projection of ν to the line ℓ_α (which was defined in Example 7.4.2). The only assumption we make is that for at least one of the mappings of $\mathcal{S}$ its linear part is a constant multiple of an orthogonal matrix of a rotation by an *irrational angle.* The set of angles α for which ν_α is singular is denoted by

$$\mathfrak{Sing} := \left\{\alpha \in [0, \pi) : \nu_\alpha \perp \mathcal{L}^1\right\}.$$

Then we have the following dichotomy: Either

(a) $\mathfrak{Sing} = \varnothing$, or
(b) $\mathfrak{Sing}$ is a dense G_δ subset of $[0, \pi)$.

Notice that this claim does not follow from Proposition 7.4.6, since projections of a planar self-similar measure with an irrational rotation are generally not self-similar.

SKETCH OF THE PROOF. The fact that $\mathfrak{Sing}$ is a G_δ set follows from Theorem 7.4.8. To see that $\mathfrak{Sing}$ is dense, if nonempty, we assume that $\alpha \in \mathfrak{Sing}$ and that the linear part of S_1 is a constant multiple of the irrational rotation by angle γ. Then the projection $\nu^\alpha_{i_1 \dots i_n}$ of the measure $(S_{i_1 \dots i_n})_*\nu$ to the line ℓ_α is also singular for all n and $(i_1, \dots, i_n)$. In particular, this holds for $(i_1, \dots, i_n) = (\underbrace{1, \dots, 1}_{n})$. Using this, an elementary argument yields that $\nu^{\alpha - n\gamma}$, the projection of ν to the line $\ell_{\alpha - n\gamma}$, is also singular. This holds for all n, and γ is an irrational angle. So, we obtain that $\mathfrak{Sing}$ is a dense subset of $[0, \pi)$. □

7.5. Planar self-similar sets of positive Lebesgue measure and empty interior

The following has been a folklore open problem for a long time:

OPEN PROBLEM. *Does there exists a self-similar set on the line of positive Lebesgue measure but with an empty interior?*

We do not have a good guess; however, in higher dimensions the analogous question is answered positively, as we show below. The following example appeared in [**CJP⁺06**]; it was inspired by an example of a planar self-affine set having similar properties, due to Jordan and Pollicott [**JP06**].

EXAMPLE 7.5.1. Let $\mathbf{t} = (t_1, t_2) \in [0, 2]^2 \subset \mathbb{R}^2$. Consider ten similarities, with the same contraction ratio $\frac{1}{3}$, given by the formulas below, where we omit the

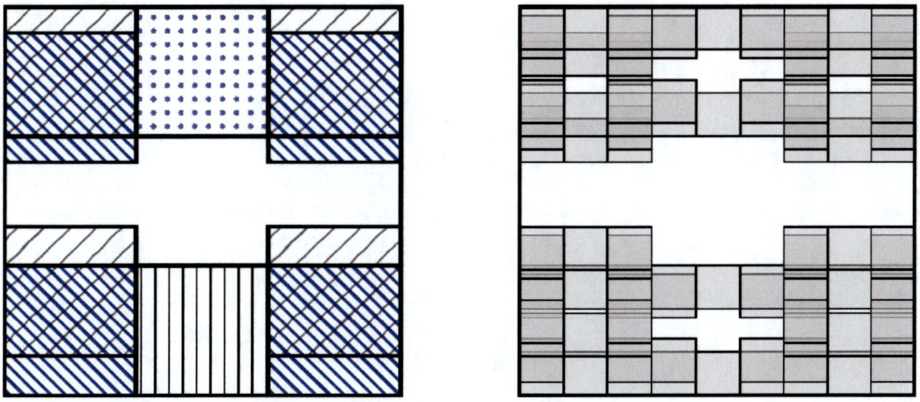

FIGURE 7.3. The first and the second approximations of the IFS of Example 7.5.1 for $t_1 = 0.3$ and $t_2 = 1.8$

variables (x, y) from notation, in order to keep it more compact:

$$\begin{array}{lll} S_0 = \frac{1}{3}(x, y) & S_4 = \frac{1}{3}(x+1, y) & S_6 = \frac{1}{3}(x+2, y) \\ S_1 = \frac{1}{3}(x, y+t_1) & & S_7 = \frac{1}{3}(x+2, y+t_1) \\ S_2 = \frac{1}{3}(x, y+t_2) & & S_8 = \frac{1}{3}(x+2, y+t_2) \\ S_3 = \frac{1}{3}(x, y+2) & S_5 = \frac{1}{3}(x+1, y+2) & S_9 = \frac{1}{3}(x+2, y+2). \end{array}$$

See Figure 7.3. Denote by $\Lambda^{\mathbf{t}} = \Lambda^{(t_1,t_2)}$ the attractor of the IFS.

THEOREM 7.5.2 ([**CJP$^+$06**]). *For all* $\mathbf{t} \in [0, 2]^2$ *the attractor* $\Lambda^{\mathbf{t}}$ *has empty interior, but for Lebesgue-a.e.* $\mathbf{t} \in [0, 2]^2$ *it has positive 2-dimensional Lebesgue measure* $\mathcal{L}^2$.

PROOF. Note that the projection of the IFS to the x-axis is simply $\{\frac{x}{3}, \frac{x+1}{3}, \frac{x+2}{3}\}$, which corresponds to the standard base-3 expansion. For $x \in [0, 1]$, we write $x = \sum_{n=1}^{\infty} a_n(x)3^{-n}$, with $a_n(x) \in \{0, 1, 2\}$, assuming for simplicity that $x \neq \frac{k}{3^n}$, $k \in \mathbb{N}$, so that the base-3 expansion is unique. Denote by $L_x^{\mathbf{t}}$ the vertical slice of $\Lambda^{\mathbf{t}}$, that is, $L_x^{\mathbf{t}} = \{y \in \mathbb{R} : (x, y) \in \Lambda^{\mathbf{t}}\}$. It is immediate from the structure of the IFS that

$$L_x^{\mathbf{t}} = \left\{ \sum_{n=1}^{\infty} b_n 3^{-n} : b_n \in \left\{ \begin{array}{ll} \{0, t_1, t_2, 2\}, & \text{if } a_n(x) \in \{0, 2\}; \\ \{0, 2\}, & \text{if } a_n(x) = 1. \end{array} \right. \right\}. \tag{7.69}$$

It follows that the slice $L_{1/2}^{\mathbf{t}}$, when $a_n(x) = 1$ for all $n \geqslant 1$, is just the classical middle-$\frac{1}{3}$ Cantor set $\mathcal{C}_{1/3}$. Similarly, if the base-3 expansion of x terminates with all 1's, then the vertical slice $L_x^{\mathbf{t}}$ is a finite union of scaled copies of $\mathcal{C}_{1/3}$. Thus $L_x^{\mathbf{t}}$ has empty interior (as a subset of $\mathbb{R}$) for a dense set of $x \in \mathbb{R}$, and hence $\Lambda^{\mathbf{t}}$ has empty interior for all $\mathbf{t}$.

It remains to prove the claim that $\mathcal{L}^2(\Lambda^{\mathbf{t}}) > 0$ for a.e. $\mathbf{t} \in [0, 2]^2$. By Fubini's theorem, it suffices to show that $\mathcal{L}^1(L_x^{\mathbf{t}}) > 0$ for $\mathcal{L}^1$-a.e. $x \in [0, 1]$, for $\mathcal{L}^2$-a.e. $\mathbf{t} \in [0, 2]^2$. To this end, fix a typical x, so that the frequency of each digit $0, 1, 2$ in the expansion $x = \sum_{n=1}^{\infty} a_n(x)3^{-n}$ is equal to $1/3$. In view of the (7.69), define the symbolic space

$$\Sigma_x = \prod_{n=1}^{\infty} \mathcal{A}_n(x), \quad \text{where } \mathcal{A}_n(x) = \begin{cases} \{0, \alpha, \beta, 2\}, & \text{if } a_n(x) \in \{0,2\}; \\ \{0,2\}, & \text{if } a_n(x) = 1. \end{cases}$$

and the "uniform" Bernoulli measure on Σ_x by

$$\mu_x = \prod_{n=1}^{\infty} \mu_x^{(n)}, \quad \text{where } \mu_x^{(n)} = \begin{cases} \{\frac14, \frac14, \frac14, \frac14\}, & \text{if } a_n(x) \in \{0,2\}; \\ \{\frac12, \frac12\}, & \text{if } a_n(x) = 1. \end{cases}$$

Define the natural projection from Σ_x onto the slice $L_x^{\mathbf{t}}$ by

$$\Pi_x^{\mathbf{t}}(\mathbf{i}) = \sum_{n=1}^{\infty} \phi^{\mathbf{t}}(\mathbf{i}_n) 3^{-n}, \quad \text{where } \mathbf{i}_n \in \mathcal{A}_n(x),$$

$$\phi^{\mathbf{t}}(0) = 0,\ \phi^{\mathbf{t}}(\alpha) = t_1,\ \phi^{\mathbf{t}}(\beta) = t_2,\ \phi^{\mathbf{t}}(2) = 2,$$

and consider the measure $\nu_x^{\mathbf{t}} = (\Pi_x^{\mathbf{t}})_* \mu_x$. Alternatively, $\nu_x^{\mathbf{t}}$ is the conditional measure on $L_x^{\mathbf{t}}$ of the self-similar measure on $\Lambda^{\mathbf{t}}$ corresponding to the probability vector $\left(\frac{1}{12}, \frac{1}{12}, \frac{1}{12}, \frac{1}{12}, \frac16, \frac16, \frac{1}{12}, \frac{1}{12}, \frac{1}{12}, \frac{1}{12}\right)$, but we will not need this. We are going to use a standard transversality argument to show that for a typical x, for $\mathcal{L}^2$-a.e. $\mathbf{t} \in [0,2]^2$, the measure $\nu_x^{\mathbf{t}}$ is absolutely continuous with respect to $\mathcal{L}^1$, which will imply the desired claim. Below we denote, for $x \in [0,1]$ and $\mathbf{i}, \mathbf{j} \in \Sigma_x$:

$$\psi_x^{\mathbf{t}}(\mathbf{i}, \mathbf{j}) = \Pi_x^{\mathbf{t}}(\mathbf{i}) - \Pi_x^{\mathbf{t}}(\mathbf{j}).$$

LEMMA 7.5.3. *There exists $C > 0$ such that for all $x \in [0,1]$ and all $\mathbf{i}, \mathbf{j} \in \Sigma_x$, we have*

$$\Delta_x(r; \mathbf{i}, \mathbf{j}) := \mathcal{L}^1 \left\{ \mathbf{t} \in [0,2]^2 : |\psi_x^{\mathbf{t}}(\mathbf{i})| \leqslant r \right\} \leqslant C 3^{|\mathbf{i} \wedge \mathbf{j}|} r, \quad \textit{for } r > 0. \tag{7.70}$$

Here, as usual, $|\mathbf{i} \wedge \mathbf{j}|$ is the length of the maximal initial segment of $\mathbf{i}$ and $\mathbf{j}$.

PROOF OF THE LEMMA. We have, denoting $n = |\mathbf{i} \wedge \mathbf{j}|$,

$$\psi_x^{\mathbf{t}}(\mathbf{i}, \mathbf{j}) = 3^{-n} \left(\phi^{\mathbf{t}}(\mathbf{i}_n) - \phi^{\mathbf{t}}(\mathbf{j}_n) + \sum_{k=1}^{\infty} 3^{-k} (\phi^{\mathbf{t}}(\mathbf{i}_{k+n}) - \phi^{\mathbf{t}}(\mathbf{j}_{k+n})) \right).$$

There are two cases. If $\{\phi^{\mathbf{t}}(\mathbf{i}_n), \phi^{\mathbf{t}}(\mathbf{j}_n)\} = \{0,2\}$, then

$$|\psi_x^{\mathbf{t}}(\mathbf{i}, \mathbf{j})| \geqslant 3^{-n} \left(2 - 2 \cdot \sum_{k=1}^{\infty} 3^{-k} \right) = 3^{-n},$$

in view of $\|\phi^{\mathbf{t}}\|_\infty = 2$, and (7.70) follows. Otherwise, there is $j \in \{1,2\}$ such that $t_j \in \{\phi^{\mathbf{t}}(\mathbf{i}_n), \phi^{\mathbf{t}}(\mathbf{j}_n)\}$. Then

$$\left| \frac{\partial \psi_x^{\mathbf{t}}(\mathbf{i}, \mathbf{j})}{\partial t_j} \right| \geqslant 3^{-n} \left(1 - \sum_{k=1}^{\infty} 3^{-k} \right) = 3^{-n} 2^{-1},$$

which also implies (7.70). □

Now, following the scheme of Mattila that was already used many times in the book, it is enough to verify that for a typical $x \in [0,1]$,

$$\mathcal{J} := \int_{[0,2]^2} \int_{L_x^{\mathbf{t}}} \underline{D}(\nu_x^{\mathbf{t}}, y) \, d\nu_x^{\mathbf{t}}(y) \, d\mathbf{t} < \infty, \tag{7.71}$$

where $\underline{D}(\nu_x^{\mathbf{t}}, y) = \liminf_{r \to 0} \frac{1}{2r} \cdot \nu_x^{\mathbf{t}}(y - r, y + r)$. Denote $\Sigma_x^n = \prod_{k=1}^n \mathcal{A}_n(x)$. As before, Fatou's Lemma, change of variables, and then Lemma 7.5.3 yield

$$\begin{aligned}
\mathcal{J} &\leqslant \liminf_{r\to 0} \frac{1}{2r} \int_{\Sigma_x} \int_{\Sigma_x} \psi_x^{\mathbf{t}}(\mathbf{i}, \mathbf{j})\, d\mu_x(\mathbf{i})\, d\mu_x(\mathbf{j}) \\
&\leqslant C \sum_{n=1}^{\infty} \sum_{j_1 \ldots j_n \in \Sigma_n^x} (\mu_x[j_1 \ldots j_n])^2 \cdot 3^n \\
&= C \sum_{n=1}^{\infty} 4^{-n_0(x)} 2^{-n_1(x)} 4^{-n_2(x)} 3^n,
\end{aligned}$$

where $n_i(x)$ denotes the number of i's in the first n terms $a_1(x), \ldots, a_n(x)$ of the base-3 expansion of x. For a typical $x \in [0, 1]$, for any $\delta > 0$ there exists $N \in \mathbb{N}$ such that for $n \geqslant N$, $n_i(x) \geqslant (\frac{1}{3} - \delta)n$ for $i = 0, 1, 2$. Then

$$\mathcal{J} \leqslant CC_1 + C \sum_{n=N}^{\infty} \left(4^{-(2/3-2\delta)} 2^{-(1/3-\delta)} 3\right)^n,$$

for some C_1 bounding the first N terms of the series, and note that the series converges for $\delta > 0$ sufficiently small. This implies the desired (typical) absolute continuity, and the proof is complete. □

7.6. Positive Hausdorff measure and the weak separation property

We have seen that for a self-similar set Λ, the Hausdorff measure is positive and finite, i.e., $0 < \mathcal{H}^{\dim_{\mathrm{H}}(\Lambda)}(\Lambda) < \infty$, if the OSC or the WSP hold, see Theorem 4.1.2 and Theorem 4.2.16. It is natural to inquire if it is possible to characterize the positivity of the Hausdorff measure at the dimension value. Combining the works of Fraser, Henderson, Olson and Robinson [**FHOR15**], and Farkas and Fraser [**FF15**], we will show that a self-similar set on the line is Ahlfors regular if and only if the WSP holds.

In order to do so, we need to understand the local structure of the set Λ. For $y \in \mathbb{R}^d$ and $r > 0$, let $M_{y,r}(x) = \frac{x-y}{r}$ be the "magnification function" centered at y with scale r, so that $M_{y,r}(B(y, r)) = B(0, 1)$. Let $E \subset \mathbb{R}^d$ be a compact set. We call the set T a *weak tangent set* of E if there exists a sequence $y_n \in E$ and a sequence $r_n > 0$ such that

$$\mathrm{dist}_H\big(M_{y_n,r_n}(E) \cap B(0, 1), T\big) \to 0 \text{ as } n \to \infty,$$

where dist_H denotes the Hausdorff metric defined in (1.37). Note that r_n does not necessarily tend to 0 as $n \to \infty$. The collection of weak tangent sets of E is denoted by $\mathrm{Tan}(E)$.

Fraser, Henderson, Olson and Robinson [**FHOR15**] showed that if a self-similar IFS on the line does not satisfy the WSP, then there exists a tangent set $T \in \mathrm{Tan}(\Lambda)$, which is a line segment, whereas simple analysis shows that every tangent set of an s-Ahlfors regular set is s-Ahlfors regular. This will allow us to show the equivalence of Ahlfors regularity and WSP in the case of $s < 1$.

To quantify the "size" of the tangent sets, we introduce another dimension concept, namely, the Assouad dimension. This dimension was introduced by Bouligand [**Bou28**] in his work on bi-Lipschitz embeddings and was further studied by Assouad in [**Ass77, Ass79**].

Let X be a nonempty bounded subset of $\mathbb{R}^d$. Similarly to Definition 1.4.1 of the box dimension, denote by $N_\delta(E)$ the smallest number of balls of radius δ which can cover E. The *Assouad dimension* of E is defined by

$$\begin{aligned}\dim_{\mathrm{A}}(E) = \inf\{s \geqslant 0 : \text{there exists } K \geqslant 0 \text{ such that} \\ N_\rho(E \cap B(x,\delta)) \leqslant K(\delta/\rho)^s \\ \text{for every } x \in E \text{ and } 0 < \rho < \delta \leqslant 1\}.\end{aligned} \tag{7.72}$$

Nowadays, the Assouad dimension is widely studied in fractal geometry, see Fraser et al. [**Fra14, FH17, FJ17**] and Mackay [**Mac11**]. Compared to the box dimension which measures how the global set can be covered optimally, the Assouad dimension finds small parts in the set which are the "hardest" to cover. No wonder that it is related to the local structure of the set. Clearly, $\overline{\dim}_{\mathrm{B}}(E) \leqslant \dim_{\mathrm{A}}(E)$. Moreover, $\dim_{\mathrm{A}}$ is finitely stable and monotone. For further properties we refer the reader to Fraser's book [**Fra20**].

The main goal of this section is to prove the following theorem due to Farkas and Fraser [**FF15**].

THEOREM 7.6.1 (Farkas, Fraser [**FF15**]). *Let $\Lambda \subset \mathbb{R}$ be a self-similar set on the real line, such that $\dim_{\mathrm{H}}(\Lambda) < 1$ and Λ is not a singleton. Then the following are equivalent:*

(i) *Λ satisfies the WSP,*
(ii) $\mathcal{H}^{\dim_{\mathrm{H}}(\Lambda)}(\Lambda) > 0$,
(iii) *Λ is Ahlfors regular,*
(iv) $\dim_{\mathrm{H}}(\Lambda) = \dim_{\mathrm{A}}(\Lambda)$,
(v) $\dim_{\mathrm{H}}(T) = \dim_{\mathrm{H}}(\Lambda)$ *for all* $T \in \mathrm{Tan}(\Lambda)$,
(vi) *there is no* $T \in \mathrm{Tan}(\Lambda)$, *which contains a line segment.*

As Farkas and Fraser [**FF15**, Proposition 3.3] pointed out, the equivalence (i) $\Longleftrightarrow$ (ii) above does not hold in higher dimensions.

EXAMPLE 7.6.2. Let Λ be the middle-third Cantor set defined by the IFS $\mathcal{S}_1 := \left(f_1 : x \mapsto \frac{x}{3}, f_2 : x \mapsto \frac{x+2}{3}\right)$. The Strong Separation Property holds for $\mathcal{S}_1$, hence $\dim_{\mathrm{H}}(\Lambda) = \frac{\log 2}{\log 3} =: t$, $0 < \mathcal{H}^t(\Lambda) < \infty$. Next, consider the IFS

$$\mathcal{S}_2 := \left(g_1 : x \mapsto \frac{x}{3}, g_2 : x \mapsto \frac{x+1}{3}, g_3 : x \mapsto \frac{x+2}{3}, g_4 : x \mapsto \frac{x}{3} + t\right)$$

where $t = 2\sum_{k=1}^{\infty} 3^{-k^2}$. Clearly, the attractor of $\mathcal{S}_2$ is $[0,1]$.

Observe that the set $\Lambda \times [0,1]$ is a self-similar set itself with respect to the product IFS

$$\mathcal{S}_1 \times \mathcal{S}_2 := (h_{i,j}(x,y) = (f_i(x), g_j(y)))_{i \in \{1,2\}, j \in \{1,2,3,4\}}\,.$$

Furthermore, $\dim_{\mathrm{H}}(\Lambda \times [0,1]) = t + 1$ by Corollary 2.2.2, and Remark 2.2.3 implies that

$$\mathcal{H}^{t+1}(\Lambda \times [0,1]) \geqslant \mathcal{H}^t(\Lambda) \cdot \mathcal{H}^1([0,1]) > 0.$$

Finally, we will demonstrate that the IFS $\mathcal{S}_1 \times \mathcal{S}_2$ does not satisfy the WSP. It is enough to show that $\mathcal{S}_2$ does not satisfy the WSP. To see that, let us define a sequence $\mathbf{i}^{(n)} = (i_1, i_2, \dots, i_{n^2})$ such that

$$i_k = \begin{cases} 1, & \text{if } k = \ell^2 \text{ for some } \ell \in \mathbb{N}, \\ 3, & \text{otherwise.} \end{cases}$$

Moreover, let $\mathbf{j}^{(n)} = (4, \overbrace{1, \ldots, 1}^{(n^2-1) \text{ times}})$. Then $g_{\mathbf{i}^{(n)}}(x) = 3^{-n^2}x + 2\sum_{k=1}^{n} 3^{-k^2}$ and $g_{\mathbf{j}^{(n)}}(x) = 3^{-n^2}x + t$. Hence, for every $x \in \mathbb{R}$

$$\left|(g_{\mathbf{i}^{(n)}})^{-1} \circ g_{\mathbf{j}^{(n)}}(x) - x\right| = \left|\sum_{k=n+1}^{\infty} \frac{2}{3^{k^2-n^2}}\right| \leqslant \frac{1}{3^{n-1}},$$

and so by Theorem 4.2.4(i), the WSP fails.

Here is the plan for the rest of the section: we present further properties of the tangent sets and the Assouad dimension and study it for overlapping self-similar sets, following Fraser, Henderson, Olson, Robinson [**FHOR15**]. Then combining this with the result on Ahlfors regularity of Farkas, Fraser [**FF15**], we show the proof of Theorem 7.6.1.

7.6.1. Assouad dimension and tangent sets. Here we study the the local structure, i.e. the tangent sets, are related to the Assouad dimension.

LEMMA 7.6.3. *Let $E \subset \mathbb{R}^d$ be an Ahlfors regular set. Then* $\dim_{\mathrm{A}}(E) = \dim_{\mathrm{H}}(E)$.

PROOF. It is enough to show that $\dim_{\mathrm{H}}(E) \geqslant \dim_{\mathrm{A}}(E)$. Denote by s the Hausdorff dimension of E. Since E is Ahlfors regular, there exists a constant $C > 0$ such that for every $0 < r$ and every $x \in E$

$$C^{-1}r^s \leqslant \mathcal{H}^s(E \cap B(x,r)) \leqslant Cr^s,$$

with $s = \dim_{\mathrm{H}}(E)$.

Let $0 < r < R$ be arbitrary. Let $P_{r/2}(E \cap B(x,R))$ be the largest number of disjoint balls of radius $r/2$ with centers in $E \cap B(x,R)$ and let $\{B(y_i, r/2)\}_i$ be such balls. Clearly, $\bigcup_i B(y_i, r/2) \subseteq B(x, 3R/2)$ and therefore,

$$\begin{aligned} C^{-1}(r/2)^s P_{r/2}(E \cap B(x,R)) &\leqslant \sum_i \mathcal{H}^s(E \cap B(y_i, r/2)) \\ &= \mathcal{H}^s(E \cap \bigcup_i B(y_i, r/2)) \\ &\leqslant \mathcal{H}^s(E \cap B(x, 3R/2)) \leqslant C(3R/2)^s. \end{aligned}$$

Hence,

$$N_r(E \cap B(x,R)) \leqslant P_{r/2}(E \cap B(x,R)) \leqslant C^2 3^s \left(\frac{R}{r}\right)^s,$$

hence $\dim_{\mathrm{A}}(E) \leqslant s$ by the definition of the Assouad dimension (7.72), as desired. In the last inequality we used that $N_r(F) \leqslant P_{r/2}(F)$ for any set $F \subset \mathbb{R}^d$. □

Käenmäki, Ojala, Rossi [**KOR18**] showed that there is a strong connection between the tangent sets of E and its Assouad dimension.

THEOREM 7.6.4 (Käenmäki, Ojala, Rossi [**KOR18**]). *Let E be a compact set of $\mathbb{R}^d$. Then*

$$\dim_{\mathrm{A}}(E) = \max\{\dim_{\mathrm{H}}(T) : T \in \mathrm{Tan}(E)\}.$$

In particular, for every $T \in \mathrm{Tan}(E)$, $\dim_{\mathrm{H}}(T) \leqslant \dim_{\mathrm{A}}(E)$, and there exists $T \in \mathrm{Tan}(E)$ such that $\dim_{\mathrm{H}}(T) = \dim_{\mathrm{A}}(E)$.

For the proof of the theorem, see [**KOR18**, Proposition 5.7]. For us, the following simpler statement is enough.

LEMMA 7.6.5. *Let E be a compact set of $\mathbb{R}^d$. Then*

$$\dim_{\mathrm{A}}(E) \geqslant \sup\{\overline{\dim}_{\mathrm{B}}(T) : T \in \mathrm{Tan}(E)\}.$$

PROOF. It suffices to show that $\overline{\dim}_{\mathrm{B}}(T) \leqslant \dim_{\mathrm{A}}(E)$ for every $T \in \mathrm{Tan}(E)$. By definition, there exists a sequence $x_n \in E$ and r_n such that $M_{x_n,r_n}(E) \cap B(0,1) \to T$ in the Hausdorff metric. Now, let $\delta > 0$ be arbitrary and choose $N \geqslant 1$ such that $\mathrm{dist}_H(M_{x_n,r_n}(E) \cap B(0,1), T) \leqslant \delta$ for every $n \geqslant N$. By the definition of Assouad dimension, for every $s > \dim_{\mathrm{A}}(E)$ there exists a constant $K > 0$ such that $N_\rho(E \cap B(y,r)) \leqslant K(r/\rho)^s$ for every $x \in E$, $0 < \rho < r \leqslant 1$. Thus, for $n \geqslant N$,

$$\begin{aligned} N_{2\delta}(T) &\leqslant N_\delta(M_{x_n,r_n}(E) \cap B(0,1)) = N_{\delta \cdot r_n}(E \cap B(x_n, r_n)) \\ &\leqslant K\left(\frac{r_n}{\delta \cdot r_n}\right)^s = K\delta^{-s}, \end{aligned}$$

which implies the claim. □

7.6.2. Local structure of self-similar sets. First we show that every tangent set of a self-similar set Λ contains a similarity image of Λ.

LEMMA 7.6.6. *Let $\Lambda \subset \mathbb{R}$ be a self-similar set. Then for every $T \in \mathrm{Tan}(\Lambda)$ there exists a similarity map $f : \mathbb{R} \to \mathbb{R}$ such that $f(\Lambda) \subseteq T$.*

PROOF. Let $\mathcal{S} = (S_i(x) = r_i x + t_i)_{i=1}^m$ be the IFS of similarities on the real line, with attractor Λ, and let $T \in \mathrm{Tan}(\Lambda)$ be arbitrary but fixed. Then there exists a sequence $y_n \in \Lambda$ and $r_n > 0$ such that

$$\mathrm{dist}_H(M_{y_n,r_n}(\Lambda) \cap B(0,1), T) \to 0 \text{ as } n \to \infty.$$

For $n \in \mathbb{N}$ there exists $\mathbf{i}_n = (i_1, \ldots, i_{k_n}) \in \Sigma^*$ such that $y_n \in f_{\mathbf{i}_n}(\Lambda)$ and $|f_{\mathbf{i}_n}(\Lambda)| < r_n \leqslant |f_{\mathbf{i}_n^-}(\Lambda)|$. Then $M_{y_n,r_n} \circ f_{\mathbf{i}_n}(x) = (r_{\mathbf{i}_n}/r_n)x + (t_{\mathbf{i}_n} - y_n)/r_n$ and

$$M_{y_n,r_n} \circ f_{\mathbf{i}_n}(\Lambda) \subset M_{y_n,r_n}(\Lambda) \cap B(0,1).$$

Since the sequences $r_{\mathbf{i}_n}/r_n$ and $(t_{\mathbf{i}_n} - y_n)/r_n$ are bounded, taking a subsequence we can make sure that $r_{\mathbf{i}_{n_k}}/r_{n_k} \to r^*$ and $(t_{\mathbf{i}_{n_k}} - y_{n_k})/r_{n_k} \to t^*$, and then by compactness of $(\mathfrak{C}, \mathrm{dist}_H)$ we obtain $f^*(\Lambda) \subset T$, where $f^*(x) = r^*x + t^*$. □

Now we prove that self-similar sets without the WSP have a line segment as a tangent set. This allows us to calculate the Assouad dimension as well. This result first appeared in Fraser, Henderson, Olson, Robinson [**FHOR15**] but we present here a simplified proof due to Rutar [**Rut**].

THEOREM 7.6.7 (Fraser et al. [**FHOR15**]). *Let $\Lambda \subset \mathbb{R}$ be a self-similar set which is not a singleton. If Λ does not satisfy the WSP, then there exists $T \in \mathrm{Tan}(\Lambda)$ which contains an interval. In particular,* $\dim_{\mathrm{A}}(\Lambda) = 1$.

PROOF. In view of Lemma 7.6.5, it is enough to show that there exists a weak tangent set T which contains an interval. This will be our goal.

Let $\mathcal{S} = (S_i(x) = r_i x + t_i)_{i=1}^m$ be the IFS of similarities on the real line with attractor Λ. Without loss of generality, by choosing a proper conjugation, we may assume that $S_1(0) = 0$ and $S_i(\overline{B}(0,1)) \subset \overline{B}(0,1)$ for every $i = 1, \ldots, m$. Moreover, by possibly considering the second iterates of the maps (i.e., considering the IFS $(S_i \circ S_j)_{i,j=1}^m$), we may also assume without loss of generality that $r_1 > 0$.

Suppose that $\mathcal{S}$ does not satisfy the WSP. By Theorem 4.2.4(iii), for any $\varepsilon > 0$ there exist $\mathbf{i}, \mathbf{j} \in \Sigma^*$ such that $S_\mathbf{i} \circ S_\mathbf{j} \neq \mathrm{Id}$, but

$$S_\mathbf{i}^{-1} \circ S_\mathbf{j}(x) = \gamma x + \delta,\ x \in \mathbb{R}, \quad \text{with } 0 \leqslant |\delta| \leqslant \varepsilon \text{ and } |1-\gamma| \leqslant \varepsilon. \tag{7.73}$$

Recall that $\mathrm{Id}(x) \equiv x$, so that $S_\mathbf{i}^{-1} \circ S_\mathbf{j} \neq \mathrm{Id}$ is equivalent to $(\gamma, \delta) \neq (1, 0)$.

FACT 7.6.8. *We may assume that $r_\mathbf{i}, r_\mathbf{j} > 0$ and $\delta > 0$ for every $\varepsilon > 0$.*

PROOF OF THE FACT. Taking $\varepsilon < 1$ we guarantee that $\gamma > 0$. If $r_\mathbf{i} < 0$ for some ε then from $0 < \gamma = r_\mathbf{i}^{-1} r_\mathbf{j}$ we have $r_\mathbf{j} < 0$, hence there exists $i \in \{1, \ldots, m\}$ with $r_i < 0$. Redefining $\mathbf{i}' := \mathbf{i}i$, $\mathbf{j}' = \mathbf{j}i$ results in $r_{\mathbf{i}'}, r_{\mathbf{j}'} > 0$, whereas the map

$$S_{\mathbf{i}'}^{-1} \circ S_{\mathbf{j}'} = S_i^{-1} \circ S_\mathbf{i}^{-1} \circ S_\mathbf{j} \circ S_i,$$

remains close to the identity Id, since it acts by

$$S_{\mathbf{i}'}^{-1} \circ S_{\mathbf{j}'}(x) = S_i^{-1}(\gamma S_i(x) + \delta) = \gamma x + \frac{S_i(0)(\gamma - 1) + \delta}{r_i} =: \gamma x + \delta',$$

with $|\delta'| \leqslant \frac{2\varepsilon}{|r_{\min}|}$, where $|r_{\min}| = \min_{i \leqslant m} |r_i|$.

This achieves the first goal, so suppose $r_\mathbf{i}, r_\mathbf{j} > 0$, but $\delta \leqslant 0$. Since Λ is not a singleton, we may assume that the fixed point of S_2 is not 0, that is, $S_2(0) = t_2 \neq 0$. Now, if $\delta = 0$, then $\gamma \neq 1$, and we can redefine $\mathbf{i}' := \mathbf{i}22$, $\mathbf{j}' := \mathbf{j}22$. The signs $r_{\mathbf{i}'}$ and $r_{\mathbf{j}'}$ remain unchanged, whereas

$$S_{\mathbf{i}'}^{-1} \circ S_{\mathbf{j}'}(x) = \gamma x + \frac{S_{22}(0)(\gamma - 1)}{r_2^2} =: \gamma x + \delta'',$$

with a small $\delta'' \neq 0$. Finally, if $\delta < 0$, exchanging the roles of $\mathbf{i}$ and $\mathbf{j}$ switches the sign of δ. □

Now we continue with the proof of the theorem. Fix a large $N \in \mathbb{N}$ and define inductively, for $\ell = 1, \ldots, N$, some words $\mathbf{i}_\ell, \mathbf{j}_\ell, \mathbf{p}_\ell, \mathbf{q}_\ell \in \Sigma^*$, and $n_\ell \in \mathbb{N} \cup \{0\}$. Using the observation (7.73), we start by fixing $\mathbf{i}_1$ and $\mathbf{j}_1$, such that

$$S_{\mathbf{i}_1}^{-1} \circ S_{\mathbf{j}_1} = \gamma_1 x + \delta_1, \quad \text{with } 0 < \delta_1 \leqslant \frac{1}{N} \text{ and } |\gamma_1 - 1| \leqslant \frac{r_1}{2N},\ \gamma_1 > 0,$$

where we recall that $S_1(x) = r_1 x$ with $r_1 > 0$. Then let $n_1 \in \mathbb{N} \cup \{0\}$ be the unique integer satisfying

$$\frac{r_1}{N} < \frac{\delta_1}{r_1^{n_1}} \leqslant \frac{1}{N}.$$

We also let $\mathbf{p}_1 = \mathbf{q}_0 = \varnothing$ (the empty word).

Then inductively, for $\ell = 2, \ldots, N$, set

$$\mathbf{p}_\ell := \mathbf{j}_{\ell-1} 1^{n_{\ell-1}} \mathbf{p}_{\ell-1}, \quad \text{and} \quad \mathbf{q}_\ell := \mathbf{i}_{\ell-1} 1^{n_{\ell-1}} \mathbf{q}_{\ell-1}.$$

(Recall that 1^n denotes the finite word $\overbrace{1 \cdots 1}^{n\text{-times}}$.) After that, choose the words $\mathbf{i}_\ell, \mathbf{j}_\ell \in \Sigma^*$, satisfying $S_{\mathbf{i}_\ell}^{-1} \circ S_{\mathbf{j}_\ell}^{-1}(x) = \gamma_\ell x + \delta_\ell$, with

$$0 < \delta_\ell \leqslant \frac{r_1 r_{\mathbf{q}_\ell}}{N} \quad \text{and} \quad |S_{\mathbf{p}_\ell}(0)(\gamma_\ell - 1)| \leqslant \frac{r_1 r_{\mathbf{q}_\ell}}{2N},\quad \gamma_\ell > 0. \tag{7.74}$$

Let $n_\ell \in \mathbb{N} \cup \{0\}$ be the unique integer satisfying

$$\frac{r_1}{N} < \frac{\delta_\ell}{r_1^{n_\ell} r_{\mathbf{q}_\ell}} \leqslant \frac{1}{N}. \tag{7.75}$$

After the (finite) inductive procedure is complete, let

$$\widetilde{\mathbf{q}}_\ell := \mathbf{i}_N 1^{n_N} \cdots \mathbf{i}_\ell 1^{n_\ell}, \quad \text{i.e.,} \quad \mathbf{q}_N = \widetilde{\mathbf{q}}_\ell \mathbf{q}_\ell, \quad \ell = 1, \ldots, N.$$

Note that, by construction, $S_{1^{n_\ell}}^{-1} \circ S_{\mathbf{i}_{k_\ell}}^{-1} \circ S_{\mathbf{j}_{k_\ell}} \circ S_{1^{n_\ell}}(x) = \gamma_{k_\ell} x + r_1^{-n_\ell} \delta_{k_\ell}$. Hence, for every $\ell = 1, \ldots, N-1$,

$$\begin{aligned} S_{\widetilde{\mathbf{q}}_{\ell+1}\mathbf{p}_{\ell+1}}(0) - S_{\widetilde{\mathbf{q}}_\ell \mathbf{p}_\ell}(0) &= S_{\widetilde{\mathbf{q}}_\ell} \circ S_{\mathbf{i}_\ell 1^{n_\ell}}^{-1} \circ S_{\mathbf{j}_\ell 1^{n_\ell}} \circ S_{\mathbf{p}_\ell}(0) - S_{\widetilde{\mathbf{q}}_\ell \mathbf{p}_\ell}(0) \\ &= r_{\widetilde{\mathbf{q}}_\ell} \left((\gamma_\ell - 1) S_{\mathbf{p}_\ell}(0) + r_1^{-n_\ell} \delta_\ell \right) =: \rho_N \cdot \Delta_\ell, \end{aligned} \tag{7.76}$$

where $\rho_N := r_{\mathbf{q}_N}$. By (7.74) and (7.75), for every $\ell = 1, \ldots, N-1$, holds

$$\frac{r_1}{2N} \leqslant \Delta_\ell = \frac{(\gamma_\ell - 1) S_{\mathbf{p}_\ell}(0) + r_1^{-n_\ell} \delta_\ell}{r_{\mathbf{q}_\ell}} \leqslant \frac{2 + r_1}{2N}. \tag{7.77}$$

In view of (7.76), we have

$$\rho_N^{-1} \left(S_{\widetilde{\mathbf{q}}_{\ell+1}\mathbf{p}_{\ell+1}}(0) - S_{\widetilde{\mathbf{q}}_1 \mathbf{p}_1}(0) \right) = \Delta_1 + \cdots + \Delta_\ell, \quad \ell = 1, \ldots, N.$$

It follows that

$$M_{y_N, \rho_N}(\Lambda) \cap \overline{B}(0,1) \supset \{0, \Delta_1, \ \Delta_1 + \Delta_2, \ldots, \Delta_1 + \cdots + \Delta_N\} =: T_N,$$

where $y_N = S_{\widetilde{\mathbf{q}}_1 \mathbf{p}_1}(0)$ (recall that the whole construction depended on N). Passing to a subsequence, we can ensure that there exists a weak tangent set T of Λ such that $M_{y_N,\rho_N}(\Lambda) \cap \overline{B}(0,1) \to T$ in the Hausdorff metric. Moreover, by (7.77), we obtain that $T_N \cap [0, r_1/2] \to [0, r_1/2]$ n the Hausdorff metric, as $N \to \infty$. Thus, $[0, r_1/2] \subset T$, which by Lemma 7.6.5 implies that $\dim_{\mathrm{A}}(\Lambda) \geqslant 1$. □

Now we are ready to prove Theorem 7.6.1, the main result of the section.

PROOF OF THEOREM 7.6.1. Theorem 4.2.16(ii-b) shows that (i) implies (ii). We saw in Theorem 3.4.3 that (ii) implies (iii). In Lemma 7.6.3, we have shown that (iii) implies (iv). Combining Lemmas 7.6.5 and 7.6.6, we get that (iv) implies (v). Since $\dim_{\mathrm{H}}(\Lambda) < 1$, clearly (v) implies (vi). Finally, since $\dim_{\mathrm{H}}(\Lambda) < 1$ and Λ contains at least two points, from (vi) it follows that Λ satisfies the WSP, otherwise it would contradict Theorem 7.6.7. □

Finally, we note that García [**Gar20**] generalised the result of Fraser, Henderson, Olson and Robinson [**FHOR15**], Theorem 7.6.7, to higher dimension. In particular, García expressed the Assouad dimension of self-similar sets, which do not satisfy the weak separation property, using orthogonal projections along special, so-called overlapping directions. Here we state only two corollaries of the main result of García [**Gar20**].

THEOREM 7.6.9 (García). *Let $\mathcal{S} = (S_i(x) = r_i O_i x + b_i)_{i=1}^m$ be a self-similar IFS on $\mathbb{R}^d$ and let Λ be the attractor of $\mathcal{S}$. Suppose that $\mathcal{S}$ does not have the WSP, moreover, the set $\{O_{\mathbf{i}} : \mathbf{i} \in \Sigma^*\}$ is dense in $O(d,\mathbb{R})$ or $SO(d,\mathbb{R})$, where $O(d,\mathbb{R})$ denotes the orthonormal group of $d \times d$ real matrices, and $SO(d,\mathbb{R})$ denotes the special linear subgroup of $O(d,\mathbb{R})$. Then $\dim_{\mathrm{A}}(\Lambda) = d$.*

On the plane, a much weaker assumption on the orthogonal matrices is sufficient.

THEOREM 7.6.10 (García). *Let $\mathcal{S} = (S_i(x) = r_i O_i x + b_i)_{i=1}^m$ be a self-similar IFS on $\mathbb{R}^2$ and let Λ be the attractor of $\mathcal{S}$. Suppose that $\mathcal{S}$ does not have the WSP, moreover, there exists an $i \in \{1, \ldots, m\}$ such that O_i is a rotation by an angle different from 0 and π. Then $\dim_A(\Lambda) = 2$.*

7.7. Examples

7.7.1. Projections of planar self-similar sets and a problem of Furstenberg. Let Λ be the attractor of the IFS (f_1, f_2, f_3) in $\mathbb{R}^2$ where

$$f_1(x,y) = \Big(\frac{x}{3}, \frac{y}{3}\Big),\ f_2(x,y) = \Big(\frac{x+1}{3}, \frac{y}{3}\Big),\ f_3(x,y) = \Big(\frac{x}{3}, \frac{y+1}{3}\Big).$$

The set Λ can be written explicitly in terms of base-3 expansions:

$$\Lambda = \Big\{ \sum_{n=1}^{\infty} a_n 3^{-n} :\ a_n \in \{(0,0), (1,0), (0,1)\} \Big\}.$$

The IFS clearly has the Strong Separation Property, hence $\dim_H(\Lambda) = \frac{\log 3}{\log 3} = 1$. It resembles the classical Sierpiński gasket, which is obtained by replacing the contraction ratio 1/3 with 1/2; thus, Kenyon [**Ken97**] called it the "one-dimensional Sierpiński gasket". The question is about *linear projections* of Λ. For $\theta \in [0, \pi)$ let Λ_θ be the orthogonal projection of S onto the line of slope $\tan(\theta)$. Note that $\Lambda_0 = C_{1/3}$, the classical middle-third Cantor set, and it is not hard to see that, e.g., $\Lambda_{-\pi/4}$ and $\Lambda_{\arctan(1/2)}$ are closed intervals (an exercise). What about other values of θ?

By Marstrand's projection theorem, $\dim_H(\Lambda_\theta) = 1$ for Lebesgue a.e. θ. In the 1970's Furstenberg conjectured that $\dim_H(\Lambda_\theta) = 1$ for all θ with $\tan\theta \notin \mathbb{Q}$. The conjecture has been open until recently, with partial results by Kenyon [**Ken97**] and Świątek–Veerman [**SV02**], and finally settled affirmatively by Hochman [**Hoc14b**]. Below we show the derivation from Hochman's Corollary 6.8.6, see [**Hoc14b**]. The argument below follows [**Hoc14b**, Th. 1.6], but is slightly more general, so as to include other examples.

Consider the homogeneous linear IFS on $\mathbb{R}$ given by

$$\mathcal{F} = (S_j(x) = \lambda x + d_j)_{j=1}^m, \tag{7.78}$$

with attractor $\Lambda_{\mathcal{F}}$. Recall that

$$\Delta_n = \Delta_n(\mathcal{F}) = \min\{|S_{\mathbf{i}}(0) - S_{\mathbf{j}}(0)| :\ |\mathbf{i}| = |\mathbf{j}| = n,\ \mathbf{i} \neq \mathbf{j}\}$$

is the minimal displacement between distinct cylinders of order n.

LEMMA 7.7.1. *Let $\mathcal{F}$ be the IFS given by (7.78), where $\lambda = 1/m$, $d_i = a_i - b_i t$, with $t \in \mathbb{R}$ and $a_i, b_i \in \mathbb{Z}$. Suppose that $\mathcal{F}$ has super-exponential concentration of cylinders, that is, $\lim_{n\to\infty}(-\frac{1}{n}\log\Delta_n) = \infty$. Then $t \in \mathbb{Q}$. More precisely, if*

$$\Delta_n \leqslant (2m^3)^{-n} \quad \textit{for all } n \geqslant n_0, \tag{7.79}$$

then $t \in \mathbb{Q}$.

Combining Lemma 7.7.1 with Theorem 6.8.5, we have that if $t \notin \mathbb{Q}$, then $\dim_H(\Lambda_{\mathcal{F}}) = 1$.

PROOF. We have, for any $\mathbf{i}, \mathbf{j}$, words of length n:

$$S_{\mathbf{i}}(0) - S_{\mathbf{j}}(0) = \sum_{k=1}^{n} (a_{i_k} - a_{j_k}) m^{-(k-1)} - t \sum_{k=1}^{n} (b_{i_k} - b_{j_k}) m^{-(k-1)} = p_n - t \cdot q_n,$$

where

$$p_n, q_n \in X_n := \Big\{ \sum_{k=1}^{n} c_k m^{-(k-1)} : c_k \in \mathbb{Z} \Big\}.$$

Thus, $\Delta_n = |p_n - t \cdot q_n|$, with $p_n, q_n \in X_n$. Suppose that (7.79) holds. If $q_n = 0$ for some $n \geqslant n_0$, then $\Delta_n = |p_n| < (2m^3)^{-n}$, but p_n is a rational with denominator not larger than m^n, hence $p_n = 0$. In that case $\Delta_n = 0$, we have exact overlaps, and $t = p_n/q_n \in \mathbb{Q}$.

Therefore, we can assume that $q_n \neq 0$ for $n \geqslant n_0$, whence $q_n \geqslant m^{-n}$. Dividing $|p_n - t \cdot q_n| \leqslant (2m^3)^{-n}$ by q_n yields $|t - p_n/q_n| < (2m^2)^{-n}$. Subtracting successive terms, by the triangle inequality we have

$$\left| \frac{p_{n+1}}{q_{n+1}} - \frac{p_n}{q_n} \right| < 2^{-n+1} m^{-2n} \text{ for all } n \geqslant n_0.$$

However, $\frac{p_{n+1}}{q_{n+1}} - \frac{p_n}{q_n}$ is a rational with denominator $\leqslant m^{2n+1}$, and so

$$\left| \frac{p_{n+1}}{q_{n+1}} - \frac{p_n}{q_n} \right| \neq 0 \Rightarrow \left| \frac{p_{n+1}}{q_{n+1}} - \frac{p_n}{q_n} \right| \geqslant m^{-(2n+1)}.$$

Since $2^{-n+1} m^{-2n} < m^{-(2n+1)}$ for large enough n, we obtain that $p_n/q_n = p_{n+1}/q_{n+1}$ for all large enough n. Therefore, there exists n_1 such that $|t - p_{n_1}/q_{n_1}| < (2m^2)^{-n}$ for $n > n_1$, and hence $t = p_{n_1}/q_{n_1}$. □

Another question is about the Lebesgue measure of Λ_θ. Kenyon [**Ken97**] proved that $\mathcal{L}^1(\Lambda_\theta) = 0$ for all lines with irrational slope. As for rational slopes, he showed that if $\tan(\theta) = p/q$ (in reduced form), then $\mathcal{L}^1(\Lambda_\theta) > 0$ if and only if $p + q = 0$ modulo 3 (as for $u = -1$ or $u = 1/2$). Note that it follows from Schief's Corollary 4.1.3 that $\mathcal{L}^1(\Lambda_\theta) > 0$ if and only if Λ_θ contains an interval, because it a self-similar subset of the line of dimension one.

Finally, one can ask what is the dimension of Λ_θ when $\tan(\theta)$ is rational. It is of course equal to one when $\mathcal{L}^1(\Lambda_\theta) > 0$, and otherwise there are exact overlaps and $\dim_{\mathrm{H}}(\Lambda_\theta) < 1$. In fact, then the IFS has the Weak Separation Property and the dimension may, in principle, be computed as in Section 4.2.3.

7.7.2. Other contraction rates. More generally, it is natural to inquire what happens when we project instead

$$\Lambda^\lambda := \Big\{ \sum_{n=1}^{\infty} a_n \lambda^n : a_n \in \{(0,0), (1,0), (0,1)\} \Big\}, \quad \lambda \in (0, 1/2).$$

Then $\dim_{\mathrm{H}}(\Lambda^\lambda) = \log 3/\log(1/\lambda)$, and we should distinguish between the "subscritical case" $\lambda < 1/3$ and the "supercritical case" $\lambda > 1/3$; whereas Furstenberg's problem was concerned with the "critical case" $\lambda = 1/3$. As before, we consider the projections $\Lambda_\theta^\lambda = \mathrm{proj}_{\mathbf{v}(\theta)}(\Lambda^\lambda)$, depending on the angle $\theta \in [0, \pi)$, where $\mathbf{v}(\theta) = (\cos(\theta), \sin(\theta))$ and $\mathrm{proj}_{\mathbf{v}}$ denotes the orthogonal projection to the line determined by the vector $\mathbf{v}$.

(i) $\lambda \in (0,1/3)$. By the Marstrand projection theorem, $\dim_H(\Lambda_\theta^\lambda) = s_\lambda := \log 3/\log(1/\lambda)$ for a.e. $\theta \in [0,\pi)$, and moreover,

$$\dim_H(\mathcal{E}_\lambda) \leqslant s_\lambda, \quad \text{where } \mathcal{E}_\lambda := \{\theta : \dim_H(\Lambda_\theta^\lambda) < s_\lambda\},$$

by classical estimates of the dimension of the "exceptional set". From the work of Hochman [**Hoc14b**] it follows that in fact $\dim_H(\mathcal{E}_\lambda) = \dim_P(\mathcal{E}_\lambda) = 0$. As in [**Hoc14b**], one can deduce that $\dim_H(\Lambda_\theta^\lambda) = s_\lambda$ for all rational $\lambda < 1/3$ and θ with irrational slope.

The next question is about the packing and Hausdorff measure of the projection Λ_θ^λ. Since the family of projections satisfies the transversality condition, by Theorem 7.1.2 we have $0 < \mathcal{P}^{s_\lambda}(\Lambda_\theta^\lambda) < \infty$ for $\mathcal{L}^1$-a.e. $\theta \in (0,\pi]$. To compute the Hausdorff measure, we turn to Section 7.3. Let $\mathcal{IP}_\lambda$ be the set of intersection parameters for Λ_θ^λ, equivalently,

$$\mathcal{IP}_\lambda = \{\theta \in [0,\pi) : \operatorname{proj}_{\mathbf{v}(\theta)} : \Lambda^\lambda \to \mathbb{R} \text{ is not 1-to-1}\}.$$

Note that $\mathcal{H}^{s_\lambda}(\Lambda_\theta^\lambda) > 0$ for all $\theta \notin \mathcal{IP}_\lambda$ since in this case Λ_θ^λ is an attractor of a self-similar IFS on the line satisfying the SSP. It is easy to see that $[0,\pi)\backslash\mathcal{IP}_\lambda$ contains intervals when $\lambda < 1/3$, but otherwise the structure of the set is not so clear. The next lemma sharpens Proposition 7.3.5 for families of projections.

LEMMA 7.7.2 ([**PSS00b**, Lem. 2.7]). *Let $\mathcal{K} \subset \mathbb{R}^2$ be a self-similar set that has the Strong Separation Property, and let $\{\mathcal{K}^\theta\}_{\theta\in[0,\pi)}$ be the family of projections $\mathcal{K}^\theta = \operatorname{proj}_{\mathbf{v}(\theta)}(\mathcal{K})$. Let $\mathcal{IP}$ be the corresponding set of intersection parameters. Then $\dim_H(\mathcal{IP}) \leqslant \dim_H(\mathcal{K}-\mathcal{K})$.*

PROOF. The Strong Separation Property implies that there exists $\eta > 0$ such that $\bigcup_{i\neq j}(\mathcal{K}_i - \mathcal{K}_j) \subset \{\mathbf{x} \in \mathbb{R}^2 : |\mathbf{x}| \geqslant \eta\}$; note that $\Upsilon(\mathbf{x}) = \mathbf{x}/|\mathbf{x}|$ is Lipschitz on the latter set. It follows that

$$\mathcal{IP} = \Big\{\theta \in [0,\pi) : (\cos\theta, \sin\theta) \in \Upsilon\Big(\bigcup_{i\neq j}(\mathcal{K}_i - \mathcal{K}_j)\Big)\Big\}.$$

Thus, $\dim_H(\mathcal{IP}) \leqslant \max_{i,j}\dim_H(\mathcal{K}_i - \mathcal{K}_j) \leqslant \dim_H(\mathcal{K}-\mathcal{K})$. □

Returning to the family of projections $\{\Lambda_\theta^\lambda\}_{\theta\in[0,\pi)}$, we have that $\Lambda^\lambda - \Lambda^\lambda$ is a self-similar set in $\mathbb{R}^2$, an attractor of a homogeneous IFS of seven maps with contraction ratio λ. Thus, Lemma 7.7.2 implies that $\mathcal{IP}_\lambda$ has Hausdorff dimension less than one and hence zero Lebesgue measure for $\lambda < 1/7$, hence $\mathcal{H}^{s(\lambda)}(\Lambda_\theta^\lambda) > 0$ for a.e. $\theta \in [0,\pi)$. To proceed further, it is convenient to pass to the family of sets

$$\mathcal{K}_u^\lambda = \Big\{\sum_{n=0}^{\infty} a_n\lambda^n : a_n \in \{0,1,u\}\Big\}, \tag{7.80}$$

which is the attractor of the IFS $(\lambda x, \lambda x + 1, \lambda x + u)$. We have

$$\mathcal{K}_{\tan\theta}^\lambda = \cos\theta \cdot \Lambda_\theta^\lambda,$$

so one can consider these families of self-similar sets interchangeably. We are going to use the claims from Remark 7.3.4. We have $\Gamma = \{0, \pm 1, \pm u, \pm(u-1)\}$. By symmetry, we can assume that $u \geqslant 2$ without loss of generality. Then $\max\Gamma = u$, $g = 1$ and $G = \max\{1, u-2\}$. It follows from Proposition 7.3.3, (7.58), and (7.59) that $\mathcal{IP}_\lambda \supset [\frac{1-\lambda}{\lambda}, \frac{2(1-\lambda)}{1-3\lambda}]$. This interval is nontrivial for $\lambda \in (\frac{1}{5}, \frac{1}{3})$. By Theorem 7.1.4, $\mathcal{H}^{s(\lambda)}(\mathcal{K}_u^\lambda) = 0$ for a.e. $u \in [\frac{1-\lambda}{\lambda}, \frac{2(1-\lambda)}{1-3\lambda}]$, given that $\lambda \in (\frac{1}{5}, \frac{1}{3})$.

Thus, for a typical parameter u in this range, the natural measure on the attractor is the packing measure rather than the Hausdorff measure. The case of $\lambda \in (\frac{1}{7}, \frac{1}{5})$ remains open, but we expect that $\mathcal{L}^1(\mathcal{IP}_\lambda) > 0$ for a.e. $\lambda \in (\frac{1}{7}, \frac{1}{5})$.

(ii) $\lambda \in (1/3, 1/2)$. By the Marstrand projection theorem, $\mathcal{L}^1(\Lambda_\theta^\lambda) > 0$ for a.e. $\theta \in [0, \pi)$. In the paper [**SS16**], it was shown that, in fact, the exceptional set of angles for this statement has Hausdorff dimension zero. The question on *interior points*, however, remains open. It is a well-known open problem whether there exists a self-similar set on the line of positive Lebesgue measure, but with empty interior. It is easy to see that for any $\lambda > 1/3$ there are intervals of θ for which the projection Λ_θ^λ is an interval, but otherwise, even "almost-all" type results are unavailable as of this writing.

7.7.3. Families with contraction rate as parameter. Here we again consider the family $\{\mathcal{K}_u^\lambda\}$ from (7.80), but now we fix u and consider λ as parameter. We will restrict ourselves to the subcritical case of $\lambda < \frac{1}{3}$. Assume that $u \geqslant 2$ (the case $u < 2$ may be reduced to this one by a linear transformation). The Open Set Condition holds for $\lambda \leqslant \frac{1}{1+u}$, so we focus on $\lambda \in (\frac{1}{1+u}, \frac{1}{3})$. The first issue is to check transversality. It follows from Definition 6.3.1 and (6.23) that a sufficient condition for transversality is absence of double zeros for the class of power series $\mathcal{B}_u$. It follows from Corollary 6.4.5(iv) that transversality in λ holds for $\lambda \in (\frac{1}{1+u}, \frac{1}{1+\sqrt{u}})$.

Case $u \in (2, 4]$. We have $\frac{1}{1+\sqrt{u}} \geqslant \frac{1}{3}$, hence transversality holds on the entire interval $\lambda \in (\frac{1}{1+u}, \frac{1}{3})$. By Proposition 7.3.3, with a little computation, we obtain that for $u \in (2, 4]$ the corresponding set of intersection parameters (in λ) contains a nonempty interval $[\frac{1}{1+u}, \frac{1}{3}]$. It follows that for a.e. $\lambda \in [\frac{1}{1+u}, \frac{1}{3}]$ we have $\mathcal{H}^{s(\lambda)}(\mathcal{K}_u^\lambda) = 0$ and $0 < \mathcal{P}^{s(\lambda)}(\mathcal{K}_u^\lambda) < \infty$, so again the natural measure is the packing measure rather than the Hausdorff measure.

Case $u > 4$. For $\lambda \in (\frac{1}{1+\sqrt{u}}, \frac{1}{3})$ the transversality condition may or may not hold, and it is not trivial to check this. However, by Hochman's Theorem 6.8.11, we know that $\dim_{\mathrm{H}}(\mathcal{K}_u^\lambda) = s(\lambda) = \frac{\log 3}{-\log \lambda}$ for all $\lambda \in (\frac{1}{1+u}, \frac{1}{3})$ outside a set of Hausdorff and packing dimension zero. As for the Hausdorff measure, we can sometimes apply Proposition 7.1.5, which does not assume transversality, but only requires that the set of intersection parameters $\mathcal{IP}$ contain an interval. We obtain, after a little computation, that for a dense set of $\lambda \in [\frac{u-2}{3u-2}, \frac{1}{3}]$ there is dimension drop and $\mathcal{H}^{s(\lambda)}(\mathcal{K}_u^\lambda) = 0$ for a dense G_δ set of λ in $[\frac{u-2}{3u-2}, \frac{1}{3}]$.

7.7.4. More on the $\{0, 1, 3\}$-problem and expansions with deleted digits. Recall that the $\{0, 1, 3\}$-problem concerns the IFS

$$\mathcal{S}_{013}^\lambda := (S_0(x) = \lambda x, \quad S_1(x) = \lambda x + 1, \quad S_3(x) = \lambda x + 3), \qquad x \in \mathbb{R},$$

so its attractor is exactly the set $\mathcal{K}_3^\lambda$ from (7.80). We have already encountered this interesting family of self-similar sets with overlaps, first in Section 1.6.1 and then in Section 6.5. We will adopt the notation

$$y_0 . y_1 \dots y_n = y_0 + y_1 \lambda + \dots + y_n \lambda^n,$$

when the parameter λ is fixed. Observe that the smallest closed interval containing $\mathcal{K}_3^\lambda$ is $I_\lambda := [0, 3.33\dots] = [0, 3/(1-\lambda)]$.

It turned out that even in the supercritical case $\lambda > 1/3$ coincidence of cylinders may cause a dimension drop. We outline one special case of this type below, following [**KSS95**].

LEMMA 7.7.3 ([**KSS95**]). *Let λ be the positive solution of the equation* $1 = 2x + 2x^2 + 2x^3$. *Then* $\dim_{\mathrm{H}}(\mathcal{K}_3^\lambda) < 1$ *and hence* $\mathcal{L}^1(\mathcal{K}_3^\lambda) = 0$.

PROOF. We will only show that the Lebesgue measure vanishes; the dimension drop claim is proved similarly, using the appropriate Hausdorff measure, and is left as an exercise.

The equation for λ shows that in base λ we have $1. = .222$, hence

$$1.111 = 0.333. \tag{7.81}$$

This implies that we can eliminate one similitude passing to the 4th level. However, this is not enough since $\lambda^{-4} \approx 72.66 < 3^4 - 1 = 80$. But we also have $0.3 = 0.2222 = 1.0002$, hence

$$0.3001 = 1.0003. \tag{7.82}$$

This implies that passing to the 5th level we can eliminate 9 cylinders (by replacing the two consecutive zeros in 03001 with arbitrary digits in $\{0, 1, 3\}$), but still this doesn't seem to be enough. So we have to use a different method, using additional information about our IFS.

Let $J = (\lambda\mathcal{K}_3^\lambda) \cap (\lambda\mathcal{K}_3^\lambda + 1)$, then

$$\begin{aligned} \mathcal{L}^1(\mathcal{K}_3^\lambda) &= \mathcal{L}^1(\lambda\mathcal{K}_3^\lambda) + \mathcal{L}^1(\lambda\mathcal{K}_3^\lambda + 1) + \mathcal{L}^1(\lambda\mathcal{K}_3^\lambda + 3) - \mathcal{L}^1(J) \\ &= 3\lambda\mathcal{L}^1(\mathcal{K}_3^\lambda) - \mathcal{L}^1(J), \end{aligned} \tag{7.83}$$

where we used that only the first two level-1 cylinders overlap. Now (7.81) implies that $1.111 + \lambda^4\mathcal{K}_3^\lambda \subset J$, and (7.82) implies that

$$1.0003 + \lambda^5\mathcal{K}_3^\lambda, \quad 1.0013 + \lambda^5\mathcal{K}_3^\lambda, \quad 1.0033 + \lambda^5\mathcal{K}_3^\lambda$$

are all subsets of J. We claim that these four subsets are mutually disjoint. Indeed, this follows from the inequalities (base λ)

$$\begin{aligned} 1.000\overline{3} &< 1.0013, \\ 1.001\overline{3} &< 1.0033, \\ 1.003\overline{3} &< 1.111, \end{aligned}$$

which are easy to verify. For instance, to check the first one we need to show that $1. > .0\overline{3}$, which is equivalent to $1 > 3\lambda^2/(1-\lambda)$, or $1 - \lambda - 3\lambda^2 > 0$. This, in turn, follows from

$$1. = .222 = .1442 > .13.$$

The other inequalities are checked similarly, and we leave the verification to the reader. Thus we obtain

$$\mathcal{L}^1(\mathcal{K}_3^\lambda) = 3\lambda\mathcal{L}^1(\mathcal{K}_3^\lambda) - \mathcal{L}^1(J) \leqslant 3\lambda\mathcal{L}^1(\mathcal{K}_3^\lambda) - \lambda^4\mathcal{L}^1(\mathcal{K}_3^\lambda) - 3\lambda^5\mathcal{L}^1(\mathcal{K}_3^\lambda).$$

If $\mathcal{L}^1(\mathcal{K}_3^\lambda) > 0$, then it follows that $1 \leqslant 3\lambda - \lambda^4 - 3\lambda^5$. But $3\lambda - 1 = 2\lambda^4$, so we obtain $2\lambda^4 > \lambda^4 + 3\lambda^5$, hence $3\lambda < 1$, a contradiction. □

7.7.5. Zero measure in the overlapping case. It follows from the discussion in Section 7.7.3 that $\mathcal{H}^{s(\lambda)}(\mathcal{K}_u^\lambda) = 0$ for $u = 3$, for Lebesgue-a.e. $\lambda \in (\frac{1}{4}, \frac{1}{3})$, that is, the attractor $\mathcal{K}_3^\lambda$ from the $\{0,1,3\}$ problem has zero Hausdorff measure in the similarity dimension, almost everywhere in the "subcritical" overlapping region; equivalently, the OSC does not hold. Here we give an elementary proof of a stronger result: the OSC fails for *all* parameters $\lambda \in (\frac{1}{4}, \frac{1}{3})$. In fact, we show it for a larger family of attractors of IFS corresponding to "expansions with deleted digits" from [**PS95**] and [**Sol98**].

PROPOSITION 7.7.4. *Let $D \subset \mathbb{N} \cup \{0\}$, with $\#D = m$, such that $D - D \supset \{0, \dots, \ell - 1\}$ for some $\ell \in \mathbb{N}$; in particular, this holds for $S = \{0, 1, 3\}$, with $m = 3$ and $\ell = 4$. Fix an arbitrary $\lambda \in (\frac{1}{\ell}, \frac{1}{m})$ and consider the set*

$$\mathcal{K}_D^\lambda := \Big\{x = \sum_{j=0}^{\infty} a_n \lambda^n : a_n \in D\Big\},$$

which is the attractor of the IFS $(S_j(x) = \lambda x + d_j)_{j \leqslant m}$, where $D = \{d_1, \dots, d_m\}$.

We claim that the OSC does not hold for this IFS; equivalently, $\mathcal{H}^{s(\lambda)}(\mathcal{K}_D^\lambda) = 0$, where $s(\lambda) = \frac{\log m}{-\log \lambda}$ is the similarity dimension.

REMARK 7.7.5. This improves [**Sol98**, Theorem 1.1(ii)], where the claim is proved for a.e. $\lambda \in (\frac{1}{\ell}, \frac{1}{m})$ under the same assumptions. Our proof here only works in this special case of integer digits, and so it does not make Theorem 7.1.4 and Proposition 7.1.5 obsolete.

PROOF. Let $\Sigma = D^{\mathbb{N}}$. As usual, for $u \in \Sigma_n = D^n$ we have

$$S_u(x) = S_{u_1} \circ \cdots \circ S_{u_n}(x) = \lambda^n x + \sum_{j=1}^{n} u_j \lambda^{j-1}.$$

By the Bandt-Graf Theorem in the form of Proposition 7.3.1, in order to show that the Hausdorff measure is zero it is enough to find, for any $\varepsilon > 0$, two distinct words $u, v \in D^n$, for some $n \in \mathbb{N}$, such that

$$|S_u(0) - S_v(0)| = \Big|\sum_{j=1}^{n} (u_j - v_j)\lambda^{j-1}\Big| \leqslant \varepsilon \lambda^n |\mathcal{K}_D^\lambda|. \tag{7.84}$$

Consider the set of numbers $\big\{\sum_{j=1}^n a_j \lambda^{j-1} : a_j \in \{0, \dots, \ell - 1\}\big\}$. There are ℓ^n such numbers, possibly with repetitions. Since all of them lie in $[0, \frac{\ell-1}{1-\lambda}]$, there exist distinct $a_1 \dots a_n$ and $b_1 \dots b_n$ in $\{0, \dots, \ell - 1\}^n$ such that

$$\Big|\sum_{j=1}^{n} (a_j - b_j)\lambda^{j-1}\Big| \leqslant c\ell^{-n}, \quad \text{where } c = \frac{\ell - 1}{1 - \lambda}.$$

Note that $a_j - b_j \in \{0, \pm 1, \dots, \pm(\ell - 1)\}$. Recall that $D - D \supset \{0, \dots, \ell - 1\}$ by assumption, hence for every $j = 1, \dots, n$ there exist $u_j, v_j \in D$ such that $a_j - b_j = u_j - v_j$, and moreover, $u = u_1 \dots u_n$ and $v = v_1 \dots v_n$ are distinct. Then

$$|S_u(0) - S_v(0)| = \Big|\sum_{j=1}^{n} (a_j - b_j)\lambda^{j-1}\Big| \leqslant c\ell^{-n},$$

which is smaller than $\varepsilon \lambda^n |\mathcal{K}_D^\lambda|$, for n sufficiently large, depending on ε, since $\lambda > \ell^{-1}$. This shows (7.84), and the proof is complete. $\square$

CHAPTER 8

Fourier-analytic and number-theoretic methods

Recall that the (infinite) Bernoulli convolution measure ν_λ is defined as the distribution of the random series $Y_\lambda = \sum_{n=0}^{\infty} a_n \lambda^n$, where a_n are i.i.d. Bernoulli random variables, with the values $\{-1,1\}$ taken with probabilities $(\frac{1}{2},\frac{1}{2})$. Equivalently, ν_λ is the self-similar measure for the IFS $(-1+\lambda x, 1+\lambda x)$, with probabilities $(\frac{1}{2},\frac{1}{2})$. As we have already seen in (6.28), the Fourier transform of ν_λ has an explicit infinite product representation:

$$(8.1) \qquad \widehat{\nu}_\lambda(\xi) = \int_{\mathbb{R}} e^{it\xi}\, d\nu_\lambda(t) = \prod_{n=0}^{\infty} \tfrac{1}{2}(\widehat{\delta_{\lambda^n} + \delta_{-\lambda^n}})(\xi) = \prod_{n=0}^{\infty} \cos(\lambda^n \xi).$$

where the convergence easily follows for $\lambda \in (0,1)$.

In this chapter we start with some classical results on Bernoulli convolutions, mostly using Fourier-analytic and Diophantine techniques. We prove singularity of ν_λ for Pisot parameters, due to Erdős (1939), and absolute continuity for Garsia parameters (defined in the chapter), due to Garsia (1962). We also give a proof of the original Erdős (1940) result, with some refinements by Kahane, showing that the Fourier transform of ν_λ has power decay at infinity for all but zero Hausdorff dimensional set of exceptional parameters. In a surprising development, this approach, which we call the "Erdős–Kahane method", has recently found many new applications, in particular, in the work of Shmerkin and collaborators. In particular, we prove a lemma, due to Shmerkin, saying that the convolution of a measure of dimension one and another measure, having a power Fourier decay, is absolutely continuous. We then use it to derive from Hochman's Theorem 6.8.11 the absolute continuity of ν_λ for all but zero Hausdorff dimensional set of exceptional parameters, following Shmerkin [**Shm14**].

In Section 8.5 the properties of ν_λ, in particular, their local dimensions, are linked to properties of β-expansions of real numbers, with $\beta = 1/\lambda$; in particular, the growth of the number of expansions is considered. In Section 8.6, we define the Garsia entropy (which is a special case of the random walk entropy for Bernoulli convolutions, see (3.14)) and prove an important result, essentially due to Garsia, saying that for Pisot parameters the Bernoulli convolution measure is not just singular, but has Hausdorff dimension smaller than one. We also discuss various results about the dimensions of ν_λ without proof. we have $\dim_{\rm H}(\nu_\lambda) = 1$.

We will return to the question of computing the dimension of a homogeneous self-similar measure on the line in Section 11.3 from another point of view. There it will be expressed as the ratio of the *projection entropy* and the Lyapunov exponent, as an application of the Feng–Hu Theorem.

8.1. Pisot–Vijayaraghavan and Garsia numbers

An algebraic integer $\theta > 1$ is called a PV number (after Charles Pisot and Tirukkannapuram Vijayaraghavan) if all its Galois conjugates (e.g., the other roots of the minimal polynomial) are less than one in absolute value. For the needed properties of PV numbers (often called simply "Pisot numbers") and a more detailed discussion, the reader is referred to Appendix D.

8.1.1. PV parameters.

THEOREM 8.1.1 (Erdős 1939, [**Erd39**]). *If $\lambda \in (0,1), \lambda \neq 1/2$, and $1/\lambda$ is a PV number, then ν_λ is singular. Moreover, the Fourier transform $\widehat{\nu}_\lambda(\xi)$ does not tend to 0 as $\xi \to \infty$.*

PROOF OF THEOREM 8.1.1. Denote $\theta = 1/\lambda$. Recall Lemma D.0.3, which says that

$$\operatorname{dist}(\theta^n, \mathbb{Z}) \leqslant (d-1)\rho^n, \quad n \in \mathbb{N}, \tag{8.2}$$

where d is the degree of θ and $\rho < 1$ is the maximal absolute value of the conjugate. By (8.1), for $N \geqslant 1$,

$$\begin{aligned} \widehat{\nu}_\lambda(\pi\theta^N) &= \cos(\pi\theta^N)\cos(\pi\theta^{N-1}) \cdot \ldots \cdot \cos(\pi)\cos(\lambda\pi)\cos(\lambda^2\pi)\ldots \\ &= \prod_{n=1}^{N} \cos(\pi\theta^n) \cdot \widehat{\nu}_\lambda(\pi). \end{aligned} \tag{8.3}$$

Since θ^n is an algebraic integer, we have that $\theta^n \neq k + \frac{1}{2}$ for $n, k \in \mathbb{N}$, see Proposition D.0.1. Also $\lambda^n \neq 1/2$ for $n \geqslant 1$, since $\lambda \neq \frac{1}{2}$ and θ is PV. It follows that $\widehat{\nu}_\lambda(\pi\theta^N) \neq 0$. Now (8.3) and (8.2) imply

$$|\widehat{\nu}_\lambda(\pi\theta^N)| \geqslant \prod_{n=1}^{\infty} |\cos((d-1)\rho^n)| \cdot |\widehat{\nu}_\lambda(\pi)| =: \delta > 0,$$

for all $N \geqslant 1$. In order to see that the infinite product converges, observe that

$$\sum_{n=1}^{\infty} 1 - |\cos((d-1)\rho^n| \leqslant \sum_{n=1}^{\infty} (d-1)\rho^n < \infty.$$

Thus, $\widehat{\nu}_\lambda(\xi) \not\to 0$ as $\xi \to \infty$. By the Riemann–Lebesgue lemma (Lemma C.0.2), ν_λ is not absolutely continuous, so ν_λ is singular in view of the Law of Pure Types (see Proposition 3.1.4). □

OPEN PROBLEM. Is ν_λ absolutely continuous for all $\lambda \in (1/2, 1)$ other than reciprocals of PV numbers?

Using Theorem D.0.5(i), Salem proved the following

THEOREM 8.1.2 (see [**Sal63**, Th.IV.II]). *If $\lambda \in (0,1)$ and $1/\lambda$ is not a PV number, then $\lim_{\xi\to\infty} \widehat{\nu}_\lambda(\xi) = 0$.*

Thus, the above Open Problem cannot be resolved just using the (non)vanishing of the Fourier transform. We sketch the proof of Theorem 8.1.2 for completeness. For short, we introduce the notation $\|t\| = \operatorname{dist}(t, \mathbb{Z})$.

Proof sketch of Theorem 8.1.2. If $\lim_{\xi\to\infty} \widehat{\nu}_\lambda(\xi) \neq 0$, then there exist $\alpha_k \in [1,\theta)$, where $\theta = \lambda^{-1}$, and increasing $m_k \in \mathbb{N}$ such that

$$\prod_{n=0}^{m_k} |\cos(\pi\alpha_k\theta^n)| \geqslant |\widehat{\nu}_\lambda(\pi\alpha_k\theta^{m_k})| \geqslant \delta > 0. \tag{8.4}$$

Without loss of generality, we can assume that $\alpha_k \to \alpha \in [1,\theta]$. Note an elementary inequality

$$|\cos(\pi t)| \leqslant 1 - \|t\|^2/4 \leqslant e^{-\|t\|^2/4}, \tag{8.5}$$

where $\|t\| = \text{dist}(t,\mathbb{Z})$. Thus (8.4) implies for all $k \geqslant 1$:

$$\sum_{n=0}^{m_k} \|\alpha_k\theta^n\|^2 \leqslant 4\log(1/\delta) =: C_\delta < \infty.$$

Then $\sum_{n=0}^{m_k} \|\alpha_s\theta^n\|^2 \leqslant C_\delta$ for $s \geqslant k$, and letting $s \to \infty$ we obtain $\sum_{n=0}^{m_k} \|\alpha\theta^n\|^2 \leqslant C_\delta$. Finally, we let $k \to \infty$ to get

$$\sum_{n=0}^{\infty} \|\alpha\theta^n\|^2 < \infty,$$

which implies that θ is a PV number by Theorem D.0.5(i). $\square$

Remark 8.1.3.

(a) The early study of Bernoulli convolutions ν_λ was related to some questions of harmonic analysis. A set $E \subset (0,2\pi)$ is a *set of uniqueness* for trigonometric series, if

$$\sum_{n=0}^{\infty} (a_n\cos(nx) + b_n\sin(nx)) = 0,\ \forall\, x \in E \implies a_n = b_n = 0,\ \forall\, n.$$

Salem and Zygmund (see [**Sal63**]) relied on Theorem 8.1.2 to show that a Cantor set with constant ratio of dissection $\lambda \in (0,1/2)$ (e.g., the support of ν_λ) is a set of uniqueness if and only if $1/\lambda$ is a PV number.

(b) Essentially the same proof as in Theorem 8.1.1 shows that the Fourier transform does not tend to zero, and hence the measure is singular, for the biased Bernoulli convolutions ν_λ^p and, more generally, for any homogeneous self-similar measure with inverse-PV contraction ratio and the digits (translation parameters) lying in $\mathbb{Z}$, up to an affine change of variable, independent of the weights.

(c) It follows from Proposition 7.4.6 that the set of "singular parameters" $S_\perp := \{\lambda \in (1/2,1) : \nu_\lambda \perp \mathcal{L}^1\}$ is a G_δ set. This implies a curious alternative: either $S_\perp$ is dense G_δ in $(\frac{1}{2},1)$, or there exists $\varepsilon > 0$ such that ν_λ is absolutely continuous for all $\lambda \in (1-\varepsilon,1)$. Indeed, if ν_λ is singular, then ν_{λ^m} is singular as well, for $m \geqslant 2$, in view of (6.29), and it is easy to see that for any sequence $\lambda_n \to 1$, $\lambda_n \in (0,1)$, the set $\{\lambda_n^m : n \geqslant 1, m \geqslant 1\}$ is dense in $[0,1]$. Note that the currently known set of singular parameters, the set of PV numbers, is countable and closed

8.1.2. Garsia parameters. We say that an algebraic integer $\theta > 1$ is a *Garsia number* (after Adriano Garsia [**Gar62**]) if if its minimal polynomial has constant term ± 2 and all its roots are greater than 1 in absolute value. These numbers include $2^{1/k}$ for $k \geqslant 1$, but also many other algebraic integers; see Appendix D for other examples and a discussion; see especially the paper by Hare and Panju

[**HP13**]. It is clear that Garsia numbers are all less than or equal to 2 (with 2 being the trivial case).

THEOREM 8.1.4 (Garsia [**Gar62**]). *Let θ be a Garsia number, and $\lambda = 1/\theta$. Then the Bernoulli convolution measure ν_λ is absolutely continuous with a bounded density.*

Recently, Yu [**Yu22**] obtained a stronger regularity result for Garsia parameters close to 1.

THEOREM 8.1.5 (Yu [**Yu22**]). *Let θ be a Garsia number such that $\theta \in (1, \sqrt{2}]$, and $\lambda = 1/\theta$. Then the Bernoulli convolution measure ν_λ is absolutely continuous with a continuous density.*

PROOF OF THEOREM 8.1.4. It is convenient to make a linear change of variable and assume that ν_λ is the self-similar measure for the IFS (S_0, S_1) with probabilities $(1/2, 1/2)$, where

$$S_j(x) = j + \lambda x, \quad j \in \{0, 1\}.$$

Then we have for any $N \geqslant 1$:

$$\nu_\lambda = 2^{-N} \sum_{\mathbf{j} \in \{0,1\}^N} (\nu_\lambda)_{\mathbf{j}}, \quad \text{where} \quad (\nu_\lambda)_{\mathbf{j}} = \nu_\lambda \circ S_{\mathbf{j}}^{-1}, \tag{8.6}$$

and

$$S_{\mathbf{j}}(x) = S_{j_1 \ldots j_N}(x) = \sum_{n=0}^{N-1} j_{n+1} \lambda^n + \lambda^N x.$$

Note that $\operatorname{spt}(\nu_\lambda) = [0, \frac{1}{1-\lambda}]$ and $\operatorname{spt}((\nu_\lambda)_{\mathbf{j}}) = [S_{\mathbf{j}}(0), S_{\mathbf{j}}(0) + \frac{\lambda^N}{1-\lambda}]$.

Now let $x \in \mathbb{R}$ and $\delta > 0$. We are going to estimate $\nu_\lambda(B(x, \delta))$ from above. Let N be the minimal positive integer such that $\lambda^N \leqslant \delta$. Then we have for $\mathbf{j} \in \{0, 1\}^N$:

$$\operatorname{spt}((\nu_\lambda)_{\mathbf{j}}) \cap B(x, \delta) \neq \varnothing \implies S_{\mathbf{j}}(0) \in \left[x - \delta - \frac{\delta}{1 - \lambda}, x + \delta\right). \tag{8.7}$$

Now we invoke Corollary D.0.11(ii), which implies that for $\mathbf{i} \neq \mathbf{j}$ in $\{0, 1\}^N$ we have

$$|S_{\mathbf{i}}(0) - S_{\mathbf{j}}(0)| \geqslant c \cdot 2^{-N}, \tag{8.8}$$

since

$$S_{\mathbf{i}}(0) - S_{\mathbf{j}}(0) = \sum_{n=0}^{N-1} (i_{n+1} - j_{n+1}) \lambda^n$$

is a polynomial in $\lambda = \theta^{-1}$ of height one and degree less than N. Combining (8.6), (8.7) and (8.8) yields that

$$\nu_\lambda(B(x, \delta)) \leqslant \frac{2\delta + \delta(1 - \lambda)^{-1}}{c \cdot 2^{-N}} \cdot 2^{-N} = \frac{2\delta + \delta(1 - \lambda)^{-1}}{2c} \cdot \mathcal{L}^1(B(x, \delta)),$$

which shows that ν_λ is absolutely continuous with bounded density, as desired. □

REMARK 8.1.6.

(a) It was crucial for the proof that the Bernoulli convolution ν_λ has uniform weights (is unbiased). In fact, for a long time not a single concrete example was known of an absolutely continuous biased Bernoulli convolution, even though they were known to be typically absolutely continuous in the relevant parameter interval (see Corollary 6.7.2). Only recently Varjú [**Var19a**] found such examples of arithmetic nature; in particular, all Garsia numbers sufficiently close to 1 have this property.

(b) One can easily generalize the above as follows: instead of ν_λ consider for $m \geqslant 2$ the self-similar measure $\nu_{\lambda,m}$ corresponding to the IFS $(S_j)_{j=0}^{m-1}$ of m maps, where $S_j(x) = j + \lambda x$, with $\mathbf{p} = (1/m, \ldots, 1/m)$. (It is important that the weights are uniform and there are no "deleted digits".) Define "m-Garsia" numbers to be algebraic integers whose minimal polynomial has constant term $\pm m$ and all zeros greater than 1 in modulus. Then $\nu_{\lambda,m}$ is absolutely continuous with a bounded density, whenever $1/\lambda$ is m-Garsia.

8.2. Fourier dimension and its variant

Before proceeding further, we introduce the notion of *Fourier dimension.* We are not trying to be comprehensive here: an excellent book by Mattila [**Mat15**] is devoted to Fourier-analytic techniques and their applications to dimension theory. We also borrow material from the paper [**EPS15**] by Ekström, Persson, and Schmeling. As pointed out in [**Mat15**, p. 54], the Fourier dimension has not been investigated systematically until [**EPS15**]. See also the survey by Ekström and Schmeling [**ES17**].

We will need a classical formula for the s-energy of a compactly supported Borel measure in $\mathbb{R}^d$ in terms of the Fourier transform, see [**Mat15**, Theorem 3.10]. Let $\mathcal{M}(A)$ denote the set of finite Borel measures on $A \subset \mathbb{R}^d$ of compact support $\mathrm{spt}(\mu) \subset A$. In addition, following [**EPS15**], we denote by $\mathcal{P}(A)$ the set of Borel probability measures on $\mathbb{R}^d$ that give full measure to A.

THEOREM 8.2.1. *Let $\mu \in \mathcal{M}(\mathbb{R}^d)$. Then for $0 < s < d$, there is a constant $c(s,d)$ such that*

$$\mathcal{E}_s(\mu) = \iint \frac{d\mu(x)\, d\mu(y)}{|x-y|^s} = c(s,d) \int_{\mathbb{R}^d} |\xi|^{s-d} |\widehat{\mu}(\xi)|^2 d\xi. \tag{8.9}$$

Below we will sometimes use the usual big-O and little-o notation. That is, for real functions g and f, we write $g = O(f)$ if $g(\xi)/f(\xi)$ is uniformly bounded, and $g = o(f)$ if $g(\xi)/f(\xi) \to 0$, as $\xi \to \infty$.

Denote

$$\mathscr{D}_d(s) = \{\nu \in \mathcal{M}(\mathbb{R}^d) : \ |\widehat{\nu}(\xi)| = O(|\xi|^{-s}), \ |\xi| \to \infty\}.$$

Thus the set $\mathscr{D}_d := \bigcup_{s>0} \mathscr{D}_d(s)$ is the set of compactly supported Borel probability measures on $\mathbb{R}^d$ with a power Fourier decay. The *Fourier dimension of a measure* μ is defined by

$$\dim_{\mathrm{F}}(\mu) = \sup\{s \in [0,d] : \ \mu \in \mathscr{D}_d(s/2)\}, \tag{8.10}$$

and the *Fourier dimension of a set* A is

$$\dim_{\mathrm{F}}(A) = \sup\{\dim_{\mathrm{F}}(\mu) : \ \mu \in \mathcal{P}(A)\}. \tag{8.11}$$

By (8.9), if $\mu \in \mathscr{D}_d(s/2)$, then $\mathcal{E}_{s-\varepsilon}(\mu) < \infty$ for all $\varepsilon > 0$, and thus we have, by Lemma 1.9.9,

$$\dim_{\mathrm{F}}(\mu) \leqslant \dim_{\mathrm{H}}(\mu), \tag{8.12}$$

and for all Borel sets $A \subset \mathbb{R}^d$,

$$\dim_{\mathrm{F}}(A) \leqslant \dim_{\mathrm{H}}(A). \tag{8.13}$$

Sometimes the Fourier dimension of a set A is defined differently, namely, instead of measures $\mu \in \mathcal{P}(A)$ one uses Borel probability measures $\mu \in \mathcal{M}(A)$. This way we obtain what is called the *compact Fourier dimension* in [**EPS15**], denoted $\dim_{\mathrm{FC}}(A)$. For a closed set A holds $\dim_{\mathrm{FC}}(A) = \dim_{\mathrm{F}}(A)$, The point here is that a probability measure μ with $\mu(A) = 1$ of fast Fourier decay may have $\mathrm{spt}(\mu) \not\subset A$.

Next we quote some remarks from [**EPS15**]; see the paper for more information.

REMARK 8.2.2.

(a) Any countable set has Fourier dimension equal to zero.

(b) A line segment has Fourier dimension one as a subset of $\mathbb{R}$, but Fourier dimension zero as a subset of $\mathbb{R}^2$.

(c) The classical middle-third Cantor set, and more generally, $\mathcal{C}_\alpha$ for $1/\alpha$ that is a PV-number, has the property that for **any** $\mu \in \mathcal{M}(\mathcal{C}_\alpha)$ we have $|\widehat{\mu}(\xi)| \not\to 0$, as $\xi \to \infty$. It follows that such $\mathcal{C}_\alpha$ have Fourier dimension zero. This, of course, is stronger than Theorem 8.1.1, and, as shown in [**Sal63**], equivalent to the claim on sets of uniqueness in Remark 8.1.3(a). For the proof, see [**Sal63**].

(d) A *Salem set* is a set A which satisfies $\dim_{\mathrm{H}}(A) = \dim_{\mathrm{F}}(A)$. Salem sets abound among *random fractals*, but they are hard to construct explicitly. An explicit example of a Salem set in $\mathbb{R}$ of any prescribed dimension is the set of α-well approximable numbers, namely

$$E(\alpha) = \bigcap_{n=1}^{\infty} \bigcup_{k=n}^{\infty} \{x \in [0,1] : \ \|xk\| \leqslant k^{-(1+\alpha)}\},$$

where $\|\cdot\|$ denotes the distance to the nearest integer. By a theorem of Jarnik and Besicovitch, $\dim_{\mathrm{H}}(E(\alpha)) = 2/(2+\alpha)$, see [**Fal85**, Theorem 8.16], and Kaufman [**Kau81**] showed that there is a measure in $\mathcal{P}(E(\alpha))$ of Fourier dimension $2/(2+\alpha)$. See also [**Blu98**].

(e) It may be derived from a theorem by Davenport, Erdős, and LeVeque [**DEL63**] (see [**QR03**]) that if a measure $\mu \in \mathcal{M}(\mathbb{R})$ has positive Fourier dimension, then μ-a.e. x is normal to every integer base $b \geqslant 2$ which means that $\{b^k x\}$ is uniformly distributed mod 1. It follows that if there exists an integer base, such that a set does not contain any normal numbers to it, then this set has Fourier dimension zero. This applies, in particular, to the middle-third Cantor set, since the base-3 expansion of its elements contains no 1's.

It turns out that the Fourier dimension is not *finitely stable*; in fact,

$$\dim_{\mathrm{F}}(A \cup B) \geqslant \max\{\dim_{\mathrm{F}}(A), \dim_{\mathrm{F}}(B)\},$$

but the inequality can be strict; and the same is true for the compact Fourier dimension $\dim_{\mathrm{FC}}$ of a set, see [**Eks15, EPS15**]. In order to rectify this and achieve finite and countable stability, Ekström, Persson, and Schmeling [**EPS15**] introduced the following.

DEFINITION 8.2.3. The modified Fourier dimension of a finite Borel measure is defined by

$$\dim_{\mathrm{FM}}(\mu) = \sup\{\dim_{\mathrm{F}}(\nu) : \ \nu \in \mathcal{P}(\mathbb{R}^d), \mu \ll \nu\},$$

and the modified Fourier dimension of a Borel set is

$$\dim_{\mathrm{FM}}(A) = \sup\{\dim_{\mathrm{F}}(\mu) : \ \mu \in \mathcal{P}(\mathbb{R}^d), \mu(A) > 0\}.$$

It is proved in [**EPS15**, Theorem 17] that the modified Fourier dimension is monotone and countably stable. The paper also gives sufficient conditions for the Fourier dimension to be countably stable.

THEOREM 8.2.4 (Ekström, Persson, and Schmeling).

(i) *Suppose that* $\{A_k\}$ *is a finite or countable collection of Borel sets in* $\mathbb{R}^d$ *such that* $A_n \cap \overline{\bigcup_{k\neq n} A_k}$ *is countable for all* n. *Then* $\dim_{\mathrm{F}}\left(\bigcup_k A_k\right) = \sup_k \dim_{\mathrm{F}}(A_k)$.

(ii) *Suppose that* μ *and* ν *are finite Borel measures on* $\mathbb{R}^d$ *whose supports are compact and disjoint. Then*

$$\dim_{\mathrm{F}}(\mu+\nu) = \min\{\dim_{\mathrm{F}}(\mu), \dim_{\mathrm{F}}(\nu)\}.$$

For the proof see [**EPS15**, Cor. 4 and Prop. 11].

8.2.1. Comparison with the Sobolev dimension. Recall the notion of Sobolev dimension introduced in (6.48) for a finite Borel measure μ in $\mathbb{R}^d$:

$$\dim_s(\mu) = \sup\left\{\alpha > 0: \int_{\mathbb{R}^d} |\widehat{\mu}(\xi)|^2 |(1+|\xi|)^{\alpha-d}\, d\xi < \infty\right\}.$$

As already mentioned, if $\sigma = \dim_s(\mu) < d$, it coincides with the correlation dimension of μ, and if $\sigma = \dim_s(\mu) > d$, then ν is absolutely continuous with the Radon–Nikodym derivative in the fractional Sobolev space $H^{\sigma-d}(\mathbb{R}^d)$.

When considering absolutely continuous measures, it is natural to consider the Fourier dimension without the restriction that it is less than or equal to d: namely, let

$$\widetilde{\dim}_{\mathrm{F}}(\mu) = \sup\{s > 0: \ \mu \in \mathscr{D}_d(s/2)\}, \tag{8.14}$$

It is immediate that

$$\dim_{\mathrm{F}}(\mu) \leqslant \widetilde{\dim}_{\mathrm{F}}(\mu) \leqslant \dim_s(\mu); \tag{8.15}$$

of course, $\dim_{\mathrm{F}}(\mu) = \widetilde{\dim}_{\mathrm{F}}(\mu)$ whenever the latter is less than or equal to d. In the opposite direction we have the following

LEMMA 8.2.5. *For any finite compactly supported Borel measure* μ *on* $\mathbb{R}^d$ *the following holds*

$$\widetilde{\dim}_{\mathrm{F}}(\mu) \geqslant \frac{\dim_s(\mu) - d}{2}. \tag{8.16}$$

Since the claim has meaning only when $\dim_s(\mu) > d$, we can assume that μ is absolutely continuous, with a density $f \in L^2(\mathbb{R}^d)$, and then $\widehat{\mu} = \widehat{f}$. The result is then immediate from the following standard fact from harmonic analysis; we borrow the proof of [**PSS00a**, Lemma 5.3].

CLAIM. *Suppose that* $f \in L^2(\mathbb{R})$ *has compact support and*

$$\|f\|_{2,\gamma}^2 = \int_{\mathbb{R}^d} |\widehat{f}(\xi)|^2 |\xi|^{2\gamma}\, d\xi < \infty.$$

Then

$$|\widehat{f}(\xi)| = o(|\xi|^{-\gamma}), \quad |\xi| \to \infty.$$

PROOF. Let $\tilde{f}(x) = f(-x)$ and let ϕ be a smooth compactly supported function on $\mathbb{R}^d$ equal to one on the support of $f * \tilde{f}$. Then $f * \tilde{f} = (f * \tilde{f})\phi$ hence

$$\begin{aligned} |\widehat{f}(\xi)|^2 &= (|\widehat{f}|^2 * \widehat{\phi})(\xi) \\ &= \int_{\mathbb{R}^d} |\widehat{f}(\xi-\eta)|^2 \widehat{\phi}(\eta)\, d\eta \\ &\leqslant C \int_{|\eta|<|\xi/2|} |\widehat{f}(\xi-\eta)|^2\, d\eta + C \int_{|\eta|\geqslant|\xi/2|} |\widehat{\phi}(\eta)|\, d\eta. \end{aligned}$$

Since $|\widehat{\phi}(\eta)|$ is rapidly decreasing, the second integral is $O(|\xi|^{-q})$ for any $q > 0$. The first integral can be estimated from above by

$$\begin{aligned} \int_{|\eta|<|\xi/2|} |\widehat{f}(\xi-\eta)|^2\, d\eta &\leqslant |\xi/2|^{-2\gamma} \int_{|\eta|<|\xi/2|} |\widehat{f}(\xi-\eta)|^2 |\xi-\eta|^{2\gamma}\, d\eta \\ &\leqslant |\xi/2|^{-2\gamma} \int_{|\tau|\geqslant|\xi/2|} |\widehat{f}(\tau)|^2 |\tau|^{2\gamma}\, d\tau = o(|\xi|^{-2\gamma}), \end{aligned}$$

and the claim follows, hence the lemma. □

8.3. Fourier decay

8.3.1. Erdős–Kahane argument. Salem proved (see Theorem 8.1.2) that $\widehat{\nu}_\lambda(\xi) \to 0$ as $\xi \to \infty$ whenever λ^{-1} is not a PV number. However, his proof does not provide any quantitative estimates for the rate of decay. Meanwhile, Erdős [**Erd40**], using a beautiful combinatorial argument, demonstrated that for a.e. $\lambda \in (0,1)$ the Fourier transform $\widehat{\nu}_\lambda$ decays at infinity at a power rate, which has many useful consequences. Later, Kahane [**Kah71**] observed that Erdős' argument actually yields power Fourier decay outside a set of zero Hausdorff dimension. Next we state the result precisely. Recall the definition from the last section in the special case of measures on the real line:

$$\mathscr{D}_1(s) = \{\nu \in \mathcal{M}(\mathbb{R}) :\ |\widehat{\nu}(\xi)| = O(|\xi|^{-s}),\ |\xi| \to \infty\}.$$

THEOREM 8.3.1 ([**Erd40**], [**Kah71**]). *For any $1 < a < b < \infty$ and $\delta > 0$ there exists $s = s(a,b,\delta) > 0$ such that*

$$\dim_{\mathrm{H}} \big(\{\lambda \in [b^{-1}, a^{-1}] :\ \nu_\lambda \notin \mathscr{D}_1(s)\}\big) < \delta. \tag{8.17}$$

COROLLARY 8.3.2 (Erdős, Kahane).

(i) *For any $\delta > 0$ and $k \in \mathbb{N}$ there exists $a(k,\delta) < 1$ such that*

$$\dim_{\mathrm{H}} \left(\left\{\lambda \in [a(k,\delta),1) :\ \frac{d\nu_\lambda}{dx} \notin C^k(\mathbb{R})\right\}\right) < \delta;$$

(ii) $\dim_{\mathrm{H}}\{\lambda \in (0,1) :\ \nu_\lambda \notin \bigcup_{s>0} \mathscr{D}_1(s)\} = 0$.

PROOF OF THE COROLLARY ASSUMING THEOREM 8.3.1. Fix $\delta > 0$ and find $s = s(\sqrt{2}, 2, \delta)$ from Theorem 8.3.1, suich that

$$\dim_{\mathrm{H}} \big(\{\lambda \in [2^{-1}, 2^{-1/2}] :\ |\widehat{\nu}_\lambda(\xi)| \neq O(|\xi|^{-s})\}\big) < \delta. \tag{8.18}$$

We recall (6.29):

$$\widehat{\nu}_\lambda(\xi) = \widehat{\nu}_{\lambda^m}(\xi)\widehat{\nu}_{\lambda^m}(\lambda\xi) \cdot \ldots \cdot \widehat{\nu}_{\lambda^m}(\lambda^{m-1}\xi) \quad \text{for } m \geqslant 2.$$

Combining this formula with (8.18) yields

$$\dim_{\mathrm{H}}\Big(\{\lambda\in[2^{-1/m},2^{-1/(2m)}]:\ |\widehat{\nu}_\lambda(\xi)|\ne O(|\xi|^{-ms})\}]\Big)<\delta.$$

If $|\widehat{\nu}_\lambda(\xi)|=O(|\xi|^{-\alpha})$ for $\alpha>1$, then $\widehat{\nu}_\lambda\in L^1(\mathbb{R})$, whence ν_λ is absolutely continuous with a density $\frac{d\nu_\lambda}{dx}\in C(\mathbb{R})$. If $|\widehat{\nu}_\lambda(\xi)|=O(|\xi|^{-\alpha})$ for $\alpha>k+1$, then $\widehat{\frac{d^k}{dx^k}(\frac{d\nu_\lambda}{dx})}=\xi^k\widehat{\nu}_\lambda(\xi)$ (in the distributional sense) is in $L^1(\mathbb{R})$, and therefore, $\frac{d\nu_\lambda}{dx}\in C^k(\mathbb{R})$, see Theorem C.0.1. This implies the assertion of the corollary for $a(k,\delta)=2^{-1/\lceil(k+1)/s\rceil}$. □

REMARK 8.3.3. Erdős and Kahane did not provide explicit estimates for $a(k,\delta)$, but it was clear that even $a(0,1)$ is rather close to 1. The paper [**PSS00a**] gives an exposition of the Erdős–Kahane argument with explicit constants $s=s(\sqrt{2},2,1)=6\cdot 10^{-4}$ and $a(0,1)=.99933$.

Before the proof of Theorem 8.3.1, as a "warm-up" we will prove Salem's Theorem, which says that

$$\{\theta>1:\ \exists\,\alpha>0,\ \text{such that}\ \|\alpha\theta^n\|\to 0\}\ \text{is countable.}\tag{8.19}$$

Recall that $\|t\|=\mathrm{dist}(t,\mathbb{Z})$.

PROOF OF (8.19). Note that in (8.19) one can assume $\alpha\in[1,\theta)$ without loss of generality. For $\theta>1$ and $\alpha\in[1,\theta)$ let

$$\alpha\theta^n=K_n+\varepsilon_n,\tag{8.20}$$

where $K_n\in\mathbb{N}$ and $|\varepsilon_n|\leqslant 1/2$, so that $\|\alpha\theta^n\|=|\varepsilon_n|$. Next observe that

$$\begin{aligned}
\Big|K_{n+2}-\frac{K_{n+1}^2}{K_n}\Big| &= K_{n+1}\Big|\frac{K_{n+2}}{K_{n+1}}-\frac{K_{n+1}}{K_n}\Big|\\
&\leqslant K_{n+1}\Big(\Big|\theta-\frac{K_{n+2}}{K_{n+1}}\Big|+\Big|\theta-\frac{K_{n+1}}{K_n}\Big|\Big)\\
&= K_{n+1}\Big(\Big|\frac{\theta\varepsilon_{n+1}-\varepsilon_{n+2}}{K_{n+1}}\Big|+\Big|\frac{\theta\varepsilon_n-\varepsilon_{n+1}}{K_n}\Big|\Big)\\
&= |\theta\varepsilon_{n+1}-\varepsilon_{n+2}|+\frac{K_{n+1}}{K_n}|\theta\varepsilon_n-\varepsilon_{n+1}|.
\end{aligned}\tag{8.21}$$

It is clear that $\lim_{n\to\infty}\frac{K_{n+1}}{K_n}=\theta$, and if $\varepsilon_n\to 0$, then the right-hand side of (8.21) will eventually become smaller than $1/2$. But in that case K_{n+2} is uniquely determined as the nearest integer to K_{n+1}^2/K_n for all n sufficiently large!

Thus, any θ in the set in (8.19) is a limit of a ratio K'_{n+1}/K'_n defined by a sequence of integers K'_n such that K'_{n+2} is the nearest integer to $\frac{(K'_{n+1})^2}{K'_n}$ for every $n\geqslant 1$ and $K'_2\geqslant K'_1$. Since there are only countably many such sequences, the countability of (8.19) follows. □

The idea of Erdős [**Erd40**] was that if *most* of the $|\varepsilon_n|$ are small, then *most* of the integers K_{n+2} are determined uniquely by the two previous ones, giving a bound for the cardinality of a good cover for the "bad" set of θ. This is implemented in the next proof.

PROOF OF THEOREM 8.3.1. Denote $\theta = \lambda^{-1}$. From (8.1) we have, for $t \in [1, \theta)$,

$$\widehat{\nu}_\lambda(\pi\theta^N t) = \prod_{n=1}^{N} \cos(\pi\theta^n t)\widehat{\nu}_\lambda(\pi t)\,. \tag{8.22}$$

As in (8.20), let $\theta^n t = K_n + \varepsilon_n$ where $K_n \in \mathbb{N}$ and $\varepsilon_n \in [-\frac{1}{2}, \frac{1}{2})$ (the dependence on θ and t is not written explicitly but should be kept in mind). Let

$$\rho := [2(b+1)(b+2)]^{-1}. \tag{8.23}$$

For $k \geqslant 2$ denote by $\mathcal{E}_N(k)$ the set of $\theta \in [a, b]$ such that there exists $t \in [1, \theta)$ for which

$$\#\{n \in [1, N] : \ |\varepsilon_n| \geqslant \rho\} \leqslant \frac{N}{k}\,.$$

Further, let $\mathcal{E}(k) = \limsup_{N\to\infty} \mathcal{E}_N(k) = \bigcap_{N=1}^{\infty} \bigcup_{M\geqslant N} \mathcal{E}_M(k)$. Since by (8.22),

$$|\widehat{\nu}_\lambda(\pi\theta^N t)| \leqslant \prod_{n=1}^{N} |\cos(\pi\varepsilon_n)|\,,$$

it is immediate that

$$\theta \notin \mathcal{E}(k) \implies \widehat{\nu}_\lambda(\xi) = O(|\xi|^{-s}),$$

where

$$s = \frac{-\log[\cos(\pi\rho)]}{k\log\theta} \geqslant \frac{-\log[\cos(\pi\rho)]}{k\log b}.$$

Thus, to prove (8.17) we only need to estimate $\dim_{\mathrm{H}}(\mathcal{E}(k))$.

LEMMA 8.3.4. *Let $1 < a < b < \infty$. The following hold for $n \geqslant n_0 := 1 + \log_a(3/2)$:*

(i) *Given K_n, K_{n+1}, there are at most $A := 1 + (b+1)(b+2)$ possibilities for K_{n+2}, independent of $\theta \in [a, b]$ and $t \in [1, \theta)$.*
(ii) *If $\max\{|\varepsilon_n|, |\varepsilon_{n+1}|, |\varepsilon_{n+2}|\} < \rho = [2(b+1)(b+2)]^{-1}$, then K_{n+2} is uniquely determined by K_n and K_{n+1}, independent of $\theta \in [a, b]$ and $t \in [1, \theta)$, namely, K_{n+2} is the nearest integer to $\frac{K_{n+1}^2}{K_n}$.*

PROOF. In view of (8.21),

$$\Big|K_{n+2} - \frac{K_{n+1}^2}{K_n}\Big| \leqslant b|\varepsilon_{n+1}| + |\varepsilon_{n+2}| + \frac{K_{n+1}}{K_n}(b|\varepsilon_n| + |\varepsilon_{n+1}|). \tag{8.24}$$

Note that

$$\theta^n \geqslant \frac{3\theta}{2} \geqslant 1 + \frac{1}{2}\theta \ \text{ for } n \geqslant n_0 = 1 + \log_a(3/2),$$

hence

$$\frac{K_{n+1}}{K_n} = \frac{\alpha\theta^{n+1} - \varepsilon_{n+1}}{\alpha\theta^n - \varepsilon_n} \leqslant \theta + 1 \leqslant b + 1 \ \text{ for } n \geqslant n_0.$$

Combining this with (8.24) yields

$$\Big|K_{n+2} - \frac{K_{n+1}^2}{K_n}\Big| \leqslant (b+1)(b+2)\max\{|\varepsilon_n|, |\varepsilon_{n+1}|, |\varepsilon_{n+2}|\}.$$

Now both (i) and (ii) are immediate since $K_{n+2} \in \mathbb{N}$. □

Note that

$$(8.25)\qquad |\theta - \frac{K_{N+1}}{K_N}| = |\frac{\theta\varepsilon_N - \varepsilon_{N+1}}{K_N}| \leqslant \frac{(b+1)/2}{a^N - 1/2} \leqslant \frac{2b}{a^N},$$

using a crude bound in the last step. Thus, we get a cover of $\mathcal{E}_N(k)$ by intervals of length $2b \cdot a^{-N}$, centered at the points K_{N+1}/K_N. The cardinality of the cover is bounded by the number of integer sequences $K_1, \ldots, K_{N+1}$ obtained from $\theta \in \mathcal{E}_N(k)$ and $t \in [1, \theta)$. There are at most $C(a, b)$ initial parts of the sequence $1, \ldots, n_0(a)$, where n_0 is from Lemma 8.3.4. For $\Gamma \subset [1, N] \cap \mathbb{N}$ consider those $\theta \in [a, b]$ for which there exists $t \in [1, \theta)$ such that $|\varepsilon_n| < \rho$ for $n \in [1, N]\backslash\Gamma$. It follows from Lemma 8.3.4 that the number of sequences $K_1, \ldots, K_{N+1}$ corresponding to such θ is bounded above by $C(a, b)(A)^{3\,\#(\Gamma)}$. Thus, the total number of sequences $K_1, \ldots, K_{N+1}$ corresponding to $\mathcal{E}_N(k)$ does not exceed

$$C(a, b) \cdot \sum_{j \leqslant \lfloor N/k \rfloor} \binom{N}{j} \cdot (A)^{3j}.$$

The number of subsets of $[1, N] \cap \mathbb{N}$ of size $\leqslant N/k$ is not greater than $\exp(O(\frac{\log k}{k} N))$, as one can deduce, e.g. using Stirling's formula. Therefore, by (8.25),

$$\dim_{\mathrm{H}} \left(\mathcal{E}(k)\right) \leqslant \lim_{N \to \infty} O\left(\frac{\frac{\log k}{k} N + \frac{3N}{k} \log A}{N \log a}\right) = O\Big(\frac{\log k}{k}\Big).$$

To complete the proof it remains to notice that the right-hand side tends to zero as $k \to \infty$. □

REMARK 8.3.5.

(a) The proof of Theorem 8.3.1 extends with minimal changes to any family of IFS $(d_j + \lambda x)_{j=1}^m$, with any probability vector (the only assumption is that there are at least two distinct digits d_j), with a parameter $\lambda \in (0, 1)$. In fact, the Hausdorff dimension zero exceptional set of λ's for *some* power decay of the Fourier transform is the same as for the family of Bernoulli convolutions ν_λ (although the rate of decay s in the quantitative statement—the analog of (8.17)—will certainly depend on the probabilities).

(b) The only specific numbers λ for which ν_λ is known to belong to $\mathscr{D}_1(s)$ for some $s > 0$ are the Garsia numbers—this was proved by Dai, Feng, and Wang [**DFW07**]. Interestingly, their proof works only for the classical unbiased Bernoulli convolutions and for their analogs corresponding to m-Garsia numbers from Remark 8.1.6(b). In fact, for the proof of the Fourier decay they need that $\cos(\theta^n t)$ sometimes gets close to zero and not just bounded away from one in absolute value. Recently, Streck [**Str23**] revisited the phenomenon behind the result of [**DFW07**]. He obtained new examples of self-similar measures with power Fourier decay (including those with λ an algebraic number, but not an algebraic integer), however no new Bernoulli convolutions.

(c) Even weaker than polynomial quantitative decay estimates of the Fourier transform for specific numbers are of interest. Logarithmic decay was proved for rational parameters by Kershner [**Ker36**] and for non-Pisot, non-Salem algebraic integers by Bufetov and Solomyak [**BS14**, Corollary A.3].

(d) The Fourier decay for *nonhomogeneous* (as well as homogeneous) self-similar measures has been actively investigated in recent years, using a variety of methods. The reader interested in this topic should consult [**LS22, AHW21, AHW22, Sol21, Bré21, VY22**].

8.3.2. Fourier decay for Salem numbers. An algebraic integer $\theta > 1$ is a *Salem number*, see Definition D.0.2, if its Galois conjugates satisfy $|\theta_j| \leqslant 1$ and at least one of the conjugates has modulus equal to one (i.e. θ is not Pisot).

LEMMA 8.3.6. *Let θ be a Salem number and $\lambda = \theta^{-1}$. Then*

$$\limsup_{\xi\to\infty} |\widehat{\nu}_\lambda(\xi)|\,|\xi|^\varepsilon = \infty \quad \text{for all } \varepsilon > 0. \tag{8.26}$$

Thus $\dim_{\mathrm{F}}(\nu_\lambda) = 0$.

Proof. Recall that $\|x\|$ denotes the distance of $x \in \mathbb{R}$ to the nearest integer. It is known (see Theorem D.0.7) that for each Salem number θ and any $\delta > 0$ one can find $t \geqslant 1$ such that

$$\|t\theta^n\| \leqslant \delta \quad \text{for all } n \geqslant 1. \tag{8.27}$$

We have by (8.1) and (8.27) for $\lambda = \theta^{-1}$:

$$\begin{aligned} |\widehat{\nu}_\lambda(\pi t\theta^n)| &\geqslant |\widehat{\nu}_\lambda(\pi t)| \prod_{k=1}^{n} \cos(\pi\|t\theta^k\|) \\ &\geqslant |\widehat{\nu}_\lambda(\pi t)|(1-c\delta^2)^n \geqslant c'|\widehat{\nu}_\lambda(\pi t)|(\pi t\theta^n)^{-\varepsilon}, \end{aligned}$$

where $\varepsilon = -\frac{\log(1-c\delta^2)}{\log\theta}$. Since $\varepsilon \to 0$ as $\delta \to 0$, it remains to show that $\widehat{\nu}_\lambda(\pi t) \neq 0$.

Suppose $\widehat{\nu}_\lambda(\pi t) = 0$, then $t = \frac{2k+1}{2}\theta^l$ for some $l \geqslant 0$. But $\{\theta^n\}_{n\geqslant 1}$ is dense mod 1 (see [**Sal63**]), hence $t\theta^n$ is dense mod $\frac{1}{2}$ which contradicts (8.27) for $\delta < \frac{1}{2}$. □

8.3.3. Fourier dimension of Bernoulli convolutions. Kahane [**Kah71**] asked whether it is true that $\dim_{\mathrm{F}}(\nu_\lambda) = s\text{-}\dim(\nu_\lambda) = \frac{\log 2}{-\log\lambda}$, for a.e. $\lambda \in (0,1)$. The upper bound $\dim_{\mathrm{F}}(\nu_\lambda) \leqslant s\text{-}\dim(\nu_\lambda)$ in the range $\lambda \in (0,\frac{1}{2})$ follows from (8.12). However, in [**PS98**] it was shown that the answer is, in general, negative. In fact, it follows from [**PS98**, Cor. 1.6] that

$$\dim_{\mathrm{F}}(\nu_\lambda) \leqslant 1/2 < \frac{\log 2}{-\log\lambda}, \quad \text{for all } \lambda \in (1/4, 3/8).$$

The reason is that $\nu_\lambda * \nu_\lambda$ does not have a density in L^2, hence $\widehat{\nu}_\lambda \notin L^4$ for $\lambda < 3/8$, see the proof of Theorem 6.5.1.

Corollary 8.3.2 implies that for all $\lambda \in (0,1)$ outside a set of Hausdorff dimension zero, the Fourier dimension of ν_λ is positive, but it seems difficult to give a good lower bound. Nevertheless, we do have the following

COROLLARY 8.3.7. *For a.e. λ in the transversality interval $J = (\frac{1}{2}, 0.668\ldots)$ for $\{-1,0,1\}$-power series, we have*

$$\dim_{\mathrm{F}}(\nu_\lambda) \geqslant \frac{1}{2}\Big(\frac{\log 2}{-\log\lambda} - 1\Big).$$

Less sharp lower bounds can be given a.e. outside of the transversality interval, using the “thinning and convolving” argument.

PROOF. We only have to combine Theorem 6.8.2(i) with Lemma 8.2.5. □

8.3.4. Sobolev dimension of Bernoulli convolutions. Recall that by Theorem 6.8.2(i) we have

$$\dim_s(\nu_\lambda) \geqslant \frac{\log 2}{-\log\lambda} \quad \text{for a.e. } \lambda \in (1/2, 0.668\ldots);$$

this was used in the proof of the above corollary (and similarly, less sharp lower bounds can be given outside of this interval). For the upper bound the following holds:

PROPOSITION 8.3.8. *We have*

$$\dim_s(\nu_\lambda) \leqslant \frac{2\log 2}{-\log\lambda} \quad \textit{for all } \lambda \in [1/2, 1). \tag{8.28}$$

Observe that this estimate is sharp for $\lambda = 2^{1/k}$, $k \in \mathbb{N}$. In particular, the Sobolev dimension of Lebesgue measure on a finite interval is equal to 2.

PROOF. This is an adaptation of the argument in [**PSS00a**, Section 5]. Fix $\lambda \in [\frac{1}{2}, 1)$ and let $\theta = \lambda^{-1}$. We have from (8.1), for all $t \in [1, \theta)$ and $n \in \mathbb{N}$:

$$|\widehat{\nu}_\lambda(\pi t\theta^n)| = |\widehat{\nu}_\lambda(\pi t)| \cdot \prod_{k=1}^{n} |\cos(\pi t\theta^k)|.$$

By Koksma's Theorem, for a.e. $t > 0$ the sequence $\{\theta^n t\}_{n\geqslant 1}$ is uniformly distributed mod 1 (see [**KN74**, Cor. I.4.3]). For such $t \in [1, \theta)$, with at most a finite number of exceptions, at which $\widehat{\nu}_\lambda(\pi t) = 0$, we have

$$\begin{aligned}
\frac{1}{n}\log|\widehat{\nu}_\lambda(\pi t\theta^n)| &= \frac{1}{n}\log|\widehat{\nu}_\lambda(\pi t)| + \frac{1}{n}\sum_{k=1}^{n}\log|\cos(\pi t\theta^k)| \\
&\to \int_0^1 \log|\cos(\pi u)|\, du = -\log 2, \quad \text{as } n \to \infty.
\end{aligned}$$

By Egorov's Theorem, we can choose a set $E \subset [1, \theta)$ of positive measure, such that the convergence for $t \in E$ is uniform. Let $\alpha > \frac{2\log 2}{\log\theta} > 1$. By the definition of Sobolev dimension, in order to prove (8.28), it suffices to show that $\int_{\mathbb{R}} |\widehat{\nu}_\lambda(\xi)|^2(1 + |\xi|)^{\alpha-1}\, d\xi = \infty$. Let $0 < \varepsilon < \frac{1}{2}\alpha\log\theta - \log 2$. By uniform convergence, there exists $N = N(\varepsilon)$ such that

$$\frac{1}{n}\log|\widehat{\nu}_\lambda(\pi t\theta^n)| \geqslant -\log 2 - \varepsilon, \quad \text{for all } n \geqslant N \text{ and for all } t \in E.$$

We have

$$\begin{aligned}
\int_{\mathbb{R}} |\widehat{\nu}_\lambda(\xi)|^2(1 + |\xi|)^{\alpha-1}\, d\xi &\geqslant \pi\sum_{n=N}^{\infty}\int_{E\theta^n} |\widehat{\nu}_\lambda(\pi\xi)|^2(1 + |\pi\xi|)^{\alpha-1}\, d\xi \\
&\geqslant \pi\sum_{n=N}^{\infty} e^{-2n(\log 2+\varepsilon)}(1 + |\pi\theta^n|)^{\alpha-1} \cdot \mathcal{L}^1(E) \cdot \theta^n \\
&\geqslant \text{const} \cdot \sum_{n=N}^{\infty} 2^{-2n}e^{-2n\varepsilon}\theta^{n\alpha} = \infty,
\end{aligned}$$

by our choice of ε, and the proof if complete. □

8.4. Applications of Fourier decay

8.4.1. Absolute continuity via convolution. The following simple lemma of Shmerkin [**Shm14**, Lemma 2.1] turned out to be very useful. Although the results below hold in $\mathbb{R}^d$, we restrict ourselves to $d = 1$ for simplicity. Recall that the *lower* Hausdorff dimension of a measure μ is defined by

$$\underline{\dim}_{\mathrm{H}}(\mu) = \inf\{\dim_{\mathrm{H}}(A) : \mu(A) > 0\} = \operatorname*{ess\,inf}_{x\sim\mu}\left(\liminf_{r\to 0}\frac{\log\mu(B(x,r))}{\log r}\right).$$

Clearly, $\dim_{\mathrm{H}}(\mu) \geqslant \underline{\dim}_{\mathrm{H}}(\mu)$.

LEMMA 8.4.1 (Shmerkin). *Let μ, ν be compactly supported positive Borel measures on the real line, and suppose that* $\dim_{\mathrm{F}}(\nu) > 0$.

(i) *Suppose that $D_2(\mu) = 1$. Then $\mu * \nu \ll \mathcal{L}^1$, and moreover, it has a density in $L^2(\mathbb{R})$.*
(ii) *Suppose that $\underline{\dim}_{\mathrm{H}}(\mu) = 1$. Then $\mu * \nu \ll \mathcal{L}^1$.*

PROOF. Positivity of the Fourier dimension means that there exists $\alpha > 0$ such that $|\widehat{\nu}(\xi)| \leqslant C|\xi|^{-\alpha}$.

(i) By the assumption $D_2(\mu) = 1$ and Proposition 1.9.8, we have $\mathcal{E}_{1-2\alpha}(\mu) < \infty$, which, in view of (8.9), yields

$$\int_{\mathbb{R}} |\xi|^{-2\alpha}|\widehat{\mu}(\xi)|^2\, d\xi < \infty.$$

Then

$$\int_{\mathbb{R}} |\widehat{\mu * \nu}(\xi)|^2\, d\xi = \int_{\mathbb{R}} |\widehat{\mu}(\xi)|^2|\widehat{\nu}(\xi)|^2\, d\xi < \infty,$$

hence $\widehat{\mu * \nu} \in L^2(\mathbb{R})$, and finally, $\mu * \nu$ is absolutely continuous with an L^2-density by Lemma C.0.7.

(ii) By the assumption $\underline{\dim}_{\mathrm{H}}(\mu) = 1$, we conclude that

$$\liminf_{r\to 0}\frac{\log\mu(B(x,r))}{\log r} \geqslant 1$$

for μ-a.e. x. By Egorov's Theorem, for every $\varepsilon > 0$ there exists a set A_ε, with $\mu(A_\varepsilon) > 1 - \varepsilon$, such that $\mu_\varepsilon := (\mu|_{A_\varepsilon})/\mu(A_\varepsilon)$ satisfies

$$\mu_\varepsilon(B(x,r)) \leqslant C_\varepsilon r^{1-\alpha} \quad \text{for all } x \in A_\varepsilon.$$

The latter implies $D_2(\mu_\varepsilon) \geqslant 1 - \alpha$ by Definition 1.9.7. Then $\mu_\varepsilon * \nu$ is absolutely continuous by the same argument as above. Letting $\varepsilon \to 0$ we obtain that $\mu * \nu \ll \mathcal{L}^1$, as desired. □

PROOF OF SHMERKIN'S THEOREM 6.8.13. We closely follow the argument of [**Shm14**]. We are given a 1-parameter family of IFS $\mathcal{F}_\lambda = (\lambda x + d_1, \ldots, \lambda x + d_m)$ is given, with $\lambda \in (0,1)$, and a probability vector $\mathbf{p} = (p_1, \ldots, p_k)$ (we omit the digit set D from the subscript to simplify the notation). Consider the IFS obtained by "skipping every kth digit":

$$\mathcal{F}_\lambda^{(k)} = \Big(\lambda^k x + \sum_{j=0}^{k-2} a_{i_{j+1}}\lambda^j\Big)_{\mathbf{i}\in\{1,\ldots,m\}^{k-1}}.$$

Notice that this is a linear homogeneous IFS with m^{k-1} maps, all having contraction ratio λ^k. To a word $\mathbf{i} = i_1 \ldots i_{k-1}$ we associate the probability $p_{\mathbf{i}}^{(k)} = p_{i_1} \cdot \ldots \cdot p_{i_k}$

and let $\mathbf{p}^{(k)} = (p_{\mathbf{i}}^{(k)})_{\mathbf{i}\in\{1,\dots,m\}^{k-1}}$. The similarity dimension of the "thinned out" IFS equals

$$s(\mathcal{F}_\lambda^{(k)}, \mathbf{p}^{(k)}) = \frac{h_{\mathbf{p}^{(k)}}}{-\log(\lambda^k)} = \frac{(k-1)h_{\mathbf{p}}}{-k\log\lambda} = \left(1 - \frac{1}{k}\right) s(\mathcal{F}_\lambda, \mathbf{p}).$$

We are going to apply Hochman's Theorem 6.8.11 to the IFS $(\mathcal{F}_\lambda^{(k)}, \mathbf{p}^{(k)})$. Observe that that the nondegeneracy assumption (6.52) is satisfied, because if $\mathbf{i} \neq \mathbf{j}$, then $\Pi_{\mathcal{F}_\lambda^{(k)}}(\mathbf{i}) - \Pi_{\mathcal{F}_\lambda^{(k)}}(\mathbf{i})$ is a nontrivial power series in λ with bounded coefficients (recall that the digits d_i are all distinct). Thus, Theorem 6.8.11 implies that there is a set E_k of Hausdorff dimension zero such that

$$\dim_{\mathrm{H}}(\nu(\mathcal{F}_\lambda^{(k)}, \mathbf{p}^{(k)})) = 1 \text{ when } \lambda \in (0,1)\backslash E_k \text{ and } s(\mathcal{F}_\lambda, \mathbf{p}) > \frac{k}{k-1}. \tag{8.29}$$

On the other hand,

$$\nu(\mathcal{F}_\lambda, \mathbf{p}) = \nu(\mathcal{F}_\lambda^{(k)}, \mathbf{p}^{(k)}) * \nu(\mathcal{F}_{\lambda^k}, \mathbf{p}).$$

This follows from the fact that $\nu(\mathcal{F}_\lambda, \mathbf{p})$ is the distribution of the random series $\sum_{n=1}^\infty d_{i_n}\lambda^{n-1}$, which we split into the sum of two independent series: $\sum_{k\nmid n}^\infty d_{i_n}\lambda^{n-1}$, whose distribution is $\nu(\mathcal{F}_\lambda^{(k)}, \mathbf{p}^{(k)})$, and $\sum_{k\mid n}^\infty d_{i_n}\lambda^{n-1}$, whose distribution is $\nu(\mathcal{F}_{\lambda^k}, \mathbf{p})$. By Corollary 8.3.2(ii), the set

$$\widetilde{E} := \{\lambda \in (0,1) : \dim_{\mathrm{F}}(\nu(\mathcal{F}_\lambda, \mathbf{p})) = 0\} \tag{8.30}$$

has Hausdorff dimension zero (as we already mentioned, the Erdős–Kahane arguments applies to the case of any set of distinct digits). Let $E_k' = \{\lambda : \lambda^k \in \widetilde{E}\}$. We deduce from (8.30) and Lemma 8.4.1(ii) that if $\lambda \in (0,1)\backslash(E_k \cup E_k')$, and $s(\mathcal{F}_\lambda, \mathbf{p}) > 1 + 1/k$, then $\nu(\mathcal{F}_\lambda, \mathbf{p})$ is absolutely continious. This yields the desired result, with exceptional set $E = \bigcup_{k=1}^\infty (E_k \cup E_k')$. □

8.4.2. L^q densities. Theorem 6.8.14 was deduced by P. Shmerkin [**Shm19**] from Theorem 6.8.17 in a similar way, splitting the homogeneous self-similar measure into a convolution, with the help of the following theorem, which generalizes Lemma 8.4.1(i).

THEOREM 8.4.2. *Let $1 < q < \infty$ and let μ, ν be positive Borel measures on $\mathbb{R}$ of compact support. If*

$$\begin{aligned} 1 - D_q(\mu) &< \dim_{\mathrm{F}}(\nu) \qquad \textit{for } q \in (1,2], \\ (q-1)(1 - D_q(\mu)) &< \dim_{\mathrm{F}}(\nu) \qquad \textit{for } q \in (2,+\infty), \end{aligned}$$

then $\mu * \nu$ is asbolutely continuous with an L^q density. In particular, this is the case when $D_q(\mu) = 1$ and $\dim_{\mathrm{F}}(\nu) > 0$.

The proof uses Littlewood–Paley decomposition, see [**SS16**, Theorem 4.4]. In [**SS16**, Theorem A] the following strengthening of Theorem 6.8.13 was obtained (we state only a special case):

THEOREM 8.4.3. *Let $m \geqslant 2$ and $d_1, \dots, d_m \in \mathbb{R}$ are distinct. Consider the self-similar IFS $\mathcal{F}_\lambda = (\lambda x + d_1, \dots, \lambda x + d_m)$. For $\lambda \in (0,1)$ and a probability vector $\mathbf{p} = (p_1, \dots, p_m)$, let $\nu(\mathcal{F}_\lambda, \mathbf{p})$ be the corresponding self-similar measure $\nu(\lambda, \mathbf{p})$. There exists a set $\mathcal{E} \subset (0,1)$, with $\dim_{\mathrm{H}}(\mathcal{E}) = 0$, such that if $\lambda \in (0,1)\backslash\mathcal{E}$ and $\mathbf{p}$ is such that the similarity dimension $s(\nu(\mathcal{F}_\lambda, \mathbf{p})) > 1$, then $\nu(\mathcal{F}_\lambda, \mathbf{p})$ is absolutely continuous, with a density in L^q for* some $q > 1$.

Even though Theorem 6.8.14 is much stronger, this earlier result had useful applications, and moreover, its proof is much simpler than that of Theorem 8.4.3. It is similar to the proof of Theorem 6.8.13 above, with the following modifications:

(a) Self-similar measures are *exact dimensional* by [**FH09**], which implies, using Theorem 2.6.4 that in (8.29) Hausdorff dimension 1 may be replaced by *entropy dimension*:

$$\dim_e(\nu(\mathcal{F}_\lambda^{(k)}, \mathbf{p}^{(k)})) = 1 \text{ when } \lambda \in (0,1)\backslash E_k \text{ and } s(\mathcal{F}_\lambda, \mathbf{p}) > \frac{k}{k-1}.$$

(b) It was proved in [**SS16**, Theorem 5.1] that

$$\dim_e(\nu) = \lim_{q \to 1+} D_q(\nu)$$

for any self-similar measure.

(c) Since $\dim_{\mathrm{F}}(\nu(\mathcal{F}_{\lambda^k}, \mathbf{p})) > 0$ for $\lambda \in (0,1)\backslash E'_k$, Theorem 8.4.2 implies that ν_λ has a density in L^q for some $q > 1$ when $s(\mathcal{F}_\lambda, \mathbf{p}) > 1 + 1/k$ and $\lambda \in E_k \cup E'_k$, so we have the same set of exceptions as in Theorem 6.8.13.

8.5. Growth rate of β-expansions and local dimensions of Bernoulli convolutions

For $\beta \in (1,2]$ and $x \in I_\beta = [0, (\beta-1)^{-1}]$, a β-expansion of x is a sequence $(\omega_n)_{n=1}^{\infty} \in \Sigma = \{0,1\}^{\mathbb{N}}$ such that

$$x = \sum_{n=1}^{\infty} \omega_n \beta^{-n}.$$

For $\beta = 2$ these are the familiar binary expansions; every $x \in I_2 = [0,1]$, except dyadic rationals in $(0,1)$, has a unique expansion, whereas the dyadic rationals have two expansions. The case $\beta \in (1,2)$ is much more complicated, which is caused by the overlapping structure of the IFS $(S_0^\beta, S_1^\beta) = (\beta^{-1}x, \beta^{-1}(x+1))$. This can be visualized in the following figure:

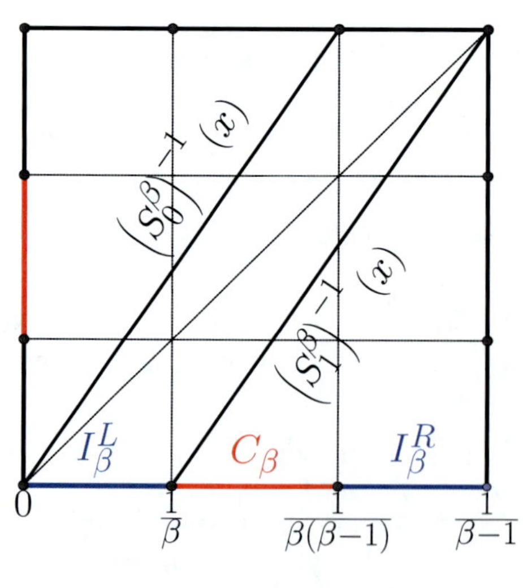

FIGURE 8.1. $1 < \beta < \frac{\sqrt{5}+1}{2}$

It shows the graphs of the inverse functions $(S_0^\beta)^{-1}(x) = \beta x$ and $(S_1^\beta)^{-1}(x) = \beta x - 1$. Consider the intervals

$$I_\beta^L := \Big[0, \frac{1}{\beta}\Big], \quad C_\beta := \Big[\frac{1}{\beta}, \frac{1}{\beta(\beta-1)}\Big], \quad I_\beta^R := \Big[\frac{1}{\beta(\beta-1)}, \frac{1}{\beta-1}\Big]$$

The interval C_β is sometimes called the "overlap" or "switch region". In order to find all expansions of $x \in I_\beta$ we start the iteration of

$$x \mapsto \begin{cases} \beta x, & \text{if } x \in I_\beta^L; \\ \beta x \ \underline{\textit{and}}\ \beta x + 1, & \text{if } x \in C_\beta; \\ \beta x + 1, & \text{if } x \in I_\beta^R. \end{cases} \tag{8.31}$$

This way we obtain a "tree" of expansions, which "branches" every time the point falls into the switch region C_β. All β-expansions arise from infinite paths in the tree obtained this way, where we write 0 or 1 depending on which map is applied.

This is, of course, the IFS corresponding to the classical Bernoulli convolution ν_λ, with $\lambda = \beta^{-1}$, up to an affine linear change of variable. For convenience, in this section we denote by $\nu_{\beta^{-1}}$ the version of Bernoulli convolution adjusted this way, that is, with the support equal to I_β. We emphasize that we consider the unbiased case, with the weights $(\frac{1}{2}, \frac{1}{2})$, throughout this section.

Let $\Pi_\beta : \Sigma \to I_\beta$ be the natural projection map:

$$\Pi_\beta(\omega) = \sum_{n=1}^\infty \omega_n \beta^{-n}.$$

In order to study the growth rate of β-expansions, we introduce some notation. Denote

$$E_\beta(x) = \Big\{(\omega_n) \in \{0,1\}^{\mathbb{N}} : \ x = \sum_{n=1}^\infty \omega_n \beta^{-n}\Big\}.$$

This is simply the "slice" of the natural projection:

$$E_\beta(x) = \Pi_\beta^{-1}(\{x\}).$$

Further, let

$$E_\beta^n(x) = \Big\{(\omega_1, \ldots, \omega_n) : \ \exists\, (\omega_{n+1}, \omega_{n+2}, \ldots) : \ x = \sum_{k=1}^\infty \omega_k \beta^{-k}\Big\}$$

be the set of prefixes of length n of the sequences from $E_\beta(x)$. Observe that

$$\omega \in E_\beta^n(x) \iff x \in I_\omega^{(\beta)} \ \text{ for } \omega \in \{0,1\}^n, \ \text{ where } \ I_\omega^{(\beta)} := S_\omega^\beta(I_\beta).$$

Denote

$$\mathcal{N}_n(x;\beta) = \#(E_\beta^n(x)).$$

By definition, the growth rate of $\mathcal{N}_n(x)$ gives the box dimension of the slice $E_\beta(x)$:

$$\underline{\dim}_B(E_\beta(x)) = \liminf_{n\to\infty} \frac{\log \mathcal{N}_n(x;\beta)}{n \log 2},$$

and similarly for $\overline{\dim}_B(E_\beta(x))$, with lim sup. Here we use the metric on Σ given by

$$\varrho(\omega, \tau) = 2^{-\sup\{n:\ \omega_1\ldots\omega_n = \tau_1\ldots\tau_n\}}.$$

8.5.1. Growth rate of β-expansions for a "typical" x. Sidorov [**Sid03**] showed that for any $\beta \in (1,2)$, for Lebesgue-almost every $x \in I_\beta$, there are uncountable many expansions. One can guess that the "expected" growth rate of $\mathcal{N}_n(x)$ for a typical x is $\log(\frac{2}{\beta})$. To justify this, note that the growth rate of $\mathcal{N}_n(x)$ corresponds to (some kind of) dimension of $E_\beta(x)$, which is the slice of the natural projection map $\Pi_\beta : \Sigma \to I_\beta$. If we use an alternative metric on Σ defined by

$$\varrho_\beta(\omega,\tau) = \beta^{-\sup\{n:\ \omega_1\ldots\omega_n=\tau_1\ldots\tau_n\}},$$

then the natural projection map is Lipschitz. In this metric Σ has dimension (box and Hausdorff) equal to $s_\beta = \frac{\log 2}{\log\beta}$, which is the similarity dimension of the IFS. Then one can expect that for a "typical" β and a "typical" $x \in I_\beta$, the dimension of the slice is $s_\beta - 1$. Adjusting for the metric on Σ, this corresponds to the typical branching rate (growth rate of $\mathcal{N}_n(x)$) equal to $(s_\beta - 1)\cdot\log\beta = \log(2/\beta)$. The following theorem makes this precise.

THEOREM 8.5.1. *Let $\beta \in (1,2)$, and let $\nu = \nu_{\beta^{-1}}$ be the natural self-similar measure for the IFS $(\beta^{-1}x, \beta^{-1}(x+1))$ (that is, ν is the classical unbiased Bernoulli convolution, up to an affine linear change of variable). Then for all $x \in I_\beta$,*

$$\underline{d}_\nu(x) + \overline{\dim}_B(E_\beta(x))\cdot\frac{\log 2}{\log\beta} = \frac{\log 2}{\log\beta},$$

$$\overline{d}_\nu(x) + \underline{\dim}_B(E_\beta(x))\cdot\frac{\log 2}{\log\beta} = \frac{\log 2}{\log\beta},$$

or, equivalently,

$$\limsup_{n\to\infty}\frac{1}{n}\log\mathcal{N}_n(x;\beta) = \log 2 - \underline{d}_\nu(x)\cdot\log\beta, \tag{8.32}$$

$$\liminf_{n\to\infty}\frac{1}{n}\log\mathcal{N}_n(x;\beta) = \log 2 - \overline{d}_\nu(x)\cdot\log\beta.$$

The proof uses an idea from the study of slices of fractal sets, especially the Sierpiński carpet, see Manning and Simon [**MS13**].

PROOF. We will prove (8.32); the other claim is proved in exactly the same way. It follows from the definitions and the decomposition of $\nu_{\beta^{-1}}$ into the average of similar copies on cylinder sets of order n that

$$\nu\big(B(x,(\beta-1)^{-1}\beta^{-n})\big) \geqslant \frac{\mathcal{N}_n(x;\beta)}{2^n}. \tag{8.33}$$

This immediately implies the upper bound $\leqslant$ in (8.32).

In order to prove the lower bound, define

$$\widetilde{E}^n_\beta(x) = \big\{\omega \in \{0,1\}^n : I^{(\beta)}_\omega \cap B(x,\beta^{-n}) \neq \varnothing,\ \text{but } x \notin I^{(\beta)}_\omega\big\},$$

$$\widetilde{\mathcal{N}}_n(x;\beta) := \#(\widetilde{E}^n_\beta(x)).$$

It is clear that

$$\nu(B(x,\beta^{-n})) \leqslant \frac{\mathcal{N}_n(x;\beta) + \widetilde{\mathcal{N}}_n(x;\beta)}{2^n}, \tag{8.34}$$

so we need to estimate the rate of growth of $\widetilde{\mathcal{N}}_n(x;\beta)$. We claim that

$$\widetilde{\mathcal{N}}_{n+1}(x;\beta) \leqslant \widetilde{\mathcal{N}}_n(x;\beta) + \mathcal{N}_n(x;\beta). \tag{8.35}$$

Indeed, note that words from $\widetilde{E}_\beta^{n+1}(x)$ can only be the "children" of words from $E_\beta^n(x) \cup \widetilde{E}_\beta^n(x)$. If $\omega \in E_\beta^n(x)$, then at least one of $\omega 0, \omega 1$ belongs to $E_\beta^{n+1}(x)$, and hence not to $\widetilde{E}_\beta^{n+1}(x)$, since

$$x \in I_\omega^{(\beta)} = I_{\omega 0}^{(\beta)} \cup I_{\omega 1}^{(\beta)}.$$

Finally, if $\omega \in \widetilde{E}_\beta^n(x)$, then at most one of $\omega 0, \omega 1$ belongs to $\widetilde{E}_\beta^{n+1}(x)$. We verify this last assertion in the case when $x < \sum_{k=1}^n \omega_k \beta^{-k}$; the other case follows by symmetry. Recall that

$$I_\omega^{(\beta)} = \left[\sum_{k=1}^n \omega_k \beta^{-k}, \sum_{k=1}^n \omega_k \beta^{-k} + (\beta-1)^{-1}\beta^{-n}\right].$$

By definition,

$$x < \sum_{k=1}^n \omega_k \beta^{-k}, \quad \omega \in \widetilde{E}_\beta^n(x) \quad \Longleftrightarrow \quad \sum_{k=1}^n \omega_k \beta^{-k} \in (x, x+\beta^{-n}).$$

But then $\sum_{k=1}^n \omega_k \beta^{-k} + \beta^{-n-1} > x + \beta^{-n-1}$, which means that $\omega 1 \notin \widetilde{E}_\beta^{n+1}(x)$. This confirms the last assertion, and the claim (8.35) is proved. Iterating (8.35) yields

$$\widetilde{\mathcal{N}}_n(x;\beta) \leqslant 1 + \sum_{k=1}^n \mathcal{N}_k(x;\beta) \leqslant 1 + n\mathcal{N}_n(x;\beta).$$

Combined with (8.34) and the definition of local dimension of ν, this implies the desired lower bound $\geqslant$ in (8.32). $\square$

Combined with the recent progress on Bernoulli convolutions, especially due to Shmerkin, we get the following.

PROPOSITION 8.5.2.

(i) *For all $\beta \in (1,2)$, outside a set of Hausdorff dimension zero, for a.e. $x \in I_\beta$,*

$$\lim_{n\to\infty} \frac{1}{n} \log \mathcal{N}_n(x;\beta) = \log\left(\frac{2}{\beta}\right). \tag{8.36}$$

(ii) *For all $\beta \in (1,\sqrt{2})$, outside a set of Hausdorff dimension zero,* (8.36) *holds for all $x \in \mathrm{Int}(I_\beta)$, and not just for almost all.*

(iii) *For all $\beta \in (\sqrt{2},2)$, outside a set of Hausdorff dimension zero, for* all *$x \in I_\beta$, holds*

$$\limsup_{n\to\infty} \frac{1}{n} \log \mathcal{N}_n(x;\beta) \leqslant \log\left(\frac{2}{\beta}\right).$$

A result in this direction was obtained by Kempton [**Kem13**], except that his result is for almost every β (Shmerkin's theorem was not yet available), but more importantly, he had lim sup instead of lim in (i), and a.e. x, instead of all x in the interior of I_β, in the analog of (ii).

PROOF.

(i) By Theorem 6.8.13, the Bernoulli convolution $\nu_{\beta^{-1}}$ is absolutely continuous for all $\beta \in (1,2)$, outside a set of Hausdorff dimension zero. Fix such a β and let $\nu = \nu_{\beta^{-1}}$. Then ν has local dimension equal to one, ν-a.e., and by the theorem of

Mauldin and Simon (see Proposition 3.1.4(ii)), this means a.e. with respect to the Lebesgue measure in I_β. It remains to apply Theorem 8.5.1.

(ii) By Theorem 6.8.14, the Bernoulli convolution $\nu_{\beta^{-1}}$ is absolutely continuous, with a density in L^2, for all $\beta \in (1,2)$, outside a set of Hausdorff dimension zero. Fix $\beta \in (1,\sqrt{2})$, such that $\nu_{\beta^{-2}}$ has a density $h(x) \in L^2$. Then $\nu = \nu_{\beta^{-1}}$ has a continuous density

$$(h * h(\beta^{-1}\cdot))(x) = \int h(t)\, h(\beta^{-1}(x-t))\, dt.$$

Since $h(t) \geqslant 0$ a.e. in I_{β^2} by the Mauldin–Simon result, we obtain that the density of ν is strictly positive (and continuous) everywhere in the interior of I_β. Thus $d_\nu(x) = 1$ for all $x \in \mathrm{Int}(I_\beta)$, and the claim follows.

(iii) By Theorem 6.8.14, for all $\beta \in (\sqrt{2}, 2)$, outside a set of zero Hausdorff dimension, $\nu_{\beta^{-1}}$ has a density in L^q for all $q > 1$. Then by Lemma 6.8.16 and (8.33) we have for all $\varepsilon > 0$ and for all $x \in \mathbb{R}$:

$$\mathcal{N}_n(x;\beta) \leqslant C_\varepsilon \cdot 2^n \beta^{-n(1-\varepsilon)}.$$

This implies the desired claim.

□

The last proposition, describing the typical growth rate of β-expansions, is for typical β. In fact, there are exceptions, namely, the Pisot numbers. Feng and Sidorov [**FS11**] showed the following:

THEOREM 8.5.3 (Feng and Sidorov). *If $\beta \in (1,2)$ is a Pisot number, then there exists $\gamma = \gamma(\beta) < \log(\frac{2}{\beta})$, such that*

$$\lim_{n\to\infty} \frac{1}{n} \log \mathcal{N}_n(x;\beta) = \gamma \quad \text{for a.e. } x \in I_\beta.$$

We conclude this discussion with a few additional comments. Let

$$f_n(x) := \frac{(\beta-1)\beta^n}{2^n} \mathcal{N}_n(x;\beta).$$

The following is an easy consequence of (8.33), apparently, first pointed out in [**PW94**]:

$$\nu_{\beta^{-1}} \text{ has a bounded density} \iff \sup_x \limsup_{n\to\infty} f_n(x) < C < \infty,$$

where $C = C(\beta)$ is independent of x. This is now known to hold for all $\beta \in (1,\sqrt{2})$ outside a set of Hausdorff dimension zero, by Theorem 6.8.14.

Kempton [**Kem13**] conjectured that if $\nu_{\beta^{-1}}$ is absolutely continuous, then $\lim_{n\to\infty} f_n(x)$ exists a.e. and is equal to the density of $\nu_{\beta^{-1}}$. He proved this claim for a.e. $\beta \in (1,\sqrt{2})$ in [**Kem16**, Prop. 5.1]. In this later paper Kempton also considers an interesting question about the Hausdorff measure of the set $E_\beta(x)$ of all β-expansions of x in the dimension $\log_2(\frac{2}{\beta})$.

THEOREM 8.5.4 (Kempton). *The set $E_\beta(x)$ of β-expansions of x has positive $\log_2(\frac{2}{\beta})$-dimensional Hausdorff measure for Lebesgue a.e.-$x \in I_\beta$ if and only if the corresponding Bernoulli convolution is absolutely continuous with bounded density.*

8.5.2. Growth of β-expansions for exceptional x and local dimensions of Bernoulli convolutions. We have seen that for a typical $\beta \in (1,2)$ and *typical* x, and even for every x for a typical $\beta \in (1,\sqrt{2})$, the rate of growth of $\mathcal{N}_n(x;\beta)$ is $\log(\frac{2}{\beta})$. But what about *exceptional* x? As we have seen above, this question is directly related to the spectrum of local dimensions of the measure $\nu_{\beta^{-1}}$. It turns out that there is a kind of "phase transition" at $\beta = \beta_g = \frac{\sqrt{5}+1}{2} = 1.618034\ldots$, the golden ratio, and another one (although in a less obvious way) at $\beta = \beta_{\mathrm{KL}} = 1.78723\ldots$, the Komornik–Loreti constant [**KL98**]. This costant is defined as the unique positive solution of the equation $1 = \sum_{n=1}^{\infty} b_n x^{-n+1}$, where $(b_n)_1^\infty$ is the Thue–Morse sequence, i.e., b_n is the parity of the number of 1's in the binary representation of n. The reason for these "phase transitions" is the existence of nontrivial x having a unique β-expansion and the structure of the set of such x.

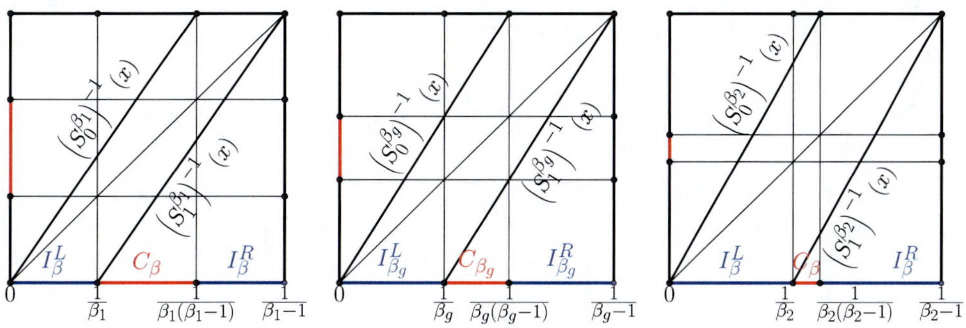

FIGURE 8.2. $1 < \beta_1 < \beta_g = \frac{\sqrt{5}+1}{2} < \beta_2 < 2$

For $\beta \in (1,2)$ let

$$\mathcal{A}_\beta = \{x \in I_\beta : \#(E_\beta(x)) = 1\}$$

be the set of x having a unique expansion in base β. Of course, 0 and $(\beta-1)^{-1}$, the endpoints of I_β, always have a unique expansion. Unique expansions have been extensively studied; the reader is referred to the survey by de Vries and Komornik [**dVK16**] for the state of the art by 2016, but many new results were obtained since then. Here are the basic results: the first one is due to Erdős, Jóo, and Komornik [**EJK90**], and the rest were obtained by Glendinning and Sidorov [**GS01**].

(i) For all $\beta \in (1,\beta_g]$ there are no nontrivial x with a unique expansion: $\mathcal{A}_\beta = \{0,(\beta-1)^{-1}\}$; in fact, for all $x \in \mathrm{Int}(I_\beta)$ the set of expansions $E_\beta(x)$ is infinite.
(ii) For all $\beta \in (\beta_g,\beta_{\mathrm{KL}})$, the set $\mathcal{A}_\beta$ is countably infinite.
(iii) For $\beta = \beta_{\mathrm{KL}}$, the set $\mathcal{A}_\beta$ is uncountable of zero Hausdorff dimension.
(iv) For $\beta \in (\beta_{\mathrm{KL}},2)$, the set $\mathcal{A}_\beta$ has positive Hausdorff dimension, which increases monotonically and tends to 1, as $\beta \uparrow 2$.

More recently, points having exactly m, for $m \geqslant 2$, or countably many expansions, have been investigated, see [**Sid09, KK19**], and shown to have a similarly complicated structure. As an immediate consequence of Theorem 8.5.1 we obtain the following.

COROLLARY 8.5.5. *Let $\beta \in (\beta_g,2)$, $x \in I_\beta$, and suppose that $E_\beta(x)$ is finite or countably infinite. Then the local dimension of $\nu_{\beta^{-1}}$ at x equals the similarity* *dimension:*

$$d_{\nu_{\beta^{-1}}}(x) = s_\beta = \frac{\log 2}{\log \beta} > 1.$$

In the next proposition, the first statement is due to Feng and Sidorov [**FS11**], and the other two are from Jordan, Shmerkin, and Solomyak [**JSS11**].

PROPOSITION 8.5.6.

(i) *For any* $\beta \in (1, \beta_g)$ *there exists* $\kappa > 0$ *such that*

$$\overline{d}_{\nu_{\beta^{-1}}}(x) \leqslant (1-\kappa) \cdot \frac{\log 2}{\log \beta} \quad \textit{for all} \quad x \in \mathrm{Int}(I_\beta).$$

(ii) *For any* $\beta \in (\beta_g, 2)$ *the set* $\left\{x \in I_\beta : d_{\nu_{\beta^{-1}}}(x) = \frac{\log 2}{\log \beta}\right\}$ *is infinite.*

(iii) *For any* $\beta \in (\beta_{\mathrm{KL}}, 2)$ *the set* $\left\{x \in I_\beta : d_{\nu_{\beta^{-1}}}(x) = \frac{\log 2}{\log \beta}\right\}$ *has positive Hausdorff dimension, which tends to 1 as* $\beta \uparrow 2$.

PROOF. (i) See [**FS11**, Cor. 1.6].

(ii) and (iii). This follows from the properties of $\mathcal{A}_\beta$ given above, combined with Corollary 8.5.5. □

We remark that in [**JSS11**] the statements (ii) and (iii) above were proved in a more complicated way, since a stronger property than uniqueness of β-expansion was used to verify that the local dimension equals the similarity dimension. On the other hand, [**JSS11**] considered also the case of biased Bernoulli convolutions, which we do not discuss here.

It should be pointed out that local dimensions of Bernoulli convolutions are still poorly understood, even for a typical β. In particular, we do not know whether $\dim_{\mathrm{H}}(\{x \in I_\beta : d_{\nu_{\beta^{-1}}}(x) = \frac{\log 2}{\log \beta}\}) = 0$ for $\beta \in (\beta_g, \beta_{\mathrm{KL}})$, and whether other values of local dimension, other than 1 and $\frac{\log 2}{\log \beta}$ are achieved for transcendental β.

We should also mention an interesting paper of Christoph Bandt [**Ban18**], in which he studied Bernoulli convolutions as a two-dimensional family, with one dimension corresponding to the interval where the IFS is defined (normalized to be $[0, 1]$) and the other dimension corresponding to the parameter. Using techniques of symbolic dynamics, Bandt studied network-like branching orbits in multivalued maps, like (8.31), in connection with the properties of the "two-dimensional density" and local dimensions of Bernoulli convolution.

8.6. Garsia entropy and dimension of Bernoulli convolutions for algebraic parameters

Recall the spectacular Theorem 6.8.9 of Varjú [**Var19b**] saying that for all transcendental parameters λ the Bernoulli convolution measure has dimension 1. Thus we focus on the case of algebraic λ below.

Let $\lambda \in (\frac{1}{2}, 1)$ and consider the sequence of finite approximations to the Bernoulli convolution ν_λ. Here we will use the the digits $\{0, 1\}$ for convenience, so that ν_λ is the self-similar measure for the IFS $(S_0^\lambda, S_1^\lambda)$, with $S_j(x) = \lambda x + j$ for $j = 0, 1$, and the probability vector $(\frac{1}{2}, \frac{1}{2})$. Define

$$Y_\lambda^{(n)} = \sum_{i=0}^{n-1} a_i \lambda^i,$$

where $a_i \in \{0,1\}$ are i.i.d. with equal probabilities, and let $\nu_\lambda^{(n)}$ be the distribution of $Y_\lambda^{(n)}$. Recall that $H(\cdot)$ denotes the Shannon entropy of a discrete measure.

DEFINITION 8.6.1. The quantity

$$h_\lambda := \lim_{n\to\infty} \frac{H(\nu_\lambda^{(n)})}{n}$$

is called the *Garsia entropy* of λ (or of $\beta = \lambda^{-1}$).

Existence of the limit easily follows from subadditivity by Fekete's Lemma 1.9.21. This quantity (of course, not the name "Garsia entropy") was introduced by Garsia [**Gar63**]. A word of warning: in some papers, the Garsia entropy is defined as $\frac{h_\lambda}{-\log\lambda}$. More explicitly, let

$$\mathcal{P}_\lambda^{(n)} = \Big\{\sum_{i=0}^{n-1} a_i\lambda^i : a_i \in \{0,1\}\Big\}$$

be the range of $Y_\lambda^{(n)}$. For $x \in \mathcal{P}_\lambda^{(n)}$ let

$$p_\lambda^{(n)}(x) = \#\Big\{a_0\ldots a_{n-1} \in \{0,1\}^n : x = \sum_{i=0}^{n-1} a_i\lambda^i\Big\}.$$

Then

$$H(\nu_\lambda^{(n)}) = -\sum_{x\in\mathcal{P}_\lambda^{(n)}} \frac{p_\lambda^{(n)}(x)}{2^n}\log\frac{p_\lambda^{(n)}(x)}{2^n}.$$

The Garsia entropy, in fact, coincides with the random walk entropy defined in (3.14). More precisely, let

$$\eta_\lambda = \frac{1}{2}(\delta^\lambda_{S_0} + \delta_{S_1^\lambda}).$$

Then

$$\nu_\lambda^{(n)} = \eta_\lambda^{*n}\cdot 0 = \sum_{|\mathbf{i}|=n} 2^{-n}\delta_{S_{\mathbf{i}}^\lambda(0)},$$

using the notation (3.12), and we have

$$H(\nu_n) = H(\eta_\lambda^{*n}),$$

since this is a homogeneous IFS on $\mathbb{R}$. Thus,

$$h_\lambda = h_{\mathrm{RW}}(\eta_\lambda) = \lim_{n\to\infty}\frac{H(\nu_\lambda^{(n)})}{n} = \inf_{n\in\mathbb{N}}\frac{H(\nu_\lambda^{(n)})}{n}.$$

The next result is essentially due to Garsia [**Gar63**], with some clarifications by Przytycki–Urbański [**PU89**]. We also use the presentation of Lalley [**Lal98**].

THEOREM 8.6.2 (Garsia). *Let $\lambda \in (\frac{1}{2},1)$ be the reciprocal of a Pisot number and ν_λ the corresponding Bernoulli convolution measure. Then*

$$\dim_{\mathrm{H}}(\nu_\lambda) < 1. \tag{8.37}$$

Note, however, that this is not a new proof of singularity of ν_λ, since singularity is actually used in the proof of theorem.

PROOF. The idea is to apply Theorem 3.2.7, which says that

$$\dim_{\mathrm{H}}(\nu_\lambda) \leqslant \frac{h_\lambda}{-\log\lambda},$$

So, in order to obtain (8.37), it suffices to find a single $n \in \mathbb{N}$ such that

$$\frac{H(\nu_\lambda^{(n)})}{-n\log\lambda} < 1.$$

Since λ is fixed, we will omit λ from notation below, and will write $\nu_n := \nu_\lambda^{(n)}$ for short. Let $\mathcal{P}_n := \operatorname{spt}(\nu_n)$. The key point, where the property of $1/\lambda$ being Pisot is used, is the well-known Garsia Lemma, see Corollary D.0.11(i), which implies that the minimal positive value of a polynomial $\sum_{i=0}^{n-1} b_i\lambda^i$, with $b_i \in \{-1,0,1\}$, is bounded below by $c\lambda^n$, where $c > 0$ is independent of $n \in \mathbb{N}$. It follows that distinct points of $\mathcal{P}_n$ are separated by a distance not less than $c\lambda^n$, and hence

$$\#\mathcal{P}_n \leqslant \frac{1}{(1-\lambda)c\lambda^n} =: C_1\lambda^{-n}. \tag{8.38}$$

Here $C_1 = c^{-1}(1-\lambda)^{-1}$.

LEMMA 8.6.3. *For any $\delta > 0$ there exists $n \in \mathbb{N}$ and $\mathcal{Q}_n \subset \mathcal{P}_n$ such that*

$$\frac{\#\mathcal{Q}_n}{\#\mathcal{P}_n} \leqslant \delta \tag{8.39}$$

and

$$\nu_n\Big(\bigcup_{F\in\mathcal{Q}_n} F\Big) \geqslant 1-\delta. \tag{8.40}$$

First we deduce the theorem from the lemma, following [**Lal98**], and then prove the lemma. We will use an elementary property of *conditional entropy*, see Section 9.5.1 in the next chapter. Let ξ_n be the partition of $\mathcal{P}_n$ into points and $\eta_n = \{\mathcal{Q}_n, \mathcal{P}_n\backslash\mathcal{Q}_n\}$. Denote $\rho := \nu_n\big(\bigcup_{F\in\mathcal{P}_n\backslash\mathcal{Q}_n} F\big)$, so that $0 < \rho \leqslant \delta$, and $\overline{\rho} = 1-\rho$. Below we consider entropy of partitions with respect to the measure ν_n. By the property **(b)** from Section 9.5.1, with α the trivial partition, we obtain, in view of (8.38):

$$\begin{aligned} H(\nu_n) = H_{\nu_n}(\xi_n) &= H_{\nu_n}(\xi_n \,|\, \eta_n) + H_{\nu_n}(\eta_n) \\ &\leqslant \overline{\rho}\log(\#\mathcal{Q}_n) + \rho\log(\#(\mathcal{P}_n\backslash\mathcal{Q}_n)) - \overline{\rho}\log\overline{\rho} - \rho\log\rho \\ &\leqslant \overline{\rho}\log(\delta C_1\lambda^{-n}) + \rho\log(C_1\lambda^{-n}) - \overline{\rho}\log\overline{\rho} - \rho\log\rho \\ &= -n\log\lambda + (\log C_1 + \overline{\rho}\log\delta) - \overline{\rho}\log\overline{\rho} - \rho\log\rho. \end{aligned}$$

In the second line we used the fact hat the probability measure that maximizes the entropy is the uniform distribution. Now, $-\overline{\rho}\log\overline{\rho} - \rho\log\rho \to 0$ as $\rho \to 0$ and $\overline{\rho}\log\delta \to -\infty$ as $\delta \to 0$, so taking $\delta > 0$ sufficiently small and using that $0 < \rho \leqslant \delta$ we can ensure that $H(\nu_n) < -n\log\lambda$, as desired. □

PROOF OF LEMMA 8.6.3. Let $\mathscr{D}_n$ be the partition of $\mathbb{R}$ into dyadic intervals $[\frac{k}{2^n}, \frac{k+1}{2^n})$. We already know that $\nu = \nu_\lambda$ is singular, hence for any $\varepsilon > 0$ there exists $N \in \mathbb{N}$ and $\mathcal{E}_N \subset \mathscr{D}_N$ such that

$$\#\mathcal{E}_N \leqslant \varepsilon\cdot 2^N \quad \text{and} \quad \nu\Big(\bigcup_{Q\in\mathcal{E}_N} Q\Big) \geqslant 1-\varepsilon.$$

Let $n \in \mathbb{N}$ be such that

$$2^{-N-1} \leqslant \frac{\lambda^n}{1-\lambda} < 2^{-N}.$$

By self-similarity,

$$\nu = \sum_{|\mathbf{i}|=n} 2^{-n}(S_{\mathbf{i}})_*\nu,$$

whereas $\nu_n = \sum_{|\mathbf{i}|=n} 2^{-n}\delta_{S_{\mathbf{i}}(0)}$. Recall that for $|\mathbf{i}| = n$,

$$\operatorname{spt}\big((S_{\mathbf{i}})_*\nu\big) = \Big[S_{\mathbf{i}}(0), S_{\mathbf{i}}(0) + \frac{\lambda^n}{1-\lambda}\Big].$$

Thus,

$$\nu(Q) \leqslant \nu_n(Q \cup Q'), \tag{8.41}$$

where $Q' \in \mathscr{D}_N$ is the dyadic interval adjacent to Q from the left, that is, for $Q = [\frac{k}{2^n}, \frac{k+1}{2^n})$ we denote $Q' = [\frac{k-1}{2^n}, \frac{k}{2^n})$. Let

$$\mathcal{E}'_N = \mathcal{E}_N \cup \{Q' : \ Q \in \mathcal{E}_N\}.$$

Then $\#\mathcal{E}'_N \leqslant 2 \cdot \#\mathcal{E}_N \leqslant 2\varepsilon \cdot 2^N$ and

$$\nu_n\Big(\bigcup_{Q\in\mathcal{E}'_N} Q\Big) \geqslant 1-\varepsilon. \tag{8.42}$$

Let $\mathcal{Q}_n$ be the set of $x \in \operatorname{spt}(\nu_n)$ which lie in one of the intervals of $\mathcal{E}'_N$. Then (8.42) says that (8.40) is satisfied for $\delta \geqslant \varepsilon$, and it remains to verify (8.39).

Since any two distinct points of $\operatorname{spt}(\nu_n)$ are separated by a distance not less than $c\lambda^n$, each interval of $\mathscr{D}_N$ contains not more than $\frac{2^{-N}}{c\lambda^n} \leqslant \frac{2}{c(1-\lambda)} = 2C_1$ atoms of ν_n. Therefore,

$$\#\mathcal{Q}_n \leqslant 2C_1 \cdot \#\mathcal{E}'_N \leqslant 2C_1 \cdot 2\varepsilon \cdot 2^N. \tag{8.43}$$

On the other hand, we claim that

$$\#\mathcal{P}_n \geqslant 2^{N-1}. \tag{8.44}$$

Indeed, $[0,1] \subset \operatorname{spt}(\nu)$, hence $\nu(Q) > 0$ for every $Q \in \mathscr{D}_N$, $Q \subset [0,1]$. By (8.41), either Q or Q' must contain an atom of ν_N, and (8.44) follows. We obtain that

$$\#\mathcal{Q}_n/\#\mathcal{P}_n \leqslant 8C_1\varepsilon,$$

which implies (8.39) for $\varepsilon > 0$ sufficiently small. This concludes the proof of the lemma and of the theorem. □

8.6.1. Computing and estimating the dimension of ν_λ. Over the years, a lot of work was done in this direction, following Garsia [**Gar63**]. This is a vast area, and we only survey some of the results, without proof. Alexander and Yorke [**AY84**] proved that for $1/\lambda$ a Pisot number, the ratio $\frac{h_\lambda}{-\log\lambda}$ is equal to the entropy dimension $\dim_e(\nu_\lambda)$; later Lalley [**Lal98**] showed that it is equal to the Hausdorff dimension $\dim_H(\nu_\lambda)$. (Of course, now that we know exact dimensionality of ν_λ, equality of these dimensions also follows from Theorem 2.6.4.)

In [**AY84**] nonrigorous numerical estimates for the case of $\lambda_g = \frac{\sqrt{5}-1}{2}$ (the inverse golden ratio) gave $\dim_H(\nu_{\lambda_g}) = 0.9957\ldots$. Later, Alexander and Zagier

[**AZ91**] obtained the same value rigorously. In fact, they proved a beautiful explicit formula

$$\dim_{\mathrm{H}}(\nu_{\lambda_g}) = \frac{1}{-\log\lambda_g}\left(1 - \frac{1}{18}\sum_{\substack{k>i>0\\ \gcd(k,i)=1}} \frac{k\log k}{4^{e(k,i)}}\right) \tag{8.45}$$

where $e(k, i)$ is the number of steps in the simple Euclidean algorithm applied to k and i (formally: $e(i, i) = 0, e(i+k, i) = e(i+k, k) = e(i, k)+1$). Sidorov and Vershik [**SV98**] gave another, ergodic-theoretic, derivation of this formula and an alternative proof of singularity of ν_{λ_g}, based on the analysis of the symbolic dynamical system which they called the "goldenshift". Independently, Feng [**Fen99, Fen05**] and Grabner, Kirschenhofer and Tichy [**GKT02**] generalized the formula (8.45) to the case of "multinacci" numbers, that is, for $\lambda_n > 0$ such that

$$\lambda_n^n + \lambda_n^{n-1} + \cdots + \lambda_n = 1, \quad n \geqslant 3,$$

(the golden ratio $\lambda_g = \lambda_2$ is sometimes called "Fibonacci number", hence the name "multinacci"). We have already seen these numbers in Section 5.2; they are Pisot reciprocals.

As was discussed already in Section 4.4, for $1/\lambda$ Pisot, the IFS defining the Bernoulli convolution ν_λ is of finite type, and similar methods to those described there to compute the dimension of the attractor, may be used to compute the dimension of the self-similar measure. This approach was used in [**Lal98**], and later, in a more sophisticated way, by Feng [**Fen05**], who computed, in particular, the dimensions of ν_λ for many "multinacci" numbers with great accuracy. Interestingly, the dimension in the case of the golden ratio is far from being the lowest among Pisot reciprocals: in particular, Feng obtained

$$\dim_{\mathrm{H}}(\nu_{\lambda_3}) = 0.9804093\ldots, \quad \dim_{\mathrm{H}}(\nu_{\lambda_4}) = 0.98692647\ldots \tag{8.46}$$

The breakthrough work of Hochman [**Hoc14b**] spurred renewed activity. He proved

THEOREM 8.6.4 (Hochman). *Let $\lambda \in (\frac{1}{2}, 1)$ be an algebraic integer. Then*

$$\dim_{\mathrm{H}}(\nu_\lambda) = \min\Big\{1, \frac{h_\lambda}{-\log\lambda}\Big\},$$

where h_λ is the Garsia entropy.

This was not explicitly stated in [**Hoc14b**], but easily follows from his results, as explained in [**BV20**, Theorem 19].

In several recent papers, the authors investigate lower bounds for the dimension of ν_λ and/or for h_λ. It is easy to see that $\dim_{\mathrm{H}}(\nu_\lambda) > \frac{1}{2}$ for all $\lambda \in (\frac{1}{2}, 1)$, see [**HS18**, Cor. 2.3]. Indeed, this follows from the convolution structure of ν_λ, using an elementary fact that the dimension of a measure does not decrease when it is convolved with another measure. Writing $\nu_\lambda = \nu_{\lambda^2} * \nu_{\lambda^2}(\lambda\cdot)$, which was already used in Chapter 6, we obtain

$$\dim_{\mathrm{H}}(\nu_\lambda) \geqslant \dim_{\mathrm{H}}(\nu_{\lambda^2}) = \frac{\log 2}{-2\log\lambda} > \frac{1}{2}, \quad \text{for } \lambda \in (2^{-1}, 2^{-1/2}),$$

since for such λ, the measure ν_{λ^2} is the self-similar for an IFS having the Strong Separation Property. Continuing in this manner, one obtains the lower bound $\dim_{\mathrm{H}}(\nu_\lambda) > \frac{1}{2}$ for all $\lambda \in (\frac{1}{2}, 1)$. Using analogous methods in a more sophisticated way, together with numerical computations, Hare and Sidorov [**HS18**] showed that $\dim_{\mathrm{H}}(\nu_\lambda) \geqslant 0.82$ for all $\lambda \in (\frac{1}{2}, 1)$.

Recently Kleptsyn, Pollicott, and Vytnova [**KPV22**] introduced a new method to get uniform lower bounds for the dimension of self-similar measures on the line, which, together with sophisticated numerical methods allowed them to obtain the bound $\dim_{\mathrm{H}}(\nu_\lambda) \geqslant 0.96399$ for all $\lambda \in (\frac{1}{2}, 1)$. Their approach is based on the connection with random processes and an estimation of correlation dimension.

Almost simultaneously, Feng and Feng [**FF22**] obtained a fantastic lower bound $\dim_{\mathrm{H}}(\nu_\lambda) \geqslant 0.9804085$, which is best possible to 5 significant digits, in view of (8.46). (In fact, the result was announced by De–Jun Feng at a conference in Cambridge, UK, in March 2019.) Their approach is very different: it is based on the expression for $\dim_{\mathrm{H}}(\nu_\lambda)$ as the ratio of the projection entropy (see Definition 11.1.1) and $-\log\lambda$, as an application of the Feng–Hu Theorem. Projection entropy and the Feng–Hu Theorem will be our topic in Section 11.3.

Of course, these latest achievements became possible thanks to advances in high-speed computing.

As far as specific algebraic numbers are concerned, showing $h_\lambda \geqslant 1$ implies that $\dim_{\mathrm{H}}(\nu_\lambda) = 1$, in view of Theorem 8.6.4. Hare and Sidorov [**HS18**] proved that $\dim_{\mathrm{H}}(\nu_\lambda) = 1$ for nontrivial roots of Pisot numbers. Akiyama, Feng, Kempton, and Persson [**AFKP20**] gave an expression for the Garsia entropy in terms of products of matrices, which moreover, provides an explicit rate of convergence. This way, they verified the equality $\dim_{\mathrm{H}}(\nu_\lambda) = 1$ for a large number of algebraic non-Pisot parameters. However, the conjecture that this holds in all non-Pisot cases, as of this writing, remains open.

As already mentioned in the introductory remarks to this chapter, we will give another expression for $\dim_{\mathrm{H}}(\nu_\lambda)$ in Section 11.3, namely as the ratio of the projection entropy (see Definition 11.1.1) and $-\log\lambda$, as an application of the Feng–Hu Theorem.

CHAPTER 9

Elements of Ergodic Theory

This chapter introduces some basic notions and facts of Ergodic Theory, which are used repeatedly, such as Ergodic Theorems, entropy, and the Shannon–McMillan–Breiman Theorem. Proofs are not given, except in the case of the Multiplicative (Oseledeč) Ergodic Theorem, for which we provide a sketch. For the proofs and additional material we refer to the books, such as [**Wal82, Pet83, PU10, EW11**].

Let $(X, \mathcal{B}, \mu)$ be a probability measure space. A measurable map $T : X \to X$ (which may only be defined almost everywhere) is called *measure-preserving* if

$$\mu(T^{-1}B) = \mu(B) \quad \text{for all } B \in \mathcal{B}.$$

The most basic examples of measure-preserving systems are given by translations and endomorphisms of the torus, equipped with the Haar measure. This includes the maps $x \mapsto x + \alpha \pmod 1$ and $x \mapsto qx \pmod 1$ with $q > 1$ on the unit circle $\mathbb{T}^1 \approx [0, 1) = \mathbb{R}/\mathbb{Z}$.

A measure-preserving system $(X, \mathcal{B}, T, \mu)$ is said to be *ergodic*, and μ is called an *ergodic measure* for T if

$$B \in \mathcal{B} \text{ and } T^{-1}B = B \implies \mu(B) \in \{0, 1\}. \tag{9.1}$$

It is easy to see that ergodicity is equivalent to the following property:

$$f : X \to \mathbb{C} \text{ is measurable and } f = f \circ T \ \mu\text{-a.e.} \implies f \equiv \text{const} \ \mu\text{-a.e.}$$

EXAMPLE 9.0.1.

(a) Let $X = [0, 1]$ and let $T : [0, 1] \to [0, 1]$ be the tent map, that is, $T(x) = 1 - 2\left|x - \frac{1}{2}\right|$. Then the Lebesgue measure is is invariant and ergodic.

(b) Let $X = [0, 1)$ and let $T(x) := x + 1/3 \pmod 1$. Then the Lebesgue measure is *not* an ergodic measure, but $\mu_\alpha := \frac{1}{3}\delta_\alpha + \frac{1}{3}\delta_{\alpha+1/3} + \frac{1}{3}\delta_{\alpha+2/3}$ is an ergodic measure for all $\alpha \in X$. More generally, for the circle rotation $x \mapsto x + \alpha \pmod 1$ the Lebesgue mesure is ergodic if and only α is irrational.

(c) Perhaps, the most important example of a measure-preserving system in fractal geometry is the following: $X = \{1, \ldots, m\}^{\mathbb{N}}$ for $m \geqslant 2$, and T is the left shift. Further, let $\mu := \mathbf{p}^{\mathbb{N}}$ be the infinite product (Bernoulli) measure for a probability vector $\mathbf{p} = (p_1, \ldots, p_m)$. Then μ is invariant and ergodic.

9.1. Ergodic theorems

Let $(X, \mathcal{B}, T, \mu)$ be a measure-preserving system. Consider the time average $f^* : X \to X$ of a function $f \in L^1(\mu)$ over the T-orbit $\{x, T(x), T^2(x), \ldots\}$ of

$x \in X$, defined by

$$f^*(x) := \lim_{n\to\infty} \frac{1}{n} \sum_{i=0}^{n-1} f(T^i(x)), \tag{9.2}$$

whenever the limit exists. The Birkhoff (pointwise) Ergodic Theorem says that the time average exists μ-a.e. and is T-invariant, hence it is equal to a constant (the space average) whenever the measure is ergodic. Moreover, the time average function f^* may be identified with the *conditional expectation* $\mathbb{E}(f|\mathcal{I})$ of f with respect to the σ-algebra of T-invariant sets:

$$\mathcal{I} = \{B \in \mathcal{B} : T^{-1}B = B\} \tag{9.3}$$

Conditional expectation is introduced in Definition 9.3.1.

THEOREM 9.1.1 (Birkhoff Ergodic Theorem). *Let $(X, \mathcal{B}, T, \mu)$ be a measure-preserving system and $f \in L^1(\mu)$. Then the time average function f^* exists for μ-almost all $x \in X$ and*

$$f^* = \mathbb{E}(f|\mathcal{I}). \tag{9.4}$$

We have

$$f^* \circ T = f^*, \quad f^* \in L^1(\mu) \text{ and } \int f^* d\mu = \int f d\mu. \tag{9.5}$$

Moreover, if μ is an ergodic measure, then

$$f^*(x) = \lim_{n\to\infty} \frac{1}{n} \sum_{k=0}^{n-1} f(T^k(x)) = \int f d\mu \quad \text{for } \mu\text{-a.e. } x. \tag{9.6}$$

In particular, for an ergodic measure μ and a measurable set $H \subset X$ we have

$$\lim_{n\to\infty} \frac{1}{n} \# \{0 \leqslant k \leqslant n-1 : T^k(x) \in H\} = \mu(H), \quad \text{for } \mu\text{-a.e. } x. \tag{9.7}$$

COROLLARY 9.1.2. *Suppose that μ_1 and μ_2 are two ergodic probability measures for the same system $(X, \mathcal{B}, T)$. Then either $\mu_1 = \mu_2$, or μ_1 and μ_2 are mutually singular.*

PROOF. If $\mu_1 \neq \mu_2$, then there exists a function $f \in L^1(X, \mathcal{B})$ such that $\int f\, d\mu_1 \neq \int f\, d\mu_2$. For $j = 1, 2$ define

$$A_j = \left\{x \in X : \lim_{n\to\infty} \frac{1}{n} \sum_{i=0}^{n-1} f(T^i(x)) = \int f\, d\mu_j\right\}.$$

Then $A_1 \cap A_2 = \varnothing$, but Birkhoff's Ergodic Theorem implies that $\mu_j(A_j) = 1$ for $j = 1, 2$. The claim follows. □

The following theorem is a generalisation of Birkhoff's Ergodic Theorem for sequences of L^1 functions.

THEOREM 9.1.3 (Maker's Ergodic Theorem). *Let μ be a T-invariant probability measure and let $f_k \in L^1(\mu)$ be a sequence which converges both in L^1 and almost everywhere to $f \in L^1(\mu)$. Then*

$$\lim_{n\to\infty} \frac{1}{n} \sum_{k=0}^{n-1} f_{n-k}(T^k(x)) = \mathbb{E}(f|\mathcal{I})(x), \tag{9.8}$$

holds almost everywhere and in L^1.

For the proof, see e.g., the book by Mañé [**Mañ12**, Corollary 1.6]. Observe that whenever μ is ergodic, (9.8) yields that for μ-a.e. x,

$$\lim_{n\to\infty}\frac{1}{n}\sum_{k=0}^{n-1} f_{n-k}(T^k(x)) = \int f d\mu. \tag{9.9}$$

We will also need the *Subadditive* Ergodic Theorem, which is extremely useful in the study of self-affine IFS.

THEOREM 9.1.4 (Kingman's Subadditive Ergodic Theorem). *Let $(\Omega,\mathcal{A},\mu)$ be a probability space and let $T\colon \Omega \mapsto \Omega$ be a measure preserving transformation. Let $\{f_n\}_{n=1}^{\infty}$ be a sequence of measurable functions $f_n\colon \Omega \mapsto \mathbb{R}\cup\{-\infty\}$ such that*

(a) $f_1^+ = \max\{0, f_1\} \in L^1(\mu)$,
(b) $f_{n+k}(x) \leqslant f_n(x) + f_k \circ T^n(x)$ *for μ-almost every x and every $n,k \geqslant 1$.*

Then there exists a measurable function $f\colon \Omega \mapsto \mathbb{R}\cup\{-\infty\}$ such that

(1) $f^+ \in L^1(\mu)$ *and $f\circ T = f$ almost everywhere,*
(2) $\lim_{n\to\infty}\frac{1}{n}f_n(x) = f(x)$ *for μ-almost every x, and*
(3) $\lim_{n\to\infty}\frac{1}{n}\int f_n d\mu = \inf_n \frac{1}{n}\int f_n d\mu = \int f d\mu$.

The proof of the theorem can be found, e.g., in [**Ste89**]. The following result was first proven by Furstenberg and Kesten [**FK60**] and is an immediate corollary of Kingman's Subadditive Ergodic Theorem.

THEOREM 9.1.5 (Furstenberg–Kesten). *Let $(\Omega,\mathcal{A},\mu)$ be a probability space and let $T\colon \Omega \mapsto \Omega$ be a measure preserving transformation. Let $\mathcal{B}$ be a Banach algebra and let $\Phi : \Omega \to \mathcal{B}$ be a measurable map such that $\log^+\|\Phi(\cdot)\|$ is integrable. Then for μ a.e. $\omega\in\Omega$ the following limit exists:*

$$\lim_{n\to\infty}\frac{1}{n}\log\left\|\prod_{i=0}^{k-1}\Phi(T^i\omega)\right\| =: \lambda(\omega) \tag{9.10}$$

and the function $\lambda(\omega)$ is T-invariant, that is, $\lambda(\omega)=\lambda(T(\omega))$. Further, $\lambda(\omega)$ takes values from $\mathbb{R}\cup\{-\infty\}$, $\lambda^+\in L^1$ and

$$\begin{aligned}\int \lambda(\omega)d\mu &= \lim_{n\to\infty}\frac{1}{n}\int \log\left\|\prod_{i=0}^{k-1}\Phi(T^i\omega)\right\|d\mu \\ &= \inf_n\left\{\frac{1}{n}\int \log\left\|\prod_{i=0}^{k-1}\Phi(T^i\omega)\right\|\right\}d\mu.\end{aligned} \tag{9.11}$$

9.2. Multiplicative Ergodic Theorem (Oseledeč Theorem)

Our aim is to introduce the Lyapunov exponents of a T-invariant ergodic measure. We write $GL(d,\mathbb{R})$ for the group of real invertible $d\times d$ matrices, and $\|\cdot\|$ denotes the Euclidean norm for vectors and matrices. For a measurable map

$$A\colon \Omega \to GL(d,\mathbb{R}), \qquad \text{with } \log^+\|A(\cdot)\|\in L^1(\mu), \tag{9.12}$$

we study the exponential growth rate of the norm of the matrix product

$$P_n(A,\omega) := A(T^{n-1}\omega)\cdots A(\omega) \tag{9.13}$$

acting on a vector. That is, we study the limit

$$\lim_{n\to\infty}\frac{1}{n}\log\|\underbrace{A(T^{n-1}\omega)\cdots A(\omega)}_{P_n(A,\omega)}\cdot\mathbf{u}\|, \tag{9.14}$$

depending on $\mathbf{u}\in\mathbb{R}^d$ and $\omega\in\Omega$ (whenever the limit exists). Oseledeč theorem says that for a T-invariant ergodic measure μ on Ω, there is a finite set $\{\lambda_1,\ldots,\lambda_s\}$ (the Lyapunov exponents of μ) such that for a μ-typical ω, for all $\mathbf{u}\in\mathbb{R}^d\setminus\{0\}$, the limit in (9.14) exists and can take values only from the finite set $\{\lambda_1,\ldots,\lambda_s\}$.

THEOREM 9.2.1 (Oseledeč Theorem). *Let μ be a T-invariant measure on Ω. Then for a measurable function A as in* (9.12), *there exists $\Omega'\subset\Omega$, with $\mu(\Omega')=1$, such that for all $\omega\in\Omega'$ we have*

(1) *The limit of the sequence of matrices*

$$\lim_{n\to\infty}\left(P_n^*(A,\omega)\cdot P_n(A,\omega)\right)^{1/2n} =: P(\omega)$$

exists, and $P(\omega)$ is a positive semidefinite matrix.

(2) *Let*

$$\exp(\lambda_1(\omega)) > \cdots > \exp(\lambda_{s(\omega)}(\omega))$$

be the distinct *eigenvalues of $P(\omega)$. Let $E_1(\omega),\ldots,E_{s(\omega)}(\omega)$ be the corresponding eigenspaces and let*

$$F_r(\omega) := E_r(\omega)\bigoplus\cdots\bigoplus E_{s(\omega)}(\omega),\quad 1\leqslant r\leqslant s(\omega);$$

furthermore, define $F_{s(\omega)+1}(\omega):=\{0\}$. Then we have

$$\lim_{n\to\infty}\frac{1}{n}\log\|P_n(A,\omega)\cdot\mathbf{u}\| = \lambda_r(\omega),\quad \mathbf{u}\in F_r(\omega)\backslash F_{r+1}(\omega),\ r=1,\ldots,s(\omega). \tag{9.15}$$

(3) *The functions*

$$\omega\mapsto\dim(F_r(\omega)\ominus F_{r+1}(\omega)),\quad \omega\mapsto\lambda_r(\omega),$$

are T-invariant.

(4) *$A(\omega)\cdot F_r(\omega)=F_r(T\omega)$ holds for $r\leqslant s(\omega)$.*

(5) *If μ is ergodic, then the following functions are constant on Ω':*

$$s(\mu):\equiv s(\omega),\ \textit{for } 1\leqslant r\leqslant s,\ \lambda_r(\mu):\equiv\lambda_r(\omega),\ d_r(\mu):\equiv\dim(F_r(\omega)\ominus F_{r+1}(\omega)).$$

DEFINITION 9.2.2. For an ergodic T-invariant measure on Ω, the numbers $\lambda_1(\mu)>\cdots>\lambda_s(\mu)$ are the Lyapunov exponents of the measure μ for the function $A\colon\Omega\mapsto GL(d,\mathbb{R})$, with multiplicities $d_1(\mu),\ldots,d_s(\mu)$ respectively. Now we define $\widetilde{\chi}_k(\mu)$ for $1\leqslant k\leqslant d$ to be the kth element of the sequence

$$\underbrace{\lambda_1(\mu),\ldots,\lambda_1(\mu)}_{d_1(\mu)},\underbrace{\lambda_2(\mu),\ldots,\lambda_2(\mu)}_{d_2(\mu)},\ldots,\underbrace{\lambda_s(\mu),\ldots,\lambda_s(\mu)}_{d_s(\mu)}.$$

That is, $\widetilde{\chi}_k(\mu)=\lambda_i(\mu)$ if $d_1(\mu)+\cdots+d_{i-1}(\mu)<k\leqslant d_1(\mu)+\cdots+d_i(\mu)$.

For a complete proof of Oseledeč theorem see [**Kre85**, p.42–48]. We give a sketch below, following the lines of that proof.

SKETCH OF THE PROOF. Equality (A.5) implies that

$$P_n(A,\omega)^{\wedge k} = A^{\wedge k}(T^{n-1}\omega)\cdots A^{\wedge k}(T\omega)A^{\wedge k}(\omega) = P_n(A^{\wedge k},\omega), \tag{9.16}$$

where $P_n(A^{\wedge k},\omega)$ is the n-fold composition of the corresponding linear operators.

First observe that by (A.7) we have $\log^+ \|A^{\wedge k}(\cdot)\| \in L^1(\mu)$. Thus we can apply both the Birkhoff theorem and the Furstenberg–Kesten theorem to get a full measure subset $\Omega' \subset \Omega$ such that for all $\omega \in \Omega'$ the following T-invariant limits exist:

$$\lim_{n\to\infty} \frac{1}{n} \sum_{i=0}^{n-1} \log^+ \|A(T^i\omega)\| \quad \in \quad \mathbb{R},$$

$$\varrho(k,\omega) := \lim_{n\to\infty} \frac{1}{n} \log \|P_n(A^{\wedge k},\omega)\| \quad \in \quad \mathbb{R} \cup \{-\infty\}. \tag{9.17}$$

Consider the Singular Value Decomposition

$$P_n(A,\omega) = U_n(\omega) D_n(\omega) V_n^*(\omega). \tag{9.18}$$

Using (A.6) for $P_n(A,\omega)$ we obtain from (9.17) that for every $1 \leqslant k \leqslant d$,

$$\lim_{n\to\infty} \frac{1}{n} \big(\log \alpha_1(P_n(A,\omega)) + \cdots + \log \alpha_k(P_n(A,\omega))\big) = \varrho(k,\omega). \tag{9.19}$$

This implies that for every $1 \leqslant k \leqslant d$ the following limit exists:

$$\widetilde{\chi}_k(\omega) := \lim_{n\to\infty} \frac{1}{n} \log \alpha_k(P_n(A,\omega)). \tag{9.20}$$

and by (A.1) we obtain that

$$\widetilde{\chi}_1(\omega) \geqslant \widetilde{\chi}_2(\omega) \geqslant \cdots \geqslant \widetilde{\chi}_d(\omega). \tag{9.21}$$

Partition the set $\{1,\ldots,d\}$:

$$\{1,\ldots,d\} = \mathcal{I}_1(\omega) \cup \cdots \cup \mathcal{I}_{s(\omega)}(\omega),$$

where $\mathcal{I}_k(\omega) := \{i_{k-1}(\omega)+1,\ldots,i_k(\omega)\}$ and the sequence $\{i_k(\omega)\}_{k=0}^{s(\omega)}$ is defined as follows: $i_0(\omega) = 0$, $i_{s(\omega)} := d$, and for $i < s(\omega)$ the increasing sequence of integers $\{i_j(\omega)\}_{j=1}^{s(\omega)-1}$ consists of those numbers k for which $\widetilde{\chi}_k(\omega) > \widetilde{\chi}_{k+1}(\omega)$. Clearly, the function $s(\omega)$ is T-invariant. So, if $k \in \mathcal{I}_i$ then $\widetilde{\chi}_k(\omega)$ takes the same value, which we denote $\lambda_i(\omega)$, that is,

$$\lambda_i(\omega) = \widetilde{\chi}_k(\omega) \iff k \in \mathcal{I}_i.$$

Define the decreasing sequence $\{G_r\}_{r=1}^{s(\omega)}$ of subspaces of $\mathbb{R}^d$:

$$G_r := \operatorname{span}\{\mathbf{e}_k\}_{k\in\mathcal{I}_r\cup\cdots\cup\mathcal{I}_{s(\omega)}},$$

where $\mathbf{e}_k$ is the kth coordinate unit vector of the natural basis of $\mathbb{R}^d$. It is proved in [**Kre85**, p. 45–47] that for every $1 \leqslant r \leqslant s(\omega)$ the following limit of subspaces exists:

$$\lim_{n\to\infty} V_n \cdot G_r =: F_r. \tag{9.22}$$

Furthermore, for every $\mathbf{u} \in F_r \backslash F_{r+1}$ we have

$$\lim_{n\to\infty} \frac{1}{n} \log \|P_n(A,\omega)\mathbf{u}\| = \lambda_r(\omega). \tag{9.23}$$

Assuming that (9.22) holds, a *very rough* idea how to verify (9.23) is as follows: Choose a large n for which the subspaces satisfy $F_r \approx V_n \cdot G_r$. Then for $\mathbf{u} \in F_r \backslash F_{r+1}$ we have $V_n^{-1}\mathbf{u} \in G_r \backslash G_{r+1}$, that is, for the vector $V_n^{-1}\mathbf{u}$ the first $d_1(\omega) + \cdots + d_{r-1}(\omega)$

components are all zero, but there is $k_0 \in \mathcal{I}_r$ such that the k_0-th component of $V_n^{-1}\mathbf{u}$ is nonzero. Using this, (9.18) and (9.20) we obtain that

$$\|P_n(A,\omega)\mathbf{u}\| = \|D_n(\omega) \cdot V_n^{-1}(\omega) \cdot \mathbf{u}\| \asymp \exp(n\lambda_r(\omega)),$$

since the first $d_1(\omega) + \cdots + d_{r-1}(\omega)$ components of the vector $D_n(\omega) \cdot V_n^{-1}(\omega) \cdot \mathbf{u}$ are all zero but the k_0th component is the k_0th component of $V_n^{-1}\mathbf{u}$ (which is nonzero) multiplied by $D_n(\omega(k_0,k_0)) \asymp \exp(n\lambda_r(\omega))$. All the other components of this vector are of less than or equal magnitude. $\square$

COROLLARY 9.2.3 (of the proof). *Suppose that $\mathcal{F} = \{A_i \cdot \mathbf{x} + \mathbf{t}_i\}_{i=1}^m$ is a contractive self-affine IFS on $\mathbb{R}^d$. Let μ be an arbitrary ergodic measure on Σ. The function $A\colon \Sigma \to GL(\mathbb{R},d)$ in (9.12) is defined by*

$$A(\mathbf{i}) := A_{i_1}^*. \tag{9.24}$$

For $i = 1,\dots,d$ let $\widetilde{\chi}_i(\mu)$ be the Lyapunov exponents of μ for the function A. Then for all $1 \leqslant k \leqslant d$ and for μ-almost all $\mathbf{i} \in \Sigma$ we have

$$\lim_{n\to\infty} \frac{1}{n} \log \alpha_k\left(A_{i_n}^* \cdots A_{i_1}^*\right) = \widetilde{\chi}_k(\mu), \tag{9.25}$$

where $\alpha_k\left(A_{i_n}^* \cdots A_{i_1}^*\right)$ is the kth largest singular value of $A_{i_n}^* \cdots A_{i_1}^*$.

Note that the Lyapunov exponents $\widetilde{\chi}_i(\mu)$ defined in (9.25) are negative, however, in our previous discussion, see for example (1.105), Theorem 3.2.2 or Section 6.2, the Lyapunov exponents were positive quantities. For consistency, in the upcoming chapters the Lyapunov exponents will be given by $\chi_i(\mu) := -\widetilde{\chi}_i(\mu)$.

9.3. Conditional expectation and martingales

For the material in this section, see e.g., the books by Parry [**Par97**, Chapter 2] or Einsiedler and Ward [**EW11**, Chapter 5].

DEFINITION 9.3.1. Let $(\Omega,\mathcal{B},\mu)$ be a probability space, and let $\mathscr{A} \subseteq \mathcal{B}$ be a sub-σ-algebra. Given $f \in L^1(\Omega,\mathcal{B},\mu)$, the *conditional expectation* of f given $\mathscr{A}$, is a function $\mathbb{E}(f|\mathscr{A}) : \Omega \to \mathbb{R}$, characterized (almost everywhere) by the following properties:

- $\mathbb{E}(f|\mathscr{A})$ is $\mathscr{A}$-measurable;
- for any $A \in \mathscr{A}$, holds $\int_A \mathbb{E}(f|\mathscr{A})\, d\mu = \int_A f\, d\mu$.

The existence of such a function can be deduced from the Radon–Nikodym Theorem.

Let $\mathcal{B}_n$ be an increasing sequence of σ-algebras converging to $\mathcal{B}$; this means that $\mathcal{B}$ is the smallest σ-algebra containing all $\mathcal{B}_n$; this is denoted by

$$\mathcal{B} = \bigvee_{n=1}^{\infty} \mathcal{B}_n.$$

A sequence of integrable $\mathcal{B}_n$-measurable functions $\{F_n\}_{n\geqslant 1}$ is called a *martingale* (with respect to the filtration $\mathcal{B}_n$) if

$$\mathbb{E}(F_{n+1}|\mathcal{B}_n) = F_n, \quad \text{for all } n \geqslant 1.$$

It follows from the definition of conditional expectation that for any $f \in L^1(\Omega,\mathcal{B},\mu)$, the sequence $F_n = \mathbb{E}(f|\mathcal{B}_n)$ is a martingale.

The following important result is a form of Doob's increasing martingale theorem.

THEOREM 9.3.2 (Increasing martingale theorem). *Let $(\Omega, \mathcal{B}, \mu)$ be a probability space. Suppose that $\{\mathscr{A}_n\}_{n\geqslant 1}$ is an increasing sequence of σ-algebras, such that $\bigvee_{n=1}^{\infty} \mathscr{A}_n = \mathscr{A} \subseteq \mathcal{B}$. Then for any $f \in L^1(\Omega, \mathcal{B}, \mu)$,*

$$\mathbb{E}(f|\mathscr{A}_n) \to E(f|\mathscr{A})$$

almost everywhere and in $L^1(\Omega, \mathcal{B}, \mu)$.

A similar result holds for a decreasing sequence of σ-algebras.

THEOREM 9.3.3 (Decreasing martingale theorem). *Let $(X, \mathcal{B}, \mu)$ be a probability space. Suppose that $\{\mathscr{A}_n\}_{n\geqslant 1}$ is a decreasing sequence of σ-algebras, such that $\bigcap_{n=1}^{\infty} \mathscr{A}_n = \mathscr{A}_\infty \subseteq \mathcal{B}$. Then for any $f \in L^1(\Omega, \mathcal{B}, \mu)$,*

$$\mathbb{E}(f|\mathscr{A}_n) \to E(f|\mathscr{A}_\infty)$$

almost everywhere and in $L^1(\Omega, \mathcal{B}, \mu)$.

In Chapter 13 we will need another result related to martingales, namely, the *Law of Large Numbers for martingale differences.* Suppose that $\mathcal{B}_n$ is an increasing sequence of σ-algebras such that $\bigvee_{n=1}^{\infty} \mathcal{B}_n = \mathcal{B}$. A sequence of integrable $\mathcal{B}$-measurable functions f_n is called a *martingale difference sequence* if

- f_n is $\mathcal{B}_n$-measurable;
- $\mathbb{E}(f_{n+1}|\mathcal{B}_n) = 0$ for all n.

It follows from the definitions that f_n is a martingale difference sequence if and only if $S_n = f_1 + \cdots + f_n$ is a martingale; hence the name.

THEOREM 9.3.4. *Let $(\Omega, \mathcal{B}, \mu)$ be a probability space and $\mathcal{B}_n$ an increasing sequence of σ-algebras converging to $\mathcal{B}$. Let $\{f_n\}_{n=0}^{\infty}$ be a martingale difference sequence such that $\sup_n \|f_n\|_{L^2(\mu)} < \infty$. Then*

$$\lim_{N\to\infty} \frac{1}{N} \sum_{n=0}^{N-1} f_n(x) = 0 \quad \mu\text{-a.e.}$$

For a proof, see [**Fel71**, Chapter VII.9, Theorem 3].

9.4. Measurable partitions, Lebesgue spaces, Rokhlin's theorem

In this section we follow the books [**PU10**, Section 2.6] and [**EW11**, Chapter 5]. The results of this section are about the existence of regular systems of conditional measures and are due to Rokhlin [**Roh52**]. In the book [**EW11**] a similar theory was developed, but instead of Lebesgue spaces the authors considered a special case of *standard Borel spaces* and instead of measurable partitions that appear in the original Rokhlin paper, the assertions in [**EW11**] are stated for countably generated sub-σ-algebras. However, one can easily handle the difficulties which arise from these differences. Since the proofs in [**EW11**, Section 5.3] are much more accessible than the proofs in the original Rokhlin's paper and the setup in [**EW11**] is general enough to cover all the applications of Rokhlin's theory that appear in this book, we devote the next subsection to sketch the Einsiedler–Ward treatment.

9.4.1. System of conditional measures for countably generated sub-σ-algebras of Borel probability spaces.

DEFINITION 9.4.1. Assume that X is a dense subset of a compact metric space $\overline{X}$, where $\overline{X}$ is the closure of X . We write $\mathcal{B}$ for the restriction of the Borel σ-algebra in $\overline{X}$ to X.

(a) The pair $(X, \mathcal{B})$ is called a Borel space.
(b) Given a probability measure μ on $\mathcal{B}$, the space $(X, \mathcal{B}, \mu)$ is called a Borel probability space.

We need the following terminology:

(i) The complement of a null set is called a conull set.
(ii) Let $(X, \mathcal{B}, \mu)$ be a Borel probability space. When we write $\mathscr{L}^1(X, \mathcal{B}, \mu)$ or $\mathscr{L}^\infty(X, \mathcal{B}, \mu)$ we refer to genuine integrable and bounded functions respectively.
(iii) Let $\mathcal{C}$ and $\mathcal{C}'$ be σ-algebras on X. We write $\mathcal{C} \overset{\mu}{\subset} \mathcal{C}'$ if for any $A \in \mathcal{C}$ there exists $A' \in \mathcal{C}'$ such that $\mu(A \triangle A') = 0$. Moreover, we write

$$\mathcal{C} \overset{\mu}{=} \mathcal{C}', \tag{9.26}$$

if both $\mathcal{C} \overset{\mu}{\subset} \mathcal{C}'$ and $\mathcal{C}' \overset{\mu}{\subset} \mathcal{C}$.

We will need the following class of σ-algebras:

DEFINITION 9.4.2. Let X be a set as in Definition 9.4.1. We say that a σ-algebra $\mathcal{A} \subset \mathcal{B}$ on X is *countably-generated* if there exists a sequence $\{A_1, A_2, \ldots\}$, with $A_k \in \mathcal{A}$, such that $\mathcal{A}$ is the smallest σ-algebra containing A_k for all k. That is, $\mathcal{A} = \sigma(\{A_1, A_2, A_3, \ldots\})$.

The fact that there are sufficiently many countably generated σ-algebras is stated next:

LEMMA 9.4.3. *If $(X, \mathcal{B}, \mu)$ is a Borel probability space and $\mathcal{F} \subset \mathcal{B}$ is a σ-algebra, then there is a countably generated σ-algebra $\mathcal{A} \subset \mathcal{F}$ such that $\mathcal{F} \overset{\mu}{=} \mathcal{A}$.*

For the proof see [**EW11**, p. 141]. The main result of this subsection is a version of Rokhlin's Theorem about the existence of regular conditional probability measures, which was stated in [**EW11**, Theorem 5.14] in the following way:

THEOREM 9.4.4 (Rokhlin). *Let $(X, \mathcal{B}, \mu)$ be a Borel probability space (as in Definition 9.4.1) and let $\mathcal{A} \subset \mathcal{B}$ be sub-σ-algebra. Then there exists a conull set $X' \subset X$ and a system of probability measures on X (called conditional measures)*

$$\{\mu_x^{\mathcal{A}} : x \in X'\} \tag{9.27}$$

with the following properties:

(a) *For all $f \in \mathscr{L}^1(X, \mathcal{B}, \mu)$ there exists a conull set such that for all x belonging to this conull set the integral $\int f(y) d\mu_x^{\mathcal{A}}(y)$ exists and on this conull set the mapping $x \mapsto \int f(y) d\mu_x^{\mathcal{A}}(y)$ is $\mathcal{A}$-measurable. Moreover, for all $A \in \mathcal{A}$ we have*

$$\int_A \int_X f(y) d\mu_x^{\mathcal{A}}(y) d\mu(x) = \int_A f d\mu. \tag{9.28}$$

This implies that for all $f \in \mathscr{L}^1(X, \mathcal{B}, \mu)$ and for μ-a.e. x we have

$$\mathbb{E}(f|\mathcal{A})(x) = \int f(y) d\mu_x^{\mathcal{A}}(y). \tag{9.29}$$

(b) *If Property **(a)** holds only for a dense countable set of functions $f \in C(\overline{X})$ instead of for all $f \in \mathscr{L}^1(X, \mathcal{B}, \mu)$, it already uniquely determines $\mu_x^{\mathcal{A}}$ for a.e. $x \in X$.*

(c) *If $\widetilde{\mathcal{A}}$ is a σ-algebra satisfying $\widetilde{\mathcal{A}} \overset{\mu}{=} \mathcal{A}$ then $\mu_x^{\mathcal{A}} = \mu_x^{\widetilde{\mathcal{A}}}$ for μ-a.e. $x \in X$.*

(d) *Assume that $\mathcal{A}$ is countable generated. For every $x \in X'$ we define the atom containing x:*

$$[x]_{\mathcal{A}} := \bigcap_{x \in A \in \mathcal{A}} A. \tag{9.30}$$

Then

(i) $[x]_{\mathcal{A}} = \bigcap_{x \in A_i} A_i \cap \bigcap_{x \notin A_i} (X \backslash A_i)$,

(ii) $\mu_x^{\mathcal{A}}([x]_{\mathcal{A}}) = 1$ *and*

(iii) $\mu_x^{\mathcal{A}} = \mu_y^{\mathcal{A}}$ *for all $x, y \in X'$ with $[x]_{\mathcal{A}} = [y]_{\mathcal{A}}$.*

We will also need the assertion [**EW11**, Proposition 5.20]

PROPOSITION 9.4.5. *Using the notation of Theorem 9.4.4 assume that $\mathcal{A}' \subset \mathcal{A}$ are countably generated σ-algebras. Then*

(a) *$[z]_{\mathcal{A}} \subset [z]_{\mathcal{A}'}$ for every $z \in X$.*

(b) *For almost every $z \in X$ and for $\mu_z^{\mathcal{A}'}$-a.e. $x \in [z]_{\mathcal{A}'}$ the conditional measure for $\mu_z^{\mathcal{A}'}$ with respect to $\mathcal{A}$ is given by*

$$\left(\mu_z^{\mathcal{A}'}\right)_x^{\mathcal{A}} = \mu_x^{\mathcal{A}}. \tag{9.31}$$

9.4.2. System of conditional measures for measurable partitions on Lebesgue spaces.

DEFINITION 9.4.6. Suppose we are given a partition ξ of the probability space $(\Omega, \mathcal{A}, \nu)$. We do not assume that ξ is countable, nor that the elements of ξ are contained in $\mathcal{A}$. Consider the family of those sets in $\mathcal{A}$ which are (not necessarily finite) unions of some elements of ξ. The σ-algebra generated by this family of sets is denoted by $\widehat{\xi}$.

Given two partitions ξ and η, we write $\xi \leqslant \eta$ if every element of ξ is the union of some elements of η and say that η is finer than ξ.

The *quotient space* $(\Omega_\xi, \mathcal{A}_\xi, \nu_\xi)$ is defined as follows:

- Ω_ξ is the set of elements of ξ.
- The canonical map $\tau : x \mapsto \xi(x)$ defines a σ-algebra $\mathcal{A}_\xi := \tau(\widehat{\xi})$ on Ω_ξ, where $\xi(x)$ denotes the element of ξ that contains x.
- Finally, $\nu_\xi := \nu \circ \tau^{-1}$.

DEFINITION 9.4.7 (Basis of a partition). Let ξ be a partition of the probability space $(\Omega, \mathcal{A}, \nu)$ as in Definition 9.4.6. We say that a sequence $\mathbf{B} = \{B_n\}_{n=1}^{\infty}$ is a *basis of* ξ if it satisfies the following conditions:

(a) $B_n \in \widehat{\xi}$ for every n.

(b) If C_1, C_2 are two distinct elements of ξ, then there exists n such that

$$\text{either } \ C_1 \subset B_n \text{ and } C_2 \subset \Omega \backslash B_n, \quad \text{or } \ C_2 \subset B_n \text{ and } C_1 \subset \Omega \backslash B_n.$$

To get an idea about what is going on, consider the following example:

EXAMPLE 9.4.8. Let $(\Omega, \mathcal{A}, \nu) := \big([0,1]^2, \mathcal{B}_2, \mathcal{L}^2\big)$, that is, the restriction of the two-dimensional Lebesgue measure to the Borel σ-algebra of the unit square. Further, let ξ be the partition of $[0,1]^2$ into vertical line segments, that is, $(x, y) \overset{\xi}{\sim} (x', y')$ if and only if $x = x'$. Then

$$\widehat{\xi} = \{B \times [0,1] : B \in \mathcal{B}_1\},$$

where $\mathcal{B}_1$ is the Borel σ-algebra on $[0,1]$. The quotient space $(\Omega_\xi, \mathcal{A}_\xi, \nu_\xi)$ can be identified with the one-dimensional Lebesgue measure restricted to $\mathcal{B}_1$. Moreover, in this case,

$$\{B_n := [0, r_n] \times [0,1]\}_{n\geqslant 1}$$

is a basis of the partition ξ, where $\{r_n\}_{n=1}^\infty$ is a dense subset of $[0,1]$.

DEFINITION 9.4.9. We say that ξ is a *measurable partition* if ξ has a basis.

Here is an equivalent definition:

DEFINITION 9.4.10. We say that ξ is a measurable partition of the probability space $(\Omega, \mathcal{A}, \nu)$ if there exists a sequence of finite partitions into measurable sets $\{\xi_n\}_{n=1}^\infty$, with $\xi_n \leqslant \mathcal{A}$, such that

$$\xi = \bigvee_{n=1}^{\infty} \xi_n := \{C_1 \cap C_2 \cap \cdots \mid C_i \in \xi_i\}. \tag{9.32}$$

It follows that every atom of a measurable partition is indeed measurable.

Partitions are often derived from the fibres (inverse branches) of measurable maps. These partitions will be measurable under mild assumptions.

PROPOSITION 9.4.11. *Let $(\Omega, \mathcal{A}, \nu)$ be a probability space. Let X be a separable metric space equipped with the Borel σ-algebra and $\phi\colon \Omega \mapsto X$ a measurable map. Let ξ be the partition defined by the preimages of points, that is, for every $\omega \in \Omega$, $\xi(\omega) := \phi^{-1}(\phi(\omega))$. Then ξ is measurable.*

PROOF. Denote $R = \{r_n\}_{n=1}^\infty$ a dense and countable subset of X. Since ϕ is measurable, $\phi^{-1}(B(r_n, 1/m)) \in \mathcal{A}$ for every $n, m \geqslant 1$. It is enough to show that $\mathbf{B} = \{\phi^{-1}(B(r_n, 1/m))\}_{n,m=1}^\infty$ is a basis of ξ.

Clearly, $\phi^{-1}(B(r_n, 1/m)) \in \widehat{\xi}$ for every $n, m \geqslant 1$, since $\phi^{-1}(B(r_n, 1/m)) = \bigcup\{\xi(\omega) : \omega \in \Omega,\ \phi(\omega) \in B(r_n, 1/m)\}$. On the other hand, if $C_1, C_2 \in \xi$ are two distinct elements, then $C_1 = \phi^{-1}(x)$ and $C_2 = \phi^{-1}(y)$ for some $x \neq y$. Thus we can choose $m \geqslant 1$ such that $d(x, y) > 2/m$. Also, there exists $r_n \in R$ such that $d(x, r_n) < 1/m$, and thus, $y \notin B(r_n, 1/m)$. Hence $C_1 \subset \phi^{-1}(B(r_n, 1/m))$ and $C_2 \subset \Omega\backslash\phi^{-1}(B(r_n, 1/m))$, concluding the proof. □

DEFINITION 9.4.12 (Basis of a probability space). We say that $\mathbf{B} = \{B_n\}_{n=1}^\infty$ is a basis of the probability space $(\Omega, \mathcal{A}, \nu)$ if the following two conditions hold:

(a) $\mathbf{B}$ is a basis for the partition of X into points.
(b) For any $A \in \mathcal{A}$ there exists $C \in \sigma(\mathbf{B})$ such that

$$A \subset C \text{ and } \nu(C\backslash A) = 0,$$

where $\sigma(\mathbf{B})$ is the σ-algebra generated by $\mathbf{B}$.

Note that if the probability space $(\Omega, \mathcal{A}, \nu)$ has a basis $\mathbf{B} = \{B_n\}_{n=1}^{\infty}$ and we write $B_n^1 := B_n$ and $B_n^{-1} := \Omega \backslash B_n$ then by **(a)** for every $\{\varepsilon_n\}_{n=1}^{\infty}$, with $\varepsilon_n \in \{-1, 1\}$ we have

$$\# \bigcap_{n=1}^{\infty} B_n^{\varepsilon_n} \leqslant 1, \tag{9.33}$$

DEFINITION 9.4.13. A probability space $(\Omega, \mathcal{A}, \nu)$ is called

- separable, if it has a basis $\mathbf{B} = \{B_n\}_{n=1}^{\infty}$.
- complete with respect to the basis $\mathbf{B}$, if for all $\{\varepsilon_n\}_{n=1}^{\infty}$, $\bigcap_{n=1}^{\infty} B_n^{\varepsilon_n} \neq \varnothing$.

The space $(\Omega, \mathcal{A}, \nu)$ is complete (mod 0) with respect to a basis $\mathbf{B}$ if $(\Omega, \mathcal{A}, \nu)$ can be included as a full measure subset into $(\overline{\Omega}, \overline{\mathcal{A}}, \overline{\nu})$ which is complete with respect to a basis $\overline{\mathbf{B}}$ whose restriction to Ω is equal to $\mathbf{B}$. If a space is complete mod 0 with respect to one basis, then it is complete (mod 0) with respect to any other basis.

EXAMPLE 9.4.14. Consider $(\Omega, \mathcal{A}, \mathcal{L}^1)$, where $\Omega = [0, 1]$, with the Lebesgue σ-algebra $\mathcal{A}$ and the Lebesgue measure $\mathcal{L}^1$. Define $B_n = \{x \in [0, 1] : x_n = 0\}$ where $x_1, x_2 \ldots$, are the digits of the dyadic expansion of x. Then $\{B_n\}_{n \geqslant 1}$ is a basis, and $(\Omega, \mathcal{A}, \mathcal{L}^1)$ is complete with respect to this basis.

Now we can introduce an important class of probability spaces called *Lebesgue spaces*.

DEFINITION 9.4.15 (Lebesgue Space). A probability space $(\Omega, \mathcal{A}, \nu)$ that is complete (mod 0) with respect to any of its bases is called a Lebesgue space.

Recall that a probability space $(\Omega, \mathcal{A}, \nu)$ is complete if every subset of every null set is measurable.

THEOREM 9.4.16.

(i) *Let Ω be a complete separable metric space and ν is an atomless probability measure on the Borel σ-algebra $\mathcal{B}(\Omega)$. Then $(\Omega, \mathcal{B}(\Omega), \nu)$, as well as its completion $(\overline{\Omega}, \overline{\mathcal{B}}(\Omega), \overline{\nu})$, is a Lebesgue space.*

(ii) *Assume that Ω is a complete separable metric space and ν is a probability measure on the σ-algebra $\mathcal{A}$ which contains $\mathcal{B}(\Omega)$, and $\mathcal{A}$ is the smallest σ-alegbra for which the measure space $(\Omega, \mathcal{A}, \nu)$ is complete. Then $(\Omega, \mathcal{A}, \nu)$ is a Lebesgue space.*

(iii) *A probability space is a Lebesgue space if and only if it is isomorphic to the union of $\big([0, 1], \mathcal{L}^1\big)$ and countably many atoms.*

For a proof of parts (i) and (iii) see [**Roy68**, Theorem 9, p. 327], and the proof of part (ii) is available in [**Roh52**, Section 2.7].

Let ξ be a measurable partition of a Lebesgue space. We say that a property holds for almost all elements of ξ if the union of atoms of ξ for which the property holds is a (measurable) set of full measure.

Now we can state a the theorem of Rokhlin on the canonical system of conditional measures with respect to a measurable partition.

THEOREM 9.4.17 (Rokhlin). *Let ξ be a measurable partition of the Lebesgue space $(\Omega, \mathcal{A}, \nu)$. Then for almost every $x \in \Omega$ there exists a probability measure ν_x^ξ, uniquely characterized by the following properties:*

(i) *For any integrable function f on $(\Omega, \mathcal{A}, \nu)$, we have*

$$\mathbb{E}\,(f|\widehat{\xi})(x) = \int f(y)\, d\nu_x^\xi(y).$$

In other words, we have that $\int f(y)\, d\nu_x^\xi(y)$ exists for a set of $x \in \Omega$ of full ν measure, on this set $x \mapsto \int f(y)\, d\nu_x^\xi(y)$ depends $\widehat{\xi}$-measurably on x, and for all $A \in \widehat{\xi}$ holds

$$\int_A \int f(y)\, d\nu_x^\xi(y)\, d\nu(x) = \int_A f\, d\nu.$$

(ii) $\operatorname{spt}(\nu_x^\xi) \subset \xi(x)$. *Moreover, $\nu_x^\xi = \nu_y^\xi$ whenever $\xi(x) = \xi(y)$.*

The family $\{\nu_x^\xi\}_x$ is called a canonical system of conditional measures; it is unique (mod 0).

Another version of the theorem can be stated using the quotient probability space $(\Omega_\xi, \mathcal{A}_\xi, \nu_\xi)$ introduced in Definition 9.4.6, see [**PU10**, Section 2.6] for details.

THEOREM 9.4.18 (Rokhlin). *Let ξ be a measurable partition of a Lebesgue space $(\Omega, \mathcal{A}, \nu)$. Then for almost every $C \in \xi$ there exists a Lebesgue space $(C, \mathcal{A}_C, \nu_C^\xi)$ such that*

(i) *If $B \in \mathcal{A}$ then $B \cap C \in \mathcal{A}_C$ holds for almost all $C \in \xi$.*
(ii) *For $B \in \mathcal{A}$ the function $C \mapsto \nu_C^\xi(B)$ is ν_ξ-measurable, and*

$$\nu(B) = \int_{\Omega_\xi} \nu_C^\xi(B)\, d\nu_\xi(C). \tag{9.34}$$

This implies that whenever $f \in L^1(\Omega, \mathcal{A}, \nu)$, holds

$$\int f d\nu = \int_\Omega \left(\int_{\xi(x)} f|_{\xi(x)} d\nu_{\xi(x)}^\xi \right) d\nu(x). \tag{9.35}$$

If the partition ξ is finite or countable, then the measure ν_C^ξ is the standard conditional measure of elementary probability: $\nu_C^\xi(E) = \nu(E \cap C)/\nu(C)$.

REMARK 9.4.19. Let $(\Omega, \mathcal{A}, \nu)$ be a Lebesgue space and $(X, \mathcal{B})$ a separable metric space with the Borel σ-algebra. Given $\phi : \Omega \to X$, a measurable map, we saw in Proposition 9.4.11 that the partition ξ of Ω defined by the preimages of points is measurable. If ϕ is onto, then the quotient probability space $(\Omega_\xi, \mathcal{A}_\xi, \nu_\xi)$ is isomorphic to $(X, \mathcal{B}, \phi_*\nu)$, see [**PU10**, Theorem 2.6.11].

Let $(\Omega, \mathcal{A}, \nu)$ be a Lebesgue space. There is a one-to-one correspondence between measurable partitions (mod 0) and complete sub-σ-algebras of $\mathcal{A}$, that is, σ-algebras $\mathcal{A}' \subseteq \mathcal{A}$, such that the restriction $\nu|_{\mathcal{A}'}$ is complete. In fact, let

$$\mathcal{A}(\xi) := \left\{ H : \exists\, U \in \widehat{\xi} \text{ such that } \nu(H \triangle U) = 0 \right\}.$$

This correspondence preserves the following operations:

$$\mathcal{A}\left(\bigvee_{\tau} \xi_\tau\right) = \bigvee_{\tau} \mathcal{A}(\xi_\tau), \quad \mathcal{A}\left(\bigwedge_{\tau} \xi_\tau\right) = \bigwedge_{\tau} \mathcal{A}(\xi_\tau), \tag{9.36}$$

where $\eta := \bigvee_{\tau} \xi_\tau$ the crudest partition that is finer than any of the partitions ξ_τ. That is,

- $\eta \geqslant \xi_\tau$ for every τ,
- if $\eta' \geqslant \xi_\tau$ for every τ and η' is measurable then $\eta' \geqslant \eta$.

To get the definition of $\bigwedge_{\tau} \xi_\tau$ we need to replace $\geqslant$ with $\leqslant$ above. On the other hand, $\bigvee_{\tau} \mathcal{A}(\xi_\tau)$ is the intersection of all σ-algebras which contain all $\mathcal{A}(\xi_\tau)$. Finally, $\bigwedge_{\tau} \mathcal{A}(\xi_\tau) := \bigcap_{\tau} \mathcal{A}(\xi_\tau)$.

We can now give an alternative definition of conditional expectation, which was introduced in Section 9.3. Let ξ be a measurable partition of a Lebesgue space $(\Omega, \mathcal{A}, \nu)$. For $f \in L^1(\Omega, \mathcal{A}, \nu)$ and ν-a.e. $x \in \Omega$ define

$$\mathbb{E}(f|\xi) = \mathbb{E}_\nu(f|\xi)(x) := \nu^{\xi}_{\xi(x)}(f) = \int_{\xi(x)} f|_{\xi(x)} d\nu^{\xi}_{\xi(x)}. \tag{9.37}$$

Then from (9.35) we obtain

$$\int f d\nu = \int \mathbb{E}(f|\xi)\, d\nu,$$

and it follows that $\mathbb{E}(f|\xi) = \mathbb{E}(f|\widehat{\xi})$, see Definition 9.3.1.

The martingale convergence theorems 9.3.2 and 9.3.3 can be restated in terms of partitions:

THEOREM 9.4.20. *Assume that either $\xi_n \uparrow \xi$; in this case let $\xi := \bigvee_{n=1}^{\infty} \xi_n$, or $\xi_n \downarrow \xi$; in this case let $\xi := \bigwedge_{n=1}^{\infty} \xi_n$. Let f be an integrable function. Then we have*

$$\mathbb{E}(f|\xi_n) \to \mathbb{E}(f|\xi) \quad \nu\text{-a.e.}$$

9.4.3. Finite partitions versus finite sub-algebras. Given a probability space $(\Omega, \mathcal{A}, \nu)$ and two finite partitions $\alpha = \{A_1, \dots, A_k\}$ and $\beta = \{B_1, \dots, B_n\}$ such that $A_i, B_j \in \mathcal{A}$ for all $i \leqslant k$, $j \leqslant n$. We write $\alpha \vee \beta$ for the common refinement. That is, the elements of the partition $\alpha \vee \beta$ are the nonempty sets of the form $A_i \cap B_j$, $i \leqslant k$, $j \leqslant n$. The finite σ-algebra $\widehat{\alpha}$, introduced in Definition 9.4.6, consists of the empty set and those elements of $\mathcal{A}$ which are unions of some elements of α. Conversely, let $\mathscr{C} = \{C_1, \dots, C_p\}$ and $\mathscr{D} = \{D_1, \dots, D_q\}$ be finite σ-algebras such that $C_i, D_j \in \mathcal{A}$ for all $i \leqslant p$, $j \leqslant q$. We write $\mathscr{C} \vee \mathscr{D}$ for the smallest sub-σ-algebra of $\mathcal{A}$ which contains both $\mathscr{C}$ and $\mathscr{D}$. Since $\mathscr{C}$ and $\mathscr{D}$ are finite,

$$\mathscr{C} \vee \mathscr{D} = \{\text{ union of sets of the form } A_i \cap B_j,\ i \leqslant p,\ j \leqslant q\}. \tag{9.38}$$

As earlier, for an $\varepsilon_i \in \{-1, 1\}$ we write

$$C_i^{\varepsilon_i} = \begin{cases} C_i, & \text{if } \varepsilon_i = 1\ ; \\ \Omega \backslash C_i, & \text{if } \varepsilon_i = -1\ . \end{cases}$$

Let

(9.39)
$$\widehat{\mathscr{C}} := \{C_1^{\varepsilon_1} \cap \cdots \cap C_k^{\varepsilon_k} : (\varepsilon_1, \ldots, \varepsilon_k) \in \{-1, 1\}^n \text{ , and } C_1^{\varepsilon_1} \cap \cdots \cap C_k^{\varepsilon_k} \neq \varnothing\}.$$

Then $\widehat{\mathscr{C}}$ is a partition of Ω whose elements are contained in $\mathcal{A}$. It is easy to see that

$$\widehat{\widehat{\alpha}} = \alpha \quad \text{and} \quad \widehat{\widehat{\mathscr{C}}} = \mathscr{C}. \tag{9.40}$$

Moreover, let $\alpha, \beta, \mathscr{C}, \mathscr{D}$ as above. Then we have

$$\widehat{\alpha \vee \beta} = \widehat{\alpha} \vee \widehat{\beta}, \quad \text{and} \quad \widehat{\mathscr{C} \vee \mathscr{D}} = \widehat{\mathscr{C}} \vee \widehat{\mathscr{D}}. \tag{9.41}$$

In this way, we have a one-to-one correspondence between the finite partitions of measurable elements, and finite sub-σ-algebras, and this correspondence preserves the $\vee$ operation.

9.4.4. Connection between the systems of conditional measures of sections 9.4.1 and 9.4.2. Let $(X, \mathcal{B})$ as in Definition 9.4.1. That is, let X be dense in the compact metric space $\overline{X}$. We assume that $(\overline{X}, \overline{\mathcal{B}}, \nu)$ is an atomless complete measure space and $\overline{\mathcal{B}}$ contains all Borel sets of $\overline{X}$ and no proper sub-σ-algebras of $\overline{\mathcal{B}}$ has these properties. Moreover, we assume that $\nu(\overline{X} \backslash X) = 0$. Let $\mathcal{B}$ be the restriction of $\overline{\mathcal{B}}$ to X. Then by Theorem 9.4.16 $(X, \mathcal{B}, \mu)$ is a Lebesgue space. Let ξ be a measurable partition of $(X, \mathcal{B}, \mu)$. Then, according to Definition 9.4.10 there exists a sequence of finite partitions $\xi_n = \{C_1^n, \ldots, C_{k_n}^n\}$, where $C_i^n \in \mathcal{B}$, such that $\xi = \bigvee_{n=1}^{\infty} \xi_n$. That is, if $\Gamma := \left\{(i_1, i_2, \ldots) \in \prod_{n=1}^{\infty} \{1, \ldots, k_n\} : \bigcap_{n=1}^{\infty} C_{i_n}^n \neq \varnothing\right\}$, then

$$\xi = \left\{\bigcap_{n=1}^{\infty} C_{i_n}^n : (i_1, i_2, \ldots) \in \Gamma\right\}. \tag{9.42}$$

Then the smallest sub-σ-algebra of $\mathcal{B}$ which contains $\widehat{\xi}_n$ for every n is $\widehat{\xi}$, that is,

$$\widehat{\xi} = \bigvee_{n=1}^{\infty} \widehat{\xi}_n. \tag{9.43}$$

Using that $\widehat{\xi}_n$ is finite for all n we get that $\widehat{\xi}$ is a countably generated sub-σ-algebra of $\mathcal{B}$. More precisely,

$$\widehat{\xi} = \sigma(\{C_{i_n}^n : i_n \in \{1, \ldots, k_n\},\ n \in \mathbb{N}\}). \tag{9.44}$$

We can apply Theorem 9.4.4 to the Borel probability space $(X, \mathcal{B}, \mu)$ letting $\mathcal{A} := \widehat{\xi}$. Using that ξ_n is a partition, for every $x \in X$ and n there exists a unique $i_n = i_n(x) \in \{1, \ldots, k_n\}$ such that $x \in C_{i_n(x)}^n$. Combining part (i) of part(d) of Theorem 9.4.4 with (9.32), we obtain that the atom containing x is

$$[x]_{\widehat{\xi}} = \bigcap_{n=1}^{\infty} C_{i_n(x)}^n = [x]_{\xi}. \tag{9.45}$$

In this way Theorem 9.4.4 implies Theorem 9.4.17 in the special case when the Lebesgue space is a Borel probability space in the sense of Definition 9.4.1. In particular, comparing the conditional measures in Theorem 9.4.4 and heorem 9.4.17 we obtain that

$$\mu_x^{\widehat{\xi}} = \mu_x^{\xi}, \quad \text{for } \mu\text{-a.e. } x \in X. \tag{9.46}$$

LEMMA 9.4.21. *Let $(X, \mathcal{B}, \mu)$ be as in Definition 9.4.1 and let ξ and P be a measurable partitions. Additionally, we assume that P has countably many (finite or countably infinite) elements, all of them with positive measure. Then for μ-almost every $x \in X$, for every $H \in \mathcal{B}$ we have*

$$\mu_x^{\xi \vee P}(H) = \frac{\mu_x^\xi(H \cap P(x))}{\mu_x^\xi(P(x))}, \tag{9.47}$$

where $P(x)$ is the element of P that contains x.

PROOF. Let $\mathcal{A}' := \widehat{\xi}$, $\mathcal{A} = \widehat{\xi \vee P} = \widehat{\xi} \vee \widehat{P}$. Then by Proposition 9.4.5, for μ-almost every $z \in X$ and for $\mu_z^{\mathcal{A}'}$-a.e. $x \in [z]_{\mathcal{A}'}$ we have $\left(\mu_z^{\mathcal{A}'}\right)_x^{\mathcal{A}} = \mu_x^{\mathcal{A}}$. Using this in the second step and (9.46) in the third step below, we get for an arbitrary $H \in \mathcal{B}$ that

$$\begin{aligned}
\mu_x^{\xi \vee P}(H) &= \mu_x^{\mathcal{A}}(H) = \left[\mu_z^{\mathcal{A}'}\right]_x^{\mathcal{A}}(H) = \left[\mu_z^\xi\right]_x^{[z]_\xi \cap P}(H) = \frac{\mu_z^\xi(H \cap P(x))}{\mu_z^\xi(P(x))} \\
&= \frac{\mu_x^\xi(H \cap P(x))}{\mu_x^\xi(P(x))}.
\end{aligned} \tag{9.48}$$

The last equality holds since we assumed that $x \in [z]_{\mathcal{A}'}$. This implies that $x \in [z]_\xi$, $[z]_\xi = [x]_\xi$ and $\mu_x^\xi = \mu_z^\xi$. Since (9.48) holds for μ-a.e. $z \in X$ and μ_z^ξ-a.e. $x \in [z]_\xi$ we obtain by Theorem 9.4.17 that (9.48) holds for μ-a.e. $x \in X$.

□

9.4.5. A theorem of Simmons. If the partition is defined as the set of fibres (inverse branches) of a measurable map $\phi\colon \Omega \mapsto X$, where X has some metric structure, it would be helpful to be able to derive the family of conditional measures using the metric properties of X. This is not always possible: in order to do so, we have to equip Ω with a metric as well.

THEOREM 9.4.22 (Simmons [**Sim12**]). *Let $(\Omega, \mathcal{A}, \nu)$ be a probability space such that Ω is a compact metric space. Let X be a locally compact separable ultrametric space or a separable Riemannian manifold equipped with the Borel σ-algebra. Let $\phi\colon \Omega \mapsto X$ be measurable and let $\xi = \{\phi^{-1}(\phi(\omega))\}_{\omega \in \Omega}$ be the measurable partition into inverse images of ϕ. Then for ν-almost every ω there exists a measure $\widehat{\nu}_\omega^\xi$ supported on $\xi(\omega)$ such that*

$$\widehat{\nu}_\omega^\xi = \lim_{\delta \to 0} \frac{\nu|_{\phi^{-1}(B(\phi(\omega),\delta))}}{\nu(\phi^{-1}(B(\phi(\omega),\delta)))},$$

where the limit is taken in weak- topology. Moreover, $\widehat{\nu}_\omega^\xi = \nu_\omega^\xi$ for ν-almost every ω, where ν_ω^ξ is the conditional measure from Theorem 9.4.17.*

9.5. Entropy

Entropy of a partition, and of a σ-invariant measure on a symbolic space, was already introduced in Section 1.9.6, where some basic facts were proven. Here we state some additional important properties and results without proof. For the material in this section, see, e.g., the book by Walters [**Wal82**] and Petersen [**Pet83**], or the book by Downarowicz [**Dow11**] for a more recent treatment.

Let $(\Omega, \mathcal{A}, \nu)$ be a Lebesgue space, and let $\alpha = \{A_1, \ldots, A_k\}$, $\xi = \{C_1, \ldots, C_k\}$, $\eta = \{D_1, \ldots D_k,\}$ be measurable partitions of Ω. Note that for finite (or countable) partitions, being measurable simply means that all the atoms are measurable, and there is no need for the complicated definition of the last section, like Definition 9.4.10. Recall that $\xi \vee \eta$ denotes the common refinement of ξ and η, that is, the elements of $\xi \vee \eta$ are subsets of Ω of the form $C_i \cap D_j$. In this section T is always a measure preserving transformation on $(\Omega, \mathcal{A}, \nu)$. The partition $\xi_k^\ell := T^{-k}\xi \vee \cdots \vee T^{-\ell}\xi$ will play an important role. The finite σ-algebras generated by α, ξ and η are denoted by $\widehat{\alpha}, \widehat{\xi}$ and $\widehat{\eta}$ respectively. Whenever $\mathcal{B}, \mathcal{F}_n$ or $\mathcal{F}$ appear in this section, they are always meant to be sub-σ-algebras of $\mathcal{A}$. We write $\nu(C|\mathcal{B}) := \mathbb{E}(\mathbb{1}_C|\mathcal{B})$.

9.5.1. Entropy and conditional entropy of partitions. First we introduce the *information function* and the *entropy* of the partition ξ:

$$\mathbf{I}_\mu(\xi) := -\sum_{i=1}^{k} \mathbb{1}_{C_i} \log \mu(C_i),$$

$$H_\mu(\xi) := \int I_\mu(\xi) d\nu = -\sum_{i=1}^{k} \mu(C_i) \log \mu(C_i).$$

We usually suppress the subscript μ if it does not lead to confusion. The meaning of entropy $H_\mu(\xi)$ is the average information gained when we are told which element of ξ contains a certain point of Ω. The *conditional entropy* of ξ given η is the average information gained when we are told which element of ξ contains a certain point of Ω, given that we know the element of η which contains this point. The *conditional information function* and conditional entropy are defined by

$$\mathbf{I}(\xi|\eta) := -\sum_{i=1}^{k} \mathbb{1}_{C_i} \log \nu(C_i|\widehat{\eta}) \quad \text{and} \quad H(\xi|\widehat{\eta}) := \int \mathbf{I}(\xi|\widehat{\eta}) d\nu. \tag{9.49}$$

This definition extends to any $\mathcal{B}$ sub-σ-algebra of $\mathcal{A}$:

$$\mathbf{I}(\xi|\mathcal{B}) := -\sum_{i=1}^{k} \mathbb{1}_{C_i} \log \nu(C_i|\mathcal{B}) \text{ and } H(\xi|\mathcal{B}) := \int \mathbf{I}(\xi|\mathcal{B}) d\nu. \tag{9.50}$$

Here are some important properties of the conditional entropy:

(a) $H(\xi|\eta) = \sum_{i=1}^{k} \nu(D_i) H(\xi_{D_i})$, where $\xi_{D_i} := \{C_1 \cap D_i, \ldots C_k \cap D_i\}$.

(b) $H(\xi \vee \eta|\alpha) = H(\eta|\alpha) + H(\xi|\alpha \vee \eta)$. Frequently, we use this with the trivial partition with only one class $\alpha = \{\Omega\}$.

(c) If $\xi \leqslant \eta$ then $H(\xi|\alpha) \leqslant H(\eta|\alpha)$.

(d) If $\xi \leqslant \eta$ then $H(\alpha|\xi) \geqslant H(\alpha|\eta)$.

(e) $H(\xi \vee \eta|\alpha) \leqslant H(\xi|\alpha) + H(\eta|\alpha)$.

(f) Let $\mathcal{F}_n$ be an increasing sequence of sub-σ-algebras of $\mathcal{A}$ and assume that $\mathcal{F}_n \uparrow \mathcal{F}$. Then $\mathbf{I}(\xi|\mathcal{F}_n)$ converges both almost everywhere and in L^1 to $\mathbf{I}(\xi|\mathcal{A})$. Hence, $H(\xi|\mathcal{F}_n) \to H(\xi|\mathcal{F})$.

(g) If m is a probability measure on Ω (not necessarily invariant for T), then $I_{m \circ T^{-1}}(\xi|\mathcal{B}) \circ \sigma = I_m(T^{-1}\xi|T^{-1}\mathcal{B})$.

(h) $H(\xi_0^{n-1}) = H(\xi) + \sum_{j=1}^{n-1} H(\xi_1^j)$.

9.5.2. The entropy of a measure preserving transformation. It easily follows from property (g) above that the nonnegative sequence $\frac{1}{n}H\left(\xi_0^{n-1}\right)_{n=1}^{\infty}$ is nonincreasing. Let $h(T,\xi) := \lim_{n\to\infty}\frac{1}{n}H(\xi_0^{n-1})$. Here is a list of some important properties of the entropy function $h(T,\xi)$:

(i) $h(T,\xi) \leqslant H(\xi)$ and $h(T,\xi) \leqslant h(T,\eta) + H(\xi|\eta)$.
(ii) $h(T,\xi \vee \eta) \leqslant h(T,\xi) + h(T,\eta)$.
(iii) If $\xi \leqslant \eta$ then $h(T,\xi) \leqslant h(T,\eta)$.
(iv) $h(T^k,\xi) = kh(T,\xi)$ and $h(T^{-1}) = h(T)$ if T is invertible.
(v) $h(T,\xi) = h(T,T^{-1}\xi) = h(T,\xi_1^{k-1})$ for any $k \geqslant 1$.
(vi) $h(T,\xi) = \lim_{n\to\infty} H(\xi|\xi_1^n) =: H(\xi|\xi_1^\infty)$.

The entropy of T is defined as

$$h_\nu(T) := \sup_{\xi \text{ finite partition}} h(T,\xi). \tag{9.51}$$

This definition is difficult to use, since it involves a supremum over all finite partitions. Fortunately, the following theorem of Kolmogorov and Sinai makes it much more manageable. See, e.g., [**Wal82**, Theorem 4.18] for a proof.

THEOREM 9.5.1 (Kolmogorov, Sinai). *Let T be a measure preserving transformation (but not necessarily invertible) of the probability space $(X,\mathcal{B},\mu)$, and let ξ be a finite generating partition, that is, $\bigvee_{i=0}^{\infty} T^{-i}\widehat{\xi} = \mathcal{B}$ (mod 0). Then $h_\mu(T) = h_\mu(T,\xi)$.*

Our main example of a measure preserving transformation is the left shift σ on the symbolic space $\Sigma = \{1,\ldots,m\}^{\mathbb{N}}$, with a σ-invariant measure μ. Since the partition into 1st order cylinder sets $\{[i] : i \leqslant m\}$ is a generating partition, we obtain the following

COROLLARY 9.5.2. *Let μ be a σ-invariant probability measure on Σ. Then*

$$h_\mu = h_\mu(\sigma) = \lim_{n\to\infty} -\frac{1}{n}\sum_{|\mathbf{i}|=n} \mu([\mathbf{i}])\log\mu([\mathbf{i}]).$$

The following theorem says that for a ν-typical x and n large enough, $\nu(\xi_0^{n-1}(x)) \sim \mathrm{e}^{-nh_\nu}$; its proof may be found, e.g., in [**Par04**, Theorem 7, p. 39] and [**Pet83**, Theorem 2.3].

THEOREM 9.5.3 (Shannon, McMillan, Breiman). *Let T be a measure preserving transformation on the probability space $(\Omega,\mathcal{A},\nu)$ and let ξ be a countable partition with $H(\xi) < \infty$. Let $\mathcal{I}$ be the σ-algebra of T-invariant sets defined in* (9.3). *Let*

$$f := \boldsymbol{I}(\xi|\xi_1^\infty). \tag{9.52}$$

Then

$$\lim_{n\to\infty} -\frac{1}{n}\log\nu\left(\xi_0^{n-1}(x)\right) = \mathbb{E}\left(f|\mathcal{I}\right) \text{ a.e. and in } L^1. \tag{9.53}$$

If the measure ν is ergodic, then the right hand side in (9.53) *is equal to the constant $h_\nu(T)$.*

Specializing to the main case used in the book, we obtain the following.

COROLLARY 9.5.4. *Let μ be an ergodic shift-invariant measure on the symbolic space Σ. Then*

$$\lim_{n\to\infty} -\frac{1}{n}\log \mu([\mathbf{i}|_n]) = h_\mu(\sigma), \quad \textit{for } \mu\textit{-a.e. } \mathbf{i},$$

where $[\mathbf{i}|_n] = [i_1 \dots i_n]$ is the cylinder set.

Indeed, this follows from Theorem 9.5.3 by noting that if ξ is the partition of Σ into the cylinder sets according to the 1st digit, then $\xi_0^{n-1} = \xi \vee \cdots \vee \sigma^{-n+1}\xi$ is the partition of Σ into the cylinder sets according to the first n digits.

CHAPTER 10

Self-affine sets and measures

In this chapter we start an in-depth investigation of self-affine IFS. This theory uses many tools and concepts from Ergodic Theory. In the previous chapter a brief introduction to the needed material was provided, with references to the books where the proofs may be found.

In Section 1.10 we considered self-affine IFS $\left(f_i(\mathbf{x}) = A_i\mathbf{x} + \mathbf{t}_i\right)_{i=1}^m$ on $\mathbb{R}^d$, with A_i linear contractions and $\mathbf{t}_i \in \mathbb{R}^d$ for all i. Theorem 1.10.4 asserts that if all contractions are smaller than $1/2$, then for almost all translates $\mathbf{t}_i$, the Hausdorff dimension of the attractor is the minimum of d and the affinity dimension $s(A_1, \ldots, A_m)$ defined in (1.127). (This was proved by Falconer [**Fal88**] under the stronger requirement $\|A_i\| < 1/3$ and improved to its present form by Solomyak [**Sol98**].) The scope of this theorem was extended by Jordan, Pollicott, and Simon [**JPS07**] using the subadditive pressure of Falconer [**Fal94**] and Barreira [**Bar96**]. In this chapter we prove that if the so-called *self-affine transversality condition* [**JPS07**] holds for a parametrized family of self-affine IFS, then for almost all parameters the Hausdorff dimension of the attractor is the minimum of the affinity dimension and d. If the affinity dimension is greater than d, then the d-dimensional Lebesgue measure of the attractor is positive. Moreover, the dimension of an ergodic stationary measure for the IFS is equal to the *Lyapunov dimension* of the measure. (All of these statements are claimed to hold for almost all parameters.) The notion of Lyapunov dimension of a self-affine measure is the natural generalization of the ratio—*entropy over the Lyapunov exponent*—which gives the dimension of a self-similar measure under strong separation, see Corollary 1.9.11.

The second part of the chapter is concerned with thermodynamic formalism, which turned out to be crucial in the study of self-affine IFS. Particularly important was the work of Käenmäki [**K04**], who proved the existence of the *equilibrium measure* for the singular value function, now called the *Käenmäki measure.* We do not prove the existence of such measures in full generality, but illustrate the result by showing it for 2×2 irreducible matrices, following Feng [**Fen09**], Feng and Lau [**FL02**] and Feng and Käenmäki [**FK11**]. We present a number of properties of Käenmäki measures and their applications, some with proofs and some with partial proofs or referring to the original papers. This is a currently active area, and we also discuss some more recent developments, including the notion of *domination.*

10.1. Introduction

Consider a self-affine IFS on $\mathbb{R}^d$:

$$\mathcal{F} = \left(f_i(\mathbf{x}) = A_i\mathbf{x} + \mathbf{t}_i\right)_{i=1}^m, \qquad \mathbf{x}, \mathbf{t}_i \in \mathbb{R}^d,\ \|A_i\| < 1. \tag{10.1}$$

Let $\Pi : \Sigma \to \mathbb{R}^d$ be the natural projection map associated with the IFS, where $\Sigma = \{1, \ldots, m\}^{\mathbb{N}}$ is the symbolic space. Given μ, an ergodic shift-invariant measure on Σ, we consider the push-forward measure $\nu = \Pi_*\mu$, called the *stationary measure* for the IFS $\mathcal{F}$ corresponding to μ. (The stationary measure is self-affine when $\mu = \mathbf{p}^{\mathbb{N}}$ is a Bernoulli measure.) We next define the *Lyapunov dimension* of ν denoted by $D(\nu)$, and sometimes by $D(\mu)$, when the linear maps of the IFS are fixed. Consider the map $C : \Sigma \to GL(d, \mathbb{R})$ given by

$$C(\mathbf{i}) = A_{i_1}^*.$$

According to Oseledeč Theorem 9.2.1 and Corollary 9.2.3, the Lyapunov exponents of the measure μ for the function C are well defined; denote them by

$$0 < \chi_1(\mu) \leqslant \chi_2(\mu) \leqslant \cdots \leqslant \chi_d(\mu).$$

Note that $\chi_k(\mu) = -\widetilde{\chi}_k(\mu)$, where $\widetilde{\chi}_k(\mu)$ is defined in Corollary 9.2.3. In particular,

$$\chi_k(\mu) = \lim_{n\to\infty} \frac{-1}{n} \log \alpha_k(A_{i_1} \cdot \ldots \cdot A_{i_n}), \quad \text{for } \mu\text{-a.e. } \mathbf{i} \in \Sigma \text{ and } k = 1, \ldots, d. \tag{10.2}$$

Let $h_\mu = h_\mu(\sigma)$ be the entropy of μ.

DEFINITION 10.1.1 (Lyapunov dimension $D(\nu)$ for $\nu = \Pi_*\mu$).

(a) If $k := k(\mu) = \max\{i : h_\mu - \chi_1(\mu) - \cdots - \chi_i(\mu) > 0\} \leqslant d$, then we let

$$D(\nu) = D(\mu) := k + \frac{h_\mu - \chi_1(\mu) - \cdots - \chi_k(\mu)}{\chi_{k+1}(\mu)};$$

(b) If $h_\mu - \chi_1(\mu) - \cdots - \chi_d(\mu) > 0$, then we let

$$D(\nu) = D(\mu) := d \cdot \frac{h_\mu}{\chi_1(\mu) + \cdots + \chi_d(\mu)}.$$

A simple consequence of the definition is the inequality

$$D(\mu) \leqslant \frac{h_\mu}{\chi_1(\mu)}. \tag{10.3}$$

Note that the Lyapunov dimension depends on the measure μ and on the linear maps of the IFS A_i, $i \leqslant m$, but not on the translations $\mathbf{t}_i$. See Figure 10.1 for a visual representation of the formula for $D(\nu) = D(\mu)$.

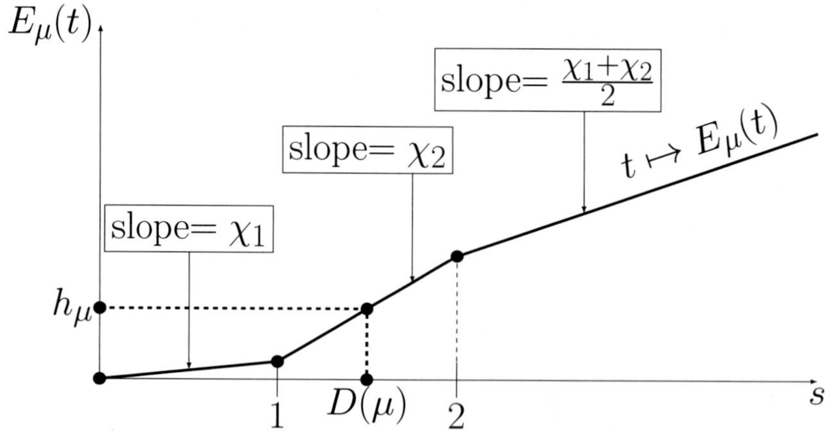

FIGURE 10.1. Definition of the Lyapunov dimension. The precise meaning of the function $E_\mu(t)$ will be given in Definition 10.5.1.

Recall that $s(A_1, \dots, A_m)$ denotes the *affinity dimension* of the IFS, see Definition 1.10.2. Now we can state the main theorem of the first part of this chapter.

THEOREM 10.1.2. *Consider the family of IFS parameterized by translations (the family $\mathcal{F}^{\mathbf{t}}$ in Theorem 1.10.4). That is, for $\mathbf{t} := (\mathbf{t}_1, \dots, \mathbf{t}_m) \in \mathbb{R}^d \times \cdots \times \mathbb{R}^d$ we consider the self-affine IFS acting on $\mathbb{R}^d$:*

$$\mathcal{F}^{\mathbf{t}} := (A_i \cdot \mathbf{x} + \mathbf{t}_i)_{i=1}^m . \tag{10.4}$$

Let $r_i := \|A_i\|$. Our principal assumption is that

$$\max_{i \neq j} \{r_i + r_j\} < 1. \tag{10.5}$$

Let $\Pi^{\mathbf{t}}$ be the natural projection, that is,

$$\Pi^{\mathbf{t}}(\mathbf{i}) := \lim_{n \to \infty} \left(\mathbf{t}_{i_1} + \sum_{k=2}^{n} A_{i_1} \cdots A_{i_{k-1}} \cdot \mathbf{t}_{i_k} \right). \tag{10.6}$$

Consider $\Lambda^{\mathbf{t}} = \Pi^{\mathbf{t}}(\Sigma)$, the attractor of the IFS. For an ergodic shift-invariant measure μ on the symbolic space Σ, let $\nu_{\mathbf{t}} = \Pi^{\mathbf{t}}_* \mu$ be the corresponding stationary measure for the IFS. Then for $\mathcal{L}^{md}$-a.e. $\mathbf{t} \in \mathbb{R}^{md}$ the following hold:

(i) $\dim_{\mathrm{H}}(\nu_{\mathbf{t}}) = \min\{D(\nu_{\mathbf{t}}), d\} = \min\{D(\mu), d\}$;
(ii) *If* $D(\nu_{\mathbf{t}}) = D(\mu) > d$ *then* $\nu_{\mathbf{t}} \ll \mathcal{L}^d$;
(iii) $\dim_{\mathrm{H}}(\Lambda^{\mathbf{t}}) = \dim_{\mathrm{B}}(\Lambda^{\mathbf{t}}) = \min\{d, s(A_1, \dots, A_m)\}$.
(iv) *If* $s(A_1, \dots, A_m) > d$ *then* $\mathcal{L}^d(\Lambda^{\mathbf{t}}) > 0$.

Actually, these results are slightly more general than those in the papers [**Sol98**] and [**JPS07**]. Namely, in these papers the stronger condition: $\max_i r_i = \max_i \|A_i\| < \frac{1}{2}$ was assumed instead of (10.5).

It is widely believed that the following holds:

CONJECTURE 10.1.3. *For a typical (in an appropriate sense) self-affine IFS the Hausdorff dimension of the limit set is equal to* $\min\{d, s(A_1, \dots, A_m)\}$.

There has been important progress towards proving this conjecture in special cases, which we will describe later in the book; however, we are still far from resolving it (and not even sure how to state it precisely). On the other hand, a result in this direction for a random perturbation of a given self-affine attractor Λ was obtained in [**JPS07**]. Fix the matrices $A_1, \dots, A_m$, about which we assume only that they are invertible and their norms are less than one, that is, we do not need assumption (10.5). We construct the randomly perturbed attractor using the formula (1.36) in such a way that with each application of the functions $f_i : \mathbb{R}^d \to \mathbb{R}^d$ from the IFS

$$\mathcal{F} := \left(f_i(\mathbf{x}) = A_i \cdot \mathbf{x} + \mathbf{t}_i\right)_{i=1}^m \tag{10.7}$$

we make a random additive error Y independently for each i and at each step. Assume that Y has an absolutely continuous distribution η with a bounded density supported on a ball B centered at the origin, which can be chosen arbitrarily small (or arbitrarily large). The errors are "coded" by $\mathbf{y} \in B^{\mathcal{T}}$, where $\mathcal{T}$ is the regular m-adic tree having m^n nodes on the nth level. The space $B^{\mathcal{T}}$ is equipped with the probability measure $\mathbb{P} = Y^{\mathcal{T}}$. For each $\mathbf{y} \in B^{\mathcal{T}}$ we get a kind of "natural projection" $\Pi^{\mathbf{y}} : \Sigma \to \mathbb{R}^d$ and the attractor $\Lambda^{\mathbf{y}} = \Pi^{\mathbf{y}}(\Sigma)$, which is no longer self-affine, but rather "statistically self-affine". For a rigorous formulation of the model, see [**JPS07**].

THEOREM 10.1.4 ([**JPS07**]). *Consider the random perturbation of a self-affine IFS* (10.7) *as described above, where the matrices* $A_1, \ldots, A_m$ *are fixed and satisfy* $\|A_i\| < 1$ *for all* i. *Let* μ *be an ergodic shift-invariant measure on* Σ. *For* $\mathbb{P}$*-a.e.* $\mathbf{y} \in B^{\mathcal{T}}$ *the following hold:*

(i) *If* $s(A_1, \ldots, A_m) \leqslant d$, *then* $\dim_{\mathrm{H}}(\Lambda^{\mathbf{y}}) = s(A_1, \ldots, A_m)$;
(ii) *If* $s(A_1, \ldots, A_m) > d$, *then* $\mathcal{L}^d(\Lambda^{\mathbf{y}}) > 0$;
(iii) $\dim_{\mathrm{H}}(\Pi^{\mathbf{y}}_* \mu) = \min\{d, D(\mu)\}$;
(iv) *if* $D(\mu) > d$, *then* $\Pi^{\mathbf{y}}_* \mu \ll \mathcal{L}^d$.

REMARK 10.1.5.

(a) There is a closely related result by Peres, Simon and Solomyak in [**PSS06**] for self-similar IFS with random multiplicative errors.

(b) In Theorem 10.1.4 it was important that the errors are sampled from a compactly supported distribution. Jordan and Jurga [**JJ14**] studied a different randomly perturbed self-affine model, with noncompactly supported errors.

(c) Instead of random IFS with errors coming from an i.i.d. process, it is sometimes important to consider a generalization when the random errors are generated by an arbitrary ergodic system. Urbański [**Urb16**] obtained formulas for the Hausdorff dimension of attractors in such a setting, for (possibly infinite) conformal IFS on the line, satisfying a transversality condition.

10.2. Singular value function

Let T be a $d \times d$ matrix with real entries. Recall (1.123) that the singular value function $\varphi^t(T)$ of T is defined by

$$\varphi^t(T) := \begin{cases} \alpha_1 \cdots \alpha_{k-1} \alpha_k^{t-(k-1)}, & \text{if } k-1 < t \leqslant k \leqslant d; \\ (\alpha_1 \cdots \alpha_d)^{t/d}, & \text{if } t > d, \end{cases} \tag{10.8}$$

where $\alpha_1 \geqslant \cdots \geqslant \alpha_d$ are the singular values of T (see Section A.1 for the definition). It is clear that for every $h > 0$ we have

$$\alpha_1^h \geqslant \frac{\varphi^{t+h}(T)}{\varphi^t(T)} \geqslant \alpha_d^h. \tag{10.9}$$

An easy calculation shows that the singular value function is sub-multiplicative: for each $t \geqslant 0$,

$$\varphi^t(T_1 T_2) \leqslant \varphi^t(T_1) \varphi^t(T_2). \tag{10.10}$$

Indeed, the sub-multiplicativity is clear for $t > d$. For $t \leqslant d$, using the property (A.6) of the norm of the exterior products, one can see that

$$\varphi^t(T) = \|T^{\wedge k}\|^{t-(k-1)} \|T^{\wedge (k-1)}\|^{k-t}, \tag{10.11}$$

where k is the unique positive integer such that $k - 1 < t \leqslant k$. Then (10.10) follows by sub-multiplicativity of the matrix norms, see Section A.2.

We write Σ^* for the set of finite words in the alphabet $\{1, \ldots, m\}$, that is, $\Sigma^* = \bigcup_{n=0}^{\infty} \{1, \ldots, m\}^n$.

For a self-affine IFS $\mathcal{F}$ of the form (10.1) choose $0 < a_{\mathcal{F}} < b_{\mathcal{F}} < 1$ such that for all $i = 1, \ldots, m$,

$$1 > b_{\mathcal{F}} \geqslant \alpha_1(A_i) \geqslant \cdots \geqslant \alpha_d(A_i) \geqslant a_{\mathcal{F}} > 0. \tag{10.12}$$

Using that $\alpha_1(A_{\mathbf{i}}) = \|A_{\mathbf{i}}\|$ and $\alpha_d(A_{\mathbf{i}}) = \|A_{\mathbf{i}}^{-1}\|^{-1}$ for every $\mathbf{i} \in \Sigma^*$, we obtain

$$b_{\mathcal{F}}^{|\mathbf{i}|} \geqslant \alpha_1(A_{\mathbf{i}}) \geqslant \cdots \geqslant \alpha_d(A_{\mathbf{i}}) \geqslant a_{\mathcal{F}}^{|\mathbf{i}|}, \tag{10.13}$$

hence

$$b_{\mathcal{F}}^{t|\mathbf{i}|} \geqslant \varphi^t(A_{\mathbf{i}}) \geqslant a_{\mathcal{F}}^{t|\mathbf{i}|}. \tag{10.14}$$

10.2.1. Connection between singular values and Lyapunov exponents. For a finite word $\mathbf{i} = (i_1, \ldots, i_n) \in \Sigma^*$ we write $\overleftarrow{\mathbf{i}} := (i_n, \ldots, i_1)$. The following lemma asserts that when computing the singular values, we can switch the order of the matrix product at the price of passing to the transposes:

LEMMA 10.2.1. *For every $\mathbf{i} \in \Sigma^*$ and for every $1 \leqslant k \leqslant d$ we have*

$$\alpha_k(A_{\mathbf{i}}) = \alpha_k\left(A^*_{\overleftarrow{\mathbf{i}}}\right). \tag{10.15}$$

PROOF. In view of

$$(A_{i_1} \cdots A_{i_n})^* \cdot (A_{i_1} \cdots A_{i_n}) = \left(A^*_{i_n} \cdots A^*_{i_1}\right) \cdot \left(A^*_{i_n} \cdots A^*_{i_1}\right)^*,$$

it suffices to prove that for any $d \times d$ matrix B, the matrices BB^* and B^*B have the same eigenvalues with the same multiplicities (apply this to $B = A^*_{i_{n-1}} \cdots A^*_{i_0}$.) However, this immediately follows from the singular value decomposition, see Section A.1. ☐

LEMMA 10.2.2. *Let μ be an arbitrary ergodic σ-invariant measure on the symbolic space Σ. We write $0 < \chi_1(\mu) \leqslant \cdots \leqslant \chi_d(\mu)$ for the Lyapunov exponents of the measure μ for the matrix cocycle $C(\mathbf{i}) := A^*_{i_1}$ (see Theorem 9.2.1 and Definition 9.2.2). Then for all $1 \leqslant k \leqslant d$ and μ-almost all $\mathbf{i} \in \Sigma$ we have*

$$\lim_{n\to\infty} \frac{-1}{n} \log \alpha_k(A_{\mathbf{i}}) = \chi_k(\mu). \tag{10.16}$$

This is immediate from Lemma 10.2.1 and Corollary 9.2.3.

Recall that the affinity dimension is defined by

$$s(A_1, \ldots, A_m) = \inf\Big\{t > 0 : \sum_{n=0}^{\infty} \sum_{|\mathbf{i}|=n} \varphi^t(A_{\mathbf{i}}) < \infty\Big\}. \tag{10.17}$$

10.2.2. Affinity dimension and finite measures. Consider the $d \times d$ nonsingular matrices $A_1, \ldots, A_m$ with $\|A_i\| < 1$.

For every $\ell > 0$ and $s > 0$ we define a set function $\mathcal{N}^s_\ell$ on the subsets of Σ by

$$\mathcal{N}^s_\ell(E) := \inf\Big\{\sum_k \varphi^s(A_{\mathbf{i}_k}) : \mathbf{i}_k \in \Sigma^*, E \subset \textstyle\bigcup_k [\mathbf{i}_k] \text{ and } |\mathbf{i}_k| \geqslant \ell\Big\},$$

where $[\mathbf{i}_k]$ is the cylinder set of $\mathbf{i} \in \Sigma$ starting with $\mathbf{i}_k$. Then the set function

$$E \mapsto \mathcal{N}^s(E) := \lim_{\ell\to\infty} \mathcal{N}^s_\ell(E) = \sup_\ell \mathcal{N}^s_\ell(E)$$

defines a regular Borel measure of Hausdorff type (called "net measure" by Falconer [**Fal88**]). This is a special case of the classical Carathéodory construction, see e.g. [**Mat95**, Theorem 4.2]. Further, we obtain from (10.9) and (10.13) that if $t < s$ and $\mathcal{N}^t(\Sigma) < \infty$ then $\mathcal{N}^s(\Sigma) = 0$. This implies that we can define $\tilde{s}(A_1, \ldots, A_m)$ by

$$\tilde{s}(A_1, \ldots, A_m) = \inf\{s : \mathcal{N}^s(\Sigma) = 0\} = \sup\{s : \mathcal{N}^s(\Sigma) = \infty\}. \tag{10.18}$$

LEMMA 10.2.3. $\widetilde{s}(A_1,\dots,A_m) = s(A_1,\dots,A_m)$.

PROOF. Fix the matrices $A_1,\dots,A_m$ and write s and $\widetilde{s}$ instead of $s(A_1,\dots,A_m)$ and $\widetilde{s}(A_1,\dots,A_n)$. It is obvious that $\widetilde{s} \leqslant s$. The opposite inequality can be proved, following [**Fal88**, p. 345], as follows: Let $t > \widetilde{s}$. Then there exists a finite set of words $\mathcal{A} \subset \Sigma^*$, such that $\{[\mathbf{i}] : \mathbf{i} \in \mathcal{A}\}$ is a cover of Σ and $\sum_{\mathbf{i}\in\mathcal{A}} \varphi^t(A_{\mathbf{i}}) < 1$. Let $p := \max_{\mathbf{i}\in\mathcal{A}} |\mathbf{i}|$. For $r > p$ let

$$\mathcal{A}_r := \{\mathbf{i}_1 \dots \mathbf{i}_q : \mathbf{i}_j \in \mathcal{A},\ |\mathbf{i}_1 \dots \mathbf{i}_q| \geqslant r \text{ but } |\mathbf{i}_1 \dots \mathbf{i}_{q-1}| < r\}.$$

Then $\{[\mathbf{i}] : \mathbf{i} \in \mathcal{A}_r\}$ is also a cover of Σ, and one can easily deduce from the sub-multiplicativity of φ^t and the fact that $\mathcal{A}_r$ is a minimal cut-set that

$$\sum_{\mathbf{i}\in\mathcal{A}_r} \varphi^t(A_{\mathbf{i}}) < 1. \tag{10.19}$$

Indeed, this follows, applying repeatedly that for every word $\mathbf{j} \in \Sigma^*$ we have $\sum_{\mathbf{i}\in\mathcal{A}} \varphi^t(A_{\mathbf{ji}}) \leqslant \sum_{\mathbf{i}\in\mathcal{A}} \varphi^t(A_{\mathbf{j}})\varphi^t(A_{\mathbf{i}}) < \varphi^t(A_{\mathbf{j}})$.

In view of (10.14), we have $\varphi^t(A_{\mathbf{i}}) \leqslant b_{\mathcal{F}}^{|\mathbf{i}|t} < 1$. By (10.10), if $\mathbf{j}'$ is a prefix of $\mathbf{j}$, i.e. $\mathbf{j} = \mathbf{j}'\mathbf{i}$ for some $\mathbf{i} \in \Sigma^*$, then $\varphi^t(A_{\mathbf{j}}) \leqslant \varphi^t(A_{\mathbf{j}'})\varphi^t(A_{\mathbf{i}}) \leqslant \varphi^t(A_{\mathbf{j}'})$. Thus

$$\sum_{|\mathbf{i}|=r+p} \varphi^t(A_{\mathbf{i}}) \leqslant m^p \sum_{\mathbf{j}\in\mathcal{A}_r} \varphi^t(A_{\mathbf{j}}) \leqslant m^p.$$

Hence,

$$\lim_{k\to\infty} \left(\sum_{|\mathbf{i}|=k} \varphi^t(A_{\mathbf{i}}) \right)^{1/k} \leqslant \lim_{k\to\infty} m^{\frac{p}{k}} = 1. \tag{10.20}$$

The inequality (10.9) yields that for every ℓ,

$$a_{\mathcal{F}}^{\ell h} \leqslant \frac{\sum_{|\mathbf{i}|=\ell} \varphi^{s+h}(A_{\mathbf{i}})}{\sum_{|\mathbf{i}|=\ell} \varphi^s(A_{\mathbf{i}})} \leqslant b_{\mathcal{F}}^{\ell h}.$$

So, the function $g(s) := \lim_{\ell\to\infty} \left(\sum_{|\mathbf{i}|=\ell} \varphi^s(A_{\mathbf{i}}) \right)^{1/\ell}$ is continuous and strictly decreasing. We know from (10.20) that $g(t) \leqslant 1$. Hence, $g(t_1) < 1$ for every $t < t_1$. This implies by the root test that $\sum_{\ell=1}^{\infty} \sum_{|\mathbf{i}|=\ell} \varphi^{t_1}(A_{\mathbf{i}}) < \infty$, that is, $s < t_1$. Using that $\widetilde{s} < t < t_1$ were arbitrary, we obtained that $s \leqslant \widetilde{s}$. □

The next lemma allows us to find a finite measure whose properties are closely related to the affinity dimension.

LEMMA 10.2.4. *If $\mathcal{N}^s(\Sigma) > 0$ for some $s > 0$, then there exists a finite positive measure μ supported on Σ and a constant c_0 such that*

$$\mu([\mathbf{i}]) \leqslant c_0 \cdot \varphi^s(A_{\mathbf{i}}) \textit{ for every } \mathbf{i} \in \Sigma^*.$$

PROOF. Fix an $s > 0$ such that $\mathcal{N}^s(\Sigma) > 0$. It follows from [**Rog70**, Theorem 54] that there exists a compact set $\Sigma' \subset \Sigma$ such that $0 < \mathcal{N}^s(\Sigma') < \infty$. We define a metric ρ on Σ' as follows:

$$\rho(\mathbf{i},\mathbf{j}) := \begin{cases} \varphi^s(A_{\mathbf{i}\wedge\mathbf{j}}), & \text{if } \mathbf{i} \neq \mathbf{j}; \\ 0, & \text{if } \mathbf{i} = \mathbf{j}. \end{cases}$$

We can see that ρ is a metric since for every $\omega \in \Sigma^*$ we have

$$\varphi^s(A_{\mathbf{i}\wedge\mathbf{j}}) \leqslant \max\{\varphi^s(A_{\mathbf{i}\wedge\omega}), \varphi^s(A_{\omega\wedge\mathbf{j}})\}.$$

Then it follows from [**Rog70**, Theorem 53] that the measure $\mathcal{N}^s$ is the 1-dimensional Hausdorff measure on the compact set Σ'. It remains to apply Frostman's Lemma in compact metric spaces, see Theorem 1.4.13(ii). □

REMARK 10.2.5. Lemma 10.2.4 can be proved more directly, using the method of *flows on trees* from Bishop and Peres [**BP17**, Section 3.1]. Recall the set-up from Section 1.3.1. Consider the set of words Σ^* as an m-regular tree, with the empty word being the root vertex. A word $\mathbf{i} = i_1 \dots i_n$ is identified with a vertex at level n. For a vertex $\mathbf{i} = i_1 \dots i_n$, with $n \geqslant 1$, we denote by $\mathbf{i}^- = i_1 \dots i_{n-1}$ its "parent" vertex. A *flow* on the tree is a nonnegative function f on the set of edges of the m-regular tree, satisfying the "inflow=outflow" property: $f(\mathbf{i}^-,\mathbf{i}) = \sum_{\mathbf{j}:\, \mathbf{i}=\mathbf{j}^-} f(\mathbf{i},\mathbf{j})$. The *norm* of the flow $\|f\|$ is defined as the sum of the flows from the root vertex to its "children". The set of infinite (nonbacktracking) paths in the tree from the root may be identified with the symbolic space Σ, and then flows correspond to nonnegative Borel measures on Σ. A *cut-set* is a set of edges that intersects every infinite path from the root, and a *minimal cut-set* is a cut-set which is minimal under the inclusion relation. Minimal cut-sets are finite by compactness; they correspond to disjoint open covers of Σ. The *norm* of the flow $\|f\|$ is defined as the sum of the flows from the root vertex to its "children". It is easy to see that $\|f\|$ is equal to $\sum_{\mathbf{e}\in\mathcal{P}} f(\mathbf{e})$ for any minimal cut-set $\mathcal{P}$.

For an edge $\mathbf{e} = (\mathbf{i}^-,\mathbf{i})$ assign a *conductance* $C(\mathbf{e}) = \varphi^s(A_{\mathbf{i}})$. A *legal flow* on the tree is a flow, such that $f(\mathbf{e}) \leqslant C(\mathbf{e})$ for all edges $\mathbf{e}$. Thus, a legal flow corresponds to a finite Borel measure μ on Σ satisfying $\mu([\mathbf{i}]) \leqslant \varphi^s(A_{\mathbf{i}})$, and so we need to show that there exists a *nonzero* legal flow. As a corollary of the MaxFlow-MinCut Theorem of Ford and Fulkerson on trees, for which [**BP17**, Theorem 3.1.5] provides a short proof, one can obtain that

$$\max_{\text{legal flows}} \|f\| = \inf_{\text{cut-sets}} C(\mathcal{P}), \quad \text{where } C(\mathcal{P}) := \sum_{\mathbf{e}\in\mathcal{P}} C(\mathbf{e}). \tag{10.21}$$

Our choice of conductance yields $\inf_{\text{cut-sets}} C(\mathcal{P}) = \inf_\ell \mathcal{N}^s_\ell(\Sigma)$, which is positive by our assumption $\mathcal{N}^s(\Sigma) > 0$. This implies that the maximizing measure in (10.21) satisfies the desired property.

The measure constructed in Lemma 10.2.4 is used to give a lower bound for the Hausdorff dimension of the self-affine attractor. Note that this measure is not shift-invariant. In Section 10.5, following Käenmäki [**K04**], we will define, and construct in a special case, a measure which has similar properties but is shift-invariant and ergodic under certain assumptions.

10.2.3. Upper bound for the box dimension. Now we return to pay one of our "debts", namely, we show that the upper box dimension of a self-affine set is bounded above by the affinity dimension. The argument is basically a small modification of the one in Section 1.10.

PROPOSITION 10.2.6. *Let $\mathcal{F} = (A_i\mathbf{x} + \mathbf{t}_i)_{i=1}^m$ be a self-affine IFS in $\mathbb{R}^d$ with attractor Λ. Then*

$$\overline{\dim}_{\mathrm{B}}(\Lambda) \leqslant \min\{d, s(A_1,\dots,A_m)\}. \tag{10.22}$$

PROOF. First, we may assume without loss of generality that $B = B(0,1)$ is mapped strictly into itself by all the maps of the IFS $\mathcal{F}$. We can also assume that $s(A_1,\dots,A_m) < d$. Let $t > s(A_1,\dots,A_m)$ be arbitrary noninteger and let $k = \lceil t \rceil$. Then by the definition (10.17), there exists K such that $\sum_{|\mathbf{i}|=K} \varphi^t(A_{\mathbf{i}}) < 1$. Using the Kth iterates we will construct a Moran-like cover of the set $\Lambda \subset B$. Having K fixed, we let, for $\delta > 0$ sufficiently small,

$$\widehat{\mathcal{M}}_\delta := \left\{ \mathbf{i} \in \bigcup_{n=0}^{\infty} \Sigma_{nK} : \alpha_k(A_{\mathbf{i}}) \leqslant \delta < \alpha_k(A_{\mathbf{i}_{-K}}) \right\},$$

where $\mathbf{i}_{-K} = (i_1,\dots,i_{(n-1)K})$ for $\mathbf{i} = (i_1,\dots,i_{nK})$.

Just as on page 51, for every $\mathbf{i} \in \widehat{\mathcal{M}}_\delta$ let $R_{\mathbf{i}}$ be the parallelepiped containing the ellipsoid $f_{\mathbf{i}}(B)$, with the sides parallel to the axes of lengths $2\alpha_1(A_{\mathbf{i}}),\dots,2\alpha_d(A_{\mathbf{i}})$. Then we cover $R_{\mathbf{i}}$ with $n_k(\mathbf{i})$-many d-dimensional cubes of side length 2δ, where

$$n_k(\mathbf{i}) = \prod_{\ell=1}^{k-1} \left\lceil \frac{\alpha_\ell(A_{\mathbf{i}})}{\delta} \right\rceil.$$

Hence, the minimal number of sets of diameter approximately δ needed to cover Λ can be estimated by

$$N_{2\delta\sqrt{d}}(\Lambda) \leqslant \sum_{\mathbf{i}\in\widehat{\mathcal{M}}_\delta} n_k(\mathbf{i}) \leqslant 2^k\delta^{-t} \sum_{\mathbf{i}\in\widehat{\mathcal{M}}_\delta} \delta^{t-(k-1)} \prod_{\ell=1}^{k-1} \alpha_\ell(A_{\mathbf{i}}) \leqslant C_K 2^k \delta^{-t} \sum_{\mathbf{i}\in\widehat{\mathcal{M}}_\delta} \varphi^t(A_{\mathbf{i}}),$$

where $C_K = \max_{\mathbf{i}\in\Sigma_K} \left(\left\| \left(A_{\mathbf{i}}^{\wedge(k-1)}\right)^{-1} \right\| \cdot \|A_{\mathbf{i}}^{\wedge k}\| \right)^{t-(k-1)}$. Here we first used that $\lceil x \rceil \leqslant x + 1 \leqslant 2x$ for every $x \geqslant 1$, then

$$\delta < \alpha_k(A_{\mathbf{i}_{-K}}) = \alpha_k(A_{\mathbf{i}} A^{-1}_{i_{(K-1)n+1}\cdots i_{Kn}}),$$

(A.8), and the inequality

$$\alpha_k(AB) = \frac{\|A^{\wedge k}B^{\wedge k}\|}{\|A^{\wedge(k-1)}B^{\wedge(k-1)}\|} \leqslant \alpha_k(A)\|B^{\wedge k}\|\|(B^{\wedge(k-1)})^{-1}\|,$$

and finally, (10.11). To conclude, as in (10.19), we have $\sum_{\mathbf{i}\in\widehat{\mathcal{M}}_\delta} \varphi^t(A_{\mathbf{i}}) < 1$, since $\widehat{\mathcal{M}}_\delta$ is a minimal cut-set, and the claim follows. □

10.3. Self-affine Generalized Projection Scheme

In Section 6.6, a generalized projection scheme was presented, which is applicable to self-similar IFS and after small modifications, to self-conformal IFS (see Chapter 14). Following [**JPS07**], in this section we construct an analogous scheme suitable for self-affine IFS and their random perturbations.

For $\mathbf{i}, \mathbf{j} \in \Sigma$, we use the notation $\alpha_k(\mathbf{i} \wedge \mathbf{j}) := \alpha_k(A_{\mathbf{i}\wedge\mathbf{j}})$ for $1 \leqslant k \leqslant d$, and let $\alpha_0(\mathbf{i} \wedge \mathbf{j}) := \infty$, $\alpha_{d+1}(\mathbf{i} \wedge \mathbf{j}) := 0$. For $1 \leqslant \ell \leqslant d$ denote

$$J_\ell := [\alpha_{\ell+1}(\mathbf{i} \wedge \mathbf{j}), \alpha_\ell(\mathbf{i} \wedge \mathbf{j})]. \tag{10.23}$$

Define a function $Z_{\mathbf{i}\wedge\mathbf{j}} : [0, \infty) \to [0, 1]$ (see Figure 10.2) by

$$\begin{aligned} Z_{\mathbf{i}\wedge\mathbf{j}}(\rho) &:= \prod_{k=1}^{d} \frac{\min\{\rho, \alpha_k(\mathbf{i} \wedge \mathbf{j})\}}{\alpha_k(\mathbf{i} \wedge \mathbf{j})} \\ &= \sum_{\ell=1}^{d} \frac{\rho^\ell}{\alpha_1(\mathbf{i} \wedge \mathbf{j}) \cdots \alpha_\ell(\mathbf{i} \wedge \mathbf{j})} \cdot \mathbb{1}_{J_\ell} + \mathbb{1}_{J_0}. \end{aligned} \tag{10.24}$$

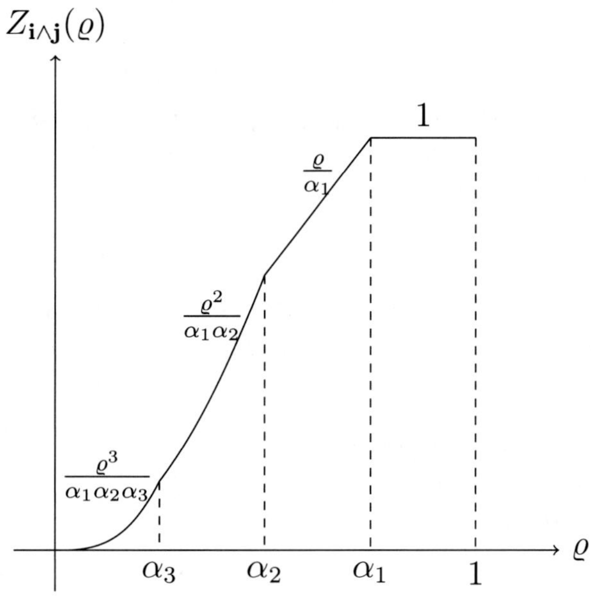

FIGURE 10.2. Definition of the function Z

In particular, if all the maps of the IFS $\mathcal{F}$ are similarities with ratios $\{r_1, \ldots, r_m\}$ then $Z_{\mathbf{i}\wedge\mathbf{j}}(\rho) = \min\left\{1, \frac{\rho^d}{r^d_{\mathbf{i}\wedge\mathbf{j}}}\right\}$. Throughout this section we always use the following:

BASIC ASSUMPTIONS AND NOTATIONS. Let U be a compact metric space (parameter space) with a finite Borel measure $\mathcal{M}$. Consider a contractive IFS $\mathcal{F}$ of the form (10.1), where the translations $\mathbf{t}_i$ depend on parameter $u \in U$ and may differ from step to step. Assume that we are also given a continuous map $\Pi : U \times \Sigma \to \mathbb{R}^d$. For any $u \in U$ and $\mathbf{i} \in \Sigma$ we write $\Pi^u := \Pi(u, \mathbf{i})$. For example in the set-up of Theorem 10.1.2 the parameter space U can be any closed ball and $\mathcal{M} = \mathcal{L}^d|_U$ and $\Pi^u = \Pi^{\mathbf{t}}$ is the natural projection from (10.6). We need the following two definitions.

DEFINITION 10.3.1 (**Self-affine Hölder condition**). We say that the self-affine Hölder condition holds if there exists a constant $K > 0$ such that for every

$u \in U$ and $\mathbf{i} \in \Omega$ and for every $n \in \mathbb{N}$ we can find an isometry $G = G(\mathbf{i}, n) : \mathbb{R}^d \to \mathbb{R}^d$ with

$$\Pi^u([\mathbf{i}|_n]) \subset K \cdot G\left([0, \alpha_1(\mathbf{i}|_n)] \times \cdots \times [0, \alpha_d(\mathbf{i}|_n)]\right), \tag{10.25}$$

where $\mathbf{i}|_n$ denotes the initial part of the sequence $\mathbf{i}$ of length n.

This condition is used only for upper bounds on the Hausdorff dimension and usually it is easy to check. For instance, it holds automatically in the set-up of Theorem 10.1.2.

DEFINITION 10.3.2 (**Self-affine transversality condition**). We say that the self-affine transversality condition holds if there exists a constant $C > 0$ (independent of $\mathbf{i}, \mathbf{j}$) such that for all $\mathbf{i}, \mathbf{j} \in \Sigma$, $\mathbf{i} \neq \mathbf{j}$, for all $\rho > 0$ we have

$$\mathcal{M}\big(\{u \in U : |\Pi^u(\mathbf{i}) - \Pi^u(\mathbf{j})| < \rho\}\big) < C \cdot Z_{\mathbf{i} \wedge \mathbf{j}}(\rho). \tag{10.26}$$

In the self-similar case this is equivalent to the transversality condition from Definition 6.3.1, and like in the self-similar case it may be difficult to check.

THEOREM 10.3.3. [**JPS07**] *If the self-affine Hölder condition and self-affine transversality condition are satisfied, then for $\mathcal{M}$-almost all $u \in U$ holds:*

(i) $\dim_{\mathrm{H}}(\Pi^u(\Sigma)) = \min\{d, s(A_1, \ldots, A_m)\}$; *and*
(ii) *if $s(A_1, \ldots, A_m) > d$, then $\mathcal{L}^d(\Pi^u(\Sigma)) > 0$.*

We also have the following result about the dimension and absolute continuity of the push-forward measure of an ergodic measure.

THEOREM 10.3.4. *Let μ be an ergodic shift-invariant measure on Σ. If the self-affine Hölder condition and self-affine transversality condition are satisfied, then for $\mathcal{M}$-almost all $u \in U$ holds:*

(i) $\dim_{\mathrm{H}}(\Pi^u_*\mu) = \min\{d, D(\mu)\}$; *and*
(ii) *if $h_\mu > \chi_1(\mu) + \cdots + \chi_d(\mu)$, then $\Pi^u_*\mu \ll \mathcal{L}^d$;*
(iii) *the measure $\Pi^u_*\mu$ is exact dimensional.*

Parts (i) and (ii) were proved in [**JPS07**, Theorem 4.3], and part (iii) was proved by E. Rossi [**Ros14**]. In particular, if the system is self-similar, then this theorem says that almost surely, the Hausdorff dimension of the push-forward measure of an ergodic measure is equal to entropy divided by the Lyapunov exponent, assuming this ratio is not greater than d. Otherwise this push-forward measure is (almost surely) absolutely continuous with respect to the Lebesgue measure, see Theorem 6.6.2.

To obtain lower bounds for the dimension in Theorems 10.3.3 and 10.3.4, we need the following proposition.

PROPOSITION 10.3.5. *Let μ be a Radon measure on Σ and $s > 0$ is such that for some $c_1 > 0$, for all finite words $\mathbf{i} \in \Sigma^*$ we have*

$$\mu([\mathbf{i}]) \leqslant c_1 \varphi^s(A_{\mathbf{i}}). \tag{10.27}$$

Then for $\mathcal{M}$-a.e. $u \in U$ holds:

(i) *If $s \leqslant d$ then $\dim_{\mathrm{H}}(\Pi^u_*\mu) \geqslant s$;*
(ii) *if $s > d$, then $\Pi^u_*\mu \ll \mathcal{L}^d$.*

To prove part (i) of Proposition 10.3.5 we need the following lemma.

LEMMA 10.3.6. *Assume that the self-affine transversality condition holds. Then for every $t \notin \mathbb{N}$, $0 < t < d$, there exists a constant $c_2 = c_2(t)$ (independent of $\mathbf{i}, \mathbf{j}$) such that for all $\mathbf{i}, \mathbf{j} \in \Sigma$, $\mathbf{i} \neq \mathbf{j}$, we have*

$$\int_U |\Pi^u(\mathbf{i}) - \Pi^u(\mathbf{j})|^{-t} \, d\mathcal{M}(u) < c_2 \cdot \left(\varphi^t(A_{\mathbf{i}\wedge\mathbf{j}})\right)^{-1}. \tag{10.28}$$

PROOF. Fix $t > 0$ and $\mathbf{i}, \mathbf{j} \in \Sigma$. We can write

$$\begin{aligned} &\int_U \quad |\Pi^u(\mathbf{i}) - \Pi^u(\mathbf{j})|^{-t} \, d\mathcal{M}(u) \\ = \quad & t \int_0^\infty \mathcal{M}\left\{u \in U : |\Pi^u(\mathbf{i}) - \Pi^u(\mathbf{j})| < \rho\right\} \rho^{-t-1} d\rho. \end{aligned}$$

Thus by (10.26) and (10.8) it is enough to prove that there exists a constant such that, denoting

$$\mathcal{I}^t_{\mathbf{i}\wedge\mathbf{j}} := \int_0^\infty Z_{\mathbf{i}\wedge\mathbf{j}}(\rho)\rho^{-t-1} d\rho, \quad t \in (k-1, k), \tag{10.29}$$

we have

$$\mathcal{I}^t_{\mathbf{i}\wedge\mathbf{j}} \leqslant \text{const} \cdot \left[\alpha_1(\mathbf{i}\wedge\mathbf{j}) \cdots \alpha_{k-1}(\mathbf{i}\wedge\mathbf{j}) \cdot \alpha_k^{t-(k-1)}(\mathbf{i}\wedge\mathbf{j})\right]^{-1}. \tag{10.30}$$

From now on we suppress $\mathbf{i}\wedge\mathbf{j}$ from notation and simply write α_k, $\mathcal{I}^t$, and $Z(\rho)$. Represent the integral in (10.29) as a sum

$$\mathcal{I}^t = \mathcal{I}_1 + \mathcal{I}_2, \tag{10.31}$$

where

$$\mathcal{I}_1 := \int\limits_{\rho=0}^{\alpha_k} Z(\rho)\rho^{-t-1} d\rho \quad \text{and} \quad \mathcal{I}_2 := \int\limits_{\rho=\alpha_k}^{\infty} Z(\rho)\rho^{-t-1} d\rho.$$

First we prove that $\mathcal{I}_1$ is bounded by the expression on the right-hand side of (10.30), and then we show that the same holds for $\mathcal{I}_2$. The proof then follows. Clearly,

$$0 < \rho < \alpha_i \quad \Longrightarrow \quad \frac{\rho^i}{\alpha_1 \cdots \alpha_i} < \frac{\rho^{i-1}}{\alpha_1 \cdots \alpha_{i-1}}.$$

Thus

$$\mathcal{I}_1 < \int\limits_{\rho=0}^{\alpha_k} \frac{\rho^k}{\alpha_1 \cdots \alpha_k} \rho^{-t-1} d\rho = \frac{1}{k-t}\left[\alpha_1 \cdots \alpha_{k-1}\alpha_k^{t-(k-1)}\right]^{-1}, \tag{10.32}$$

whence (10.30) holds with the constant $\frac{1}{k-t}$. We now recall (10.23) and write

$$\mathcal{I}_2 = \sum_{\ell=0}^{k-1} \int_{J_\ell} Z(\rho)\rho^{-t-1} d\rho, \tag{10.33}$$

where we set $\alpha_0^{-t} = 0$. Observe that for $\ell \geqslant 0$ we have

$$\int_{J_\ell} Z(\rho)\rho^{-t-1} d\rho = \frac{1}{t-\ell}(\alpha_1 \cdots \alpha_\ell)^{-1}\left[\alpha_{\ell+1}^{\ell-t} - \alpha_\ell^{\ell-t}\right].$$

Using that

$$\frac{1}{t-(\ell+1)}(\alpha_1 \cdots \alpha_{\ell+1})^{-1}\alpha_{\ell+1}^{\ell+1-t} > \frac{1}{t-\ell}(\alpha_1 \cdots \alpha_\ell)^{-1}\alpha_{\ell+1}^{\ell-t},$$

one can see that the quantity which we add in the sum (10.33) for each ℓ is smaller than that which we subtract for $\ell+1$. Therefore,

$$\mathcal{I}_2 < \frac{1}{t-(k-1)} \cdot \left(\alpha_1 \cdots \alpha_{k-1}\alpha_k^{t-(k-1)}\right)^{-1},$$

which implies (10.30) with the constant $\frac{1}{t-(k-1)}$. Together with (10.31) and (10.32), this implies (10.30) for $\mathcal{I}^t$ with the constant $\frac{1}{t-(k-1)} + \frac{1}{k-t}$, and the proof of the lemma is complete. □

Proof of Proposition 10.3.5. We first prove part (i) and assume that $s \leqslant d$. Fix $t < s$. In view of Lemma 1.9.9, if we prove that for $\mathcal{M}$-a.e. $u \in U$ holds $\mathcal{E}_t(\Pi^u_*\mu) < \infty$, then $\dim_{\mathrm{H}}(\Pi^u_*\mu) \geqslant t$. For this it suffices to show that

$$\mathcal{I}^t := \int\limits_U \iint\limits_{\Omega\times\Omega} |\Pi^u(\mathbf{i}) - \Pi^u(\mathbf{j})|^{-t}\, d\mu(\mathbf{i})\, d\mu(\mathbf{j})\, d\mathcal{M}(u) < \infty.$$

Using Fubini's Theorem, Lemma 10.3.6 and (10.27) we obtain

$$\begin{aligned} \mathcal{I}^t &< c_2 \sum_{k=0}^{\infty} \sum_{|\omega|=k} \iint\limits_{\mathbf{i}\wedge\mathbf{j}=\omega} \left(\varphi^t(A_\omega)\right)^{-1} d\mu(\mathbf{i})\, d\mu(\mathbf{j}) \\ &\leqslant c_2 c_1 \sum_{k=0}^{\infty} \sum_{|\omega|=k} \mu([\omega])\varphi^s(A_\omega) \left(\varphi^t(A_\omega)\right)^{-1}. \end{aligned} \tag{10.34}$$

It follows from (10.9) and (10.13) that

$$\varphi^s(A_\omega) \cdot \left(\varphi^t(A_\omega)\right)^{-1} \leqslant b_{\mathcal{F}}^{k\cdot(s-t)}.$$

Using this and (10.34) we obtain that

$$\mathcal{I}^t \leqslant c_1 c_2 \sum_{k=0}^{\infty} b_{\mathcal{F}}^{k\cdot(s-t)} \underbrace{\sum_{|\omega|=k} \mu([\omega])}_{=1} < \infty.$$

Since t can be chosen arbitrarily close to s, this completes the proof of part (i).

We next prove part (ii), so that now $s > d$. It follows from the definition (10.24) of $Z_{\mathbf{i}\wedge\mathbf{j}}(\rho)$ that for distinct $\mathbf{i}, \mathbf{j} \in \Sigma$ and $\rho > 0$ we have the bound

$$Z_{\mathbf{i}\wedge\mathbf{j}}(\rho) \leqslant \frac{\rho^d}{\varphi^d(A_{\mathbf{i}\wedge\mathbf{j}})}.$$

Thus by the self-affine transversality condition (10.26), the inequality

$$\mathcal{M}\big(\{u \in U : |\Pi^u(\mathbf{i}) - \Pi^u(\mathbf{j})| < \rho\}\big) < C \cdot \frac{\rho^d}{\varphi^d(A_{\mathbf{i}\wedge\mathbf{j}})} \tag{10.35}$$

holds for all $\mathbf{i} \neq \mathbf{j}$ in Σ and $\rho > 0$. Furthermore, from (10.14) we obtain that for $\mathbf{i} \in \Sigma^*$,

$$\frac{\mu([\mathbf{i}])}{\varphi^d(A_{\mathbf{i}})} \leqslant c_1 \cdot \frac{\varphi^s(A_{\mathbf{i}})}{\varphi^d(A_{\mathbf{i}})} \leqslant b_{\mathcal{F}}^{|\mathbf{i}|(s-d)}. \tag{10.36}$$

To verify absolute continuity of $\Pi_*^u\mu$ for $\mathcal{M}$-a.e. u we use the method of Mattila Theorem B.2.5 which has already appeared in the book: see the proofs of Theorem 2.3.1(ii), Theorem 6.3.3(ii), and Theorem 6.6.2(iv). Namely, it is enough to show that

$$I := \iint \liminf_{r\to 0} \frac{\Pi_*^u\mu(B(x,r))}{(2r)^d}\, d\Pi_*^u\mu(x)\, d\mathcal{M}(u) < \infty.$$

Applying Fatou's Lemma, Fubini's Theorem and (10.35), we obtain

$$\begin{aligned} I &\leqslant \liminf_{r\to 0} \frac{1}{(2r)^d} \iiint \mathbb{1}_{\{(\mathbf{i},\mathbf{j}):|\Pi^u(\mathbf{i})-\Pi^u(\mathbf{j})|\leqslant r\}}\, d\mu(\mathbf{i})\, d\mu(\mathbf{j})\, d\mathcal{M}(u) \\ &= \liminf_{r\to 0} \frac{1}{(2r)^d} \iint \mathcal{M}\{u : |\Pi^u(\mathbf{i}) - \Pi^u(\mathbf{j})| \leqslant r\}\, d\mu(\mathbf{i})\, d\mu(\mathbf{j}) \\ &\leqslant C \iint \frac{1}{\varphi^d(A_{\mathbf{i}\wedge\mathbf{j}})}\, d\mu(\mathbf{i})\, d\mu(\mathbf{j}). \end{aligned}$$

Finally, by (10.36) we get

$$\begin{aligned} I &\leqslant C \sum_{k=0}^{\infty} \sum_{|\mathbf{i}|=k} \mu([\mathbf{i}])^2 (\varphi^d(A_{\mathbf{i}}))^{-1} \\ &\leqslant c_1 C + C \sum_{k=0}^{\infty} b_{\mathcal{F}}^{k(s-d)} \sum_{|\mathbf{i}|=k} \mu([\mathbf{i}]) < \infty. \end{aligned}$$

This completes the proof of the proposition. □

PROOF OF THEOREM 10.3.3. We first prove part (i). The upper bound follows from (10.25), and it holds for all parameters. In the case of actual self-affine IFS, like in Theorem 10.1.2, the upper bound for the dimension follows from Proposition 10.2.6. To get the lower bound, fix an arbitrary $t < \min\{d, s(A_1,\ldots,A_m)\}$. Then $\mathcal{N}^t(\Sigma) = \infty$ by (10.18) and Lemma 10.2.3. Using Lemma 10.2.4 we can find a measure μ which is supported on Σ and satisfies (10.27), i.e.,

$$\mu([\mathbf{i}]) \leqslant c_1\varphi^t(A_{\mathbf{i}}) \text{ for } \mathbf{i} \in \Sigma^*. \tag{10.37}$$

We then apply Proposition 10.3.5 to deduce that for $\mathcal{M}$-a.e. u we have $\dim_{\mathrm{H}}(\Pi^u(\Sigma)) \geqslant t$. This completes the proof of part (i) of the theorem.

Next we turn to the proof of part (ii). Since by assumption $d < s(A_1,\ldots,A_m)$ we can choose

$$d < t < s(A_1,\ldots,A_m).$$

As in the proof of part (i), there exists a measure μ supported on Σ which satisfies (10.37). Thus we can apply part (ii) of Proposition 10.3.5 to deduce that for $\mathcal{M}$-a.e. $u \in U$ we have $\Pi_*^u\mu \ll \mathcal{L}^d$ and thus $\mathcal{L}^d(\Pi^u(\Sigma)) > 0$. □

The proof of Theorem 10.3.4 is somewhat similar to that of Theorem 6.7.1(iii)—it also makes use of Egorov's Theorem and the Shannon–McMillan–Breiman Theorem—but it is rather technical, so we refer the reader to [**JPS07**]. In [**JPS07**] it is also proved that the self-affine transversality condition holds in the randomly perturbed case, and this yields Theorem 10.1.4.

10.3.1. Lower bound for the Lyapunov dimension. Here we prove a lower bound for the Lyapunov dimension, which shows that the inequality "$\leqslant$" in Theorem 10.3.4(a), even the stronger one for the packing dimension of the self-affine measure, holds always, and not just for almost every parameter.

PROPOSITION 10.3.7. *Let μ be an ergodic shift-invariant measure on Σ. Then*

$$\dim_{\mathrm{H}}(\Pi^{\mathbf{t}}_*\mu) \leqslant \dim_{\mathrm{P}}(\Pi^{\mathbf{t}}_*\mu) \leqslant \min\{d, D(\mu)\},$$

where $\Pi^{\mathbf{t}}$ is the natural projection (10.6) *with respect to the affine IFS* (10.7).

PROOF. We may assume without loss of generality that $D(\mu) < d$. Let $\varepsilon > 0$ be arbitrary but fixed. We show that for μ-almost every $\mathbf{i} \in \Sigma$,

$$\overline{d}_{\Pi^{\mathbf{t}}_*\mu}(\Pi^{\mathbf{t}}(\mathbf{i})) \leqslant D(\mu), \tag{10.38}$$

and then the statement follows by Theorem 1.9.5.

Let B be a ball centered at the origin such that $f_i(B) \subseteq B$ for every $i = 1, \ldots, m$. Since for every $\mathbf{i} \in \Sigma$, $f_{\mathbf{i}|_n}(B)$ is an ellipsoid, by the definition of singular values, it can be enclosed in a box (parallelepiped) with side lengths $K\alpha_1(\mathbf{i}|_n), \ldots, K\alpha_d(\mathbf{i}|_n)$, where $K > 0$ depends only on the radius of B. Without loss of generality, we may assume that $K \geqslant 1$. Denote the box by $R_{\mathbf{i}|_n}$. Fix $k \leqslant d$ and divide $R_{\mathbf{i}|_n}$ into

$$N(\mathbf{i}|_n) := \left\lceil \frac{\alpha_1(\mathbf{i}|_n)}{\alpha_{k+1}(\mathbf{i}|_n)} \right\rceil \cdots \left\lceil \frac{\alpha_k(\mathbf{i}|_n)}{\alpha_{k+1}(\mathbf{i}|_n)} \right\rceil$$

many boxes with side lengths

$$\overbrace{K\alpha_{k+1}(\mathbf{i}|_n), \ldots, K\alpha_{k+1}(\mathbf{i}|_n)}^{k+1}, K\alpha_{k+2}(\mathbf{i}|_n) \ldots, K\alpha_d(\mathbf{i}|_n),$$

so that each of the boxes has diameter $\leqslant C\alpha_{k+1}(\mathbf{i}|_n)$, with $C = K\sqrt{d}$. Denote by $P_n(\mathbf{i})$ the box which contains $\Pi^{\mathbf{t}}(\mathbf{i})$. Let

$$\begin{aligned} Q_n(\mathbf{i}) &:= \{\mathbf{j} \in \Sigma : \mathbf{j} \in [\mathbf{i}|_n] \text{ and } \Pi^{\mathbf{t}}(\mathbf{j}) \in P_n(\mathbf{i})\} \\ &\subseteq \{\mathbf{j} \in \Sigma : |\Pi^{\mathbf{t}}(\mathbf{j}) - \Pi^{\mathbf{t}}(\mathbf{i})| \leqslant C\alpha_{k+1}(\mathbf{i}|_n)\}. \end{aligned} \tag{10.39}$$

Observe that $Q_n(\mathbf{i}) \subseteq (\Pi^{\mathbf{t}})^{-1}(P_n(\mathbf{i}))$, but because of possible overlaps the equality need not hold. Let

$$A_n = \left\{\mathbf{i} \in \Sigma : \mu(Q_n(\mathbf{i})) > \frac{1}{n^2} \cdot \frac{\mu([\mathbf{i}|_n])}{N(\mathbf{i}|_n)}\right\}.$$

By definition, for every $n \geqslant 1$,

$$\mu\left((A_n)^c\right) = \sum_{|\mathbf{i}|=n} \mu\left([\mathbf{i}] \cap (A_n)^c\right) \leqslant \sum_{|\mathbf{i}|=n} N(\mathbf{i}) \frac{1}{n^2} \frac{\mu([\mathbf{i}])}{N(\mathbf{i})} = \frac{1}{n^2}.$$

Thus, for the set $A = \bigcap_{k=1}^{\infty} \bigcup_{n=k}^{\infty} (A_n)^c$, we have $\mu(A) = 0$ by the Borel–Cantelli Lemma.

Finally, for μ-almost every $\mathbf{i} \in \Sigma \backslash A$, so that $\mathbf{i} \in A_n$ for all sufficiently large n, we have

$$\begin{aligned}
\overline{d}_{\Pi^{\mathbf{t}}_* \mu}(\Pi^{\mathbf{t}}(\mathbf{i})) &= \limsup_{n\to\infty} \frac{\log \Pi^{\mathbf{t}}_* \mu(B(\Pi^{\mathbf{t}}(\mathbf{i}), C\alpha_{k+1}(\mathbf{i}|_n)))}{\log(C\alpha_{k+1}(\mathbf{i}|_n))} \\
&\leqslant \limsup_{n\to\infty} \frac{\log \mu(Q_n(\mathbf{i}))}{\log(C\alpha_{k+1}(\mathbf{i}|_n))} \qquad \text{by (10.39)} \\
&\leqslant \lim_{n\to\infty} \frac{-2\log n + \log \mu([\mathbf{i}|_n]) - \log N(\mathbf{i}|_n)}{\log C + \log \alpha_{k+1}(\mathbf{i}|_n)} \\
&= \lim_{n\to\infty} \frac{-\frac{1}{n}\log \mu([\mathbf{i}|_n]) + \frac{1}{n}\log \alpha_1(\mathbf{i}|_n) + \cdots + \frac{1}{n}\log \alpha_k(\mathbf{i}|_n)}{-\frac{1}{n}\log \alpha_{k+1}(\mathbf{i}|_n)} + k \\
&= k + \frac{h_\mu - \chi_1(\mu) - \cdots - \chi_k(\mu)}{\chi_{k+1}(\mu)},
\end{aligned}$$

by the Shannon–McMillan–Breiman Theorem 9.5.3 and (10.2), so that (10.38) follows by Definition 10.1.1 of the Lyapunov dimension. □

10.4. Self-affine transversality and the proof of Theorem 10.1.2

In order to deduce Theorem 10.1.2, we need the following transversality result. The idea of the proof is due to Falconer [**Fal88**] who assumed that $\max_i \|A_i\| < 1/3$. This condition was relaxed in[**Sol98**] to $\max_i \|A_i\| < 1/2$, and our proposition is a further generalization in this direction.

PROPOSITION 10.4.1. *Suppose that we are given $d \times d$ matrices $A_1, \dots, A_m$ satisfying:*

$$r := \max_{i \neq j} \{\|A_i\| + \|A_j\|\} < 1. \tag{10.40}$$

Let $U := B(\mathbf{u}_0, \delta)$ be an arbitrary ball in $\mathbb{R}^{md}$ and let $\mathcal{M}$ be the measure $\mathcal{L}^{md}$ restricted to U. For $\mathbf{u} = (u_1, \dots, u_m) \in U$ consider the following family of IFS on $\mathbb{R}^d$:

$$\mathcal{F}^{\mathbf{u}} = (f_i^{\mathbf{u}}(x) = A_i \cdot x + u_i).$$

Let $\Pi^{\mathbf{u}}$ be the natural projection, that is, $\Pi^{\mathbf{u}}(\mathbf{i}) := \lim_{n\to\infty} f_{i_1}^{\mathbf{u}} \circ \cdots \circ f_{i_n}^{\mathbf{u}}(0)$. Then the self-affine transversality condition (10.26) holds.

PROOF. Fix arbitrary distinct $\mathbf{i}, \mathbf{j} \in \Sigma := \{1, \dots, m\}^{\mathbb{N}}$. It is an immediate consequence of the definition of $\Pi^{\mathbf{u}}$ that for $\mathbf{i} \neq \mathbf{j}$ in Σ with $k := |\mathbf{i} \wedge \mathbf{j}|$ we have

$$\begin{aligned}
&\Pi^{\mathbf{u}}(\mathbf{i}) - \Pi^{\mathbf{u}}(\mathbf{j}) \\
= \;& A_{\mathbf{i}\wedge\mathbf{j}}\big(u_{i_{k+1}} - u_{j_{k+1}} + (A_{i_{k+1}} u_{i_{k+2}} + A_{i_{k+1}} A_{i_{k+2}} u_{i_{k+3}} + \cdots) \\
- \;& (A_{j_{k+1}} u_{j_{k+2}} + A_{j_{k+1}} A_{j_{k+2}} u_{j_{k+3}} + \cdots)\big) \\
= \;& A_{\mathbf{i}\wedge\mathbf{j}}(u_{i_{k+1}} - u_{j_{k+1}} + \underbrace{E_1 \cdot u_1 + \cdots + E_m \cdot u_m}_{E(\mathbf{u})}),
\end{aligned} \tag{10.41}$$

where $E_1, \dots, E_m$ are $d \times d$ matrices satisfying

$$\|E_i\| < \frac{2\eta}{1-\eta}, \tag{10.42}$$

where $\eta := \max_{1\leqslant i\leqslant m} \|A_i\|$. Since $u_{i_{k+1}} \neq u_{j_{k+1}}$, without loss of generality we may assume that

$$i_{k+1} = 1 \text{ and } j_{k+1} = 2.$$

LEMMA 10.4.2. *We have*

$$\min\{\|E_1\|, \|E_2\|\} \leqslant r_1 + r_2 \leqslant r. \tag{10.43}$$

PROOF OF THE LEMMA. One can easily see that (10.43) follows from

$$\|E_1\|(1-r_1) + \|E_2\|(1-r_2) \leqslant r_1 + r_2, \tag{10.44}$$

in view of the assumption $r_1 + r_2 < 1$. Recall that $r_i = \|A_i\|$ and observe that (10.41) implies

$$E_\ell = \sum_{n\geqslant k+1:\ i_{n+1}=\ell} A_{i_{k+1}} \cdots A_{i_n} - \sum_{n\geqslant k+1:\ j_{n+1}=\ell} A_{j_{k+1}} \cdots A_{j_n},$$

hence

$$\|E_\ell\| \leqslant \sum_{n\geqslant k+1:\ i_{n+1}=\ell} r_{i_{k+1}} \cdots r_{i_n} + \sum_{n\geqslant k+1:\ j_{n+1}=\ell} r_{j_{k+1}} \cdots r_{j_n}. \tag{10.45}$$

Applying this for $\ell = 1, 2$ we obtain

$$\begin{aligned}
&\|E_1\|(1-r_1) + \|E_2\|(1-r_2) \\
&\leqslant \sum_{n\geqslant k+1:\ i_{n+1}=1} r_{i_{k+1}} \cdots r_{i_n}(1-r_{i_{n+1}}) + \sum_{n\geqslant k+1:\ j_{n+1}=1} r_{j_{k+1}} \cdots r_{j_n}(1-r_{j_{n+1}}) \\
&\quad + \sum_{n\geqslant k+1:\ i_{n+1}=2} r_{i_{k+1}} \cdots r_{i_n}(1-r_{i_{n+1}}) + \sum_{n\geqslant k+1: j_{n+1}=2} r_{j_{k+1}} \cdots r_{j_n}(1-r_{j_{n+1}}) \\
&\leqslant \sum_{n=k+1}^{\infty} r_{i_{k+1}} \cdots r_{i_n}(1-r_{i_{n+1}}) + \sum_{n=k+1}^{\infty} r_{j_{k+1}} \cdots r_{j_n}(1-r_{j_{n+1}}) \\
&= r_1 + r_2,
\end{aligned}$$

confirming (10.44), as desired. □

Suppose that $\|E_1\| \leqslant r$. We claim that the following linear transformation is invertible: $\mathcal{T} : \mathbb{R}^{md} \to \mathbb{R}^{md}$, given by

$$\mathcal{T} : (u_1, u_2 \ldots, u_m) \to (u_1 - u_2 + E(\mathbf{u}), u_2, \ldots, u_m), \tag{10.46}$$

and moreover, $\|\mathcal{T}^{-1}\|$ is bounded from above uniformly, independent of $\mathbf{i} \neq \mathbf{j}$ fixed above.

If $\|E_1\| > r$, then by Lemma 10.4.2 we have $\|E_2\| \leqslant r$, and then the proof is analogous, but instead of $\mathcal{T}$ we consider the linear transformation

$$(u_1, u_2 \ldots, u_m) \mapsto (u_1, u_1 - u_2 + E(\mathbf{u}), u_3, \ldots, u_m).$$

In order to verify the claim about $\mathcal{T}^{-1}$ (assuming $\|E_1\| \leqslant r$), note that

$$\mathcal{T}^{-1}(v_1, \ldots, v_m) = \Big((I - E_1)^{-1}\Big(v_2 - \sum_{i=2}^{m} E_i v_i\Big), v_2, \ldots, v_m\Big),$$

hence

$$\|\mathcal{T}^{-1}\| \leqslant \frac{1}{1-r} \cdot \left(1 + (m-1)\frac{2\eta}{1-\eta}\right) + m - 1 =: K, \tag{10.47}$$

as desired. Clearly, this implies that

$$\det(\mathcal{T}^{-1}) \leqslant K^{md}.$$

Next observe that for a nonsingular matrix A with singular values $\alpha_1, \ldots, \alpha_m$, and for any positive numbers δ', ρ we have

$$\mathcal{L}^d\Big(\{x \in B(\mathbf{0}, \delta') : \|Ax\| \leqslant \rho\}\Big) \leqslant \prod_{i=1}^{d} 2 \cdot \min\Big\{\frac{\rho}{\alpha_i}, \delta'\Big\}. \tag{10.48}$$

This is because $A^{-1}B(\mathbf{0}, \rho)$ is a d-dimensional ellipsoid, contained in a box with sides of length $\frac{2\rho}{\alpha_i}$, aligned in the directions of the column vectors of the orthogonal matrix U_A which appears in the singular value decomposition (A.3) of the matrix A, i.e., the axes of the ellipsoid.

Our goal is to prove that there exists $C = C(U)$, with $U = B(\mathbf{u}_0, \delta)$, such that for any $\rho > 0$,

$$L := \mathcal{L}^{md}\Big(\{\mathbf{u} \in U : | \underbrace{\Pi^{\mathbf{u}}(\mathbf{i}) - \Pi^{\mathbf{u}}(\mathbf{j})}_{A_{\mathbf{i}\wedge\mathbf{j}} \cdot (u_1 - u_2 + E(\mathbf{u}))} | \leqslant \rho\}\Big) \leqslant C \cdot Z_{\mathbf{i}\wedge\mathbf{j}}(\rho). \tag{10.49}$$

Let $\delta_1 := |U| + \delta$. Then $U \subset B(\mathbf{0}, \delta_1)$. By (10.42) and (10.46), if $\mathbf{u} \in U$, then

$$\|u_1 - u_2 + E(\mathbf{u})\| \leqslant \delta_2 \quad \text{and} \quad \|\mathcal{T}(\mathbf{u})\| \leqslant \delta_3, \tag{10.50}$$

where

$$\delta_2 := 2\delta_1 + m \cdot \frac{2\eta}{1-\eta} \quad \text{and} \quad \delta_3 := \max\{1, \delta_2 + (m-1)\delta_1\}.$$

Hence denoting $\mathbf{v} = (v_1, \ldots, v_m) = \mathcal{T}\mathbf{u} \in \underbrace{\mathbb{R}^d \times \cdots \times \mathbb{R}^d}_{m}$, we have

$$\begin{aligned}
L &\leqslant \mathcal{L}^{md}(\{\|\mathbf{u}\| < \delta_1 : \|A_{\mathbf{i}\wedge\mathbf{j}} \cdot (u_1 - u_2 + E(\mathbf{u}))\| < \rho\}) \\
&\leqslant \det\left(\mathcal{T}^{-1}\right) \cdot \mathcal{L}^{md}(\{\|\mathbf{v}\| < \delta_3 : \|A_{\mathbf{i}\wedge\mathbf{j}} \cdot v_1)\| < \rho\}) \\
&\leqslant \underbrace{K^{md}(2\delta_3)^{d(m-1)}}_{C_1} \cdot \mathcal{L}^d(\{v_1 \in B(\mathbf{0}, \delta_3) : \|A_{\mathbf{i}\wedge\mathbf{j}} \cdot v_1\| < \rho\}) \\
&\leqslant C_1 \prod_{\ell=1}^{d} \min\left\{\frac{\rho}{\alpha_\ell(A_{\mathbf{i}\wedge\mathbf{j}})}, \delta_3\right\} \leqslant \underbrace{C_1 \delta_3^d}_{C_2} \prod_{\ell=1}^{d} \min\left\{\frac{\rho}{\alpha_\ell(A_{\mathbf{i}\wedge\mathbf{j}})}, 1\right\} \\
&= C_2 \prod_{\ell=1}^{d} \frac{\min\{\rho, \alpha_\ell(A_{\mathbf{i}\wedge\mathbf{j}})\}}{\alpha_\ell(A_{\mathbf{i}\wedge\mathbf{j}})} = C_2\, Z_{\mathbf{i}\wedge\mathbf{j}}(\rho),
\end{aligned}$$

in view of (10.24), where in the fourth displayed line we used (10.48). We have verified (10.49), and the proposition is proved completely. □

PROOF OF THEOREM 10.1.2. Parts (iii) and (iv) of Theorem 10.1.2 immediately follow from Theorem 10.3.3 and Proposition 10.4.1. Parts (i) and (ii) follow from Theorem 10.3.4, whose proof may be found in [**JPS07**]. □

10.5. Thermodynamic formalism

10.5.1. Subadditive pressure. Consider a self-affine IFS $\mathcal{F} := \left(f_i(\mathbf{x}) = A_i\mathbf{x} + \mathbf{t}_i\right)_{i=1}^{m}$ on $\mathbb{R}^d$. In this section we study the limiting behavior of the sequence $\frac{1}{n} \cdot \log \sum\limits_{|\mathbf{i}|=n} \varphi^t(A_{\mathbf{i}})$, since it gives the exponential growth rate of the sum in the definition of the Hausdorff dimension corresponding to the natural covering system (1.127),(10.17). First we show that for all $t \geqslant 0$ the limit

$$\lim_{n\to\infty} \frac{1}{n} \cdot \log \sum_{|\mathbf{i}|=n} \varphi^t(A_{\mathbf{i}}) \text{ exists and is finite.} \tag{10.51}$$

For the existence of the limit in (10.51), observe that

$$\sum_{|\mathbf{i}|=q+r} \varphi^s(A_{\mathbf{i}}) = \sum_{|\mathbf{i}|=q} \sum_{|\mathbf{j}|=r} \varphi^s(A_{\mathbf{i}} \cdot A_{\mathbf{j}}). \tag{10.52}$$

By (10.10), this implies that the sequence $\left\{\log \sum\limits_{|\mathbf{i}|=n} \varphi^s(A_{\mathbf{i}})\right\}_{n=1}^{\infty}$ is subadditive, hence the limit exists by Fekete's Lemma 1.9.21. On the other hand, (10.14) implies that

$$\log\left(ma_{\mathcal{F}}^t\right) \leqslant \frac{1}{n} \cdot \log \sum_{|\mathbf{i}|=n} \varphi^t(A_{\mathbf{i}}) \leqslant \log\left(mb_{\mathcal{F}}^t\right). \tag{10.53}$$

which ensures finiteness of the limit in (10.51).

Now we are ready to define the subadditive pressure function $P : \mathbb{R}_+ \to \mathbb{R}$ (introduced by Falconer [**Fal94**] and in a different way by Barreira [**Bar96**]) by

$$P(t) := \lim_{n\to\infty} \frac{1}{n} \cdot \log \sum_{|\mathbf{i}|=n} \varphi^t(A_{\mathbf{i}}), \quad t \geqslant 0. \tag{10.54}$$

Observe that putting together (10.9) and (10.13) yields

$$\log a_{\mathcal{F}} \leqslant \frac{\frac{1}{n} \cdot \log \sum\limits_{|\mathbf{i}|=k} \varphi^{t+h}(A_{\mathbf{i}}) - \frac{1}{n} \cdot \log \sum\limits_{|\mathbf{i}|=k} \varphi^t(A_{\mathbf{i}})}{h} \leqslant \log b_{\mathcal{F}} \tag{10.55}$$

for every $h \in \mathbb{R}$, and hence $P(t)$ is a continuous and strictly decreasing function on $\mathbb{R}_+$. An immediate computation yields that

$$\log m = P(0) > 0 \text{ and } P(t) < 0 \text{ for all } t > \frac{\log m}{\log \frac{1}{b_{\mathcal{F}}}}. \tag{10.56}$$

Thus there is a unique positive zero of the pressure function $P(t)$, namely $t_0 \in (0, \log m/\log \frac{1}{b_{\mathcal{F}}})$ such that $P(t_0) = 0$. Recall again the affinity dimension (1.127) and (10.17). Using the root test, it is easy to see that

$$\text{if } e^{P(t)} = \lim_{n\to\infty} \Big(\sum_{|\mathbf{i}|=n} \varphi^t(A_{\mathbf{i}})\Big)^{1/n} < 1, \text{ then } \sum_{n=0}^{\infty} \sum_{|\mathbf{i}|=n} \varphi^t(A_{\mathbf{i}}) < \infty; \text{ and}$$

$$\text{if } e^{P(t)} = \lim_{n\to\infty} \Big(\sum_{|\mathbf{i}|=n} \varphi^t(A_{\mathbf{i}})\Big)^{1/n} > 1, \text{ then } \sum_{n=0}^{\infty} \sum_{|\mathbf{i}|=n} \varphi^t(A_{\mathbf{i}}) = \infty.$$

Hence, the affinity dimension $s(A_1, \ldots, A_m)$ is equal to the unique zero of $P(t)$.

10.5.2. Energy. For $\mathbf{i} \in \Sigma \cup \Sigma^*$, such that $|\mathbf{i}| \geqslant n$, we write $\mathbf{i}|_n := (i_1, \dots, i_n)$. Put

$$\psi_n^t(\mathbf{i}) := \log \varphi^t(A_{\mathbf{i}|_n}). \tag{10.57}$$

It follows from (10.14) and (10.10) that for all n, k, $|\mathbf{i}| \geqslant n+k$, holds:

$$nt \log a_{\mathcal{F}} \leqslant \psi_n^t(\mathbf{i}) \leqslant nt \log b_{\mathcal{F}}, \tag{10.58}$$

$$\psi_{n+k}^t(\mathbf{i}) \leqslant \psi_n^t(\mathbf{i}) + \psi_k^t(\sigma^n \mathbf{i}) \tag{10.59}$$

By the Subadditive Ergodic Theorem 9.1.4, for an arbitrary σ-invariant ergodic measure μ on Σ, the following limit exists and is constant μ-a.e.:

$$\lim_{n\to\infty} \frac{1}{n}\psi_n^t(\mathbf{i}) = \lim_{n\to\infty} \frac{1}{n} \int_\Sigma \psi_n^t(\mathbf{j})\, d\mu(\mathbf{j}) \quad \text{for } \mu\text{-a.e. } \mathbf{i} \in \Sigma. \tag{10.60}$$

DEFINITION 10.5.1. For an ergodic σ-invariant measure μ the *energy function* is defined by

$$E_\mu(t) := \lim_{n\to\infty} \frac{-1}{n} \int \psi_n^t(\mathbf{j})\, d\mu(\mathbf{j}) = \lim_{n\to\infty} \frac{-1}{n} \sum_{|\mathbf{i}|=n} \mu([\mathbf{i}]) \log \varphi^s(A_{\mathbf{i}}).$$

When $s = d$, by (1.124) and by the multiplicativity of the determinant we get

$$E_\mu(d) = -\int \log|\det(A_{i_1})| d\mu(\mathbf{i}) = \sum_{k=1}^d \chi_k(\mu). \tag{10.61}$$

COROLLARY 10.5.2. *Let $k < s \leqslant k+1$. Then*

$$E_\mu(s) = \begin{cases} \chi_1(\mu) + \cdots + \chi_k(\mu) + (s-k)\,\chi_{k+1}(\mu), & \text{if } s < d; \\ \frac{s}{d}\left(\chi_1(\mu) + \cdots + \chi_d(\mu)\right), & \text{if } s \geqslant d. \end{cases} \tag{10.62}$$

See Figure 10.1.

PROOF. It follows from (10.60) that for μ-a.e. $\mathbf{i}$ we have $E_\mu(s) = \lim_{n\to\infty} \frac{1}{n}\psi_n^s(\mathbf{i})$. On the other hand, Lemma 10.2.2 and (10.8) imply that for μ-a.e. $\mathbf{i}$ the limit $\lim_{n\to\infty} \frac{1}{n}\psi_n^s(\mathbf{i})$ is equal to the right-hand side of (10.62). □

10.5.3. Variational principle and the Käenmäki measure.

THEOREM 10.5.3 (Käenmäki [**K04**]). *Consider a self-affine IFS $\mathcal{F} := \left(f_i(\mathbf{x}) = A_i\mathbf{x} + \mathbf{t}_i\right)_{i=1}^m$ on $\mathbb{R}^d$, where A_i are nonsingular matrices.*

(i) *For any $s > 0$ and any ergodic σ-invariant measure μ, we have*

$$P(s) \geqslant h_\mu - E_\mu(s). \tag{10.63}$$

(ii) *Moreover, for any $s > 0$ there is an ergodic σ-invariant measure μ_K^s (called an equilibrium measure), such that*

$$P(s) = h_{\mu_K^s} - E_{\mu_K^s}(s). \tag{10.64}$$

In particular, there exists an ergodic measure μ_K on Σ such that

$$0 = P(s(A_1, \dots, A_m)) = h_{\mu_K} - E_{\mu_K}(s(A_1, \dots, A_m)). \tag{10.65}$$

DEFINITION 10.5.4. The measure μ_K satisfying (10.65) is called a *Käenmäki measure* for the self-affine IFS.

Corollary 10.5.2 and Definition 10.1.1 of the Lyapunov dimension $D(\mu)$ yield

LEMMA 10.5.5. *For every ergodic σ-invariant measure μ we have*

$$h_\mu - E_\mu(D(\mu)) = 0. \tag{10.66}$$

Putting together Theorem 10.5.3 and Lemma 10.5.5, we obtain:

$$s(A_1, \ldots, A_m) = D(\mu_K) = \sup\{D(\mu) : \mu \text{ is an ergodic } \sigma\text{-invariant measure}\}. \tag{10.67}$$

We start by showing (10.63), which is rather simple. Part (ii), which is the main part of the theorem, will be proven later, and only in a special case.

PROOF OF THEOREM 10.5.3(**i**). Since the logarithm is concave, we have for every $n \geqslant 1$ and every ergodic measure μ:

$$\begin{aligned} \frac{1}{n} \cdot \log \sum_{|\mathbf{i}|=n} \varphi^s(A_\mathbf{i}) &= \frac{1}{n} \cdot \log \sum_{|\mathbf{i}|=n} \mu([\mathbf{i}]) \frac{\varphi^s(A_\mathbf{i})}{\mu([\mathbf{i}])} \\ &\geqslant \frac{-1}{n} \cdot \sum_{|\mathbf{i}|=n} \mu([\mathbf{i}]) \log \mu([\mathbf{i}]) + \frac{1}{n} \cdot \sum_{|\mathbf{i}|=n} \mu([\mathbf{i}]) \log \varphi^s(A_\mathbf{i}). \end{aligned}$$

Then the inequality (10.63) follows by the definition of the energy (Definition 10.5.1) and Corollary 9.5.2. □

10.5.4. On the uniqueness of the Käenmäki measure. It is a natural question whether the Käenmäki measure is unique. First, we present an example of Käenmäki and Vilppolainen [**KV10**], showing that the Käenmäki measure need not be unique. Consider the IFS $\big(f_1(x) = A_1 x, f_2(x) = A_2 x + \mathbf{t}\big)$, where

$$A_1 = \begin{pmatrix} \frac{2}{3} & 0 \\ 0 & \frac{1}{3} \end{pmatrix}, \; A_2 = \begin{pmatrix} \frac{1}{3} & 0 \\ 0 & \frac{2}{3} \end{pmatrix} \text{ and } \mathbf{t} = \begin{pmatrix} \frac{2}{3} \\ \frac{1}{3} \end{pmatrix}.$$

Then $\|A_\mathbf{i}\| = 3^{-|\mathbf{i}|} \max\{2^{\#_1 \mathbf{i}}, 2^{\#_2 \mathbf{i}}\}$ for every $\mathbf{i} \in \Sigma^*$. Hence, $1 \leqslant \sum_{|\mathbf{i}|=n} \|A_\mathbf{i}\| \leqslant 2$ and $s(A_1, A_2) = 1$. For $i = 1, 2$, let $\mu_i = \{p_{i,1}, p_{i,2}\}^{\mathbb{N}}$ be Bernoulli measures on Σ with probabilities $p_{1,1} = p_{2,2} = 1/3$ and $p_{1,2} = p_{2,1} = 2/3$. Thus, $h_{\mu_i} = \log 3 - \frac{2}{3} \log 2$. Moreover,

$$\begin{aligned} E_{\mu_i}(1) &= \lim_{n\to\infty} \frac{-1}{n} \sum_{|\mathbf{i}|=n} p_{i,1}^{\#_1 \mathbf{i}} p_{i,2}^{\#_2 \mathbf{i}} \log \|A_\mathbf{i}\| \\ &= \lim_{n\to\infty} \frac{-1}{n} \sum_{k=0}^{n} \binom{n}{k} p_{i,1}^k p_{i,2}^{n-k} \log \max\left\{\frac{2^k}{3^n}, \frac{2^{n-k}}{3^n}\right\} \\ &= \log 3 + \lim_{n\to\infty} \frac{-\log 2}{n 3^n} \sum_{k=0}^{n} \binom{n}{k} 2^k \max\{k, n-k\} \\ &\leqslant \log 3 + \lim_{n\to\infty} \frac{-\log 2}{n 3^n} \sum_{k=0}^{n} \binom{n}{k} 2^k k \\ &= \log 3 + \lim_{n\to\infty} \frac{-\log 2}{3^n} \sum_{k=1}^{n} \binom{n-1}{k} 2^k = \log 3 - \frac{2}{3} \log 2. \end{aligned}$$

Therefore, $h_{\mu_i} - E_{\mu_i}(1) \geqslant 0 = P(s(A_1, A_2))$ and thus, both μ_1 and μ_2 are equilibrium Käenmäki measures.

Observe that in this example the matrices were diagonal and thus both preserve a nontrivial subspace. Recently, Morris and Sert [**MS21**] showed that nonuniqueness of the equilibrium measure might happen even in the case when the set of matrices is strongly irreducible (for the definition of strong irreducibility see Definition 1.12.2). Therefore, in order to have a hope for uniqueness of the Käenmäki measure for self-affine systems, we need to assume an even stronger irreducibility condition.

DEFINITION 10.5.6. We say that a collection $(A_1, \dots, A_m)$ of $d \times d$ invertible matrices is *k-irreducible* if there is no proper subspace V of $\bigwedge^k \mathbb{R}^d$ such that V is preserved by all the matrices, i.e. $A_i^{\wedge k}(V) = V$ for $i = 1, \dots, m$. We say that the collection is *k-strongly irreducible* if there is no finite collection $V_1, \dots, V_n$ of proper subspaces of $\bigwedge^k \mathbb{R}^d$ such that $(A_i^{\wedge k})(\bigcup_{j=1}^n V_j) = \bigcup_{j=1}^n V_j$ for all $i = 1, \dots, m$.

Note that for $k = 0$ and $k = d$, every tuple of nonsingular matrices is trivially k-strongly irreducible. (Strong) irreducibility introduced in Definition 1.12.2 is exactly (strong) 1-irreducibility. The following result is a combination of [**KM18**, Th. 3] and [**KM18**, Remark 4.2], and generalizes earlier work [**FK11**].

THEOREM 10.5.7 (Käenmäki, Morris [**KM18**]). *Let $k \in \{0, \dots, d-1\}$ and $k < s < k+1$. Suppose that the m-tuple $(A_1, \dots, A_m)$ of $d \times d$ invertible matrices is k- and $(k+1)$-irreducible and either k-strongly irreducible or $(k+1)$-strongly irreducible. Then there exists a unique ergodic shift-invariant measure μ such that*

$$P(s) = h_\mu - E_\mu(s).$$

Moreover, there exists a uniform constant $C \geqslant 1$ such that for every finite word $\mathbf{i} \in \Sigma^*$,

$$C^{-1} \leqslant \frac{\mu([\mathbf{i}])}{e^{-|\mathbf{i}|P(s)}\varphi^s(A_\mathbf{i})} \leqslant C.$$

In particular, this measure is fully supported.

Note that in the case $s \geqslant d$ the uniqueness of the equilibrium measure is standard, by [**Bow75**], and for $s = k$, in the k-strongly irreducible case, it follows by [**FL09**].

We do not give complete proofs of Theorems 10.5.3 and 10.5.7. To demonstrate the main ideas behind the theorems, we prove only a special case, following Feng [**Fen09**], Feng and Lau [**FL02**] and Feng and Käenmäki [**FK11**].

THEOREM 10.5.8. *Let $(A_1, \dots, A_m)$ be a 1-irreducible tuple of 2×2 invertible matrices. Then for every $s \geqslant 0$ there exist a unique fully supported ergodic shift-invariant measure $\mu = \mu^s$ and a constant $C \geqslant 1$ such that for every finite word $\mathbf{i} \in \Sigma^*$*

$$C^{-1} \leqslant \frac{\mu([\mathbf{i}])}{e^{-|\mathbf{i}|P(s)}\varphi^s(A_\mathbf{i})} \leqslant C. \tag{10.68}$$

This is the φ^s-equilibrium measure, that is, $P(s) = h_\mu - E_\mu(s)$.

Before we prove the theorem, we need an important lemma, which guarantees that the product of the matrices in the tuple is quasi-multiplicative.

LEMMA 10.5.9. *The 1-irreducible tuple $(A_1, \dots, A_m)$ of 2×2 invertible matrices is quasi-multiplicative, namely, there exist $C \geqslant 1$ and $K \geqslant 1$ such that for every $\mathbf{i}, \mathbf{j} \in \Sigma^*$ there exists $\omega \in \Sigma^*$, with $|\omega| \leqslant K$, such that*

$$\|A_\mathbf{i} A_\omega A_\mathbf{j}\| \geqslant C^{-1} \|A_\mathbf{i}\| \|A_\mathbf{j}\|.$$

PROOF. We argue by contradiction, namely, suppose that for every $C \geqslant 1$ and $K \geqslant 1$ there exist $\mathbf{i}, \mathbf{j} \in \Sigma^*$ such that for every $\omega \in \Sigma^*$ with $|\omega| \leqslant K$,

$$\|A_{\mathbf{i}} A_\omega A_{\mathbf{j}}\| < C^{-1} \|A_{\mathbf{i}}\| \|A_{\mathbf{j}}\|.$$

Let $u_{\mathbf{i}}$ and $v_{\mathbf{j}}$ be unit vectors such that $\|A_{\mathbf{i}}^T u_{\mathbf{i}}\| = \|A_{\mathbf{i}}\|$ and $\|A_{\mathbf{j}} v_{\mathbf{j}}\| = \|A_{\mathbf{j}}\|$, and consider $\widetilde{u}_{\mathbf{i}} = \frac{A_{\mathbf{i}}^T u_{\mathbf{i}}}{\|A_{\mathbf{i}}\|}$ and $\widetilde{v}_{\mathbf{j}} = \frac{A_{\mathbf{j}} v_{\mathbf{j}}}{\|A_{\mathbf{j}}\|}$. Thus, $|\langle A_\omega \widetilde{u}_{\mathbf{j}}, \widetilde{u}_{\mathbf{i}} \rangle| < C^{-1}$ by our assumption. Letting $C \to \infty$ and $K \to \infty$ and choosing $\widetilde{u}$ and $\widetilde{v}$ to be accumulation points of $\{\widetilde{u}_{\mathbf{i}}\}$ and $\{\widetilde{v}_{\mathbf{j}}\}$, we get that for every $\omega \in \Sigma^*$,

$$\langle A_\omega \widetilde{v}, \widetilde{u} \rangle = 0.$$

Consider the subspace $V = \operatorname{span}\{A_\omega \widetilde{v} : \omega \in \Sigma^*\}$. By definition, $V \neq \{0\}$ and $A_i V = V$ for every $i = 1, \dots, m$. Moreover, since $\widetilde{u} \perp V$, V is a proper subspace, which contradicts our assumption. □

To simplify notation, denote $Z_n^s = \sum_{|\mathbf{i}|=n} \varphi^s(A_{\mathbf{i}})$.

LEMMA 10.5.10. *The sequence Z_n^s is almost multiplicative; that is, there exists a constant $D > 0$ such that for every $n, p \geqslant 0$,*

$$Z_{n+p}^s \leqslant Z_n^s Z_p^s \leqslant D Z_{n+p}^s. \tag{10.69}$$

Moreover, for every $n \geqslant 1$,

$$e^{nP(s)} \leqslant Z_n^s \leqslant D e^{nP(s)}. \tag{10.70}$$

PROOF. The first inequality in (10.69) holds by sub-multiplicativity (10.10), so we focus on the second one. For any invertible 2×2 matrix A we have by (10.8):

$$\varphi^s(A) = \begin{cases} \|A\|^s, & \text{if } 0 \leqslant s < 1 \\ \|A\|^{2-s} |\det(A)|^{s-1}, & \text{if } 1 \leqslant s < 2 \\ |\det(A)|^{2/s} & \text{if } s \geqslant 2. \end{cases} \tag{10.71}$$

Thus, by Lemma 10.5.9, for every $\mathbf{i}, \mathbf{j} \in \Sigma^*$ there exists ω with $|\omega| \leqslant K$, such that

$$\varphi^s(A_{\mathbf{i}}) \varphi^s(A_{\mathbf{j}}) \leqslant C' \varphi^s(A_{\mathbf{i}} A_\omega A_{\mathbf{j}}) \leqslant C' \sum_{|\omega| \leqslant K} \varphi^s(A_{\mathbf{i}} A_\omega A_{\mathbf{j}}). \tag{10.72}$$

Therefore,

$$\begin{aligned} Z_{n+p}^s \leqslant Z_n^s Z_p^s &\leqslant C' \sum_{|\mathbf{i}|=n} \sum_{|\omega| \leqslant K} \sum_{|\mathbf{j}|=p} \varphi^s(A_{\mathbf{i}} A_\omega A_{\mathbf{j}}) \\ &= C' \sum_{|\mathbf{i}|=n} \sum_{|\omega| \leqslant K} \sum_{|\mathbf{j}|=p} \varphi^s(A_{\mathbf{i}} A_{\mathbf{j}} A_\omega) \\ &\leqslant Z_{n+p}^s C' \sum_{k=0}^{K} Z_k^s. \end{aligned}$$

Thus, taking $D := C' \sum_{k=0}^K Z_k^s$ we get (10.69). Furthermore, by (10.69),

$$\left(Z_{n \cdot p}^s\right)^{1/p} \leqslant Z_n^s \leqslant D \left(Z_{n \cdot p}^s\right)^{1/p},$$

and letting $p \to \infty$, we get (10.70) by the definition of pressure (10.54). □

PROOF OF THEOREM 10.5.8. We need to show the existence of an equilibrium measure satisfying (10.68). For a finite word $\mathbf{j} \in \Sigma^*$, denote by $\delta_{\mathbf{j}}$ the Dirac mass on $\mathbf{j}1^\infty$, where $1^\infty = (1, 1, \ldots)$. Let

$$\nu_n := (Z_n^s)^{-1} \sum_{|\mathbf{j}|=n} \varphi^s(A_{\mathbf{j}})\delta_{\mathbf{j}} \quad \text{and} \quad \mu_n^s = \frac{1}{n}\sum_{\ell=0}^{n-1} \nu_{2n} \circ \sigma^{-\ell}.$$

Let μ^s be an arbitrary accumulation point of the sequence $(\mu_n^s)_{n\geqslant 1}$ in the weak* topology and consider n_k such that $\mu_{n_k}^s \to \mu^s$. It is easy to see that μ^s is σ-invariant.

Next, for a finite word $\mathbf{i} \in \Sigma^*$,

$$\begin{aligned} \mu^s([\mathbf{i}]) &= \lim_{k\to\infty} \frac{1}{n_k} \sum_{\ell=0}^{n_k-1} \sum_{|\tau|=\ell} \nu_{2n_k}([\tau\mathbf{i}]) \\ &= \lim_{k\to\infty} \frac{1}{n_k} \sum_{\ell=0}^{n_k-1} \sum_{|\tau|=\ell} \sum_{|w|=2n_k-|\mathbf{i}|-\ell} \nu_{2n_k}([\tau\mathbf{i}w]). \end{aligned} \tag{10.73}$$

Hence, by the definition of ν_{2n_k} and subadditivity of φ^s,

$$\begin{aligned} \mu^s([\mathbf{i}]) &\leqslant \varphi^s(A_{\mathbf{i}}) \lim_{k\to\infty} \frac{1}{n_k} \sum_{\ell=0}^{n_k-1} \frac{Z_\ell^s Z_{2n_k-|\mathbf{i}|-\ell}^s}{Z_{2n_k}^s} \\ &\leqslant \varphi^s(A_{\mathbf{i}}) \cdot D \cdot \lim_{k\to\infty} \frac{Z_{2n_k-|\mathbf{i}|}^s}{Z_{2n_k}^s} \leqslant \varphi^s(A_{\mathbf{i}}) \cdot D^2 e^{-|\mathbf{i}|P(s)}, \end{aligned}$$

where we used (10.69) and then (10.70). This gives the upper bound in (10.68), and now we turn to showing the lower bound. We again start with (10.73), use submultiplicativity of φ^s, and denote $D' := \sum_{k=0}^K Z_k^s$, to obtain

$$\begin{aligned} \mu^s([\mathbf{i}]) &\geqslant \frac{1}{(D')^2} \liminf_{k\to\infty} \frac{1}{n_k} \sum_{\ell=0}^{n_k-1} \sum_{|\tau|=\ell} \sum_{|\tau'|\leqslant K} \sum_{|\omega|=2n_k-|\mathbf{i}|-\ell} \sum_{|\omega'|\leqslant K} \frac{\varphi^s(A_{\tau'\tau\mathbf{i}\omega\omega'})}{Z_{2n_k}^s} \\ &= \frac{1}{(D')^2} \liminf_{k\to\infty} \frac{1}{n_k} \sum_{\ell=0}^{n_k-1} \sum_{|\tau|=\ell} \sum_{|\tau'|\leqslant K} \sum_{|\omega|=2n_k-|\mathbf{i}|-\ell} \sum_{|\omega'|\leqslant K} \frac{\varphi^s(A_{\tau\tau'\mathbf{i}\omega'\omega})}{Z_{2n_k}^s}. \end{aligned}$$

Hence, by (10.72), applied twice, and then (10.70),

$$\begin{aligned} \mu^s([\mathbf{i}]) &\geqslant \frac{(C')^2}{(D')^2} \liminf_{k\to\infty} \frac{1}{n_k} \sum_{\ell=0}^{n_k-1} \sum_{|\tau|=\ell} \sum_{|\omega|=2n_k-|\mathbf{i}|-\ell} \frac{\varphi^s(A_\tau)\varphi^s(A_{\mathbf{i}})\varphi^s(A_\omega)}{Z_{2n_k}^s} \\ &= \frac{(C')^2}{(D')^2} \varphi^s(A_{\mathbf{i}}) \liminf_{k\to\infty} \frac{1}{n_k} \sum_{\ell=0}^{n_k-1} \frac{Z_\ell^s Z_{2n_k-|\mathbf{i}|-\ell}^s}{Z_{2n_k}^s} \\ &\geqslant \frac{(C')^2}{(D')^2} \varphi^s(A_{\mathbf{i}}) \liminf_{k\to\infty} \frac{Z_{2n_k-|\mathbf{i}|}^s}{Z_{2n_k}^s} \\ &\geqslant \frac{(C')^2}{D(D')^2} \varphi^s(A_{\mathbf{i}})\, e^{-|\mathbf{i}|P(s)}, \end{aligned}$$

concluding the proof of (10.68).

The property $h_{\mu^s} - E_{\mu^s}(s) = P(s)$ immediately follows from (10.68), Definition 10.5.1 and Corollary 9.5.2.

Finally, we prove the uniqueness and ergodicity of μ^s. It is enough to show that any measure which satisfies (10.68) is ergodic. The uniqueness will follow from Corollary 9.1.2 and the fact that any two measures satisfying (10.68) are mutually absolutely continuous.

So let μ be a shift-invariant measure satisfying (10.68). Let $\mathbf{i}, \mathbf{j} \in \Sigma^*$. Then for any $n > |\mathbf{i}|$,

$$\sum_{j=0}^{K} \mu([\mathbf{i}] \cap \sigma^{-n-j}[\mathbf{j}]) = \sum_{j=0}^{K} \sum_{|v|=n+j-|\mathbf{i}|} \mu([\mathbf{i}v\mathbf{j}])$$
$$\geqslant C^{-1} \sum_{j=0}^{K} e^{-(n+j+|\mathbf{j}|)P(s)} \sum_{|v|=n+j-|\mathbf{i}|} \varphi^s(A_{\mathbf{i}} A_v A_{\mathbf{j}})$$

then writing $v = \omega u$, $|\omega| = j \leqslant K$, $|u| = n - |\mathbf{i}|$, and using (10.72),

$$\geqslant (C')^{-1} C^{-1} e^{-(n+|\mathbf{j}|)P(s)} \min\{1, e^{-KP(s)}\} \sum_{|u|=n-|\mathbf{i}|} \varphi^s(A_{\mathbf{i}}) \varphi^s(A_u A_{\mathbf{j}})$$
$$\geqslant (C')^{-1} C^{-3} \min\{1, e^{-KP(s)}\} \mu([\mathbf{i}]) \mu(\sigma^{-(n-|\mathbf{i}|)}[\mathbf{j}])$$
$$= (C')^{-1} C^{-3} \min\{1, e^{-KP(s)}\} \mu([\mathbf{i}]) \mu([\mathbf{j}]).$$

Approximating arbitrary measurable $A, B \subset \Sigma$ by disjoint unions of cylinder sets, we obtain for a positive constant:

$$\liminf_{n\to\infty} \sum_{j=0}^{K} \mu(A \cap \sigma^{-n-j} B) \geqslant \text{const} \cdot \mu(A)\mu(B).$$

If μ is not ergodic, then there exists $B = \sigma^{-1}B$ such that $0 < \mu(B) < 1$. But then

$$0 < \text{const} \cdot \mu(B)\mu(B^c) \leqslant \liminf_{n\to\infty} \sum_{j=0}^{K} \mu(B \cap \sigma^{-n-j} B^c) = K\mu(B \cap B^c) = 0,$$

which is a contradiction. □

10.6. Dominated matrices

In this section, we discuss briefly a special property of matrix tuples, which has several nice features and important applications. We restrict ourselves to the planar case for simplicity.

DEFINITION 10.6.1. We say that the tuple $(A_1, \ldots, A_m)$ of 2×2 invertible matrices is *dominated* if there exist $C_1 > 0$ and $0 < \tau < 1$ such that for every $\mathbf{i} \in \Sigma^*$,

$$\frac{\alpha_2(A_{\mathbf{i}})}{\alpha_1(A_{\mathbf{i}})} \leqslant C_1 \tau^{|\mathbf{i}|}.$$

The notion of domination is closely related to the action of the matrix group $GL(2, \mathbb{R})$ on the projective line. We say that $\mathcal{C} \subset \mathbb{PR}^1$ is a *multicone* if it is a proper subset and a finite union of closed projective intervals. Sometimes we refer to closed projective intervals as *cones*.

The angle $\sphericalangle(V, W)$ between two subspaces $V, W \in \mathbb{PR}^1$ is defined by

$$\sin \sphericalangle(V, W) := \frac{\|v \wedge w\|}{\|v\| \|w\|}, \text{ where } v \in V, \ w \in W, \tag{10.74}$$

are arbitrary nonzero vectors and $\wedge$ denotes the exterior product.

THEOREM 10.6.2 (Avila, Bochi, Yoccoz [**ABY10**]). *The tuple $(A_1, \ldots, A_m)$ of 2×2 invertible matrices is dominated if and only if there exists a strongly invariant multicone $\mathcal{C}$, that is, $\bigcup_{i=1}^m A_i\mathcal{C} \subset \mathcal{C}^o$, where $\mathcal{C}^o$ denotes the interior of $\mathcal{C}$.*

By Theorem 10.6.2, a simple example of a dominated matrix tuple is when all entries of all matrices are strictly positive, in which case the positive quadrant of the plane is strongly invariant.

Clearly, if the tuple $(A_1, \ldots, A_m)$ of 2×2 invertible matrices has a strongly invariant multicone $\mathcal{C}$ then the multicone $\mathcal{D} = \overline{\mathbb{PR}^1 \backslash \mathcal{C}}$ is strongly invariant with respect to the matrices $(A_1^{-1}, \ldots, A_m^{-1})$. Recall that a nonsingular matrix A acts on the projective line $\mathbb{PR}^1$ by $\operatorname{span}\{\mathbf{x}\} \mapsto \operatorname{span}\{A\mathbf{x}\}$. We will abuse the notation slightly by writing $\theta \mapsto A\theta$ for the induced action. We will also write $A|_V$ for the restriction of A to the line corresponding to $V \in \mathbb{PR}^1$ and denote by $\|A|_V\|$ the norm of A restricted to V. Note that, by definition, for any two matrices A, B and a subspace V,

$$\|AB|_V\| = \|A|_{BV}\| \cdot \|B|_V\|. \tag{10.75}$$

An important property of dominated matrix tuples is the following lemma, which essentially states that the action of the matrices over the subspaces in $\mathcal{C}$ is uniformly comparable in norm to the largest singular value.

LEMMA 10.6.3 (Bochi, Morris [**BM15**]). *Let $\mathcal{B}, \mathcal{C} \subset \mathbb{PR}^1$ be multicones such that $\mathcal{B} \subset \mathcal{C}^o$. Then there exists a constant $C_2 > 1$ such that for every 2×2 invertible matrix A with $A\mathcal{C} \subset \mathcal{B}$ and $V \in \mathcal{B}$ we have $\|A|_V\| \geqslant C_2^{-1}\|A\|$.*

The proof is left as an exercise; the extension to higher dimensions may be found in [**BM15**, Lemma 2.2]. Now suppose that $(A_1, \ldots, A_m)$ is a dominated matrix tuple, and let $\mathcal{C}$ be the corresponding strongly invariant multicone. Then using the definition of the angle $\sphericalangle$ (10.74), Lemma 10.6.3, and the equalities $|\det(A_{\mathbf{i}})| = \alpha_1(A_{\mathbf{i}})\alpha_2(A_{\mathbf{i}})$, $\|A_{\mathbf{i}}\| = \alpha_1(\mathbf{i})$, we get that for every $V, W \in \mathcal{C}$,

$$\begin{aligned}\sin \sphericalangle(A_{\mathbf{i}}V, A_{\mathbf{i}}W) &= \frac{|\det(A_{\mathbf{i}})|}{\|A_{\mathbf{i}}|_V\|\|A_{\mathbf{i}}|_W\|} \sin \sphericalangle(V, W) \\ &\leqslant C_2^2 \cdot \frac{\alpha_2(A_{\mathbf{i}})}{\alpha_1(A_{\mathbf{i}})} \sin \sphericalangle(V, W) \leqslant C_1C_2^2 \cdot \tau^{|\mathbf{i}|} \sin \sphericalangle(V, W).\end{aligned}$$

This implies that for every $\mathbf{i} \in \Sigma$ the diameter of $A_{\mathbf{i}|_n}\mathcal{C}$ converges to 0 as n tends to infinity, and thus we can define a natural projection $V \colon \Sigma \to \mathbb{PR}^1$ as

$$\{V(\mathbf{i})\} := \bigcap_{n=1}^{\infty} A_{\mathbf{i}|_n}\mathcal{C}.$$

Notice that, by definition, for any $\mathbf{i} \in \Sigma$ we have $V(\mathbf{i}) = A_{i_1}V(\sigma\mathbf{i})$. It is easy to see that V is Hölder continuous with respect to the usual metric on the symbolic space and $\sphericalangle$. Similarly, we can define a Hölder continuous map $W \colon \Sigma \to \mathbb{PR}^1$ by

$$\{W(\mathbf{i})\} := \bigcap_{n=1}^{\infty} A_{i_1}^{-1} \cdots A_{i_n}^{-1}\mathcal{D}, \text{ where } \mathcal{D} = \overline{\mathbb{PR}^1 \backslash \mathcal{C}}. \tag{10.76}$$

From the Hölder continuity of V it follows that the map $f(\mathbf{i}) := \log \|A_{i_1}|_{V(\sigma\mathbf{i})}\|$ is Hölder continuous. The map $g(\mathbf{i}) := \log |\det(A_{i_1})|$ is clearly Hölder continuous as well. Let $S_n f(\mathbf{i}) = \sum_{k=0}^{n-1} f(\sigma^k \mathbf{i})$, and similarly define $S_n g$. Then by Lemma 10.6.3 and (10.75), for every $\mathbf{i} \in \Sigma$ and $n \geqslant 1$ we have that for $0 \leqslant s \leqslant 1$,

$$e^{s \cdot S_n f(\mathbf{i})} \leqslant \varphi^s(A_{\mathbf{i}|_n}) = \|A_{\mathbf{i}|_n}\|^s \leqslant C_2 \|A_{\mathbf{i}|_n}|_{V(\sigma^n \mathbf{i})}\|^s = C_2 e^{s \cdot S_n f(\mathbf{i})}, \tag{10.77}$$

and for $1 \leqslant s \leqslant 2$,

$$\begin{aligned} e^{(2-s)\cdot S_n f(\mathbf{i}) + (s-1)\cdot S_n g(\mathbf{i})} &\leqslant \varphi^s(A_{\mathbf{i}|_n}) = \|A_{\mathbf{i}|_n}\|^{2-s} |\det(A_{\mathbf{i}|_n})|^{s-1} \\ &\leqslant C_2 \|A_{\mathbf{i}|_n}|_{V(\sigma^n \mathbf{i})}\|^{2-s} |\det(A_{\mathbf{i}|_n})|^{s-1} \\ &= C_2 e^{(2-s)\cdot S_n f(\mathbf{i}) + (s-1)\cdot S_n g(\mathbf{i})}. \end{aligned} \tag{10.78}$$

The inequality (10.77), together with Lemma 10.2.2, implies that the Lyapunov exponent can be expressed as the integral of a Hölder continuous map, i.e.,

$$\chi_1(\mu) = -\int \log \|A_{i_1}|_{V(\sigma \mathbf{i})}\| \, d\mu(\mathbf{i}). \tag{10.79}$$

Similarly, there exists a constant $C > 0$ such that $\|(A^{-1})_{\mathbf{i}|_n}\| \leqslant C \|(A^{-1})_{\mathbf{i}|_n}|W(\sigma^n \mathbf{i})\|$ and hence,

$$\chi_2(\mu) = \int \log \|A_{i_1}^{-1}|_{W(\sigma \mathbf{i})}\| \, d\mu(\mathbf{i}). \tag{10.80}$$

Inequalities (10.77) and (10.78) also imply that φ^s is almost-multiplicative for dominated matrix tuples, that is, for every $\mathbf{i}, \mathbf{j} \in \Sigma^*$, holds

$$\varphi^s(A_{\mathbf{i}})\varphi^s(A_{\mathbf{j}}) \geqslant \varphi^s(A_{\mathbf{i}}A_{\mathbf{j}}) \geqslant C^{-1}\varphi^s(A_{\mathbf{i}})\varphi^s(A_{\mathbf{j}}).$$

Combining this with (10.68) we get that for every $\mathbf{i}, \mathbf{j} \in \Sigma^*$

$$\begin{aligned} \mu([\mathbf{ij}]) \geqslant C^{-1} e^{-(|\mathbf{i}|+|\mathbf{j}|)P(s)} \varphi^s(A_{\mathbf{i}}A_{\mathbf{j}}) &\geqslant C^{-2} e^{-(|\mathbf{i}|+|\mathbf{j}|)P(s)} \varphi^s(A_{\mathbf{i}})\varphi^s(A_{\mathbf{j}}) \\ &\geqslant C^{-4}\mu([\mathbf{i}])\mu([\mathbf{j}]). \end{aligned}$$

So equations (10.77) and (10.78), similarly to the proof of Theorem 10.5.8, imply the following statement:

PROPOSITION 10.6.4. *Suppose that $(A_1, \ldots, A_m)$ is a dominated matrix tuple of 2×2 invertible matrices. Then the φ^s-equilibrium measure, defined in Theorem 10.5.3, is a Gibbs measure with respect to a Hölder continuous potential for $0 < s < 2$.*

Bárány, Käenmäki and Morris [**BKM20**] proved that the converse statement holds as well, namely, if for $0 < s < 2$ the equilibrium measure μ is a Gibbs measure with respect to a Hölder continuous potential, then the tuple of matrices $(A_1, \ldots, A_m)$ is dominated or $\left(\sqrt{|\det(A_1^{-1})|}A_1, \ldots, \sqrt{|\det(A_m^{-1})|}A_m\right)$ is contained in a compact subgroup of $SL^{\pm}(2, \mathbb{R})$.

REMARK 10.6.5. One may wonder: how does Proposition 10.6.4 compare to Theorem 10.5.8? First, note that the conditions are different: domination does not imply 1-irreducibility and 1-irreducibility certainly does not imply domination. Second, although Theorem 10.5.8 does provide an ergodic equilibrium measure $\mu = \mu^s$, the potential ϕ^s need not be Hölder continuous, as was demonstrated to be the case under the domination assumption. Hölder continuity of the potential gives a crucial advantage in further investigations.

10.7. A survey of some related results

10.7.1. Smoothness of the affinity dimension. Domination plays an important role in other issues in the theory of affine iterated function systems as well. For instance, Feng and Shmerkin [**FS14**] proved that the affinity dimension $s(A_1, \ldots, A_m)$ of the IFS $\mathcal{F} = \left(f_i(\mathbf{x}) = A_i\mathbf{x} + \mathbf{t}_i\right)_{i=1}^m$ on $\mathbb{R}^d$ depends continuously on the tuple $(A_1, \ldots, A_m)$. The proof relies, in particular, on subsystems with dominated linear parts. Recently, Jurga and Morris [**JM20**] showed that the dependence of $s(A_1, \ldots, A_m)$ on $(A_1, \ldots, A_m)$ is analytic in the planar case (i.e., s depends locally analytically on all the elements of the 2×2 matrices), if $(A_1, \ldots, A_m)$ is dominated and the invariant multicone $\mathcal{C}$ is a simple cone.

10.7.2. Approximation of the affinity dimension. In general, it is a nontrivial problem to estimate the numerical value of the affinity dimension for specific iterated function systems. The naive approximation is very weak. Namely, let s_n be the unique solution of the equation $\sum_{|\mathbf{i}|=n} \varphi^{s_n}(A_\mathbf{i}) = 1$. Then $|s - s_n| = O(1/n)$.

Pollicott and Vytnova [**PV15**] gave a much more effective approximation method for planar invertible affine contractions with positive linear parts under the conditions of Hueter–Lalley Theorem 1.12.1, quoted from [**HL95**]. The idea is to reformulate the problem on the projective line and use powerful methods of thermodynamic formalism in the setting of real analytic expanding maps, which turns out to be possible thanks to the positivity of the matrices.

Morris [**Mor22**] generalized their result recently for higher dimensional self-affine systems. Here we state the special case of the result from Morris [**Mor22**] for planar self-affine IFS. Denote by $\lambda_1(A)$ the largest in modulus eigenvalue of a matrix A.

THEOREM 10.7.1 (Morris [**Mor22**]). *Let $A_1, \ldots, A_m$ be planar matrices such that A_i has strictly positive entries for every $i = 1, \ldots, m$. For each $n \geqslant 1$ and $s \in (0, 2)$, let*

$$t_n^1(s) = \sum_{|\mathbf{i}|=n} \frac{|\lambda_1(A_\mathbf{i})|^{s+2}}{|\lambda_1(A_\mathbf{i})|^2 - \det(A_\mathbf{i})}, \tag{10.81}$$

$$t_n^2(s) = \sum_{|\mathbf{i}|=n} \frac{|\lambda_1(A_\mathbf{i})|^{4-s} |\det(A_\mathbf{i})|^{s-1}}{|\lambda_1(A_\mathbf{i})|^2 - \det(A_\mathbf{i})}. \tag{10.82}$$

Moreover, for $\ell = 1, 2$ let

$$a_n^\ell(s) := \frac{(-1)^n}{n!} \det \begin{pmatrix} t_1^\ell(s) & n-1 & 0 & \cdots & 0 & 0 \\ t_2^\ell(s) & t_1^\ell(s) & n-2 & 0 & \cdots & 0 \\ t_3^\ell(s) & t_2^\ell(s) & t_1^\ell(s) & \ddots & 0 & 0 \\ \vdots & \vdots & \vdots & \ddots & \ddots & \vdots \\ t_{n-1}^\ell(s) & t_{n-2}^\ell(s) & t_{n-3}^\ell(s) & \cdots & t_1^\ell(s) & 1 \\ t_n^\ell(s) & t_{n-1}^\ell(s) & t_{n-2}^\ell(s) & \cdots & t_2^\ell(s) & t_1^\ell(s) \end{pmatrix}.$$

Suppose that $s(A_1, \ldots, A_m) \in (\ell - 1, \ell)$ for some $\ell \in \{1, 2\}$. For each $n \geqslant 1$ let $s_n^\ell \in \mathbb{R}$ denote the smallest positive real number such that $\sum_{i=1}^n a_i^\ell(s_n^\ell) = -1$. Then there exists $\gamma > 0$ such that

$$|s(A_1, \ldots, A_m) - s_n^\ell| = O(e^{-\gamma n^2}).$$

10.7.3. Dimension gap phenomenon. It follows from Theorems 10.1.2 and 10.5.3 that for a "typical" self-affine IFS, satisfying the norm bound (10.5), the dimension of the attractor is equal to the affinity dimension, which is equal to the Lyapunov dimension of the Käenmäki measure, which is also equal to the Hausdorff dimension of the push-forward of the Käenmäki measure. This suggests the following questions:

QUESTION 10.7.2. Let $\mathcal{F}$ be a self-affine IFS (10.1), with attractor Λ and natural projection $\Pi : \Sigma \to \Lambda$. Is it true that

$$\dim_{\mathrm{H}}(\Lambda) = \sup\{\dim_{\mathrm{H}}(\Pi_*\mu) : \mu \text{ is an ergodic measure on } \Sigma\}? \tag{10.83}$$

QUESTION 10.7.3. Assuming that the answer to the previous question is positive, does there exist a dimension-maximizing measure, i.e., can sup in (10.83) be replaced by max?

Analogous questions have been extensively studied in a more general context of dynamical systems. A long-standing open problem has been the analog of Question 10.7.2 for *expanding repellers*, when the invariant measure can be considered on the fractal itself, see e.g. [**GP97**, **GP96**]. We do not give a general definition, but note that the attractor of an IFS can be represented as an expanding repeller whenever the IFS has the Strong Separation Property. There is a large literature with miscellaneous results, giving affirmative answers in special cases, for instance, for many classes of self-affine carpets (self-affine carpets will be the topic of Section 11.5). However, Das and Simmons [**DS17**] demonstrated that the answer to Question 10.7.2 is negative, by constructing a "Barański sponge", a 3-dimensional generalization of Barański carpet, which has a "dimension gap", namely,

$$\exists\, c > 0 : \dim_{\mathrm{H}}(\Lambda) - \dim_{\mathrm{H}}(\Pi_*\mu) \geqslant c \text{ for every ergodic } \mu \text{ on } \Sigma.$$

The paper [**DS17**] contains a thorough review of the literature on Question 10.7.2, and we refer the reader to it for additional information.

10.7.4. Bernoulli measures and equilibrium states. Another natural question is whether it is possible that the φ^s-equilibrium measure is not just quasi-Bernoulli, but Bernoulli? It turns out that this may happen only for a very restrictive family of self-affine IFS. Let $\mathcal{F} = \big(f_i(\mathbf{x}) = A_i\mathbf{x} + \mathbf{t}_i\big)_{i=1}^m$ be an IFS of affine maps on $\mathbb{R}^d$. Morris and Sert [**MS19a**, Theorem 4] showed that if the m-tuple of matrices $(A_1, \dots, A_m)$ is irreducible (i.e., there is no proper subspace, which is preserved by all the matrices), then the equilibrium measure μ_K^s defined in Theorem 10.5.3 is not Bernoulli for $0 < s < d$, unless there is a change of coordinates $\mathbf{y} = B\mathbf{x}$ such that $Bf_i(B^{-1}\mathbf{x}) = BA_iB^{-1}\mathbf{x} + B\mathbf{t}_i$ are similarity maps for all $i \leqslant m$. Thus, if the matrices are irreducible and the equilibrium measure is Bernoulli, then the self-affine IFS is essentially self-similar.

Actually, Morris and Sert [**MS19a**] proved more. Namely, they showed that if the tuple of matrices $(A_1, \dots, A_m)$ is irreducible, the IFS is not self-similar up to a linear change of coordinates, and $0 < s(A_1, \dots, A_m) < d$, then for every self-affine measure $\nu = \sum_{i=1}^m p_i (f_i)_*\nu$ we have $\dim_{\mathrm{H}}(\nu) < s(A_1, \dots, A_m)$.

The case when the tuple $(A_1, \dots, A_m)$ is reducible is poorly understood in higher dimensions. Up to a linear change of coordinates, we may assume that all the matrices are upper-block triangular. In the case $d = 2$, it is not hard to prove that the φ^s-equilibrium measures are Bernoulli for every $s > 0$, see for example

[**BKM20**, Proposition 2.7]. For $d \geqslant 3$, Käenmäki and Morris [**KM18**] showed that the equilibrium measures depend only on the block-diagonal parts of the matrices. For further information, see [**KM18**].

10.7.5. Drop of the affinity dimension. There was a folklore conjecture in the theory of self-affine IFS, stating that the affinity dimension is always strictly reduced when any one of the maps is removed. The analog for self-similar IFS is, of course, trivial. This was settled by Bochi and Morris [**BM18**, Theorem 2].

THEOREM 10.7.4 (Bochi and Morris). *Let $m \geqslant 2$ and let $A_1, \ldots, A_m : \mathbb{R}^d \to \mathbb{R}^d$ be invertible affine contractions. Then $s(A_1, \ldots, A_{m-1}) < s(A_1, \ldots, A_m)$.*

This was derived from the description of equilibrium (Käenmäki) measures for the potential φ^s, specifically, from the claim that all of them are fully supported. The latter was shown in [**FK11**] for $d = 2$, extended to $d = 3$ in [**KM18**], and finally, extended to all d in [**BM18**, Theorem 1] (note that [**KM18**] was written earlier, but published simultaneously with [**BM18**]).

CHAPTER 11

Diagonally self-affine IFS

We introduced self-affine iterated function systems in Section 1.2. Here we consider the special case when the linear part of every mapping in the self-affine IFS is a diagonal matrix.

DEFINITION 11.0.1 (Diagonally self-affine IFS in $\mathbb{R}^d$). We say that $\mathcal{S}$ is a diagonally self-affine IFS in $\mathbb{R}^d$ if

$$\mathcal{S} := (S_i(\mathbf{x}) = A_i \cdot \mathbf{x} + \mathbf{t}_i)_{i=1}^m, \tag{11.1}$$

where A_i is a $d \times d$ diagonal matrix for $i = 1, \ldots, m$:

$$A_i = \mathrm{Diag}(r_{i,1}, \ldots, r_{i,d}) \text{ and } \mathbf{t}_i = (t_{i,1}, \ldots, t_{i,d}). \tag{11.2}$$

All the diagonal elements satisfy $r_{i,j} \in (-1,1)\backslash\{0\}$.

Without loss of generality, in this chapter we always assume that

$$S_i([0,1]^d) \subset [0,1]^d. \tag{11.3}$$

We emphasize that in the general case we do not assume any separation conditions.

First in Section 11.1, we present the Feng–Hu Theorem [**FH09**, Theorem 2.11] in full generality which asserts that the dimension of projections of ergodic measures can be expressed by the Ledrappier–Young formula. In fact, Ledrappier and Young [**LY85a**] in 1985 expressed the dimension of invariant measures of C^2 diffeomorphisms in terms of entropy and Lyapunov exponents. Feng and Hu established the appropriate version for diagonally self-affine IFS. To do so, they needed to introduce the concept of *projection entropy*. This concept is rather technical, and it may be hard to see its relevance at first. In Section 11.2, we present a typical application of the Feng–Hu Theorem, which does not use the projection entropy explicitly, because in this case the IFS is in $\mathbb{R}^2$ and strong separation is assumed.

To clarify the meaning of projection entropy, we discuss some of its properties in Sections 11.1.4 and 11.3.1. In addition, in Section 11.3 we consider the special case when $d = 1$; then the self-affine IFS is a self-similar IFS on the line with possible overlaps. Even here we confine ourselves to the homogeneous case (when all the contraction ratios are the same). In that special case we present a complete proof, which contains the idea of the proof in the most general case and shows the importance of projection entropy.

Then we continue the study of diagonally self-affine IFS and their box-counting dimension in Section 11.4 under a special separation condition, called rectangular open set condition. Finally, in Section 11.5, we compute the Hausdorff and box-counting dimension of self-affine sets of several important classes. The Przytycki–Urbański-sets and the Bedford–McMullen carpets are analyzed in detail. We include the upper bound for the Hausdorff dimension for completeness, although it does not rely on the Feng–Hu Theorem. We then survey other classes of diagonally self-affine sets, such as Gatzouras–Lalley carpets, Barański carpets and discuss the recent work extending the results to the case of lower-triangular self-affine IFS.

We should note that Kenyon and Peres [**KP96**] used the Ledrappier–Young formula to express the dimension of invariant measures for self-affine IFS on Bedford–McMullen carpets and their higher-dimensional generalizations.

11.1. The Feng–Hu Theorem in full generality

In order to define the projection entropy we need to introduce some terminology. A diagonally self-affine IFS $\mathcal{S} = (S_1, \ldots, S_m)$ on $\mathbb{R}^d$ can be viewed as a *direct product IFS* $\mathcal{S} = \Phi_1 \times \cdots \times \Phi_d$, where $\{\Phi_j\}_{j=1}^d$ are self-similar IFS's on the line:

$$\mathcal{S} = (S_k(z_1, \ldots, z_d))_{k=1}^m := \left((S_{k,1}(z_1), \ldots, S_{k,d}(z_d))\right)_{k=1}^m, \tag{11.4}$$

where

$$\Phi_j := (S_{i,j}(x) := r_{i,j}x + t_{i,j})_{i=1}^m, \quad j = 1, \ldots, d. \tag{11.5}$$

The expression "direct product of IFS's" is slightly misleading. For example, if we take the product of two IFS's on the line, each of which consists of the same number, say m, functions, then the direct product of these IFS's on the plane still consists of the same number of functions: m and not m^2.

We write $\Lambda \subset \mathbb{R}^d$ for the attractor of $\mathcal{S}$, and $\Sigma := \{1, \ldots, m\}^{\mathbb{N}}$ for the symbolic space. Further, $\mathcal{M}_\sigma(\Sigma)$ (resp. $\mathcal{E}_\sigma(\Sigma)$) is the set of σ-invariant (resp. ergodic) measures on Σ. The natural projection from Σ to Λ is denoted by Π. Similarly, the first $1 \leqslant \ell \leqslant d$ components of $\mathcal{S}$ determine a self-affine IFS on $\mathbb{R}^\ell$ denoted by

$$\mathcal{S}^\ell := \Phi_1 \times \cdots \times \Phi_\ell = \left((S_{k,1}(z_1), \ldots, S_{k,\ell}(z_\ell))\right)_{k=1}^m. \tag{11.6}$$

The attractor of $\mathcal{S}^\ell$ is denoted by $\Lambda_\ell \subset \mathbb{R}^\ell$.

11.1.1. Three different projections for diagonally self-affine IFS.

We use three different projections $\Pi_\ell : \Sigma \to \mathbb{R}^\ell$, $\Gamma_k : \Sigma \to \mathbb{R}$ and $\text{proj}_k : [0,1]^k \to [0,1]^{k-1}$, and introduce some measurable partitions corresponding to these projections. Below we collect notation used in this section. Figure 11.1 helps visualize it.

NOTATION 4.

γ: *Borel σ-algebra of $\mathbb{R}$.*

$\mathcal{P}$: *partition of Σ according to the first symbol:*
$$\mathcal{P} := \{[j] : 1 \leqslant j \leqslant m\}, \text{ where } [j] := \{\mathbf{i} \in \Sigma : i_1 = j\}.$$

μ, ν: *μ is an ergodic measure on Σ, $\nu := \Pi_*\mu$.*

Π_ℓ: *natural projection from Σ to Λ_ℓ, the attractor of the IFS $\mathcal{S}^\ell$.*

ν_ℓ: *push-forward measure of μ by Π_ℓ.*

χ_k: *for $k = 1, \ldots, d$ and $\mathbf{i} \in \Sigma$ we set $\psi_k(\mathbf{i}) := r_{i_1,k}$. Then $\chi_k = \chi_k(\mathcal{S}, \mu) := -\int_\Sigma \log \psi_k(\mathbf{i}) d\mu(\mathbf{i})$.*

η_ℓ: *measurable partition of Σ corresponding to Π_ℓ; that is,*
$$\eta_\ell := \left\{\Pi_\ell^{-1}(\mathbf{a}) : \mathbf{a} \in \Lambda_\ell\right\} \subset \Sigma.$$

$\{\mu_{\mathbf{i}}^{\eta_\ell}\}_{\mathbf{i} \in \Sigma}$: *the system of conditional measures (on Σ) corresponding to the measurable partition η_ℓ.*

Γ_k: *the natural projection from Σ to the attractor of the IFS Φ_k; that is, $\Pi_\ell(\mathbf{i}) = (\Gamma_1(\mathbf{i}), \ldots, \Gamma_\ell(\mathbf{i}))$.*

proj_k: *the projection from* $[0,1]^k$ *to the first* $k-1$ *coordinates; that is,* $\mathrm{proj}_k : [0,1]^k \to [0,1]^{k-1}$, $\mathrm{proj}_k(x_1,\dots,x_k) := (x_1,\dots,x_{k-1})$, $k = 2,\dots,d$.

ξ^k: *the corresponding measurable partition; that is,*

$$\xi^k(x_1,\dots,x_k) := \{\mathbf{y} \in [0,1]^k : x_1 = y_1,\dots,x_{k-1} = y_{k-1}\}.$$

$(\nu_k)^{\xi^k}_{\mathbf{x}}$: *the conditional measures of* ν_k *with respect to the partition element of* ξ^k *which contains* $\mathbf{x}$.

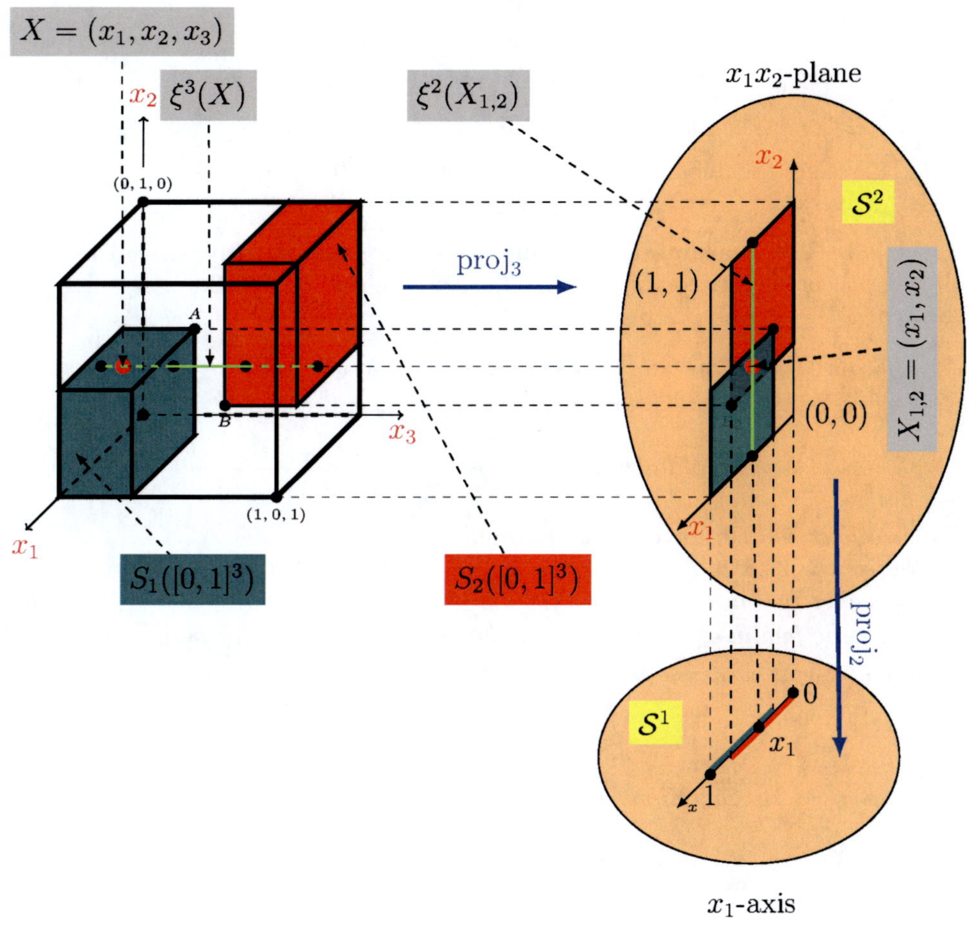

FIGURE 11.1. IFS's $\mathcal{S}^2, \mathcal{S}^1$ and partitions ξ^3 and ξ^2.

By conjugating the IFS with a proper permutation matrix, we may assume without loss of generality that

$$0 < \chi_1(\mu) \leqslant \chi_2(\mu) \leqslant \cdots \leqslant \chi_d(\mu). \tag{11.7}$$

11.1.2. Projection entropy. Following some ideas from the papers of Ledrappier and Young [**LY85a**] and [**LY85b**], Feng and Hu [**FH09**] introduced *projection entropy*. Below we use some facts about conditional information and conditional entropy (denoted by $I_\mu(\xi|\mathcal{B})$ and $H_\mu(\xi|\mathcal{B})$ respectively) of a countable measurable partition ξ given a σ-algebra $\mathcal{B}$. The definitions of $I_\mu(\xi|\mathcal{B})$ and $H_\mu(\xi|\mathcal{B})$

are given in (9.50), their most important properties are collected in Chapter 9. Further details are available, for example, in [**PU10**].

DEFINITION 11.1.1. Let $\mu \in \mathcal{M}_\sigma(\Sigma)$.

- The *projection entropy* of μ under Π with respect to $\mathcal{S}$ is

$$h_\Pi := h_\Pi(\mu) := H_\mu(\mathcal{P}|\sigma^{-1}\Pi^{-1}\gamma) - H_\mu(\mathcal{P}|\Pi^{-1}\gamma). \tag{11.8}$$

- The *local projection entropy* of μ at $\mathbf{i} \in \Sigma$ is

$$h_\Pi(\mu, \mathbf{i}) := \mathbb{E}_\mu\left(f|\mathcal{I}\right)(\mathbf{i}), \tag{11.9}$$

where $\mathcal{I}$ stands for the σ-algebra of σ-invariant sets (see (9.3)) and

$$f := I_\mu(\mathcal{P}|\sigma^{-1}\Pi^{-1}\gamma) - I_\mu(\mathcal{P}|\Pi^{-1}\gamma).$$

- We get the definition of h_{Π_ℓ} if we replace $\mathcal{S}$ and Π with $\mathcal{S}^\ell$ and Π_ℓ respectively.

11.1.3. The statement of Feng–Hu Theorem in full generality. The following theorem is a combination of [**FH09**, Theorem 2.11, Corollary 4.16 and Theorem 6.2].

THEOREM 11.1.2 (Feng, Hu [**FH09**]). *Let $\mu \in \mathcal{E}_\sigma(\Sigma)$ and $\nu := \Pi_*\mu$. Then ν is exact dimensional and for ν-a.e. $\mathbf{x}$ we have*

$$d_\nu(\mathbf{x}) = \dim_{\mathrm{H}}(\nu) = \frac{h_{\Pi_1}(\mu)}{\chi_1} + \sum_{j=2}^{d} \frac{h_{\Pi_j}(\mu) - h_{\Pi_{j-1}}(\mu)}{\chi_j}. \tag{11.10}$$

This is called the Ledrappier–Young formula.

Moreover, for $\mathbf{x} = (x_1, \dots, x_d) \in \Lambda$ and $k = 2, \dots, d$, let $\mathbf{x}_k = (x_1, \dots, x_k) \in \Lambda_k$. Consider the measure ν_k on Λ_k, the projection $\mathrm{proj}_k : \Lambda_k \to \Lambda_{k-1}$, *and the corresponding family of conditional measures $(\nu_k)^{\xi^k}_{\mathbf{x}_k}$, defined for ν_k-a.e. $\mathbf{x}_k$. Then for ν_k-a.e. $\mathbf{x}_k$ the measure $(\nu_k)^{\xi^k}_{\mathbf{x}_k}$ is exact dimensional, and*

$$\dim_{\mathrm{H}}(\nu_k)^{\xi^k}_{\mathbf{x}_k} = \frac{h_{\Pi_k}(\mu) - h_{\Pi_{k-1}}(\mu)}{\chi_k}. \tag{11.11}$$

The measure ν_1 is exact dimensional as well, and

$$\dim_{\mathrm{H}}(\nu_1) = \frac{h_{\Pi_1}(\mu)}{\chi_1}. \tag{11.12}$$

Thus,

$$\dim_{\mathrm{H}}(\nu) = \dim_{\mathrm{H}}(\nu_1) + \sum_{k=2}^{d} \dim_{\mathrm{H}}(\nu_k)^{\xi^k}_{\mathbf{x}_k}, \tag{11.13}$$

where $\nu_d = \nu$.

Now we explain the meaning of the formula (11.10) when $d = 3$, as in Figure 11.1. Take a ν-typical $\mathbf{x} = (x_1, x_2, x_3)$. Consider $\xi^3(\mathbf{x})$, (the green segment on the left side of Figure 11.1). The corresponding conditional measure is $(\nu)^{\xi^3}_{\mathbf{x}}$. Now we are in the top right of Figure 11.1. Consider the projection $\mathbf{x}_2 = (x_1, x_2)$ of $\mathbf{x}$ by proj_3. Let $\xi^2(\mathbf{x}_2)$ be the element of the partition ξ^2 which contains $\mathbf{x}_2$. (This is the vertical green segment in the top right of Figure 11.1.) The corresponding conditional measure is $(\nu_2)^{\xi^2}_{\mathbf{x}_2}$. By Theorem 11.1.2,

$$\dim_{\mathrm{H}}\left((\nu_3)^{\xi^3}_{\mathbf{x}}\right) = \frac{h_{\Pi_3}(\mu) - h_{\Pi_2}(\mu)}{\chi_3}, \tag{11.14}$$

and

$$\dim_{\mathrm{H}}\left((\nu_2)^{\xi^2}_{\mathbf{x}_2}\right) = \frac{h_{\Pi_2}(\mu) - h_{\Pi_1}(\mu)}{\chi_2}, \quad \dim_{\mathrm{H}}(\nu_1) = \frac{h_{\Pi_1}}{\chi_1}. \tag{11.15}$$

Moreover,

$$d_\nu(\mathbf{x}) = \dim_{\mathrm{H}}(\nu) = \dim_{\mathrm{H}}(\nu_1) + \dim_{\mathrm{H}}(\nu_2)^{\xi^2}_{\mathbf{x}_2} + \dim_{\mathrm{H}}(\nu_3)^{\xi^3}_{\mathbf{x}}. \tag{11.16}$$

for ν-almost every $\mathbf{x}$.

11.1.4. Projection entropy and conditional measures. Here we show the relation between projection entropy and conditional measures in case of self-affine measures. Let $\mathbf{p} = (p_1, \dots, p_m)$ be a probability vector and let $\mu_{\mathbf{p}}$ be the Bernoulli measure on Σ with probabilities $\mathbf{p}$. Recall that the entropy of the system is given by

$$h_{\mu_{\mathbf{p}}} = -\sum_{i=1}^{m} p_i \log p_i.$$

Now, we provide a simple formula for the projection entropy $h_{\Pi_\ell}(\mu)$ in this special case.

LEMMA 11.1.3. *Let $\mathcal{S}$ and $\mathcal{S}^\ell$ be the IFS's defined in* (11.4) *and* (11.6), *and let Π_ℓ and η_ℓ be the natural projection of $\mathcal{S}^\ell$ and the corresponding partition of Σ respectively, see Notation* 4. *Then*

$$h_{\Pi_\ell}(\mu_{\mathbf{p}}) = h_{\mu_{\mathbf{p}}} + \int_\Sigma \log(\mu^{\eta_\ell}_{\mathbf{i}}([i_1]))\, d\mu_{\mathbf{p}}(\mathbf{i}) \tag{11.17}$$

for all $\ell = 1, \dots, d$.

Lemma 11.1.3 will follow from the more general Proposition 12.3.8, but we give here an alternative proof for this special case. Recall that, by definition,

$$h_{\Pi_\ell}(\mu_{\mathbf{p}}) = H_{\mu_{\mathbf{p}}}(\mathcal{P}|\sigma^{-1}\Pi_\ell^{-1}\gamma) - H_{\mu_{\mathbf{p}}}(\mathcal{P}|\Pi_\ell^{-1}\gamma),$$

where γ denotes the Borel σ-algebra on $\mathbb{R}$. Lemma 11.1.3 follows from this equality, combined with the following lemma.

LEMMA 11.1.4. *Using the notation above, we have*

$$H_{\mu_{\mathbf{p}}}(\mathcal{P}|\sigma^{-1}\Pi_\ell^{-1}\gamma) = -\sum_{k=1}^{m} p_k \log(p_k). \tag{11.18}$$

$$H_{\mu_{\mathbf{p}}}(\mathcal{P}|\Pi_\ell^{-1}\gamma) = -\int_\Sigma \log(\mu^{\eta_\ell}_{\mathbf{i}}([i_1]))\, d\mu_{\mathbf{p}}(\mathbf{i}), \tag{11.19}$$

for every $\ell = 1, \dots, d$.

PROOF. Let $[k] \in \mathcal{P}$, then the σ-algebra generated by the function $\mathbb{1}_{[k]}$ is $\mathcal{C}_{\mathbb{1}_{[k]}} = \{\varnothing, [k], \Sigma\backslash[k], \Sigma\} \subseteq \widehat{\mathcal{P}}$, where $\widehat{\mathcal{P}}$ is the σ-algebra generated by the partition $\mathcal{P}$. One can easily check that $\widehat{\mathcal{P}}$ and $\sigma^{-1}\Pi_\ell^{-1}\gamma$ are independent σ-algebras. Indeed, any set $A \in \widehat{\mathcal{P}}$ can be written as $A = \bigcup_k [i_k]$ and so for every $B \in \Pi_\ell^{-1}\gamma$, $A \cap \sigma^{-1} B = \bigcup_k \{\mathbf{j} : j_1 = i_k, \sigma\mathbf{j} \in B\}$, and so

$$\mu_{\mathbf{p}}(A \cap \sigma^{-1}B) = \sum_k \mu_{\mathbf{p}}(\{\mathbf{j} : j_1 = i_k, \sigma\mathbf{j} \in B\}) = \sum_k p_{i_k}\mu_{\mathbf{p}}(B) = \mu_{\mathbf{p}}(A)\mu_{\mathbf{p}}(B).$$

Let $\varphi \in L^1(\Sigma, \mathcal{C}, \mu)$ and denote by $\mathcal{C}_\varphi$ the σ-algebra generated by the map φ. If $\mathcal{C}_\varphi$ and $\mathcal{G}$ are independent σ-algebras, then $\mathbb{E}_\mu(\varphi|\mathcal{G}) = \mathbb{E}_\mu(\varphi) = \int \varphi \, d\mu$. Thus, with the choice $\varphi = \mathbb{1}_{[k]}$, we obtain

$$\mathbb{E}_{\mu_{\mathbf{p}}}(\mathbb{1}_{[k]}|\sigma^{-1}\Pi_\ell^{-1}\gamma) = \mu_{\mathbf{p}}([k]) = p_k.$$

So by (9.50), we can conclude that

$$\begin{aligned} H_{\mu_{\mathbf{p}}}(\mathcal{P}|\sigma^{-1}\Pi_\ell^{-1}\gamma) &= -\int \sum_k \mathbb{1}_{[k]} \log \mathbb{E}_{\mu_{\mathbf{p}}}(\mathbb{1}_{[k]}|\sigma^{-1}\Pi_\ell^{-1}\gamma) \, d\mu_{\mathbf{p}}(\mathbf{i}) \\ &= -\sum_{k=1}^m p_k \log(p_k). \end{aligned}$$

On the other hand, an application of (9.37) and (9.50) gives

$$\begin{aligned} H_{\mu_{\mathbf{p}}}(\mathcal{P}|\Pi_\ell^{-1}\gamma) &= -\sum_{k=1}^m \int_\Sigma \mathbb{1}_{[k]}(\mathbf{i}) \log \mathbb{E}_{\mu_{\mathbf{p}}}(\mathbb{1}_{[k]}|\Pi_\ell^{-1}\gamma)(\mathbf{i}) \, d\mu_{\mathbf{p}}(\mathbf{i}) \\ &= -\sum_{k=1}^m \int_\Sigma \mathbb{1}_{[k]}(\mathbf{i}) \log \mu_{\mathbf{i}}^{\eta_\ell}([k]) \, d\mu_{\mathbf{p}}(\mathbf{i}) \\ &= -\int_\Sigma \log(\mu_{\mathbf{i}}^{\eta_\ell}([i_1])) \, d\mu_{\mathbf{p}}(\mathbf{i}). \end{aligned}$$

□

A simple corollary of Lemma 11.1.3 is the following.

Corollary 11.1.5. *If $\mathcal{S}^\ell$ satisfies the Strong Separation Property, then $h_{\Pi_\ell}(\mu_{\mathbf{p}}) = h_{\mathbf{p}}$.*

Proof. If the SSP holds, then $\eta_\ell(\mathbf{i}) = \Pi_\ell^{-1}(\Pi_\ell(\mathbf{i})) = \{\mathbf{i}\}$ for every $\mathbf{i} \in \Sigma$, hence $\mu_{\mathbf{i}}^{\eta_\ell} = \delta_{\mathbf{i}}$ and in particular, $\mu_{\mathbf{i}}^{\eta_\ell}([i_1]) = 1$. □

For further discussion, see Section 11.3.1.

11.2. Feng–Hu Theorem in $\mathbb{R}^2$ assuming separation

Here we consider box-like self-affine IFS like the one in Figure 11.2. More precisely, assume that we are given horizontal and vertical contractions: $(r_{i,1})_{i=1}^m$ and $(r_{i,2})_{i=1}^m$ respectively, with $|r_{i,j}| \in (0,1)$, and translations $(\mathbf{t}_i = (t_{i,1}, t_{i,2}))_{i=1}^m$, and consider the diagonally self-affine IFS in $\mathbb{R}^2$:

$$\mathcal{S} := \left(S_i(\mathbf{x}) = \begin{pmatrix} r_{i,1} & 0 \\ 0 & r_{i,2} \end{pmatrix} \cdot \begin{pmatrix} x_1 \\ x_2 \end{pmatrix} + \begin{pmatrix} t_{i,1} \\ t_{i,2} \end{pmatrix} \right)_{i=1}^m . \tag{11.20}$$

In this section we always assume that the so-called *Rectangular Strong Separation Property* (RSSP) holds:

$$S_i([0,1]^2) \cap S_j([0,1]^2) = \varnothing, \quad \forall i \neq j \in \{1, \ldots, m\}, \tag{11.21}$$

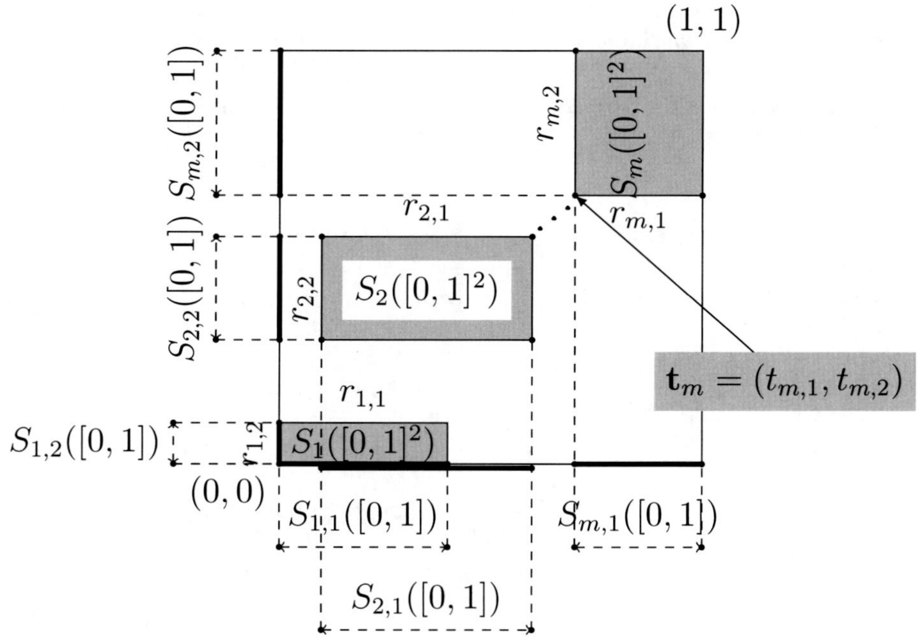

FIGURE 11.2. The IFS's Φ_1 and Φ_2 are the horizontal and vertical projections of $\mathcal{S} = (S_1, S_2, \ldots, S_m)$ respectively.

whereas, as we mentioned earlier,

$$S_i([0,1]^2) \subset [0,1]^2, \qquad \forall i \in \{1, \ldots, m\}$$

is always assumed as well.

As usual, we write Λ for the attractor of $\mathcal{S}$. The projections of $\mathcal{S}$ to the horizontal and vertical axes are the self-similar IFS Φ_1, Φ_2 respectively. That is,

$$\Phi_1 = (S_{i,1} = r_{i,1}x_1 + t_{i,1})_{i=1}^{m} \text{ and } \Phi_2 = (S_{i,2} = r_{i,2}x_2 + t_{i,2})_{i=1}^{m}. \tag{11.22}$$

In contrast with $\mathcal{S}$, it can happen that Φ_1 and Φ_2 satisfy no separation conditions, as can be seen in Figure 11.2.

Next we introduce the notation used in this section.

NOTATION 5. *As usual, we write $\Sigma := \{1, \ldots, m\}^{\mathbb{N}}$ for the symbolic space. The symbols Π_{Φ_1}, Π_{Φ_2} and $\Pi_{\mathcal{S}}$ stand for the natural projections which correspond to the IFS's Φ_1, Φ_2 and $\mathcal{S}$ respectively. Let $\mathbf{p} := (p_1, \ldots, p_m)$ be a probability vector. The push-forward measures of the infinite product measure $\mu_{\mathbf{p}} := \mathbf{p}^{\mathbb{N}}$ by Π_{Φ_1}, Π_{Φ_2} and $\Pi_{\mathcal{S}}$ are denoted by ν_1, ν_2 and ν respectively. The Lyapunov exponents of ν are χ_1, χ_2, which are actually the Lyapunov exponents of the self-similar measures ν_1 and ν_2 respectively, that is,*

$$\chi_j(\mathbf{p}) := -\sum_{i=1}^{m} p_i \log |r_{i,j}| > 0, \quad j = 1, 2.$$

Recall that the entropy of $\mu_{\mathbf{p}}$ is $h_{\mathbf{p}} = -\sum_{k=1}^{m} p_i \log p_i$.

Usually the Lyapunov exponents are enumerated in increasing order. In contrast with that, here we label them according to the axis to which the self-affine system is projected to, but without loss of generality (possibly applying the conjugation exchanging the coordinates $f(x,y) = (y,x)$), we may assume that the *direction x-dominates*, that is,

$$0 < \chi_1(\mathbf{p}) \leqslant \chi_2(\mathbf{p}). \tag{11.23}$$

For example, if all horizontal contractions are 1/3 and all vertical contractions are 1/9, then clearly the direction x-dominates: $\chi_1(\mathbf{p}) = \log 3 < \log 9 = \chi_2(\mathbf{p})$.

According to Definition 10.1.1, the Lyapunov dimension of the self-affine measure ν is

$$D(\nu) := \begin{cases} \dfrac{h_{\mathbf{p}}}{\chi_1(\mathbf{p})}, & \text{if } h_{\mathbf{p}} \leqslant \chi_1(\mathbf{p}); \\ 1 + \dfrac{h_{\mathbf{p}} - \chi_1(\mathbf{p})}{\chi_2(\mathbf{p})}, & \text{if } \chi_1(\mathbf{p}) < h_{\mathbf{p}} \leqslant \chi_1(\mathbf{p}) + \chi_2(\mathbf{p}); \\ 2 \cdot \dfrac{h_{\mathbf{p}}}{\chi_1(\mathbf{p}) + \chi_2(\mathbf{p})}, & \text{if } h_{\mathbf{p}} > \chi_1(\mathbf{p}) + \chi_2(\mathbf{p}). \end{cases} \tag{11.24}$$

The Feng–Hu Theorem 11.1.2 implies

THEOREM 11.2.1. *Assume that $\mathcal{S} := (S_1, \ldots, S_m)$ is a diagonally self-affine IFS, $\mathbf{p} = (p_1, \ldots, p_m)$ is a probability vector, the RSSP holds, and direction x-dominates: $0 < \chi_1(\mathbf{p}) \leqslant \chi_2(\mathbf{p})$. Then*

$$\dim_{\mathrm{H}}(\nu) = \frac{h_{\mathbf{p}}}{\chi_2(\mathbf{p})} + \left(1 - \frac{\chi_1(\mathbf{p})}{\chi_2(\mathbf{p})}\right) \cdot \dim_{\mathrm{H}}(\nu_1). \tag{11.25}$$

In particular,

$$D(\nu) = \dim_{\mathrm{H}}(\nu) \Longleftrightarrow \dim_{\mathrm{H}}(\nu_1) = \min\left\{1, \frac{h_{\mathbf{p}}}{\chi_1(\mathbf{p})}\right\}. \tag{11.26}$$

Theorem 11.1.2 implies Theorem 11.2.1 since in the latter we assumed strong separation, whence $h_{\Pi_2}(\mu) = h_\mu = h_{\mathbf{p}}$, see Corollary 11.1.5. Furthermore,

$$\begin{aligned} \dim_{\mathrm{H}}(\nu) &= \dim_{\mathrm{H}}(\nu_1) + \dim_{\mathrm{H}}(\nu_2)^{\xi^2}_{\mathbf{x}} \\ &= \frac{h_{\Pi_1}(\mu)}{\chi_1(\mathbf{p})} + \frac{h_{\Pi_2}(\mu) - h_{\Pi_1}(\mu)}{\chi_2(\mathbf{p})} \\ &= \frac{h_{\mathbf{p}}}{\chi_2(\mathbf{p})} + \left(1 - \frac{\chi_1(\mathbf{p})}{\chi_2(\mathbf{p})}\right) \frac{h_{\Pi_1}(\mu)}{\chi_1(\mathbf{p})}, \end{aligned}$$

and $\dim_{\mathrm{H}}(\nu_1) = \frac{h_{\Pi_1}(\mu)}{\chi_1(\mathbf{p})}$.

An alternative proof follows by Theorem 12.3.2 in the next section.

11.2.1. Feng–Wang formula. An important consequence of Theorem 11.2.1 is the following: in order to calculate the Hausdorff dimension of a diagonally self-affine measure ν under the RSSP it is enough to determine the dimension of its orthogonal projection onto the main axis, which is a self-similar measure. Needless to say that such reduction is essential in the understanding of geometric properties. Feng and Wang [**FW05**] obtained similar result for the L^q dimension and L^q-spectrum for diagonally self-affine measures with the RSSP.

Recall the definition of L^q-dimension and L^q-spectrum, see Definition 2.6.1 and (2.28) for $q > 0$. The symbol $\mathscr{D}_n$ stands for the partition of $\mathbb{R}^d$ into grid boxes $\prod_{i=1}^d [k_i 2^{-n}, (k_i+1)2^{-n})$, with $k_i \in \mathbb{Z}$. For $q \neq 0$ denote

$$\tau_n^{(q)}(\nu) = \sum_{Q \in \mathscr{D}_n} (\nu(Q))^q.$$

The L^q spectrum of ν is defined by

$$\tau(\nu, q) = \lim_{n\to\infty} \frac{\log \tau_n^{(q)}(\nu)}{-n \log 2} \quad \text{for } q > 0, \tag{11.27}$$

if the limit exists. By Lemma 2.6.6, the L^q dimension of ν is $D_q(\nu) = \frac{\tau(\nu,q)}{q-1}$ for $q > 0$ and $q \neq 1$.

Let $\mathcal{S}$ be the diagonally self-affine IFS of the form (11.20). Consider two closed subsets of the simplex $\Delta = \{(p_1, \dots, p_m) \in \mathbb{R}^m : p_i \geqslant 0 \text{ and } \sum_{i=1}^m p_i = 1\}$ with respect to the direction with the dominating contraction:

$$\Gamma_1 = \{\mathbf{p} \in \Delta : \chi_1(\mathbf{p}) \geqslant \chi_2(\mathbf{p})\}$$
$$\Gamma_2 = \{\mathbf{p} \in \Delta : \chi_1(\mathbf{p}) \leqslant \chi_2(\mathbf{p})\}.$$

The *relative entropy* $h(\mathbf{q}|\mathbf{p})$ from the probability vector $\mathbf{p}$ to $\mathbf{q}$ is defined by

$$h(\mathbf{q}|\mathbf{p}) = \sum_{i=1}^m q_i \log \frac{q_i}{p_i}. \tag{11.28}$$

It is also known as the *Kullback–Leibler divergence* in the literature. Using that $\log(x) \leqslant x - 1$, one can show that $h(\mathbf{q}|\mathbf{p}) \geqslant 0$, and $h(\mathbf{q}|\mathbf{p}) = 0$ implies $\mathbf{q} = \mathbf{p}$. The following theorem is due to Feng and Wang [**FW05**, Theorem 1].

THEOREM 11.2.2 (Feng and Wang). *Assume that $\mathcal{S} := (S_1, \dots, S_m)$ is a diagonally self-affine IFS, $\mathbf{p} = (p_1, \dots, p_m)$ is a probability vector, and the RSSP holds. Let $\nu = (\Pi_{\mathcal{S}})_* \mu_{\mathbf{p}}$, and let $\nu_1 = (\Pi_{\Phi_1})_* \mu_{\mathbf{p}}$, $\nu_2 = (\Pi_{\Phi_2})_* \mu_{\mathbf{p}}$. Consider the following quantities:*

$$\theta_1(\mathbf{p}, q) := \inf_{\mathbf{q} \in \Gamma_1} \frac{q h(\mathbf{q}|\mathbf{p}) - (1-q) h_{\mathbf{q}}}{\chi_1(\mathbf{q})} + \tau(\nu_2, q)\left(1 - \frac{\chi_2(\mathbf{q})}{\chi_1(\mathbf{q})}\right),$$
$$\theta_2(\mathbf{p}, q) := \inf_{\mathbf{q} \in \Gamma_2} \frac{q h(\mathbf{q}|\mathbf{p}) - (1-q) h_{\mathbf{q}}}{\chi_2(\mathbf{q})} + \tau(\nu_1, q)\left(1 - \frac{\chi_1(\mathbf{q})}{\chi_2(\mathbf{q})}\right).$$

Then $\tau(\nu, q) = \min\{\theta_1(\mathbf{p}, q), \theta_2(\mathbf{p}, q)\}$ for all $q > 0$.

In the special case when $|r_{i,2}| \leqslant |r_{i,1}|$ for every $i = 1, \dots, m$, one can show by a simple calculation that $\tau(\nu, q)$ is the unique root of the equation

$$\sum_{i=1}^m |r_{i,2}|^{\tau(\nu_1,q) - \tau(\nu,q)} |r_{i,1}|^{-\tau(\nu_1,q)} p_i^q = 1.$$

11.2.2. An application of Feng–Hu formula: Typical translations. Let us show a simple application of the Feng–Hu formula to planar diagonally self-affine sets. Recall that the affinity dimension $\dim_{\mathrm{aff}}(\mathcal{S})$ was introduced in Definition 1.10.2 as the most natural guess for the value of the Hausdorff and box dimension of the attractor Λ of $\mathcal{S}$, namely,

$$\dim_{\mathrm{aff}}(\mathcal{S}) = \inf\Big\{t \geqslant 0 : \sum_{n=0}^{\infty} \sum_{|\mathbf{i}|=n} \varphi^t(A_{\mathbf{i}}) < \infty\Big\}.$$

Since the matrices A_i for $i = 1, \ldots, m$ are diagonal, we get that

$$\varphi^t(A_{\mathbf{i}}) = \begin{cases} \max\{|r_{\mathbf{i},1}|, |r_{\mathbf{i},2}|\}^t, & \text{if } 0 \leqslant t \leqslant 1; \\ \max\{|r_{\mathbf{i},1}|, |r_{\mathbf{i},2}|\} \min\{|r_{\mathbf{i},1}|, |r_{\mathbf{i},2}|\}^{t-1}, & \text{if } 1 < t \leqslant 2; \\ \left(|r_{\mathbf{i},1}||r_{\mathbf{i},2}|\right)^{t/2}, & \text{if } t > 2. \end{cases} \tag{11.29}$$

The affinity dimension $\dim_{\mathrm{aff}}(\mathcal{S})$ of $\mathcal{S}$ can be computed explicitly as follows: First we compute the similarity dimensions s_j of the IFS's Φ_j, $j = 1, 2$. Let $\widehat{s}_j := \min\{s_j, 1\}$. We define d_1, d_2 as the solutions of

$$\sum_{i=1}^{m} |r_{i,1}|^{\widehat{s}_1} \cdot |r_{i,2}|^{d_1 - \widehat{s}_1} = 1 \quad \text{and} \quad \sum_{i=1}^{m} |r_{i,2}|^{\widehat{s}_2} \cdot |r_{i,1}|^{d_2 - \widehat{s}_2} = 1. \tag{11.30}$$

Then an easy calculation shows that if $\mathcal{S}$ satisfies the RSSP, then

$$\dim_{\mathrm{aff}}(\mathcal{S}) = \max\{d_1, d_2\}. \tag{11.31}$$

In order to explain the meaning of the formula (11.31), note that $\sum_{i=1}^{m} |r_{i,j}|^{s_j} = 1$, for $j = 1, 2$, hence

$$s_j \leqslant 1 \implies \widehat{s}_j = s_j \implies d_j = s_j.$$

On the other hand, if $s_j > 1$, then $\sum_{i=1}^{m} |r_{i,1}| \cdot |r_{i,2}|^{d_j - 1} = 1$. Observe also that $\sum_{i=1}^{m} |r_{i,1}| \cdot |r_{i,2}| < 1$ by the RSSP (11.21), hence $\widehat{s}_j \leqslant d_j < 2$ for $j = 1, 2$.

In order to verify (11.31), note that (11.29) yields

$$\begin{aligned} &\max\left\{\left(\sum_{i=1}^{m} |r_{i,1}|^{\min\{1,t\}} |r_{i,2}|^{\max\{1,t\}-1}\right)^n, \left(\sum_{i=1}^{m} |r_{i,2}|^{\min\{1,t\}} |r_{i,1}|^{\max\{1,t\}-1}\right)^n\right\} \\ &\quad \leqslant \sum_{|\mathbf{i}|=n} \varphi^t(A_{\mathbf{i}}) \\ &\quad \leqslant \left(\sum_{i=1}^{m} |r_{i,1}|^{\min\{1,t\}} |r_{i,2}|^{\max\{1,t\}-1}\right)^n + \left(\sum_{i=1}^{m} |r_{i,2}|^{\min\{1,t\}} |r_{i,1}|^{\max\{1,t\}-1}\right)^n. \end{aligned}$$

for every $n \geqslant 1$ and $t \in [0, 2]$. Then (11.31) follows by considering the cases $t \leqslant 1$ and $t \in (1, 2)$ separately.

Let us remark that

$$\max\{d_1, d_2\} \leqslant \widehat{s}_1 + \widehat{s}_2. \tag{11.32}$$

If $\max\{\widehat{s}_1, \widehat{s}_2\} < 1$ then $\widehat{s}_i = s_i$ and by the definition (11.30), $d_i = s_i$, and so, (11.32) follows. If $\min\{\widehat{s}_1, \widehat{s}_2\} = 1$ then by $\max\{d_1, d_2\} \leqslant 2$, (11.32) follows again. Hence, we may assume that $\min\{\widehat{s}_1, \widehat{s}_2\} < 1 = \max\{\widehat{s}_1, \widehat{s}_2\}$. Let us say that $\widehat{s}_2 < 1 = \widehat{s}_1$. Then $\widehat{s}_2 = s_2$, and so $d_2 = s_2$, and

$$\sum_{i=1}^{m} |r_{i,1}| \cdot |r_{i,2}|^{s_2} \leqslant \max_i\{|r_{i,1}|\} < 1,$$

which implies that $d_1 \leqslant 1 + s_2 \leqslant \widehat{s}_1 + \widehat{s}_2$.

The following theorem shows that for typical translations the dimension equals the affinity dimension if the RSSP holds, without the extra assumption of Proposition 10.4.1, $\max_{i \neq j} \{\max\{|r_{i,1}|, |r_{i,2}|\} + \max\{|r_{j,1}|, |r_{j,2}|\}\} < 1$. Moreover, “typical” is understood in a stronger sense, since we get an estimate of the packing dimension

of the exceptional set. We consider the same IFS as in (11.20), but emphasize the parameter dependence in notation:
(11.33)

$$\mathcal{S}^{\mathbf{t}} := \left(S_i^{\mathbf{t}}(\mathbf{x}) = \begin{pmatrix} r_{i,1} & 0 \\ 0 & r_{i,2} \end{pmatrix} \cdot \begin{pmatrix} x_1 \\ x_2 \end{pmatrix} + \begin{pmatrix} t_{i,1} \\ t_{i,2} \end{pmatrix} \right)_{i=1}^{m}, \quad \mathbf{t} = (\mathbf{t}_1, \ldots, \mathbf{t}_m) \in \mathbb{R}^{2m}.$$

The attractor of $\mathcal{S}^{\mathbf{t}}$ will be denoted by $\Lambda^{\mathbf{t}}$, and the push-forward self-affine/self-similar measures, as in Notation 5, by $\nu^{\mathbf{t}}$ and $\nu_j^{\mathbf{t}}$, $j = 1, 2$.

THEOREM 11.2.3. *Let $\mathcal{S}$ be the IFS of the form in* (11.33), *and let*

$$\mathcal{T} = \{\mathbf{t} = (\mathbf{t}_1, \ldots, \mathbf{t}_m) \in \mathbb{R}^{2m} : \text{ the RSSP holds for } \mathcal{S}^{\mathbf{t}}\}.$$

Then there exists a set $E \subset \mathcal{T}$ such that $\dim_{\mathrm{P}}(E) \leqslant 2m - 1$ *and for every* $(\mathbf{t}_1, \ldots, \mathbf{t}_m) \in \mathcal{T} \backslash E$ *we have*

$$\dim_{\mathrm{H}}(\Lambda^{\mathbf{t}}) = \dim_{\mathrm{B}}(\Lambda^{\mathbf{t}}) = \dim_{\mathrm{aff}}(\mathcal{S}^{\mathbf{t}}).$$

PROOF. By Proposition 10.2.6 and (11.31) we have $\overline{\dim}_{\mathrm{B}}(\Lambda^{\mathbf{t}}) \leqslant \max\{d_1, d_2\}$ for all $\mathbf{t}$, where d_1, d_2 are defined in (11.30), so it is enough to show that $\dim_{\mathrm{H}}(\Lambda^{\mathbf{t}}) \geqslant \max\{d_1, d_2\}$ for "typical" $\mathbf{t} \in \mathcal{T}$.

Suppose that $d_1 \geqslant d_2$. Let

$$\mathbf{p} = \left(|r_{i,1}|^{\hat{s}_1} |r_{i,2}|^{d_1 - \hat{s}_1} \right)_{i=1}^{m} \quad \text{and} \quad \mathbf{q} = \left(|r_{i,2}|^{\hat{s}_2} |r_{i,1}|^{d_2 - \hat{s}_2} \right)_{i=1}^{m}.$$

By the positivity of the relative entropy $h(\mathbf{p}|\mathbf{q})$ defined in (11.28), we have

$$\begin{aligned} 0 \leqslant h(\mathbf{p}|\mathbf{q}) &= \sum_{i=1}^{m} |r_{i,1}|^{\hat{s}_1} |r_{i,2}|^{d_1 - \hat{s}_1} \log \left(\frac{|r_{i,1}|^{\hat{s}_1} |r_{i,2}|^{d_1 - \hat{s}_1}}{|r_{i,2}|^{\hat{s}_2} |r_{i,1}|^{d_2 - \hat{s}_2}} \right) \\ &= (\hat{s}_1 + \hat{s}_2 - d_1)\chi_2(\mathbf{p}) - (\hat{s}_1 + \hat{s}_2 - d_2)\chi_1(\mathbf{p}). \end{aligned}$$

Since $\hat{s}_1 + \hat{s}_2 \geqslant d_1$ by (11.32), the assumption $d_1 \geqslant d_2$ implies that $\chi_1(\mathbf{p}) \leqslant \chi_2(\mathbf{p})$.

On the other hand, by Corollary 6.8.12 of Hochman, there exists a set $E_1 \subset \mathbb{R}^m$ such that $\dim_{\mathrm{P}}(E_1) \leqslant m - 1$ and $\dim_{\mathrm{H}}(\nu_1^{\mathbf{t}}) = \min\{1, \frac{h_{\mathbf{p}}}{\chi_1(\mathbf{p})}\}$ for every $(t_{1,1}, \ldots, t_{m,1}) \in \mathbb{R}^m \backslash E_1$. Hence, by Theorem 11.2.1, and in particular (11.26), for every $(t_{1,1}, \ldots, t_{m,1}, t_{1,2}, \ldots, t_{m,2}) \in \mathcal{T} \backslash (E_1 \times \mathbb{R}^m)$ holds

$$\begin{aligned} \dim_{\mathrm{H}}(\Lambda^{\mathbf{t}}) \geqslant \dim_{\mathrm{H}}(\nu^{\mathbf{t}}) = D(\nu^{\mathbf{t}}) &= \min\left\{ \frac{h_{\mathbf{p}}}{\chi_1(\mathbf{p})}, 1 + \frac{h_{\mathbf{p}} - \chi_1(\mathbf{p})}{\chi_2(\mathbf{p})} \right\} \\ &= \min\left\{ \hat{s}_1 + (d_1 - \hat{s}_1)\frac{\chi_2(\mathbf{p})}{\chi_1(\mathbf{p})}, 1 + (d_1 - \hat{s}_1) + (\hat{s}_1 - 1)\frac{\chi_1(\mathbf{p})}{\chi_2(\mathbf{p})} \right\} = d_1. \end{aligned}$$

Finally, (2.17) implies that $\dim_{\mathrm{P}}(E_1 \times \mathbb{R}^m) \leqslant 2m - 1$. □

11.3. Dimension of homogeneous self-similar measures

As we have already mentioned, we will not prove the Feng–Hu Theorem 11.1.2 in complete generality. However, we will prove an important corollary of their result, using the methods of [**FH09**]. Namely, we will show that homogeneous self-similar measures on $\mathbb{R}$ are exact dimensional. Let us first introduce the notation used in this section.

Let Λ be the attractor of the self-similar IFS $\mathcal{H} = (h_i(x) := r_i x + u_i)_{i=1}^{m}$ on $\mathbb{R}$. The symbolic space is $\Sigma = \{1, 2, \ldots, m\}^{\mathbb{N}}$ and the natural projection is $\Pi(\mathbf{i}) := \lim_{n \to \infty} h_{\mathbf{i}|n}(0)$ for $\mathbf{i} \in \Sigma$. For a probability vector $\mathbf{q} = (q_1, \ldots, q_m)$ we write

$\mu_{\mathbf{q}} := \mathbf{q}^{\mathbb{N}}$ for the infinite product measure on Σ, and its push-forward is denoted by $\nu_{\mathbf{q}}$. To measure how significant is the overlap between the cylinders $\Lambda_i := h_i(\Lambda)$, we consider the following partition of Σ:

$$\xi(\mathbf{i}) := \Pi^{-1}\Pi(\mathbf{i}).$$

Observe that ξ is a measurable partition of Σ by Proposition 9.4.11. We write $\widehat{\xi}$ for the σ-algebra generated by ξ. Thus, the elements of $\widehat{\xi}$ are unions of the elements of ξ. We can consider the family of conditional measures with respect to ξ, and Theorem 9.4.18 implies that there exists a subset $\widehat{\Sigma} \subset \Sigma$, with $\mu_{\mathbf{q}}(\widehat{\Sigma}) = 1$, such that for all $\mathbf{i} \in \widehat{\Sigma}$ there exists a probability measure $\mu_{\mathbf{i}}^{\xi}$ defined on $\xi(\mathbf{i})$ with the properties:

- for every Borel set $A \subset \Sigma$, the mapping $\mathbf{i} \mapsto \mu_{\mathbf{i}}^{\xi}(A)$ is $\widehat{\xi}$-measurable;
- for every Borel set $U \subset \Sigma$ we have

$$\mu_{\mathbf{q}}(U) = \int \mu_{\mathbf{i}}^{\xi}(U) d\mu_{\mathbf{q}}(\mathbf{i}). \tag{11.34}$$

As we have seen in Theorem 11.1.2, projection entropy, defined by formula (11.8), plays a central role in the dimension theory of box-like self-affine sets. In this section we demonstrate how the projection entropy appears naturally. Recall that the entropy and Lyapunov exponent of $\mu_{\mathbf{q}}$ are defined by

$$h_{\mathbf{q}} = -\sum_{i=1}^{m} q_i \log q_i. \text{ and } \chi(\mu_{\mathbf{q}}) = -\sum_{i=1}^{m} q_i \log r_i.$$

THEOREM 11.3.1 (Feng and Hu). *Let $\nu_{\mathbf{q}} = \Pi_*(\mu_{\mathbf{q}})$ be a self-similar measure for the self-similar IFS $\mathcal{H}$ on the line. Then $\nu_{\mathbf{q}}$ is exact dimensional and*

$$\dim_{\mathrm{H}}(\nu_{\mathbf{q}}) = \frac{h_{\Pi}(\mu_{\mathbf{q}})}{\chi(\mu_{\mathbf{q}})}, \tag{11.35}$$

where $h_{\Pi}(\mu_{\mathbf{q}})$ is the projection entropy introduced in Definition 11.1.1.

By Lemma 11.1.3, we obtain that

$$\dim_{\mathrm{H}}(\nu_{\mathbf{q}}) = \frac{h_{\mu_{\mathbf{q}}} - H}{\chi(\mu_{\mathbf{q}})}, \quad \text{where } H = -\int \log \mu_{\mathbf{i}}^{\xi}([i_1])\, d\mu_{\mathbf{q}}(\mathbf{i}) \geqslant 0. \tag{11.36}$$

In the proof we will assume for simplicity that $\mathcal{H}$ is a *homogeneous* self-similar IFS on the line:

$$\mathcal{S} := (S_i(x) = \rho x + t_i)_{i=1}^{m}, \quad 0 < \rho < 1, \tag{11.37}$$

and so $\chi(\mu_{\mathbf{q}}) = -\log \rho$. In order to better visualize the steps of the proof, let us "draw" the symbolic space Σ (see Figure 11.3) by identifying the elements of Σ with the elements of the attractor $\widetilde{\Lambda}$ of a self-similar IFS $\widetilde{S}$ on the plane. We can guarantee the existence of such an IFS $\widetilde{S}$ only if $\rho < \frac{1}{m}$. However, in the case when $\rho \geqslant \frac{1}{m}$, the steps of the proof are identical; the IFS $\widetilde{S}$ is used for a visual representation only, and the entire proof happens in the symbolic space.

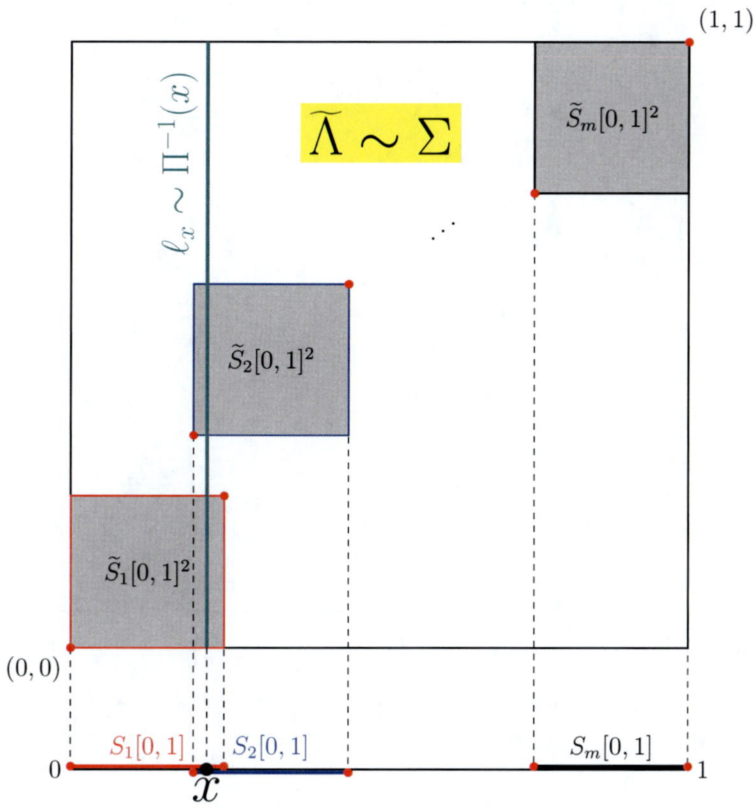

FIGURE 11.3. Symbolic space Σ is identified with the attractor $\widetilde{\Lambda}$ of $\widetilde{\mathcal{S}} = \left(\widetilde{S}_1, \dots, \widetilde{S}_m\right)$

Assuming that $\rho < \frac{1}{m}$, consider the self-similar IFS $\widetilde{\mathcal{S}}$ on $[0,1]^2$, given by

$$\widetilde{\mathcal{S}} := \left(\widetilde{S}_i(x_1, x_2) := \rho \cdot (x_1, x_2) + (t_i, t'_i)\right)_{i=1}^{m},$$

where we choose t'_i, $i = 1, \dots, m$, in such a way that $\widetilde{S}_i([0,1]^2) \cap \widetilde{S}_j([0,1]^2) = \varnothing$ if $i \neq j$. Then the attractor $\widetilde{\Lambda}$ of $\widetilde{\mathcal{S}}$ is bi-Lipschitz equivalent to $\Sigma := \{1, \dots, m\}^{\mathbb{N}}$, with the metric: $\operatorname{dist}(\mathbf{i}, \mathbf{j}) := \rho^{|\mathbf{i} \wedge \mathbf{j}|}$, via the canonical identification

$$\widetilde{\Pi}(\mathbf{i}) := \lim_{n \to \infty} \widetilde{S}_{i_1 \dots i_n}(0,0).$$

Let Λ be the attractor of $\mathcal{S}$ and pick $\mathbf{i} \in \Sigma$. Consider $x := \Pi(\mathbf{i})$, and let ℓ_x be the vertical line corresponding to x (see Figure 11.3). With the identification $\Sigma \cong \widetilde{\Lambda}$, the set $\Pi^{-1}(x) \subset \Sigma$ corresponds to $\ell_x \cap \widetilde{\Lambda}$. Below we write $B(x, r) := (x - r, x + r)$ and carry out the proof referring to Figure 11.4. With a reference to the corresponding regions in Figure 11.4, we consider the following subsets of Σ:

- $\operatorname{dotted}(\mathbf{i}, n) = \Pi^{-1}(B(\Pi(\mathbf{i}), \rho^n))$,
- $\operatorname{blue}(\mathbf{i}, n) = \Pi^{-1}(B(\Pi(\sigma\mathbf{i}), \rho^{n-1}))$,
- $\operatorname{orange}(\mathbf{i}, n) = \sigma^{-1}\Pi^{-1}B(\Pi(\sigma\mathbf{i}), \rho^{n-1})$,

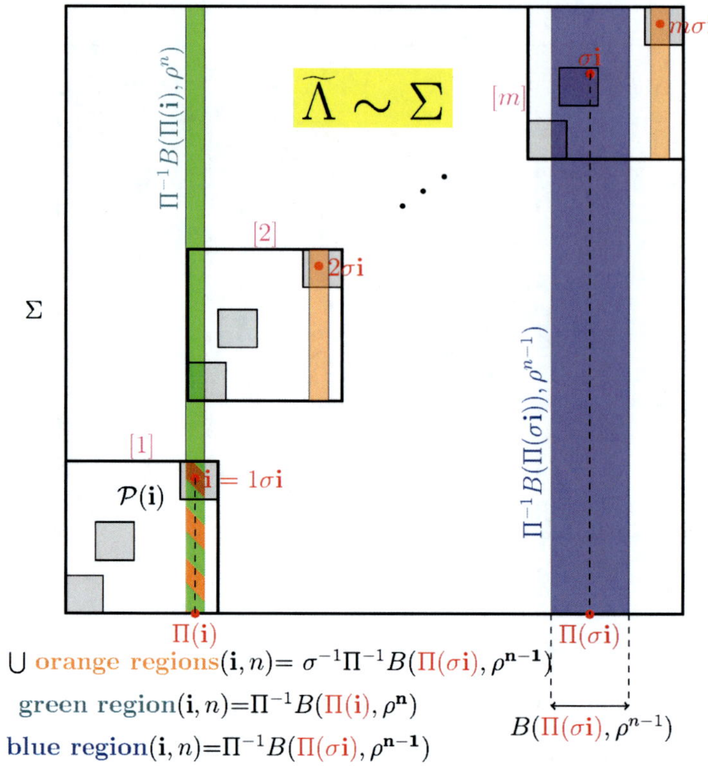

FIGURE 11.4. To the proof of Theorem 11.3.1

Clearly, for every $0 \leqslant \ell \leqslant n-1$ we have

$$\text{blue}(\sigma^\ell \mathbf{i}, n-\ell) = \text{dotted}(\sigma^{\ell+1}\mathbf{i}, n-(\ell+1)). \tag{11.38}$$

By the shift invariance of the measure μ, we have

$$\mu(\sigma^{-1}\Pi^{-1}B(\Pi(\sigma\mathbf{i}), \rho^{n-1})) = \mu\left(\Pi^{-1}(B(\Pi(\sigma\mathbf{i}), \rho^{n-1}))\right). \tag{11.39}$$

To estimate the local dimension of ν at x, we consider

$$\log \nu(B(x, \rho^n)) = \log \mu\left(\Pi^{-1}B(x, \rho^n)\right).$$

The first "trick" is the observation that

$$\log \nu(B(x, \rho^n)) = \sum_{k=0}^{n-1} \log \frac{\mu(\Pi^{-1}B(\Pi(\sigma^k\mathbf{i}), \rho^{n-k}))}{\mu(\Pi^{-1}B(\Pi(\sigma^{k+1}\mathbf{i}), \rho^{n-(k+1)}))}. \tag{11.40}$$

Indeed, by (11.38) the right-hand side is a telescopic sum. After cancellations only the first and the last terms remain. These are $\log \nu(B(x, \rho^n))$ and $-\log \underbrace{\mu(\Pi^{-1}B(\Pi(\sigma^n\mathbf{i}), \rho^0))}_{=1} = 0$ respectively. Let

$$H_n(\mathbf{i}) = \log \frac{\mu\left(\Pi^{-1}B(\Pi(\mathbf{i}), \rho^n)\right)}{\mu\left(\Pi^{-1}B(\Pi(\sigma\mathbf{i}), \rho^{n-1})\right)}. \tag{11.41}$$

With this notation, we obtain from (11.40) that

$$\log \nu(B(x,\rho^n)) = \sum_{k=0}^{n-1} H_{n-k}(\sigma^k \mathbf{i}). \tag{11.42}$$

The second "trick" is that using (11.39), we present H_n in a form which is seemingly more complicated:

$$H_n(\mathbf{i}) = \underbrace{\log \frac{\mu(\Pi^{-1}(B(\Pi(\mathbf{i}),\rho^n)) \cap \sigma^{-1}\Pi^{-1}(B(\Pi(\sigma\mathbf{i}),\rho^{n-1})))}{\mu(\sigma^{-1}\Pi^{-1}B(\Pi(\sigma\mathbf{i}),\rho^{n-1}))}}_{=:W_n(\mathbf{i})} - \underbrace{\log \frac{\mu(\Pi^{-1}(B(\Pi(\mathbf{i}),\rho^n)) \cap \sigma^{-1}\Pi^{-1}B(\Pi(\sigma\mathbf{i}),\rho^{n-1}))}{\mu(\Pi^{-1}(B(\Pi(\mathbf{i}),\rho^n)))}}_{=:G_n(\mathbf{i})}. \tag{11.43}$$

To write down $W_n(\mathbf{i})$ and $G_n(\mathbf{i})$ more compactly, we define $B^{\Pi}(\mathbf{i},r), B^{\Pi\sigma}(\mathbf{i},r) \subset \Sigma$ as follows:

$$B^{\Pi}(\mathbf{i},r) := \Pi^{-1}B(\Pi(\mathbf{i}),r) \quad \text{and} \quad B^{\Pi\sigma}(\mathbf{i},r) := (\Pi\sigma)^{-1}B(\Pi\sigma(\mathbf{i}),r). \tag{11.44}$$

Recall that we defined the natural partition $\mathcal{P}$ of Σ (the partition according to the first digit) in Notation 4. We write $\mathcal{P}(\mathbf{i})$ for the class of the partition $\mathcal{P}$ containing $\mathbf{i}$. Observe that

$$\begin{aligned} B^{\Pi\sigma}(\mathbf{i},\rho^{n-1}) \cap \mathcal{P}(\mathbf{i}) &= \sigma^{-1}\Pi^{-1}B(\Pi(\sigma\mathbf{i}),\rho^{n-1}) \cap \mathcal{P}(\mathbf{i}) \\ &= \Pi^{-1}(B(\Pi(\mathbf{i}),\rho^n)) \cap \sigma^{-1}\Pi^{-1}B(\Pi(\sigma\mathbf{i}),\rho^{n-1}) \\ &= \Pi^{-1}(B(\Pi(\mathbf{i}),\rho^{n-1})) \cap \mathcal{P}(\mathbf{i}) \\ &= B^{\Pi}(\mathbf{i},\rho^n) \cap \mathcal{P}(\mathbf{i}). \end{aligned} \tag{11.45}$$

With this notation, the formal definitions of $W_n(\mathbf{i}), G_n(\mathbf{i})$ are:

$$W_n(\mathbf{i}) := \log \frac{\mu\left(B^{\Pi\sigma}(\mathbf{i},\rho^{n-1}) \cap \mathcal{P}(\mathbf{i})\right)}{\mu\left(B^{\Pi\sigma}(\mathbf{i},\rho^{n-1})\right)} \tag{11.46}$$

and

$$G_n(\mathbf{i}) := \log \frac{\mu\left(B^{\Pi}(\mathbf{i},\rho^n) \cap \mathcal{P}(\mathbf{i})\right)}{\mu\left(B^{\Pi}(\mathbf{i},\rho^n)\right)}. \tag{11.47}$$

That is, by (11.42) we have

$$\frac{1}{n}\log \nu(B(\Pi(\mathbf{i}),\rho^n)) = \frac{1}{n}\sum_{k=0}^{n-1} W_{n-k}(\sigma^k\mathbf{i}) - \frac{1}{n}\sum_{k=0}^{n-1} G_{n-k}(\sigma^k\mathbf{i}). \tag{11.48}$$

If we recall that the local dimension $d_\nu(x)$ is the limit, as n tends to infinity, of the left hand side of (11.48), and if we also recall the definition of the projection entropy h_Π given in (11.8) then we can see that Theorem 11.3.1 (in our homogeneous case) immediately follows from the next proposition:

PROPOSITION 11.3.2. *For μ-a.a. $\mathbf{i} \in \Sigma$ we have*

$$\lim_{n\to\infty} \frac{1}{n} \sum_{k=0}^{n-1} G_{n-k}(\sigma^k \mathbf{i}) = H_\mu(\mathcal{P}|\Pi^{-1}\gamma) \tag{11.49}$$

and

$$\lim_{n\to\infty} \frac{1}{n} \sum_{k=0}^{n-1} W_{n-k}(\sigma^k \mathbf{i}) = H_\mu(\mathcal{P}|\sigma^{-1}\Pi^{-1}\gamma). \tag{11.50}$$

To verify this we need two assertions. The first one is a special case of Maker's Ergodic theorem, see Theorem 9.1.3.

PROPOSITION 11.3.3. *Let $\mathfrak{m}$ be a σ-invariant measure and let $F_k \in L^1(\Sigma, \mathfrak{m})$ be a sequence which converges both in L^1 and almost everywhere to $F \in L^1(\Sigma, \mathfrak{m})$. Then*

$$\lim_{n\to\infty} \frac{1}{n} \sum_{i=0}^{n-1} F_{n-i}(\sigma^i(x)) = \mathbb{E}_{\mathfrak{m}}\left(F|\mathcal{I}\right)(x), \tag{11.51}$$

almost everywhere and in L^1.

Whenever $\mathfrak{m}$ is ergodic, (11.51) yields that for $\mathfrak{m}$-a.e. $\mathbf{j} \in \Sigma$ we have

$$\lim_{n\to\infty} \frac{1}{n} \sum_{k=0}^{n-1} F_{n-k}(\sigma^k(\mathbf{j})) = \int F(\mathbf{i})\, d\mathfrak{m}(\mathbf{i}). \tag{11.52}$$

The other ingredient of the proof of Proposition 11.3.2 is

LEMMA 11.3.4. *Let $\Phi : \Sigma \to \Lambda$ be a continuous map and let ξ be a countable partition of Σ with $H_\mu(\xi) < \infty$. Denote*

$$B^\Phi(\mathbf{i}, r) := \Phi^{-1}B(\Phi(\mathbf{i}), r). \tag{11.53}$$

Then for μ-a.e. $\mathbf{i} \in \Sigma$ we have

$$\lim_{r\to 0} \log \frac{\mu\left(B^\Phi(\mathbf{i}, r) \cap \xi(\mathbf{i})\right)}{\mu\left(B^\Phi(\mathbf{i}, r)\right)} = -I_\mu\left(\xi|\Phi^{-1}\gamma\right)(\mathbf{i}). \tag{11.54}$$

Moreover, for the nonnegative function

$$g(\mathbf{i}) := -\inf_{r>0} \log \frac{\mu\left(B^\Phi(\mathbf{i}, r) \cap \xi(\mathbf{i})\right)}{\mu\left(B^\Phi(\mathbf{i}, r)\right)}, \tag{11.55}$$

we have

$$g \in L^1\left(\Sigma, \mu\right). \tag{11.56}$$

By the Lebesgue Dominated Convergence Theorem this lemma implies that

$$\lim_{r\to 0} \log \frac{\mu\left(B^\Phi(\mathbf{i}, r) \cap \xi(\mathbf{i})\right)}{\mu\left(B^\Phi(\mathbf{i}, r)\right)} = -I_\mu\left(\xi|\Phi^{-1}\gamma\right)(\mathbf{i}) \text{ in } L^1 \text{ and a.e.} \tag{11.57}$$

Combining these assertions, we obtain the proof of Theorem 11.3.1 as follows.

PROOF OF THEOREM 11.3.1 ASSUMING LEMMA 11.3.4. Applying Lemma 11.3.4 first with $\Phi = \Pi$, and then with $\Phi = \Pi \circ \sigma$, for G_n and W_n respectively (see (11.47) and (11.46)), yields

$$G_n \to G := -I_\mu\left(\mathcal{P}|\Pi^{-1}\gamma\right) \text{ and } W_n \to W := -I_\mu\left(\mathcal{P}|\sigma^{-1}\Pi^{-1}\gamma\right) \tag{11.58}$$

both in L^1 and almost everywhere. Below we use Proposition 11.3.3 twice: in the first step in (11.59) with $\Phi = \Pi$ and then in (11.60) with $\Phi = \Pi \circ \sigma$. In both cases in the second step we use the ergodicity of μ to get that for μ-a.e. $\mathbf{i} \in \Sigma$:

$$\frac{1}{n}\sum_{k=0}^{n-1} G_{n-k}(\sigma^k \mathbf{i}) \to \mathbb{E}_\mu\left(G|\mathcal{I}\right)(\mathbf{i}) = \int G(\mathbf{i})\, d\mu(\mathbf{i}) = H_\mu\left(\mathcal{P}|\Pi^{-1}\gamma\right), \tag{11.59}$$

and

$$\frac{1}{n}\sum_{k=0}^{n-1} W_{n-k}(\sigma^k \mathbf{i}) \to \mathbb{E}_\mu\left(W|\mathcal{I}\right)(\mathbf{i}) = \int W(\mathbf{i})\, d\mu(\mathbf{i}) = H_\mu\left(\mathcal{P}|\sigma^{-1}\Pi^{-1}\gamma\right). \tag{11.60}$$

Substitute these into (11.48), divide by $\log \rho$ on both sides, and take limit on both sides to get the assertion of Theorem 11.3.1. □

So, to complete the proof of Theorem 11.3.1 the only thing left is to prove Lemma 11.3.4.

PROOF OF LEMMA 11.3.4. In order to prove (11.54), it is enough to verify that for every $A \in \xi$ we have

$$\lim_{r\to 0} \frac{\mu\left(B^\Phi(\mathbf{i},r) \cap A\right)}{\mu\left(B^\Phi(\mathbf{i},r)\right)} = \mathbb{E}_\mu\left(\mathbb{1}_A|\Phi^{-1}\gamma\right)(\mathbf{i}) \quad \text{for } \mu\text{-a.e. } \mathbf{i} \in \Sigma. \tag{11.61}$$

Fix an arbitrary $A \in \xi$ and define the measure $\widetilde{\nu}_A$ on $[0,1]$ by

$$\widetilde{\nu}_A(H) := \mu\left(A \cap \Phi^{-1}H\right)$$

for $H \subset [0,1]$ Borel. Then

$$\widetilde{\nu}_A(\Phi(\mathbf{i}) - r, \Phi(\mathbf{i}) + r) = \mu\left(B^\Phi(\mathbf{i},r) \cap A\right) \text{ and } \widetilde{\nu}(\Phi(\mathbf{i}) - r, \Phi(\mathbf{i}) + r) = \mu\left(B^\Phi(\mathbf{i},r)\right). \tag{11.62}$$

Clearly, $\widetilde{\nu}_A \ll \widetilde{\nu}$. Since both $\widetilde{\nu}$ and $\widetilde{\nu}_A$ are Radon measures, we can apply the Radon–Nykodim Theorem (see Theorem B.2.5 or [**Mat95**, Theorem 2.12]), which yields that the derivative

$$\mathfrak{D}_A(x) := D(\widetilde{\nu}_A, \widetilde{\nu}, x) = \lim_{r\to 0} \frac{\widetilde{\nu}_A(x-r, x+r)}{\widetilde{\nu}(x-r, x+r)} \tag{11.63}$$

exists $\widetilde{\nu}$-almost everywhere, where $D(\widetilde{\nu}_A, \widetilde{\nu}, x)$ corresponds to the notation in Definition B.2.4, and $\mathfrak{D}_A(x)$ is a Borel function (see [**Mat95**, p. 20]). Moreover,

$$\int_H \mathfrak{D}_A(x)\, d\widetilde{\nu}(x) = \widetilde{\nu}_A(H). \tag{11.64}$$

Note that in Definition B.2.4 we defined the derivative of measures in higher generality with a slightly different notation. Putting together (11.62) and (11.63), we obtain that

$$\mathfrak{D}_A(\Phi(\mathbf{i})) = \lim_{r\to 0} \frac{\mu\left(B^\Phi(\mathbf{i},r) \cap A\right)}{\mu\left(B^\Phi(\mathbf{i},r)\right)} \quad \text{for } \mu\text{-a.e. } \mathbf{i} \in \Sigma. \tag{11.65}$$

To complete the proof of (11.61) the only thing left is to verify that

$$\mathfrak{D}_A(\Phi(\mathbf{i})) = \mathbb{E}_\mu\left(\mathbb{1}_A|\Phi^{-1}\gamma\right)(\mathbf{i}) \text{ for } \mu\text{-a.e. } \mathbf{i} \in \Sigma. \tag{11.66}$$

By the definition of the conditional expectation, this holds if

(a) $\mathfrak{D}_A(\Phi(\mathbf{i}))$ is $\Phi^{-1}\gamma$ measurable and
(b) for each $B \in \Phi^{-1}\gamma$, we have $\int_B \mathbb{1}_A = \int_B \mathfrak{D}_A \circ \Phi$.

The claim (a) holds because $\mathfrak{D}_A(x)$ is a Borel function, so $\mathfrak{D}_A \circ \Phi$ is $\Phi^{-1}\gamma$ measurable. The claim (b) holds since for an arbitrary $H \subset \mathbb{R}$,

$$\int\limits_{\Phi^{-1}H} \mathbb{1}_A(\mathbf{i})\, d\mu(\mathbf{i}) = \mu(\Phi^{-1}(H) \cap A) = \widetilde{\nu}_A(H) = \int\limits_{\Phi^{-1}H} \mathfrak{D}_A(\Phi(\mathbf{i}))\, d\mu(\mathbf{i}),$$

where in the last step we used (11.64). This completes the proof of (11.61) which implies that (11.54) holds.

Now we verify (11.56). The key of the proof is the following observation: Recall that $A \in \xi$. For a $\lambda > 0$ set

$$H_A(\lambda) := \left\{ x \in \Lambda : \inf_{r>0} \frac{\widetilde{\nu}_A(x-r, x+r)}{\widetilde{\nu}(x-r, x+r)} < \lambda \right\}.$$

Then there exists a constant $Q > 0$ such

$$\widetilde{\nu}_A(H_A(\lambda)) \leqslant Q\lambda. \tag{11.67}$$

Indeed, for every $x \in H_A(\lambda)$ choose $r = r(x)$ such that for the open interval $B = B(x,r) = (x-r, x+r)$ holds $\widetilde{\nu}_A(B) < \lambda\widetilde{\nu}(B)$. Let $\mathcal{B}$ be the collection of these balls. Then by Besicovitch's Covering Theorem (see Theorem B.3.2), there exists a constant Q (which does not depend on $\mathcal{B}$) such that we can partition $\mathcal{B}$ into subfamilies $\mathcal{B}_1, \dots, \mathcal{B}_Q$, such that the balls in each of these subfamilies are pairwise disjoint and the union of all balls in these families together cover $H_A(\lambda)$. Hence for each $j = 1, \dots, Q$,

$$\widetilde{\nu}_A\left(\bigcup_{B \in \mathcal{B}_j} B \right) \leqslant \lambda \cdot \widetilde{\nu}\left(\bigcup_{B \in \mathcal{B}_j} B \right) \leqslant \lambda.$$

This concludes the proof of (11.67).

Observe that $g(\mathbf{i})$, defined in (11.55), can be expressed as

$$g(\mathbf{i}) = -\sum_{A \in \xi} \mathbb{1}_A(\mathbf{i}) \log g^A(\mathbf{i}), \tag{11.68}$$

where

$$g^A(\mathbf{i}) := \inf_{r>0} \frac{\mu\left(B^\Phi(\mathbf{i}, r) \cap A\right)}{\mu\left(B^\Phi(\mathbf{i}, r)\right)}.$$

Hence, using (11.68) in the second and (11.67) in the third step we obtain

$$\begin{aligned}
\int g\, d\mu &= \int_0^\infty \mu\{g > t\}\, dt = \int_0^\infty \sum_{A \in \xi} \mu(A \cap \{g^A < \mathrm{e}^{-t}\})\, dt \\
&\leqslant \sum_{A \in \xi} \int_0^\infty \min\{\mu(A), Q\mathrm{e}^{-t}\}\, dt = \sum_{A \in \xi} (-\mu(A) \log \mu(A) + \mu(A) + \mu(A) \log Q) \\
&= H_\mu(\xi) + 1 + \log Q < \infty.
\end{aligned}$$

□

11.3.1. No dimension drop is equivalent to weak almost unique coding. Now, using Theorem 11.3.1 and in particular (11.36), we can relate the drop of the dimension of self-similar measures to the injectivity of the natural projection Π. The following conditions appeared and were used first by Kolossváry and Simon [**KS19**] to express that the overlap between the cylinders is not too significant.

Recall that $\mathcal{H} = (h_i(x) := r_i x + u_i)_{i=1}^m$ is a self-similar IFS on $\mathbb{R}$ and ξ is a partition of Σ defined by $\xi(\mathbf{i}) = \Pi^{-1}\Pi(\mathbf{i})$. We write $\mathbf{q} > 0$ for a probability vector if all of its components are strictly positive.

DEFINITION 11.3.5.

NDD: We say that $\mathcal{H}$ has the *No Dimension Drop (NDD)* property if for all probability vectors $\mathbf{q} > 0$ such that $\dim_S(\nu_\mathbf{q}) = \frac{h_\mathbf{q}}{\chi_{\mu_\mathbf{q}}} \leqslant 1$, we have

$$\dim_H(\nu_\mathbf{q}) = \frac{h_\mathbf{q}}{\chi_{\mu_\mathbf{q}}}.$$

AUC: We say that $\mathcal{H}$ satisfies the *Almost Unique Coding (AUC)* condition if for all probability vectors $\mathbf{q} > 0$ such that $\frac{h_\mathbf{q}}{\chi_{\mu_\mathbf{q}}} \leqslant 1$, there is an exceptional set $\mathcal{R}_\mathcal{H} = \mathcal{R}_\mathcal{H}(\mathbf{q}) \subset \Sigma$ such that

$$\mu_\mathbf{q}(\mathcal{R}_\mathcal{H}) = 0 \text{ and } \#\xi(\mathbf{i}) = 1, \quad \text{for every } \mathbf{i} \in \Sigma \backslash \mathcal{R}_\mathcal{H}. \tag{11.69}$$

WAUC: We say that $\mathcal{H}$ has the *Weak Almost Unique Coding (WAUC)* property if for all probability vectors $\mathbf{q} > 0$ such that $\frac{h_\mathbf{q}}{\chi_{\mu_\mathbf{q}}} \leqslant 1$, there exists a set $\mathcal{B}_\mathcal{H} = \mathcal{B}_\mathcal{H}(\mathbf{q}) \subset \Sigma$ for which

$$\mu_\mathbf{q}(\mathcal{B}_\mathcal{H}) = 0 \text{ and } \#(\xi(\mathbf{i}) \backslash \mathcal{B}_\mathcal{H}) = 1, \quad \text{for every } \mathbf{i} \in \Sigma \backslash \mathcal{B}_\mathcal{H}.$$

Note that WAUC means that there exists a set of full measure $\Sigma \backslash \mathcal{B}_\mathcal{H}$ in the symbolic space, on which the natural projection $\Pi|_{\Sigma \backslash \mathcal{B}_\mathcal{H}}$ is injective.

Let $\delta_\mathbf{i}$ denote the Dirac-delta measure at the point $\mathbf{i} \in \Sigma$.

PROPOSITION 11.3.6. *For any self-similar IFS on the line the conditions NDD and WAUC are equivalent. More precisely,*

$$\text{NDD} \iff \left(\forall\, \mathbf{q} > 0, \ \frac{h_\mathbf{q}}{\chi_{\mu_\mathbf{q}}} \leqslant 1 \implies \mu_\mathbf{i}^\xi = \delta_\mathbf{i} \textit{ for } \mu_\mathbf{q}\textit{-a.e. } \mathbf{i} \in \Sigma\right) \iff \text{WAUC}. \tag{11.70}$$

We remark that the same holds for arbitrary ergodic measures, not just for Bernoulli, according to [**Fen23**, Corollary 4.7].

We prove the assertion in two steps.

PROOF OF THE FIRST EQUIVALENCE IN (11.70). In fact, the proof works for each $\mathbf{q} > 0$, satisfying $\frac{h_\mathbf{q}}{\chi_{\mu_\mathbf{q}}} \leqslant 1$, individually, so we fix such a $\mathbf{q} > 0$ for the duration of the proof. By (11.36) and the definition of NDD, we obtain:

$$\text{NDD} \iff H = 0 \iff \mu_\mathbf{i}^\xi([i_1]) = 1 \text{ for } \mu_\mathbf{q}\text{-a.e. } \mathbf{i} \in \Sigma.$$

Thus it suffices to show that

$$\mu_\mathbf{i}^\xi([i_1]) = 1 \text{ for } \mu_\mathbf{q}\text{-a.e. } \mathbf{i} \in \Sigma \iff \mu_\mathbf{i}^\xi = \delta_\mathbf{i} \text{ for } \mu_\mathbf{q}\text{-a.e. } \mathbf{i} \in \Sigma. \tag{11.71}$$

The direction $\Longleftarrow$ in (11.71) is obvious. In the other direction, we will show that for $\mu_\mathbf{q}$-a.e. $\mathbf{i} \in \Sigma$,

$$\mu_\mathbf{i}^\xi([i_1]) = 1 \implies \mu_\mathbf{i}^\xi([i_1 \ldots i_n]) = 1 \text{ for all } n \implies \mu_\mathbf{i}^\xi = \delta_\mathbf{i}.$$

Observe that the measure $\sigma_*(\mu_{\mathbf{i}}^{\xi\vee\mathcal{P}})$ is supported on the partition element $\xi(\sigma\mathbf{i})$, and using the σ-invariance of $\mu_{\mathbf{q}}$ we obtain

$$\int \sigma_*(\mu_{\mathbf{i}}^{\xi\vee\mathcal{P}})\, d\mu_{\mathbf{q}}(\mathbf{i}) = \sigma_*(\mu_{\mathbf{q}}) = \mu_{\mathbf{q}} = \int \mu_{\mathbf{i}}^{\xi}\, d\mu_{\mathbf{q}}(\mathbf{i}) = \int \mu_{\sigma\mathbf{i}}^{\xi}\, d\mu_{\mathbf{q}}(\mathbf{i}).$$

Using the uniqueness of conditional measures, we get that $\sigma_*(\mu_{\mathbf{i}}^{\xi\vee\mathcal{P}}) = \mu_{\sigma\mathbf{i}}^{\xi}$ for $\mu_{\mathbf{q}}$-a.e. $\mathbf{i}$. By an analogous argument we get that for every $\ell \geqslant 1$ and for $\mu_{\mathbf{q}}$-a.e. $\mathbf{i}$,

$$\sigma_*^{\ell}\mu_{\mathbf{i}}^{\xi\vee\mathcal{P}\vee\sigma^{-1}\mathcal{P}\vee\cdots\vee\sigma^{-(\ell-1)}\mathcal{P}} = \mu_{\sigma^{\ell}\mathbf{i}}^{\xi}. \tag{11.72}$$

Thus, for every $n \geqslant 1$,

$$\begin{aligned}
\mu_{\mathbf{i}}^{\xi}([i_1\ldots i_n]) &= \mu_{\mathbf{i}}^{\xi}([i_1]) \cdot \prod_{k=1}^{n-1} \frac{\mu_{\mathbf{i}}^{\xi}([i_1\ldots i_{k+1}])}{\mu_{\mathbf{i}}^{\xi}([i_1\ldots i_k])} \\
&= \mu_{\mathbf{i}}^{\xi}([i_1]) \cdot \prod_{k=1}^{n-1} \mu_{\mathbf{i}}^{\xi\vee\mathcal{P}\vee\cdots\vee\sigma^{-(k-1)}\mathcal{P}}(\sigma^{-k}[i_{k+1}]) \\
&= \prod_{k=0}^{n-1} \mu_{\sigma^k\mathbf{i}}^{\xi}([i_{k+1}]).
\end{aligned}$$

In fact, the first equality above is obvious and the last one follows from (11.72). To see that the second equality holds observe that for $P = \mathcal{P} \vee \sigma^{-1}\mathcal{P} \vee \cdots \vee \sigma^{k-1}\mathcal{P}$

$$\begin{aligned}
\frac{\mu_{\mathbf{i}}^{\xi}([i_1\ldots i_{k+1}])}{\mu_{\mathbf{i}}^{\xi}([i_1\ldots i_k])} &= \frac{\mu_{\mathbf{i}}^{\xi}([i_1\ldots i_k] \cap \sigma^{-k}[i_{k+1}])}{\mu_{\mathbf{i}}^{\xi}([i_1\ldots i_k])} = \frac{\mu_{\mathbf{i}}^{\xi}(P(\mathbf{i}) \cap \sigma^{-k}[i_{k+1}])}{\mu_{\mathbf{i}}^{\xi}(P(\mathbf{i}))} \\
&= \mu_{\mathbf{i}}^{\xi\vee P}(\sigma^{-k}[i_{k+1}]) = \mu_{\mathbf{i}}^{\xi\vee\mathcal{P}\vee\sigma^{-1}\mathcal{P}\vee\cdots\vee\sigma^{k-1}\mathcal{P}}(\sigma^{-k}[i_{k+1}]),
\end{aligned}$$

where the last but one equality follows from Lemma 9.4.21. Now the second implication in the 1st equivalence follows from the Monotone Convergence Theorem:

$$1 = \lim_{n\to\infty} \mu_{\mathbf{i}}^{\xi}([i_1\ldots i_n]) = \mu_{\mathbf{i}}^{\xi}(\{\mathbf{i}\}) \implies \mu_{\mathbf{i}}^{\xi} = \delta_{\mathbf{i}} \text{ for } \mu_{\mathbf{q}}\text{-a.e. } \mathbf{i} \in \Sigma.$$

□

PROOF OF THE SECOND EQUIVALENCE IN (11.70).

$\Longrightarrow$ **direction:** Let $\widehat{\Sigma}$ be the full measure subset of Σ for which $\mu_{\mathbf{i}}^{\xi}$ exists. We claim that the conditions in the definition of WAUC are satisfied with

$$\mathcal{B}_{\mathcal{H}} = \mathcal{B} := (\Sigma\backslash\widehat{\Sigma}) \bigcup \left\{\mathbf{i} \in \widehat{\Sigma} : \mu_{\mathbf{i}}^{\xi} \neq \delta_{\mathbf{i}}\right\}.$$

By assumption, $\mu_{\mathbf{q}}(\mathcal{B}) = 0$. Moreover, for any $\mathbf{i} \in \Sigma\backslash\mathcal{B}$ we have $\mathbf{i} \in \widehat{\Sigma}$, so the probability measure $\mu_{\mathbf{i}}^{\xi}$ exists and $\mu_{\mathbf{i}}^{\xi} = \delta_{\mathbf{i}}$. If $\mathbf{j} \in \xi(\mathbf{i})\backslash\mathcal{B}$, then $\xi(\mathbf{j}) = \xi(\mathbf{i})$, thus

$$\delta_{\mathbf{i}} = \mu_{\mathbf{i}}^{\xi} = \mu_{\mathbf{j}}^{\xi} = \delta_{\mathbf{j}},$$

hence $\mathbf{i} = \mathbf{j}$. We showed that for every $\mathbf{i} \in \Sigma\backslash\mathcal{B}$ holds $\xi(\mathbf{i})\backslash\mathcal{B} = \{\mathbf{i}\}$, as desired.

$\Longleftarrow$ **direction:** Clearly, WAUC is equivalent to the existence of $\mathcal{B}$ with $\mu_{\mathbf{q}}(\mathcal{B}) = 0$ such that

$$\text{if } \mathbf{i} \notin \mathcal{B} \text{ then } \xi(\mathbf{i})\backslash\mathcal{B} = \{\mathbf{i}\}\,. \tag{11.73}$$

Using (11.34) for $\mathcal{B}$ we obtain that the set

$$\widehat{\Sigma}_{\mathcal{H}} := \left\{\mathbf{i} \in \widehat{\Sigma} : \mu_{\mathbf{i}}^{\xi}(\mathcal{B}) = 0\right\}$$

has full measure:

$$\mu_{\mathbf{q}}(\widehat{\Sigma}_{\mathcal{H}}) = 1, \tag{11.74}$$

Assume that $\mathbf{i} \in \widehat{\Sigma}_{\mathcal{H}}$. Then

$$\mu_{\mathbf{i}}^{\xi}\left(\xi(\mathbf{i})\backslash\mathcal{B}\right) = 1 \quad \text{and} \quad \xi(\mathbf{i})\backslash\mathcal{B}_{\mathcal{H}} = \{\mathbf{i}\},$$

in view of (11.73). That is, $\mu_{\mathbf{i}}^{\xi}(\{\mathbf{i}\}) = 1$ whenever $\mathbf{i} \in \widehat{\Sigma}_{\mathcal{H}}$. Combining this with (11.74), we get that for a $\mu_{\mathbf{q}}$-full measure set of $\mathbf{i}$ we have $\mu_{\mathbf{i}}^{\xi} = \delta_{\mathbf{i}}$. □

Finally, we show here that the Almost Unique Coding condition holds under the Strong Open Set Condition, see Definition 1.5.2.

LEMMA 11.3.7. *Let $\mathcal{S}$ be an IFS of contractive maps and let Π be the natural projection. Suppose that $\mathcal{S}$ satisfies the strong open set condition (SOSC), i.e. there exists an open bounded set U such that*

$$S_k(U) \subseteq U,\ S_k(U) \cap S_j(U) = \varnothing \text{ for every } k \neq j \text{ and } \Lambda \cap U \neq \varnothing.$$

Then the almost unique coding (AUC) holds and in particular, $h_{\Pi}(\mu_{\mathbf{p}}) = h_{\mathbf{p}}$.

PROOF. It is enough to show that the set

$$\mathcal{R} = \{\mathbf{i} \in \Sigma : \text{ there exist } \mathbf{j} \neq \mathbf{i} \text{ such that } \Pi(\mathbf{i}) = \Pi(\mathbf{j})\}$$

has zero measure. Since $S_k(U) \cap S_j(U) = \varnothing$ for every $k \neq j$, if $\Pi(\mathbf{i}) = \Pi(\mathbf{j})$ with $i_1 \neq j_1$ then $\Pi(\mathbf{i}) = \Pi(\mathbf{j}) \in \partial U = \overline{U}\backslash U$. Therefore $\mathcal{R} \subset \bigcup_{k=0}^{\infty} \sigma^{-k}\Pi^{-1}(\partial U)$. Hence, it is enough to show that $\mu_{\mathbf{p}}(\Pi^{-1}(\partial U)) = 0$.

To this end, we need to study the set $\mathcal{C} = \{\mathbf{i} \in \Sigma : \Pi(\mathbf{i}) \in U\}$. Since $\bigcup_{k=1}^{m} S_k(U) \subseteq U$ we get that $\sigma^{-1}\mathcal{C} \subset \mathcal{C}$ and therefore, either $\mu_{\mathbf{p}}(\mathcal{C}) = 0$ or $\mu_{\mathbf{p}}(\mathcal{C}) = 1$ by the ergodicity of $\mu_{\mathbf{p}}$. But observe that since $\Lambda \cap U \neq \varnothing$, there exists $\mathbf{i} \in \Sigma^*$ such that $\Pi([\mathbf{i}]) \subset U$. Hence, $\mu_{\mathbf{p}}(\mathcal{C}) \geqslant \mu_{\mathbf{p}}([\mathbf{i}]) > 0$ and by ergodicity $\mu(\mathcal{C}) = 1$, as required.

Finally, AUC obviously implies WAUC, which is equivalent to $h_{\Pi}(\mu_{\mathbf{p}}) = h_{\mathbf{p}}$ by Proposition 11.3.6. □

A simple corollary of Theorem 11.3.1 and Lemma 11.3.7 says that the dimension of self-similar measures does not drop under the Strong Open Set Condition, which strengthens Corollary 1.9.11 and Theorem 3.2.1.

COROLLARY 11.3.8. *Let $\mathcal{S}$ be a self-similar IFS on $\mathbb{R}^d$ with contraction ratios given by the vector $\mathbf{r} = (r_1, \ldots, r_m)$, and let ν be the self-similar measure for $\mathcal{S}$ with the probability vector $\mathbf{p} = (p_1, \ldots, p_m)$. If $\mathcal{S}$ satisfies the SOSC, then $\dim_{\mathrm{H}}(\nu) = \frac{-\sum_{i=1}^{m} p_i \log p_i}{-\sum_{i=1}^{m} p_i \log r_i} = \frac{h_{\mathbf{p}}}{\chi_{\mathbf{r}}^{\mathbf{p}}}$.*

11.4. Box dimension of self-affine sets with aligned cylinders

We continue the study the geometry of attractors of diagonally self-affine IFS in $\mathbb{R}^2$ and compute the box dimension under the Rectangular Open Set Condition. The box dimension was computed for several special classes of IFS, which will be discussed in Section 11.5, but Feng and Wang [**FW05**] gave a dimension formula for the box dimension in the general case, which depends on the contraction ratios and the box dimension of the projections. In this section we present their result, but we follow the proof in Bárány [**Bár12**], which is a generalization of the box dimension result of Barański [**Bar07**].

As before, consider

$$\mathcal{S} := \left(S_i(\mathbf{x}) = \begin{pmatrix} r_{i,1} & 0 \\ 0 & r_{i,2} \end{pmatrix} \cdot \begin{pmatrix} x_1 \\ x_2 \end{pmatrix} + \begin{pmatrix} t_{i,1} \\ t_{i,2} \end{pmatrix} \right)_{i=1}^{m}, \tag{11.75}$$

where $|r_{i,j}| \in (0,1)$, and let $A_i = \begin{pmatrix} r_{i,1} & 0 \\ 0 & r_{i,2} \end{pmatrix}$.

DEFINITION 11.4.1. We say that the IFS $\mathcal{S}$ in (11.75) satisfies the Rectangular Open Set Condition (ROSC) if

$$S_i([0,1]^2) \subset [0,1]^2 \text{ and } S_i((0,1)^2) \cap S_j((0,1)^2) = \emptyset, \quad \forall i \neq j \in \{1,\dots,m\}. \tag{11.76}$$

Compare this with the RSSP, see (11.21).

As usual, denote by Λ the attractor of $\mathcal{S}$. The projections of $\mathcal{S}$ to the horizontal and vertical axes are the self-similar IFS Φ_1, Φ_2 respectively, namely,

$$\Phi_1 = (S_{i,1} = r_{i,1}x_1 + t_{i,1})_{i=1}^{m} \text{ and } \Phi_2 = (S_{i,2} = r_{i,2}x_2 + t_{i,2})_{i=1}^{m}. \tag{11.77}$$

Denote the attractor of Φ_j by Λ_j, for $j = 1,2$. Since these sets are self-similar, the box dimension of Λ_i exists for $i = 1,2$, see Theorem 3.3.1. The statement of the next theorem is from [**Bár12**, Theorem 4.1], but it can be shown that it is equivalent to [**FW05**, Corollary 1].

THEOREM 11.4.2 (Feng and Wang). *Assume that $\mathcal{S} := (S_1,\dots,S_m)$ is a diagonally self-affine IFS with attractor Λ and suppose that the ROSC holds. Then the box dimension of Λ exists and* $\dim_B(\Lambda) = \max\{d_1,d_2\}$, *where d_1 and d_2 are the unique solutions of the equations*

$$\sum_{i=1}^{m} |r_{i,1}|^{\dim_B(\Lambda_1)} |r_{i,2}|^{d_1-\dim_B(\Lambda_1)} = 1 \text{ and } \sum_{i=1}^{m} |r_{i,2}|^{\dim_B(\Lambda_2)} |r_{i,1}|^{d_2-\dim_B(\Lambda_2)} = 1 \tag{11.78}$$

respectively. Moreover, if $|r_{i,1}| \geqslant |r_{i,2}|$ for every $i = 1,\dots,m$, then $\dim_B(\Lambda) = d_1$.

These formulas resemble (11.31) and (11.30) for the affinity dimension, which were shown under the RSSP, although the proof works under the ROSC as well. In fact, the formulas agree when $\dim_B(\Lambda_j) = \min\{1, s_j\}$, where s_j is the similarity dimension of Φ_j.

Before we prove Theorem 11.4.2, let us introduce some further notation. First, define two collections of rectangles in $[0,1]^2$ as follows: for $K \in \mathbb{N}$, let

$$\mathfrak{R}_K^1 = \left\{ \left[\frac{j-1}{K}, \frac{j}{K}\right] \times [0,1] : j = 1,\dots,K \text{ s.t. } \left[\frac{j-1}{K}, \frac{j}{K}\right] \cap \Lambda_1 \neq \emptyset \right\} \text{ and}$$

$$\mathfrak{R}_K^2 = \left\{ [0,1] \times \left[\frac{j-1}{K}, \frac{j}{K}\right] : j = 1,\dots,K \text{ s.t. } \left[\frac{j-1}{K}, \frac{j}{K}\right] \cap \Lambda_2 \neq \emptyset \right\}.$$

By the definition of the box dimension,

$$\#\mathfrak{R}_K^1 = K^{\dim_B(\Lambda_1)+o(1)} \text{ and } \#\mathfrak{R}_K^2 = K^{\dim_B(\Lambda_2)+o(1)}. \tag{11.79}$$

For $\rho > 0$, define a partition of Σ by

$$\Delta_\rho := \{\mathbf{i} \in \Sigma^* : \min\{|r_{\mathbf{i},1}|, |r_{\mathbf{i},2}|\} \leqslant \rho < \min\{|r_{\mathbf{i}_-,1}|, |r_{\mathbf{i}_-,2}|\}\}. \tag{11.80}$$

The set Δ_ρ is analogous to the Moran minimal cut-set introduced in (1.73), but here we consider not the diameter of the axis parallel cylinder $S_{\mathbf{i}}([0,1]^2)$, but its shortest side. Now, let

$$\Delta_\rho^1 = \{\mathbf{i} \in \Delta_\rho : |r_{\mathbf{i},1}| \geqslant |r_{\mathbf{i},2}|\} \text{ and } \Delta_\rho^2 = \{\mathbf{i} \in \Delta_\rho : |r_{\mathbf{i},1}| < |r_{\mathbf{i},2}|\}.$$

For $\rho > 0$ and $\mathbf{i} \in \Delta_\rho$, denote by $K = K(\mathbf{i},\rho) \geqslant 1$ the unique integer such that

$$\frac{\max\{|r_{\mathbf{i},1}|,|r_{\mathbf{i},2}|\}}{K} \leqslant \rho < \frac{\max\{|r_{\mathbf{i},1}|,|r_{\mathbf{i},2}|\}}{K-1}.$$

Proof of Theorem 11.4.2. For every $\rho > 0$, $i \in \{1,2\}$, $\mathbf{i} \in \Delta_\rho^i$, and $R \in \mathfrak{R}^i_{K(\mathbf{i},\rho)}$, the rectangle $S_{\mathbf{i}}(R)$ has side-lengths $\frac{\max\{|r_{\mathbf{i},1}|,|r_{\mathbf{i},2}|\}}{K(\mathbf{i},\rho)}$ and $\min\{|r_{\mathbf{i},1}|,|r_{\mathbf{i},2}|\}$. It follows from the definitions that the side-lengths of $S_{\mathbf{i}}(R)$ are approximately equal to ρ, so that there exists a constant $C > 0$ independent of $\rho, \mathbf{i}$ and R, such that the ratio of the side-lengths of $S_{\mathbf{i}}(R)$ is in the interval $[C^{-1}, C]$. Indeed, suppose that $i = 1$ (the case $i = 2$ is analogous). Then $\mathbf{i} \in \Delta_\rho^1$ implies that $|r_{\mathbf{i},1}| \geqslant |r_{\mathbf{i},2}|$, so $\rho \approx |r_{\mathbf{i},2}|$ and $K = K(\mathbf{i},\rho) \approx \frac{|r_{\mathbf{i},1}|}{\rho}$. The side lengths of $R \in \mathfrak{R}^1_{K(\mathbf{i},\rho)}$ are K^{-1} and 1, hence the side lengths of $S_{\mathbf{i}}(R)$ are $\frac{|r_{\mathbf{i},1}|}{K} \approx \rho$ and $|r_{\mathbf{i},2}| \approx \rho$. We call these sets *approximate squares* of side-length ρ. With a slight abuse of notation, for $\mathbf{i} \in \Delta_\rho^i$ let $S_{\mathbf{i}}(\mathfrak{R}^i_{K(\mathbf{i},\rho)}) = \left\{S_{\mathbf{i}}(R) : R \in \mathfrak{R}^i_{K(\mathbf{i},\rho)}\right\}$. Consider

$$\mathfrak{P}_\rho := \bigcup_{i\in\{1,2\}} \bigcup_{\mathbf{i}\in\Delta_\rho^i} S_{\mathbf{i}}(\mathfrak{R}^i_{K(\mathbf{i},\rho)}).$$

By construction and the ROSC, $R^o \cap (R')^o = \varnothing$ for every $R \neq R'$ in $\mathfrak{P}_\rho$. From the facts listed above, it clearly follows that there exists $C > 0$ such that $C^{-1} \leqslant \frac{\#\mathfrak{P}_\rho}{N_\rho(\Lambda)} \leqslant C$, where $N_\rho(\Lambda)$ is minimal number of sets of diameter ρ needed to cover Λ, see Definition 1.4.1. Hence,

$$\underline{\dim}_{\mathrm{B}}(\Lambda) = \liminf_{\rho\to 0} \frac{\log \#\mathfrak{P}_\rho}{-\log\rho} \text{ and } \overline{\dim}_{\mathrm{B}}(\Lambda) = \limsup_{\rho\to 0} \frac{\log \#\mathfrak{P}_\rho}{-\log\rho}. \tag{11.81}$$

First, we show the upper bound $\overline{\dim}_{\mathrm{B}}(\Lambda) \leqslant \max\{d_1,d_2\}$. Let $\tau_j = \max_i \frac{\log|r_{i,j}|}{\log|r_{i,j}|}$, $j = 1,2$, and let d_1, d_2 be as in (11.78). By the definition of $\mathfrak{P}_\rho$ and (11.79),

$$\begin{aligned}\#\mathfrak{P}_\rho &= \sum_{\mathbf{i}\in\Delta_\rho^1} \#\mathfrak{R}^1_{K(\mathbf{i},\rho)} + \sum_{\mathbf{i}\in\Delta_\rho^2} \#\mathfrak{R}^2_{K(\mathbf{i},\rho)} \\ &= \sum_{\mathbf{i}\in\Delta_\rho^1} K(\mathbf{i},\rho)^{\dim_{\mathrm{B}}(\Lambda_1)+o(1)} + \sum_{\mathbf{i}\in\Delta_\rho^2} K(\mathbf{i},\rho)^{\dim_{\mathrm{B}}(\Lambda_2)+o(1)}.\end{aligned}$$

Then by the definition of $K(\mathbf{i},\rho)$ and Δ_ρ, we have

$$\begin{aligned}
\#\mathfrak{P}_\rho &\leqslant C_1 \sum_{\mathbf{i}\in\Delta_\rho^1} \left(\frac{|r_{\mathbf{i},1}|}{\rho}\right)^{\dim_B(\Lambda_1)+o(1)} + C_1 \sum_{\mathbf{i}\in\Delta_\rho^2} \left(\frac{|r_{\mathbf{i},2}|}{\rho}\right)^{\dim_B(\Lambda_2)+o(1)} \\
&\leqslant C_2 \sum_{\mathbf{i}\in\Delta_\rho^1} \left(\frac{|r_{\mathbf{i},1}|}{|r_{\mathbf{i},2}|}\right)^{\dim_B(\Lambda_1)+o(1)} + C_2 \sum_{\mathbf{i}\in\Delta_\rho^2} \left(\frac{|r_{\mathbf{i},2}|}{|r_{\mathbf{i},1}|}\right)^{\dim_B(\Lambda_2)+o(1)} \\
&\leqslant C_3 \rho^{-\tau_1|o(1)|-d_1} \sum_{\mathbf{i}\in\Delta_\rho^1} \left(\frac{|r_{\mathbf{i},1}|}{|r_{\mathbf{i},2}|}\right)^{\dim_B(\Lambda_1)} |r_{\mathbf{i},2}|^{d_1} \\
&\quad + C_3 \rho^{-\tau_2|o(1)|-d_2} \sum_{\mathbf{i}\in\Delta_\rho^2} \left(\frac{|r_{\mathbf{i},2}|}{|r_{\mathbf{i},1}|}\right)^{\dim_B(\Lambda_2)} |r_{\mathbf{i},1}|^{d_2} \\
&\leqslant 2C_3 \rho^{-\max\{d_1+\tau_1|o(1)|, d_2+\tau_2|o(1)|\}}.
\end{aligned}$$

This, together with (11.81), implies the desired upper bound. In the last step we used the following

FACT 11.4.3. *For any $\rho > 0$ holds*

$$\sum_{\mathbf{i}\in\Delta_\rho} |r_{\mathbf{i},1}|^{\dim_B(\Lambda_1)} |r_{\mathbf{i},2}|^{d_1-\dim_B(\Lambda_1)} = 1 \quad \textit{and} \quad \sum_{\mathbf{i}\in\Delta_\rho} |r_{\mathbf{i},2}|^{\dim_B(\Lambda_2)} |r_{\mathbf{i},1}|^{d_2-\dim_B(\Lambda_2)} = 1.$$

PROOF. We can apply Lemma 1.3.3 about flows on trees. Define a function f on the edges of the m-regular tree, identified with Σ^*, as follows:

$$f(\mathbf{i}_-, \mathbf{i}) = |r_{\mathbf{i},1}|^{\dim_B(\Lambda_1)} |r_{\mathbf{i},2}|^{d_1-\dim_B(\Lambda_1)}.$$

It follows from (11.78) that f is a flow of norm $\|f\| = 1$, and since Δ_ρ is a minimal cut-set, that is, $\{[\mathbf{i}] : \mathbf{i} \in \Delta_\rho\}$ is a partition of Σ, we obtain from Lemma 1.3.3 that

$$\sum_{\mathbf{i}\in\Delta_\rho} f(\mathbf{i}_-, \mathbf{i}) = 1.$$

This proves the first equality, and the second one is proved analogously. □

Now, we continue with the proof of the lower bound $\underline{\dim}_B(\Lambda) \geqslant \max\{d_1, d_2\}$. Similarly to the above, one can easily show that there exists $C > 0$ such that for all $\rho > 0$,

$$\#\mathfrak{P}_\rho \geqslant C \sum_{\mathbf{i}\in\Delta_\rho^1} \left(\frac{|r_{\mathbf{i},1}|}{|r_{\mathbf{i},2}|}\right)^{\dim_B(\Lambda_1)+o(1)} + C \sum_{\mathbf{i}\in\Delta_\rho^2} \left(\frac{|r_{\mathbf{i},2}|}{|r_{\mathbf{i},1}|}\right)^{\dim_B(\Lambda_2)+o(1)}. \tag{11.82}$$

Before we continue, we need to show the following fact.

FACT 11.4.4. *Under the assumptions above,* $\sum_{i=1}^m |r_{i,1}|^{\dim_B(\Lambda_1)} |r_{i,2}|^{\dim_B(\Lambda_2)} \leqslant 1$. *In particular,* $\max\{d_1, d_2\} \leqslant \dim_B(\Lambda_1) + \dim_B(\Lambda_2)$.

PROOF. Denote $\alpha_j = \dim_B(\Lambda_j)$ for $j = 1, 2$, and argue by contradiction, that is, suppose that

$$\sum_{i=1}^m |r_{i,1}|^{\alpha_1} |r_{i,2}|^{\alpha_2} = \gamma > 1.$$

We claim that this implies

$$\liminf_{\rho\to 0}\frac{\log\sum_{\mathbf{i}\in\Delta_\rho}|r_{\mathbf{i},1}|^{\alpha_1}|r_{\mathbf{i},2}|^{\alpha_2}}{-\log\rho}>0. \tag{11.83}$$

We again apply Lemma 1.3.3 about flows on trees. Define a function f on the edges of the m-regular tree, identified with Σ^*, as follows:

$$f(\mathbf{i}_-,\mathbf{i})=\frac{|r_{\mathbf{i},1}|^{\alpha_1}|r_{\mathbf{i},2}|^{\alpha_2}}{\left(\sum_{i=1}^m|r_{i,1}|^{\alpha_1}|r_{i,2}|^{\alpha_2}\right)^{|\mathbf{i}|}}=\frac{|r_{\mathbf{i},1}|^{\alpha_1}|r_{\mathbf{i},2}|^{\alpha_2}}{\gamma^{|\mathbf{i}|}}.$$

It is clear that f is a flow, and we obtain by Lemma 1.3.3 that $\sum_{\mathbf{i}\in\Delta_\rho}f(\mathbf{i}_-,\mathbf{i})=1$, since Δ_ρ is a minimal cut-set. By the definition of Δ_ρ we have $|\mathbf{i}|\geqslant -c\log\rho$ for $c>0$, and (11.83) follows. Using (11.82), we get

$$\begin{aligned}\#\mathfrak{P}_\rho &\geqslant C_2\sum_{\mathbf{i}\in\Delta_\rho^1}\left(\frac{|r_{\mathbf{i},1}|}{|r_{\mathbf{i},2}|}\right)^{\dim_{\mathrm{B}}(\Lambda_1)}\rho^{\tau_1|o(1)|}+C_2\sum_{\mathbf{i}\in\Delta_\rho^2}\left(\frac{|r_{\mathbf{i},2}|}{|r_{\mathbf{i},1}|}\right)^{\dim_{\mathrm{B}}(\Lambda_2)}\rho^{\tau_2|o(1)|}\\ &\geqslant C_3\rho^{\max\{\tau_1,\tau_2\}|o(1)|-\dim_{\mathrm{B}}(\Lambda_1)-\dim_{\mathrm{B}}(\Lambda_2)}\sum_{\mathbf{i}\in\Delta_\rho}|r_{\mathbf{i},1}|^{\dim_{\mathrm{B}}(\Lambda_1)}|r_{\mathbf{i},2}|^{\dim_{\mathrm{B}}(\Lambda_2)}.\end{aligned}$$

Hence, by (11.81) and (11.83) we obtain that $\underline{\dim}_{\mathrm{B}}(\Lambda)>\dim_{\mathrm{B}}(\Lambda_1)+\dim_{\mathrm{B}}(\Lambda_2)$. On the other hand, since $\Lambda\subseteq\Lambda_1\times\Lambda_2$, we get that $\overline{\dim}_{\mathrm{B}}(\Lambda)\leqslant\dim_{\mathrm{B}}(\Lambda_1)+\dim_{\mathrm{B}}(\Lambda_2)$ by Proposition 2.2.4, which is a contradiction. □

Now, let us assume first that $d_1>d_2$. Then there exists $C>0$ such that for every $\rho>0$ and for every $\mathbf{i}\in\Delta_\rho^2$,

$$\begin{aligned}\frac{|r_{\mathbf{i},2}|^{\dim_{\mathrm{B}}(\Lambda_2)}|r_{\mathbf{i},1}|^{d_2-\dim_{\mathrm{B}}(\Lambda_2)}}{|r_{\mathbf{i},1}|^{\dim_{\mathrm{B}}(\Lambda_1)}|r_{\mathbf{i},2}|^{d_1-\dim_{\mathrm{B}}(\Lambda_1)}} &= \left(\frac{|r_{\mathbf{i},2}|}{|r_{\mathbf{i},1}|}\right)^{\dim_{\mathrm{B}}(\Lambda_2)+\dim_{\mathrm{B}}(\Lambda_1)}|r_{\mathbf{i},1}|^{d_2}|r_{\mathbf{i},2}|^{-d_1}\\ &= \left(\frac{|r_{\mathbf{i},2}|}{|r_{\mathbf{i},1}|}\right)^{\dim_{\mathrm{B}}(\Lambda_2)+\dim_{\mathrm{B}}(\Lambda_1)-d_1}|r_{\mathbf{i},1}|^{d_2-d_1}\geqslant C\rho^{d_2-d_1}.\end{aligned}$$

Therefore,

$$\begin{aligned}\sum_{\mathbf{i}\in\Delta_\rho^2}|r_{\mathbf{i},1}|^{\dim_{\mathrm{B}}(\Lambda_1)}|r_{\mathbf{i},2}|^{d_1-\dim_{\mathrm{B}}(\Lambda_1)} &\leqslant C\rho^{d_1-d_2}\sum_{\mathbf{i}\in\Delta_\rho^2}|r_{\mathbf{i},2}|^{\dim_{\mathrm{B}}(\Lambda_2)}|r_{\mathbf{i},1}|^{d_2-\dim_{\mathrm{B}}(\Lambda_2)}\\ &\leqslant C\rho^{d_1-d_2}.\end{aligned}$$

Thus, by (11.82),

$$\begin{aligned}\#\mathfrak{P}_\rho &\geqslant C\sum_{\mathbf{i}\in\Delta_\rho^1}\left(\frac{|r_{\mathbf{i},1}|}{|r_{\mathbf{i},2}|}\right)^{\dim_{\mathrm{B}}(\Lambda_1)+o(1)}\\ &\geqslant C_2\rho^{\tau_1|o(1)|}\rho^{-d_1}\sum_{\mathbf{i}\in\Delta_\rho^1}|r_{\mathbf{i},1}|^{\dim_{\mathrm{B}}(\Lambda_1)}|r_{\mathbf{i},2}|^{d_1-\dim_{\mathrm{B}}(\Lambda_1)}\\ &\geqslant C_2\rho^{\tau_1|o(1)|}\rho^{-d_1}(1-C\rho^{d_1-d_2}),\end{aligned}$$

where we used that $\sum_{\mathbf{i}\in\Delta_\rho}|r_{\mathbf{i},1}|^{\dim_{\mathrm{B}}(\Lambda_1)}|r_{\mathbf{i},2}|^{d_1-\dim_{\mathrm{B}}(\Lambda_1)}=1$ by Fact 11.4.3. Hence, by (11.81) $\underline{\dim}_{\mathrm{B}}(\Lambda)\geqslant d_1$.

Now, assume that $d:=d_1=d_2$. Then for every $\mathbf{i}\in\Delta_\rho^1$

$$\frac{|r_{\mathbf{i},1}|^{\dim_{\mathrm{B}}(\Lambda_1)}|r_{\mathbf{i},2}|^{d-\dim_{\mathrm{B}}(\Lambda_1)}}{|r_{\mathbf{i},2}|^{\dim_{\mathrm{B}}(\Lambda_2)}|r_{\mathbf{i},1}|^{d-\dim_{\mathrm{B}}(\Lambda_2)}}=\left(\frac{|r_{\mathbf{i},1}|}{|r_{\mathbf{i},2}|}\right)^{\dim_{\mathrm{B}}(\Lambda_2)+\dim_{\mathrm{B}}(\Lambda_1)-d}\geqslant 1.$$

Thus,

$$\sum_{\mathbf{i}\in\Delta^1_\rho} |r_{\mathbf{i},1}|^{\dim_B(\Lambda_1)}|r_{\mathbf{i},2}|^{d-\dim_B(\Lambda_1)} + \sum_{\mathbf{i}\in\Delta^2_\rho} |r_{\mathbf{i},2}|^{\dim_B(\Lambda_2)}|r_{\mathbf{i},1}|^{d-\dim_B(\Lambda_2)} \geqslant 1,$$

and by (11.82), $\#\mathfrak{P}_\rho \geqslant C\rho^{\max\{\tau_1,\tau_2\}|o(1)|-d}$, which implies that $\underline{\dim}_B(\Lambda) \geqslant d$. The case $d_1 < d_2$ is identical to $d_1 > d_2$.

Now, we prove the last claim of the theorem. Suppose that $|r_{i,1}| \geqslant |r_{i,2}|$ for every $i = 1, \dots, m$. Then

$$\frac{|r_{i,1}|^{\dim_B(\Lambda_1)}|r_{i,2}|^{d_1-\dim_B(\Lambda_1)}}{|r_{i,2}|^{\dim_B(\Lambda_2)}|r_{i,1}|^{d_1-\dim_B(\Lambda_2)}} = \left(\frac{|r_{i,1}|}{|r_{i,2}|}\right)^{\dim_B(\Lambda_2)+\dim_B(\Lambda_1)-d_1} \geqslant 1,$$

where in the last inequality we used Fact 11.4.4. Hence,

$$\sum_{i=1}^{m} |r_{i,2}|^{\dim_B(\Lambda_2)}|r_{i,1}|^{d_1-\dim_B(\Lambda_2)} \leqslant 1,$$

and so $d_2 \leqslant d_1$, whence $\dim_B(\Lambda) = \max\{d_1, d_2\} = d_1$. □

11.5. Applications: Self-affine carpets

Here we consider the so-called Przytycki–Urbański sets and their generalizations, as well as Bedford–McMullen carpets, with complete details. Then we survey some other classes of self-affine sets with diagonal linear parts, in particular, Gatzouras–Lalley carpets. Finally, we discuss more recent results on self-affine sets with lower triangular linear parts.

11.5.1. Generalized Przytycki–Urbański sets. Here we consider the attractors of the two self-affine IFSs of Figure 11.5.

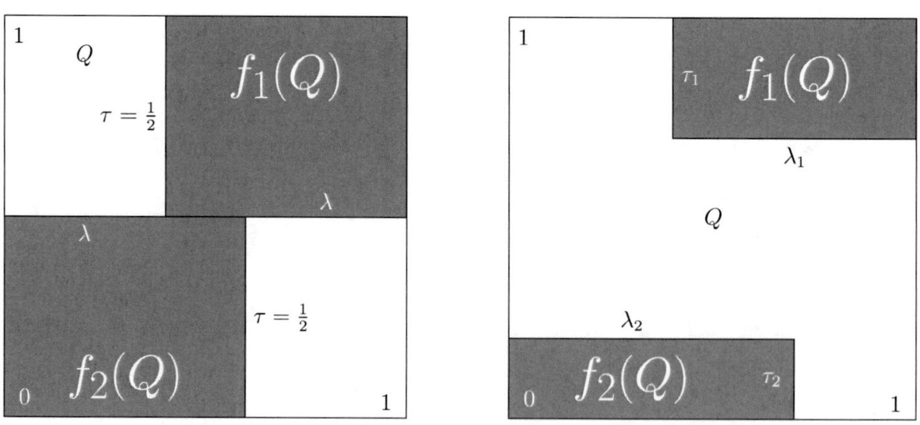

FIGURE 11.5. From left to right: PU-set, Generalized PU-set

11.5.1.1. *Definitions of the three IFSs.* Now we give the formal definitions of PU-, generalized PU- and generalized four corner sets.

EXAMPLE 11.5.1 (PU-set). Przytycki and Urbański introduced a one-parameter family of self-affine IFS $\{\mathcal{F}_\lambda\}_{\lambda\in(0.5,1)}$ on the plane (see the left of Figure 11.5) as the attractor Λ^λ of

$$\mathcal{F}_\lambda := \left(f_i^{(\lambda)}(x) = A^{(\lambda)} \cdot x + t_i^{(\lambda)}\right)_{i=1}^{2}, \tag{11.84}$$

where $\lambda \in \left(\frac{1}{2}, 1\right)$ and

$$A^{(\lambda)} = \begin{bmatrix} \lambda & 0 \\ 0 & \frac{1}{2} \end{bmatrix},\ t_1^{(\lambda)} = (0,0),\ t_2^{(\lambda)} = \left(1-\lambda, \frac{1}{2}\right).$$

That is, for each parameter $\lambda \in \left(\frac{1}{2}, 1\right)$ the fixed point of $f_1^{(\lambda)}$ and $f_2^{(\lambda)}$ are the opposite corners $(0,0)$ and $(1,1)$ of the unit square $Q = [0,1]^2$ as shown on Figure 11.5. Its attractor Λ^λ is called the PU-set.

Pollicott–Weiss [**PW94**], Bedford–Urbański [**BU90**], and Neunhäuserer [**Neu01**] considered a generalization of this system that we call a generalized PU-system:

EXAMPLE 11.5.2 (Generalized PU-set). See the right of Figure 11.5. Let

$$\mathcal{F}_{\boldsymbol{\lambda},\boldsymbol{\tau}} := \left(f_1(x) = \begin{bmatrix} \lambda_1 & 0 \\ 0 & \tau_1 \end{bmatrix} \cdot x,\ f_2(x) = \begin{bmatrix} \lambda_2 & 0 \\ 0 & \tau_2 \end{bmatrix} \cdot x + \begin{bmatrix} 1-\lambda_2 \\ 1-\tau_2 \end{bmatrix}\right) \tag{11.85}$$

where $\boldsymbol{\lambda} := (\lambda_1, \lambda_2)$, $\boldsymbol{\tau} := (\tau_1, \tau_2)$ satisfy $0 < \tau_i < \lambda_i$, $i = 1, 2$, and also $\lambda_1 + \lambda_2 > 1$, $\tau_1 + \tau_2 \leqslant 1$.

THEOREM 11.5.3. *Let $\Lambda^{\boldsymbol{\lambda},\boldsymbol{\tau}}$ be the attractor of the IFS $\mathcal{F}_{\boldsymbol{\lambda},\boldsymbol{\tau}}$ given by* (11.85), *and let s be the unique solution of the equation*

$$\lambda_1 \tau_1^{s-1} + \lambda_2 \tau_2^{s-1} = 1.$$

Then

$$\dim_{\mathrm{B}}(\Lambda^{\boldsymbol{\lambda},\boldsymbol{\tau}}) = s. \tag{11.86}$$

Moreover, if the self-similar measure $\nu_{\boldsymbol{\lambda}}^{\boldsymbol{p}}$ of the IFS $(\lambda_1 x, \lambda_2 x + 1 - \lambda_2)$ with probabilities $\boldsymbol{p} = (\lambda_1 \tau_1^{s-1}, \lambda_2 \tau_2^{s-1})$ has Hausdorff dimension 1, then

$$\dim_{\mathrm{H}}(\Lambda^{\boldsymbol{\lambda},\boldsymbol{\tau}}) = s. \tag{11.87}$$

The measure $\nu_{\boldsymbol{\lambda}}^{\boldsymbol{p}}$ has dimension 1 for Lebesgue almost every $\boldsymbol{\lambda}$, see Theorem 6.8.11 (an earlier Theorem 6.7.5 gives a partial result).

PROOF. The proof easily follows from Theorem 11.4.2 and Theorem 11.2.1 and is left for the reader. □

In particular, the PU-set considered in Example 11.5.1 has dimension $2 + \frac{\log\lambda}{\log 2}$ for a.e. $\lambda \in \left(\frac{1}{2}, 1\right)$. For this set, we can give a more delicate analysis on the dimension of vertical slices, due to Theorem 8.5.1. Denote by $\mathrm{proj}\colon \mathbb{R}^2 \to \mathbb{R}$ the orthogonal projection onto the x-axis. Further, for $x \in [0,1]$ let $L_x = \mathrm{proj}^{-1}(x) \cap \Lambda$.

THEOREM 11.5.4.

(i) *For every $\lambda \in (1/2, 1)$ outside of a set of Hausdorff dimension 0, holds*

$$\dim_{\mathrm{H}}(L_x) = \dim_{\mathrm{B}}(L_x) = 1 + \frac{\log\lambda}{\log 2} \quad \textit{for Lebesgue a.e. } x \in (0,1). \tag{11.88}$$

(ii) *For every* $\lambda \in (2^{-1/2}, 1)$ *outside of a set of Hausdorff dimension 0, holds*

$$\dim_P(L_x) = \dim_B(L_x) = 1 + \frac{\log \lambda}{\log 2} \quad \textit{for all } x \in (0,1). \tag{11.89}$$

PROOF. (i) Let $\nu_\lambda = \nu_\lambda^{1/2}$ be the Bernoulli convolution with probabilities $(\frac{1}{2}, \frac{1}{2})$. Further, denote by $\Pi_1 : \{0,1\}^{\mathbb{N}} \to [0,1]$ the natural projection with respect to the IFS $(g_0(x) = \lambda x,\ g_1(x) = \lambda x + (1-\lambda))$. Applying Theorem 8.5.1 for $\beta = \lambda^{-1}$ we obtain that for every $x \in (0,1)$,

$$\dim_B(L_x) = \dim_B(E_\beta(x)) = \frac{\log \beta}{\log 2} \cdot d_{\nu_\lambda}(x) + 1.$$

Theorem 6.8.14 implies that ν_λ is absolutely continuous with respect to the Lebesgue measure for all $\lambda \in (1/2, 1)$ outside of a set of Hausdorff dimension 0. Fix such a point λ; we have $d_{\nu_\lambda}(x) = 1$ for Lebesgue-a.e. $x \in (0,1)$ and thus,

$$\dim_B(L_x) = 1 + \frac{\log \lambda}{\log 2} \quad \text{for } \mathcal{L}^1\text{-a.e. } x \in (0,1).$$

Now let us consider the family of conditional measures $(\nu_\lambda^{1/2})_x^{\mathrm{proj}^{-1}}$ with respect of the inverse branches $\mathrm{proj}^{-1}(\mathrm{proj}(x))$ for ν_λ-almost every x. By Feng–Hu's formula (11.15),

$$\begin{aligned} \dim_H((\nu_\lambda^{1/2})_x^{\mathrm{proj}^{-1}}) &= \frac{h_{\Pi_2} - h_{\Pi_1}}{\log 2} \quad \text{for } \nu_\lambda\text{-almost every } x, \\ \dim_H(\nu_\lambda) &= \frac{h_{\Pi_1}}{-\log \lambda}, \end{aligned}$$

where Π_2 is the natural projection with respect to the IFS (11.84) and then $\Pi_1 = \mathrm{proj} \circ \Pi_2$. Since the IFS $\mathcal{F}_\lambda$ satisfies the Strong Separation Condition, we have $h_{\Pi_2} = \log 2$. Moreover, since ν_λ is absolutely continuous, we conclude that $h_{\Pi_1} = -\log \lambda$. Thus,

$$\dim_H(L_x) \geqslant \dim_H((\nu_\lambda^{1/2})_x^{\mathrm{proj}^{-1}}) = 1 + \frac{\log \lambda}{\log 2}$$

for ν_λ-almost every x, which implies that the equality (11.88) holds for Lebesgue a.e. x as well.

(ii) Theorem 6.8.14 also implies that there exists a set $\mathcal{E}$ with $\dim_H(\mathcal{E}) = 0$ such that ν_λ is absolutely continuous with respect to the Lebesgue measure with a density in L^q for all $q > 1$ for every $\lambda \in (2^{-1/2}, 1) \backslash \mathcal{E}$. Similarly to Proposition 8.5.2(ii), one can show that for every $\lambda \in (2^{-1/2}, 1)$ outside of a set of Hausdorff dimension 0,

$$\dim_B(L_x) = 1 + \frac{\log \lambda}{\log 2} \quad \text{for every } x \in (0,1). \tag{11.90}$$

Thus, it remains to verify that $\dim_P(L_x) = 1 + \frac{\log \lambda}{\log 2}$ for all $x \in (0,1)$ and for every $\lambda \in (0,1)$ outside of a set of zero Hausdorff dimension.

It is easy to see that if L_x contains an isolated point in the topology of the line $\mathrm{proj}^{-1}(\mathrm{proj}(x))$, then there must be a finite sequence $\mathbf{i} \in \{0,1\}^*$ such that either $x = g_{\mathbf{i}}(0)$ or $x = g_{\mathbf{i}}(1)$. Thus, taking

$$\mathcal{E}' = \big\{\lambda : \exists \mathbf{i} \neq \mathbf{j} \in \{0,1\}^* \text{ and } k, l \in \{0,1\} \text{ such that } S_{\mathbf{i}}(k) = S_{\mathbf{j}}(l)\big\},$$

we have that $\dim_H(\mathcal{E}') = 0$ and for every $\lambda \in (2^{-1/2}, 1) \backslash \mathcal{E}'$, the set L_x contains at most one isolated point for every $x \in (0,1)$.

Fix $\lambda \in (2^{-1/2},1)\backslash(\mathcal{E}\cup\mathcal{E}')$. For $x \in (0,1)$ let $\widetilde{L}_x$ be the subset of the line segment L_x obtained after removing the possible isolated point. Thus, $\widetilde{L}_x$ is compact. By Proposition 1.4.18, for every $\varepsilon > 0$ there exists a cover $\{F_i\}_i$ of $\widetilde{L}_x$ by closed sets such that

$$\dim_{\mathrm{P}}(\widetilde{L}_x) \geqslant \sup_i \overline{\dim}_{\mathrm{B}}(F_i) - \varepsilon.$$

By Baire's Category Theorem, there exists j such that F_j is open in the topology of $\widetilde{L}_x$. Hence there exists $\mathbf{i} \in \{0,1\}^*$ such that $f_{\mathbf{i}}([0,1]^2) \cap \widetilde{L}_x \subset F_j$ and $x \in g_{\mathbf{i}}((0,1))$. Therefore,

$$\overline{\dim}_{\mathrm{B}}(F_j) \geqslant \overline{\dim}_{\mathrm{B}}\big(f_{\mathbf{i}}([0,1]^2) \cap \widetilde{L}_x\big) = \overline{\dim}_{\mathrm{B}}\big([0,1]^2 \cap f_{\mathbf{i}}^{-1}(\widetilde{L}_x)\big) = \overline{\dim}_{\mathrm{B}}\big(\widetilde{L}_{g_{\mathbf{i}}^{-1}(x)}\big).$$

By (11.90), $\overline{\dim}_{\mathrm{B}}\big(\widetilde{L}_{g_{\mathbf{i}}^{-1}(x)}\big) = 1 + \frac{\log\lambda}{\log 2}$. Thus, $1 + \frac{\log\lambda}{\log 2} \geqslant \dim_{\mathrm{P}}(\widetilde{L}_x) \geqslant 1 + \frac{\log\lambda}{\log 2} - \varepsilon$ for every $\varepsilon > 0$, and the claim follows. □

Returning to the discussion of the relation of the almost unique coding (AUC), weak almost unique coding (WAUC) and no dimension drop (NDD), see Definition 11.3.5, we mention here that a consequence of Theorem 11.5.4 is that the WAUC does not imply the AUC. (Clearly, the AUC implies the WAUC.) Here we present an example, due to De–Jun Feng [**Fen**].

EXAMPLE 11.5.5. Consider the IFS $\mathcal{S} = (g_0(x) = \lambda x, g_1(x) = \lambda x + 1 - \lambda)$ with the natural projection $\Pi_1 \colon \{0,1\}^{\mathbb{N}} \to [0,1]$, as in the proof of Theorem 11.5.4. Let $E_1 \subseteq (2^{-1/2}, 0.9)$ be the set such that (11.89) holds for every $\lambda \in (2^{-1/2}, 0.9)\backslash E_1$ and $\dim_{\mathrm{H}}(E_1) = 0$. Clearly, (11.89) implies by the Lipschitz property of the natural projection that $\xi(\mathbf{i}) = \Pi_1^{-1}(\Pi_1(\mathbf{i}))$ is uncountable for every $\mathbf{i} \in \Sigma\backslash\{(0,0,\ldots),(1,1\ldots)\}$, hence the AUC does not hold.

For $p \in (0,1)$ let ν_λ^p be the corresponding Bernoulli convolution. By Theorem 6.8.11, there exists a set $E_2 \subseteq (2^{-1/2}, 0.9)$ with $\dim_{\mathrm{H}}(E_2) = 0$ such that $\dim_{\mathrm{H}}(\nu_\lambda^p) = \frac{p\log p + (1-p)\log(1-p)}{\log\lambda} \leqslant 1$ for ALL $p \in (0,1)$ be such that $-p\log p - (1-p)\log(1-p) \leqslant -\log 0.9$ and for all $\lambda \in (2^{-1/2}, 0.9)\backslash E_2$. Then for every $\lambda \in (2^{-1/2}, 0.9)\backslash(E_1 \cup E_2)$ the IFS $\mathcal{S}$ satisfies the WAUC (see Proposition 11.3.6), but not the AUC, see (11.69).

11.5.2. Bedford–McMullen carpets. Let $a > b \geqslant 2$ be integers and let $A \subseteq \{0,\ldots,b-1\} \times \{0,\ldots,a-1\}$. The attractor of the IFS

$$\left(f_{(i,j)}\begin{pmatrix} x \\ y \end{pmatrix} = \begin{pmatrix} \frac{1}{b} & 0 \\ 0 & \frac{1}{a} \end{pmatrix}\begin{pmatrix} x \\ y \end{pmatrix} + \begin{pmatrix} i/b \\ j/a \end{pmatrix}\right)_{(i,j)\in A} \tag{11.91}$$

is called a *Bedford–McMullen carpet.* Before we turn to the calculation of the dimension, let us introduce some notation. Denote

$$\begin{aligned} A^X &:= \{i : \exists j \text{ s.t. } (i,j) \in A\}, \\ T_i^X &:= \{j : (i,j) \in A\}. \end{aligned}$$

See Figure 11.6.

One can easily see from the definition of the sub-additive pressure function (10.54) or formula (11.31), that the affinity dimension (the unique zero of the pressure function) is equal to

$$s_0 = \min\left\{\frac{\log\#(A)}{\log b},\ 1 + \frac{\log\#(A) - \log b}{\log a}\right\}. \tag{11.92}$$

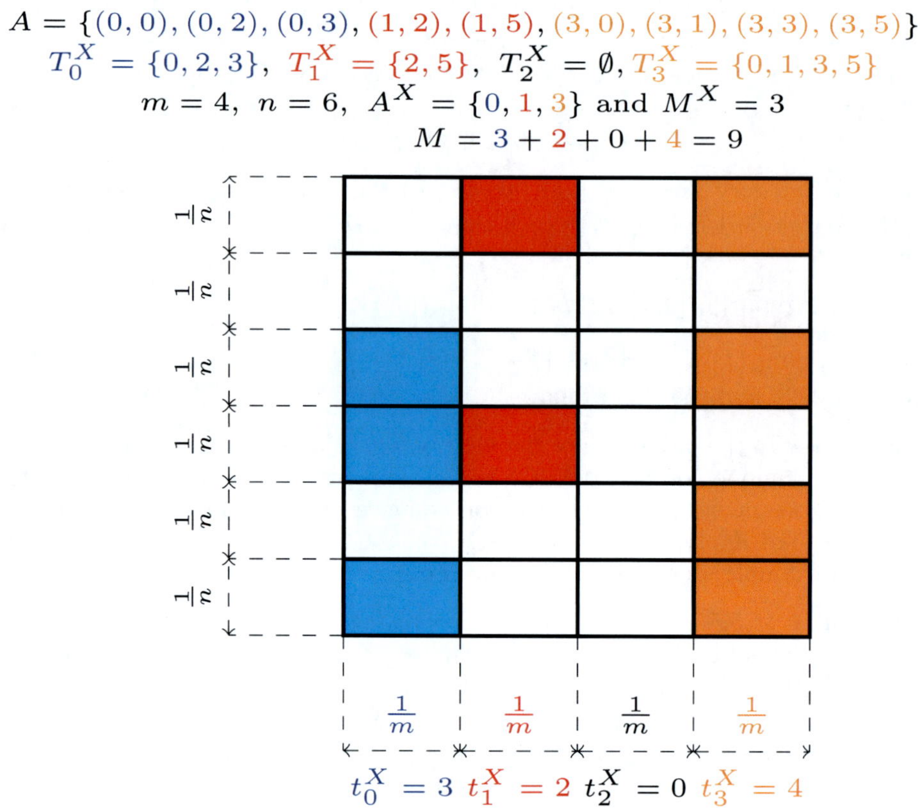

FIGURE 11.6. Bedford–McMullen carpet.

As usual, let $\Sigma = A^{\mathbb{N}}$ be the symbolic space, and denote by Π_2 the natural projection, which can be written explicitly: for $\mathbf{i} = ((i_1, j_1), (i_2, j_2), \ldots)$,

$$\Pi_2(\mathbf{i}) = \left(\sum_{k=1}^{\infty} \frac{i_k}{b^k}, \sum_{k=1}^{\infty} \frac{j_k}{a^k} \right).$$

Observe that the projection of Λ to the horizontal axis, denoted by $\text{proj}(\Lambda)$, is a self-similar set on the line. In fact, $\text{proj}(\Lambda)$ is the attractor of the IFS

$$\left(g_i : x \mapsto \frac{x+i}{b} \right)_{i \in A^X}. \tag{11.93}$$

Clearly, the IFS in (11.93) satisfies the Open Set Condition with respect to the interval $(0,1)$. Consider the natural projection from Σ to $[0,1]$ denoted by Π_1, that is, for $\mathbf{i} = ((i_1, j_1), (i_2, j_2), \ldots)$, let $\Pi_1(\mathbf{i}) = \sum_{k=1}^{\infty} \frac{i_k}{b^k}$. For short, let $t_i^X := \#T_i^X$.

Bedford and McMullen independently computed the Hausdorff dimension of the attractor Λ.

THEOREM 11.5.6 (Bedford [**Bed84**], McMullen [**McM84**]). *The following formulas hold:*

$$\dim_{\mathrm{H}}(\Lambda) = \frac{\log\left(\sum_{i\in A^X} (t_i^X)^{\frac{\log b}{\log a}}\right)}{\log b}; \tag{11.94}$$

$$\dim_{\mathrm{B}}(\Lambda) = \frac{\log \#(A^X)}{\log b} + \frac{\log\left(\#(A)/\#(A^X)\right)}{\log a}. \tag{11.95}$$

PROOF. The proof of (11.95) follows easily from Theorem 11.4.2. Since $a \geqslant b$, we get that $\dim_{\mathrm{B}}(\Lambda)$ is the unique solution of

$$\#(A)\left(\frac{1}{b}\right)^{\dim_{\mathrm{B}}(\mathrm{proj}(\Lambda))}\left(\frac{1}{a}\right)^{d-\dim_{\mathrm{B}}(\mathrm{proj}(\Lambda))} = 1.$$

Since $\dim_{\mathrm{B}}(\mathrm{proj}(\Lambda)) = \frac{\log \#(A^X)}{\log b}$, the claim follows.

Now we turn to the proof of (11.94). We begin with the lower bound, which relies on the Feng–Hu Theorem. (For a direct proof, see [**BP17**, Section 4.2].)

Let μ be the Bernoulli measure on Σ with the probability vector $(p_{(i,j)})_{(i,j)\in A}$, and let $\nu = (\Pi_2)_*\mu$. Since $\dim_{\mathrm{H}}(\Lambda) \geqslant \dim_{\mathrm{H}}(\nu)$, by maximizing the value of $\dim_{\mathrm{H}}(\nu)$ we will get a lower bound. First, let us compute the exact value of the Hausdorff dimension of the measure ν using the Feng–Hu formula (11.10). By (11.15) and (11.16), hold

$$\dim_{\mathrm{H}}(\nu) = \frac{h_{\Pi_1}(\mu)}{\log b} + \frac{h_{\Pi_2}(\mu) - h_{\Pi_1}(\mu)}{\log a} \quad \text{and} \quad \dim_{\mathrm{H}}(\mathrm{proj}_*\,\nu) = \frac{h_{\Pi_1}(\mu)}{\log b}.$$

Since the IFS (11.91) satisfies the Strong Open Set Condition, we have $h_{\Pi_2}(\mu) = h(\mu) = -\sum_{(i,j)\in A} p_{i,j}\log p_{i,j}$, see Lemma 11.3.7. On the other hand, the measure $\mathrm{proj}_*\,\mu$ is a self-similar measure with respect to the probabilities $\{\sum_{j\in T_i^X} p_{i,j},\ i \in A^X\}$ for the IFS (11.93). Thus, by Corollary 1.9.11,

$$\dim_{\mathrm{H}}(\mathrm{proj}_*\,\nu) = \frac{-\sum_{i\in A^X}\sum_{j\in T_i^X} p_{i,j}\log\sum_{j\in T_i^X} p_{i,j}}{\log b}.$$

Therefore,

$$\dim_{\mathrm{H}}(\nu) = \frac{-\sum_{(i,j)\in A} p_{i,j}\log p_{i,j}}{\log a} + \left(1 - \frac{\log b}{\log a}\right)\frac{-\sum_{i\in A^X}\sum_{j\in T_i^X} p_{i,j}\log\sum_{j\in T_i^X} p_{i,j}}{\log b}. \tag{11.96}$$

Taking the probabilities

$$p_{i,j} = \frac{(t_i^X)^{\frac{\log b}{\log a}-1}}{\sum_{k\in A^X}(t_k^X)^{\frac{\log b}{\log a}}} \quad \text{and noting that} \quad \sum_{j\in T_i^X} p_{i,j} = q_i := \frac{(t_i^X)^{\frac{\log b}{\log a}}}{\sum_{k\in A^X}(t_k^X)^{\frac{\log b}{\log a}}}, \tag{11.97}$$

we obtain, after a short computation:

$$\dim_{\mathrm{H}}(\Lambda) \geqslant \frac{\log\left(\sum_{i\in A^X}(t_i^X)^{\frac{\log b}{\log a}}\right)}{\log b}. \tag{11.98}$$

Now we turn to the proof of the upper bound. Our claim is that the lower bound in (11.98) is sharp. The usual way to show that is to find an optimal cover for the set Λ. However, our natural cover, which is related to the affinity dimension, is not optimal unless certain conditions hold, see later. Therefore, we use here Theorem 1.9.5, a variant of the mass distribution principle. That is, we will show that for the self-affine measure ν with probability vector defined in (11.97),

$$\liminf_{r\to 0+}\frac{\log\nu(B_r(\Pi_2(\mathbf{i})))}{\log r}\leqslant\frac{\log\left(\sum_{i\in A^X}(t_i^X)^{\frac{\log m}{\log n}}\right)}{\log b}\quad\text{for }\underline{\textbf{every}}\ \mathbf{i}\in\Sigma. \tag{11.99}$$

Theorem 1.9.5 then implies that $\dim_{\mathrm{H}}(\Lambda)\leqslant\frac{1}{\log b}\log\left(\sum_{i\in A^X}(t_i^X)^{\frac{\log b}{\log a}}\right)$, as desired.

For $l\in\mathbb{N}$, let $k=k(l)\geqslant 1$ be the smallest integer such that $b^{-k}\leqslant a^{-l}$, i.e. $k(l)=\left\lceil l\frac{\log a}{\log b}\right\rceil$. Let $\mathbf{i}=((i_1,j_1),(i_2,j_2),\ldots)$. Denote by $C_l(\mathbf{i})$ the following "approximate square"

$$C_l(\mathbf{i})=\left\{\left(\sum_{r=1}^{\infty}\frac{i'_r}{b^r},\ \sum_{r=1}^{\infty}\frac{j'_r}{a^r}\right)\in\Lambda : i_p=i'_p\text{ for }p=1,\ldots,\left\lceil l\frac{\log a}{\log b}\right\rceil\right.$$
$$\left.\text{and }j_q=j'_q\text{ for }q=1,\ldots,l\right\}.$$

In other words, $C_l(\mathbf{i})$ is the union of all $k(l)$th level cylinder sets $\mathcal{C}_{k(l)}$ such that $\Pi_1(\mathbf{i})\in\operatorname{proj}\left(f_{(i_1,j_1)}\circ\cdots\circ f_{(i_l,j_l)}(\Lambda)\cap\mathcal{C}_{k(l)}\right)$.

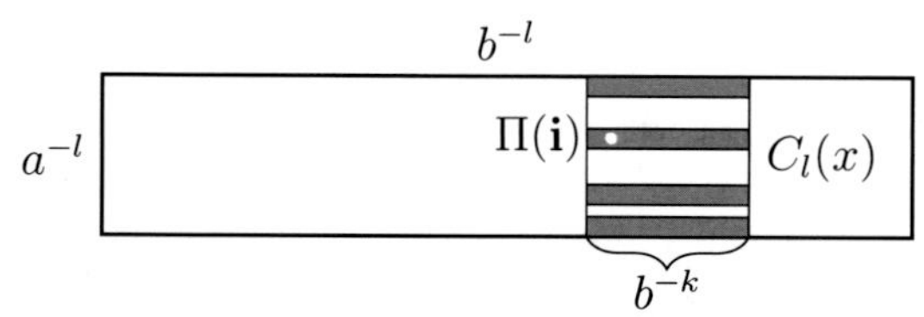

It is not hard to see that

$$C_l(\mathbf{i})=f_{(i_1,j_1)}\circ\cdots\circ f_{(i_l,j_l)}\left(\operatorname{proj}^{-1}\left[\sum_{r=1}^{k(l)-l}\frac{i_{l+r}}{b^r},\ \sum_{r=1}^{k(l)-l}\frac{i_{l+r}}{b^r}+\frac{1}{b^{k(l)-l}}\right]\right).$$

Therefore, by the definition of measure ν,

$$\nu\left(B_{\sqrt{2}/a^l}(\Pi_2(\mathbf{i}))\right)\geqslant\nu(C_l(\mathbf{i}))=$$
$$\frac{(t_{i_1}^X)^{\frac{\log b}{\log a}-1}\cdots(t_{i_l}^X)^{\frac{\log b}{\log a}-1}\cdot(t_{i_{l+1}}^X)^{\frac{\log b}{\log a}}\cdots(t_{i_{k(l)}}^X)^{\frac{\log b}{\log a}}}{\left(\sum_{i'\in A^X}(t_{i'}^X)^{\frac{\log b}{\log a}}\right)^{k(l)}}.$$

Thus,

$$\frac{\log\nu\left(B_{\sqrt{2}/a^l}(\Pi_2(\mathbf{i}))\right)}{-l\log a}\leqslant\frac{k(l)\log b}{l\log a}\cdot\frac{\log\sum_{i'\in A^X}(t_{i'}^X)^{\frac{\log b}{\log a}}}{\log b}+$$
$$+\frac{-1}{\log a}\left(\frac{1}{k(l)}\sum_{r=1}^{k(l)}\log t_{i_r}^X-\frac{1}{l}\sum_{r=1}^{l}\log t_{i_r}^X\right).$$

Since $k(l)/l \to \log a/\log b$ as $l \to \infty$, it is enough to show that

$$\limsup_{l\to\infty} \frac{\left(\prod_{r=1}^{k(l)} t_{i_r}^X\right)^{1/k(l)}}{\left(\prod_{r=1}^{l} t_{i_r}^X\right)^{1/l}} \geqslant 1, \tag{11.100}$$

which implies (11.99). In order to verify (11.100), we need the following simple lemma:

LEMMA 11.5.7. *Let $\{a_n\}$ be a sequence of positive real numbers and let $c > 1$. If* $\limsup_{n\to\infty} \frac{a_{[cn]}}{a_n} < 1$ *then* $\liminf_{n\to\infty} a_n = 0$.

The proof of the lemma is left as an exercise. Now, for every sequence $i_1, i_2, \ldots, i_r, \ldots$ and every $l \geqslant 1$

$$a_l := \left(\prod_{r=1}^{l} t_{i_r}^X\right)^{1/l} \geqslant 1,$$

hence (11.99) holds by the lemma. □

REMARK 11.5.8.

(a) Comparing (11.94) with (11.92), we obtain by a simple algebraic calculation:

$$s_0 = \dim_{\mathrm{H}} \Lambda \iff \#(A^X) = \#(A) \quad \text{or} \quad t_i^X = \frac{\#(A)}{a} \text{ for all } i \in Q. \tag{11.101}$$

(b) Using the formulas (11.92) and (11.95), one can easily check that

$$s_0 = \dim_{\mathrm{B}} \Lambda \iff \#(A^X) = \#(A) \quad \text{or} \quad \#(A^X) = b.$$

In other words, the box dimension is equal to the affinity dimension if and only if

$$\dim_{\mathrm{B}}\big(\mathrm{proj}(\Lambda)\big) = \min\{1, \dim_{\mathrm{B}} \Lambda\}.$$

(c) We say that the carpet has *uniform vertical fibers* if there is an equal number of chosen rectangles in every nonempty column. The formulas show that the Hausdorff dimension of the Bedford–McMullen carpet is equal to the box dimension if and only the carpet has uniform vertical fibers.

(d) Gatzouras–Lalley [**LG92**] and Bishop–Peres [**BP17**, Theorem 4.3.2] showed that for *any gauge function* ϕ, the Hausdorff measure $\mathcal{H}^\phi$ of the Bedford–McMullen carpet is either zero or infinity, provided the carpet has nonuniform vertical fibers. Peres [**Per94**] proved that the Hausdorff measure $\mathcal{H}^s$ (where s is the Hausdorff dimension) is infinite and found more precise information in terms of gauge functions.

(e) Motivated by the article of McMullen [**McM84**] which just appeared then, Przytycki and Urbański [**PU89**], in a note added to their paper after it was completed, indicated how their method can be used to construct a Hölder-continuous function from $[0, 1]$ to $\mathbb{R}$, which is Hölder-continuous of order $\alpha = \log m/\log n$ for some integers $n > m > 1$, such that the Hausdorff dimension of its graph is an arbitrary number between 1 and $2 - \alpha$.

EXAMPLE 11.5.9. Let $\Lambda^l, \Lambda^m, \Lambda^r$ be the Bedford–McMullen carpets generated by the IFS on the left, middle and right sub-figure of Figure 11.7 respectively. Then

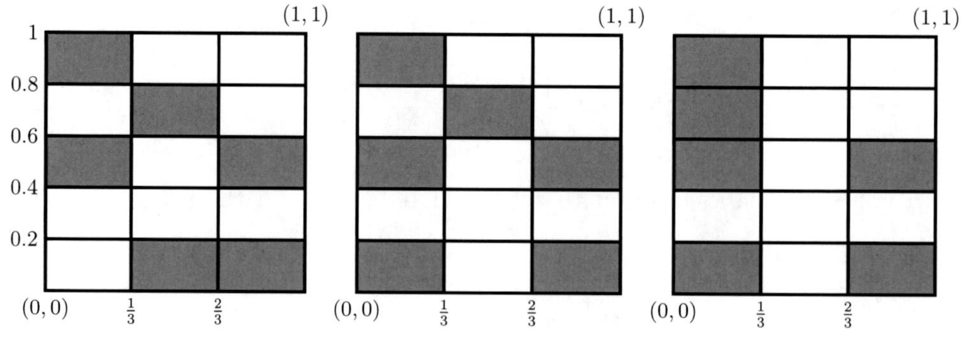

FIGURE 11.7. Three Bedford–McMullen IFS

left: $s := \dim_{\mathrm{H}} \Lambda^l = \dim_{\mathrm{B}} \Lambda^l = \dim_{\mathrm{aff}} \Lambda$ and $0 < \mathcal{H}^s(\Lambda^l) < \infty$,
middle: $s := \dim_{\mathrm{H}} \Lambda^m < \dim_{\mathrm{B}} \Lambda^m = \dim_{\mathrm{aff}} \Lambda^m$ and $\mathcal{H}^s(\Lambda^m) = \infty$,
right: $s := \dim_{\mathrm{H}} \Lambda^r < \dim_{\mathrm{B}} \Lambda^r < \dim_{\mathrm{aff}} \Lambda^r$ and $\mathcal{H}^s(\Lambda^r) = \infty$.

11.5.3. Gatzouras–Lalley carpets. In 1992 Gatzouras an Lalley [**LG92**] introduced a family of self-affine carpets, more general than Bedford–McMullen carpets. Below we use the same notation as in [**KS19**].

DEFINITION 11.5.10. The diagonal self-affine IFS

$$\mathcal{F} = (f_i(\underline{x}) := A_i\underline{x} + t_i)_{i=1}^m,\ A_i = \begin{pmatrix} b_i & 0 \\ 0 & a_i \end{pmatrix},\ 0 < a_i < b_i < 1,\ t_i = \begin{pmatrix} t_{i,1} \\ t_{i,2} \end{pmatrix}, \tag{11.102}$$

is a Gatzouras–Lalley (GL) IFS if the first order cylinder rectangles

$$R_i := f_i(R) = t_i + [0, b_i] \times [0, a_i] \tag{11.103}$$

(which are axes parallel rectangles) satisfy

(1) $R_i \subset R$,
(2) $R_i^\circ \cap R_j^\circ = \varnothing$ if $i \neq j$.
(3) Either $\mathrm{proj}_x(R_i^\circ) \cap \mathrm{proj}_x(R_j^\circ) = \varnothing$ or $\mathrm{proj}_x(R_i) = \mathrm{proj}_x(R_j)$. In the latter case we say that R_i and R_j are in the same column.

Since all of the parameters of f_i are determined by R_i, we can represent $\mathcal{F}$ graphically by drawing R_i, $i = 1, \ldots, m$ inside R (see Figure 11.8). We can label the elements of $\mathcal{F}$ so that the labels of those cylinder rectangles which are in one column form an interval in $[m] = \{1, \ldots, m\}$. More precisely, we write $M \leqslant m$ for the number of columns which contain at least one R_i and we partition $\mathcal{I} := [m]$ as $\mathcal{I} = \bigsqcup_{\widehat{i}=1}^{M} \mathcal{I}_{\widehat{i}}$ where $\#\mathcal{I}_{\widehat{i}} = N_{\widehat{i}}$, $\widehat{i} = 1, \ldots, M$, and

$$\mathcal{I}_1 = \{1, \ldots, N_1\}, \quad \mathcal{I}_{\widehat{i}} = \{N_1 + \ldots + N_{\widehat{i}-1} + 1, \ldots, N_1 + \ldots + N_{\widehat{i}}\}, \tag{11.104}$$

where $k, \ell \in \mathcal{I}_{\widehat{i}}$ means that R_k, R_ℓ are in the $\widehat{i}$-th column:

$$k, \ell \in \mathcal{I}_{\widehat{i}} \iff \begin{cases} b_k = b_\ell =: r_{\widehat{i}}, \\ t_{k,1} = t_{\ell,1} =: u_{\widehat{i}}. \end{cases} \tag{11.105}$$

This induces a mapping

$$\phi : \{1, 2, \ldots, m\} \to \{1, 2, \ldots, M\},\ \ \phi(i) := \widehat{i}\ \text{ if } i \in \mathcal{I}_{\widehat{i}}. \tag{11.106}$$

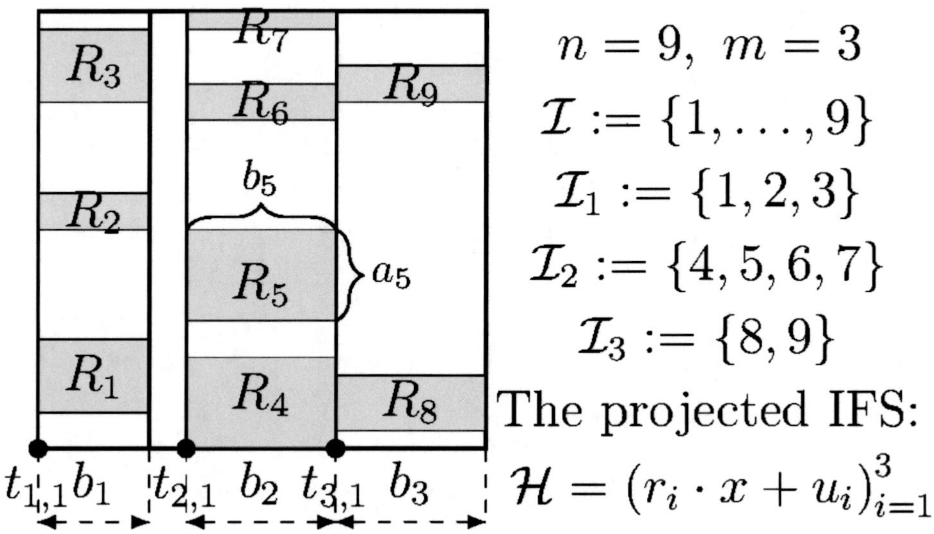

FIGURE 11.8. A Gatzouras–Lalley IFS

Projecting $\mathcal{F}$ to the x-axis we get the "horizontal" self-similar IFS $\mathcal{H}$:

$$\mathcal{H} = \{h_{\widehat{i}}(x) := r_{\widehat{i}} \cdot x + u_{\widehat{i}}\}_{\widehat{i}=1}^{M}, \tag{11.107}$$

Without loss of generality we may assume that

$$0 = \min_{1 \leqslant \widehat{i} \leqslant M} \left\{ \mathrm{Fix}(h_{\widehat{i}}) \right\}, \qquad 1 = \max_{1 \leqslant \widehat{i} \leqslant M} \left\{ \mathrm{Fix}(h_{\widehat{i}}) \right\}, \tag{11.108}$$

where $\mathrm{Fix}(h_{\widehat{i}})$ is the fixed point of $h_{\widehat{i}}$. We write Λ and $\widehat{\Lambda}$ for the attractor of $\mathcal{F}$ and $\mathcal{H}$ respectively and we denote the corresponding symbolic spaces by $\Sigma := [m]^{\mathbb{N}}$, $\Sigma_{\mathcal{H}} := [M]^{\mathbb{N}}$. The natural projections from Σ and from $\Sigma_{\mathcal{H}}$ to the attractors Λ and $\Lambda_{\mathcal{H}}$ are denoted by Π and $\Pi_{\mathcal{H}}$ respectively. The mapping ϕ defined in (11.106) extends naturally component-wise to $\Phi : \Sigma \to \Sigma_{\mathcal{H}}$. Then the following diagram commutes:

$$\begin{array}{ccc} \Sigma & \xrightarrow{\ \Phi\ } & \Sigma_{\mathcal{H}} \\ {\scriptstyle \Pi}\downarrow & & \downarrow{\scriptstyle \Pi_{\mathcal{H}}} \\ \Lambda & \xrightarrow[\mathrm{proj}_x]{} & \Lambda_{\mathcal{H}} \end{array} \tag{11.109}$$

For every probability vector $\mathbf{p} = (p_i)_{i=1}^m$ we associate another probability vector, which is the "horizontal projection" of $\mathbf{p}$, as follows:

$$\mathbf{q} = \mathbf{q_p} := (q_j)_{j=1}^M, \quad \text{where } q_j := \sum_{k \in \mathcal{I}_j} p_k. \tag{11.110}$$

Consider the infinite product measures corresponding to $\mathbf{p}$ and $\mathbf{q} = \mathbf{q_p}$:

$$\mu_{\mathbf{p}} := \mathbf{p}^{\mathbb{N}} \text{ and } \mu_{\mathbf{q}} := \mathbf{q}^{\mathbb{N}} \tag{11.111}$$

on Σ and $\Sigma_{\mathcal{H}}$ respectively and observe that

$$\Phi_* \mu_{\mathbf{p}} = \mu_{\mathbf{q}}. \tag{11.112}$$

Using this and that the diagram in (11.109) commutes, we get that

$$\nu_{\mathbf{q}} := (\Pi_{\mathcal{H}})_* \mu_{\mathbf{q}} = (\mathrm{proj}_x)_* \nu_{\mathbf{p}}, \tag{11.113}$$

where $\nu_{\mathbf{p}} := \Pi_* \mu_{\mathbf{p}}$. By the Feng–Hu Theorem (11.25) we obtain

$$\begin{aligned} \dim_{\mathrm{H}}(\nu_{\mathbf{p}}) &= \frac{h_{\mu_{\mathbf{p}}}}{\chi_2(\nu_{\mathbf{p}})} + \left(1 - \frac{\chi_1(\nu_{\mathbf{p}})}{\chi_2(\nu_{\mathbf{p}})}\right) \cdot \frac{h_{\mu_{\mathbf{q}}}}{\chi_{\mu_{\mathbf{q}}}} \\ &= \frac{\sum_{i=1}^{m} p_i \log p_i}{\sum_{i=1}^{m} p_i \log a_i} + \left(1 - \frac{\sum_{i=1}^{m} p_i \log b_i}{\sum_{i=1}^{m} p_i \log a_i}\right) \frac{\sum_{i=1}^{M} q_i \log q_i}{\sum_{i=1}^{M} q_i \log b_i}. \end{aligned} \tag{11.114}$$

Recall that $\mathcal{P}_m$ is the set of all probability vectors $(p_1, \ldots, p_m)$, with all $p_i > 0$.

THEOREM 11.5.11 (Gatzouras and Lalley [**LG92**]). *There exists $\mathbf{p}^* \in \mathcal{P}_m$ such that*

$$\alpha^* := \dim_{\mathrm{H}}(\Lambda) = \dim_{\mathrm{H}}(\nu_{\mathbf{p}^*}) = \sup_{\mathbf{p} \in \mathcal{P}_m} \dim_{\mathrm{H}}(\nu_{\mathbf{p}}). \tag{11.115}$$

Gatzouras and Lalley also computed the box dimension of Λ.

THEOREM 11.5.12 (Gatzouras and Lalley [**LG92**]). *Let s_x be the similarity dimension of the IFS $\mathcal{H}$, that is,*

$$\sum_{k=1}^{M} r_i^{s_x} = 1. \tag{11.116}$$

(By construction, $s_x \leqslant 1$ always holds.) Then

$$\dim_{\mathrm{P}}(\Lambda) = \dim_{\mathrm{B}}(\Lambda) = s, \quad \text{where } s \text{ is the solution of } \sum_{k=1}^{m} b_i^{s_x} a_i^{s - s_x} = 1, \tag{11.117}$$

in particular, the box dimension exists.

In general $\dim_{\mathrm{H}}(\Lambda) < \dim_{\mathrm{B}}(\Lambda)$. This holds for example for the GL IFS that appears in Figure 11.8. Gatzouras and Lalley pointed out that

$$\begin{aligned} \dim_{\mathrm{H}}(\Lambda) = \dim_{\mathrm{B}}(\Lambda) &\iff 0 < \mathcal{H}^{\alpha^*}(\Lambda) < \infty \\ &\iff \sum_{j \in \mathcal{I}_{\widehat{i}}} a_j^{s - s_x} = 1, \ \forall \widehat{i} \in [M]. \end{aligned} \tag{11.118}$$

where $\alpha^* = \dim_{\mathrm{H}}(\Lambda)$ and $s = \dim_{\mathrm{B}}(\Lambda)$. If the last of the equivalent condition in (11.118) holds, then we say that Λ has uniform vertical structure.

An important special case is *diagonally homogeneous case.* This is when there exist $a, b \in (0,1)$, $a < b$, such that

$$b_i \equiv b, \ a_i \equiv a, \ \forall i \in [m]. \tag{11.119}$$

In this case the uniform vertical structure condition becomes:

$$N_{\widehat{i}} \equiv m/M, \ \forall \widehat{i} \in [M]. \tag{11.120}$$

THEOREM 11.5.13. *In the diagonally homogeneous case, i.e., when (11.119) holds, we have*

$$\dim_{\mathrm{H}}(\Lambda) = \frac{1}{-\log b} \log \sum_{\widehat{j}=1}^{M} N_{\widehat{j}}^{\frac{\log b}{\log a}}, \tag{11.121}$$

$$\dim_{\mathrm{B}}(\Lambda) = \frac{\log m}{-\log a} + \left(1 - \frac{\log b}{\log a}\right) \frac{\log M}{-\log b}.$$

The optimal $\mathbf{p}^* = (p_k^*)_{k=1}^m \in \mathcal{P}_m$*, for which*

$$\dim_{\mathrm{H}}(\nu_{\mathbf{p}^*}) = \dim_{\mathrm{H}}(\Lambda) \tag{11.122}$$

is given by:

$$p_k^* := N_{\widehat{i}}^{\frac{\log b}{\log a} - 1} \cdot \Big(\sum_{\widehat{j}=1}^{M} N_{\widehat{j}}^{\frac{\log b}{\log a}} \Big)^{-1}, \quad \text{if } k \in \mathcal{I}_{\widehat{i}} \tag{11.123}$$

In particular, $\dim_{\mathrm{H}}(\Lambda) = \dim_{\mathrm{B}}(\Lambda)$ *if and only if* $N_{\widehat{i}} = m/M$ *for all* $\widehat{i} \in [M]$.

The optimal $\mathbf{p}^* = (p_k^*)_{k=1}^m \in \mathcal{P}_m$ in the theorem is obtained by a routine use of the method of Lagrange multipliers.

Finally, we compute the affinity dimension (see Definition 1.10.2) of Λ, or rather of the GL IFS $\mathcal{F}$ from (11.102). Define $\widetilde{s}_x$ as the unique real solution of

$$\sum_{i=1}^{m} b_i^{\widetilde{s}_x} = 1. \tag{11.124}$$

(compare with the definition of s_x in (11.116)). Then

$$\dim_{\mathrm{aff}}(\Lambda) = s_A \quad \text{where} \quad \sum_{i=1}^{m} b_i^{\min\{\widetilde{s}_x,1\}} a_i^{s_A - \min\{\widetilde{s}_x,1\}} = 1 \tag{11.125}$$

Hence,

- if $\widetilde{s}_x < 1$ then $s_A = \widetilde{s}_x$,
- if $s_x \geqslant 1$ then s_A is the solution of the equation $\sum_{i=1}^{m} b_i a_i^{s_A - 1} = 1$.

This implies that

$$\dim_{\mathrm{B}}(\Lambda) = \dim_{\mathrm{aff}}(\Lambda) \iff N_{\widehat{i}} = 1 \text{ for all } \widehat{i} \in [M] \text{ or } \Lambda_{\mathcal{H}} = [0,1]. \tag{11.126}$$

11.5.4. Barański carpets. Another generalization of the Bedford–McMullen carpets was introduced by Barański [**Bar07**]. Compared to the Gatzouras–Lalley carpets, the Barański carpet does not have a common strong contracting direction, but it keeps the grid structure. Let us now describe it in detail.

Let $N, M \geqslant 2$ be integers and let $a_1, \ldots, a_N$ and $b_1, \ldots, b_M$ be positive reals such that $\sum_{i=1}^N a_i = \sum_{j=1}^M b_j = 1$. Let $A \subseteq \{1, \ldots, N\} \times \{1, \ldots, M\}$ be nonempty with at least two elements. For $(i,j) \in A$, let

$$f_{(i,j)}(\underline{x}) = \begin{pmatrix} a_i & 0 \\ 0 & b_j \end{pmatrix} \underline{x} + \begin{pmatrix} \sum_{k=1}^{i-1} a_k \\ \sum_{k=1}^{j-1} b_k \end{pmatrix} \text{ and } \mathcal{F} = (f_{(i,j)})_{(i,j) \in A}. \tag{11.127}$$

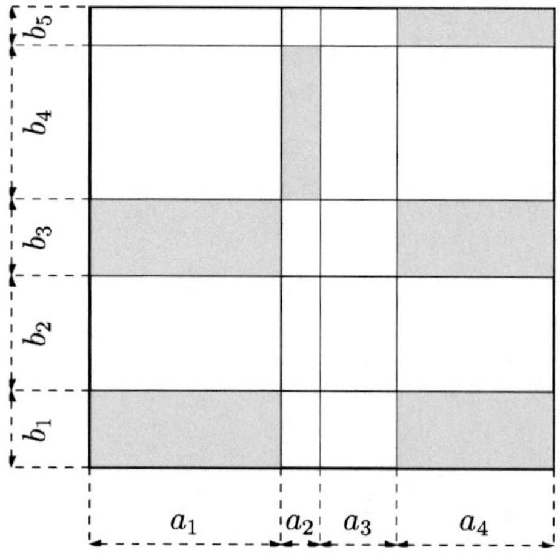

$$N = 4, \; M = 5$$

$$A = \{(1,1),(4,1),(1,3),(4,3),(2,4),(4,5)\}$$

$$A^X = \{1,2,4\} \qquad A^Y = \{1,3,4,5\}$$

FIGURE 11.9. A Barański IFS

By construction, the unit square is mapped into itself by all the maps of $\mathcal{F}$ defined in (11.127). For an example of $\mathcal{F}$, see Figure 11.9. As usual, denote by Λ the attractor of $\mathcal{F}$, and let $\text{proj}_x(\Lambda)$ and $\text{proj}_y(\Lambda)$ be the orthogonal projections of Λ to the x- and y-axis respectively. It is easy to see that $\mathcal{F}$ satisfies the ROSC. Let us define

$$\begin{aligned} A^X &:= \{i \in \{1,\dots,N\} : \exists j \text{ such that } (i,j) \in A\}, \\ A^Y &:= \{j \in \{1,\dots,M\} : \exists i \text{ such that } (i,j) \in A\}. \end{aligned}$$

Clearly, the projections $\text{proj}_x(\Lambda)$ and $\text{proj}_y(\Lambda)$ are self-similar sets satisfying the OSC, and

$$\begin{aligned} &\dim_{\mathrm{H}}(\text{proj}_x(\Lambda)) = \dim_{\mathrm{B}}(\text{proj}_x(\Lambda)) = s_x, \text{ where } \sum_{i\in A^X} a_i^{s_x} = 1, \text{ and} \\ &\dim_{\mathrm{H}}(\text{proj}_y(\Lambda)) = \dim_{\mathrm{B}}(\text{proj}_y(\Lambda)) = s_y, \text{ where } \sum_{j\in A^Y} b_j^{s_y} = 1. \end{aligned} \tag{11.128}$$

The following theorem is a simple corollary of Theorem 11.4.2.

THEOREM 11.5.14 (Barański). *For the IFS defined in* (11.127), $\dim_{\mathrm{B}}(\Lambda) = \max\{d_1, d_2\}$, *where* d_1 *and* d_2 *are the unique solution of the equations*

$$\sum_{(i,j)\in A} a_i^{s_x} b_j^{d_1 - s_x} = 1 \text{ and } \sum_{(i,j)\in A} b_j^{s_y} a_i^{d_2 - s_y} = 1,$$

where s_x *and* s_y *are defined in* (11.128).

For $i \in A^X$ and $j \in A^Y$, define

$$T_i^X := \{j \in \{1,\dots,M\} : (i,j) \in A\}, \text{ and } T_j^Y := \{i \in \{1,\dots,N\} : (i,j) \in A\}$$

respectively. With a probability vector $\mathbf{p} = (p_{(i,j)})_{(i,j)\in A}$ we associate another two probability vectors $\mathbf{q}^X := (q_i^X)_{i\in A^X}$ and $\mathbf{q}^Y := (q_j^Y)_{j\in A^Y}$, which are the "horizontal" and "vertical" projections of $\mathbf{p}$, as follows:

$$q_i^X := \sum_{j\in T_i^X} p_{(i,j)} \text{ and } q_j^Y := \sum_{i\in T_j^Y} p_{(i,j)}. \tag{11.129}$$

Consider the infinite product measures corresponding to $\mathbf{p}$, $\mathbf{q}^X$ and $\mathbf{q}^Y$:

$$\mu_{\mathbf{p}} := \mathbf{p}^{\mathbb{N}}, \mu_{\mathbf{q}^X} := (\mathbf{q}^X)^{\mathbb{N}} \text{ and } \mu_{\mathbf{q}^Y} := (\mathbf{q}^Y)^{\mathbb{N}} \tag{11.130}$$

on $A^{\mathbb{N}}$, $(A^X)^{\mathbb{N}}$ and $(A^Y)^{\mathbb{N}}$ respectively. Simple calculations show that

$$\chi_1(\mathbf{p}) = \chi(\mathbf{q}^X) = -\sum_{i \in A^X} q_i^X \log a_i \text{ and } \chi_2(\mathbf{p}) = \chi(\mathbf{q}^Y) = -\sum_{j \in A^Y} q_j^Y \log b_j.$$

Using Feng–Hu Theorem 11.2.1, it is easy to show that

$$\dim_{\mathrm{H}}(\nu_{\mathbf{p}}) = \begin{cases} \dfrac{h_{\mathbf{p}}}{\chi_2(\mathbf{p})} + \left(1 - \dfrac{\chi_1(\mathbf{p})}{\chi_2(\mathbf{p})}\right) \dfrac{h_{\mathbf{q}^X}}{\chi_1(\mathbf{p})}, & \text{if } \chi_1(\mathbf{p}) \leqslant \chi_2(\mathbf{p}); \\ \dfrac{h_{\mathbf{p}}}{\chi_1(\mathbf{p})} + \left(1 - \dfrac{\chi_2(\mathbf{p})}{\chi_1(\mathbf{p})}\right) \dfrac{h_{\mathbf{q}^Y}}{\chi_2(\mathbf{p})}, & \text{otherwise.} \end{cases} \tag{11.131}$$

The following theorem was obtained by Barański [**Bar07**, Theorem A].

THEOREM 11.5.15 (Barański). *For the IFS defined in* (11.127), $\dim_{\mathrm{H}}(\Lambda) = \max_{\mathbf{p} \in \mathcal{P}} \dim_{\mathrm{H}}(\nu_{\mathbf{p}})$.

11.5.5. Lower triangular self-affine IFS. Here we describe, without proof, some more recent developments, on self-affine sets with lower-triangular linear parts. We start with a result of Bárány, Rams, and Simon [**BRS19**] and then discuss "Triangular Gatzouras–Lalley carpets" studied by Kolossváry and Simon [**KS19**].

DEFINITION 11.5.16. We say that $\mathcal{F}$ is a lower triangular self-affine IFS if

$$\mathcal{F} = (f_i(\underline{x}) := A_i \underline{x} + t_i)_{i=1}^N, A_i = \begin{pmatrix} b_i & 0 \\ d_i & a_i \end{pmatrix}, \; a_i, b_i \in (0,1), t_i = \begin{pmatrix} t_{i,1} \\ t_{i,2} \end{pmatrix}. \tag{11.132}$$

Without loss of generality we may assume that $R_i := f_i(R) \subset R$, where $R := [0,1]^2$. The attractor of $\mathcal{F}$ is denoted by Λ, and we write $\Sigma := \{1, \ldots, N\}^{\mathbb{N}}$ and $\Pi : \Sigma \to \Lambda$ for the natural projection. If $\mathcal{F}$ is an IFS of the form (11.137), then the linear part of $f_{i_1 \ldots i_n} := f_{i_1} \circ \cdots \circ f_{i_n}$ is

$$A_{i_1 \ldots i_n} := A_{i_1} \cdots A_{i_n} := \begin{pmatrix} b_{i_1 \ldots i_n} & 0 \\ \widetilde{d}_{i_1 \ldots i_n} & a_{i_1 \ldots i_n} \end{pmatrix} \tag{11.133}$$

where

$$\widetilde{d}_{i_1 \ldots i_n} = \sum_{\ell=1}^{n} d_{i_\ell} \cdot \prod_{k<\ell} a_{i_k} \cdot \prod_{r=\ell+1}^{n} b_{i_r}. \tag{11.134}$$

The proof of the following Fact is contained in [**Arn98**, p.129].

FACT 11.5.17. *Let* $\mu_{\mathbf{p}} := \mathbf{p}^{\mathbb{N}}$ *for a probability vector* $\mathbf{p} \in \mathcal{P}_N$. *Then the Lyapunov exponents of the measure* $\nu_{\mathbf{p}} := \Pi_* \mu_{\mathbf{p}}$ *are*

$$\chi_1(\nu_{\mathbf{p}}) \leqslant \chi_2(\nu_{\mathbf{p}}) \in \left\{ -\sum_{i=1}^{N} p_i \log a_i, -\sum_{i=1}^{N} p_i \log b_i \right\}. \tag{11.135}$$

Bárány, Rams and Simon [**BRS19**] gave dimension estimates of the attractors of lower triangular IFS in the diagonally homogeneous case (the diagonal entries of all linear parts are the same) with possible overlaps between the cylinders. See Figure 11.10 for two typical examples included.

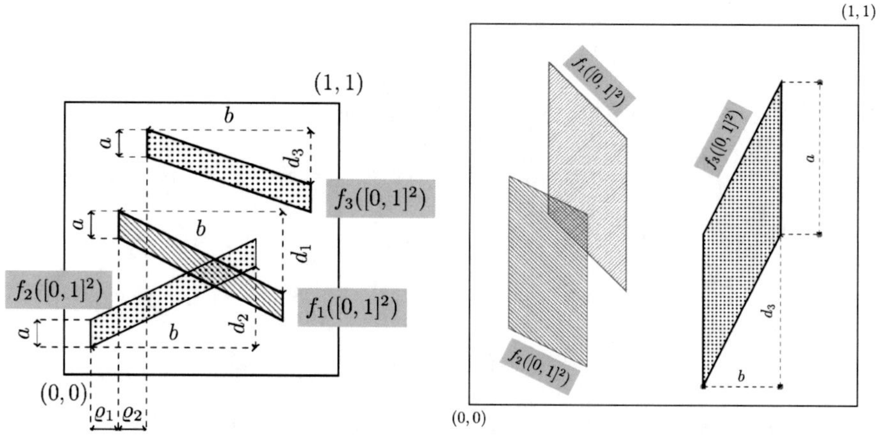

FIGURE 11.10. Figures from [**BRS19**]. The diagonal entries of all linear parts are a and b.

We say that direction x dominates if

$$b_i > a_i \text{ for all } i \in [N]. \tag{11.136}$$

For example, in the left side of Figure 11.10 dimension x dominates, but this is not the case in the right side.

11.5.5.1. *Lower triangular self-affine IFS when direction x dominates.*

DEFINITION 11.5.18. We say that $\mathcal{F}$ is a lower triangular self-affine IFS with the first direction dominating if

$$\mathcal{F} = (f_i(\underline{x}) := A_i\underline{x}+t_i)_{i=1}^N, A_i = \begin{pmatrix} b_i & 0 \\ d_i & a_i \end{pmatrix}, 0 < a_i < b_i < 1, t_i = \begin{pmatrix} t_{i,1} \\ t_{i,2} \end{pmatrix}. \tag{11.137}$$

Without loss of generality we may assume that $R_i := f_i(R) \subset R$, where $R := [0,1]^2$. The attractor of $\mathcal{F}$ is denoted by Λ.

Then $R_{i_1\dots i_n} := f_{i_1\dots i_n}(R)$ is a parallelogram having two vertical sides for every n and $(i_1,\dots,i_n)$. Let us write $\gamma_{i_1\dots i_n}$ for the angle between any of the nonvertical sides and the x-axis (see Figure 11.11). It is important that this angle is uniformly bounded away from $\pi/2$ (in n and $i_1,\dots,i_n$).

A simple calculation shows that

$$\tan\gamma_{i_1\dots i_n} = \frac{d_{i_1}}{b_{i_1}} + \sum_{\ell=2}^{n} \frac{d_{i_\ell}}{b_{i_\ell}} \cdot \prod_{k=1}^{\ell-1} \frac{a_{i_k}}{b_{i_k}}. \tag{11.138}$$

CLAIM. *Suppose that direction x dominates. Then*

(i) *There exists $C < \infty$ such that for every n and $(i_1,\dots,i_n) \in [N]^n$ we have*

$$\tan\gamma_{i_1\dots i_n} < C. \tag{11.139}$$

(ii) *The Lyapunov exponents of the self-similar measure $\nu_{\mathbf{p}}$ are*

$$\chi_1(\nu_{\mathbf{p}}) = -\sum_{i=1}^{N} p_i \log b_i \text{ and } \chi_2(\nu_{\mathbf{p}}) = -\sum_{i=1}^{N} p_i \log a_i. \tag{11.140}$$

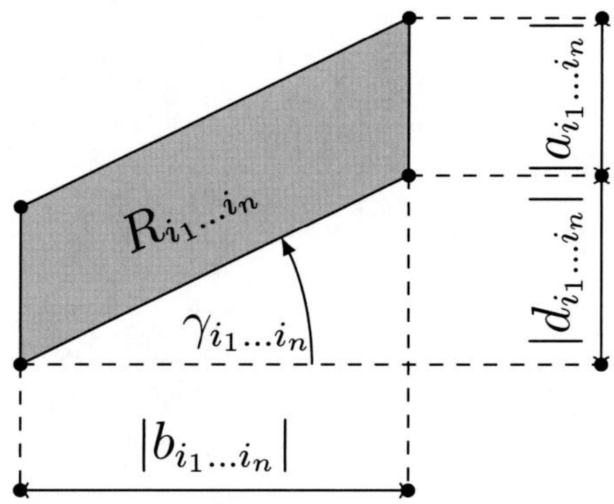

FIGURE 11.11. Typical level-n cylinder

Part (i) is immediate from (11.138) and the assumption that x-direction dominates. Part (ii) comes from(11.135).

Using this claim and [**Bár15**, Theorem 2.3] we obtain that

$$\dim_{\mathrm{H}}(\nu_{\mathbf{p}}) = \frac{\sum_{i=1}^{N} p_i \log p_i}{\sum_{i=1}^{N} p_i \log a_i} + \frac{\sum_{i=1}^{N} p_i \log a_i - \sum_{i=1}^{N} p_i \log b_i}{\sum_{i=1}^{N} p_i \log a_i} \cdot \dim_{\mathrm{H}}\left((\mathrm{proj}_x)_* \nu_{\mathbf{p}}\right). \tag{11.141}$$

The Hausdorff dimension of the attractor Λ is

$$\dim_{\mathrm{H}}(\Lambda) = \sup_{\mathbf{p}\in\mathcal{P}_N} \dim_{\mathrm{H}}(\nu_{\mathbf{p}}), \tag{11.142}$$

where $\mathcal{P}_N$ is the collection of probability vectors of N components, all of which are positive.

11.5.6. Triangular Gatzouras–Lalley carpets.

DEFINITION 11.5.19. We say that a lower triangular self-affine IFS

$$\mathcal{F} = (f_i(\underline{x}) := A_i\underline{x} + t_i)_{i=1}^{N}, \quad A_i = \begin{pmatrix} b_i & 0 \\ d_i & a_i \end{pmatrix}, \; 0 < a_i < b_i < 1, t_i = \begin{pmatrix} t_{i,1} \\ t_{i,2} \end{pmatrix}, \tag{11.143}$$

is a Triangular Gatzouras–Lalley (TGL) carpet if the parallelograms $R_i := f_i(R)$, where $R := [0,1]^2$ satisfy

(1) $R_i \subset R$,
(2) $R_i^\circ \cap R_j^\circ = \varnothing$ if $i \neq j$.
(3) Either $\mathrm{proj}_x(R_i^\circ) \cap \mathrm{proj}_x(R_i^\circ) = \varnothing$ or $\mathrm{proj}_x(R_i) = \mathrm{proj}_x(R_i)$. In the latter case we say that R_i and R_j are in the same column.

Barański [**Bar08**] proved that the same formulas hold for the various dimensions for TGL carpets as for GL carpets. Kolossváry and Simon [**KS19**] investigated TGL carpets with overlaps (see e.g. Figure 11.13) and gave conditions under which the dimension formulas for the Gatzouras–Lalley carpets remain valid.

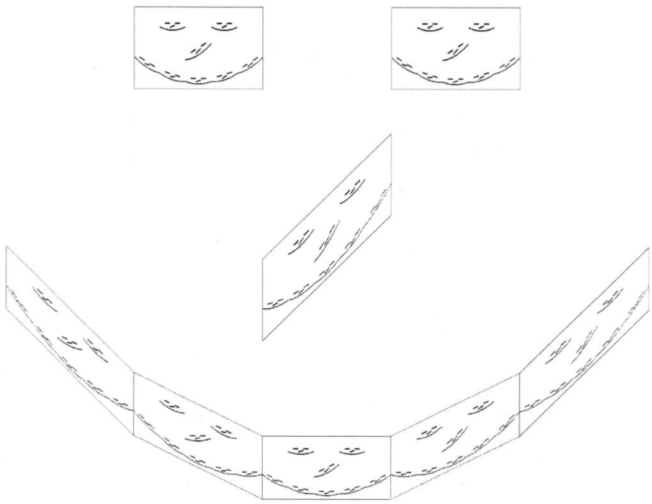

FIGURE 11.12. The "self-affine smiley", for which $\dim_{\mathrm{H}}(\Lambda) = 1.20665\ldots < 1.21340\ldots = \dim_{\mathrm{B}}(\Lambda) = \dim_{\mathrm{aff}}(\Lambda)$. The Figure is from [**KS19**] where the exact definitions are available.

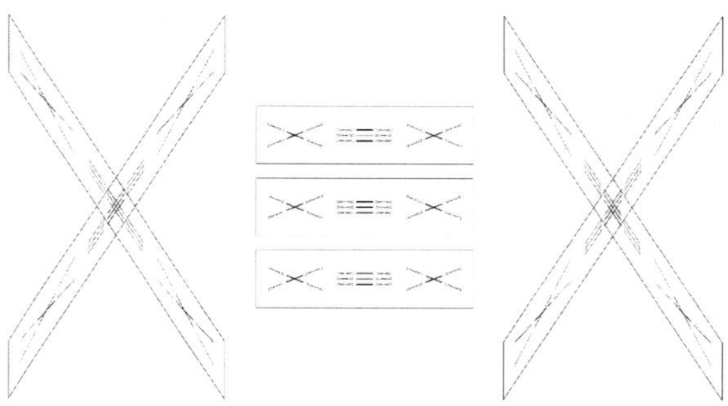

FIGURE 11.13. Example "$X \equiv X$" for which $\dim_{\mathrm{H}}(\Lambda) = 1.13259\ldots < \dim_{\mathrm{B}}(\Lambda) = 1.13626\ldots < \dim_{\mathrm{aff}}(\Lambda) = 1.2170\ldots$. The Figure is from [**KS19**] where the exact definitions are available.

CHAPTER 12

Exact dimensionality and dimension conservation

This chapter is devoted to important phenomena in the theory of planar self-similar and self-affine systems, namely the *dimension conservation.* It can be deduced from a more general result of Furstenberg [**Fur08**, Theorem 3.1] that if an IFS $(S_1, \ldots, S_m)$ consists of similarities on $\mathbb{R}^d$ of the form $S_i(x) = r_i x + b_i$, then for every self-similar measure ν and for every k-dimensional subspace $\mathbf{v}$ of $\mathbb{R}^d$, we have

$$\dim(\nu_x^{\mathbf{v}}) + \dim(\operatorname{proj}_{\mathbf{v}} \nu) = \dim(\nu), \quad \text{for } \operatorname{proj}_{\mathbf{v}} \nu\text{-a.e. } x, \tag{12.1}$$

where $\nu_x^{\mathbf{v}}$ are the conditional measures defined on the slices $\operatorname{proj}_{\mathbf{v}}^{-1}(x)$. Moreover, all the measures in the last formula are exact dimensional. In this case we simply write $\dim(\nu)$ without a subscript.

Later Bárány [**Bár15**] and Bárány and Käenmäki [**BK17**] extended the result to the case of self-affine measures. To be more precise, they showed that (12.1) holds for a self-affine measure ν, for almost every proper subspace of the plane. However, both papers required extra assumptions on the matrices. The most recent result on this topic is due to Feng [**Fen23**], who extended the dimension conservation result to the case of quasi-Bernoulli measures without any further assumptions on the matrices.

In this chapter, we study the dimension conservation phenomena for planar self-similar and self-affine systems with Strong Separation Property. Many of the results remain true in higher dimensions and without separation conditions, but the proofs are more technical. We highlight the use of two important tools: the *Furstenberg measure* and the *Ledrappier–Young formula.* Furstenberg measure is, in general, the stationary measure for the random walk on the Grassmanian manifold, induced by the matrices of the IFS. The Ledrappier–Young formula was introduced for diagonally self-affine IFS in the previous chapter, and here it is extended to more general self-affine IFS in the plane.

12.1. Random walk on the projective line and the Furstenberg measure

One can consider $\mathbb{PR}^1$ as the unit circle modulo the identification $\theta \sim -\theta$. Recall that a nonsingular matrix A acts on the projective line $\mathbb{PR}^1$ by $\operatorname{span}\{\mathbf{x}\} \mapsto \operatorname{span}\{A\mathbf{x}\}$. We will abuse the notation slightly by writing $\theta \mapsto A\theta$ for the induced action and will write $A|_\theta$ for the restriction of A to the line corresponding to $\theta \in \mathbb{PR}^1$. We will use some standard notation and denote by $SL^{\pm}(2, \mathbb{R})$ the group of 2×2 real matrices with absolute value of the determinant equal to one. Further, $O(2, \mathbb{R})$ is the orthogonal group of 2×2 real matrices, and $SO(2, \mathbb{R})$ is the special linear group—the subgroup of $O(2, \mathbb{R})$ of matrices with determinant equal to one.

Let $\mathcal{F} = (f_1, \ldots, f_m)$ be an IFS of planar affine transformations of the form $f_i(x) = A_i x + t_i$. As usual, we assume that A_i are nonsingular, contracting matrices.

Let $B_i = \sqrt{|\det(A_i)|}A_i^{-1}$, and let $\mathcal{B} = (B_1, \ldots, B_m)$. Note that $B_i \in SL^{\pm}(2,\mathbb{R})$. Let $\mathbf{p} = (p_1, \ldots, p_m)$ be a probability vector with all $p_i > 0$ and let $\mu = \mathbf{p}^{\mathbb{N}}$. The IFS induces a random walk on the projective line. That is, one can define a random walk $X_1, X_2, \ldots$ on $\mathbb{PR}^1$ by

$$\mathbb{P}(X_{k+1} = W | X_k = V) = \sum_{i: B_i V = W} p_i.$$

Note that we use the convention that the empty sum is zero. We call a measure η on $\mathbb{PR}^1$ $(\mathbf{p}, \mathcal{B})$-*invariant (or* μ-*stationary)*, if

$$\int g(\theta) d\eta(\theta) = \int \sum_{k=1}^{m} p_k g(B_k \theta) d\eta(\theta), \tag{12.2}$$

for every integrable, real-valued function g. Alternatively, for every Borel set $C \subset \mathbb{PR}^1$,

$$\eta(C) = \sum_{k=1}^{m} p_k \eta(B_k^{-1} C) = \sum_{k=1}^{m} p_k \eta(A_k C).$$

Consider the *skew product transformation* on $\Sigma \times \mathbb{PR}^1$ defined by

$$T(\mathbf{i}, \theta) = (\sigma \mathbf{i}, B_{i_1} \theta),$$

where σ is the usual left shift on Σ.

LEMMA 12.1.1. *The measure* η *is* μ-*stationary, where* $\mu = \mathbf{p}^{\mathbb{N}}$ *if and only if the measure* $\mu \times \eta$ *is invariant with respect to the skew product transformation* T.

For a proof, see [**Via14**, Proposition 5.5]. We will say that the measure η is $(\mathbf{p}, \mathcal{B})$-*ergodic* if it is μ-stationary and the measure $\mu \times \eta$ is an invariant ergodic measure for the skew product transformation T. In the next section, when we want to emphasize the matrices of the IFS, we will refer to the same object as a $(\mathbf{p}, (A_i^{-1})_{i=1}^m)$-invariant ergodic measure.

Denote the smallest closed semigroup containing $\mathcal{B}$ by $SG_{\mathcal{B}}$ and the smallest closed subgroup by $G_{\mathcal{B}}$. That is,

$$SG_{\mathcal{B}} = \overline{\{B_{\mathbf{i}} : \mathbf{i} \in \Sigma^*\}} \text{ and}$$

$$G_{\mathcal{B}} = \overline{\{B_{i_1}^{j_1} \cdots B_{i_n}^{j_n} : n \geqslant 0,\ j_1, \ldots, j_n \in \{-1, 1\},\ i_1, \ldots, i_n \in \{1, \ldots, m\}\}}.$$

Recall from Definition 1.12.2 that $G_{\mathcal{B}}$ (respectively $SG_{\mathcal{B}}$) is *reducible* if there exists $\theta \in \mathbb{PR}^1$ such that $M\theta = \theta$ for every $M \in G_{\mathcal{B}}$ (resp. $M \in SG_{\mathcal{B}}$), otherwise, it is *irreducible*. We call $G_{\mathcal{B}}$ (respectively $SG_{\mathcal{B}}$) *strongly irreducible* if there is no finite collection of subspaces $\theta_1, \ldots, \theta_k \in \mathbb{PR}^1$ such that $M(\bigcup_{i=1}^k \theta_i) = \bigcup_{i=1}^k \theta_i$ for every $M \in G_{\mathcal{B}}$ (respectively $M \in SG_{\mathcal{B}}$). Note that $G_{\mathcal{B}}$ being reducible corresponds, up to conjugation, to the case when all of the matrices of $\mathcal{B}$ are lower triangular.

Next we state the main result of this section, due to Furstenberg [**Fur63**]. Recall that $\chi_j(\mu)$, for $j = 1, \ldots, d$, are the Lyapunov exponents of the self-affine measure $\nu = \Pi_* \mu$ in $\mathbb{R}^d$, see Theorem 9.2.1 and Lemma 10.2.2. Since we only consider the planar case here, there are two Lyapunov exponents $\chi_2(\mu) \geqslant \chi_1(\mu) > 0$, where

$$\chi_j(\mu) = \lim_{n \to \infty} \frac{-1}{n} \log \alpha_j(A_{i_1} \cdots A_{i_n}) \text{ for } \mu\text{-almost every } (i_1, i_2, \ldots) \in \Sigma.$$

Similar results hold in any dimension.

THEOREM 12.1.2 (Existence of Furstenberg measure). *Let $(A_k)_{k=1}^m$ be a finite set of nonsingular, contracting matrices, let $\mathbf{p} = (p_1, \ldots, p_m)$ be a probability vector, and let $\mu = \mathbf{p}^{\mathbb{N}}$. Then there exists a measure ν_F on the projective line such that*

(1) $\int f(\theta) d\nu_F(\theta) = \int \sum_{k=1}^m p_k f(A_k^{-1}\theta)\, d\nu_F(\theta)$ *for every measurable, real-valued function $f \geqslant 0$, i.e., ν_F is μ-stationary,*
(2) *$\mu \times \nu_F$ is $T : (\mathbf{i}, \theta) \mapsto (\sigma\mathbf{i}, A_{i_1}^{-1}\theta)$ invariant and ergodic,*
(3)
$$\lim_{n\to\infty} \frac{1}{n} \log \|A_{i_n}^{-1} \cdots A_{i_1}^{-1}|_\theta\| = \sum_{k=1}^m p_k \int \log \|A_k^{-1}|_\theta\|\, d\nu_F(\theta) = \chi_2(\mu)$$
for $\mu \times \nu_F$-a.e. $(\mathbf{i}, \theta)$.

Moreover, if $(A_k)_{k=1}^m$ is strongly irreducible, then ν_F is unique.

We call the measure ν_F the *Furstenberg measure.* This measure plays an essential role in our further analysis. In the rest of this section, we provide a partial proof of Theorem 12.1.2 in our special planar case, following the book of Bougerol and Lacroix [**BL85**]. Several lemmas stated below are exercises in [**BL85**], however, for completeness we provide the proofs. The reader familiar with the Furstenberg measure may wish to skip directly to Section 12.2.

12.1.1. Compact group case. First, we study the Furstenberg measure in the case when $G_{\mathcal{B}}$ is compact. There are two sub-cases: when $G_{\mathcal{B}}$ is finite and when it is infinite. Roughly speaking, if $G_{\mathcal{B}}$ is finite then it contains only rational rotations and if $G_{\mathcal{B}}$ is infinite then there is an irrational rotation in $G_{\mathcal{B}}$.

LEMMA 12.1.3. *If $G_{\mathcal{B}}$ is compact, then it is conjugate to a compact subgroup of $O(2,\mathbb{R})$. Namely, there exists an invertible matrix M such that $MG_{\mathcal{B}}M^{-1} = \{MBM^{-1} : B \in G_{\mathcal{B}}\} \subset O(2,\mathbb{R})$.*

Furthermore, if $G_{\mathcal{B}}$ is compact and infinite then either $MG_{\mathcal{B}}M^{-1} = O(2,\mathbb{R})$ or $MG_{\mathcal{B}}M^{-1} = SO(2,\mathbb{R})$.

PROOF. Since $G_{\mathcal{B}}$ is a compact subgroup of $GL(2,\mathbb{R})$, there exists a right-invariant Haar probability measure m on $G_{\mathcal{B}}$ (See Definition 2.3.4). Let $\langle\cdot,\cdot\rangle$ be the standard inner product on $\mathbb{R}^2$. It is easy to see that the formula $\langle u, v\rangle' := \int\langle Au, Av\rangle\, dm(A)$ defines an inner product on $\mathbb{R}^2$ such that $\langle u, v\rangle' = \langle Bu, Bv\rangle'$ for every $B \in G_{\mathcal{B}}$. Every inner product on $\mathbb{R}^2$ is related to the standard one by a change of basis, so there exists $M \in GL(2,\mathbb{R})$ such that $\langle u, v\rangle' = \langle Mu, Mv\rangle$ for all $u, v \in \mathbb{R}^2$. In particular, $\langle MAM^{-1}u, MAM^{-1}v\rangle = \langle AM^{-1}u, AM^{-1}v\rangle' = \langle M^{-1}u, M^{-1}v\rangle' = \langle u, v\rangle$ for all $u, v \in \mathbb{R}^2$ and $A \in G_{\mathcal{B}}$, which yields $G_{\mathcal{B}} \subset M^{-1}O(2,\mathbb{R})M$.

Let us turn to the second claim of the lemma. By the first claim, we may assume without loss of generality that $\mathcal{B} \subset O(2,\mathbb{R})$ and $G_{\mathcal{B}}$ is itself a compact and infinite subgroup of $O(2,\mathbb{R})$. The statement will follow from the fact that in that case there exists a word $\mathbf{i}$ of length $|\mathbf{i}| \geqslant 2$ such that $B_{\mathbf{i}}$ is an irrational rotation on the unit circle, and hence either $SG_{\mathcal{B}} = G_{\mathcal{B}} = SO(2,\mathbb{R})$ (if $\mathcal{B}$ contains no reflections) or $SG_{\mathcal{B}} = G_{\mathcal{B}} = O(2,\mathbb{R})$ (if $\mathcal{B}$ contains a reflection), whence $G_{\mathcal{B}}$ is strongly irreducible. Thus, the proof can be finished by the next claim. □

CLAIM 12.1.4. *If $G_{\mathcal{B}}$ is infinite, then there exists a word $\mathbf{i}$ of length $|\mathbf{i}| \geqslant 2$ such that $B_{\mathbf{i}}$ is an irrational rotation on the unit circle.*

PROOF OF THE CLAIM. This follows from the fact that either $\mathcal{B} \cap SO(2,\mathbb{R})$ contains an irrational rotation, or there are two reflections $B_i, B_j \in \mathcal{B}$ (i.e., $B_i, B_j \in \mathcal{B} \cap (O(2,\mathbb{R})\backslash SO(2,\mathbb{R}))$) such that B_iB_j is an irrational rotation. Suppose to the contrary that this is not the case, so $\mathcal{B} \cap SO(2,\mathbb{R})$ contains only rational rotations and the pairwise products of reflections in $\mathcal{B}$ are rational rotations as well. Let q be largest common denominator of the rotation angles (mod π). Then, using that for every $O \in O(2,\mathbb{R})$ with rotation angle $2\pi p/q$ and for every $R \in O(2,\mathbb{R})\backslash SO(2,\mathbb{R})$, we have $RO = O'R$, where $O' \in O(2,\mathbb{R})$ is a rotation with angle $(q-p)2\pi/q$, we get that every element of $\{B_{\mathbf{i}}\}_{\mathbf{i}\in\Sigma^*}$ is either a rational rotation with denominator q or a product of a rational rotation of denominator q with a reflection $B \in \mathcal{B}$. This set is clearly finite and so we get a contradiction. □

LEMMA 12.1.5. *If $G_{\mathcal{B}}$ is finite, then for every $\theta \in \mathbb{PR}^1$ there exists a measure ν_F on $\mathbb{PR}^1$ such that $\nu_F(\{M\theta : M \in G_{\mathcal{B}}\}) = 1$ and the measure $\mu \times \nu_F$ is T-invariant and ergodic.*

PROOF. Fix $\theta \in \mathbb{PR}^1$ and let $\Theta = \{M\theta : M \in G_{\mathcal{B}}\}$. Further, consider the following Markov chain $\{X_n\}$ on Θ:

$$\mathbb{P}(X_{n+1} = K \mid X_n = L) = \sum_{i:B_iL=K} p_i \text{ for } K, L \in \Theta.$$

It is easy to see that Θ is an irreducible Markov chain, and it has finitely many states. Hence, it has a unique stationary measure ν_F supported on Θ, which is ergodic, see [**Dur12**, Theorem 1.23]. □

The next two lemmas are standard, but we couldn't find a good reference, and so we provide a proof for the reader's convenience. They concern the case when $G_{\mathcal{B}}$ is compact and infinite, hence it is conjugate to $O(2,\mathbb{R})$ or $SO(2,\mathbb{R})$ by Lemma 12.1.3. The case of $SO(2,\mathbb{R})$ is simpler, since it is a compact Abelian group; in fact, it is isomorphic to the rotation group of the circle. In particular, the proof of the next lemma in that case can be easily deduced from [**Pet83**, Theorem 2.4.8].

LEMMA 12.1.6. *Suppose that $G_{\mathcal{B}}$ is compact and infinite. Let $\widetilde{T}$ be the skew product transformation on $\Sigma \times G_{\mathcal{B}}$ given by*

$$\widetilde{T}(\mathbf{i}, M) = (\sigma\mathbf{i}, B_{i_1}M).$$

Then the measure $\mu \times m$ is $\widetilde{T}$-invariant and ergodic, where m is the right-invariant Haar measure on $G_{\mathcal{B}}$.

PROOF. Since $(B)_*m = m$ for every $B \in G_{\mathcal{B}}$, the measure $\mu \times m$ is $\widetilde{T}$-invariant. Using a result from Parry [**Par97**, Corollary 4.5], it is enough to show that there exists a point $(\mathbf{i}, M) \in \Sigma \times G_{\mathcal{B}}$ having a dense orbit, in order to verify that $\mu \times m$ is ergodic. Since $SG_{\mathcal{B}} = G_{\mathcal{B}}$ and so for every $M, M' \in G_{\mathcal{B}}$, $M'M^{-1} \in G_{\mathcal{B}} = SG_{\mathcal{B}}$, we have that for every $M, M' \in G_{\mathcal{B}}$ and $\delta > 0$ there exists a finite length word $\mathbf{i}(M, M', \delta)$ such that $\|B_{\mathbf{i}(M,M',\delta)} - M'M^{-1}\| = \|B_{\mathbf{i}(M,M',\delta)}M - M'\| < \delta$. Let $(\mathbf{y}_1, M_1), (\mathbf{y}_2, M_2), \ldots$ be a dense sequence in $\Sigma \times G_{\mathcal{B}}$. Now we define $\mathbf{j} \in \Sigma$ inductively. Let M be arbitrary, $n_1 = |\mathbf{i}(M, M_1, 1)| + 1$, and let $\mathbf{i}_1'$ be the word $\mathbf{i}(M, M_1, 1)$ in the reverse order. We define $\mathbf{j}|_{n_1} = \mathbf{i}_1'\mathbf{y}_1|_1$. Now, suppose that n_k is defined and $\mathbf{j}$ is defined up to the n_kth symbol. Let $\mathbf{j}_k'$ denote the word $\mathbf{j}|_{n_k}$ in the reverse order. Then let $\mathbf{i}_k = \mathbf{i}(B_{\mathbf{j}_k'}M, M_{k+1}, 1/(k+1))$,

$$n_{k+1} := n_k + |\mathbf{i}(B_{\mathbf{j}_k'}M, M_{k+1}, 1/(k+1))| + k + 1,$$

and let $\mathbf{j}|_{n_{k+1}}$ be defined as the juxtaposition

$$\mathbf{j}|_{n_{k+1}} := \mathbf{j}|_{n_k}\mathbf{i}'_k\mathbf{y}_{k+1}|_{k+1},$$

where $\mathbf{i}'_k$ is the word $\mathbf{i}_k$ in reversed order. This way $\mathbf{j}$ is well defined, moreover, $\widetilde{T}^{n_k-k}(\mathbf{j}, M)$ is in the $1/(k+1)$-neighborhood of $(\mathbf{y}_k, M_k)$, hence the orbit of $(\mathbf{j}, M)$ is dense. □

LEMMA 12.1.7. *If $G_{\mathcal{B}}$ is compact and infinite, then there exists a unique $(\mathbf{p}, \mathcal{B})$-ergodic measure η. Furthermore, the measure η is equivalent to the Lebesgue measure.*

PROOF. Let $\theta \in \mathbb{PR}^1$ be arbitrary but fixed, and let $p : G_{\mathcal{B}} \to \mathbb{PR}^1$ be such that $p(B) = B\theta$. The map p naturally extends to a mapping from $\Sigma \times G_{\mathcal{B}}$ to $\Sigma \times \mathbb{PR}^1$, which we still denote by p, with a slight abuse of notation. It is easy to see that $p \circ \widetilde{T} = T \circ p$, hence, the measure $\mu \times p_*m$ is T-invariant and ergodic by Lemma 12.1.6. By Lemma 12.1.3, there exists $M \in GL(d, \mathbb{R})$ such that $MG_{\mathcal{B}}M^{-1}$ is $O(2, \mathbb{R})$ or $SO(2, \mathbb{R})$, so $p_*m = (M^{-1})_*\mathcal{L}$, where $\mathcal{L}$ denotes the Lebesgue measure on $\mathbb{PR}^1$, i.e. the unique uniformly distributed Borel regular probability measure on $\mathbb{PR}^1$.

It remains to show that p_*m is the unique $(\mathbf{p}, \mathcal{B})$-invariant measure. Without loss of generality, we may assume that $G_{\mathcal{B}}$ is $O(2, \mathbb{R})$ or $SO(2, \mathbb{R})$, and so $p_*m = \mathcal{L}$. Clearly, $G_{\mathcal{B}^*} = G_{\mathcal{B}}$, where $\mathcal{B}^* = \{B^* : B \in \mathcal{B}\}$. Hence, by Lemma 12.1.6 the measure $\mu \times m$ is $\hat{T}$-invariant and ergodic, where $\hat{T}\colon (\mathbf{i}, M) \in \Sigma \times G_{\mathcal{B}^*} \to (\sigma\mathbf{i}, B^*_{i_1}M) \in \Sigma \times G_{\mathcal{B}^*}$.

Let $\delta > 0$ and let $M \in G_{\mathcal{B}^*}$ be arbitrary but fixed. Since $\mu \times m$ is ergodic and $(\mu \times m)(\Sigma \times B_\delta(M)) > 0$, we get that for $\mu \times m$-a.e. $(\mathbf{i}, A) \in \Sigma \times B_\delta(\mathrm{Id})$, we have $\hat{T}^n(\mathbf{i}, A) \in \Sigma \times B_\delta(M)$ for infinitely many n. Since $\|BA - BA'\| = \|A - A'\|$ for every $B, A, A' \in G_{\mathcal{B}}$, we get that $\hat{T}^n(\mathbf{i}, \mathrm{Id}) \in \Sigma \times B_{2\delta}(M)$ for μ-a.e. $\mathbf{i}$ and for infinitely many n. Let $\mathcal{P}_{\delta,M} = \{\mathbf{i} \in \Sigma^* : \|B^*_{i_n} \cdots B^*_{i_1} - M\| < 2\delta\}$. Since $\mu(\bigcup_{\mathbf{i}\in\mathcal{P}_{\delta,M}}[\mathbf{i}]) = 1$, there exists a $\hat{\mathcal{P}}_{\delta,M} \subset \mathcal{P}_{\delta,M}$ such that $[\mathbf{i}] \cap [\mathbf{j}] = \varnothing$ for any $\mathbf{i} \neq \mathbf{j} \in \hat{\mathcal{P}}_{\delta,M}$ and $\mu(\bigcup_{\mathbf{i}\in\hat{\mathcal{P}}_{\delta,M}}[\mathbf{i}]) = 1$. Then for any $(\mathbf{p}, \mathcal{B})$-invariant measure η,

$$\int f(\theta)\, d\eta(\theta) = \sum_{\mathbf{i}\in\hat{\mathcal{P}}_{\delta,M}} p_{\mathbf{i}} \int f(B_{i_1} \cdots B_{i_n}\theta)\, d\eta(\theta).$$

Therefore, for any continuous function $f : \mathbb{PR}^1 \to \mathbb{R}$ with $|f(\theta_1) - f(\theta_2)| \leqslant o_f(\delta) \to 0$ as $\sphericalangle(\theta_1, \theta_2) \leqslant \delta \to 0$, we have

$$\begin{aligned}
&\left|\int f(\theta)\, d\eta(\theta) - \int f(M^*\theta)\, d\eta(\theta)\right| = \\
&\qquad \left|\sum_{\mathbf{i}\in\hat{\mathcal{P}}_{\delta,M}} p_{\mathbf{i}} \int f(B_{i_1} \cdots B_{i_n}\theta)\, d\eta(\theta) - \int f(M^*\theta)\, d\eta(\theta)\right| \leqslant \\
&\qquad\qquad \sum_{\mathbf{i}\in\hat{\mathcal{P}}_{\delta,M}} p_{\mathbf{i}} \int \left|f(B_{i_1} \cdots B_{i_n}\theta) - \int f(M^*\theta)\right| d\eta(\theta).
\end{aligned}$$

Since for every $\mathbf{i} \in \hat{\mathcal{P}}_{\delta,M}$,

$$\sin\left(\sphericalangle(B_{i_1} \cdots B_{i_n}\theta, M^*\theta)\right) \leqslant \|B_{i_1} \cdots B_{i_n} - M^*\| = \|B^*_{i_n} \cdots B^*_{i_1} - M\| < 2\delta,$$

we get $\left|\int f(\theta)\, d\eta(\theta) - \int f(M^*\theta)\, d\eta(\theta)\right| \leqslant o_f(2\delta)$. Since $\delta > 0$ was arbitrary, we obtain that $\int f(\theta)\, d\eta(\theta) = \int f(M^*\theta)\, d\eta(\theta)$ for every continuous function. Since $M^* \in G_{\mathcal{B}}$ was arbitrary, by the uniqueness of uniformly distributed Borel regular probability measures, $\eta = \mathcal{L} = p_* m$. □

12.1.2. Noncompact group case. Now we study the Furstenberg measure in the case when $G_{\mathcal{B}}$ is noncompact. This is equivalent to the existence of a sequence $M_1, M_2, \ldots \in \mathcal{B}$ such that $\|M_n \cdots M_1\| \to \infty$ as $n \to \infty$.

LEMMA 12.1.8. *If $G_{\mathcal{B}}$ is noncompact, then one of the following three alternatives must hold:*

(1) *$G_{\mathcal{B}}$ is strongly irreducible; or*
(2) *$G_{\mathcal{B}}$ is conjugate to a subgroup of*

$$\left\{\begin{pmatrix} a^{-1} & 0 \\ 0 & \pm a \end{pmatrix} : a \neq 0\right\} \bigcup \left\{\begin{pmatrix} 0 & b^{-1} \\ \pm b & 0 \end{pmatrix} : b \neq 0\right\}, \quad \textit{or}$$

(3) *$G_{\mathcal{B}}$ is conjugate to a subgroup of*

$$\left\{\begin{pmatrix} a^{-1} & b \\ 0 & \pm a \end{pmatrix} : a \neq 0,\ b \in \mathbb{R}\right\}.$$

Note that $G_{\mathcal{B}}$ is irreducible but not strongly irreducible in the case (2), whereas $G_{\mathcal{B}}$ is reducible corresponds to case (3).

PROOF OF LEMMA 12.1.8. If $G_{\mathcal{B}}$ (i.e. $\mathcal{B}$) is not irreducible, then (3) clearly follows. So we may assume that $G_{\mathcal{B}}$ is not strongly irreducible, but irreducible. Thus there exist proper subspaces $V_1, \ldots, V_n$ such that $B_i\left(\bigcup_{j=1}^n V_j\right) = \bigcup_{j=1}^n V_j$ for every $B_i \in \mathcal{B}$. It follows that the action of $B_i \in \mathcal{B}$ on the tuple $(V_1, \ldots, V_n)$ corresponds to a permutation. By noncompactness, there exists a sequence $M_1, M_2, \ldots \in \mathcal{B}$ such that $\|M_n \cdots M_1\| \to \infty$ as $n \to \infty$. By the pigeon-hole principle, there exists a permutation $(V_{i_1}, \ldots, V_{i_n})$ and a subsequence $\{n_k\}$ of $\mathbb{N}$ such that $M_{n_k} \cdots M_1 V_j = V_{i_j}$ for every $j = 1, \ldots, n$ and $k = 1, 2, \ldots$ But then $M_{n_k} \cdots M_{n_1+1} V_j = V_j$ for every $j = 1, \ldots, n$ and $k \geqslant 2$. If $n \geqslant 3$, this is only possible when $M_{n_k} \cdots M_{n_1+1} = \pm I$ for every $k \geqslant 2$, contradicting our noncompactness assumption. Indeed, we obtain that each of V_j is the span of an eigenvector of $M_{n_k} \cdots M_{n_1+1}$, and having three linearly independent eigenvectors for a matrix in $SL^{\pm}(2, \mathbb{R})$ implies that it is the matrix I or $-I$. Hence $n = 2$, and with a change of basis, we are clearly in the case (2). □

Now we describe the Furstenberg measure in the cases listed in Lemma 12.1.8. But first we need some auxiliary statements whose proofs may be found in Bougerol and Lacroix [**BL85**].

LEMMA 12.1.9. *Suppose that $G_{\mathcal{B}}$ is not compact. If there exists a nonatomic $(\mathbf{p}, \mathcal{B})$-invariant measure η on $\mathbb{PR}^1$ satisfying (12.2), then for μ-almost every $\mathbf{i} = (i_1, i_2, \ldots)$ there exists $V(\mathbf{i}) \in \mathbb{PR}^1$ such that the sequence $(B_{i_1} \ldots B_{i_n})_* \eta$ converges to the measure $\delta_{V(\mathbf{i})}$ in the weak* sense. Moreover, $d\eta(W) = d\delta_{V(\mathbf{i})}(W) d\mu(\mathbf{i})$, η is the unique nonatomic $(\mathbf{p}, \mathcal{B})$-invariant measure, and there exists $\gamma > 0$ such that*

$$\lim_{n \to \infty} \frac{1}{n} \log \|B_{i_1} \cdots B_{i_n}\| = \gamma \textit{ for } \mu\textit{-almost every } \mathbf{i}.$$

Furthermore, for every $V, W \in \mathbb{PR}^1$ *and for* μ*-almost every* $\mathbf{i}$, $\sphericalangle(B^*_{i_n} \cdots B^*_{i_1} V, B^*_{i_n} \cdots B^*_{i_1} W) \to 0$ *as* $n \to \infty$, *and*

$$\lim_{n\to\infty} \frac{\|B^*_{i_n} \cdots B^*_{i_1}|_W\|}{\|B^*_{i_n} \cdots B^*_{i_1}\|} = |\cos(\sphericalangle(W, V(\mathbf{i}))|.$$

The proof may be found in [**BL85**, Proposition II.3.1, Proposition II.3.3 and Theorem II.3.6].

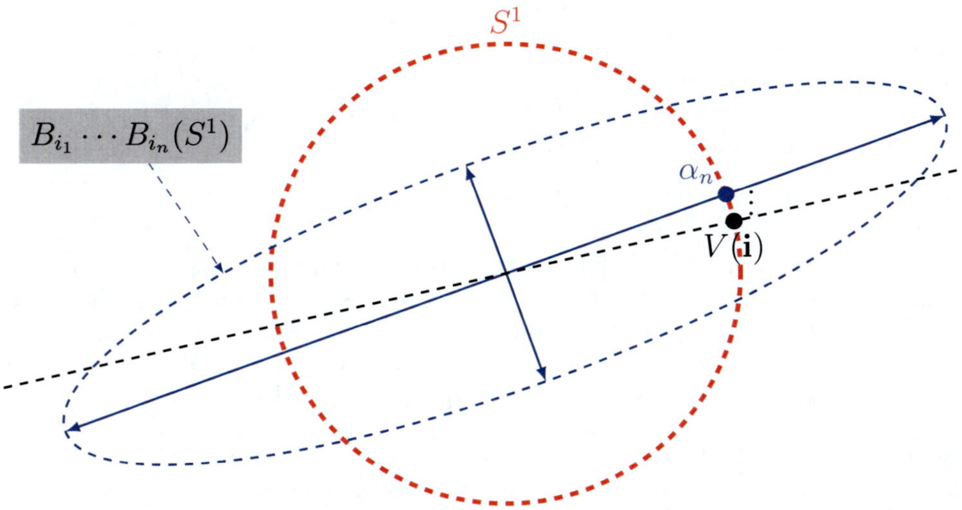

FIGURE 12.1. The convergence of $(B_{i_1} \cdots B_{i_n})_*\eta \to \delta_{V(\mathbf{i})}$ in the weak* topology can be visualised as the main axis of $B_{i_1} \cdots B_{i_n}(S^1)$ converging to $V(\mathbf{i})$.

LEMMA 12.1.10. *Let* $\mathcal{B} = \{B_1, \ldots, B_m\}$, *and suppose that* $G_{\mathcal{B}}$ *is noncompact, in the form of case* (2) *in Lemma* 12.1.8. *Let* θ_1 *and* θ_2 *be the commonly preserved subspaces by the matrices in* $G_{\mathcal{B}}$. *Let* $\mathbf{p} = (p_1, \ldots, p_m) > 0$.

(i) *If every* $B_i \in \mathcal{B}$ *has eigespaces* θ_1 *and* θ_2*; or equivalently, if all* B_i *are simultaneously diagonalizable, then for every* $0 \leqslant \alpha \leqslant 1$ *the measure* $\alpha\delta_{\theta_1} + (1-\alpha)\delta_{\theta_2}$ *is* $(\mathbf{p}, \mathcal{B})$*-invariant and* δ_{θ_1} *and* δ_{θ_2} *are the* $(\mathbf{p}, \mathcal{B})$*-ergodic measures.*

(ii) *If there exists* $B_i \in \mathcal{B}$ *for which* θ_1 *and* θ_2 *are not eigenspaces; or equivalently, if* $B_i\theta_1 = \theta_2$ *and* $B_i\theta_2 = \theta_1$, *then* $\nu_F = \frac{1}{2}\delta_{\theta_1} + \frac{1}{2}\delta_{\theta_2}$ *is the unique* $(\mathbf{p}, \mathcal{B})$*-invariant and ergodic measure.*

PROOF. The proof of part (i) is straightforward, so we turn to case (ii). Let η be a stationary measure satisfying (12.2). First, we show that η is discrete.

Suppose to the contrary that this is not the case. Decompose η into a discrete η_d and a nonatomic component η_c. It is easy to see that both η_d and η_c are stationary measures. By Lemma 12.1.9, for every V, W and for $\mu = \mathbf{p}^{\mathbb{N}}$ almost every $\mathbf{i}$ we have $\lim_{n\to\infty} \sphericalangle(B^*_{i_n} \cdots B^*_{i_1} V, B^*_{i_n} \cdots B^*_{i_1} W) = 0$. But this is impossible with the choice $V = \theta_1^\perp$ and $W = \theta_2^\perp$. Indeed, $B_i\theta_1 = \theta_2$ is equivalent to $B_i^*\theta_2^\perp = \theta_1^\perp$, and since $p_i > 0$, a μ-almost every $\mathbf{i}$ will contain infinitely many i's, causing the subspaces to "flip" infinitely often. Thus $\eta = \eta_d$.

Suppose that $\eta = \sum_{i\in\mathcal{I}} q_i\delta_{V_i}$, where $\mathcal{I}$ is a countable index set and $q_i > 0$ for every $i \in \mathcal{I}$, ordered by magnitude: $q_1 \geqslant q_2 \geqslant \cdots$. Since η_d is stationary, we have

$$\sum_{i\in\mathcal{I}} q_i\delta_{V_i} = \eta = \sum_{j=1}^{m} p_j(B_j)_*\eta = \sum_{j=1}^{m} p_j \sum_{\ell\in\mathcal{I}} q_\ell\delta_{B_jV_\ell}.$$

Therefore,

$$q_i = \sum_{\ell\in\mathcal{I}} q_\ell \sum_{j:B_jV_\ell=V_i} p_j, \quad \text{for all } i \in \mathcal{I}. \tag{12.3}$$

Since B_j are invertible, for each V_i and B_j there is at most one subspace V_ℓ such that $B_jV_\ell = V_i$. Consider the equality (12.3) for $i = 1$. We claim that for every $j = 1,\dots,m$, there exists exactly one subspace V_ℓ such that $B_jV_\ell = V_i$, and moreover, $q_\ell = q_1$. Indeed, we have from (12.3)

$$q_1 = \sum_{j:\exists\ell\in\mathcal{I}\text{ s.t. } B_jV_\ell=V_1} p_jq_\ell \leqslant q_1 \cdot \sum_{j:\exists\ell\in\mathcal{I}\text{ s.t. } B_jV_\ell=V_1} p_j \leqslant q_1,$$

where we used that $q_\ell \leqslant q_1$. It follows that both inequalities above must be equalities, which implies the claim, that is, all $j \leqslant m$ must be present in the last sum and all the corresponding q_ℓ are equal to q_1.

Now, for each q_ℓ, such that $q_\ell = q_1$, we can repeat the above argument to obtain that $q_k = q_\ell = q_1$ for each k satisfying $B_jV_k = V_\ell$ for some $k \in I$ and $j \leqslant m$. This process cannot continue indefinitely, since η is a probability measure, and so we obtain a finite subset $\mathcal{J} \subseteq \mathcal{I}$, containing 1, such that the corresponding union of subspaces $\bigcup_{\ell\in\mathcal{J}} V_\ell$ is invariant under $\mathcal{B}$, and $q_\ell = q_1$ for every $\ell \in \mathcal{J}$. By assumption, $\bigcup_{\ell\in\mathcal{J}} V_\ell \subseteq \{\theta_1,\theta_2\}$, and actually, $\bigcup_{\ell\in\mathcal{J}} V_\ell = \{\theta_1,\theta_2\}$, since in part (ii) there is a B_i switching the two subspaces. It remains to note that $\mathcal{I} = \mathcal{J}$. Indeed, If $\mathcal{I}\backslash\mathcal{J} \neq \varnothing$, then using the invariance of $\bigcup_{\ell\in\mathcal{J}} V_\ell$, we obtain that $\eta_d' = \sum_{i\in\mathcal{I}\backslash\mathcal{J}} \frac{q_i}{1-q_1-q_2}\delta_{V_i}$ is again a stationary measure, and repeating the previous argument, we get a contradiction. Thus $\nu_F = \frac{1}{2}\delta_{\theta_1} + \frac{1}{2}\delta_{\theta_2}$ is the unique stationary measure. $\square$

LEMMA 12.1.11. *Suppose that the matrices $B_i \in \mathcal{B}$ have the form*

$$B_i = \begin{pmatrix} a_i & b_i \\ 0 & c_i^{-1} \end{pmatrix},$$

so that $|a_i| = |c_i|$. Moreover, suppose that there is no common eigenspace other than e_1 and $G_{\mathcal{B}}$ is noncompact.

(i) *If $\sum_{i=1}^m p_i \log|a_i| < 0$ then there exists a measurable map $P\colon \Sigma \to \mathbb{PR}^1$ such that*

$$P(\mathbf{i}) = \operatorname{span}\left\{\begin{pmatrix} \sum_{n=1}^{\infty} a_{i_1}c_{i_1}\cdots a_{i_{n-1}}c_{i_{n-1}}b_{i_n} \\ 1 \end{pmatrix}\right\}$$

for ν-almost every $\mathbf{i}$. Moreover, the measure $\nu_F = \mu\circ P^{-1}$ is a $(\mathbf{p},\mathcal{B})$-invariant and ergodic measure.

(ii) *If $\sum_{i=1}^m p_i \log|a_i| \geqslant 0$ then δ_{e_1} is the unique $(\mathbf{p},\mathcal{B})$-invariant and $(\mathbf{p},\mathcal{B})$-ergodic measure.*

PROOF. In case (i), it is easy to see that the series $\sum_{n=1}^{\infty} a_{i_1}c_{i_1}\cdots a_{i_{n-1}}c_{i_{n-1}}b_{i_n}$ is absolutely convergent for μ-almost every $\mathbf{i}$, moreover, $P(\mathbf{i}) = B_{i_1}P(\sigma\mathbf{i})$. Therefore,

ν_F is indeed a stationary measure. Ergodicity follows from the fact that the left shift is ergodic on the two-sided shift space with respect to the Bernoulli measure μ and T is conjugated to σ by $\pi(\mathbf{i},\mathbf{j}) := (\mathbf{i}, P(\mathbf{j}))$.

Now let us turn to case (ii). First we show that any stationary measure is discrete, arguing by contradiction. Suppose that there exists a nonatomic stationary measure η. Then by Lemma 12.1.9, for μ-almost every $\mathbf{i}$, we have weak* convergence $(B_{i_1}\cdots B_{i_n})_*\eta \to \delta_{V(\mathbf{i})}$. We will show that $V(\mathbf{i}) = e_1$ for almost every $\mathbf{i}$, which contradicts the assumption that η is nonatomic. Suppose that $V(\mathbf{i}) \neq e_1$ on a positive measure set $K \subset \Sigma$. Then by Lemma 12.1.9, for μ-a.e. $\mathbf{i} \in K$,

$$\lim_{n\to\infty} \frac{|a_{i_1}\cdots a_{i_n}|^{-1}}{\|B_{i_1}\cdots B_{i_n}\|} = \lim_{n\to\infty} \frac{\|B^*_{i_n}\cdots B^*_{i_1}|_{e_1^\perp}\|}{\|B^*_{i_n}\cdots B^*_{i_1}\|} = |\cos(\measuredangle(e_1^\perp, V(\mathbf{i}))| > 0. \tag{12.4}$$

Denote $\varphi := \measuredangle(e_1^\perp, V(\mathbf{i}))$. Thus, for μ-a.e. $\mathbf{i} \in K$ and $n \geqslant 1$ sufficiently large,

$$\log(|\cos\varphi|/2) \leqslant -\log|a_{i_1}\cdots a_{i_n}| - \log\|B_{i_1}\cdots B_{i_n}\| \leqslant \log(2|\cos\varphi|),$$

which implies that $\frac{1}{n}\log\|B_{i_1}\cdots B_{i_n}\| = -\sum_{i=1}^m p_i \log|a_i|$ for μ-a.e. $\mathbf{i} \in K$, by the ergodicity of μ. But by Lemma 12.1.9 again, we have $0 < \lim_{n\to\infty} \frac{1}{n}\log\|B_{i_1}\cdots B_{i_n}\| = -\sum_{i=1}^m p_i \log|a_i| \leqslant 0$, which is a contradiction.

Thus any stationary measure can be written as $\eta = \sum_{i\in\mathcal{I}} q_i\delta_{V_i}$ with $q_1 \geqslant q_2 \geqslant \cdots$. In view of (12.2),

$$\sum_{i\in\mathcal{I}} q_i\delta_{V_i} = \sum_{i\in\mathcal{I}} \left(\sum_{\ell\in\mathcal{I}} \sum_{j:B_jV_\ell=V_i} p_j q_\ell \right) \delta_{V_i}.$$

By an argument similar to the proof of Lemma 12.1.10, one can show that $\eta = \delta_{e_1}$. □

PROOF OF THEOREM 12.1.2. If $G_{\mathcal{B}}$ is strongly irreducible and noncompact, then the proof may be found in [**BL85**, Theorem II.4.1], as well as in the original paper of Furstenberg [**Fur63**]. If $G_{\mathcal{B}}$ is noncompact but not strongly irreducible, then the claim follows by Lemma 12.1.8, and $\mathcal{B}$ is conjugated to either (2) or (3) in Lemma 12.1.8. In case (2), the existence of Furstenberg measure follows by Lemma 12.1.10 with the proper choice of the Dirac masses. In case (3), the claim follows by Lemma 12.1.11. If $G_{\mathcal{B}}$ is compact, then the assertion follows by a combination of Lemma 12.1.6 and Lemma 12.1.7. □

Note that in case (2) in Lemma 12.1.10, we have $\chi_1(\mu) = \chi_2(\mu)$. Indeed, by (10.61) and Theorem 12.1.2(3) one can easily show that

$$\chi_2(\mu) = \sum_{k=1}^m p_k \int \log\|A_k^{-1}|_\theta\|\, d\nu_F(\theta) = -\frac{1}{2}\sum_{k=1}^m p_k \log|\det(A_k)| = \frac{\chi_1(\mu)+\chi_2(\mu)}{2}.$$

LEMMA 12.1.12.

(i) *If $G_{\mathcal{B}}$ is compact, then there is a constant $C > 0$ such that for every $\theta, \theta' \in \mathbb{PR}^1$, $B \in G_{\mathcal{B}}$*

$$C^{-1}\measuredangle(\theta,\theta') \leqslant \measuredangle(B\theta, B\theta') \leqslant C\measuredangle(\theta,\theta').$$

(ii) *If $G_{\mathcal{B}}$ is not compact and strongly irreducible, then for every $\theta, \theta' \in \mathbb{PR}^1$ and μ-almost every* $\mathbf{i}$

$$\lim_{n\to\infty} \measuredangle(B_{i_n}\cdots B_{i_1}\theta, B_{i_n}\cdots B_{i_1}\theta') = 0.$$

PROOF. (i) If $G_{\mathcal{B}}$ is compact, then taking the conjugation matrix M in Lemma 12.1.3, we have

$$\begin{aligned}\measuredangle(B\theta, B\theta') &\leqslant C_1 \measuredangle(MBM^{-1}M\theta, MBM^{-1}M\theta') \\ &= C_1 \measuredangle(M\theta, M\theta') \\ &\leqslant C_2 \measuredangle(\theta, \theta').\end{aligned}$$

The other inequality is proved in exactly the same way.

To prove part (ii), observe that if $G_{\mathcal{B}}$ is strongly irreducible and noncompact, then $G_{\mathcal{B}^*}$ is also strongly irreducible and noncompact, where $\mathcal{B}^* = \{B_1^*, \ldots, B_m^*\}$. Thus, part (ii) follows by applying Theorem 12.1.2 and Lemma 12.1.9 for $\mathcal{B}^*$. □

12.2. Slices and dimension conservation

Here we make a digression and prove the Slicing Theorem 2.4.3. This will be an opportunity to introduce techniques which will be used in the study of dimension conservation phenomena in the next section.

Recall the statement of the theorem for convenience. For $\mathbf{v} \in \mathbb{S}^{d-1}$ we denote by $\mathrm{proj}_{\mathbf{v}} : \mathbb{R}^d \to \mathbb{R}$ the orthogonal projection from $\mathbb{R}^d$ to the line in the direction of $\mathbf{v}$ defined as in (2.18), and, given a Borel set E, write $S_{\mathbf{v}}(x)$ for the slice

$$S_{\mathbf{v}}(x) = \mathrm{proj}_{\mathbf{v}}^{-1}(x) \cap E.$$

Recall that for convenience, we identify the subspace defined by $\mathbf{v} \in \mathbb{S}^{d-1}$ with $\mathbb{R}$.

THEOREM 2.4.3 *Let $t > 1$ and let $E \subseteq \mathbb{R}^d$ be a Borel set with $0 < \mathcal{H}^t(E) < \infty$. Then for σ_{d-1}-a.e. $\mathbf{v} \in \mathbb{S}^{d-1}$,*

$$\mathcal{L}^1(\{x : \dim_{\mathrm{H}}(S_{\mathbf{v}}(x)) = t - 1\}) > 0.$$

Before the proof, we recall the decomposition of a measure on E into an integral of conditional measures on the slices. In the proof of the theorem, the measure will be $\mathcal{H}^t|_E$.

Let μ be a Radon probability measure on $\mathbb{R}^d$. Then by Rokhlin's Theorem (see Theorem 9.4.18) for every $\mathbf{v} \in \mathbb{S}^{d-1}$ there exists a canonical system of conditional measures, i.e., for $\mu_{\mathbf{v}} = (\mathrm{proj}_{\mathbf{v}})_* \mu$-almost every x there exists a measure $\mu_x^{\mathbf{v}}$ supported on $\mathrm{proj}_{\mathbf{v}}^{-1}(x)$ such that for every measurable set $A \subseteq \mathbb{R}^d$ the function $x \mapsto \mu_x^{\mathbf{v}}(A)$ is measurable, and

$$\mu(A) = \int \mu_x^{\mathbf{v}}(A)\, d\mu_{\mathbf{v}}(x).$$

The conditional measures are uniquely defined up to a set of zero measure. For $\mathbf{v} \in \mathbb{S}^{d-1}$ define the transversal ball centered at $x \in \mathbb{R}$ of radius r:

$$B_{\mathbf{v}}^T(x, r) = \mathrm{proj}_{\mathbf{v}}^{-1} B(x, r),$$

where $B(x, r)$ is the open Euclidean ball centered at x of radius r. By Theorem 9.4.22,

$$\mu_x^{\mathbf{v}} = \lim_{r \to 0+} \frac{\mu|_{B_{\mathbf{v}}^T(x,r)}}{\mu(B_{\mathbf{v}}^T(x, r))} = \lim_{r \to 0+} \frac{\mu|_{B_{\mathbf{v}}^T(x,r)}}{\mu_{\mathbf{v}}(B(x, r))}. \tag{12.5}$$

for $\mu_{\mathbf{v}}$-almost every x, where the limit is taken in weak* sense.

Equivalently, for any nonnegative compactly supported continuous function ϕ we have $\int \phi\, d\mu_x^{\mathbf{v}} = \lim_{r \to 0+} \frac{\int_{B_{\mathbf{v}}^T(x,r)} \phi\, d\mu}{\mu_{\mathbf{v}}(B(x,r))}$ for $\mu_{\mathbf{v}}$-almost every x. Thus, for any lower semi-continuous function $g : \mathbb{R}^d \to [0, \infty]$, writing it as a limit of a nondecreasing sequence of nonnegative compactly supported continuous functions, we obtain

$$\int g \, d\mu_x^{\mathbf{v}} \leqslant \liminf_{r \to 0^+} \frac{\int_{B_{\mathbf{v}}^T(x,r)} g \, d\mu}{\mu_{\mathbf{v}}(B(x,r))}. \tag{12.6}$$

Analogously to Proposition 2.4.2, one can prove the following slicing theorem for the local dimension of general Radon probability measures. Recall that for a measure μ, we denote by $\underline{d}_\mu(x)$ and $\overline{d}_\mu(x)$ the lower and upper local dimensions at x, see Definition 1.9.1.

PROPOSITION 12.2.1 (Ledrappier, Young). *Let μ be a compactly supported Radon measure on $\mathbb{R}^d$ and let $\mathbf{v} \in \mathbb{S}^{d-1}$. If $\underline{d}_{\mu^{\mathbf{v}}_{\mathrm{proj}_{\mathbf{v}}(\mathbf{x})}}(\mathbf{x}) \geqslant \alpha$ for μ-almost all $\mathbf{x}$, then*

$$\overline{d}_\mu(\mathbf{x}) \geqslant \alpha + \overline{d}_{\mu_{\mathbf{v}}}(\mathrm{proj}_{\mathbf{v}}(\mathbf{x})) \quad \text{and} \quad \underline{d}_\mu(\mathbf{x}) \geqslant \alpha + \underline{d}_{\mu_{\mathbf{v}}}(\mathrm{proj}_{\mathbf{v}}(\mathbf{x})) \tag{12.7}$$

for μ-almost every $\mathbf{x}$.

To the best of our knowledge, the proposition first appeared in Ledrappier and Young [**LY85b**, Lemma 11.3.1] in a more general form. Here we give a simplified version.

PROOF. By Egorov's Theorem, for every $\varepsilon > 0$ there exists a set $J_1 \subset \mathbb{R}^d$ and $C > 0$ such that $\mu(J_1) \geqslant 1 - \varepsilon$ and

$$\mu^{\mathbf{v}}_{\mathrm{proj}_{\mathbf{v}}(\mathbf{x})}(B(\mathbf{x}, 2e^{-m})) \leqslant Ce^{-m(\alpha-\varepsilon)}$$

for every $\mathbf{x} \in J_1$ and every sufficiently large $m \geqslant 1$.

By the analog of Lebesgue's density theorem B.4.2(i), which is also called as Besicovitch's density theorem, and Egorov's Theorem, there exists a set $J_2 \subset J_1$ such that $\mu(J_2) > 1 - 2\varepsilon$ and

$$\mu(J_1 \cap B(\mathbf{x}, e^{-m})) \geqslant \frac{1}{2}\mu(B(\mathbf{x}, e^{-m}))$$

for every $\mathbf{x} \in J_2$ and every sufficiently large $m \geqslant 1$. Fix m sufficiently large, so that the last two inequalities hold.

So for every $\mathbf{x} \in J_2$ and $\mathbf{y} \in B_{\mathbf{v}}^T(\mathrm{proj}_{\mathbf{v}}(\mathbf{x}), e^{-m})$ there are two possibilities. If there exists $\mathbf{y}' \in \mathrm{proj}_{\mathbf{v}}^{-1}(\mathrm{proj}_{\mathbf{v}}(\mathbf{y})) \cap J_1 \cap B(\mathbf{x}, e^{-m})$, then

$$\begin{aligned}\mu^{\mathbf{v}}_{\mathrm{proj}_{\mathbf{v}}(\mathbf{y})}(J_1 \cap B(\mathbf{x}, e^{-m})) &\leqslant \mu^{\mathbf{v}}_{\mathrm{proj}_{\mathbf{v}}(\mathbf{y})}(J_1 \cap B(\mathbf{y}', 2e^{-m})) \\ &= \mu^{\mathbf{v}}_{\mathrm{proj}_{\mathbf{v}}(\mathbf{y}')}(J_1 \cap B(\mathbf{y}', 2e^{-m})) \\ &\leqslant \mu^{\mathbf{v}}_{\mathrm{proj}_{\mathbf{v}}(\mathbf{y}')}(B(\mathbf{y}', 2e^{-m})) \leqslant Ce^{-m(\alpha-\varepsilon)}.\end{aligned}$$

Otherwise, $\mathrm{proj}_{\mathbf{v}}^{-1}(\mathrm{proj}_{\mathbf{v}}(\mathbf{y})) \cap J_1 \cap B(\mathbf{x}, e^{-m}) = \varnothing$, and then the bound is trivial since $\mu^{\mathbf{v}}_{\mathrm{proj}_{\mathbf{v}}(\mathbf{y})}(J_1 \cap B(\mathbf{x}, e^{-m})) = 0$. Hence, for every $\mathbf{x} \in J_2$ we have

$$\begin{aligned}\mu(B(\mathbf{x}, e^{-m})) &\leqslant 2\mu(J_1 \cap B(\mathbf{x}, e^{-m})) \\ &= 2\int \mu_y^{\mathbf{v}}(J_1 \cap B(\mathbf{x}, e^{-m})) d\mu_{\mathbf{v}}(y) \\ &\leqslant 2Ce^{-m(\alpha-\varepsilon)}\mu_{\mathbf{v}}(B(\mathrm{proj}_{\mathbf{v}}(\mathbf{x}), e^{-m})).\end{aligned}$$

Note that here we used Rokhlin's Theorem in the form (9.34), in view of Remark 9.4.19. Since $\varepsilon > 0$ was arbitrary small and m arbitrary sufficiently large, both inequalities in (12.7) follow. □

Finally, we turn to the proof of Theorem 2.4.3, following [**Mat95**, Ch. 10].

PROOF OF THEOREM 2.4.3. By Theorem 2.4.1, $\dim_{\mathrm{H}}(S_{\mathbf{v}}(x)) \leqslant t-1$ for every $\mathbf{v} \in \mathbb{S}^{d-1}$ and $\mathcal{L}^1$-a.e. x, thus it suffices to prove the lower bound. It is enough to show that for σ_{d-1}-a.e. $\mathbf{v}$ there exists a set $E_{\mathbf{v}}$ with $\mathcal{L}^1(E_{\mathbf{v}}) > 0$ such that for any $s < t$,

$$\dim_{\mathrm{H}}(S_{\mathbf{v}}(x)) \geqslant s-1 \text{ for a.e. } x \in E_{\mathbf{v}}. \tag{12.8}$$

(Taking a sequence $s_n \uparrow t$ we can find exceptional sets in $\mathbb{S}^{d-1}$ and $E_{\mathbf{v}}$ independent of s.)

Let $\nu = \mathcal{H}^t|_E$ and $1 < s < t$ be arbitrary but fixed. Then by Theorem 2.1.1 and Egorov's Theorem, there exists $E' \subset E$ with $\mathcal{H}^t(E') > 0$ and $C > 0$ such that $\mathcal{H}^t(E \cap B(x,r)) < Cr^{(s+t)/2}$ for every $x \in E'$ and $r > 0$. Hence, taking $\mu = \mathcal{H}^t|_{E'}$, we have $\mu(B(x,r)) \leqslant 2Cr^{(s+t)/2}$ for all $x \in \mathbb{R}^d$ and $r > 0$. It follows that $\mathcal{E}_s(\mu) < \infty$. By Theorems 2.3.1 and 2.3.3, $\mathrm{proj}_{\mathbf{v}}(E')$ has positive Lebesgue measure and $\mu_{\mathbf{v}} = (\mathrm{proj}_{\mathbf{v}})_*\mu \ll \mathcal{L}^1$ for σ^{d-1}-a.e. $\mathbf{v}$. Let $\nu_x^{\mathbf{v}} = \underline{D}(\mu_{\mathbf{v}}, x) \cdot \mu_x^{\mathbf{v}}$, where $\mu_{\mathbf{v}}$ is absolutely continuous and $\{\mu_x^{\mathbf{v}}\}$ is the family of conditional measures on the slices. We can now define

$$E_{\mathbf{v}} = \{x : \ \mu_x^{\mathbf{v}}(\mathbb{R}^d) > 0\}.$$

Note that $\underline{D}(\mu_{\mathbf{v}}, x)$ is the density of an absolutely continuous $\mu_{\mathbf{v}}$ for a.e. x (see Definition B.2.4 and Theorem B.2.5), and we obtain from (12.6):

$$\int g \, d\nu_x^{\mathbf{v}} \leqslant \liminf_{r \to 0^+} (2r)^{-1} \int_{B_{\mathbf{v}}^T(x,r)} g \, d\mu, \tag{12.9}$$

where g is a nonnegative lower semicontinuous function. Note that $\nu_x^{\mathbf{v}}$ is a finite and positive Borel measure on $S_{\mathbf{v}}(x)$ for $\mathcal{L}^1$-a.e. $x \in E_{\mathbf{v}}$, hence it will follow by Lemma 1.4.16 that if

$$\mathcal{J} := \int_{\mathbb{S}^{d-1}} \int_{E_{\mathbf{v}}} \mathcal{E}_{s-1}(\nu_z^{\mathbf{v}}) \, dz \, d\sigma_{d-1}(\mathbf{v}) < \infty \quad \text{for all } 1 < s < t,$$

then (12.8) holds. Observe that $|\mathbf{y}-\mathbf{x}|^{1-s}$ is a nonnegative lower semicontinuous function. Thus, using Fatou's Lemma, (12.9), the definition of conditional measures, and Fubini's Theorem, we get

$$\begin{aligned}
\mathcal{J} &= \iiiint \frac{1}{\|\mathbf{y}-\mathbf{x}\|^{s-1}}\, d\nu_z^{\mathbf{v}}(\mathbf{y}) d\nu_z^{\mathbf{v}}(\mathbf{x})\, dz\, d\sigma_{d-1}(\mathbf{v}) \\
&\leqslant \liminf_{r\to 0} \frac{1}{2r} \iiint \int_{B_{\mathbf{v}}^T(z,r)} \frac{1}{\|\mathbf{y}-\mathbf{x}\|^{s-1}}\, d\mu(\mathbf{y})\, d\nu_z^{\mathbf{v}}(\mathbf{x})\, dz\, d\sigma_{d-1}(\mathbf{v}) \\
&= \liminf_{r\to 0} \frac{1}{2r} \iiint \int_{B_{\mathbf{v}}^T(z,r)} \frac{1}{\|\mathbf{y}-\mathbf{x}\|^{s-1}}\, d\mu(\mathbf{y})\, d\mu_z^{\mathbf{v}}(\mathbf{x})\, d\mu_{\mathbf{v}}(z)\, d\sigma_{d-1}(\mathbf{v}) \\
&= \liminf_{r\to 0} \frac{1}{2r} \iint \int_{B_{\mathbf{v}}^T(\mathrm{proj}_{\mathbf{v}}(\mathbf{x}),r)} \frac{1}{\|\mathbf{y}-\mathbf{x}\|^{s-1}}\, d\mu(\mathbf{y})\, d\mu(\mathbf{x})\, d\sigma_{d-1}(\mathbf{v}) \\
&= \liminf_{r\to 0} \frac{1}{2r} \iint \frac{\sigma_{d-1}(\{\mathbf{v} : |\,\mathrm{proj}_{\mathbf{v}}(\mathbf{x}-\mathbf{y})| < r\})}{\|\mathbf{y}-\mathbf{x}\|^{s-1}}\, d\mu(\mathbf{y})\, d\mu(\mathbf{x}).
\end{aligned}$$

By (2.23), there exists a constant $C > 0$ such that

$$\begin{aligned}
\liminf_{r\to 0} \frac{1}{2r} \iint & \frac{\sigma^{d-1}(\{\mathbf{v} : |\,\mathrm{proj}_{\mathbf{v}}(\mathbf{x}-\mathbf{y})| < r\})}{\|\mathbf{y}-\mathbf{x}\|^{s-1}} d\mu(\mathbf{y})d\mu(\mathbf{x}) \\
&\leqslant \liminf_{r\to 0} \frac{1}{2r} \iint \frac{C\|\mathbf{y}-\mathbf{x}\|^{-1} r}{\|\mathbf{y}-\mathbf{x}\|^{s-1}} d\mu(\mathbf{y})d\mu(\mathbf{x}) \\
&= C \iint \frac{1}{\|\mathbf{y}-\mathbf{x}\|^{s}} d\mu(\mathbf{y})d\mu(\mathbf{x}) < \infty,
\end{aligned}$$

as desired. □

12.3. The Ledrappier–Young formula

This section is devoted to the proof of dimension conservation phenomena and the Ledrappier–Young formula for self-similar and self-affine measures. Let

$$\Psi = (f_i(\mathbf{x}) = A_i\mathbf{x} + \mathbf{t}_i)_{i=1}^m$$

be an IFS on $\mathbb{R}^2$ such that A_i are nonsingular, contracting matrices for every $i = 1, \dots, m$. Assume that Ψ has the Strong Separation Property. Let $\mathbf{p}$ be a probability vector and let ν be the corresponding self-affine measure. In this and the next chapter it is convenient to consider the orthogonal projection *in the direction* of $\theta \in \mathbb{PR}^1$, rather than *onto* the line θ. To do so, for each $\theta \in \mathbb{PR}^1$ we choose a unit vector $\mathbf{u}(\theta) \in \theta$ such that $\langle \mathbf{u}, (1,0)\rangle > 0$, and in the case when θ is parallel to the y-axis we set $\mathbf{u} = (0,1)$. Thus, we write

$$\mathrm{Proj}_\theta := \mathrm{proj}_{\mathbf{u}(\theta^\perp)},$$

using our earlier notation from (2.18). That is, Proj_θ denotes the orthogonal projection along the subspace $\theta \in \mathbb{PR}^1$. We denote by $\nu_{\mathbf{x}}^{\xi_\theta}$ the family of conditional measures with respect to this projection, defined ν-a.e. (see Theorem 9.4.18 for the definition of conditional measures).

THEOREM 12.3.1 (Dimension conservation). *Let Ψ be a self-affine IFS on $\mathbb{R}^2$ such that Ψ has the Strong Separation Property, and let $\mathbf{p}$ be a probability vector. Let ν_F be a $(\mathbf{p}, \{A_i^{-1}\})$-ergodic Furstenberg measure on the projective line, defined in Theorem 12.1.2. Moreover, let $\nu = \Pi_*\mu$, where μ is the Bernoulli measure corresponding to the probability vector $\mathbf{p}$, and Π denotes the natural projection. Then $\nu_{\mathbf{x}}^{\xi_\theta}$ and $(\mathrm{Proj}_\theta)_*\nu$ are exact dimensional for $\nu \times \nu_F$-a.e. $(\mathbf{x}, \theta)$. Moreover,*

$$\dim(\nu_{\mathbf{x}}^{\xi_\theta}) + \dim\big((\mathrm{Proj}_\theta)_*\nu\big) = \dim(\nu), \quad \textit{for } \nu \times \nu_F\textit{-a.e. } (\mathbf{x}, \theta).$$

This formulation of the theorem includes the results Furstenberg [**Fur08**] and Bárány and Käenmäki [**BK17**] for planar systems. Recall that we write simply "dim", without indicating which dimension it is, when the measure is exact dimensional, so $\dim(\nu)$ is simultaneously the Hausdorff dimension and almost everywhere local dimension of the measure ν.

We note that the Furstenberg measure plays a crucial role in the dimension conservation phenomena. Interestingly, Rapaport [**Rap17**] showed that there is a planar self-similar IFS, which does not satisfy the dimension conservation in a residual set of directions. More precisely, there exists self-similar IFS $\Phi = (f_i(\mathbf{x}) = \lambda R\mathbf{x} + \mathbf{t}_i)_{i=1}^m$, which has the strong separation property, and where R is an irrational rotation and the self-similar $\nu = \sum_{i=1}^m \frac{1}{m}(f_i)_*\nu$ satisfies $\dim(\nu) > 1$, such that both sets

$$\mathbf{B}_1 = \{\theta \in \mathbb{PR}^1 : (\mathrm{Proj}_\theta)_*\nu \text{ is singular}\} \text{ and}$$

$$\mathbf{B}_2 = \{\theta \in \mathbb{PR}^1 : \nu_{\mathbf{x}}^{\xi_\theta} \text{ is discrete for } \nu\text{-almost every } \mathbf{x}\}$$

contain a dense G_δ subset of $\mathbb{PR}^1$. This demonstrates that although in a measure theoretical sense almost every direction is dimension conserving, it might happen that in topological sense, almost every direction is not dimension conserving, since for every $\theta \in \mathbf{B}_2$

$$\dim(\nu_{\mathbf{x}}^{\xi_\theta}) + \dim\big((\mathrm{Proj}_\theta)_*\nu\big) \leqslant 1 < \dim(\nu).$$

It would be useful to express the dimension of the conditional measures by the Lyapunov exponents, entropy and the dimension of the projected measure, and the following theorem states that it is possible. Thus, when studying the dimension of strongly separated self-affine measure, we can restrict ourselves to the analysis of the dimension of orthogonal projections, which is simpler in some cases.

THEOREM 12.3.2 (Ledrappier–Young formula). *Under the assumptions of Theorem 12.3.1,*

$$\dim(\nu) = \frac{h_\mu}{\chi_2(\mu)} + \left(1 - \frac{\chi_1(\mu)}{\chi_2(\mu)}\right)\dim\big((\mathrm{Proj}_\theta)_*\nu\big), \quad \textit{for } \nu_F\textit{-a.e. } \theta.$$

We recall that $\chi_2(\mu) \geqslant \chi_1(\mu) > 0$ and $h_\mu = h_{\mathbf{p}}$. The importance of this theorem is that it gives a characterization of the dimension drop for self-affine measures.

COROLLARY 12.3.3. *Under the assumptions of Theorem 12.3.1 and assuming $\chi_1(\mu) < \chi_2(\mu)$,*

$$\dim(\nu) = D(\mu) \iff \dim\big((\mathrm{Proj}_\theta)_*\nu\big) = \min\{1, \dim(\nu)\} \quad \textit{for } \nu_F\textit{-a.e. } \theta,$$

where $D(\mu)$ is the Lyapunov dimension of μ.

The derivation of the corollary is a simple algebraic manipulation, left to the reader. The Lyapunov dimension of a self-affine measure was introduced in Definition 10.1.1, see also (11.24) for the planar case. This corollary should be compared with (11.26).

12.3.1. An application: Hueter–Lalley systems. Before we prove the Ledrappier–Young formula, we show that it implies Hueter and Lalley's [**HL95**] Theorem 1.12.1 and then discuss some recent developments in this area without proof. Actually, we state a slightly more general theorem, from which Theorem 1.12.1 follows.

THEOREM 12.3.4. *Let $(A_1, \dots, A_m)$ be a tuple of 2×2 invertible matrices such that the following hold:*

(i) ***1-bunched property:*** $\alpha_1(A_i)^2 < \alpha_2(A_i)$ *for every* $i = 1, \dots, m$;
(ii) ***domination:*** *there is a multicone $\mathcal{D}$ on $\mathbb{PR}^1$ which is strongly invariant with respect to* $(A_1^{-1}, \dots, A_m^{-1})$;
(iii) ***separation:*** $A_i^{-1}\mathcal{D} \cap A_j^{-1}\mathcal{D} = \varnothing$ *for all* $i \neq j$.

Let $\Psi = (A_i x + t_i)_{i=1}^m$ be a self-affine IFS such that Ψ has the Strong Separation Property. Then for every self-affine measure $\nu = \pi_* \mu$,

$$\dim(\nu) = D(\mu) < 1 \text{ and } \dim_{\mathrm{H}}(\Lambda) = \dim_{\mathrm{B}}(\Lambda) = s(A_1, \dots, A_m) < 1.$$

We note that the condition of domination in Theorem 12.3.4 is equivalent to the definition of dominated matrix tuples in Definition 10.6.1 in view of Theorem 10.6.2. Before we prove Theorem 12.3.4, we state a lemma, which is a straightforward generalization of Marstrand's Theorem 2.3.1 and Hu and Taylor's Theorem 2.3.3, so we leave the proof as an exercise.

LEMMA 12.3.5. *Let μ be a finite Borel measure on $\mathbb{R}^2$ and let λ be a finite Borel measure on $\mathbb{PR}^1$ for which there exist constants $C, s > 0$ such that for every cone I of diameter $|I|$ with respect to the metric $\measuredangle$ defined in* (10.74)*, holds $\lambda(I) \leqslant C|I|^s$. Then*

$$\dim_{\mathrm{H}}\big((\mathrm{Proj}_\theta)_*\mu\big) \geqslant \min\{1, \dim_{\mathrm{H}}(\mu), s\} \text{ for } \lambda\text{-a.e. } \theta \in \mathbb{PR}^1.$$

PROOF SKETCH OF THEOREM 12.3.4. We are going to use some facts from the theory of *nonlinear self-conformal IFS* on the line, which is the topic of Chapter 14 below. In many aspects, such systems behave like self-similar IFS, which we studied in depth earlier. Let $g_i \colon \mathcal{D} \to \mathcal{D}$ be the mappings given by $g_i(\theta) = A_i^{-1}\theta$. Then $(g_i)_{i=1}^m$ is a self-conformal IFS on $\mathcal{D}$, and by our separation assumption, it satisfies the Strong Separation Condition. Then the Furstenberg measure ν_F, whose existence follows from Theorem 12.1.2, is the unique measure satisfying $\nu_F = \sum_{i=1}^m p_i (g_i)_* \nu_F = W_* \mu$, where W is defined in (10.76). By Theorem 14.2.3, which is a generalization of Theorem 3.2.2 to the nonlinear self-conformal case,

$$\dim_{\mathrm{H}}(\nu_F) = \frac{h_\mu}{-\sum_{i=1}^m p_i \int \log |g_i'(\theta)| \, d\nu_F(\theta)} =: s < 1,$$

and ν_F is exact dimensional. Using (10.74), we obtain

$$|g_i'(\theta)| = \frac{|\det(A_i^{-1})|}{\|A_i^{-1}|_\theta\|^2}. \tag{12.10}$$

Indeed, by (10.74),

$$\sin \sphericalangle(g_i(\theta), g_i(\theta')) = \frac{\|A_i^{-1}\mathbf{u}(\theta) \wedge A_i^{-1}\mathbf{u}(\theta')\|}{\|A_i^{-1}|_\theta\| \|A_i^{-1}|_{\theta'}\|} = \frac{|\det(A_i^{-1})|}{\|A_i^{-1}|_\theta\| \|A_i^{-1}|_{\theta'}\|} \sin \sphericalangle(\theta, \theta').$$

Letting $\theta' \to \theta$, we get the desired claim. Now, by (12.10),

$$-\sum_{i=1}^{m} p_i \int \log |g_i'| \, d\nu_F = \sum_{i=1}^{m} p_i \log |\det(A_i)| + 2 \sum_{i=1}^{m} p_i \int \log \|A_i^{-1}|_\theta\| \, d\nu_F(\theta),$$

and so by (10.61) and (10.80),

$$-\sum_{i=1}^{m} p_i \int \log |g_i'| \, d\nu_F = \chi_2(\mu) - \chi_1(\mu).$$

Thus, $s = \frac{h_\mu}{\chi_2(\mu) - \chi_1(\mu)}$. Hence, in view of Corollary 12.3.3 and Lemma 12.3.5, it is enough to show that $\dim_{\mathrm{H}}(\nu) \leqslant s = \frac{h_\mu}{\chi_2(\mu) - \chi_1(\mu)}$. However, by Proposition 10.3.7 and (10.3),

$$\dim_{\mathrm{H}}(\nu) \leqslant D(\mu) \leqslant \frac{h_\mu}{\chi_1(\mu)}.$$

Using the 1-bunched condition, we have that $\chi_2(\mu) < 2\chi_1(\mu)$, and so $\frac{h_\mu}{\chi_1(\mu)} < \frac{h_\mu}{\chi_2(\mu) - \chi_1(\mu)} < 1$, as desired.

Finally, the dimension result for the set follows by taking higher iterates Ψ^n of Ψ. It is easy to see that all the assumptions of the theorem are inherited for higher iterates. Define a sequence of self-affine measures ν_n of Ψ^n with probability vector $\{\alpha_1(A_{\mathbf{i}})^{s_n}\}_{\mathbf{i} \in \{1,\ldots,m\}^n}$, where s_n is the unique solution of $\sum_{\mathbf{i}} \alpha_1(A_{\mathbf{i}})^{s_n} = 1$. Then it is not hard to see that $s_n \to s(A_1, \ldots, A_m)$; moreover,

$$\dim_{\mathrm{H}}(\Lambda) \geqslant \dim_{\mathrm{H}}(\nu_n) = \frac{h_{\mu_n}}{\chi_1(\mu_n)} \geqslant s_n + O\Big(\frac{1}{n}\Big),$$

which implies the desired claim. □

12.3.2. Further developments. The Ledrappier–Young formula has been extensively used recently in the development of the dimension theory of self-affine sets and measures, see, for example, Bárány and Rams [**BR18**], Bárány, Käenmäki and Koivusalo [**BKK18**], Morris and Shmerkin [**MS19b**] and Rapaport [**Rap18**].

Bárány, Hochman and Rapaport [**BHR19**] showed that the dimension of every self-affine measure equals its Lyapunov dimension under mild conditions.

THEOREM 12.3.6 (Bárány, Hochman and Rapaport [**BHR19**]). *Let*

$$\mathcal{F} = (f_i(\mathbf{x}) = A_i\mathbf{x} + \mathbf{t}_i)_{i=1}^m$$

be a planar self-affine IFS such that $\mathcal{F}$ satisfies the strong open set condition and the collection of matrices $\mathcal{A} = (A_1, \ldots, A_m)$ is strongly irreducible. Then for every fully supported self-affine measure $\nu = \Pi_*\mu$, where μ is the corresponding Bernoulli measure, holds

$$\dim(\nu) = D(\mu).$$

The idea of the proof was based on Corollary 12.3.3 and on the entropy growth theorem of Hochman [**Hoc14b**]. In fact, in [**BHR19**] it was shown that the projection of a self-affine measure is sufficiently well approximable by a measure with a kind of convolution structure. This observation made it possible to apply a version of the entropy growth theorem of Hochman [**Hoc14b**]. It essentially says that the dimension of the projection might drop only at most over a set of directions of zero packing dimension, like in Theorem 6.8.11. Finally, since the Furstenberg measure has positive dimension, the claim follows. For the actual proof, which is quite technical, see [**BHR19**].

Theorem 1.12.3 follows from Theorem 12.3.6, although we have seen in Section 10.7.4 that the dimension of every self-affine measure is strictly smaller than the affinity dimension in the strongly irreducible and essentially nonself-similar case. However, Morris and Shmerkin [**MS19b**] showed that the affinity dimension of the IFS can be approximated by the Lyapunov dimension of n-step self-affine measures on the plane. That is, for every $\varepsilon > 0$ there exist $n \geqslant 1$ and $\mathcal{M} \subset \{1, \ldots, m\}^n$ and probabilities $(p_{\mathbf{i}})_{\mathbf{i} \in \mathcal{M}}$ such that the Lyapunov dimension of the self-affine measure of the IFS $\Psi = (f_{\mathbf{i}})_{\mathbf{i} \in \mathcal{M}}$ with probabilities $(p_{\mathbf{i}})_{\mathbf{i} \in \mathcal{M}}$ is ε-close to the affinity dimension $s(A_1, \ldots, A_m)$. In their work, Morris and Shmerkin were using dominated subsystems, see Definition 10.6.1.

Hochman and Rapaport [**HR22**] extended Theorem 12.3.6 by weakening the SOSC to exponential separation. This is a nontrivial extension, unfortunately, also quite technical: whereas the paper [**BHR19**] relied on a version of Hochman's theory in $\mathbb{R}$, the paper [**HR22**] builds on a version of Hochman's theory in $\mathbb{R}^2$. Even more recently, Rapaport [**Rap22a**] partially extended the results of [**BHR19**] to higher dimensions; with the case of $\mathbb{R}^3$ being the most complete. In higher dimensions, stronger irreducibility assumptions are required.

In his recent paper, Feng [**Fen23**] showed that the push-forward measure of every σ-invariant ergodic measure is exact dimensional; moreover, it satisfies a formula similar to (11.10) from the Feng–Hu Theorem 11.1.2. In addition, Feng [**Fen23**] extended the dimension conservation result to the case of push-forwards of quasi-Bernoulli measures. The method of Feng [**Fen23**] relies more on the original ideas of Ledrappier and Young [**LY85a, LY85b**], whereas the method of Bárány and Käenmäki [**BK17**], whose simplified proof will be presented next, follows the ideas of Furstenberg [**Fur08**].

12.3.3. Proof of Theorems 12.3.1 and 12.3.2. Now we return to the proof of the main theorems of the section. Let us first introduce some notation. Recall the dynamics $T\colon \Sigma \times \mathbb{PR}^1 \to \Sigma \times \mathbb{PR}^1$

$$T(\mathbf{i}, \theta) = (\sigma \mathbf{i}, A_{i_1}^{-1}\theta).$$

Let as also recall that $\Pi\colon \Sigma \to \mathbb{R}^2$ denotes the natural projection from the symbolic space to the self-affine set Λ. We extend the map Π to $\Sigma \times \mathbb{PR}^1$ by $\pi(\mathbf{i}, \theta) := \Pi(\mathbf{i})$.

Define the natural partition $\mathcal{P}$ on $\Sigma \times \mathbb{PR}^1$, namely, let $\mathcal{P}(\mathbf{i}, \theta) = [i_1] \times \mathbb{PR}^1$. Further, let ξ denote the measurable partition along subspaces in $\Sigma \times \mathbb{PR}^1$, that is,

$$\xi(\mathbf{i}, \theta) = \Pi^{-1} \operatorname{Proj}_\theta^{-1} \operatorname{Proj}_\theta(\Pi(\mathbf{i})) \times \{\theta\}.$$

In other words, $\xi(\mathbf{i}, \theta)$ is the set of all pairs $(\mathbf{j}, \theta)$ for $\mathbf{j} \in \Sigma$, such that $\Pi(\mathbf{i}) - \Pi(\mathbf{j})$ is parallel to the direction θ. We denote the conditional measures of $\mu \times \nu_F$ with respect to ξ by $\mu^{\xi}_{(\mathbf{i},\theta)}$. It is easy to see from the definitions that

$$\pi_* \left(\mu^{\xi}_{(\mathbf{i},\theta)}\right) = \nu^{\xi_\theta}_{\Pi(\mathbf{i})}.$$

Denote by $\mathcal{T}_k : (\mathbf{i}, \theta) \to (k\mathbf{i}, A_k\theta)$ the kth local inverse of T. That is,

$$T \circ \mathcal{T}_k(\mathbf{i}, \theta) = (\mathbf{i}, \theta) \text{ and } T^{-1}(E) = \bigcup_{k=1}^{m} \mathcal{T}_k(E), \ E \in \Sigma \times \mathbb{PR}^1.$$

Since $\mathcal{T}_{i_1}(\xi(T(\mathbf{i}, \theta))) = \xi(\mathbf{i}, \theta) \cap \mathcal{P}(\mathbf{i}, \theta)$ and since $T_*\mu^{\xi\vee\mathcal{P}}_{(\mathbf{i},\theta)}(E) = \mu^{\xi\vee\mathcal{P}}_{(\mathbf{i},\theta)}(\mathcal{T}_{i_1}(E))$ for every measurable set A, we have that $T_*\mu^{\xi\vee\mathcal{P}}_{(\mathbf{i},\theta)}$ is supported on $\xi(T(\mathbf{i}, \theta))$. On the other hand,

$$\begin{aligned}\int T_*(\mu^{\xi\vee\mathcal{P}}_{(\mathbf{i},\theta)})\, d\mu(\mathbf{i})\, d\nu_F(\theta) &= T_*(\mu \times \nu_F) = \mu \times \nu_F \\ &= \int \mu^{\xi}_{(\mathbf{i},\theta)}\, d\mu(\mathbf{i})\, d\nu_F(\theta) = \int \mu^{\xi}_{T(\mathbf{i},\theta)}\, d\mu(\mathbf{i})\, d\nu_F(\theta),\end{aligned}$$

where we used T-invariance of $\mu \times \nu_F$ twice. Since the conditional measures are unique up to a set of zero measure, we have

$$T_*(\mu^{\xi\vee\mathcal{P}}_{(\mathbf{i},\theta)}) = \mu^{\xi}_{T(\mathbf{i},\theta)} \text{ for } \mu \times \nu_F\text{-a.e. } (\mathbf{i}, \theta). \tag{12.11}$$

We divide the proof of Theorem 12.3.1 into three propositions.

PROPOSITION 12.3.7. *For $\mu \times \nu_F$-a.e. $(\mathbf{i}, \theta)$, the measure $\nu^{\xi_\theta}_{\Pi(\mathbf{i})}$ is exact dimensional and*

$$\dim(\nu^{\xi_\theta}_{\Pi(\mathbf{i})}) = \frac{-\int \log \mu^{\xi}_{(\mathbf{j},\vartheta)}(\mathcal{P}(\mathbf{j}, \vartheta))\, d\mu(\mathbf{j})\, d\nu_F(\vartheta)}{\chi_2(\mu)}. \tag{12.12}$$

PROPOSITION 12.3.8. *For ν_F-a.e. θ, the measure $(\mathrm{Proj}_\theta)_*\nu$ is exact dimensional and*

$$\dim((\mathrm{Proj}_\theta)_*\nu) = \frac{h_\mu + \int \log \mu^{\xi}_{(\mathbf{i},\vartheta)}(\mathcal{P}(\mathbf{i}, \vartheta))\, d\mu(\mathbf{i})\, d\nu_F(\vartheta)}{\chi_1(\mu)}. \tag{12.13}$$

PROPOSITION 12.3.9. *For $\mu \times \nu_F$-a.e. $(\mathbf{i}, \theta)$,*

$$\overline{d}_\nu(\Pi(\mathbf{i})) = \limsup_{r\to 0} \frac{\log \nu(B(\Pi(\mathbf{i}), r))}{\log r} \leqslant \dim(\nu^{\xi_\theta}_{\Pi(\mathbf{i})}) + \dim((\mathrm{Proj}_\theta)_*\nu). \tag{12.14}$$

Note that the numerator in (12.13) is a version of projection entropy, see Lemma 11.1.4. Thus in the special case of diagonally self-affine IFS in $\mathbb{R}^2$, Propositions 12.3.7 and 12.3.8 are exactly the assertions of (11.11) and (11.12).

Theorem 12.3.1 follows by combining the three propositions above, as well as Proposition 12.2.1, which gives the inequality opposite to (12.14) for $\underline{d}_\nu(\pi(\mathbf{i}))$ by the exact dimensionality of conditional measures. Theorem 12.3.2 is then obtained by a simple algebraic manipulation, so it remains to prove the three propositions above. Before we do so, we show in the next corollary that Proposition 12.3.8 yields a generalization of the Feng–Hu Theorem 11.3.1 on exact dimensionality to the case of general self-similar measures (not necessarily homogeneous) on the line.

COROLLARY 12.3.10. *Let $(g_i(x) = \lambda_i x + t_i)_{i=1}^m$ be a contracting IFS on the real line and let $\mathbf{p} = (p_1, \ldots, p_m)$ be a probability vector. Then the self-similar measure $\nu = \sum_{i=1}^m p_i (g_i)_*\nu$ is exact dimensional with*

$$\dim(\nu) = \frac{h_{\mathbf{p}} + \int \log \mu^{\Pi^{-1}}_{\mathbf{i}}([i_1]) d\mu(\mathbf{i})}{-\sum_{i=1}^m p_i \log |\lambda_i|} = \frac{h_\Pi(\mu)}{-\sum_{i=1}^m p_i \log |\lambda_i|},$$

where $\mu = \mathbf{p}^{\mathbb{N}}$, Π *is the natural projection,* $\mu_{\mathbf{i}}^{\Pi^{-1}}$ *are the conditional measures of* μ *along the inverse fibers* $\Pi^{-1}(\Pi(\mathbf{i}))$, *and* h_Π *is the projection entropy, see Definition* 11.1.1.

PROOF. Let $0 < \rho < \min\{1/m, \min_i |\lambda_i|\}\}$, let $\tau_1, \dots, \tau_m$ be such that the planar self-affine IFS $(f_i(x,y) = (g_i(x), \rho y + \tau_i))_{i=1}^m$ has the Strong Separation Property and let $\widehat{\nu}$ be the corresponding self-affine measure with probabilities $\mathbf{p}$. Denote the natural projection of $(f_i)_{i=1}^m$ by π. Clearly, $\mathrm{Proj}_{\theta_y} \widehat{\nu} = \nu$ for the subspace θ_y parallel to the y-axis and $\xi(\mathbf{i}, \theta_y) = \Pi^{-1} \mathrm{Proj}_{\theta_y}^{-1} \mathrm{Proj}_{\theta_y}(\Pi(\mathbf{i})) \times \{\theta_y\}$.

By Lemma 12.1.11, $\nu_F = \delta_{\theta_y}$, and thus, Proposition 12.3.8 implies that ν is exact dimensional and

$$\dim(\nu) = \frac{h_{\mathbf{p}} + \int \log \mu_{\mathbf{i}}^{\Pi^{-1}}([i_1]) d\nu(\mathbf{i})}{-\sum_{i=1}^m p_i \log |\lambda_i|}.$$

To see that $h_{\mathbf{p}} + \int \log \mu_{\mathbf{i}}^{\Pi^{-1}}([i_1]) d\mu(\mathbf{i}) = h_\Pi$, it is enough to note that $-\int \log \mu_{\mathbf{i}}^{\Pi^{-1}}([i_1]) d\mu(\mathbf{i}) = H_\mu(\mathcal{P}|\Pi^{-1}\gamma)$, see Lemma 11.1.4. □

Before we prove Proposition 12.3.7, we show why the formula (12.12) in the statement is well defined.

LEMMA 12.3.11. *The measurable map* $(\mathbf{i}, \theta) \mapsto -\log \mu^{\xi}_{(\mathbf{i},\theta)}(\mathcal{P}(\mathbf{i}, \theta))$ *is in* $L^1(\mu \times \nu_F)$.

PROOF. Let $I = -\int \log \mu^{\xi}_{(\mathbf{i},\theta)}(\mathcal{P}(\mathbf{i}, \theta))\, d\mu(\mathbf{i})\, d\nu_F(\theta)$. Writing the integral in terms of the distribution function (see Theorem B.5.1) yields

$$\begin{aligned} I &= \int_0^\infty \mu \times \nu_F \left(\left\{ (\mathbf{i}, \theta) : -\log \mu^{\xi}_{(\mathbf{i},\theta)}(\mathcal{P}(\mathbf{i}, \theta)) \geqslant t \right\} \right) dt \\ &= \int_0^\infty \mu \times \nu_F \left(\left\{ (\mathbf{i}, \theta) : \mu^{\xi}_{(\mathbf{i},\theta)}(\mathcal{P}(\mathbf{i}, \theta)) \leqslant e^{-t} \right\} \right) dt \\ &= \int_0^\infty \sum_{i=1}^m \mu \times \nu_F \left([i] \times \mathbb{PR}^1 \bigcap \left\{ (\mathbf{i}, \theta) : \mu^{\xi}_{(\mathbf{i},\theta)}(\mathcal{P}(\mathbf{i}, \theta)) \leqslant e^{-t} \right\} \right) dt. \end{aligned}$$

Observe that $\mu \times \nu_F \left([i] \times \mathbb{PR}^1 \bigcap \left\{ (\mathbf{i}, \theta) : \mu^{\xi}_{(\mathbf{i},\theta)}(\mathcal{P}(\mathbf{i}, \theta)) \leqslant e^{-t} \right\} \right) =$

$\int \mu^{\xi}_{(\mathbf{j},\tau)} \left([i] \times \mathbb{PR}^1 \bigcap \left\{ (\mathbf{i}, \theta) : \mu^{\xi}_{(\mathbf{i},\theta)}(\mathcal{P}(\mathbf{i}, \theta)) \leqslant e^{-t} \right\} \right) d\mu(\mathbf{j})\, d\nu_F(\tau)$. So it is enough to show that for $\mu \times \nu_F$-almost every $(\mathbf{j}, \tau)$,

$$\mu^{\xi}_{(\mathbf{j},\tau)} \left([i] \times \mathbb{PR}^1 \bigcap \left\{ (\mathbf{i}, \theta) : \mu^{\xi}_{(\mathbf{i},\theta)}(\mathcal{P}(\mathbf{i}, \theta)) \leqslant e^{-t} \right\} \right) \leqslant e^{-t}. \tag{12.15}$$

Indeed, this would imply that $I \leqslant \int_0^\infty m e^{-t} dt < \infty$.

To show (12.15), we distinguish two cases. If

$$\xi(\mathbf{j}, \tau) \cap [i] \times \mathbb{PR}^1 \bigcap \left\{ (\mathbf{i}, \theta) : \mu^{\xi}_{(\mathbf{i},\theta)}(\mathcal{P}(\mathbf{i}, \theta)) \leqslant e^{-t} \right\} = \varnothing,$$

then clearly

$$\mu^{\xi}_{(\mathbf{j},\tau)} \left([i] \times \mathbb{PR}^1 \bigcap \left\{ (\mathbf{i}, \theta) : \mu^{\xi}_{(\mathbf{i},\theta)}(\mathcal{P}(\mathbf{i}, \theta)) \leqslant e^{-t} \right\} \right) = 0 \leqslant e^{-t}.$$

If $\xi(\mathbf{j},\tau) \cap [i] \times \mathbb{PR}^1 \bigcap \left\{(\mathbf{i},\theta) : \mu^\xi_{(\mathbf{i},\theta)}(\mathcal{P}(\mathbf{i},\theta)) \leqslant e^{-t}\right\} \neq \varnothing$, then let $(\mathbf{k},\vartheta)$ be an element of this set. Then $\mu^\xi_{(\mathbf{j},\tau)} = \mu^\xi_{(\mathbf{k},\vartheta)}$ by the definition of the conditional measures, and thus,

$$\begin{aligned}
&\mu^\xi_{(\mathbf{j},\tau)}\left([i] \times \mathbb{PR}^1 \bigcap \left\{(\mathbf{i},\theta) : \mu^\xi_{(\mathbf{i},\theta)}(\mathcal{P}(\mathbf{i},\theta)) \leqslant e^{-t}\right\}\right) \\
&\quad = \mu^\xi_{(\mathbf{k},\vartheta)}\left([i] \times \mathbb{PR}^1 \bigcap \left\{(\mathbf{i},\theta) : \mu^\xi_{(\mathbf{i},\theta)}(\mathcal{P}(\mathbf{i},\theta)) \leqslant e^{-t}\right\}\right) \\
&\quad = \mu^\xi_{(\mathbf{k},\vartheta)}\left(\mathcal{P}(\mathbf{k},\vartheta) \bigcap \left\{(\mathbf{i},\theta) : \mu^\xi_{(\mathbf{i},\theta)}(\mathcal{P}(\mathbf{i},\theta)) \leqslant e^{-t}\right\}\right) \\
&\quad \leqslant \mu^\xi_{(\mathbf{k},\vartheta)}\left(\mathcal{P}(\mathbf{k},\vartheta)\right) \leqslant e^{-t}.
\end{aligned}$$

□

Denote by $\mathcal{P}_1^n$ the partition of $\Sigma \times \mathbb{PR}^1$ into cylinders of length n in the first component, that is, $\mathcal{P}_1^n(\mathbf{i},\theta) = [i_1 \dots i_n] \times \mathbb{PR}^1$.

PROOF OF PROPOSITION 12.3.7. To prove the statement, observe that it is enough to show that

$$\lim_{n\to\infty} \frac{1}{n} \log \mu^\xi_{(\mathbf{i},\theta)}(\mathcal{P}_1^n(\mathbf{i})) = \int \log \mu^\xi_{(\mathbf{j},\vartheta)}(\mathcal{P}(\mathbf{j},\vartheta))\, d\mu(\mathbf{j})\, d\nu_F(\vartheta) \tag{12.16}$$

for $\mu \times \nu_F$-almost every $(\mathbf{i},\theta)$.

Indeed, $\nu^{\xi_\theta}_{\Pi(\mathbf{i})}$ is a measure on $\theta(\Pi(\mathbf{i}))$, where $\theta(\mathbf{x})$ denotes the line in the direction θ passing through $\mathbf{x}$. Moreover, $\nu^{\xi_\theta}_{\Pi(\mathbf{i})} = \nu^{\xi_\theta}_{\pi(\mathbf{i}')}$ for almost all $(\mathbf{i}',\theta) \in \xi(\mathbf{i},\theta)$. Thus, for ν_F-a.e. θ, the local dimension of $\nu^{\xi_\theta}_{\Pi(\mathbf{i})}$ at a typical point equals

$$\lim_{r\to 0} \frac{\log \nu^{\xi_\theta}_{\Pi(\mathbf{i})}\left(B_r(\Pi(\mathbf{i})) \cap \theta(\Pi(\mathbf{i}))\right)}{\log r},$$

assuming this limit exists. Now, by the definition of the natural projection Π and the Strong Separation Property, there exists a constant $c > 1$ such that for every $\mathbf{i} \in \Sigma$, every $n \in \mathbb{N}$ and every $\theta \in \mathbb{PR}^1$,

$$\begin{aligned}
B_{c^{-1}\|A^{-1}_{\mathbf{i}|_n}|_\theta\|^{-1}}(\Pi(\mathbf{i})) \cap \theta(\Pi(\mathbf{i})) \cap \Lambda \quad &\subseteq \quad \Pi(\mathcal{P}_1^n(\mathbf{i})) \cap \theta(\Pi(\mathbf{i})) \\
&\subseteq \quad B_{c\|A^{-1}_{\mathbf{i}|_n}|_\theta\|^{-1}}(\Pi(\mathbf{i})) \cap \theta(\Pi(\mathbf{i})).
\end{aligned}$$

Recall that $\Pi_*\left(\mu^\xi_{(\mathbf{i},\theta)}\right) = \nu^{\xi_\theta}_{\Pi(\mathbf{i})}$. The Strong Separation Property means that the map Π is 1-to-1, and thus

$$\nu^{\xi_\theta}_{\Pi(\mathbf{i})}\left(B_{c^{-1}r}(\Pi(\mathbf{i})) \cap \theta(\Pi(\mathbf{i}))\right) \leqslant \mu^\xi_{(\mathbf{i},\theta)}(\mathcal{P}_1^n(\mathbf{i})) \leqslant \nu^{\xi_\theta}_{\Pi(\mathbf{i})}\left(B_{cr}(\Pi(\mathbf{i})) \cap \theta(\Pi(\mathbf{i}))\right),$$

where $r = \|A^{-1}_{\mathbf{i}|_n}|_\theta\|^{-1}$. Finally, recall that $\lim_{n\to\infty} \log \|A^{-1}_{\mathbf{i}|_n}|_\theta\|^{-1} = \chi_2(\mu)$ for $\mu \times \nu_F$-a.e. $(\mathbf{i},\theta)$ by part (3) of Theorem 12.1.2, and so (12.16) implies (12.12).

Now, in order to show (12.16), observe that we have

$$\begin{aligned}
\mu^\xi_{(\mathbf{i},\theta)}(\mathcal{P}_1^n(\mathbf{i},\theta)) &= \mu^\xi_{(\mathbf{i},\theta)}\left(\bigcap_{\ell=0}^{n-1} T^{-\ell}\mathcal{P}(T^\ell(\mathbf{i},\theta))\right) \\
&= \mu^\xi_{(\mathbf{i},\theta)}(\mathcal{P}(\mathbf{i},\theta)) \prod_{k=1}^{n-1} \frac{\mu^\xi_{(\mathbf{i},\theta)}\left(\bigcap_{\ell=0}^{k} T^{-\ell}\mathcal{P}(T^\ell(\mathbf{i},\theta))\right)}{\mu^\xi_{(\mathbf{i},\theta)}\left(\bigcap_{\ell=0}^{k-1} T^{-\ell}\mathcal{P}(T^\ell(\mathbf{i},\theta))\right)}.
\end{aligned}$$

Further, by Lemma 9.4.21 and (12.11),

$$\frac{\mu^{\xi}_{(\mathbf{i},\theta)}\left(\bigcap_{\ell=0}^{k} T^{-\ell}\mathcal{P}(T^{\ell}(\mathbf{i},\theta)\right)}{\mu^{\xi}_{(\mathbf{i},\theta)}\left(\bigcap_{\ell=0}^{k-1} T^{-\ell}\mathcal{P}(T^{\ell}(\mathbf{i},\theta)\right)} = \mu^{\xi\vee\mathcal{P}_1^k}_{(\mathbf{i},\theta)}\left(T^{-k}(\mathcal{P}(T^k(\mathbf{i},\theta)))\right)$$

$$= \mu^{\xi}_{T^k(\mathbf{i},\theta)}\left(\mathcal{P}(T^k(\mathbf{i},\theta))\right).$$

The proof can be finished by an application of Birkhoff's Ergodic Theorem 9.1.1 to the ergodic system $(\Sigma\times\mathbb{PR}^1, \mu\times\nu_{\mathrm{F}}, T)$, with the function $(\mathbf{i},\theta)\mapsto -\log\mu^{\xi}_{(\mathbf{i},\theta)}(\mathcal{P}(\mathbf{i},\theta))$, which is integrable by Lemma 12.3.11. □

Before we turn to the proof of the next proposition, we state here for the reader's convenience a special case of Oseledeč's Multiplicative Ergodic Theorem 9.2.1 that we require. As above, for $\mathbf{i}\in\Sigma$ and $n\in\mathbb{N}$, we write $A^{-1}_{\mathbf{i}|_n} := (A_{\mathbf{i}|_n})^{-1} = A^{-1}_{i_n}\cdots A^{-1}_{i_1}$.

THEOREM 12.3.12 (Oseledeč' theorem). *If the semigroup $SG_{\mathcal{B}}$ generated by $\mathcal{B}$ is noncompact, then there exists a measurable function $E\colon\Sigma\to\mathbb{PR}^1$ such that for μ-almost every $\mathbf{i}$,*

(1) $A^{-1}_{i_1}E(\mathbf{i}) = E(\sigma\mathbf{i})$,
(2) $\lim_{n\to\infty}\frac{1}{n}\log\|A^{-1}_{\mathbf{i}|_n}|_{E(\mathbf{i})}\| = \chi_1(\mu)$,
(3) $\lim_{n\to\infty}\frac{1}{n}\log\sin\sphericalangle(E(\sigma^n\mathbf{i}), A^{-1}_{\mathbf{i}|_n}\theta) = 0$ *for ν_F-a.e. θ.*

Since we do not assume the noncompactness of $G_{\mathcal{B}}$, we need to consider two cases. If $G_{\mathcal{B}}$ is compact, then by Lemma 12.1.3, it is conjugate to a subgroup of $O(2,\mathbb{R})$. Let M denote the conjugation matrix, for which $M^{-1}AM\in O(2,\mathbb{R})$ for every $A\in G_{\mathcal{B}}$. Then define $V\colon\Sigma\times\mathbb{PR}^1\to\mathbb{PR}^1$ as follows

$$V(\mathbf{i},\theta) = \begin{cases} E(\mathbf{i}), & \text{if } G_{\mathcal{B}} \text{ is noncompact,}\\ M(M^{-1}\theta)^{\perp}, & \text{if } G_{\mathcal{B}} \text{ is compact with conjugation matrix } M. \end{cases}$$

LEMMA 12.3.13. *For $\mu\times\nu_F$-almost every $(\mathbf{i},\theta)$,*

(1) $A^{-1}_{i_1}V(\mathbf{i},\theta) = V(T(\mathbf{i},\theta))$,
(2) $\lim_{n\to\infty}\frac{1}{n}\log\|A^{-1}_{\mathbf{i}|_n}|_{V(\mathbf{i},\theta)}\| = \iint\log\|A^{-1}_{i_1}|_{V(\mathbf{i},\theta)}\|d\mu(\mathbf{i})d\nu_F(\theta) = \chi_1(\mu)$,
(3) $\lim_{n\to\infty}\frac{1}{n}\log\sin\sphericalangle(V(T^n(\mathbf{i},\theta)), A^{-1}_{\mathbf{i}|_n}\theta) = 0$.

PROOF. Since the case of noncompact $G_{\mathcal{B}}$ is covered by Theorem 12.3.12, we may assume that $G_{\mathcal{B}}$ is compact. For every $A\in G_{\mathcal{B}}$ we have $M^{-1}AM\in O(2,\mathbb{R})$, hence

$$\begin{aligned} V(T(\mathbf{i},\theta)) &= M(M^{-1}A^{-1}_{i_1}\theta)^{\perp} = M(M^{-1}A^{-1}_{i_1}MM^{-1}\theta)^{\perp}\\ &= MM^{-1}A^{-1}_{i_1}M(M^{-1}\theta)^{\perp} = A^{-1}_{i_1}V(\mathbf{i},\theta). \end{aligned}$$

Moreover, since $G_{\mathcal{B}}$ is compact, we get that $\chi_1(\mu) = \chi_2(\mu)$ trivially for every Bernoulli measure μ.

Finally, since the map $A\mapsto\sphericalangle(M(M^{-1}A\theta)^{\perp}, A\theta)$ is continuous on $G_{\mathcal{B}}$, by compactness we get that either there exists $\kappa>0$ such that $\sphericalangle(M(M^{-1}A\theta)^{\perp}, A\theta) > \kappa$ for every $A\in G_{\mathcal{B}}$ or there exists $A\in G_{\mathcal{B}}$ such that $\sphericalangle(M(M^{-1}A\theta)^{\perp}, A\theta) = 0$. Since $A^{-1}_{\mathbf{i}|_n}\in G_{\mathcal{B}}$ for every $\mathbf{i}\in\Sigma$ and $n\geqslant 1$, the first case immediately implies (3). So we only have to show that the second case is impossible. And in fact, if $0 = \sphericalangle(M(M^{-1}A\theta)^{\perp}, A\theta) = \sphericalangle(AM(M^{-1}\theta)^{\perp}, A\theta)$ then using that A and M are invertible, we get $\sphericalangle((M^{-1}\theta)^{\perp}, M^{-1}\theta) = 0$, which is a contradiction. □

By Lemma 12.3.13(3), we have that

$$\mu \times \nu_F\big(\{(\mathbf{i},\theta) : V(\mathbf{i},\theta) = \theta\}\big) = 0. \tag{12.17}$$

Indeed, if this were not the case, then by Birkhoff's Ergodic theorem $V(T^n(\mathbf{i},\theta)) = A_{\mathbf{i}|_n}^{-1}\theta$ for $\mu \times \nu_F$-a.e. $(\mathbf{i},\theta)$ for infinitely many $n \in \mathbb{N}$, which would contradict Lemma 12.3.13(3). This allows us to view $(\mathrm{Proj}_\theta)_*(\nu)$ as a measure on $V(\mathbf{i},\theta)$, which is the equivariant direction.

Proof of Proposition 12.3.8. Recall that the IFS $\mathcal{F} = (f_i(\mathbf{x}) = A_i\mathbf{x} + \mathbf{t}_i)_{i=1}^m$ satisfies the SSP and its attractor is denoted by Λ. Consider the "transversal ball"

$$B_\theta^{\mathrm{tr}}(\mathbf{x},\delta) = \mathrm{Proj}_\theta^{-1} B(\mathrm{Proj}_\theta \mathbf{x},\delta),$$

so that $(\mathrm{Proj}_\theta)_*\nu\big(B(\mathrm{Proj}_\theta \mathbf{x},\delta)\big) = \nu\big(B_\theta^{\mathrm{tr}}(\mathbf{x},\delta)\big)$. Next, define the "symbolic" transversal ball by

$$B^t\big((\mathbf{i},\theta),\delta\big) = \Pi^{-1} B_\theta^{\mathrm{tr}}\big(\Pi(\mathbf{i}),\delta \sin \sphericalangle(V(\mathbf{i},\theta),\theta)\big).$$

Observe that $B_\theta^{\mathrm{tr}}\big(\pi(\mathbf{i}),\delta\sin\sphericalangle(V(\mathbf{i},\theta),\theta)\big)$ is a strip of width $2\delta\sin\sphericalangle(V(\mathbf{i},\theta),\theta)$ with the center-line $\theta(\Pi(\mathbf{i}))$, so its width is 2δ *in the direction of* $V(\mathbf{i},\theta)$, which is the equivariant direction by Lemma 12.3.13(1). Furthermore, by (12.17) the symbolic transversal ball $B^t((\mathbf{i},\theta),\delta)$ has positive width for $\mu\times\nu_F$-a.e. $(\mathbf{i},\theta)$. For a visualization, see Figure 12.2.

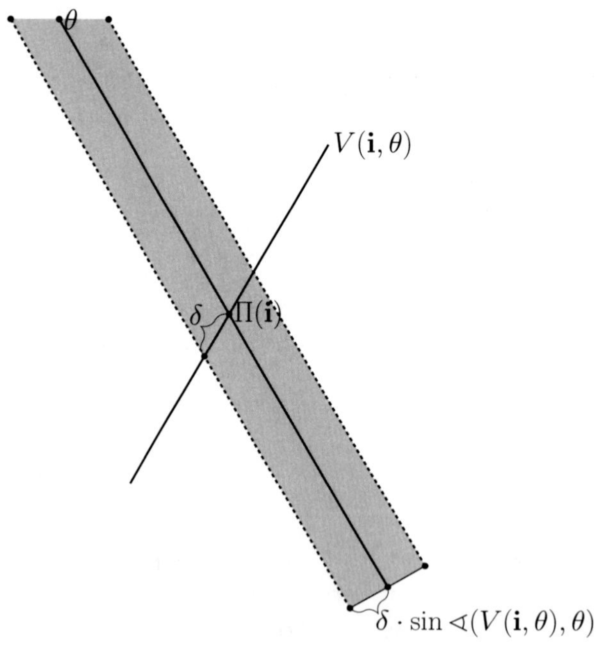

Figure 12.2. The symbolic transversal ball $B^t((\mathbf{i},\theta),\delta)$ represented on the plane, which has width $2\delta\sin\sphericalangle(V(\mathbf{i},\theta),\theta)$; the component along the direction $V(\mathbf{i},\theta)$ has length 2δ.

By the self-affinity of $\nu = \Pi_*\mu$,

$$\begin{aligned}\mu&\left(B^t((\mathbf{i},\theta),\delta) \cap \mathcal{P}(\mathbf{i},\theta)\right)\\&= \nu\left(B^{\mathrm{tr}}_\theta(\Pi(\mathbf{i}),\delta\sin\measuredangle(V(\mathbf{i},\theta),\theta)) \cap f_{i_1}(\Lambda)\right)\\&= \sum_{\ell=1}^m p_\ell\nu\Big(f_\ell^{-1}\Big(B^{\mathrm{tr}}_\theta(\Pi(\mathbf{i}),\delta\sin\measuredangle(V(\mathbf{i},\theta),\theta)) \cap f_{i_1}(\Lambda)\Big)\Big).\end{aligned}$$

Clearly, by the SSP there exists a unique ℓ for which the inverse image above is not an empty set, namely, $\ell = i_1$. By the definition of the dynamics, $f_{i_1}^{-1}(\Pi(\mathbf{i})) = \Pi(\sigma\mathbf{i})$, and the direction of the transversal ball changes to $A_{i_1}^{-1}\theta$. Moreover, the action of $A_{i_1}^{-1}\colon V(\mathbf{i},\theta) \to V(T(\mathbf{i},\theta))$ has norm $\|A_{i_1}^{-1}|_{V(\mathbf{i},\theta)}\| = \|A_{i_1}|_{V(T(\mathbf{i},\theta))}\|^{-1}$, and so the radius $\delta\sin\measuredangle(V(\mathbf{i},\theta),\theta)$ of the transversal ball becomes $\delta\|A_{i_1}|_{V(T(\mathbf{i},\theta))}\|^{-1}\sin\measuredangle(V(T(\mathbf{i},\theta)),A_{i_1}^{-1}\theta)$ under the action of $f_{i_1}^{-1}$. (This follows by considering how the width of the strip changes in the equivariant direction.) Thus,

$$\begin{aligned}\mu&\left(B^t((\mathbf{i},\theta),\delta) \cap \mathcal{P}(\mathbf{i},\theta)\right)\\&= p_{i_1}\nu\Big(f_{i_1}^{-1}\Big(B^{\mathrm{tr}}_\theta(\Pi(\mathbf{i}),\delta\sin\measuredangle(V(\mathbf{i},\theta),\theta)) \cap f_{i_1}(\Lambda)\Big)\Big)\\&= p_{i_1}\nu\left(B^{\mathrm{tr}}_{A_{i_1}^{-1}\theta}(\Pi(\sigma\mathbf{i}),\delta\|A_{i_1}|_{V(T(\mathbf{i},\theta))}\|^{-1}\sin\measuredangle(V(T(\mathbf{i},\theta)),A_{i_1}^{-1}\theta))\right)\\&= p_{i_1}\mu\left(B^t(T(\mathbf{i},\theta),\delta\|A_{i_1}|_{V(T(\mathbf{i},\theta))}\|^{-1})\right).\end{aligned}\tag{12.18}$$

In order to verify (12.13), it enough to show that $\lim_{\delta\to0}\log\mu\big(B^t((\mathbf{i},\theta),\delta)\big)/(\log\delta)$ exists and equals the right-hand side of (12.13) for $\mu\times\nu_F$-a.e. $(\mathbf{i},\theta)$. Let

$$w_\delta(\mathbf{i},\theta) = \frac{\mu\big(B^t((\mathbf{i},\theta),\delta) \cap \mathcal{P}(\mathbf{i},\theta)\big)}{\mu\big(B^t((\mathbf{i},\theta),\delta)\big)}.$$

It is not hard to see that $\sup_{\delta>0}|\log w_\delta| \in L^1$. Indeed, this is proved the same way as (11.56) in Lemma 11.3.4; see [**Bár15**, Lemma 3.6] for more details. Further, let $\delta_n(\mathbf{i},\theta) = \|A_{\mathbf{i}|_n}|_{V(T^n(\mathbf{i},\theta))}\|$, with $\delta_0 \equiv 1$. By the definition of V and Theorem 12.3.12, since $V(\mathbf{i},\theta)$ are one-dimensional equivariant subspaces, we have $\delta_n(\mathbf{i},\theta) = \|A_{i_1}|_{V(T(\mathbf{i},\theta))}\|\cdot\delta_{n-1}(T(\mathbf{i},\theta))$ for all n. So using (12.18) we get

$$\begin{aligned}\mu&\left(B^t(T^k(\mathbf{i},\theta),\delta_{n-k}(T^k(\mathbf{i},\theta))) \cap \mathcal{P}(T^k(\mathbf{i},\theta))\right)\\&= p_{i_{k+1}}\mu\left(B^t(T^{k+1}(\mathbf{i},\theta),\delta_{n-k-1}(T^{k+1}(\mathbf{i},\theta)))\right).\end{aligned}$$

for every $0 \leqslant k \leqslant n$. Hence, we get a "telescoping formula":

$$\begin{aligned}\mu&\big(B^t((\mathbf{i},\theta),\delta_n(\mathbf{i},\theta))\big)\\&= \mu\big(B^t(T^n(\mathbf{i},\theta),1)\big)\cdot\prod_{k=0}^{n-1}\frac{\mu\left(B^t(T^k(\mathbf{i},\theta),\delta_{n-k}(T^k(\mathbf{i},\theta)))\right)}{\mu\left(B^t(T^{k+1}(\mathbf{i},\theta),\delta_{n-k-1}(T^{k+1}(\mathbf{i},\theta)))\right)}\\&= \mu\big(B^t(T^n(\mathbf{i},\theta),1)\big)\cdot\left(\prod_{k=0}^{n-1}\frac{\mu\left(B^t(T^k(\mathbf{i},\theta),\delta_{n-k}(T^k(\mathbf{i},\theta))) \cap \mathcal{P}(T^k(\mathbf{i},\theta))\right)}{p_{i_{k+1}}\mu\left(B^t(T^k(\mathbf{i},\theta),\delta_{n-k}(T^k(\mathbf{i},\theta)))\right)}\right)^{-1}\\&= \mu\big(B^t(T^n(\mathbf{i},\theta),1)\big)\cdot\prod_{k=1}^n p_{i_k}\cdot\left(\prod_{k=0}^{n-1}w_{\delta_{n-k}(T^k(\mathbf{i},\theta))}(T^k(\mathbf{i},\theta))\right)^{-1}.\end{aligned}$$

By the density theorem for conditional measures (12.5), see also the proof of (11.54) in Lemma 11.3.4 for a special case, we have

$$\lim_{n\to\infty} \log w_{\delta_n(\mathbf{i},\theta)}(\mathbf{i},\theta) = \log \mu^{\xi}_{(\mathbf{i},\theta)}(\mathcal{P}(\mathbf{i},\theta)),$$

in L^1 and almost everywhere, and by Maker's Theorem 9.1.3,

$$\lim_{n\to\infty} \frac{1}{n} \log \prod_{k=0}^{n-1} w_{\delta_{n-k}(T^k(\mathbf{i},\theta))}(T^k(\mathbf{i},\theta)) = \int \log \mu^{\xi}_{(\mathbf{i},\theta)}(\mathcal{P}(\mathbf{i},\theta))\, d\mu(\mathbf{i})\, d\nu_F(\theta)$$

for $\mu \times \nu_F$-a.e. $(\mathbf{i},\theta)$. On the other hand, trivially,

$$\begin{aligned} 1 &\geqslant \mu\left(B^t(T^n(\mathbf{i},\theta),1)\right) \\ &\geqslant \nu\left(B\big(\Pi(\sigma^n\mathbf{i}), \sin\sphericalangle(V(T^n(\mathbf{i},\theta)), A^{-1}_{\mathbf{i}|_n}\theta)\big)\right). \end{aligned}$$

Note that

$$\nu(B(\Pi(\mathbf{j}),r)) \geqslant p_{j_1}\cdot\ldots\cdot p_{j_\ell} \geqslant p^{\ell}_{\min} \quad \text{if} \quad \|A_{\mathbf{j}|_\ell}\| \leqslant (1-\varepsilon)^\ell < r.$$

Hence, by Lemma 12.3.13(3),

$$\lim_{n\to\infty} \frac{1}{n} \log \mu\big(B^t(T^n(\mathbf{i},\theta),1)\big) = 0 \quad \text{for } \mu\times\nu_F\text{-a.e. } (\mathbf{i},\theta).$$

The proof can be finished by applying Birkhoff's Ergodic Theorem 9.1.1:

$$\begin{aligned} \lim_{n\to\infty} \frac{1}{n} \log \delta_n(\mathbf{i},\theta) &= \lim_{n\to\infty} \frac{1}{n} \sum_{k=1}^{n} \log \|A_{i_k}|_{V(T^k(\mathbf{i},\theta))}\| \\ &= \iint \log \|A_{i_1}|_{V(T(\mathbf{i},\theta))}\|\, d\mu(\mathbf{i})\, d\nu_F(\theta) = \chi_1(\mu) \end{aligned}$$

and $\lim_{n\to\infty} \frac{1}{n} \log \prod_{k=1}^{n} p_{i_k} = h_\mu$ for $\mu \times \nu_F$-almost every $(\mathbf{i},\theta)$. □

Finally, we turn to the proof of Proposition 12.3.9. For simplicity, let us denote the typical value of the local dimension of $\nu^{\xi_\theta}_{\Pi(\mathbf{i})}$, which exists by Proposition 12.3.7, byd_1 and the entropy $-\int \log \mu^{\xi}_{(\mathbf{i},\theta)}(\mathcal{P}(\mathbf{i},\theta))\, d\mu(\mathbf{i})\, d\nu_F(\theta)$ by H.

PROOF OF PROPOSITION 12.3.9. Applying Egorov's theorem to (12.12) and to (12.16), and the Shannon–McMillan–Breiman Theorem, for every $\varepsilon > 0$ we can choose $J_1 \subseteq \Sigma \times \mathbb{PR}^1$ and a constant $M = M(\varepsilon, J_1)$ such that $\mu \times \nu_F(J_1) > 1-\varepsilon$ and

$$\nu^{\xi_\theta}_{\Pi(\mathbf{i})}\left(B(\Pi(\mathbf{i}), e^{-m(\chi_1(\mu)-\varepsilon)})\right) \geqslant e^{-m(\chi_1(\mu)-\varepsilon)(d_1+\varepsilon)}; \tag{12.19}$$

$$\mu^{\xi}_{(\mathbf{i},\theta)}(\mathcal{P}^m_1(\mathbf{i},\theta)) \leqslant e^{-m(H-\varepsilon)}; \tag{12.20}$$

$$\mu(\mathcal{P}^m_1(\mathbf{i},\theta)) \geqslant e^{-m(h_\mu+\varepsilon)}; \tag{12.21}$$

$$\pi(\mathcal{P}^m_1(\mathbf{i},\theta)) \subseteq B(\Pi(\mathbf{i}), e^{-m(\chi_1(\mu)-\varepsilon)}), \tag{12.22}$$

for every $(\mathbf{i},\theta) \in J_1$ and every $m \geqslant M$. Below we write

$$r_m := e^{-m(\chi_1(\mu)-\varepsilon)} \tag{12.23}$$

to simplify the notation. By the analog of Lebesgue's density theorem in $\Sigma \times \mathbb{PR}^1$ (Theorem B.4.2(ii)) and Egorov's theorem, there exist $J_2 \subseteq J_1$ and $M' \geqslant M$ such that $\mu \times \nu_F(J_2) > 1 - 2\varepsilon$ and

$$\frac{\nu^{\xi_\theta}_{\Pi(\mathbf{i})}\big(\pi(J_1) \cap B(\Pi(\mathbf{i}), r_m)\big)}{\nu^{\xi_\theta}_{\Pi(\mathbf{i})}\big(B(\Pi(\mathbf{i}), r_m)\big)} \geqslant \frac{1}{2} \tag{12.24}$$

for every $(\mathbf{i}, \theta) \in J_2$ and $m \geqslant M'$. Denote, for $(\mathbf{i}, \theta) \in J_2$,

$$\mathcal{Q}_0^{m-1}(\mathbf{i}, \theta) := \Big\{P \in \mathcal{P}_1^m : \Pi(P \cap \xi(\mathbf{i}, \theta) \cap J_1) \cap B(\Pi(\mathbf{i}), r_m) \neq \varnothing\Big\} \tag{12.25}$$

Note that for $P \in \mathcal{Q}_1^m$ we have $P \cap \xi(\mathbf{i}, \theta) \cap J_1 \neq \varnothing$, hence $P = \mathcal{P}_1^m(\mathbf{j}, \theta)$ for some $\mathbf{j} \in J_1$, with $\xi(\mathbf{j}, \theta) = \xi(\mathbf{i}, \theta)$. But then $\mu^\xi_{(\mathbf{i},\theta)} = \mu^\xi_{(\mathbf{j},\theta)}$, and we can apply (12.20), with $\mathbf{j}$ instead of $\mathbf{i}$, to get

$$\mu^\xi_{(\mathbf{i},\theta)}(P) \leqslant e^{-m(H-\varepsilon)}, \quad \text{for } P \in \mathcal{Q}_1^m. \tag{12.26}$$

Further, $\pi(P) \subset B(\Pi(\mathbf{j}), r_m)$ by (12.22), and $\pi(P) \cap B(\mathbf{i}, r_m) \neq \varnothing$, hence

$$\pi(P) \subset B(\mathbf{i}, 2r_m), \quad \text{for } P \in \mathcal{Q}_1^m. \tag{12.27}$$

Thus, for $(\mathbf{i}, \theta) \in J_2$, by (12.19),

$$\begin{aligned} \frac{1}{2} r_m^{d_1+\varepsilon} &\leqslant \nu^{\xi_\theta}_{\Pi(\mathbf{i})}\big(\pi(J_1) \cap B(\Pi(\mathbf{i}), r_m)\big) \\ &= \mu^\xi_{(\mathbf{i},\theta)}\Big(\pi^{-1}\big(\pi(J_1) \cap B(\Pi(\mathbf{i}), r_m)\big)\Big) \\ &\leqslant \#\mathcal{Q}_1^m \cdot \max\{\mu^\xi_{(\mathbf{i},\theta)}(P) : P \in \mathcal{Q}_1^m\} \\ &\leqslant \#\mathcal{Q}_1^m \cdot e^{-m(H-\varepsilon)}, \end{aligned} \tag{12.28}$$

using (12.26) in the last step. On the other hand, in view of (12.27), for $(\mathbf{i}, \theta) \in J_2$:

$$\begin{aligned} \nu\big(B(\Pi(\mathbf{i}), 2r_m)\big) &\geqslant \#\mathcal{Q}_1^m \cdot \min\{\mu(P) : P \in \mathcal{Q}_1^m\} \\ &\geqslant \#\mathcal{Q}_1^m \cdot e^{-m(h_\nu+\varepsilon)} \\ &\geqslant r_m^{d_1+\varepsilon} \cdot e^{m(H-\varepsilon)-m(h_\nu+\varepsilon)}, \end{aligned}$$

where we used (12.21) and (12.28). Thus, recalling (12.23), we obtain

$$\overline{d}_\nu(\Pi(\mathbf{i})) \leqslant d_1 + \varepsilon + \frac{h_\nu - H + 2\varepsilon}{\chi_1(\mu) - \varepsilon}$$

for every $(\mathbf{i}, \theta) \in J_2$. Since $\varepsilon > 0$ was arbitrary, the statement follows. □

12.4. Dimension conservation for the natural measure

The dimension conservation phenomenon presented in Theorem 12.3.1 is necessarily of "almost every" type, i.e., it holds for almost every slice. However, in the special case of the natural measure for a self-similar IFS, one can deduce a formula which holds for every slice, connecting the local dimension of the projection and the box dimension of the slice. Results of this kind first appeared in Manning and Simon [**MS13**]. We start by recalling the notation.

Let $\mathcal{S} = (S_i(\mathbf{x}) = \lambda_i \mathbf{x} + \mathbf{t}_i)_{i=1}^m$, with $\lambda_i \in (0, 1)$, be a self-similar IFS on the plane with attractor $\Lambda \subset \mathbb{R}^2$, such that the Open Set Condition holds. Let μ be the natural measure on $\Sigma = \{1, \ldots, m\}^{\mathbb{N}}$, that is,

$$\mu([i_1, \ldots, i_n]) = \prod_{k=1}^n |\lambda_{i_k}|^s \tag{12.29}$$

for every cylinder set $[i_1, \ldots, i_n]$, where s is the similarity dimension of $\mathcal{S}$. Let $\Pi\colon \Sigma \to \Lambda$ be the usual natural projection. Moreover, let $\mathrm{proj} : \mathbb{R}^2 \to \mathbb{R}$ be the orthogonal projection to the x-axis. By conjugating the IFS via a rotation, one can easily see that the following statements hold also for orthogonal projections in any direction. It is also clear that since $\mathcal{S}$ is rotation free, the map $\pi := \mathrm{proj} \circ \Pi\colon \Sigma \to \mathrm{Proj}(\Lambda)$ is a natural projection with respect to the IFS

$$\widetilde{\mathcal{S}} = (f_i(x) = \lambda_i x + \mathrm{proj}(t_i))_{i=1}^m; \tag{12.30}$$

moreover, $\mathrm{proj}(S_i(\mathbf{x})) = f_i(\mathrm{proj}(\mathbf{x}))$ for every $\mathbf{x} \in \mathbb{R}^2$. For convenience, below we denote

$$\widetilde{\Lambda} := \mathrm{proj}(\Lambda), \quad \widetilde{\Lambda}_{\mathbf{i}} := f_{\mathbf{i}}(\widetilde{\Lambda}) \quad \text{for } \mathbf{i} \in \Sigma.$$

Without loss of generality, we may assume that $\widetilde{\Lambda} \subseteq [0,1]$ and $0, 1 \in \widetilde{\Lambda}$. Further, let $L(x) := \Lambda \cap \mathrm{proj}^{-1}(x)$ be the slice over $x \in \widetilde{\Lambda}$, and let $\nu = \Pi_* \mu$.

PROPOSITION 12.4.1. *Let $\mathcal{S}$ be an IFS as above satisfying the open set condition. Then for every $x \in \widetilde{\Lambda}$,*

$$\overline{d}_{\mathrm{proj}_* \nu}(x) + \underline{\dim}_{\mathrm{B}}(L(x)) \leqslant s \quad \text{and} \quad \underline{d}_{\mathrm{proj}_* \nu}(x) + \overline{\dim}_{\mathrm{B}}(L(x)) \leqslant s, \tag{12.31}$$

where $\sum_{i=1}^m |\lambda_i|^s = 1$. Moreover, if $\widetilde{\Lambda} = [0,1]$ and there exist unique $i, j \in \{1, \ldots, m\}$ such that $0 \in f_i([0,1])$, $1 \in f_j([0,1])$, then for every $x \in \widetilde{\Lambda}$,

$$\overline{d}_{\mathrm{proj}_* \nu}(x) + \underline{\dim}_{\mathrm{B}}(L(x)) = s \quad \text{and} \quad \underline{d}_{\mathrm{proj}_* \nu}(x) + \overline{\dim}_{\mathrm{B}}(L(x)) = s. \tag{12.32}$$

The proof of Proposition 12.4.1 follows the strategy similar to that of Theorem 8.5.1. Let $\widetilde{\mathcal{M}}_\delta = \{\mathbf{i} \in \Sigma^* : |\lambda_{\mathbf{i}}| \leqslant \delta < |\lambda_{\mathbf{i}_-}|\}$ be the usual Moran minimal cut-set, so that $\{[\mathbf{i}] : \mathbf{i} \in \widetilde{\mathcal{M}}_\delta\}$ is a partition for every $\delta \in (0,1)$. We also let $\widetilde{\mathcal{M}}_1 = \{1, \ldots, m\}$. Further, for $x \in \widetilde{\Lambda}$ let

$$\mathcal{N}(x, \delta) = \{\mathbf{i} \in \widetilde{\mathcal{M}}_\delta : x \in \widetilde{\Lambda}_{\mathbf{i}}\}. \tag{12.33}$$

It is clear that $\#\mathcal{N}(x,\delta)$ is nonincreasing in δ. Denote the largest and the smallest absolute value of the contracting ratios by $\lambda_{\max} := \max\{|\lambda_i|\}$ and $\lambda_{\min} := \min\{|\lambda_i|\}$ respectively.

LEMMA 12.4.2. *Let $\mathcal{S}$ be a self-similar IFS with OSC as above. Then for every $x \in \widetilde{\Lambda}$,*

$$\underline{\dim}_{\mathrm{B}}(L(x)) = \liminf_{\delta \to 0} \frac{\log \#\mathcal{N}(x,\delta)}{-\log \delta} \quad \text{and} \quad \overline{\dim}_{\mathrm{B}}(L(x)) = \limsup_{\delta \to 0} \frac{\log \#\mathcal{N}(x,\delta)}{-\log \delta}.$$

PROOF. Fix an arbitrary $x \in \widetilde{\Lambda}$ and denote by $N_\delta(L(x))$ the minimal number of intervals of length δ needed to cover $L(x)$. If $\mathbf{i} \in \mathcal{N}(x,\delta)$ then $S_{\mathbf{i}}(\Lambda) \cap \mathrm{proj}^{-1}(x)$ is nonempty and has diameter at most $\delta|\Lambda|$. Extending each of these sets to an interval yields

$$N_{\delta|\Lambda|}(L(x)) \leqslant \#\mathcal{N}(x,\delta).$$

On the other hand, let $\{I_j\}_{j=1}^{N_\delta(L(x))}$ be the collection of intervals that cover $L(x)$ optimally. We will show that each I_j can intersect at most finitely many sets from $\{S_{\mathbf{i}}(\Lambda) : \mathbf{i} \in \mathcal{N}(x,\delta)\}$, with a uniform upper bound.

Let $\mathcal{O}$ be the open set such that $S_i(\mathcal{O}) \subset \mathcal{O}$ and $S_i(\mathcal{O}) \cap S_j(\mathcal{O}) = \varnothing$. Without loss of generality, we may assume that $\mathrm{proj}(\mathcal{O}) \subseteq (0,1)$. Indeed, one can take the intersection of $\mathcal{O}$ with $\mathrm{proj}^{-1}((0,1))$, which clearly remains invariant and its images remain disjoint with different maps. Consider I_j as a subset of $L(x)$ and take the $\delta|\mathcal{O}|$-neighborhood of I_j in $\mathbb{R}^2$, which we denote $B(I_j, \delta|\mathcal{O}|)$. Clearly,

$$\mathcal{L}^2\big(B(I_j, \delta|\mathcal{O}|\big) \leqslant 2\delta^2|\mathcal{O}| \cdot (2|\mathcal{O}| + 1).$$

On the other hand, for every $\mathbf{i} \in \mathcal{N}(x,\delta)$, we have $\mathcal{L}^2(S_{\mathbf{i}}(\mathcal{O})) \geqslant \lambda_{\min}^2 \delta^2 \mathcal{L}^2(\mathcal{O})$. If $S_{\mathbf{i}}(\Lambda) \cap I_j \neq \varnothing$, then $S_{\mathbf{i}}(\mathcal{O}) \subset B(I_j, \delta|\mathcal{O}|)$ but since $\{S_{\mathbf{i}}(\mathcal{O})\}_{\mathbf{i}\in\mathcal{N}(x,\delta)}$ are disjoint we get that I_j can intersect at most $2|\mathcal{O}| \cdot \big(2|\mathcal{O}| + 1\big)/\big(\lambda_{\min}^2 \mathcal{L}^2(\mathcal{O})\big)$-many sets from $\{S_{\mathbf{i}}(\Lambda)\}_{\mathbf{i}\in\mathcal{N}(x,\delta)}$. Thus,

$$N_\delta(L(x)) \geqslant \frac{\lambda_{\min}^2 \mathcal{L}^2(\mathcal{O})}{2|\mathcal{O}| \cdot \big(2|\mathcal{O}| + 1\big)} \cdot \#\mathcal{N}(x,\delta),$$

and the claim of the lemma follows. □

Proof of Proposition 12.4.1. It is easy to see that $\operatorname{proj}_* \nu(B(x,\delta)) \geqslant \mathcal{N}(x,\delta) \cdot \delta^s \lambda_{\min}^s$, and therefore, (12.31) follows from Lemma 12.4.2.

We turn to the proof of (12.32) and recall that we assumed in addition that the attractor of the projected IFS $(f_i)_{i=1}^m$, given by (12.30), is $\widetilde{\Lambda} = [0,1]$. Let i_0 be the unique symbol such that $0 \in f_{i_0}([0,1])$ and let i_1 be the unique symbol such that $1 \in f_{i_1}([0,1])$, and let

$$\kappa := \min\left\{\min_{k\neq i_0} \operatorname{dist}\big(0, f_k([0,1])\big), \min_{k\neq i_1} \operatorname{dist}\big(1, f_k([0,1])\big)\right\} > 0.$$

In order to get a good upper bound for $\operatorname{proj}_* \nu(B(x,\delta))$ we need to control the number of cylinder sets $\widetilde{\Lambda}_{\mathbf{i}}$, for $\mathbf{i} \in \widetilde{\mathcal{M}}_\delta$, which do not contain x, but nevertheless intersect $B(x,\delta)$. With this in mind, define

$$\mathcal{B}(x,\delta) = \left\{\mathbf{i} \in \widetilde{\mathcal{M}}_\delta : x \notin f_{\mathbf{i}}([0,1]),\ \text{but}\ \ f_{\mathbf{i}}([0,1]) \cap B(x, \delta\kappa \cdot \lambda_{\min}) \neq \varnothing\right\}.$$

Since $f_{\mathbf{i}}([0,1])$ is connected, if $\mathbf{i} \in \mathcal{B}(x,\delta)$ then either $f_{\mathbf{i}}([0,1]) \subset [0,x)$ or $f_{\mathbf{i}}([0,1]) \subset (x,1]$.

A general remark will be helpful below: if $\mathbf{i} \in \widetilde{\mathcal{M}}_\delta$, then either $\mathbf{i} \in \widetilde{\mathcal{M}}_{\delta\lambda_{\max}}$, in which case none of the $\mathbf{i}j$ is in $\widetilde{\mathcal{M}}_{\delta\lambda_{\max}}$, or else $\mathbf{i}j \in \widetilde{\mathcal{M}}_{\delta\lambda_{\max}}$, for every $j \in \{1,\ldots,m\}$, but never longer words.

Claim. *For every $\mathbf{i} \in \mathcal{B}(x,\delta)$ there exists at most one symbol $j \in \{1,\ldots,m\}$ such that $\mathbf{i}j \in \mathcal{B}(x,\delta\lambda_{\max})$ and then necessarily $j \in \{i_0, i_1\}$.*

Indeed, if $\mathbf{i} \in \widetilde{\mathcal{M}}_{\delta\lambda_{\max}}$, the claim trivially holds, since then $\mathbf{i}j \notin \widetilde{\mathcal{M}}_{\delta\lambda_{\max}}$. Suppose otherwise, with $f_{\mathbf{i}}([0,1]) \subset (x,1]$; the case when $f_{\mathbf{i}}([0,1]) \subset [0,x)$ is similar. Let $f_{\mathbf{i}}(y)$ be the left endpoint of $f_{\mathbf{i}}([0,1])$, where $y \in \{0,1\}$. Then for every $k \neq i_y$,

$$\operatorname{dist}(x, f_{\mathbf{i}k}([0,1])) \geqslant \operatorname{dist}(f_{\mathbf{i}}(y), f_{\mathbf{i}k}([0,1])) = |\lambda_{\mathbf{i}}| \cdot \operatorname{dist}(y, f_k([0,1])) \geqslant \delta\kappa \cdot \lambda_{\min}.$$

Thus $f_{\mathbf{i}k}([0,1]) \cap B(x, \delta\kappa \cdot \lambda_{\min}) = \varnothing$, and the Claim follows.

Thus, for every $n \geqslant 1$,

$$\#\mathcal{B}(x, \lambda_{\max}^{n+1}) \leqslant \#\mathcal{B}(x, \lambda_{\max}^n) + m \cdot \#\mathcal{N}(x, \lambda_{\max}^n).$$

(The factor m can be easily improved to $m-1$, but this is irrelevant for what follows.) By induction,

$$\#\mathcal{B}(x, \lambda_{\max}^n) \leqslant m \cdot \sum_{k=1}^{n} \#\mathcal{N}(x, \lambda_{\max}^k) + \#\mathcal{B}(x,1) \leqslant 2mn \cdot \#\mathcal{N}(x, \lambda_{\max}^n),$$

using a crude bound for convenience. It is easy to see that

$$\begin{aligned}\operatorname{proj}_* \nu\big(B(x, \kappa\lambda_{\min}\lambda_{\max}^n)\big) &\leqslant \big(\#\mathcal{N}(x, \lambda_{\max}^n) + \#\mathcal{B}(x, \lambda_{\max}^n)\big) \cdot \lambda_{\max}^{sn}\\ &\leqslant (2mn+1) \cdot \#\mathcal{N}(x, \lambda_{\max}^n) \cdot \lambda_{\max}^{sn}.\end{aligned}$$

Now Lemma 12.4.2 implies the desired (12.32). □

Proposition 12.4.1 has several interesting corollaries. First, note that Proposition 12.4.1 can be formulated purely symbolically with essentially the same proof.

PROPOSITION 12.4.3. *Let $\mathcal{S} = (S_i(x) = \lambda_i x + t_i)_{i=1}^m$ be an IFS on the real line such that the attractor Λ is contained in $[0,1]$. Furthermore, let s be the similarity dimension, let μ be the natural measure defined in (12.29), let $\Pi\colon \Sigma \to \Lambda$ be the natural projection, and let $\nu = \Pi_*\mu$ be the natural measure on Λ. Then for every $x \in \Lambda$,*

$$\underline{d}_\nu(x) + \overline{\dim}_{\mathrm{B}}\big(\{\mathbf{i} \in \Sigma : \Pi(\mathbf{i}) = x\}\big) \leqslant s \text{ and } \overline{d}_\nu(x) + \underline{\dim}_{\mathrm{B}}\big(\{\mathbf{i} \in \Sigma : \Pi(\mathbf{i}) = x\}\big) \leqslant s,$$

where we equip Σ with the metric $d(\mathbf{i},\mathbf{j}) = |\Lambda_{\mathbf{i}\wedge\mathbf{j}}|$.

Furthermore, if $\Lambda = [0,1]$ and $\#\Pi^{-1}(0) = \#\Pi^{-1}(1) = 1$, then for every $x \in \Lambda$

$$\underline{d}_\nu(x) + \overline{\dim}_{\mathrm{B}}\big(\{\mathbf{i} \in \Sigma : \Pi(\mathbf{i}) = x\}\big) = s \text{ and } \overline{d}_\nu(x) + \underline{\dim}_{\mathrm{B}}\big(\{\mathbf{i} \in \Sigma : \Pi(\mathbf{i}) = x\}\big) = s.$$

The proof is essentially identical to the proof of Proposition 12.4.1, and we leave it to the reader.

COROLLARY 12.4.4. *Let $(f_i(x) = \lambda_i x + t_i)_{i=1}^m$ be a self-similar IFS on the real line with the natural projection $\Pi\colon \Sigma \to \mathbb{R}$, such that $\sum_{i=1}^m |\lambda_i| \leqslant 1$. Then*

$$\dim_{\mathrm{H}}(\Pi_*\mu) = s \iff \lim_{\delta\to 0} \frac{\log \#\{\mathbf{j} \in \widetilde{\mathcal{M}}_\delta : x \in \Pi([\mathbf{j}])\}}{-\log\delta} = 0$$

for $\nu = \Pi_*\mu$-a.e. x, where s is the similarity dimension and μ is the Bernoulli measure with probabilities $\{|\lambda_i|^s\}_{i=1}^m$, so that ν is the natural self-similar measure.

PLAN OF THE PROOF. The direction "$\Longleftarrow$" follows by Theorem 12.3.1 (or Theorem 11.3.1) and Lemma 12.4.2, whereas the direction "$\Longrightarrow$" is a simple corollary of Proposition 12.4.3. Both directions use the exact dimensionality of ν, see Theorem 11.3.1. The details are left to the reader. □

COROLLARY 12.4.5. *Let $\mathcal{S} = (S_i(\mathbf{x}) = \lambda_i\mathbf{x} + \mathbf{t}_i)_{i=1}^m$ be an IFS on the plane such that the OSC holds. Then for μ-almost every $\mathbf{i}$,*

$$\dim_{\mathrm{B}}\big(L(\operatorname{proj}(\Pi(\mathbf{i})))\big) = \dim_{\mathrm{H}}\big(L(\operatorname{proj}(\Pi(\mathbf{i})))\big) = s - \dim_{\mathrm{H}}\big(\operatorname{proj}_* \Pi_*\mu\big),$$

where μ is the natural measure.

PROOF. The upper bound $\dim_{\mathrm{B}}\big(L(\operatorname{proj}(\Pi(\mathbf{i})))\big) \leqslant s - \dim_{\mathrm{H}}\big(\operatorname{proj}_* \Pi_*\mu\big)$ follows by (12.31) and the exact dimensionality of $\operatorname{proj}_* \Pi_*\mu$. The lower bound $\dim_{\mathrm{H}}\big(L(\operatorname{proj}(\Pi(\mathbf{i})))\big) \geqslant s - \dim_{\mathrm{H}}\big(\operatorname{proj}_* \Pi_*\mu\big)$ follows from Theorem 12.3.1. □

Let us note that surprisingly the second claim of Proposition 12.4.1 implies the equality (7.36) in Remark 7.2.7, which was first observed by Simon and Solomyak [**SS98**].

COROLLARY 12.4.6. *Let $\mathcal{S} = (\lambda x + t_i)_{i=1}^m$ be a homogeneous IFS with contraction ratio $\lambda \in (0,1)$ such that $\widetilde{\Lambda} := \Lambda - \Lambda$ is an interval and $t_1 < t_j < t_m$ for every $j = 2, \ldots, m-1$. Furthermore, let μ be the uniform measure on Σ and $\nu = \Pi_* \mu$, where $\Pi \colon \Sigma \to \Lambda$ is the natural projection of the IFS $\mathcal{S}$. Then*

$$D_2(\nu) = 2s - \overline{\dim}_{\mathrm{B}}\big(\{(\mathbf{i},\mathbf{j}) \in \Sigma \times \Sigma : \Pi(\mathbf{i}) = \Pi(\mathbf{j})\}\big),$$

where $s = \frac{\log m}{-\log \lambda}$ is the similarity dimension, and Σ is equipped with the metric $d(\mathbf{i},\mathbf{j}) = |\Lambda_{\mathbf{i}\wedge\mathbf{j}}| = \lambda^{|\mathbf{i}\wedge\mathbf{j}|} \cdot |\Lambda|$.

PROOF. Let $\widetilde{\mathcal{S}} = (\lambda x + t_i - t_j)_{i,j=1}^m$. Clearly, the attractor of $\widetilde{\mathcal{S}}$ is $\widetilde{\Lambda} = \Lambda - \Lambda$. Let $\widetilde{\Sigma} = \Sigma \times \Sigma$ and $\widetilde{\Pi} \colon \widetilde{\Sigma} \to \Lambda - \Lambda$ be the natural projection of the IFS $\mathcal{S}$. Furthermore, let $\widetilde{\mu} = \mu \times \mu$ be the uniform measure on $\widetilde{\Sigma}$, and $\widetilde{\nu} = \nu * \nu = \widetilde{\Pi}_*(\mu \times \mu)$.

It is easy to see from Definition 1.9.7 and Theorem 3.6.1, that

$$D_2(\nu) = \lim_{r \to 0} \frac{\log \int \nu(B(x,r))\, d\nu(x)}{\log r} = \lim_{r \to 0} \frac{\log \iint \mathbb{1}_{\{(x,y):|x-y|<r\}}\, d\nu(y)\, d\nu(x)}{\log r} = d_{\widetilde{\nu}}(0).$$

Furthermore, clearly $\widetilde{\Pi}^{-1}(0) = \{(\mathbf{i},\mathbf{j}) \in \Sigma \times \Sigma : \Pi(\mathbf{i}) = \Pi(\mathbf{j})\}$.

Thus, in order to apply Proposition 12.4.3, it is enough to check that the endpoints of $\widetilde{\Lambda}$ are uniquely coded. But

$$\widetilde{\Pi}(\mathbf{i},\mathbf{j}) = \Pi(\mathbf{i}) - \Pi(\mathbf{j}) = \sum_{k=1}^{\infty} (t_{i_k} - t_{j_k})\lambda^k \geqslant \sum_{k=1}^{\infty} (t_1 - t_m)\lambda^k = \widetilde{\Pi}((11\ldots),(mm\ldots)),$$

and the inequality above is strict if there exists $(i_k, j_k) \neq (1,m)$, thus, the endpoint $\widetilde{\Pi}((11\ldots),(mm\ldots))$ is indeed uniquely coded. $\square$

CHAPTER 13

Local entropy averages and projections of self-affine sets and measures

This chapter is devoted to the study of orthogonal projections of self-affine and self-similar sets and measures. Marstrand's Theorem 2.3.1 and Theorem 2.3.3 of Hu and Taylor guarantee that the dimension of the orthogonal projection of a planar self-similar or self-affine set (respectively measure) does not drop for a typical direction, that is, it equals to the minimum of the dimension of the set (respectively measure) and 1. In Section 7.7.1, we have seen an example of a planar self-similar set for which we have much better estimates on the exceptional set, i.e. on the dimension of the set of directions for which the dimension of the projection is strictly smaller than the natural upper bound. A natural question is: can we improve this bound for general self-similar and self-affine sets?

Again, we restrict ourselves to the plane. In the case of self-similar measures generated by the IFS $\mathcal{S} = (S_i(x) = \lambda_i O_i x + t_i)_{i=1}^m$, if the orthogonal matrices $(O_1, \ldots, O_m)$ do not contain an irrational rotation, then the orthogonal projection of a self-similar set corresponds to a graph-directed self-similar set, see Section 1.7. Thanks to the work of Hochman [**Hoc14b**], see Theorem 6.8.5, one can show that for such systems the set of exceptional directions has packing dimension at most 0.

However, if the orthonormal matrices $(O_1, \ldots, O_m)$ do contain an irrational rotation, then there are no exceptional directions at all. This was shown by Hochman and Shmerkin [**HS12**], where they introduce an important tool, namely, the method of *local entropy averages*. Later, Falconer and Kempton [**FK17**] and Bárány, Hochman and Rapaport [**BHR19**] extended it, so that it applies to self-affine IFS under the assumption of SOSC and strong irreducibility.

In this chapter, we first introduce the method of local entropy averages in the symbolic setting, which contains the "heart" of the method. After that, we turn to a specific realization and prove a simplified version of the local entropy averages theorem, following Falconer and Kempton [**FK17**]. Then we sketch the proof of the theorem asserting that for self-affine sets with SOSC and strong irreducibility, there are no exceptional directions of projections. An important step in this proof is an application of Proposition 12.3.8 saying that the orthogonal projection of the self-affine measure under these conditions is exact-dimensional almost everywhere with respect to the Furstenberg measure.

13.1. Local entropy averages: Symbolic setting

In this section we introduce the method of Hochman and Shmerkin [**HS12**] in the symbolic setting, to avoid technical difficulties and focus on the main ideas. In the next section, it is applied in a more specific situation. We are grateful to

Pablo Shmerkin for his kind permission to use some exerpts of his Lecture Notes (unpublished).

13.1.1. Local dimension from local entropy averages. Let I be a finite alphabet. Here we consider the symbolic space $I^{\mathbb{N}}$, equipped with the metric

$$\varrho(\omega,\tau) = 2^{-|\omega\wedge\tau|} \tag{13.1}$$

and the Borel σ-algebra, which is generated by cylinder sets. Denote by $I^* = \bigcup_{n=0}^{\infty} I^n$ the set of finite words in the alphabet I, where $I^0 = \{\varnothing\}$ and $\varnothing$ is the empty word. Further, let $\mathcal{P}_k$ be the partition of $I^{\mathbb{N}}$ into cylinders of length k, so that

$$\omega \in [\omega|_k] \in \mathcal{P}_k \quad \text{for } \omega \in I^{\mathbb{N}}.$$

By convention, $\mu_{[\varnothing]} = \mu$. Given a probability measure μ on $I^{\mathbb{N}}$ and a finite word $v \in I^*$, denote by $\mu_{[v]}$ the usual conditional measure:

$$\mu_{[v]} = \frac{\mu|_{[v]}}{\mu([v])}.$$

Recall also (see Definition 1.9.15) that, given a probability measure μ on X and a (finite or countable) measurable partition $\mathcal{P}$ of X, the entropy of μ with respect to $\mathcal{P}$ is given by

$$H(\mu,\mathcal{P}) = -\sum_{D\in\mathcal{P}} \mu(D)\log\mu(D).$$

Now we can state a very useful "local entropy averages" lemma.

LEMMA 13.1.1. *Let μ be a probability measure on $I^{\mathbb{N}}$. Then for μ-a.e. $\omega \in I^{\mathbb{N}}$,*

$$\lim_{n\to\infty}\left(\frac{-\log\mu([\omega|_n])}{n} - \frac{1}{n}\sum_{k=0}^{n-1} H(\mu_{[\omega|_k]},\mathcal{P}_{k+1})\right) = 0.$$

Before the proof we point out the following corollary; recall that $\underline{d}_\mu(\omega)$ and $\overline{d}_\mu(\omega)$ are the lower and upper local dimensions of μ at ω respectively, see Section 1.9.

COROLLARY 13.1.2. *Let μ be a probability measure on $I^{\mathbb{N}}$. Then for μ-a.e. $\omega \in I^{\mathbb{N}}$,*

$$\underline{d}_\mu(\omega) = \liminf_{n\to\infty}\frac{1}{n\log 2}\sum_{k=0}^{n-1} H(\mu_{[\omega|_k]},\mathcal{P}_{k+1}),$$

and similarly, for $\overline{d}_\mu(\omega)$ with the lim sup. *In particular,* $\underline{\dim}_{\mathrm{H}}(\mu)$ *is the μ-essential infimum of the right-hand side above, see* (1.102).

The corollary is immediate by the definition of the local dimension (recall that we use the metric (13.1) on $I^{\mathbb{N}}$).

PROOF OF LEMMA 13.1.1. Define the kth conditional information function by

$$I_k(\omega) = -\log\mu_{[\omega|_{k-1}]}([\omega|_k]) = -\log\frac{\mu([\omega|_k])}{\mu([\omega|_{k-1}])},$$

so that

$$-\log\mu([\omega|_n]) = \sum_{k=1}^{n} I_k(\omega). \tag{13.2}$$

The partitions $\mathcal{P}_k$ form an increasing sequence of σ-algebras converging to the Borel σ-algebra on $I^{\mathbb{N}}$ (a filtration). We claim that

$$f_k(\omega) := I_k(\omega) - H(\mu_{[\omega|_{k-1}]}, \mathcal{P}_k)$$

is a sequence of martingale differences with respect to the filtration $\{\mathcal{P}_k\}$. It is clear that f_k is measurable with respect to $\mathcal{P}_k$, so it remains to check that

$$\mathbb{E}\,(f_{k+1}|\mathcal{P}_k)(\omega) \equiv 0, \tag{13.3}$$

see Section 9.3. Recall that, given a probability space $(X, \mathcal{B}, \nu)$ and a finite measurable partition $\mathcal{P}$, the conditional expectation of $f \in L^1(X, \nu)$ is given by

$$\mathbb{E}\,(f|\mathcal{P})(x) = \frac{1}{\nu([D(x)])} \int_{[D(x)]} f\, d\nu, \quad \text{where } x \in D(x) \in \mathcal{P}.$$

Thus,

$$\begin{aligned}
\mathbb{E}\,(I_{k+1}|\mathcal{P}_k)(\omega) &= \mathbb{E}\,\big(-\log \mu_{[\omega|_k]}([\omega|_{k+1}]) \,|\, \mathcal{P}_k\big)\,(\omega) \\
&= \frac{-1}{\mu([\omega|_k])} \int_{[\omega|_k]} \log \mu_{[\omega|_k]}([\omega|_{k+1}])\, d\mu \\
&= \frac{-1}{\mu([\omega|_k])} \sum_{i\in I} \mu([\omega|_k i]) \cdot \log \mu_{[\omega|_k]}([\omega|_k i]) \\
&= -\sum_{i\in I} \mu_{[\omega|_k]}([\omega|_k i]) \cdot \log \mu_{[\omega|_k]}([\omega|_k i]) \\
&= H(\mu_{[\omega|_k]}, \mathcal{P}_{k+1}).
\end{aligned}$$

This confirms (13.3) and the claim that $\{f_k\}$ is a sequence of martingale differences. Next, we show that $\{f_k\}$ is uniformly bounded in $L^2(I^{\mathbb{N}}, \mu)$. Since $H(\mu_{[\omega|_k]}, \mathcal{P}_{k+1}) = \mathbb{E}\,(I_{k+1}|\mathcal{P}_k)$ and conditional expectation does not increase L^2-norm, it is enough to check that $\{I_k\}$ are uniformly bounded. We have

$$\begin{aligned}
\mathbb{E}\,(I_k^2) &= \sum_{v\in I^k} \mu([v])\big(\log \mu_{[v|_{k-1}]}([v])\big)^2 \\
&= \sum_{w\in I^{k-1}} \sum_{i\in I} \big(\log \mu_{[w]}([wi])\big)^2 \cdot \mu([wi]) \\
&= \sum_{w\in I^{k-1}} \mu([w]) \sum_{i\in I} \mu_{[w]}([wi]) \cdot \big(\log \mu_{[w]}([wi])\big)^2 \\
&\leqslant \sum_{w\in I^{k-1}} \mu([w]) \cdot |I| \cdot 4e^{-2} = |I| \cdot 4e^{-2}.
\end{aligned}$$

In the last inequality we used that the map $x \mapsto x(\log x)^2$ attains its unique maximum on $[0,1]$ at e^{-2}. We have verified that $\{f_k\}$ is a sequence of martingale differences uniformly bounded in $L^2(I^{\mathbb{N}}, \mu)$, and so by Theorem 9.3.4,

$$\lim_{n\to\infty} \frac{1}{n} \sum_{k=1}^{n} f_k = 0 \quad \mu\text{-almost surely.}$$

This concludes the proof of the lemma, in view of (13.2). □

13.1.2. Tree morphisms and mass decay. Now suppose that we have two finite alphabets I and J and the corresponding symbolic spaces $I^{\mathbb{N}}$ and $J^{\mathbb{N}}$ (both are equipped with the same metric ϱ from (13.1). We have already used the representation of I^* as a tree, see Section 1.3.1. It is an $|I|$-regular tree, with the root vertex $\varnothing$ and each vertex $v \in I^n$ having the "children" vi, $i \in I$.

A *tree morphism* is a map $f : I^* \to J^*$ that preserves the tree structure. This means that for any $n \geqslant 1$ holds $f(I^n) \subset J^n$, and

$$\text{for any } \ v \in I^* \ \text{ and } \ i \in I, \ \text{ there exists } \ j = f^v(i), \ \text{ such that } f(vi) = f(v)j.$$

The function $f^v : I \to J$ is called the (natural) *induced map*, corresponding to v. Furthermore, a tree morphism f induces a map between the symbolic spaces, which we denote (in a slight abuse of notation) by the same letter: $f : I^{\mathbb{N}} \to J^{\mathbb{N}}$, defined by

$$f([v]) = [f(v)], \quad \text{for } \ v \in I^{\mathbb{N}}.$$

Let μ be a Borel probability measure on $I^{\mathbb{N}}$. For each $v \in I^*$ there is a conditional measure $\mu_{[v]}$ on the cylinder $[v]$, but here we also use a closely related *discrete measure* μ^v on the alphabet I defined by

$$\mu^v(\{i\}) = \frac{\mu([vi])}{\mu([v])} \quad \text{for all } i \in I.$$

This is a discrete measure, and we can consider its Shannon entropy: $H(\mu^v) = -\sum_{i \in I} \mu^v(\{i\}) \log \mu^v(\{i\})$. Note that, by construction,

$$H(\mu^v) = H(\mu_{[v]}, \mathcal{P}_{n+1}) \quad \text{for } v \in I^n.$$

We can also consider the push-forward measure $(f^v)_*\mu^v$ and its Shannon entropy

$$H((f^v)_*\mu^v) = H(f_*\mu_{[v]}, \mathcal{P}_{n+1}).$$

The following is a key result; it is [**HS12**, Theorem 4.4] (with a slightly different notation).

THEOREM 13.1.3 (Hochman and Shmerkin). *Let $f : I^* \to J^*$ be a tree morphism and μ a probability measure on $I^{\mathbb{N}}$. If for μ-a.e. $\omega \in I^N$,*

$$\liminf_{n\to\infty} \frac{1}{n} \sum_{i=0}^{n-1} H((f^{\omega|_i})_*\mu^{\omega|_i}) = \liminf_{n\to\infty} \frac{1}{n} \sum_{i=0}^{n-1} H((f_*\mu_{[\omega|_i]}, \mathcal{P}_{i+1}) \geqslant \alpha, \tag{13.4}$$

then

$$\underline{\dim}_{\mathrm{H}}(f_*\mu) \geqslant \alpha.$$

The significance of this theorem is that gives a global lower bound for the dimension of the image of the measure in terms of local quantities.

PROOF. Our goal is to construct a random probability measure ν on $I^{\mathbb{N}}$, such that

(a) $\mathbb{E}(\nu(A)) = \mu(A)$ for any Borel $A \subset I^{\mathbb{N}}$; and
(b) f is injective on $\mathrm{spt}(\nu)$.

First we describe the construction informally. A Borel probability measure on $I^{\mathbb{N}}$ is uniquely determined by the conditional measures on the edges of the tree. Each vertex $w \in I^*$ (more precisely, the cylinder set $[w]$) receives its ν-measure according to a random procedure, inductively on $|w|$, starting with $|w| = 1$. Let $v \in I^*$ be a vertex, such that $\mu([v]) > 0$. We need to decide how to split the mass $\nu([v])$ between the vertices $vi, i \in I$, or equivalently, how to choose the conditional measures $\nu^v(i)$. If $f(vi) = f(v)j$ and such an i is unique, that is, $f(vi') \neq f(v)j$ for $i' \neq i$, then we simply put $\nu^v(i) = \mu^v(i)$. If, however, several edges $vi_1, \ldots vi_\ell$ get mapped by f into the same edge vj, then we choose one of the i_k's randomly, with the probability proportional to μ^{i_k}, and assign to it the total μ^v-mass of the fiber. All these choices are made independently of each other. It is clear from this construction that f will be injective on $\mathrm{spt}(\nu)$. The first property, $\mathbb{E}(\nu) = \mu$, will be verified once we describe the procedure more formally.

More precisely, for each $v \in I^*$ and $j \in J$, let

$$F_{v,j} = (f^v)^{-1}(j)$$

be the corresponding fiber. For each $v \in I^*$ with $\mu([v]) > 0$ and each $j \in J$ such that $F_{v,j} \neq \varnothing$, consider the probability distribution $\mathbb{P}_{v,j}$ on the fiber $F_{v,j}$ given by

$$\mathbb{P}_{v,j}(i) = \frac{\mu^v(i)}{\mu^v(F_{v,j})}.$$

The random measure ν is defined inductively on cylinders of length n. Starting with the root $\varnothing$, select one element i_j from each nonempty fiber $F_{\varnothing,j}$ randomly from the distribution $\mathbb{P}_{\varnothing,j}$ and set $\nu^{\varnothing}(i_j) = \mu^{\varnothing}(F_{\varnothing,j})$ (the other elements of the fiber receive zero ν-mass). For each of the vertices v that received positive mass, we continue inductively in the same way, choosing elements from the fibers $F_{v,j}$ randomly according to $\mathbb{P}_{v,j}$, and passing all the mass of the fiber to the chosen element, all the choices being made independently of each other.

This process constructs a random probability measure on $I^{\mathbb{N}}$. Moreover, by construction, if $\nu([v]) > 0$, then there exists a set $J'_v \subset I$ of the same cardinality as J, such that $\nu^v = (f^v)_*\mu^v$, after identifying elements of J'_v with the corresponding elements of J.

CLAIM 13.1.4. *We have $\mathbb{E}(\nu(A)) = \mu(A)$ for any $A \subset I^{\mathbb{N}}$ Borel.*

It is enough to check the claim for cylinders $[v]$, $v \in I^*$. If $\mu([v]) = 0$, then $\nu([v]) = 0$ by construction. Otherwise,

$$\begin{aligned}
\mathbb{E}(\nu[v]) &= \mathbb{P}(\nu[v] > 0) \cdot \mathbb{E}(\nu[v] \mid \nu[v] > 0) \\
&= \prod_{i=1}^{|v|-1} \frac{\mu^{v|_i}(v_{i+1})}{\mu^{v|_i}(F_{v|_i, f^{v|_i}(v_{i+1})})} \cdot \prod_{i=1}^{|v|-1} \mu^{v|_i}(F_{v|_i, f^{v|_i}(v_{i+1})}) \\
&= \prod_{i=1}^{|v|-1} \mu^{v|_i}(v_{i+1}) \\
&= \mu([v]),
\end{aligned}$$

as desired.

Now we conclude the proof of the theorem. We argue by contradiction. By the definition of the lower Hausdorff dimension of a measure (1.101), if $\underline{\dim}_{\mathrm{H}}(f_*\mu) < \alpha$, then there exists a Borel set $E \subset I^{\mathbb{N}}$, such that $\mu(E) > 0$ and $\dim_{\mathrm{H}}(f(E)) < \alpha$. Let G be the set of all ω for which (13.4) holds. By assumption, $\mu(G) = 1$, hence $\nu(G) = 1$ almost surely. On the other hand, $\mathbb{E}(\nu(E)) > 0$, since $\mu(E) > 0$, hence there exists a realization of ν such that $\nu(G) = 1$ and $\nu(E) > 0$; fix such a realization.

Lemma 13.1.1, together with the facts that $\nu(G) = 1$ and $H(\nu^v) = H((f^v)_*\mu^v)$ for all v, implies $\underline{\dim}_{\mathrm{H}}(\nu) \geqslant \alpha$. But then all sets of positive ν-measure have Hausdorff dimension at least α, by (1.102). In particular, this is true for $E \cap \mathrm{spt}(\nu)$.

But f is injective on $\mathrm{spt}(\nu)$ and hence an isometry, so it preserves the Hausdorff dimension, and we obtain that $\dim_{\mathrm{H}}(f(E)) \geqslant \dim_{\mathrm{H}}(E \cap \mathrm{spt}(\nu)) \geqslant \alpha$. This is a contradiction with our assumption, and the theorem is proved. □

REMARK 13.1.5. Another proof of the theorem (using convexity and not requiring the random measure) was found by Hochman, see [**Hoc14a**, Theorem 8.8], where it is stated in a slightly different language, for maps between Euclidean spaces.

13.1.3. Lifting tree maps to Euclidean spaces. The original map f may be from a symbolic space to the Euclidean space, or between Euclidean spaces. It is possible to construct a tree morphism which represents f "faithfully", in the sense that the entropy and dimension are preserved up to a small error. However, both the statements and the proofs are rather technical. The first construction of "faithful maps" was given in [**HS12**, Section 5]. One of the complications in this procedure is due to the fact that, in order to make the error arbitrarily small, the size of the alphabet for the symbolic space should tend to infinity.

13.2. Local entropy averages: Concrete realization

Here we focus on a concrete realization of the method of local entropy averages, following Falconer and Kempton [**FK17**], with an emphasis on the "lifting" part of the argument, mentioned at the end of the last section.

First we introduce some notation. We write $\Sigma = \{1, \ldots, m\}^{\mathbb{N}}$ as usual, and let $\Sigma^* = \bigcup_{n=0}^{\infty} \{1, \ldots, m\}^n$. As before, we use the metric $d(\mathbf{i}, \mathbf{j}) = 2^{-\min\{k \geqslant 0:\ i_k \neq j_k\}}$. In this metric, the balls are the cylinder sets, i.e., they have the form

$$[i_1 \ldots i_k] = \{\mathbf{j} \in \Sigma : i_\ell = j_\ell \text{ for } \ell = 1, \ldots, k\}.$$

Let $f \colon \Sigma \to [-1, 1]$ be a continuous function. Suppose that there exists a "double" sequence of subsets $\mathcal{Q}_n^N \subset \Sigma^*$, for $n, N \geqslant 1$, satisfying the following properties:

(Q.1) for every $n, N \geqslant 1$, $\{[\mathbf{i}]\}_{\mathbf{i} \in \mathcal{Q}_n^N}$ is a partition of Σ, i.e., $\mathcal{Q}_n^N$ is a minimal cut-set, see Section 1.3.1;

(Q.2) there exists a constant $C > 1$ such that for every $n, N \geqslant 1$,

$$C^{-1} 2^{-Nn} \leqslant |f([\mathbf{i}])| \leqslant 2^{-Nn-1} \quad \text{for all } \mathbf{i} \in \mathcal{Q}_n^N;$$

(Q.3) for every N, and $n \geqslant m$, $\mathcal{Q}_n^N$ is a refinement of $\mathcal{Q}_m^N$;

(Q.4) for sufficiently large N, we have $\mathcal{Q}_n^N \cap \mathcal{Q}_m^N = \varnothing$ for $n \neq m$ as subsets of Σ^*.

In the next section, it is shown how to build the sequence $\{\mathcal{Q}_n^N\}_{n,N \geqslant 1}$ in a concrete situation. In the simplest case, when f is the natural projection map for a homogeneous self-similar IFS of contraction ratio $\lambda \in (0, 1)$, we can simply take $\mathcal{Q}_n^N = \mathcal{P}_{Nn\lceil \log_2(1/\lambda) \rceil}$, where $\mathcal{P}_k$ is the partition of Σ into k-level cylinders.

For $\mathbf{i} \in \Sigma$, denote by $\mathcal{Q}_n^N(\mathbf{i})$ the unique element of $\mathcal{Q}_n^N$ such that $\mathbf{i} \in [\mathcal{Q}_n^N(\mathbf{i})]$. For a measure $\mu \in \mathcal{P}(\Sigma)$ and $\mathbf{i} \in \Sigma^*$, let $\mu_{[\mathbf{i}]}$ be the conditional measure of μ on $[\mathbf{i}]$, that is, $\mu_{[\mathbf{i}]} = \frac{\mu|_{[\mathbf{i}]}}{\mu([\mathbf{i}])}$.

Recall (1.116) that $\mathscr{D}_n$ denotes the partition of the real line into dyadic intervals of length 2^{-n}, such that one of the boundary points is 0.

THEOREM 13.2.1. *Let μ be a probability measure on $\Sigma = \{1, \ldots, m\}^{\mathbb{N}}$ and let $f \colon \Sigma \to [-1, 1]$ be a continuous function. Suppose that we are given a sequence of* *minimal cut-sets* $\{\mathcal{Q}_n^N\}_{n,N\geqslant 1}$ *satisfying the properties (Q.1)–(Q.4) and there exists* $\alpha > 0$ *such that*

$$\lim_{N\to\infty} \liminf_{n\to\infty} \frac{1}{N\log 2} \cdot \frac{1}{n}\sum_{i=1}^{n} H(f_*\mu_{[\mathcal{Q}_i^N(\mathbf{i})]}, \mathscr{D}_{N(i+1)}) \geqslant \alpha \text{ for } \mu\text{-a.e. } \mathbf{i}. \tag{13.5}$$

Then $\dim_{\mathrm{H}}(f_*\mu) \geqslant \alpha$.

In order to prove the theorem, a rather lengthy preparation will be necessary. In the next subsection, we show some properties of entropy, complementing our earlier discussion in Sections 1.9, 2.6, and 3.7. After that, we build a particular family of "faithful maps", which were defined by Hochman and Shmerkin [**HS12**] in greater generality. Instead, following Falconer and Kempton [**FK17**], we construct a particular system, which has the desired properties and can be used in many situations. Finally, in Section 13.2.3, we analyze the dimension-theoretic properties of that system and conclude the proof sketch of the theorem.

13.2.1. Properties of entropy. Recall the definition of entropy dimension from (2.27) (it will play an important role in the next section). For a probability measure μ on a metric space X, the lower entropy dimension is given by

$$\underline{\dim}_e\mu = \liminf_{r\to 0} \frac{H_r(\mu)}{-\log r} = \liminf_{r\to 0} \frac{\int \log \mu(B(x,r))d\mu(x)}{\log r}.$$

Just like in Lemma 1.9.3, one can show that the lim inf can be taken along exponential sequences. Furthermore, we have seen in Proposition 2.6.11 that the entropy dimension can be calculated using the entropy with respect to the partition $\mathscr{D}_n$, i.e.,

$$\underline{\dim}_e\mu = \liminf_{n\to\infty} \frac{H(\mu,\mathscr{D}_n)}{n\log 2}, \tag{13.6}$$

and similarly for the upper entropy dimension.

Recall also Lemma 2.6.12, which in our case ($d = 1$) asserts that for every $n \geqslant 1$,

$$0 \leqslant H(\mu,\mathscr{D}_n) - H_{2^{-n}}(\mu) \leqslant 3, \tag{13.7}$$

and for every $n, k \geqslant 1$,

$$0 \leqslant H(\mu,\mathscr{D}_{n+k}) - H(\mu,\mathscr{D}_n) \leqslant 2k. \tag{13.8}$$

Let us now consider how the entropy with respect to the dyadic partition changes under continuous maps.

LEMMA 13.2.2. *There exists a uniform constant* $C > 0$ *such that the following hold:*

(i) *For every* f *and* g*, continuous functions from a metric space* X *to* $\mathbb{R}$*, for every probability measure* μ *on* X*,*

$$\sup_{x\in X} |f(x) - g(x)| \leqslant 2^{-n} \implies |H(g_*\mu,\mathscr{D}_n) - H(f_*\mu,\mathscr{D}_n)| \leqslant C.$$

(ii) *For any linear map* $S_{a,b}(x) = ax + b$ *and a probability measure* μ *on* $\mathbb{R}$*,*

$$|H\big((S_{a,b})_*\mu,\mathscr{D}_n\big) - H(\mu,\mathscr{D}_{n+\lceil\log_2 a\rceil})| \leqslant C. \tag{13.9}$$

PROOF. (i) Observe that for every $x \in \mathbb{R}$,

$$f^{-1}(B(x, 2^{-n})) \subseteq g^{-1}(B(x, 2 \cdot 2^{-n})).$$

Thus, using (13.7) and (13.8) and the fact that $r \mapsto H_r(\mu)$ is monotone decreasing, we get

$$\begin{aligned} H(g_*\mu, \mathscr{D}_n) &\leqslant H_{2^{-n}}(g_*\mu) + 3 \leqslant H_{2^{-n-1}}(f_*\mu) + 3 \\ &\leqslant H(f_*\mu, \mathscr{D}_{n+1}) + 6 \leqslant H(f_*\mu, \mathscr{D}_n) + 8. \end{aligned}$$

The other inequality follows by switching f and g.

(ii) We will use Fact 2.6.13(ii), which says that if there are two finite partitions $\mathcal{P}$ and $\mathcal{Q}$, such that every element of $\mathcal{Q}$ intersects at most K elements of $\mathcal{P}$, then $H(\mu, \mathcal{P}) - H(\mu, \mathcal{Q}) \leqslant K$. This immediately implies (13.9) for a translation $S_{1,b}(x) = x + b$, with $C = 2$. It remains to check (13.9) for a dilation $S_{a,0}(x) = ax$. Note that $S_{2^\ell,0}^{-1}(\mathscr{D}_n) = \mathscr{D}_{n+\ell}$, with $\ell \in \mathbb{Z}$, hence $H\big((S_{2^\ell,0})_*\mu, \mathscr{D}_n\big) = H(\mu, \mathscr{D}_{n+\ell})$. Then (13.9) for a dilation $S_{a,0}(x) = ax$ with $2^\ell < a < 2^{\ell+1}$ follows by another application of Fact 2.6.13(ii). The proof of the lemma is complete. □

For simplicity, below we will use the notation $\mathscr{D}_a := \mathscr{D}_{\lceil a \rceil}$ for $a > 0$.

13.2.2. Faithful maps. Before we continue with the proof of Theorem 13.2.1, let us define another collection of symbols and symbolic spaces. Let $X_N := \{-2^{N+1}, \ldots, 2^{N+1} - 2\}$ and $\Delta_N := (X_N)^{\mathbb{N}}$. Denote the partition of Δ_N into cylinder sets of length n by $\mathcal{C}_n^N$. We equip Δ_N with the metric

$$d_N(\omega, \eta) = 2^{-N \min\{k \geqslant 0:\ \omega_k \neq \eta_k\}}. \tag{13.10}$$

Thus for a finite word $\omega \in X_N^n$ of length n, the corresponding cylinder set $[\omega]$ has diameter $\|[\omega]\| = 2^{-Nn}$. For every $i \in X_N$, let S_i be the similarity map such that $S_i([-1,1])$ is the interval of length 2^{-N+1} with the left endpoint $i/2^{N+1}$, that is, $S_i(x) = 2^{-N}x + (i+2)2^{-N-1}$. It is easy to see that the attractor of the self-similar IFS $(S_i)_{i \in X_N}$ is the interval $[-1,1]$; moreover, each point is contained in at most 3 cylinders of the form $S_i([-1,1])$. Let $g_N : \Delta_N \to [-1,1]$ be the natural projection map for this IFS, that is,

$$g_N(\omega) := \lim_{n \to \infty} S_{\omega_1} \circ \cdots \circ S_{\omega_n}(0), \quad \text{for } \omega \in \Delta_N. \tag{13.11}$$

Consider a probability measure μ on Δ_N. We do not assume that μ is shift invariant, so we cannot use the measure-theoretic entropy of μ, since it is not defined. However, we will use the lower Hausdorff dimension of μ with respect to the metric (13.10), see (1.101) and (1.102), that is,

$$\underline{\dim}_{\mathrm{H}}(\mu) = \operatorname{ess}\inf_{\omega \in \Delta_N} \liminf_{n \to \infty} \frac{\log \mu([\omega_1, \ldots, \omega_n])}{-nN \log 2}.$$

By Egorov's Theorem, for every $\varepsilon > 0$ there exists a set E with $\mu(E) > 1 - \varepsilon$ and a constant $C > 0$ such that

$$\mu([\omega|_n]) \leqslant C 2^{-Nn(\underline{\dim}_{\mathrm{H}}(\mu) - \varepsilon)} \text{ for every } n \geqslant 1 \text{ and } \omega \in E. \tag{13.12}$$

LEMMA 13.2.3. *Let $g_N : \Delta_N \to [-1,1]$ be the function as in* (13.11). *Then for every $n, N \geqslant 1$ and $\omega \in X_N^n$, no point in $g_N([\omega])$ is covered by more than 3 sets $g_N([\omega j])$, $j \in X_N$. Moreover, for every probability measure $\mu \in \mathcal{P}(\Delta_N)$,*

$$\underline{\dim}_{\mathrm{H}}((g_N)_*\mu) \geqslant \underline{\dim}_{\mathrm{H}}(\mu) - \frac{\log 3}{N \log 2}.$$

PROOF. The first assertion of the lemma follows trivially by the definition of $(S_i)_{i\in X_N}$, so we turn to the proof of the inequality.

It is easy to see that for any $x \in [-1,1]$ and $n, N \geqslant 1$, there exist at most 3 symbols, $\omega_1, \omega_2, \omega_3 \in X_N$, such that $B(x, 2^{-nN}) \cap g_N([\omega_i]) \neq \varnothing$. By the first assertion of the lemma, for each ω_i, there are at most 3 symbols $\omega_{i,1}, \ldots, \omega_{i,3} \in X_N$ such that $B(x, 2^{-nN}) \cap g_N([\omega_i\, \omega_{i,j}]) \neq \varnothing$. Similarly, by induction, there are at most 3^n finite words ω in X_N^n for which $g_N([\omega]) \cap B(x, 2^{-nN}) \neq \varnothing$. Thus, by (13.12),

$$\begin{aligned} &(g_N)_*\mu|_E\big(B(x,2^{-nN})\big)\\ \leqslant\ & 3^n \max\left\{\mu([\omega]) :\ \omega \in X_N^n,\ [\omega] \cap E \cap B(x, 2^{-nN}) \neq \varnothing\right\}\\ \leqslant\ & 3^n C 2^{-Nn(\underline{\dim}_{\mathrm{H}}(\mu)-\varepsilon)}, \end{aligned}$$

where E is the set from (13.12). Therefore,

$$\underline{\dim}_{\mathrm{H}}((g_N)_*\mu|_E) \geqslant \underline{\dim}_{\mathrm{H}}(\mu) - \varepsilon - \frac{\log 3}{N \log 2}.$$

Now, suppose that contrary to the claim of the lemma, $\underline{\dim}_{\mathrm{H}}\big((g_N)_*\mu\big) < \underline{\dim}_{\mathrm{H}}(\mu) - \frac{\log 3}{N\log 2}$ and choose $\varepsilon' > 0$ such that $\underline{\dim}_{\mathrm{H}}\big((g_N)_*\mu\big) < \underline{\dim}_{\mathrm{H}}(\mu) - \frac{\log 3}{N\log 2} - \varepsilon'$. By the definition of lower Hausdorff dimension (1.102), there exists a set $F \subset \mathbb{R}$ with $(g_N)_*\mu(F) > 0$ for which

$$\underline{d}_{(g_N)_*\mu}(x) < \underline{\dim}_{\mathrm{H}}(\mu) - \frac{\log 3}{N \log 2} - \varepsilon'$$

for every $x \in F$. So $\underline{\dim}_{\mathrm{H}}\big((g_N)_*\mu|_F\big) < \underline{\dim}_{\mathrm{H}}(\mu) - \frac{\log 3}{N\log 2} - \varepsilon'$ still holds. Let $F' = g_N^{-1}(F)$. Choose an arbitrary $\varepsilon < \min\{\varepsilon', \mu(F')\}$, and let E be the set corresponding to ε in (13.12). Then $E \cap F' \neq \varnothing$ and thus,

$$\underline{d}_{(g_N)_*\mu}(x) \geqslant \underline{\dim}_{\mathrm{H}}(\mu) - \varepsilon - \frac{\log 3}{N \log 2},$$

for $x \in g_N(E \cap F')$, which is a contradiction. □

LEMMA 13.2.4. *There exists a constant $C > 0$ such that for every $N \geqslant 1$ and any probability measure μ on Δ_N,*

$$\left|H(\mu, \mathcal{C}_1^N) - H\big((g_N)_*\mu, \mathscr{D}_N\big)\right| \leqslant C. \tag{13.13}$$

PROOF. The argument is similar to the proof of Fact 2.6.13. Recall that $\mathcal{C}_1^N$ is the partition of X_N with respect to cylinder sets of the first level. For every $\omega \in \Delta_N$, $\mathscr{D}_{N+1}(g_N(\omega)) \subseteq g_N(\mathcal{C}_1^N(\omega)) = [\frac{\omega_1}{2^{N+1}}, \frac{\omega_1+4}{2^{N+1}}]$, thus

$$H(\mu, \mathcal{C}_1^N) \leqslant H((g_N)_*\mu, \mathscr{D}_{N+1}) \leqslant H((g_N)_*\mu, \mathscr{D}_N) + 2,$$

where for the last inequality we used Lemma 2.6.12(i).

For each $D \in \mathscr{D}_{N+1}$ there exist exactly 4 words $\omega_1, \ldots, \omega_4 \in X_N$ such that $S_{\omega_i}([0,1]) \cap D \neq \varnothing$. Hence,

$$\begin{aligned} H((g_N)_*\mu, \mathscr{D}_N) - H(\mu, \mathcal{C}_1^N) &\leqslant H((g_N)_*\mu, \mathscr{D}_{N+1}) - H(\mu, \mathcal{C}_1^N) \\ &= \int \log \frac{\mu(\mathcal{C}_1^N(\omega))}{\mu(g_N^{-1}\mathscr{D}_{N+1}(g_N(\omega)))}\, d\mu(\omega) \\ &\leqslant \int \frac{\mu(\mathcal{C}_1^N(\omega))}{\mu(g_N^{-1}\mathscr{D}_{N+1}(g_N(\omega)))}\, d\mu(\omega) \\ &= \sum_{D \in \mathscr{D}_{N+1}} \sum_{\substack{\omega: \\ S_\omega([-1,1]) \cap D \neq \varnothing}} \frac{\mu([\omega])}{\mu(g_N^{-1}(D))} \mu(g_N^{-1}(D)) \leqslant 4. \end{aligned}$$

□

The following result will be useful in the proof of Theorem 13.2.1. It is here that we realize the idea of lifting "faithfully" a map from the symbolic space to $\mathbb{R}$ to a (version of) tree morphism.

PROPOSITION 13.2.5. *Let $f\colon \Sigma \to [-1,1]$ be a continuous function, and suppose that we are given a sequence of minimal cut-sets $\{\mathcal{Q}_n^N\}_{n,N \geqslant 1}$ satisfying the properties (Q.1)–(Q.4). Then for every $N \geqslant 1$ there exist maps $h_N : \Sigma \to \Delta_N$ and $g_N : \Delta_N \to [-1,1]$ such that*

(i) $g_N \circ h_N = f$,
(ii) *for every $n, N \geqslant 1$ and $\mathbf{i} \in \Sigma$, there exists $\omega \in X_N^n$ such that $h_N([\mathcal{Q}_n^N(\mathbf{i})]) \subseteq [\omega]$.*

Note that h_N are analogous to tree morphisms from Section 13.1.2, but they preserve the tree structure in Σ induced by the sequence of minimal cut-sets $\{\mathcal{Q}_n^N\}$, rather than the original tree structure. The proof of Proposition 13.2.5 follows the argument of Falconer and Kempton [**FK17**, Theorem 4.3].

PROOF. We define $h_N^{(n)} : \mathcal{Q}_n^N \to X_N^n$ inductively. By the assumption (Q.2), for every $\mathbf{i} \in \mathcal{Q}_1^N$ we have $|f([\mathbf{i}])| \leqslant 2^{-N-1}$, hence there exists $j \in X_N$ such that $f([\mathbf{i}]) \subset g_N([j])$. If there exist more than one such symbol j, we fix one arbitrarily and define $h_N^{(1)} : \mathcal{Q}_1^N \to X_N$ so that $h_N^{(1)}(\mathbf{i}) := j$.

By induction, suppose that $h_N^{(n)} : \mathcal{Q}_n^N \to X_N^n$ is already defined. By the assumption (Q.3), for every $\mathbf{i} \in \mathcal{Q}_{n+1}^N$ there exists a unique $\mathbf{j} \in \mathcal{Q}_n^N$ such that $[\mathbf{i}] \subset [\mathbf{j}]$. Moreover, since $|f([\mathbf{i}])| \leqslant 2^{-(n+1)N-1}$ there exists $j \in X_N$ such that $f[\mathbf{i}] \subset g_N([h_N^{(n)}(\mathbf{j})j])$. Let us define $h_N^{(n+1)}(\mathbf{i}) = h_N^{(n)}(\mathbf{j})j$. Finally, for $\mathbf{i} \in \Sigma$ let $h_N(\mathbf{i}) = \bigcap_{n=1}^\infty [h_N^{(n)}(\mathcal{Q}_n^N(\mathbf{i}))]$. By construction, for every $n, N \geqslant 1$ and $\mathbf{i} \in \Sigma$, we have $h_N([\mathcal{Q}_n^N(\mathbf{i})]) \subset [\omega]$ for some $\omega \in X_N^n$.

On the other hand, for every $\mathbf{i} \in \Sigma$,

$$\begin{aligned} \{f(\mathbf{i})\} = \bigcap_{n=1}^\infty f([\mathcal{Q}_n^N(\mathbf{i})]) &\subseteq \bigcap_{n=1}^\infty g_N([h_N^{(n)}(\mathcal{Q}_n^N(\mathbf{i}))]) \\ &= \bigcap_{n=1}^\infty g_N([h_N(\mathbf{i})|_n]) = \{g_N(h_N(\mathbf{i}))\}. \end{aligned}$$

The proposition is proved. □

13.2.3. Dimension properties of faithful maps. This section is devoted to the proof of the following main proposition.

PROPOSITION 13.2.6. *Let* $h_N : \Sigma \to \Delta_N$ *be the function as in Proposition* 13.2.5 *and* μ *a probability measure on* Σ. *If*

$$\liminf_{n\to\infty} \frac{1}{N\log 2}\cdot\frac{1}{n}\sum_{i=1}^{n} H((h_N)_*\mu_{[\mathcal{Q}_i^N(\mathbf{i})]}, \mathcal{C}_{i+1}^N) \geqslant \beta_N \tag{13.14}$$

for μ*-almost every* $\mathbf{i}$, *then* $\dim_{\mathrm{H}}\big((h_N)_*\mu\big) \geqslant \beta_N$.

Before we prove the proposition, we show how to deduce Theorem 13.2.1.

DERIVATION OF THEOREM 13.2.1 FROM PROPOSITION 13.2.6. By Proposition 13.2.5(i) and Lemma 13.2.3, for every $N \geqslant 1$, $\dim_{\mathrm{H}}(f_*\mu) \geqslant \dim_{\mathrm{H}}\big((h_N)_*\mu\big) - \frac{\log 3}{N\log 2}$. Therefore,

$$\dim_{\mathrm{H}}(f_*\mu) \geqslant \limsup_{N\to\infty} \dim_{\mathrm{H}}\big((h_N)_*\mu\big).$$

Thus, by Proposition 13.2.6, if

$$\lim_{N\to\infty}\liminf_{n\to\infty} \frac{1}{N\log 2}\cdot\frac{1}{n}\sum_{i=1}^{n} H((h_N)_*\mu_{[\mathcal{Q}_i^N(\mathbf{i})]}, \mathcal{C}_{i+1}^N) \geqslant \alpha \text{ for } \mu\text{-a.e. } \mathbf{i}, \tag{13.15}$$

then $\dim_{\mathrm{H}}(f_*\mu) \geqslant \alpha$. Hence, it is enough to show that (13.5) implies (13.15).

By Proposition 13.2.5(ii), for every $\mathbf{i} \in \Sigma$ and $i, N \in \mathbb{N}$ there exists a cylinder set $[\omega] \in \mathcal{C}_i^N$ such that $h_N([\mathcal{Q}_i^N(\mathbf{i})]) \subseteq [\omega]$. Let us define a probability measure ν on Δ_N as follows:

$$\nu([\eta]) = \frac{(h_N)_*\mu_{[\mathcal{Q}_i^N(\mathbf{i})]}([\omega\eta])}{(h_N)_*\mu_{[\mathcal{Q}_i^N(\mathbf{i})]}([\omega])} = (h_N)_*\mu_{[\mathcal{Q}_i^N(\mathbf{i})]}([\omega\eta]), \quad \text{for } \eta \in X_N^*.$$

Then

$$f_*\mu_{[\mathcal{Q}_i^N(\mathbf{i})]} = (g_N)_*(h_N)_*\mu_{[\mathcal{Q}_i^N(\mathbf{i})]} = (S_\omega)_*(g_N)_*\nu,$$

where $S_\omega = S_{\omega_1}\circ\ldots\circ S_{\omega_i}$ has contraction ratio 2^{-iN}.

Thus, by Lemma 13.2.2(ii) there exists a constant $C > 0$ such that

$$\begin{aligned} &\Big|H(f_*\mu_{[\mathcal{Q}_i^N(\mathbf{i})]}, \mathscr{D}_{N(i+1)}) - H((g_N)_*\nu, \mathscr{D}_N)\Big| \\ &\qquad = \big|H((S_\omega)_*(g_N)_*\nu, \mathscr{D}_{N(i+1)}) - H((g_N)_*\nu, \mathscr{D}_N)\big| \leqslant C. \end{aligned} \tag{13.16}$$

Moreover, by (13.13),

$$\big|H(\nu, \mathcal{C}_1^N) - H((g_N)_*\nu, \mathscr{D}_N)\big| \leqslant C. \tag{13.17}$$

On the other hand, by the definition of the measure ν,

$$H((h_N)_*\mu_{[\mathcal{Q}_i^N(\mathbf{i})]}, \mathcal{C}_{i+1}^N) = H(\nu, \mathcal{C}_1^N). \tag{13.18}$$

Combining (13.16), (13.17) and (13.18) we get

$$\begin{aligned} H((h_N)_*\mu_{[\mathcal{Q}_i^N(\mathbf{i})]}, \mathcal{C}_{i+1}^N) &= H(\nu, \mathcal{C}_1^N) \geqslant H((g_N)_*\nu, \mathscr{D}_N) - C \\ &\geqslant H(f_*\mu_{[\mathcal{Q}_i^N(\mathbf{i})]}, \mathscr{D}_{N(i+1)}) - 2C \end{aligned}$$

and thus, (13.5) implies (13.15). ☐

ABOUT THE PROOF OF PROPOSITION 13.2.6. This is essentially Theorem 13.1.3, using the maps h_N as versions of tree morphism. The idea is to construct a random measure ν on Σ such that

(i) $\mathbb{E}(\nu) = \mu$ and,

(ii) $\liminf_{n\to\infty} \dfrac{\log (h_N)_*\nu([\mathbf{i}|_n])}{-nN\log 2} \geqslant \beta_N$ for $(h_N)_*\nu$-a.e. $\mathbf{i}$ almost surely.

The proof also uses a lemma on local entropy averages, which is analogous to [**HS12**, Lemma 4.3]. It is a special case of Lemma 13.1.1, with the symbolic space Δ_N instead of $I^{\mathbb{N}}$.

LEMMA 13.2.7. *For any probability measure ν on Δ_N,*

$$\lim_{n\to\infty}\left(\frac{-\log \nu([\mathbf{i}|_n])}{n} - \frac{1}{n}\sum_{i=0}^{n-1} H(\nu_{[\mathbf{i}|_i]}, \mathcal{C}_{i+1}^N)\right) = 0, \ \nu\text{-a.e. } \mathbf{i}.$$

The rest of the proof is left as an exercise to the reader. □

13.3. Orthogonal projections of self-affine measures

Theorem 13.2.1 on local entropy averages can be applied to study the orthogonal projections of iterated function systems. Let $\Psi = (f_i(\mathbf{x}) = A_i\mathbf{x} + \mathbf{t}_i)_{i=1}^m$ be a contracting IFS of affine transformations on $\mathbb{R}^2$, and let Λ denote its attractor. Let $\mathbf{p} = (p_1, \ldots, p_m)$ be a probability vector and let μ be the Bernoulli measure on $\Sigma = \{1, \ldots, m\}^{\mathbb{N}}$ corresponding to $\mathbf{p}$. Let $\nu = \sum_{i=1}^m p_i (f_i)_*\nu = \pi_*\mu$ be the invariant measure associated to the probability vector $\mathbf{p}$ and the IFS Ψ, where $\pi\colon \Sigma \to \mathbb{R}^2$ is the natural projection. Further, let ν_F be the associated Furstenberg measure defined in Theorem 12.1.2. Finally, as before, denote by Proj_θ the orthogonal projection along the subspace $\theta \in \mathbb{PR}^1$. That is, $\mathrm{Proj}_\theta = \mathrm{proj}_{\mathbf{u}(\theta^\perp)}$, where the unit vector $\mathbf{u}(\theta^\perp) \in \theta^\perp$ is chosen in such a way that $\langle \mathbf{u}, (1,0)\rangle > 0$, and in the case when $\theta^\perp$ is parallel to the y-axis we set $\mathbf{u} = (0,1)$.

THEOREM 13.3.1. *Let $\Psi = (f_i(\mathbf{x}) = A_i\mathbf{x} + \mathbf{t}_i)_{i=1}^m$ be such that the Strong Separation Condition (SSC) holds and the set of matrices is strongly irreducible. Moreover, suppose that the attractor Λ is not contained in a line. Then for every $\theta \in \mathbb{PR}^1$ and every fully supported self-affine measure ν,*

$$\dim_{\mathrm{H}}\big((\mathrm{Proj}_\theta)_*\nu\big) \geqslant \int \dim_{\mathrm{H}}\big((\mathrm{Proj}_\theta)_*\nu\big)\, d\nu_F(\theta).$$

In this form, the theorem appeared in [**BHR19**, Theorem 7.1], which was an improvement over [**FK17**, Theorem 3.1], where it was assumed that the matrices A_i have strictly positive entries.

Note that we do not assume that the matrices $(A_i)_{i=1}^m$ form a strictly affine system. In fact, each matrix may be a similitude. The theorem above has two important corollaries. One is a special case of Hochman and Shmerkin [**HS12**, Theorem 1.6 and Proposition 9.4] (they allow an arbitrary $d \geqslant 2$ and nonlinear maps instead of projections).

COROLLARY 13.3.2. *Let $(S_1, \ldots, S_N)$ be an IFS of similarities, satisfying the SSC and such that the generated orthogonal group is dense in $SO(2,\mathbb{R})$, i.e., it contains an irrational rotation. Then for every $\theta \in \mathbb{PR}^1$ and every self-similar measure ν,*

$$\dim_{\mathrm{H}}\big((\mathrm{Proj}_\theta)_*\nu\big) = \min\{1, \dim_{\mathrm{H}}(\nu)\}.$$

Moreover, for every $\theta \in \mathbb{PR}^1$ *the attractor* Λ *satisfies*

$$\dim_{\mathrm{H}}(\mathrm{Proj}_\theta \Lambda) = \min\{1, \dim_{\mathrm{H}}(\Lambda)\}.$$

PROOF. By Lemma 12.1.7, the Furstenberg measure ν_F is equivalent to the Lebesgue measure on $\mathbb{PR}^1$. By Theorem 2.3.3, $\dim_{\mathrm{H}}((\mathrm{Proj}_\theta)_*\nu)) = \min\{1, \dim_{\mathrm{H}}(\nu)\}$ for Lebesgue almost every $\theta \in \mathbb{PR}^1$. Since for every $\theta \in \mathbb{PR}^1$, $\dim_{\mathrm{H}}((\mathrm{Proj}_\theta)_*\mu) \leqslant \min\{1, \dim_{\mathrm{H}}(\nu)\}$, the first claim follows by Theorem 13.3.1. The second claim follows by taking the natural self-similar measure, so that $\dim_{\mathrm{H}}(\nu) = \dim_{\mathrm{H}}(\Lambda)$. □

The second corollary shows that Marstrand's Projection Theorem 2.3.1 holds in a strong sense for self-affine sets and measures, if the dimension of the measure is as large as it is possible.

COROLLARY 13.3.3. *Let* $\Psi = (f_i(\mathbf{x}) = A_i\mathbf{x} + \mathbf{t}_i)_{i=1}^m$ *be an IFS on* $\mathbb{R}^2$ *such that* A_i *are nonsingular, contracting matrices for* $i = 1, \ldots, m$, *the IFS satisfies the SSC and is strongly irreducible, and the attractor* Λ *is not contained in a line. Then for every fully supported Bernoulli measure* μ *on the symbolic space, the corresponding self-affine measure satisfies*

$$\dim_{\mathrm{H}}(\nu) = D(\mu)$$

if and only if

$$\dim_{\mathrm{H}}((\mathrm{Proj}_\theta)_*\nu) = \min\{1, \dim_{\mathrm{H}}(\nu)\} \textit{ for every } \theta \in \mathbb{PR}^1.$$

Here $D(\nu)$ *denotes the Lyapunov dimension, see Definition* 10.1.1. *In particular, if* $\dim_{\mathrm{H}}(\Lambda) = s(A_1, \ldots, A_m)$ *then*

$$\dim_{\mathrm{H}}((\mathrm{Proj}_\theta)_*\Lambda) = \min\{1, \dim_{\mathrm{H}}(\Lambda)\} \textit{ for every } \theta \in \mathbb{PR}^1,$$

where $s(A_1, \ldots, A_m)$ *is the affinity dimension, see Definition* 1.10.2.

The proof follows from Corollary 12.3.3 and Theorem 13.3.1, and is left to the reader.

Next we sketch the proof of Theorem 13.3.1; some technical details will be omitted. Without loss of generality, we may assume that $|\Lambda| \leqslant 1$, and so for every $\theta \in \mathbb{PR}^1$ holds $|\mathrm{Proj}_\theta \Lambda| \leqslant 1$.

Observe that for any subspace $\theta \in \mathbb{PR}^1$ and any invertible matrix A, the lines $A^*\theta$ and $A^{-1}\theta^\perp$ are orthogonal:

$$\sphericalangle(A^*\theta, A^{-1}\theta^\perp) = \frac{\pi}{2}. \tag{13.19}$$

Indeed, let $\mathbf{u} \in \theta$ and $\mathbf{v} \in \theta^\perp$ be arbitrary vectors, then

$$0 = \langle \mathbf{u}, \mathbf{v} \rangle = \langle \mathbf{u}, AA^{-1}\mathbf{v} \rangle = \langle A^*\mathbf{u}, A^{-1}\mathbf{v} \rangle.$$

Thus, for any affine transformation $f(\mathbf{x}) = A\mathbf{x} + \mathbf{t}$,

$$\begin{aligned}\mathrm{Proj}_\theta(f(\mathbf{x})) = \langle \mathbf{u}(\theta^\perp), A\mathbf{x} + \mathbf{t} \rangle &= \|A^*\mathbf{u}(\theta^\perp)\| \left\langle \frac{A^*\mathbf{u}(\theta^\perp)}{\|A^*\mathbf{u}(\theta^\perp)\|}, \mathbf{x} \right\rangle + \langle \mathbf{u}(\theta^\perp), \mathbf{t} \rangle \\ &= \pm \|A^*|_{\theta^\perp}\| \cdot \mathrm{Proj}_{A^{-1}\theta}(\mathbf{x}) + \mathrm{Proj}_\theta(f(0)),\end{aligned}$$

where the sign is chosen according to $\pm \frac{A^*\mathbf{u}(\theta^\perp)}{\|A^*\mathbf{u}(\theta^\perp)\|} = \mathbf{u}(A^*\theta^\perp)$.

Let $\mathbf{j} \in \Sigma^*$. Applying the last formula to $f_{\mathbf{j}}$ yields

$$\mathrm{Proj}_\theta(f_{\mathbf{j}}(\mathbf{x})) = \pm \|A_{\mathbf{j}}^*|_{\theta^\perp}\| \cdot \mathrm{Proj}_{A_{\mathbf{j}}^{-1}\theta}(\mathbf{x}) + \mathrm{Proj}_\theta(f_{\mathbf{j}}(0)), \tag{13.20}$$

where the sign $\pm$ depends on the orientation of $A_{\mathbf{j}}^* \mathbf{u}(\theta^\perp)$. Therefore, by Lemma 13.2.2(ii) we obtain, for any $n \in N$:

$$\left| H\big((\mathrm{Proj}_\theta)_*(f_{\mathbf{j}})_*\nu, \mathscr{D}_{n+\ell}\big) - H\big((\mathrm{Proj}_{A_{\mathbf{j}}^{-1}\theta})_*\nu, \mathscr{D}_n\big) \right| \leqslant C, \tag{13.21}$$

where

$$\ell = -\lceil \log_2 \|A_{\mathbf{j}}^*|_{\theta^\perp}\| \rceil,$$

for a uniform constant $C > 0$.

Without loss of generality (possibly passing to higher iterates), we may assume that $1 < -\log_2 \|A_i^*|_{\theta^\perp}\|$ for every $1 \leqslant i \leqslant m$ and $\theta \in \mathbb{PR}^1$. Consider a family of minimal cut-sets $\mathcal{Q}_n^{N,t}$ for Σ, with

$$0 \leqslant t \leqslant \max_{1 \leqslant i \leqslant m} \big(-\log_2 \|A_i^*\|\big),$$

adapted to the function $F := \mathrm{Proj}_\theta \circ \Pi \colon \Sigma \to \mathbb{R}$, defined as follows:

$$\mathcal{Q}_n^{N,t} := \{\mathbf{j} = (j_1, \dots, j_k) \in \Sigma^* : \|A_{\mathbf{j}}^*|_{\theta^\perp}\| \leqslant 2^{-Nn-t} < \|A_{j_{k-1}}^* \cdots A_{j_1}^*|_{\theta^\perp}\|\}.$$

It is easy to see that the assumptions (Q.1)–(Q.4) hold, with F instead of f.

Note that by self-affinity of ν we have

$$\Pi_*\mu_{[\mathbf{j}]} = (f_{\mathbf{j}})_*\nu \quad \text{for any } \mathbf{j} \in \Sigma^*. \tag{13.22}$$

Let $\mathbf{i} \in \Sigma$. Combining (13.21) with (13.22) for $N, i \in \mathbb{N}$ and $\mathbf{j} = \mathcal{Q}_i^{N,t}(\mathbf{i})$ (i.e., $\mathbf{j}$ is the unique element of $\mathcal{Q}_i^{N,t}$ such that $\mathbf{i} \in [\mathbf{j}]$), we get

$$\left| H\big((\mathrm{Proj}_\theta)_*\Pi_*\mu_{[\mathcal{Q}_i^{N,t}(\mathbf{i})]}, \mathscr{D}_{N(i+1)}\big) - H\Big(\big(\mathrm{Proj}_{A^{-1}_{\mathcal{Q}_i^{N,t}(\mathbf{i})}\theta}\big)_*\nu, \mathscr{D}_N\Big) \right| \leqslant C. \tag{13.23}$$

We claim that in order to prove Theorem 13.3.1, it is enough to verify the following lemma.

LEMMA 13.3.4. *Under the assumptions of Theorem* 13.3.1, *there exist* $C > 0$ *and* $t \geqslant 1$ *such that for every* $\theta \in \mathbb{PR}^1$ *and* μ*-a.e.* $\mathbf{i}$, *for all* $N \in \mathbb{N}$,

(13.24)
$$\liminf_{n \to \infty} \frac{1}{n} \sum_{i=0}^{n-1} H\Big(\big(\mathrm{Proj}_{A^{-1}_{\mathcal{Q}_i^{N,t}(\mathbf{i})}\theta}\big)_*\nu, \mathscr{D}_N\Big) \geqslant \int H\big((\mathrm{Proj}_\theta)_*\nu, \mathscr{D}_N\big)\, d\nu_F(\theta) - C;$$

(13.25)
$$\limsup_{n \to \infty} \frac{1}{n} \sum_{i=0}^{n-1} H\Big(\big(\mathrm{Proj}_{A^{-1}_{\mathcal{Q}_i^{N,t}(\mathbf{i})}\theta}\big)_*\nu, \mathscr{D}_N\Big) \leqslant \int H\big((\mathrm{Proj}_\theta)_*\nu, \mathscr{D}_N\big)\, d\nu_F(\theta) + C;$$

First we prove Theorem 13.3.1, assuming this lemma.

DERIVATION OF THEOREM 13.3.1. We wish to apply Theorem 13.2.1 with the choice of $f = F := \mathrm{Proj}_\theta \circ \Pi \colon \Sigma \to \mathbb{R}$. Thus it is enough to show that

$$\lim_{N \to \infty} \liminf_{n \to \infty} \frac{1}{N \log 2} \cdot \frac{1}{n} \sum_{i=1}^{n} H\big((\mathrm{Proj}_\theta)_*\Pi_*\mu_{[\mathcal{Q}_i^{N,t}(\mathbf{i})]}, \mathscr{D}_{N(i+1)}\big)$$
$$\geqslant \int \dim_{\mathrm{H}}\big((\mathrm{Proj}_\theta)_*\nu\big)\, d\nu_F(\theta).$$

In fact, we will show that this is equality. Using (13.23) and (13.24), we get that

$$\liminf_{N\to\infty}\liminf_{n\to\infty}\frac{1}{N\log 2}\cdot\frac{1}{n}\sum_{i=1}^{n}H\big((\mathrm{Proj}_\theta)_*\Pi_*\mu_{[\mathcal{Q}_i^{N,t}(\mathbf{i})]},\mathscr{D}_{N(i+1)}\big)\geqslant$$

$$\liminf_{N\to\infty}\frac{1}{N\log 2}\left(\liminf_{n\to\infty}\frac{1}{n}\sum_{i=1}^{n}H\Big(\big(\mathrm{Proj}_{A^{-1}_{Q_i^{N,t}(\mathbf{i})}\theta}\big)_*\nu,\mathscr{D}_N\Big)-C\right)$$

$$\geqslant\liminf_{N\to\infty}\frac{1}{N\log 2}\int H\big((\mathrm{Proj}_\theta)_*\nu,\mathscr{D}_N\big)\,d\nu_F(\theta).$$

Applying Fatou's lemma and the formula for the lower entropy dimension (13.6), we obtain

$$\liminf_{N\to\infty}\frac{1}{N\log 2}\int H((\mathrm{Proj}_\theta)_*\nu,\mathscr{D}_N)\,d\nu_F(\theta)\geqslant\int\underline{\dim}_e\big((\mathrm{Proj}_\theta)_*\nu\big)\,d\nu_F(\theta).$$

Similarly, using (13.23) and (13.25), we obtain

$$\limsup_{N\to\infty}\liminf_{n\to\infty}\frac{1}{N\log 2}\cdot\frac{1}{n}\sum_{i=1}^{n}H\big((\mathrm{Proj}_\theta)_*\Pi_*\mu_{[\mathcal{Q}_i^{N,t}(\mathbf{i})]},\mathscr{D}_{N(i+1)}\big)\leqslant$$

$$\leqslant\limsup_{N\to\infty}\frac{1}{N\log 2}\int H\big((\mathrm{Proj}_\theta)_*\nu,\mathscr{D}_N\big)\,d\nu_F(\theta).$$

Recall that $|\,\mathrm{Proj}_\theta\,\Lambda|\leqslant 1$ for all θ, hence $H\big((\mathrm{Proj}_\theta)_*\nu,\mathscr{D}_N\big)\leqslant N\log 2$ and we can use Fatou's Lemma for lim sup to obtain

$$\limsup_{N\to\infty}\frac{1}{N\log 2}\int H((\mathrm{Proj}_\theta)_*\nu,\mathscr{D}_N)\,d\nu_F(\theta)\leqslant\int\overline{\dim}_e\big((\mathrm{Proj}_\theta)_*\nu\big)\,d\nu_F(\theta).$$

By Proposition 12.3.8, we have that $(\mathrm{Proj}_\theta)_*\nu$ is exact dimensional for ν_F-almost every θ and hence, by Theorem 2.6.4, $\overline{\dim}_e\big((\mathrm{Proj}_\theta)_*\nu\big)=\underline{\dim}_e\big((\mathrm{Proj}_\theta)_*\nu\big)$ for ν_F-a.e. θ, whence

$$\int\dim_e\big((\mathrm{Proj}_\theta)_*\nu\big)\,d\nu_F(\theta)=\int\dim_{\mathrm{H}}\big((\mathrm{Proj}_\theta)_*\nu\big)\,d\nu_F(\theta),$$

concluding the proof. □

The proof of Lemma 13.3.4 follows Falconer and Kempton [**FK17**, Proposition 4.5]. Since it requires more sophisticated methods from Ergodic Theory and suspension flows, we only sketch it here.

SKETCH OF THE PROOF OF LEMMA 13.3.4. The idea of the proof is based on ergodicity of the measure $\mu\times\nu_F$ for the skew product transformation $T:(\mathbf{i},\theta)\mapsto(\sigma\mathbf{i},A_{i_1}^{-1}\theta)$ on $\Sigma\times\mathbb{PR}^1$, see Theorem 12.1.2. Consider the function

$$g(\mathbf{i},\theta)=g_N(\mathbf{i},\theta)=H\big((\mathrm{Proj}_\theta)_*\nu,\mathscr{D}_N\big).$$

It is clear that g is bounded and measurable; in fact, it is not hard to see that g is continuous, since ν is nonatomic and every line has zero mass, see Proposition 3.1.1. Thus, we can apply Birkhoff's Ergodic Theorem to obtain that for $\mu\times\nu_F$-almost every $(\mathbf{i},\theta)$

$$\lim_{n\to\infty}\frac{1}{n}\sum_{k=0}^{n-1}H\Big(\big(\mathrm{Proj}_{A_{i_k}^{-1}\cdots A_{i_1}^{-1}\theta}\big)_*\nu,\mathscr{D}_N\Big)=\int H\big((\mathrm{Proj}_\theta)_*\nu,\mathscr{D}_N\big)d\nu_F(\theta)$$

(notice that g does not depend on the variable $\mathbf{i}$, so integration with respect to μ can be omitted). One can see resemblance with the claim of Lemma 13.3.4, but unfortunately, $A_{i_k}^{-1}\cdots A_{i_1}^{-1}\theta$ is not the same as $A^{-1}_{Q_n^{N,t}(\mathbf{i})}\theta$. In fact, $A^{-1}_{Q_n^{N,t}(\mathbf{i})}\theta$ is a rare subsequence of $A_{i_k}^{-1}\cdots A_{i_1}^{-1}\theta$. In order to handle this difficulty, we need to introduce a suspension semiflow over the skew product, and then its time-N map.

A *suspension (semi)flow*, sometimes also called *special (semi)flow*, is a standard Ergodic Theory construction. Given a measure-preserving transformation $(X, \mathcal{B}, T, \mathfrak{m})$ and a measurable "roof function" $\phi : X \to [0, \infty)$, we define the space

$$\mathscr{X}_\phi := \{(x,t) \in X \times \mathbb{R} : \ 0 \leqslant t \leqslant \phi(x)\},$$

with the points $(x, \phi(x))$ and $(Tx, 0)$ identified. The suspension semiflow is defined for $s \geqslant 0$ by $\psi_s(x,t) = (x, s+t)$ for $t + s \leqslant \phi(x)$ and extended using this identification. It is easy to verify that $\{\psi_s\}_{s\geqslant 0}$ is an action of the semigroup $\mathbb{R}_+$, which preserves the measure $\mathfrak{m} \times \mathcal{L}^1$ restricted to $\mathscr{X}_\phi$. Moreover, if the "base transformation" $(X, \mathcal{B}, T, \mathfrak{m})$ is ergodic, then the suspension semiflow is ergodic, see, e.g., [**EW11**, Lemma 9.2.4].

In the problem at hand we construct the suspension flow of the skew product transformation $T : (\mathbf{i}, \theta) \mapsto (\sigma\mathbf{i}, A_{i_1}^{-1}\theta)$ on $\Sigma \times \mathbb{PR}^1$, with the roof function

$$\phi(\mathbf{i}, \theta) = -\log_2 \|A_{i_1}^*|_{\theta^\perp}\|.$$

To be explicit, we consider the space

$$W = \{(\mathbf{i}, \theta, t) \in \Sigma \times \mathbb{PR}^1 \times \mathbb{R} : 0 \leqslant t \leqslant -\log_2 \|A_{i_1}^*|_{\theta^\perp}\|\},$$

where the points $(\mathbf{i}, \theta, -\log_2 \|A_{i_1}^*|_{\theta^\perp}\|)$ and $(\sigma\mathbf{i}, A_{i_1}^{-1}\theta, 0)$ are identified. The suspension semiflow is defined by $\psi_s(\mathbf{i}, \theta, t) = (\mathbf{i}, \theta, t+s)$ for $0 \leqslant t+s \leqslant -\log_2 \|A_{i_1}^*|\theta^\perp\|$ and extended this using the identification. The measure $m := (\mu \times \nu_F \times \mathcal{L}^1)|_W$ is ψ_s-invariant, and ψ_s is ergodic. Why did we choose such a roof function? Notice that

$$\psi_s(\mathbf{i}, \theta, t) = (\sigma^k\mathbf{i}, \theta', t') \in W$$

for some $k \geqslant 0$, if

$$\|A_{\mathbf{i}|_k}^*|_{\theta^\perp}\| \leqslant 2^{-t-s} < \|A_{i_{k-1}}^* \cdots A_{i_1}^*|_{\theta^\perp}\| \text{ and } \theta' = A_{\mathbf{i}}^{-1}\theta.$$

This is a consequence of the next equality, where A, B are arbitrary invertible matrices:

$$\|B^*A^*|_{\theta^\perp}\| = \|B^*|_{(A^{-1}\theta)^\perp}\| \cdot \|A^*|_{\theta^\perp}\|,$$

which holds in view of

$$\mathbf{u}(A^*\theta^\perp) = \pm\mathbf{u}\big((A^{-1}\theta)^\perp\big),$$

see (13.19). It follows that if Π_2 denotes the projection $\Pi_2(\mathbf{i}, \theta, t) = \theta$, then for every $n \geqslant 0$ we have

$$\Pi_2 \circ (\psi_N)^n(\mathbf{i}, \theta, t) = A^{-1}_{Q_n^{N,t}(\mathbf{i})}\theta.$$

The transformation ψ_N is called the *time-N map* for the semiflow ψ_s. Falconer and Kempton [**FK17**, Lemma 5.3] proved that $\psi_N : W \to W$ is ergodic; however, technical details for this proof are out of the scope of this book. Taking it for granted, we can apply the Ergodic Theorem to obtain

$$\lim_{n\to\infty} \frac{1}{n} \sum_{i=0}^{n-1} g(A^{-1}_{Q_i^{N,t}(\mathbf{i})}\theta) = \int g(\theta) d\nu_F(\theta) \text{ for } m\text{-almost every } (\mathbf{i}, \theta, t), \tag{13.26}$$

where $g(\mathbf{i},\theta) = H\big((\mathrm{Proj}_\theta)_*\nu, \mathscr{D}_N\big)$, as above.

The next difficulty we need to handle is that we want to consider every direction $\theta \in \mathbb{PR}^1$ and not just almost every. However, by Lemma 12.1.12, for μ-a.e. $\mathbf{i} \in \Sigma$, if $\theta, \theta' \in \mathbb{PR}^1$ are close to each other, then the angle $\sphericalangle(A_{\mathbf{i}|_n}^{-1}\theta, A_{\mathbf{i}|_n}^{-1}\theta')$ remains close to zero for all $n \geqslant 1$. Thus, for a typical $\mathbf{i} \in \Sigma$, for every $\theta \in \mathbb{PR}^1$ we can choose a sufficiently close "ν_F-typical" θ' for which (13.26) holds, and then the proof can be finished using the weak*-continuity of $\theta \mapsto (\mathrm{Proj}_\theta)_*\nu$ and Lemma 13.2.2. In more details:

By Lemma 12.1.12, we have that for every $\theta \in \mathbb{PR}^1$ and for almost every $\mathbf{i} \in \Sigma$ and $t \geqslant 1$ there exists $\theta' \in \mathbb{PR}^1$ such that $(\mathbf{i}, \theta', t)$ satisfies (13.26) and

$$\limsup_{i\to\infty} \left\| \mathrm{Proj}_{A^{-1}_{Q_i^{N,t}(\mathbf{i})}\theta} - \mathrm{Proj}_{A^{-1}_{Q_i^{N,t}(\mathbf{i})}\theta'} \right\| \leqslant 2^{-N-1}.$$

As already noted above, the map $g(\theta) = H(\mathrm{Proj}_\theta \nu, \mathscr{D}_N)$ is continuous, and hence, by Lemma 13.2.2 there exists a constant $C > 0$ such that for μ-almost every $\mathbf{i}$ and for all sufficiently large i we have

$$\left| H\big((\mathrm{Proj}_{A^{-1}_{Q_i^{N,t}(\mathbf{i})}\theta'})_*\nu, \mathscr{D}_N\big) - H\big((\mathrm{Proj}_{A^{-1}_{Q_i^{N,t}(\mathbf{i})}\theta})_*\nu, \mathscr{D}_N\big) \right| \leqslant C.$$

Thus,

$$\begin{aligned}
\liminf_{n\to\infty} \frac{1}{n} \sum_{i=0}^{n-1} & H\big((\mathrm{Proj}_{A^{-1}_{Q_i^{N,t}(\mathbf{i})}\theta})_*\nu, \mathscr{D}_N\big) \\
&\geqslant \lim_{n\to\infty} \frac{1}{n} \sum_{i=0}^{n-1} H\big((\mathrm{Proj}_{A^{-1}_{Q_i^{N,t}(\mathbf{i})}\theta'})_*\nu, \mathscr{D}_N\big) - C \\
&= \int H\big((\mathrm{Proj}_\theta)_*\nu, \mathscr{D}_N\big)\, d\nu_F(\theta) - C.
\end{aligned}$$

This confirms (13.24), and (13.25) is proved similarly, replacing lim inf by lim sup and $-C$ by $+C$. □

13.3.1. Projection of self-affine carpets. Finally, we state here, without proof, a result of Ferguson, Jordan and Shmerkin [**FJS10**] on the orthogonal projection of self-affine carpets. Similarly to Corollary 13.3.2 or Corollary 13.3.3, they strengthened the estimate on the exceptional set of Marstrand's projection theorem 2.3.1. In particular, they showed that under certain irrationality conditions there are no exceptional directions except the natural axes-parallel directions. Let us recall Gatzouras–Lalley carpets from Definition 11.5.10 and Barański carpets from Section 11.5.4.

The diagonal self-affine IFS

(13.27)
$$\mathcal{F}_{\mathrm{GL}} = (f_i(\underline{x}) := A_i\underline{x} + t_i)_{i=1}^m, A_i = \begin{pmatrix} b_i & 0 \\ 0 & a_i \end{pmatrix},\ 0 < a_i < b_i < 1, t_i = \begin{pmatrix} t_{i,1} \\ t_{i,2} \end{pmatrix},$$

is called a Gatzouras–Lalley (GL) IFS if the first order cylinder rectangles

$$R_i := f_i(R) = t_i + [0, b_i] \times [0, a_i]$$

satisfy

(1) $R_i \subset R$,
(2) $R_i^\circ \cap R_j^\circ = \varnothing$ if $i \neq j$.

(3) Either $\text{proj}_x(R_i^\circ) \cap \text{proj}_x(R_j^\circ) = \varnothing$ or $\text{proj}_x(R_i) = \text{proj}_x(R_j)$. In the latter case we say that R_i and R_j are in the same column.

We call the attractor of $\mathcal{F}_{GL}$ the Gatzouras–Lalley carpets.

DEFINITION 13.3.5. We say that the Gatzouras–Lalley carpet (13.27) is of irrational type if there exists $i \in \{1,\ldots,m\}$ such that $\frac{\log a_i}{\log b_i} \notin \mathbb{Q}$ or if there exist $i,j \in \{1,\ldots,m\}$ such that $\frac{\log a_i}{\log b_j} \notin \mathbb{Q}$ and $\frac{\log a_i}{\log b_i}$ is not constant for $i \in \{1,\ldots,m\}$.

Next let us recall the definition of Barański carpets from (11.127). Let $N, M \geqslant 2$ be integers and let $a_1,\ldots,a_N$ and $b_1,\ldots,b_M$ be positive reals such that $\sum_{i=1}^N a_i = \sum_{j=1}^M b_j = 1$. Let $A \subseteq \{1,\ldots,N\} \times \{1,\ldots,M\}$ be nonempty with at least two elements. For $(i,j) \in A$, let

$$f_{(i,j)}(\underline{x}) = \begin{pmatrix} a_i & 0 \\ 0 & b_j \end{pmatrix} \underline{x} + \begin{pmatrix} \sum_{k=1}^{i-1} a_k \\ \sum_{k=1}^{j-1} b_k \end{pmatrix} \text{ and } \mathcal{F}_B = (f_{(i,j)})_{(i,j)\in A}. \tag{13.28}$$

We call the attractor of $\mathcal{F}_B$ the Barański carpet.

DEFINITION 13.3.6. We say that the Barański carpet (13.28) is of irrational type if there exists $(i,j) \in A$ such that $\frac{\log a_i}{\log b_j} \notin \mathbb{Q}$ or if there exists $(i,j) \notin A$ such that $\frac{\log a_i}{\log b_j} \notin \mathbb{Q}$ and $\frac{\log a_i}{\log b_j}$ is not constant for $(i,j) \in A$.

THEOREM 13.3.7 (Ferguson, Jordan and Shmerkin [**FJS10**]). *Let Λ be a Gatzouras–Lalley carpet of irrational type or a Barański carpet of irrational type. Then for every $\mathbf{v} \in \mathcal{S}^1 \backslash \{(1,0),(-1,0),(0,1),(0,-1)\}$ holds*

$$\dim_{\mathrm{H}}(\text{proj}_{\mathbf{v}}(\Lambda)) = \min\{1, \dim_{\mathrm{H}}(\Lambda)\}.$$

Since the Bedford–McMullen carpets are special cases of Gatzouras–Lalley carpets, we get that a Bedford–McMullen carpet is of irrational type if $\frac{\log b}{\log a} \notin \mathbb{Q}$ where a, b are defined in (11.91).

CHAPTER 14

Nonlinear conformal iterated functions systems

Although the primary focus in this book is on self-similar and self-affine iterated function systems, we use the opportunity to show how the transversality method developed in this book can be applied in the nonlinear case. The natural generalization of a self-similar IFS is a *conformal* IFS. For a connected open set $U \subseteq \mathbb{R}^d$, a function $f : U \to \mathbb{R}^d$ of class $\mathcal{C}^{1+\theta}$ is called conformal if its differential $Df(\mathbf{x})$ is a nondegenerate similitude (similarity mapping) on $\mathbb{R}^d$ for every $\mathbf{x}$ such that $\mathbf{x} \mapsto Df(\mathbf{x})$ is Hölder continuous with exponent θ. For $d = 1$, any $C^{1+\theta}$-map without singularities will be automatically a conformal map, but in higher dimensions the situation is much more rigid: in $\mathbb{R}^2 \approx \mathbb{C}$ a conformal function is either holomorphic or antiholomorphic (see [**PU10**, p. 236]), with nonvanishing derivative, and in higher dimensions it is even more rigid and has to be a composition of isometries, dilations or inversions (see [**dC92**, Theorem 5.2 (Liouville) in Chapter 8]). For simplicity we restrict ourselves to the $d = 1$ case, but most of the results can be extended to the higher-dimensional conformal case with little difficulty.

First, however, we prove in the next section an existence and uniqueness theorem of invariant measures for general iterated function systems with countably many maps.

14.1. Existence and uniqueness of invariant measure in a general case

In Proposition 1.8.2 we pointed out that for a self-similar IFS $\mathcal{S} = (S_i)_{i=1}^m$ and a probability vector $\mathbf{p} = (p_1, \dots, p_m)$ there exists a unique Borel probability measure ν satisfying

$$\nu(A) = p_1\nu(S_1^{-1}(A)) + \cdots + p_m\nu(S_m^{-1}(A)), \text{ for all Borel set } A. \tag{14.1}$$

Now we prove that this holds not only on $\mathbb{R}^d$ but on an arbitrary complete metric space and not only for a finite list of contracting similarities but for all probability vectors of countable dimension and for all countable sets of contractions whose fixed points are not too far from each other in the sense of condition (14.3) below.

Recall that $M(X)$ denotes the set of all Borel probability measures satisfying $\int f d\mu < \infty$ for all Lipschitz function $f \in \mathrm{Lip}(X)$. The dual Lipschitz metric L on $M(X)$ is defined by

$$L(\mu, \mu') := \sup_{\mathrm{Lip}(g) \leqslant 1} \left| \int g d\mu - \int g d\mu' \right|, \tag{14.2}$$

where $\mathrm{Lip}(g)$ is the Lipschitz constant of g.

THEOREM 14.1.1 (Kravchenko). *Let $\mathcal{S} = (f_i)_{i \in \mathbb{N}}$ be a countable IFS on a complete metric space (X, ρ) with $\mathrm{Lip}(f_i) < 1$ for all $i \in \mathbb{N}$ and $\mathbf{p} = (p_1, p_2, \dots)$ a* *probability vector. Let x_i be the fixed point of f_i. We assume that*

$$\sum_{i=1}^{\infty} p_i \rho(x_1, x_i) < \infty. \tag{14.3}$$

Then there exists a unique measure $\nu \in M(X)$ such that

$$\nu(A) = \sum_{i=1}^{\infty} p_i \nu\left(f_i^{-1}(A)\right), \quad \textit{for all } \nu\textit{-measurable } A. \tag{14.4}$$

PROOF. Our first goal is to prove that for $\mathfrak{F}(\nu) := \sum_{i=1}^{\infty} p_i (f_i)_* \nu$ we have

$$\mathfrak{F} : M(X) \to M(X). \tag{14.5}$$

Then we show that

$$\mathrm{Lip}(\mathfrak{F}) < 1, \quad \text{on the metric space } (M(X), L). \tag{14.6}$$

After we proved these, the assertion of the theorem follows from Banach Fixed Point Priniciple and Theorem 1.8.7.

Proof of (14.5)**:** First observe that for a Borel probability measure μ,

$$\begin{aligned} \mu \in M(X) \quad &\Longleftrightarrow \quad \forall\, x_0 \in X \int \rho(x, x_0) d\mu(x) < \infty \\ &\Longleftrightarrow \quad \exists\, x_0 \in X \text{ such that } \int \rho(x, x_0) d\mu(x) < \infty. \end{aligned} \tag{14.7}$$

Indeed, the first implication " $\Longrightarrow$ " is clear, since $x \mapsto \rho(x, x_0)$ is Lipschitz. For the other direction, let $x_0 \in X$ be arbitrary. For any $g \in \mathrm{Lip}(X)$ we have $g(x) \leqslant g(x_0) + \mathrm{Lip}(g) \cdot \rho(x, x_0)$, hence $\int g d\mu \leqslant g(x_0) + \mathrm{Lip}(g) \cdot \int \rho(x, x_0)\, d\mu(x)$.

Let $\phi(x) := \rho(x, x_1)$. Using (14.7), to verify (14.5) we only need to prove that for a Borel probability measure μ we have

$$\int \phi(x)\, d\mu(x) < \infty \Longrightarrow \int \phi(x)\, d(\mathfrak{F}\mu)(x) = \int \sum_{i=1}^{\infty} p_i \phi(f_i(x))\, d\mu(x) < \infty. \tag{14.8}$$

Note that

$$\rho(f_i(x), x_1) \leqslant \rho(f_i(x), x_i) + \rho(x_i, x_1) \leqslant \rho(x, x_i) + \rho(x_i, x_1). \tag{14.9}$$

By assumption, $K := \sum_{i=1}^{\infty} p_i \rho(x_1, x_i) < \infty$, so

$$\sum_{i=1}^{\infty} p_i \phi(f_i(x)) \leqslant \sum_{i=1}^{\infty} p_i \rho(x, x_i) + K. \tag{14.10}$$

By the triangular inequality, $\rho(x, x_i) \leqslant \rho(x_i, x_1) + \rho(x, x_1)$, hence

$$\begin{aligned} \int \sum_{i=1}^{\infty} p_i \rho(x, x_i)\, d\mu(x) \quad &\leqslant \quad \int \sum_{i=1}^{\infty} p_i \rho(x_i, x_1)\, d\mu(x) + \int \sum_{i=1}^{\infty} p_i \rho(x, x_1)\, d\mu(x) \\ &\leqslant \quad K + \int \phi(x)\, d\mu(x) < \infty. \end{aligned} \tag{14.11}$$

Now, given $\mu \in M(X)$, we have $\int \phi(x) d\mu(x) < \infty$. Putting together (14.10) and (14.11) yields (14.8), and (14.5) follows.

Proof of (14.6) Let $r_i := \mathrm{Lip}(f_i)$. Then for every $g \in \mathrm{Lip}(X)$,

$$\mathrm{Lip}(g) \leqslant 1 \Longrightarrow \mathrm{Lip}\left(r_i^{-1} g \circ f_i\right) \leqslant 1. \tag{14.12}$$

Using this and the definition of $\mathfrak{F}$, for any $\mu_1, \mu_2 \in M(X)$ we have

$$\begin{aligned} L\left(\mathfrak{F}\mu_1, \mathfrak{F}\mu_1\right) &= \sup_{\mathrm{Lip}(g)\leqslant 1} \left\{\left|\int g\, d(\mathfrak{F}\mu_1) - \int g\, d(\mathfrak{F}\mu_2)\right|\right\} \\ &= \sup_{\mathrm{Lip}(g)\leqslant 1} \left\{\left|\sum_{i=1}^{\infty} p_i \int (g \circ f_i)\, d\mu_1 - \sum_{i=1}^{\infty} p_i \int (g \circ f_i)\, d\mu_2\right|\right\} \\ &\leqslant \sum_{i=1}^{\infty} p_i r_i \sup_{\mathrm{Lip}(g)\leqslant 1} \left\{\left|\int (r_i^{-1} g \circ f_i)\, d\mu_1 - \int (r_i^{-1} g \circ f_i)\, d\mu_2\right|\right\} \\ &\leqslant \sum_{i=1}^{\infty} p_i r_i \sup_{\mathrm{Lip}(g)\leqslant 1} \left\{\left|\int g\, d\mu_1 - \int g\, d\mu_2\right|\right\} \\ &= \underbrace{\sum_{i=1}^{\infty} p_i r_i}_{<1} \cdot L(\mu_1, \mu_2). \end{aligned}$$

This completes the proof of (14.6) and so the proof of Theorem 14.1.1. □

14.2. Hyperbolic IFS on the line

Now we turn our attention to finite iterated function systems with smooth maps on the real line. That is, let $\mathcal{F} = (f_1, \dots, f_m)$ be a contracting iterated function system on a closed interval $I \subset \mathbb{R}$ and let $\Lambda_{\mathcal{F}}$ be its attractor. As usual, consider the symbolic space $\Sigma = \{1, \dots, m\}^{\mathbb{N}}$, the set of words of length n, denoted $\Sigma_n = \{1, \dots, m\}^n$, the compositions $f_{\mathbf{i}} = f_{i_1} \circ \cdots \circ f_{i_n}$ for $\mathbf{i} = (i_1, \dots, i_n) \in \Sigma_n$, and the cylinder intervals $I_{\mathbf{i}}^{\mathcal{F}} = f_{\mathbf{i}}(I)$. If we want to estimate the Hausdorff dimension of $\Lambda_{\mathcal{F}}$, the first thing that comes to mind is to consider the sequence $s_n > 0$ defined by

$$\sum_{\mathbf{i} \in \Sigma_n} |I_{\mathbf{i}}^{\mathcal{F}}|^{s_n} = 1, \tag{14.13}$$

and hope that $\lim_{n\to\infty} s_n$ is the dimension. The equality does hold in the important special case when the cylinder intervals I_j are mutually disjoint and the functions f_j are monotone, with Hölder continuous derivatives, bounded away from zero and one. In this case the attractor is known as a hyperbolic "cookie-cutter" set; then it is more common to view $\Lambda_{\mathcal{F}}$ as the *repeller* of an expanding map $\phi : \bigcup_{j=1}^m I_j \to I$ defined by $\phi(x) = f_j^{-1}(x)$ for $x \in I_j := f_j(I)$.

DEFINITION 14.2.1. Suppose that $I \subset \mathbb{R}$ is a closed interval and the functions $f_j \in \mathcal{C}^{1+\theta}(I \to \mathrm{int}(I))$, for $j = 1, \dots, m$, satisfy the conditions

$$0 < |f_j'(x)| < 1 \quad \text{for all } x \in I,\ j \leqslant m. \tag{14.14}$$

Then $\mathcal{F} = (f_1, \dots, f_m)$ is called a hyperbolic $\mathcal{C}^{1+\theta}$-smooth IFS on I.

THEOREM 14.2.2. *Suppose that $\mathcal{F} = (f_1, \dots, f_m)$ is a hyperbolic $\mathcal{C}^{1+\theta}$-smooth IFS on $I \subset \mathbb{R}$. Let $\Lambda_{\mathcal{F}}$ be the attractor of the IFS $\mathcal{F}$ and consider $s_n > 0$, defined by* (14.13). *Then there exists a limit $s = \lim_{n\to\infty} s_n$ and* $\dim_{\mathrm{H}}(\Lambda_{\mathcal{F}}) \leqslant \min\{s, 1\}$. *If the IFS has the Strong Separation Property (SSP), then* $\dim_{\mathrm{H}}(\Lambda_{\mathcal{F}}) = s$.

The separation condition may be weakened to the OSC. In the next section we will show that the same formula holds *typically* and *generically* for overlapping families of IFS, assuming an appropriate transversality condition holds.

Alternatively, the dimension may be expressed by *Bowen's formula* [**Bow75**, **Rue04**]:

$$\dim_{\mathrm{H}}(\Lambda_{\mathcal{F}}) = s(\mathcal{F}), \quad \text{where} \quad P_{\mathcal{F}}(s(\mathcal{F})) = 0. \tag{14.15}$$

Here $P_{\mathcal{F}}(t)$ is the pressure function, which can be defined by

$$P_{\mathcal{F}}(t) = \lim_{n\to\infty} \frac{1}{n} \log \sum_{\mathbf{i}\in\Sigma_n} \|f_{\mathbf{i}}'\|^t, \tag{14.16}$$

where $\|\cdot\|$ is the supremum norm on I. Note that the limit exists by sub-multiplicativity: $\|f_{\mathbf{ij}}'\| \leqslant \|f_{\mathbf{i}}'\| \cdot \|f_{\mathbf{j}}'\|$. We will call the number $s(\mathcal{F})$ the *conformal similarity dimension* of the IFS.

We will also be interested in invariant measures for the IFS. Although it is most common to consider measures associated with a "random Chaos Game", where the maps of the IFS are taken i.i.d., it is useful to consider more general push-forward measures from the symbolic space. As usual, let σ be the left shift on Σ. Given a hyperbolic IFS, there is a natural projection map $\Pi_{\mathcal{F}} : \Sigma \to \Lambda_{\mathcal{F}}$ defined by

$$\Pi_{\mathcal{F}}(\mathbf{i}) = \lim_{n\to\infty} f_{i_1\dots i_n}(x_0), \tag{14.17}$$

where $x_0 \in I$ is arbitrary. Let μ be an ergodic shift-invariant probability measure on Σ with positive entropy h_μ. We consider the projection of μ onto the limit set:

$$\nu_{\mathcal{F},\mu} = (\Pi_{\mathcal{F}})_*\mu = \mu \circ \Pi_{\mathcal{F}}^{-1}.$$

The *Lyapunov exponent* of the IFS $\mathcal{F}$ with respect to the measure μ is defined by

$$\chi_\mu(\mathcal{F}) = -\int_\Sigma \log |f_{i_1}'(\Pi_{\mathcal{F}}(\sigma\mathbf{i}))| \, d\mu(\mathbf{i}). \tag{14.18}$$

In the special case when μ is Bernoulli, the Lyapunov exponent can be written in terms of ν:

$$\mu = (p_1, \dots, p_m)^{\mathbb{N}} \implies \chi_\mu(\mathcal{F}) = -\sum_{i=1}^m p_i \int \log |f_i'(x)| \, d\nu(x).$$

THEOREM 14.2.3. *Suppose that $\mathcal{F} = (f_1, \dots, f_m)$ is a hyperbolic $C^{1+\theta}$-smooth IFS on $I \subset \mathbb{R}$. Let μ be an ergodic shift-invariant probability measure on Σ with positive entropy h_μ. Then*

$$\dim_{\mathrm{H}}(\nu_{\mathcal{F},\mu}) \leqslant \min\Big\{\frac{h_\mu}{\chi_\mu(\mathcal{F})}, 1\Big\}.$$

If, in addition, $\mathcal{F}$ satisfies the SSP, then $\dim_{\mathrm{H}}(\nu_{\mathcal{F},\mu}) = h_\mu/\chi_\mu(\mathcal{F})$ and $\nu_{\mathcal{F},\mu}$ is exact dimensional.

This is also classical; the separation condition may be weakened to OSC.

We will next sketch the proof of the theorems above since what follows builds on them and along the way we prepare the background for the overlapping case. A crucial role is played by bounded distortion estimates and some basic thermodynamic formalism.

14.2.1. Preliminaries. Let $I \subset \mathbb{R}$ be a compact interval, and let $\Theta(I)$ be the set of all IFS $\mathcal{F} = (f_1, \ldots, f_m)$ such that for each $1 \leqslant i \leqslant m$,

$$f_i : I \to \operatorname{int}(I), \ \ f_i \in \mathcal{C}^{1+\theta}, \ \ 0 < |f'(x)| < 1 \text{ on } I.$$

The number $\theta \in (0,1)$ is fixed throughout the section. Further, denote for $0 < \gamma_1 < \gamma_2 < 1$:

$$\Theta_{\gamma_1,\gamma_2}(I) = \{\mathcal{F} \in \Theta(I) : \ \gamma_1 \leqslant |f_i'(x)| \leqslant \gamma_2 \text{ for } x \in I, \ i \leqslant m\}.$$

For $\mathcal{F} \in \Theta(I)$, let

$$L(\mathcal{F}) = \sup_{x,y\in I, \ i\leqslant m} \frac{|f_i'(x) - f_i'(y)|}{|x-y|^\theta}.$$

For $h \in \mathcal{C}^{1+\theta}(I)$ define

$$\|h\|_{1+\theta} := \|h\|_\infty + \|h'\|_\infty + \sup_{x\neq y\in I} \frac{|h'(x) - h'(y)|}{|x-y|^\theta},$$

and let

$$\rho(\mathcal{F}, \mathcal{G}) := \max_{1\leqslant i\leqslant m} \|f_i - g_i\|_{1+\theta} \tag{14.19}$$

for $\mathcal{F} = (f_1, \ldots, f_m)$ and $\mathcal{G} = (g_1, \ldots, g_m)$ in $\Theta_{\gamma_1,\gamma_2}(I)$. It is not hard to see that $(\Theta_{\gamma_1,\gamma_2}(I), \rho)$ is a complete metric space.

LEMMA 14.2.4. *Let* $0 < \gamma_1 < \gamma_2 < 1$.

(i) *There exists* $C_1 = C_1(\gamma_1, \gamma_2) > 1$ *such that for all* $\mathcal{F} \in \Theta_{\gamma_1,\gamma_2}(I)$ *and* $\mathbf{i} \in \Sigma^*$ *and* $x, y \in I$,

$$C_1^{-L(\mathcal{F})} < \frac{|f_{\mathbf{i}}'(x)|}{|f_{\mathbf{i}}'(y)|} < C_1^{L(\mathcal{F})}. \tag{14.20}$$

(ii) *For any* $\mathcal{F} \in \Theta_{\gamma_1,\gamma_2}(I)$ *there is a constant* $C_2 = C_2(L(\mathcal{F}), \gamma_1, \gamma_2) > 0$ *such that for all* $\mathcal{G} \in \Theta_{\gamma_1,\gamma_2}(I)$ *satisfying* $\rho(\mathcal{F}, \mathcal{G}) \leqslant 1$ *and any* $\mathbf{i} \in \Sigma^*$,

$$\exp[-|\mathbf{i}| \cdot C_2 \cdot \rho(\mathcal{F}, \mathcal{G})^\theta] < \frac{|f_{\mathbf{i}}'(0)|}{|g_{\mathbf{i}}'(0)|} < \exp[|\mathbf{i}| \cdot C_2 \cdot \rho(\mathcal{F}, \mathcal{G})^\theta], \tag{14.21}$$

where $|\mathbf{i}|$ *denotes the length of* $\mathbf{i}$.

The property (14.20) in part (i) of the lemma is usually called *bounded distortion*; it plays a crucial role in the study of nonlinear conformal IFS.

PROOF. (i) By symmetry, it is enough to check one inequality. Let $\mathbf{i} = (i_1, \ldots, i_n)$, then we have

$$\begin{aligned}
\log \frac{|f'_{\mathbf{i}}(x)|}{|f'_{\mathbf{i}}(y)|} &= \sum_{k=1}^{n} \log \left| \frac{f'_{i_k}(f_{i_{k+1}\ldots i_n}(x))}{f'_{i_k}(f_{i_{k+1}\ldots i_n}(y))} \right| \\
&\leqslant \sum_{k=1}^{n} \left| \frac{f'_{i_k}(f_{i_{k+1}\ldots i_n}(x)) - f'_{i_k}(f_{i_{k+1}\ldots i_n}(y))}{f'_{i_k}(f_{i_{k+1}\ldots i_n}(y))} \right| \\
&\leqslant \frac{L(\mathcal{F})}{\gamma_1} \sum_{k=1}^{n} \left| f_{i_{k+1}\ldots i_n}(x) - f_{i_{k+1}\ldots i_n}(y) \right|^{\theta} \\
&\leqslant \frac{L(\mathcal{F})}{\gamma_1} \sum_{k=1}^{n} \gamma_2^{(n-k)\theta} \cdot |x-y|^{\theta} \\
&< L(\mathcal{F}) \cdot \frac{|I|^{\theta}}{\gamma_1(1-\gamma_2^{\theta})},
\end{aligned}$$

and (14.20) follows. In the 2nd line of the display above we used the inequality

$$\log\left|\frac{u}{v}\right| \leqslant \left|\frac{u-v}{v}\right|, \tag{14.22}$$

and in the 4th line we applied the Mean Value Theorem which yields

$$|f_{\tau_1\ldots\tau_k}(x) - f_{\tau_1\ldots\tau_k}(y)| \leqslant \gamma_2 \cdot |f_{\tau_2\ldots\tau_k}(x) - f_{\tau_2\ldots\tau_k}(y)|.$$

(ii) Observe that for all $p \geqslant 1$ and $x \in I$,

$$\begin{aligned}
|f_{i_1\ldots i_p}(x) - g_{i_1\ldots i_p}(x)| &\leqslant |f_{i_1}(f_{i_2\ldots i_p}(x)) - g_{i_1}(f_{i_2\ldots i_p}(x))| + \\
&\quad\; |g_{i_1}(f_{i_2\ldots i_p}(x)) - g_{i_1}(g_{i_2\ldots i_p}(x))| \\
&\leqslant \rho(\mathcal{F},\mathcal{G}) + \gamma_2 |f_{i_2\ldots i_p}(x) - g_{i_2\ldots i_p}(x)| \\
&\leqslant \ldots < \rho(\mathcal{F},\mathcal{G})(1-\gamma_2)^{-1}. \qquad (14.23)
\end{aligned}$$

Further,

$$\begin{aligned}
|f'_{i_k}(f_{i_{k+1}\ldots i_n}(0)) - g'_{i_k}(g_{i_{k+1}\ldots i_n}(0))| &\leqslant |f'_{i_k}(g_{i_{k+1}\ldots i_n}(0)) - g'_{i_k}(g_{i_{k+1}\ldots i_n}(0))| + \\
&\quad\; |f'_{i_k}(f_{i_{k+1}\ldots i_n}(0)) - f'_{i_k}(g_{i_{k+1}\ldots i_n}(0))| \\
&\leqslant \rho(\mathcal{F},\mathcal{G}) + L(\mathcal{F}) \cdot |f_{i_{k+1}\ldots i_n}(0) - g_{i_{k+1}\ldots i_n}(0)|^{\theta} \\
&\leqslant \rho(\mathcal{F},\mathcal{G}) + L(\mathcal{F}) \cdot \rho(\mathcal{F},\mathcal{G})^{\theta} \cdot (1-\gamma_2)^{-\theta} \\
&\leqslant \left[1 + L(\mathcal{F}) \cdot (1-\gamma_2)^{-\theta}\right] \cdot \rho(\mathcal{F},\mathcal{G})^{\theta}, \qquad (14.24)
\end{aligned}$$

using that $\rho(\mathcal{F},\mathcal{G}) \leqslant 1$ in the last step. Using (14.22) again, we obtain that for each $k \geqslant 1$,

$$\log \left| \frac{f'_{i_k}(f_{i_{k+1}\ldots i_n}(0))}{g'_{i_k}(g_{i_{k+1}\ldots i_n}(0))} \right| \leqslant \frac{\left[1 + L(\mathcal{F}) \cdot (1-\gamma_2)^{-\theta}\right] \cdot \rho(\mathcal{F},\mathcal{G})^{\theta}}{\gamma_1},$$

whence

$$\begin{aligned}
\frac{1}{n} \log \left| \frac{f'_{i_1\ldots i_n}(0)}{g'_{i_1\ldots i_n}(0)} \right| &= \frac{1}{n} \sum_{k=1}^{n} \log \left| \frac{f'_{i_k}(f_{i_{k+1}\ldots i_n}(0))}{g'_{i_k}(g_{i_{k+1}\ldots i_n}(0))} \right| \\
&\leqslant \gamma_1^{-1} \cdot \left[1 + L(\mathcal{F}) \cdot (1-\gamma_2)^{-\theta}\right] \cdot \rho(\mathcal{F},\mathcal{G})^{\theta}.
\end{aligned}$$

This implies (14.21) and completes the proof of the lemma. □

In the next corollary we collect some useful consequences of Lemma 14.2.4. Without loss of generality we can assume that $I = [0,1]$. Recall the notation $I_{\mathbf{i}}^{\mathcal{F}} = f_{\mathbf{i}}(I)$ for $\mathbf{i} \in \Sigma^*$.

COROLLARY 14.2.5. *Let $0 < \gamma_1 < \gamma_2 < 1$ and $\mathcal{F} \in \Theta_{\gamma_1\gamma_2}(I)$.*

(i) *For any $\mathbf{i} \in \Sigma^*$,*

$$C_1^{-L(\mathcal{F})} < \frac{|I_{\mathbf{i}}^{\mathcal{F}}|}{|f_{\mathbf{i}}'(0)|} < C_1^{L(\mathcal{F})}. \tag{14.25}$$

(ii) *For all $t > 0$,*

$$P_{\mathcal{F}}(t) = \lim_{n\to\infty} \frac{1}{n} \log \sum_{\mathbf{i}\in\Sigma_n} |f_{\mathbf{i}}'(0)|^t.$$

(iii) *For all $\mathbf{i} \neq \mathbf{j} \in \Sigma$, if $\mathbf{i} \wedge \mathbf{j} = (i_1, \dots i_n)$, then*

$$|\Pi_{\mathcal{F}}(\mathbf{i}) - \Pi_{\mathcal{F}}(\mathbf{j})| \geqslant C_1^{-L(\mathcal{F})} |f_{i_1\dots i_n}'(0)| \cdot |\Pi_{\mathcal{F}}(\sigma^n\mathbf{i}) - \Pi_{\mathcal{F}}(\sigma^n\mathbf{j})|. \tag{14.26}$$

PROOF. All the claims are immediate from Lemma 14.2.4 and the Mean Value Theorem, which implies, in particular, the following useful equality:

$$|\Pi_{\mathcal{F}}(\mathbf{i}) - \Pi_{\mathcal{F}}(\mathbf{j})| = |f_{\mathbf{i}\wedge\mathbf{j}}'(c)| \cdot |\Pi_{\mathcal{F}}(\sigma^n\mathbf{i}) - \Pi_{\mathcal{F}}(\sigma^n\mathbf{j})| \tag{14.27}$$

for some $c \in I$ and $n = |\mathbf{i} \wedge \mathbf{j}|$. □

COROLLARY 14.2.6. *Let $0 < \gamma_1 < \gamma_2 < 1$ and $\mathcal{F} \in \Theta_{\gamma_1,\gamma_2}(I)$. Then there exists $C_3 = C_3(\mathcal{F}) > 1$ such that for all $\mathbf{i} \neq \mathbf{j}$,*

$$|\Pi_{\mathcal{F}}(\mathbf{i}) - \Pi_{\mathcal{F}}(\mathbf{j})| \leqslant C_3 \cdot |I_{\mathbf{i}\wedge\mathbf{j}}^{\mathcal{F}}|. \tag{14.28}$$

If, moreover, $\mathcal{F}$ satisfies the SSP, then

$$C_3^{-1} \cdot |I_{\mathbf{i}\wedge\mathbf{j}}^{\mathcal{F}}| \leqslant |\Pi_{\mathcal{F}}(\mathbf{i}) - \Pi_{\mathcal{F}}(\mathbf{j})|. \tag{14.29}$$

PROOF. Consider (14.27). Now the upper bound (14.28) follows from Lemma 14.2.4 and Corollary 14.2.5(i). For the lower bound (14.29), observe that if $n = |\mathbf{i} \wedge \mathbf{j}|$, then $i_{n+1} \neq j_{n+1}$, and hence

$$|\Pi_{\mathcal{F}}(\sigma^n\mathbf{i}) - \Pi_{\mathcal{F}}(\sigma^n\mathbf{j})| \geqslant \delta > 0,$$

where

$$\delta := \min_{i\neq j\in\mathcal{A}} \operatorname{dist}(f_i(\Lambda_{\mathcal{F}}), f_j(\Lambda_{\mathcal{F}}))$$

is positive by the SSP. □

DEFINITION 14.2.7 (Metric associated with the IFS). Given a $\mathcal{C}^{1+\theta}$-smooth IFS $\mathcal{F}$ define an associated metric on Σ by

$$\varrho_{\mathcal{F}}(\mathbf{i},\mathbf{j}) := |I_{\mathbf{i}\wedge\mathbf{j}}^{\mathcal{F}}|, \quad \mathbf{i},\mathbf{j} \in \Sigma.$$

The axioms of a metric are easy to check. In fact, this is an *ultrametric*, that is,

$$\varrho_{\mathcal{F}}(\mathbf{i},\mathbf{j}) \leqslant \max\{\varrho_{\mathcal{F}}(\mathbf{i},\mathbf{k}), \varrho_{\mathcal{F}}(\mathbf{j},\mathbf{k})\} \quad \text{for } \mathbf{i},\mathbf{j},\mathbf{k} \in \Sigma.$$

It follows from Corollary 14.2.6 that the natural projection map Π is Lipschitz from the metric $\varrho_{\mathcal{F}}$ to the Euclidean metric, and it is bi-Lipschitz under the SSP.

14.2.2. Pressure function and Bowen's formula. Let $\mathcal{F} \in \Theta_{\gamma_1,\gamma_2}(I)$ for $0 < \gamma_1 < \gamma_2 < 1$. We next indicate some properties of the pressure function and show that the "conformal similarity dimension" $s(\mathcal{F})$ from (14.15) is well-defined and equals $\lim_{n\to\infty} s_n$, where s_n are from (14.13).

It follows from Corollary 14.2.5(ii) that

$$\log \gamma_1 \leqslant \frac{P_{\mathcal{F}}(t+u) - P_{\mathcal{F}}(t)}{u} \leqslant \log \gamma_2 \tag{14.30}$$

for all positive t and u. Consequently, $t \mapsto P_{\mathcal{F}}(t)$ is strictly decreasing and continuous. Clearly, $P_{\mathcal{F}}(0) = \log m > 0$, so (14.30) implies that the pressure function has a unique zero, which we denote $s(\mathcal{F})$.

LEMMA 14.2.8. *Suppose that $\mathcal{F} \in \Theta_{\gamma_1,\gamma_2}(I)$ for $0 < \gamma_1 < \gamma_2 < 1$.*

(i) *The solutions s_n of the equation* (14.13) *satisfy $s(\mathcal{F}) = \lim_{n\to\infty} s_n$;*

(ii) $\dim_{\mathrm{H}}(\Lambda_{\mathcal{F}}) \leqslant s(\mathcal{F})$.

PROOF OF LEMMA 14.2.8.

(i) Let

$$Q_n := \frac{1}{n} \log \sum_{\mathbf{i}\in\Sigma_n} |I_{\mathbf{i}}^{\mathcal{F}}|^s, \quad \text{where } s = s(\mathcal{F}).$$

By Lemma 14.2.4(i), (ii), and the definition of s as a solution of the Bowen equation $P_{\mathcal{F}}(s) = 0$, we have $\lim_{n\to\infty} Q_n = 0$. Now write

$$|I_{\mathbf{i}}^{\mathcal{F}}|^s = |I_{\mathbf{i}}^{\mathcal{F}}|^{s_n} \cdot |I_{\mathbf{i}}^{\mathcal{F}}|^{s-s_n}$$

and use the estimates $\gamma_1^n \leqslant |I_{\mathbf{i}}^{\mathcal{F}}| \leqslant \gamma_2^n$ (recall that $I = [0,1]$ by assumption), where $|\mathbf{i}| = n$, to obtain

$$|I_{\mathbf{i}}^{\mathcal{F}}|^s \in \left(\gamma_1^{n(s-s_n)} \cdot |I_{\mathbf{i}}^{\mathcal{F}}|^{s_n}, \ \gamma_2^{n(s-s_n)} \cdot |I_{\mathbf{i}}^{\mathcal{F}}|^{s_n}\right),$$

where we abuse the notation slightly, in the sense that the interval in the formula above has the endpoints indicated there, but which one is the right endpoint depends on the sign of $s - s_n$. Recall that $\sum_{\mathbf{i}\in\Sigma_n} |I_{\mathbf{i}}^{\mathcal{F}}|^{s_n} = 1$ by (14.13). Thus,

$$Q_n \in \left((s - s_n)\log\gamma_1, (s - s_n)\log\gamma_2\right),$$

hence $s - s_n \in \left(\frac{Q_n}{\log\gamma_1}, \frac{Q_n}{\log\gamma_2}\right)$, and so $s_n \to s$, as desired.

(ii) Fix $\varepsilon > 0$. Then by part (i) we have $s_n < s(\mathcal{F}) + \frac{\varepsilon}{2}$ for all n sufficiently large. Thus,

$$\sum_{\mathbf{i}\in\Sigma_n} |I_{\mathbf{i}}^{\mathcal{F}}|^{s(\mathcal{F})+\varepsilon} < \sum_{\mathbf{i}\in\Sigma_n} |I_{\mathbf{i}}^{\mathcal{F}}|^{s_n+\frac{\varepsilon}{2}} \leqslant \gamma_2^{\frac{n\varepsilon}{2}} \to 0, \quad \text{as n} \to \infty.$$

Since $\{I_{\mathbf{i}}^{\mathcal{F}} : \mathbf{i} \in \Sigma_n\}$ is a covering of the attractor $\Lambda_{\mathcal{F}}$, we obtain $\mathcal{H}^{s(\mathcal{F})+\varepsilon}(\Lambda_{\mathcal{F}}) = 0$, and $\varepsilon > 0$ being arbitrary, we conclude that $\dim_{\mathrm{H}}(\Lambda_{\mathcal{F}}) \leqslant s$.

□

14.2.3. Thermodynamic formalism and the proof of Theorems 14.2.2 and 14.2.3. In order to proceed, we will need some basic results on thermodynamic formalism; see [**Bow75**] for details. We remark that they can be also deduced from Section 10.5 with some straightforward modifications.

For a continuous function $\varphi : \Sigma \to \mathbb{R}$, the *topological pressure* is

$$(14.31)\qquad P(\varphi) := \lim_{n\to\infty} \frac{1}{n} \log \sum_{\mathbf{i}\in\Sigma_n} \exp\left[\sup_{\mathbf{j}\in[\mathbf{i}]} \sum_{k=0}^{n-1} \varphi(\sigma^k \mathbf{j})\right].$$

Given an IFS $\mathcal{F} \in \Theta_{\gamma_1,\gamma_2}(I)$ let

$$(14.32)\qquad \varphi_{\mathcal{F}}(\mathbf{i}) = \log\left|f'_{i_1}(\Pi_{\mathcal{F}}(\sigma\mathbf{i}))\right|, \quad \mathbf{i} \in \Sigma.$$

It is immediate from the Chain Rule that the pressure function defined in (14.16) is, in fact,

$$P_{\mathcal{F}}(t) = P(t \cdot \varphi_{\mathcal{F}}),\ t \in \mathbb{R}.$$

Consider the usual metric on Σ defined by $\mathrm{dist}(\mathbf{i},\mathbf{j}) = 2^{-|\mathbf{i}\wedge\mathbf{j}|}$.

THEOREM 14.2.9 (Bowen (1975)). *Let $\varphi : \Sigma \to \mathbb{R}$ be a Hölder-continuous function (using the metric* dist*). Then there exists a Borel probability measure μ supported on Σ and a constant $c_0 > 1$ such that for all $\mathbf{i} \in \Sigma^*$ we have*

$$(14.33)\qquad c_0^{-1} \leqslant \frac{\mu([\mathbf{i}])}{\exp(-|\mathbf{i}|P(\varphi) + \sum_{j=0}^{n-1}\varphi(\sigma^j\mathbf{j}))} \leqslant c_0$$

for all $\mathbf{j} \in [\mathbf{i}]$. Moreover, μ may be chosen σ-invariant and ergodic.

The measure μ is called the *Gibbs measure* for the potential φ.

PROOF OF THEOREM 14.2.2. In view of Lemma 14.2.8 it remains to prove that $\dim_{\mathrm{H}}(\Lambda_{\mathcal{F}}) \geqslant s(\mathcal{F})$. Consider the function $\varphi_{\mathcal{F}} : \Sigma \to \mathbb{R}$ defined by (14.32). Since $\mathcal{F}$ is $\mathcal{C}^{1+\theta}$-smooth, one can easily check that $\varphi_{\mathcal{F}}$ is Hölder continuous. Therefore, there exists a Gibbs measure μ for the potential $s(\mathcal{F})\cdot\varphi_{\mathcal{F}}$. Since $P_{\mathcal{F}}(s(\mathcal{F})\cdot\varphi_{\mathcal{F}}) = 0$, it follows from (14.33) that for all $\mathbf{j} \in \Sigma$,

$$\mu[\mathbf{j}|_n] \in \left(C_4^{-1}\left|f'_{\mathbf{j}|_n}(\Pi_{\mathcal{F}}(\sigma^n\mathbf{j}))\right|^{s(\mathcal{F})}, C_4\left|f'_{\mathbf{j}|_n}(\Pi_{\mathcal{F}}(\sigma^n\mathbf{j}))\right|^{s(\mathcal{F})}\right),$$

for some $C_4 = C_4(\mathcal{F}) > 1$. In view of Corollary 14.2.5, there exists $C_5 = C_5(\mathcal{F}) > 1$ such that

$$(14.34)\qquad \mu[\mathbf{j}|_n] \in \left(C_5^{-1}\left|I^{\mathcal{F}}_{\mathbf{j}|_n}\right|^{s(\mathcal{F})}, C_5\left|I^{\mathcal{F}}_{\mathbf{j}|_n}\right|^{s(\mathcal{F})}\right).$$

We claim that μ has exact local dimension $s(\mathcal{F})$ at every point $\omega \in \Sigma$, with respect to the metric $\varrho_{\mathcal{F}}$. For $\mathbf{j} \in \Sigma$ let $r_n^{\mathbf{j}} := \left|I^{\mathcal{F}}_{\mathbf{j}|_n}\right|$. We have $\gamma_1 \leqslant \frac{r^{\mathbf{j}}_{n+1}}{r^{\mathbf{j}}_n} < 1$, hence it suffices to consider closed balls $\overline{B}(\mathbf{j}, r_n^{\mathbf{j}})$ in the metric $\varrho_{\mathcal{F}}$ for the purpose of computing the local dimension at $\mathbf{j}$. By definition, $\overline{B}(\mathbf{j}, r_n^{\mathbf{j}}) = [\mathbf{j}|_n]$, hence

$$d_\mu(\mathbf{j}) = \lim_{n\to\infty} \frac{\log \mu[\mathbf{j}|_n]}{\log r_n^{\mathbf{j}}} = \lim_{n\to\infty} \frac{\log \mu[\mathbf{j}|_n]}{\log\left|I^{\mathcal{F}}_{\mathbf{j}|_n}\right|} = s(\mathcal{F}),$$

in view of (14.34). Thus $\dim_{\mathrm{H}}(\Sigma) = s(\mathcal{F})$, in the associated metric $\varrho_{\mathcal{F}}$. In the case of SSP, the attractor $\Lambda_{\mathcal{F}}$ is bi-Lipschitz with $(\Sigma, \varrho_{\mathcal{F}})$, hence $\dim_{\mathrm{H}}(\Lambda_{\mathcal{F}}) \geqslant s(\mathcal{F})$, completing the proof. $\square$

PROOF OF THEOREM 14.2.3. Now μ is an arbitrary ergodic measure on Σ. We again utilize the metric $\varrho_{\mathcal{F}}$ and compute the local dimension of μ almost everywhere: for $\mathbf{i} \in \Sigma$ we have

$$d_\mu(\mathbf{i}) = \lim_{n\to\infty} \frac{\log \mu[\mathbf{i}|_n]}{\log\left|I^{\mathcal{F}}_{\mathbf{i}|_n}\right|} = \frac{\lim_{n\to\infty} \frac{1}{n}\log \mu[\mathbf{i}|_n]}{\lim_{n\to\infty} \frac{1}{n}\log\left|I^{\mathcal{F}}_{\mathbf{i}|_n}\right|}, \tag{14.35}$$

assuming the limits in the right-hand side exist. Now,

$$\lim_{n\to\infty} \frac{1}{n}\log \mu[\mathbf{i}|_n] = -h_\mu \quad \text{for } \mu\text{-a.e. } \mathbf{i} \in \Sigma,$$

by the Shannon–McMillan–Breiman Theorem. As for the denominator of (14.35), we have for μ-a.e. $\mathbf{i}$:

$$\lim_{n\to\infty} \frac{1}{n}\log\left|I^{\mathcal{F}}_{\mathbf{i}|_n}\right| = \lim_{n\to\infty} \frac{1}{n}\sum_{j=1}^{n} \log f'_{i_j}(\Pi_{\mathcal{F}}(\sigma^j \mathbf{i})) = -\chi_\mu(\mathcal{F}),$$

by Corollary 14.2.5, the Chain Rule, and Birkhoff's Ergodic Theorem, see (14.18). It follows from (14.35) that the local dimension $d_\mu(\mathbf{i})$ exists as a limit and equals to $h_\mu/\chi_\mu(\mathcal{F})$ for μ-a.e. $\mathbf{i}$, and hence this is also the value of $\dim_{\mathrm{H}}(\nu(\mathcal{F},\mu))$. It remains to note again that $\Pi_{\mathcal{F}}$ is Lipschitz from $(\Sigma, \varrho_{\mathcal{F}})$ to $\Lambda_{\mathcal{F}}$ and bi-Lipschitz under the SSP assumption. The proof is complete. □

We note here that, just as for self-similar sets, the box dimension of self-conformal sets equals the Hausdorff dimension, regardless of overlaps. Moreover, the Hausdorff measure of the appropriate dimension, of any self-conformal set is finite, again, regardless of overlaps. This was first shown by Falconer [**Fal89**].

THEOREM 14.2.10. *Suppose that $\mathcal{F} = (f_1, \dots, f_m)$ is a hyperbolic $\mathcal{C}^{1+\theta}$-smooth IFS on $I \subset \mathbb{R}$. Let $\Lambda_{\mathcal{F}}$ be the attractor of the IFS $(f_1, \dots, f_m)$. Then*

$$\dim_{\mathrm{B}}(\Lambda_{\mathcal{F}}) = \dim_{\mathrm{H}}(\Lambda_{\mathcal{F}}) \quad \textit{and} \quad \mathcal{H}^{\dim_{\mathrm{H}}(\Lambda_{\mathcal{F}})}(\Lambda_{\mathcal{F}}) < \infty.$$

The proof, essentially a straightforward adaptation of the proofs of Theorem 3.3.1 and Theorem 3.4.1, using the bounded distortion Corollary 14.2.5 and Theorem 14.2.2, is left to the reader. Alternatively, we refer to Jonker and Veerman [**JV02**, Corollary C] for another proof, or see [**Fal97**, Theorem 3.2]. The same holds for a conformal contracting IFS in $\mathbb{R}^d$ for $d > 1$.

14.3. Equivalent conditions for OSC and WSP for nonlinear Iterated Function Systems

The OSC and SOSC (Open set Condition and Strong Open Set Condition respectively) were introduced in Definition 1.5.2 for general IFS. In contrast with that, the Weak Separation Property (WSP) was introduced in Definition 4.2.3 only for self-similar IFS. The extension of WSP from the self-similar to hyperbolic IFS was developed by Lau, Ngai and Wang [**LNW09**]. Then Angelevska, Käenmäki and Troscheit [**AKT20**] modified the original definition of WSP and provided equivalent conditions for it. In the second part of this section we present these results. In the first part, we give equivalent conditions for the OSC, based on [**PRSS01**]. For the proofs, we refer to the original papers.

14.3.1. Equivalent conditions for OSC for hyperbolic $\mathcal{C}^{1+\theta}$ IFS on the line. Let $\mathcal{F} = (f_1, \dots, f_m)$ be a hyperbolic $\mathcal{C}^{1+\theta}$-smooth IFS on the interval I (as in Definition 14.2.1). We say that the *Bandt–Graf condition* holds if there exists $\varepsilon > 0$ such that for any distinct $\mathbf{i}, \mathbf{j} \in \Sigma^*$ we can find $x \in \Lambda$ such that

$$|f_{\mathbf{i}}(x) - f_{\mathbf{j}}(x)| > \varepsilon \min \{|\Lambda_{\mathbf{i}}|, |\Lambda_{\mathbf{j}}|\} . \tag{14.36}$$

Note that the same condition was introduced in Section 4.1 (there we called it Condition BG2) for self-similar IFS.

The following theorem is a special case of [**PRSS01**, Theorem 1.1]

THEOREM 14.3.1 (Peres, Rams, Simon and Solomyak). *Let $\mathcal{F}$ be as above. Then the following are equivalent:*

(i) *the OSC holds;*
(ii) $0 < \mathcal{H}^s(\Lambda) < \infty$, *with* $s = P^{-1}(0)$, *where P is the topological pressure defined in* (14.31)*;*
(iii) *the Bandt–Graf condition holds;*
(iv) *the SOSC holds.*

This theorem was stated in [**PRSS01**] for self-conformal IFS on $\mathbb{R}^d$. Similarly, some of the results in the following subsection are stated in their original reference in a higher dimensional set-up, but we restrict ourselves to the special case of hyperbolic $\mathcal{C}^{1+\theta}$ IFS on the line.

14.3.2. WSP for hyperbolic $\mathcal{C}^{1+\theta}$ IFS on the line. In this subsection we always assume that $\mathcal{F} = (f_1, \dots, f_m)$ is a hyperbolic $\mathcal{C}^{1+\theta}$-smooth IFS on a compact interval $I \subset \mathbb{R}$. We present the version of WSP introduced by Angelevska, Käenmäki and Troscheit [**AKT20**]. To do so, recall that for a finite word $\mathbf{i} = (i_1, \dots, i_n) \in \Sigma^*$ with $n \geqslant 1$ we write $\mathbf{i}^- := (i_1, \dots, i_{n-1})$, and $\Lambda_{\mathbf{i}} := f_{\mathbf{i}}(\Lambda)$.

DEFINITION 14.3.2 (WSP). Let

$$\Phi(x, r) := \{f_{\mathbf{i}}|_\Lambda : |\Lambda_{\mathbf{i}}| \leqslant r < |\Lambda_{\mathbf{i}^-}| \text{ and } \Lambda_{\mathbf{i}} \cap B(x, r) \neq \varnothing\} . \tag{14.37}$$

We say that the WSP holds if

$$\sup \{\#\Phi(x, r) : x \in \Lambda \text{ and } r > 0\} < \infty. \tag{14.38}$$

We remark that WSP for the nonlinear case was introduced by Lau, Wang and Ngai [**LNW09**] in a similar way. Namely, their definition is exactly like (14.38), if we replace $\Phi(x, r)$ by

$$\Phi^*(x, r) := \{f_{\mathbf{i}} : |\Lambda_{\mathbf{i}}| \leqslant r < |\Lambda_{\mathbf{i}^-}| \text{ and } \Lambda_{\mathbf{i}} \cap B(x, r) \neq \varnothing\} . \tag{14.39}$$

The difference is as follows: in $\Phi^*(x, r)$ we consider two functions distinct if they differ somewhere on I, whereas in $\Phi(x, r)$ two functions are distinct if they differ somewhere on the attractor. Clearly, $\Phi(x, r) \leqslant \Phi^*(x, r)$, so the notion of WSP due to Angelevska, Käenmäki and Troscheit is a weaker condition than the original one given by Lau, Wang and Ngai. In particular, there exists a conformal IFS on $\mathbb{R}$ formed by three maps such that it satisfies the WSP in the sense of Definition 14.3.2 but $\sup\{\#\Phi^*(x, r) : x \in \Lambda \text{ and } r > 0\} = \infty$, see [**AKT20**, Example 7.7].

The following separation condition was introduced in [**AKT20**].

DEFINITION 14.3.3 (Identity Limit Criterion). The Identity Limit Criterion (ILC) holds if

$$\inf\left\{\|f_{\mathbf{i}}'\|_\infty^{-1}\cdot\sup_{x\in\Lambda}|f_{\mathbf{i}}(x)-f_{\mathbf{j}}(x)| : \mathbf{i},\mathbf{j}\in\Sigma^* \text{ such that } f_{\mathbf{i}}|_\Lambda\neq f_{\mathbf{j}}|_\Lambda\right\}>0. \tag{14.40}$$

The equivalence of these two conditions is established in [**AKT20**, Theorem 3.2].

THEOREM 14.3.4 (Angelevska, Käenmäki and Troscheit). *If the attractor Λ contains at least two points, then conditions WSP and ILC are equivalent.*

It was pointed out in [**AKT20**, Theorem 4.1] that the failure of these conditions implies that the Assouad dimension of the attractor is maximal. Recall that the Assouad dimension of a set A is denoted by $\dim_{\mathrm{A}}(A)$ and its definition was given in (7.72).

THEOREM 14.3.5 (Angelevska, Käenmäki and Troscheit). *Assume that Λ is not a singleton. If WSP does not hold, then* $\dim_{\mathrm{A}}(\Lambda)=1$.

Moreover, the following theorem ([**AKT20**, Corollary 4.2]) generalizes the corresponding results about self-similar IFS (see Section 7.6) to hyperbolic $\mathcal{C}^{1+\theta}$ IFS's.

THEOREM 14.3.6 (Angelevska, Käenmäki and Troscheit). *Let $\mathcal{F}=(f_1,\dots,f_m)$ be a hyperbolic $\mathcal{C}^{1+\theta}$-smooth IFS on the interval I (as in Definition 14.2.1) and let Λ be its attractor. We assume that Λ is not a singleton and $s:=\dim_{\mathrm{H}}(\Lambda)<1$. Then the following conditions are equivalent:*

(i) *the WSP holds,*
(ii) $0<\mathcal{H}^s(\Lambda)<\infty$,
(iii) *Λ is s-Ahlfors regular,*
(iv) $\dim_{\mathrm{A}}(\Lambda)=s$.
(v) *the ILC holds.*

14.3.3. The exact overlap conjecture in the hyperbolic settings. Let $\mathcal{F}=(f_1,\dots,f_m)$ be a hyperbolic $\mathcal{C}^{1+\theta}$-smooth IFS on the interval I. As usual, we write Λ for its attractor. Recall that we say that $\mathcal{F}$ has an exact overlap if there exist distinct $\mathbf{i}\neq\mathbf{j}\in\Sigma^*$ with $f_{\mathbf{i}}|_\Lambda=f_{\mathbf{j}}|_\Lambda$.

CONJECTURE 14.3.7 (dimension drop conjecture). *Either $\mathcal{F}$ has an exact overlap, or* $\dim_{\mathrm{H}}(\Lambda)=\min\{1,P^{-1}(0)\}$, *where P is the topological pressure.*

The next result [**AKT20**, Theorem 4.3] verifies this conjecture for those hyperbolic attractors on the line which have positive Hausdorff measure in its dimension.

THEOREM 14.3.8 (Angelevska, Käenmäki and Troscheit). *Suppose that $s=\dim_{\mathrm{H}}(\Lambda)<1$ and $\mathcal{H}^s(\Lambda)>0$. Then $s=P^{-1}(0)$ if and only if there is no exact overlap.*

14.4. Families of nonlinear overlapping IFS's

Our focus is on parametrized families of overlapping IFS. Let $U\subset\mathbb{R}^d$ be an open set, and suppose that we have a family of hyperbolic $\mathcal{C}^{1+\gamma}$-smooth IFS $\mathcal{F}^{\mathbf{t}}=(f_1^{\mathbf{t}},\dots,f_m^{\mathbf{t}})$ on $I\subset\mathbb{R}$ for $\mathbf{t}\in U$. We will say that the family is continuous if the maps $\mathbf{t}\mapsto f_i^{\mathbf{t}}$ are continuous from U to $\mathcal{C}^{1+\theta}(I\to I)$ with the norm $\|\cdot\|_{1+\theta}$. Let $\Pi_{\mathbf{t}}$ be the natural projection associated with the IFS $\mathcal{F}^{\mathbf{t}}$. We say that the family

$\{\mathcal{F}^{\mathbf{t}}\}_{\mathbf{t}\in U}$ satisfies the *transversality condition* if there exists $C_{\text{tr}} > 0$ such that for all $r > 0$,

$$\forall\, \mathbf{i}, \mathbf{j} \in \Sigma,\ i_1 \neq j_1 \implies \mathcal{L}^d\big(\{\mathbf{t} \in U : |\Pi_{\mathbf{t}}(\mathbf{i}) - \Pi_{\mathbf{t}}(\mathbf{j})| \leqslant r\}\big) \leqslant C_{\text{tr}} r. \tag{14.41}$$

THEOREM 14.4.1. *Let $\{\mathcal{F}^{\mathbf{t}}\}_{\mathbf{t}\in U}$ be a continuous family of hyperbolic $C^{1+\theta}$-smooth IFS satisfying the transversality condition (14.41). Let $\Lambda_{\mathbf{t}} = \Lambda_{\mathcal{F}^{\mathbf{t}}}$ be the attractor of the IFS. Then*

(i) $\dim_{\mathrm{H}}(\Lambda_{\mathbf{t}}) = \min\{s(\mathbf{t}), 1\}$ *for $\mathcal{L}^d$-a.e. $\mathbf{t} \in U$ where $P_{\mathcal{F}^{\mathbf{t}}}(s(\mathbf{t})) = 0$.*
(ii) $\mathcal{L}^1(\Lambda_{\mathbf{t}}) > 0$ *for $\mathcal{L}^d$-a.e. $\mathbf{t} \in U$ such that $s(\mathbf{t}) > 1$.*

THEOREM 14.4.2. *Let $\{\mathcal{F}^{\mathbf{t}}\}_{\mathbf{t}\in U}$ be a continuous family of hyperbolic $C^{1+\theta}$-smooth IFS satisfying the transversality condition (14.41). Let μ be an ergodic shift-invariant probability measure on Σ with positive entropy h_μ and $\nu_{\mathbf{t}} = \mu \circ \Pi_{\mathbf{t}}^{-1}$. Then*

(i) $\dim_{\mathrm{H}}(\nu_{\mathbf{t}}) = \min\left\{\frac{h_\mu}{\chi_\mu(\mathcal{F}^{\mathbf{t}})}, 1\right\}$ *for $\mathcal{L}^d$-a.e. $\mathbf{t} \in U$.*
(ii) $\nu_{\mathbf{t}} \ll \mathcal{L}^1$ *for $\mathcal{L}^d$-a.e. $\mathbf{t} \in U$ such that $h_\mu/\chi_\mu(\mathcal{F}^{\mathbf{t}}) > 1$.*

The following lemma deals with perturbations of a hyperbolic IFS.

LEMMA 14.4.3. *Let $0 < \gamma_1 < \gamma_2 < 1$.*

(i) *$\mathcal{F} \mapsto s(\mathcal{F})$ is continuous, as a function from $(\Theta_{\gamma_1,\gamma_2}(I), \rho)$ to $\mathbb{R}$.*
(ii) *Let μ be a probability measure on Σ. Then $\mathcal{F} \mapsto \chi_\mu(\mathcal{F})$ is continuous, as a function from $(\Theta_{\gamma_1,\gamma_2}(I), \rho)$ to $\mathbb{R}$.*

PROOF. (i) We will prove that

$$|P_{\mathcal{G}}(s(\mathcal{F}))| \leqslant C_2 \cdot s(\mathcal{F}) \cdot \rho(\mathcal{F}, \mathcal{G})^\theta, \tag{14.42}$$

where C_2 comes from Lemma 14.2.4(ii). Then the desired claim follows easily. Indeed, by (14.30),

$$|P_{\mathcal{G}}(s(\mathcal{F}))| = |P_{\mathcal{G}}(s(\mathcal{F})) - P_{\mathcal{G}}(s(\mathcal{G}))| \geqslant |\log \gamma_2| \cdot |s(\mathcal{F}) - s(\mathcal{G})|.$$

Thus, by (14.42),

$$|s(\mathcal{F}) - s(\mathcal{G})| \leqslant \frac{C_2 \cdot s(\mathcal{F}) \cdot \rho(\mathcal{F}, \mathcal{G})^\theta}{|\log \gamma_2|},$$

which implies continuity at $\mathcal{F}$. To complete the proof it remains to verify (14.42). Using Corollary 14.2.5(ii) and Lemma 14.2.4(ii), we obtain

$$\begin{aligned} |P_{\mathcal{G}}(s(\mathcal{F}))| &= |P_{\mathcal{G}}(s(\mathcal{F})) - P_{\mathcal{F}}(s(\mathcal{F}))| \\ &= \lim_{n\to\infty} \frac{1}{n} \left| \log \frac{\sum_{\mathbf{i}\in\Sigma_n} |f_{\mathbf{i}}'(0)|^{s(\mathcal{F})}}{\sum_{\mathbf{i}\in\Sigma_n} |g_{\mathbf{i}}'(0)|^{s(\mathcal{F})}} \right| \\ &\leqslant C_2 \cdot s(\mathcal{F}) \cdot \rho(\mathcal{F}, \mathcal{G})^\theta, \end{aligned}$$

as desired.

(ii) Letting $p \to \infty$ in (14.23) and then taking $k = 1$ and $n \to \infty$ in (14.24) yields, for $\mathcal{F}, \mathcal{G} \in \Theta_{\gamma_1,\gamma_2}(I)$ with $\rho(\mathcal{F}, \mathcal{G}) \leqslant 1$:

$$|f_{i_1}'(\Pi_{\mathcal{F}}(\sigma\mathbf{i})) - g_{i_1}'(\Pi_{\mathcal{G}}(\sigma\mathbf{i}))| \leqslant \big[1 + L(\mathcal{F}) \cdot (1 - \gamma_2)^{-\theta}\big] \cdot \rho(\mathcal{F}, \mathcal{G})^\theta,$$

hence

$$|\chi_\mu(\mathcal{F}) - \chi_\mu(\mathcal{G})| \leqslant \big[1 + L(\mathcal{F}) \cdot (1 - \gamma_2)^{-\theta}\big] \cdot \rho(\mathcal{F}, \mathcal{G})^\theta,$$

by the definition (14.18). □

The next lemma makes use of the transversality condition (14.41).

LEMMA 14.4.4. *Let $\{\mathcal{F}^{\mathbf{t}}\}_{\mathbf{t}\in U}$ be a continuous family of hyperbolic $C^{1+\theta}$-smooth IFS on I satisfying the transversality condition* (14.41). *Then for every $\alpha \in (0,1)$, for all $\mathbf{i},\mathbf{j} \in \Sigma$ with $i_1 \neq j_1$, there exists $C_3 = C_3(\alpha)$ such that*

$$\int_U \frac{d\mathbf{t}}{|\Pi_{\mathbf{t}}(\mathbf{i}) - \Pi_{\mathbf{t}}(\mathbf{j})|^{\alpha}} \leqslant C_3.$$

PROOF. In view of (14.41), we can estimate as follows:

$$\begin{aligned}\int_U \frac{d\mathbf{t}}{|\Pi_{\mathbf{t}}(\mathbf{i}) - \Pi_{\mathbf{t}}(\mathbf{j})|^{\alpha}} &= \int_0^{\infty} \mathcal{L}^d\Big\{\mathbf{t} \in U: \frac{1}{|\Pi_{\mathbf{t}}(\mathbf{i}) - \Pi_{\mathbf{t}}(\mathbf{j})|^{\alpha}} \geqslant x\Big\}\,dx \\ &= \alpha\int_0^{\infty} \mathcal{L}^d\{\mathbf{t} \in U: |\Pi_{\mathbf{t}}(\mathbf{i}) - \Pi_{\mathbf{t}}(\mathbf{j})| \leqslant r\}\,r^{-\alpha-1}\,dr \\ &= \alpha\Big(\int_0^{|I|} + \int_{|I|}^{\infty}\Big) \leqslant \frac{C_{\mathrm{tr}}\alpha}{1-\alpha}|I|^{1-\alpha} + \mathcal{L}^d(U)\cdot|I|^{-\alpha}. \qquad \square\end{aligned}$$

We next turn to the proof of Theorem 14.4.1. It essentially uses the Generalized Projection Scheme, however, there are some minor distinctions, and it is easier to adjust the argument to the setting at hand than to make the GPS "all-encompassing". We will use the symbols $\lesssim$ and $\gtrsim$ to indicate inequalities which hold up to a multiplicative constant.

PROOF OF THEOREM 14.4.1. (i) The upper bound has already been verified in Lemma 14.2.8(ii). For the lower bound, in view of Lemma 14.4.3(i), it suffices to prove that for any $\mathbf{t}_0 \in U$ and any $\varepsilon > 0$ there exists a neighborhood $B_\delta(\mathbf{t}_0)$ such that

$$\dim_{\mathrm{H}}(\Lambda_{\mathbf{t}}) \geqslant \min\{s(\mathbf{t}_0), 1\} - \varepsilon \quad \text{for } \mathcal{L}^d\text{-a.e. } \mathbf{t} \in B_\delta(\mathbf{t}_0).$$

Denote $\mathcal{F} := \mathcal{F}_{\mathbf{t}_0}$ and $s = s(\mathbf{t}_0)$. Assume that $s \leqslant 1$; the remaining case will follow from part (ii). As in the proof of Theorem 14.2.2, there exists a Gibbs measure μ for the potential $s(\mathcal{F})\cdot\phi_{\mathcal{F}}$, which satisfies (14.34). Denote the product measure $\mu\times\mu$ on $\Sigma\times\Sigma$ by μ_2. By the potential-theoretic characterization of the Hausdorff dimension, it is enough to show that

$$R(\mathbf{t}) := \iint_{\Sigma\times\Sigma} \frac{d\mu_2(\mathbf{i},\mathbf{j})}{|\Pi_{\mathbf{t}}(\mathbf{i}) - \Pi_{\mathbf{t}}(\mathbf{j})|^{s-\varepsilon}} < \infty$$

for $\mathcal{L}^d$-a.e. $\mathbf{t} \in B_\delta(\mathbf{t}_0)$ for some $\delta > 0$. Indeed, this is equivalent to finiteness of the $(s-\varepsilon)$-energy of the push-forward measure $\mu\circ\Pi_{\mathbf{t}}^{-1}$.

Since the family of IFS is continuous, in view of (14.21), we can find $\delta > 0$ such that for all $\mathbf{i} \in \Sigma^*$,

$$\big|(f_{\mathbf{i}}^{\mathbf{t}})'(0)\big|^{s-\varepsilon} \geqslant \big|f_{\mathbf{i}}'(0)\big|^{s-\varepsilon/2} \quad \text{for all } \mathbf{t} \in B_\delta(\mathbf{t}_0). \tag{14.43}$$

It is enough to prove that $\int_{B_\delta(\mathbf{t}_0)} R(\mathbf{t})\,d\mathbf{t} < \infty$. By Fubini's Theorem,

$$\int\limits_{B_\delta(\mathbf{t}_0)} R(\mathbf{t})\,d\mathbf{t} = \sum_{n\geqslant 0}\sum_{\mathbf{k}\in\Sigma_n} \iint_{A_{\mathbf{k}}}\Big(\int_{B_\delta(\mathbf{t}_0)} \frac{d\mathbf{t}}{|\Pi_{\mathbf{t}}(\mathbf{i}) - \Pi_{\mathbf{t}}(\mathbf{j})|^{s-\varepsilon}}\Big)\,d\mu_2(\mathbf{i},\mathbf{j}),$$

where $A_{\mathbf{k}} = \{(\mathbf{i},\mathbf{j}) \in \Sigma \times \Sigma : \mathbf{i} \wedge \mathbf{j} = \mathbf{k}\}$. Now, for $\mathbf{i},\mathbf{j} \in \Sigma$ with $|\mathbf{i} \wedge \mathbf{j}| = n \geqslant 0$ and $\mathbf{t} \in B_\delta(\mathbf{t}_0)$ we have

$$\begin{aligned} |\Pi_{\mathbf{t}}(\mathbf{i}) - \Pi_{\mathbf{t}}(\mathbf{j})|^{s-\varepsilon} &\geq \left|(f^{\mathbf{t}}_{\mathbf{i}\wedge\mathbf{j}})'(0)\right|^{s-\varepsilon} \cdot |\Pi_{\mathbf{t}}(\sigma^n\mathbf{i}) - \Pi_{\mathbf{t}}(\sigma^n\mathbf{j})|^{s-\varepsilon} \\ &\geqslant \left|f'_{\mathbf{i}\wedge\mathbf{j}}(0)\right|^{s-\varepsilon/2} \cdot |\Pi_{\mathbf{t}}(\sigma^n\mathbf{i}) - \Pi_{\mathbf{t}}(\sigma^n\mathbf{j})|^{s-\varepsilon}, \end{aligned}$$

by (14.26) and (14.43). Thus, we have by Lemma 14.4.4, using that $(\sigma^n\mathbf{i})_1 \neq (\sigma^n\mathbf{j})_1$:

$$\begin{aligned} \int_{B_\delta(\mathbf{t}_0)} \frac{d\mathbf{t}}{|\Pi_{\mathbf{t}}(\mathbf{i}) - \Pi_{\mathbf{t}}(\mathbf{j})|^{s-\varepsilon}} &\leq \int_{B_\delta(\mathbf{t}_0)} \frac{\left|f'_{\mathbf{i}\wedge\mathbf{j}}(0)\right|^{-(s-\varepsilon/2)} d\mathbf{t}}{|\Pi_{\mathbf{t}}(\sigma^n\mathbf{i}) - \Pi_{\mathbf{t}}(\sigma^n\mathbf{j})|^{s-\varepsilon}} \\ &\leqslant C_3 \left|f'_{\mathbf{i}\wedge\mathbf{j}}(0)\right|^{-(s-\varepsilon/2)}. \end{aligned}$$

Therefore,

$$\begin{aligned} \int_{B_\delta(\mathbf{t}_0)} R(\mathbf{t})\, d\mathbf{t} &\leq \sum_{n\geqslant 0} \sum_{\mathbf{k}\in\Sigma_n} \iint_{A_{\mathbf{k}}} \left|f'_{\mathbf{i}\wedge\mathbf{j}}(0)\right|^{-(s-\varepsilon/2)} d\mu_2(\mathbf{i},\mathbf{j}) \\ &= \sum_{n\geqslant 0} \sum_{\mathbf{k}\in\Sigma_n} \left|f'_{\mathbf{k}}(0)\right|^{-(s-\varepsilon/2)} \mu_2(A_{\mathbf{k}}) \\ &\leqslant \sum_{n\geqslant 0} \sum_{\mathbf{k}\in\Sigma_n} \left|f'_{\mathbf{k}}(0)\right|^{-(s-\varepsilon/2)} \mu([\mathbf{k}])^2. \end{aligned}$$

By (14.34) and (14.25), we have $\mu([\mathbf{k}]) \preceq |f'_{\mathbf{k}}(0)|^s$, and $|f'_{\mathbf{k}}(0)| \leqslant \gamma_2^n$, so finally,

$$\begin{aligned} \int_{B_\delta(\mathbf{t}_0)} R(\mathbf{t})\, d\mathbf{t} &\preceq \sum_{n\geqslant 0} \sum_{\mathbf{k}\in\Sigma_n} \left|f'_{\mathbf{k}}(0)\right|^{-(s-\varepsilon/2)} \mu([\mathbf{k}])^2 \\ &= \sum_{n\geqslant 0} \gamma_2^{n\varepsilon/2} \sum_{\mathbf{k}\in\Sigma_n} \mu([\mathbf{k}]) = \sum_{n\geqslant 0} \gamma_2^{n\varepsilon/2} < \infty. \end{aligned}$$

This completes the proof.

(ii) Let $\mathbf{t}_0 \in U$ be such that $s = s(\mathbf{t}_0) > 1$; denote $\mathcal{F} = \mathcal{F}_{\mathbf{t}_0}$. As in the proof of part (i), consider the Gibbs measure μ for potential $s(\mathcal{F}) \cdot \phi_{\mathcal{F}}$, which satisfies (14.34), and the push-forward measure

$$\nu_{\mathbf{t}} = \mu \circ \Pi_{\mathbf{t}}^{-1}$$

on the attractor $\Lambda_{\mathbf{t}}$. Choose $\varepsilon > 0$ such that $s - 1 > \varepsilon$, and then $\delta > 0$ so that

$$\left|(f^{\mathbf{t}}_{\mathbf{i}})'(0)\right| \geqslant \left|f'_{\mathbf{i}}(0)\right|^{1+\varepsilon} \quad \text{for all } \mathbf{t} \in B_\delta(\mathbf{t}_0) \text{ and } \mathbf{i} \in \Sigma_*. \tag{14.44}$$

It suffices to show that $\nu_{\mathbf{t}}$ is absolutely continuous with respect to the Lebesgue measure for a.e. $\mathbf{t} \in B_\delta(\mathbf{t}_0)$, for some $\delta > 0$. We apply the scheme of Mattila, which we have used repeatedly (see, e.g., the proof of Theorem 6.3.3) with appropriate modifications. The goal is to verify that

$$\mathcal{I} := \int_{B_\delta(\mathbf{t}_0)} \int_{\mathbb{R}} \underline{D}(\nu_{\mathbf{t}}, x)\, d\nu_{\mathbf{t}}(x), d\mathbf{t} < \infty, \tag{14.45}$$

where

$$\underline{D}(\nu_{\mathbf{t}}, x) = \liminf_{r\to 0} (2r)^{-1} \nu_{\mathbf{t}}[x - r, x + r]$$

is the lower derivative of the measure $\nu_{\mathbf{t}}$ at the point x. By Fatou's Lemma,

$$\mathcal{I} \leqslant \liminf_{r\to 0}(2r)^{-1}\int_{B_\delta(\mathbf{t}_0)}\int_{\mathbb{R}} \nu_{\mathbf{t}}[x-r,x+r]\,d\nu_{\mathbf{t}}(x)\,d\mathbf{t}.$$

Next we write $\nu_{\mathbf{t}}[x-r,x+r]$ as an integral of the indicator function and change the variable to obtain

$$\int_{\mathbb{R}} \nu_{\mathbf{t}}[x-r,x+r]\,d\nu_{\mathbf{t}}(x) = \iint_{\Sigma\times\Sigma} \mathbf{1}_{\{\mathbf{j}:\ |\Pi_{\mathbf{t}}(\mathbf{i})-\Pi_{\mathbf{t}}(\mathbf{j})|\leqslant r\}}\,d\mu(\mathbf{i})\,d\mu(\mathbf{j}).$$

Applying Fubini's Theorem and then decomposing $\Sigma\times\Sigma$ into the disjoint union according to the common initial word, yields

$$\mathcal{I} \leqslant \liminf_{r\to 0}(2r)^{-1}\sum_{n\geqslant 0}\sum_{\mathbf{k}\in\Sigma_n}\iint_{A_{\mathbf{k}}} \mathcal{L}^d\{\mathbf{t}\in B_\delta(\mathbf{t}_0):\ |\Pi_{\mathbf{t}}(\mathbf{i})-\Pi_{\mathbf{t}}(\mathbf{j})|\leqslant r\}\,d\mu_2(\mathbf{i},\mathbf{j}).$$

For $\mathbf{i}\neq\mathbf{j}\in\Sigma$ with $|\mathbf{i}\wedge\mathbf{j}|=n$ and $\mathbf{t}\in B_\delta(\mathbf{t}_0)$ we have

$$\begin{aligned}|\Pi_{\mathbf{t}}(\mathbf{i})-\Pi_{\mathbf{t}}(\mathbf{j})| &\succeq \left|(f^{\mathbf{t}}_{\mathbf{i}\wedge\mathbf{j}})'(0)\right|\cdot|\Pi_{\mathbf{t}}(\sigma^n\mathbf{i})-\Pi_{\mathbf{t}}(\sigma^n\mathbf{j})| \\ &\geqslant \left|f'_{\mathbf{i}\wedge\mathbf{j}}(0)\right|^{1+\varepsilon}\cdot|\Pi_{\mathbf{t}}(\sigma^n\mathbf{i})-\Pi_{\mathbf{t}}(\sigma^n\mathbf{j})|,\end{aligned}$$

by (14.26) and (14.44). Thus we can estimate, using the transversality condition (14.41),

$$\mathcal{L}^d\{\mathbf{t}\in B_\delta(\mathbf{t}_0):\ |\Pi_{\mathbf{t}}(\mathbf{i})-\Pi_{\mathbf{t}}(\mathbf{j})|\leqslant r\}\preceq C_{\mathrm{tr}}\,r\cdot\left|f'_{\mathbf{i}\wedge\mathbf{j}}(0)\right|^{-(1+\varepsilon)}.$$

Substituting this into the last inequality for the integral $\mathcal{I}$, we obtain, almost as in the proof of part (i):

$$\begin{aligned}\mathcal{I} &\preceq \sum_{n\geqslant 0}\sum_{\mathbf{k}\in\Sigma_n}\iint_{A_{\mathbf{k}}}\left|f'_{\mathbf{i}\wedge\mathbf{j}}(0)\right|^{-(1+\varepsilon)}\,d\mu_2(\mathbf{i},\mathbf{j}) \\ &\leqslant \sum_{n\geqslant 0}\sum_{\mathbf{k}\in\Sigma_n}\left|f'_{\mathbf{k}}(0)\right|^{-(1+\varepsilon)}\mu([\mathbf{k}])^2.\end{aligned}$$

Again we can estimate $\mu([\mathbf{k}])\preceq|f'_{\mathbf{k}}(0)|^s$, and $|f'_{\mathbf{k}}(0)|\leqslant\gamma_2^n$, hence

$$\mathcal{I}\preceq\sum_{n\geqslant 0}\sum_{\mathbf{k}\in\Sigma_n}\gamma_2^{n(s-1-\varepsilon)}\mu([\mathbf{k}])=\sum_{n\geqslant 0}\gamma_2^{n(s-1-\varepsilon)}<\infty,$$

using that $s>1+\varepsilon$. We have confirmed (14.45), and the proof is complete. □

About the proof of Theorem 14.4.2. It is analogous to the proof of Theorem 6.6.2 from the Generalized Projection Scheme. In appropriate places one needs to apply the bounded distortion estimates and the nonlinear transversality condition. The details are left to the reader; we only make a few comments:

(i) The upper bound is from Theorem 14.2.3, for all parameters $\mathbf{t}$. In order to prove the lower bound, we first restrict the measure μ to a subset of large measure by Egorov's Theorem, where the convergence in the Shannon–McMillan–Breiman Theorem is uniform (it is here that we need ergodicity). The restricted measure is exact dimensional, with the local dimension proportional to its entropy h_μ. The Hölder exponent of the natural projection is essentially equal to the Lyapunov exponent, and so the argument similar to the proof of Theorem 6.6.2(i) goes through without much difficulty.

(ii) In order to prove absolute continuity of the push-forward measure, we apply Egorov's Theorem as in part (i), and for the projection of the restricted measure one can show that there is L^2-density for a.e. $\mathbf{t}$ in the appropriate region, similarly to the proof of Theorem 6.6.2(iv). □

14.5. Translation families and topologically generic IFS with overlaps

Recall the definitions of the space $\Theta(I)$ and $\Theta_{\gamma_1,\gamma_2}(I)$ from Section 14.2.1. For a hyperbolic IFS $\mathcal{F} = (f_1, \ldots, f_m) \in \Theta(I)$ we consider the translation family

$$\mathcal{F}^{\mathbf{t}} := (f_1(x) + t_1, \ldots, f_m(x) + t_m), \;\; \mathbf{t} = (t_1, \ldots, t_m).$$

It is easy to see that $Q(\mathcal{F}) := \{\mathbf{t} \in \mathbb{R}^m : \mathcal{F}^{\mathbf{t}} \in \Theta(I)\}$ is a nonempty open parallelepiped in $\mathbb{R}^m$ containing the origin and

$$Q(\mathcal{F}) \subset [-|I|, |I|]^m. \tag{14.46}$$

Now we introduce notation that we need in order to state an assumption which guarantees that the transversality condition holds.

DEFINITION 14.5.1. Let $0 < \gamma_1 < \gamma_2 < 1$ and $\mathcal{F} = (f_1, \ldots, f_m) \in \Theta_{\gamma_1,\gamma_2}(I)$. We define $r_i := \max_{x \in I} \{|f_i'(x)|\}$ and write $r_{\max}(\mathcal{F}) := \max\{r_i\}_{i=1}^m$. Assume that $r_j = \max\{r_i\}_{i=1}^m$. Then we define $r_{\max^-}(\mathcal{F}) := \max\{r_i : 1 \leqslant i \leqslant m \;\&\; i \neq j\}$.

THEOREM 14.5.2. *Let $\mathcal{F}$ be an IFS as in Definition 14.5.1. Assume that*

$$r := r_{max} + r_{max^-} < 1. \tag{14.47}$$

Then $\{\mathcal{F}^{\mathbf{t}}\}_{\mathbf{t} \in Q(\mathcal{F})}$ satisfies the transversality condition. In particular, for $\mathcal{L}^m$-a.e. $\mathbf{t} \in Q(\mathcal{F})$, the following hold:

(i) $\dim_{\mathrm{H}}(\Lambda_{\mathcal{F}^{\mathbf{t}}}) = \min\{s(\mathcal{F}^{\mathbf{t}}), 1\}$;
(ii) *if* $s(\mathcal{F}^{\mathbf{t}}) > 1$, *then* $\mathcal{L}^1(\Lambda_{\mathcal{F}^{\mathbf{t}}}) > 0$.

The proof is similar to the proof of Proposition 10.4.1.

PROOF. It suffices to verify the transversality condition (14.41); then the claims (i),(ii) hold by Theorem 14.4.1. Denote by $\Pi_{\mathbf{t}}$ the natural projection associated with the IFS $\mathcal{F}^{\mathbf{t}}$. We have for all $\mathbf{i} = (i_1, i_2, \ldots) \in \Sigma$,

$$\Pi_{\mathbf{t}}(\mathbf{i}) = \lim_{n \to \infty} f^{\mathbf{t}}_{i_1 \ldots i_n}(0) = t_{i_1} + f_{i_1}\big(t_{i_2} + f_{i_2}(t_{i_3} + \cdots)\big).$$

Fix $\mathbf{i}, \mathbf{j} \in \Sigma$ with $i_1 \neq j_1$. We need to analyse

$$\Delta(\mathbf{t}) := \Delta_{\mathbf{i},\mathbf{j}}(\mathbf{t}) := \Pi_{\mathbf{t}}(\mathbf{i}) - \Pi_{\mathbf{t}}(\mathbf{j}).$$

LEMMA 14.5.3. *There are open sets $Q^k(\mathcal{F})$ for $k = i_1, j_1$, such that $Q(\mathcal{F}) \subset Q^{i_1}(\mathcal{F}) \cup Q^{j_1}(\mathcal{F})$ and*

$$\left|\frac{\partial \Delta}{\partial t_k}(\mathbf{t})\right| > 1 - r \quad \textit{if } \mathbf{t} \in Q^k(\mathcal{F}). \tag{14.48}$$

where r was defined in (14.47).

PROOF OF LEMMA 14.5.3. Observe that for all $k \leqslant m$,

$$\begin{aligned}\frac{\partial \Pi_{\mathbf{t}}(\omega)}{\partial t_k} &= \delta_{i_1,k} + f'_{i_1}(\Pi_{\mathbf{t}}(\sigma \mathbf{i}))\big[\delta_{i_2,k} + f'_{i_2}(\Pi_{\mathbf{t}}(\sigma^2 \mathbf{i}))(\delta_{i_3,k} + \cdots)\big] \\ &= \delta_{k,i_1} + \sum_{\substack{n\geqslant 1\\ i_{n+1}=k}} \prod_{k=1}^{n} f'_{i_k}\left(\Pi_{\mathbf{t}}(\sigma^k \mathbf{i})\right) \qquad (14.49) \\ &= \delta_{k,i_1} + \sum_{\substack{n\geqslant 1\\ i_{n+1}=k}} f'_{\mathbf{i}|_n}\left(\Pi_{\mathbf{t}}(\sigma^n \mathbf{i})\right), \qquad (14.50)\end{aligned}$$

where $\delta_{i,k}$ is Dirac's delta. Hence,

$$\frac{\partial \Delta}{\partial t_k}(\mathbf{t}) = \delta_{k,i_1} - \delta_{k,j_1} + E_k(\mathbf{t}),$$

where the continuous function $E_k : Q(\mathcal{F}) \to \mathbb{R}$ is defined by

$$E_k(\mathbf{t}) := \sum_{\substack{n\geqslant 1\\ i_{n+1}=k}} f'_{\mathbf{i}|_n}\left(\Pi_{\mathbf{t}}(\sigma^p \mathbf{i})\right) - \sum_{\substack{n\geqslant 1\\ j_{n+1}=k}} f'_{\mathbf{j}|_n}\left(\Pi_{\mathbf{t}}(\sigma^p \mathbf{j})\right).$$

FACT 14.5.4. *For all* $\mathbf{t} \in Q(\mathcal{F})$ *we have*

$$\min\{|E_{i_1}(\mathbf{t})|, |E_{j_1}(\mathbf{t})|\} < r < 1.$$

PROOF OF FACT 14.5.4. Clearly, for every $\mathbf{t} \in Q(\mathcal{F})$ we have

$$|E_k(\mathbf{t})| \leqslant \sum_{\substack{n\geqslant 1\\ i_{n+1}=k}} r_{\mathbf{i}|_n} + \sum_{\substack{n\geqslant 1\\ j_{n+1}=k}} r_{\mathbf{j}|_n}.$$

Hence, using in the last step that $i_1 \neq j_1$, we get

$$\begin{aligned}\sum_{k\in\{i_1,j_1\}} |E_k(\mathbf{t})|(1-r_k) &\leqslant \underbrace{\sum_{n=1}^{\infty} r_{\mathbf{i}|_n}(1-r_{i_{n+1}})}_{=r_{i_1}} + \underbrace{\sum_{n=1}^{\infty} r_{\mathbf{j}|_n}(1-r_{j_{n+1}})}_{=r_{j_1}} \\ &\leqslant r < 1.\end{aligned}$$

The Fact follows since we have the strict inequality $(1-r_{i_1}) + (1-r_{j_1}) > 1$. □

Continuity of the function $E_k(\cdot)$ implies that the following sets are open:

$$Q^k(\mathcal{F}) := \{\mathbf{t} \in Q(\mathcal{F}) : |E_k(\mathbf{t})| < r\}.$$

Moreover, by Fact 14.5.4 we get that

$$Q(\mathcal{F}) \subseteq Q^{i_1}(\mathcal{F}) \cup Q^{j_1}(\mathcal{F}). \qquad (14.51)$$

Assume that $\mathbf{t} \in Q^{i_1}(\mathcal{F})$, then

$$\left|\frac{\partial \Delta}{\partial t_{i_1}}(\mathbf{t})\right| = \left|1 - \left(1 - \frac{\partial \Delta}{\partial t_{i_1}}(\mathbf{t})\right)\right| \geqslant 1 - |-E_{i_1}(\mathbf{t}))| > 1 - r.$$

The case when $\mathbf{t} \in Q^{j_1}(\mathcal{F})$ can be settled in the same way. This completes the proof of Lemma 14.5.3. □

For $k \in \{1, \ldots, m\}$ we define

$$A_\rho^k := \left\{\mathbf{t} \in Q^k(\mathcal{F}) : |\Delta(\mathbf{t})| \leqslant \rho\right\} \text{ and } A_\rho := \{\mathbf{t} \in Q(\mathcal{F}) : |\Delta(\mathbf{t})| \leqslant \rho\}.$$

Clearly,

$$A_\rho = A_\rho^{i_1} \cup A_\rho^{j_1}. \tag{14.52}$$

Without loss of generality, we may assume that $i_1 = 1$ and $j_1 = 2$. It is enough to prove that there exists $C_{\mathrm{tr}}^1 > 0$ (independent of $\mathbf{i}, \mathbf{j} \in \Sigma$ if $i_1 \neq j_1$), such that for all $\rho > 0$,

$$\mathcal{L}^m\left(A_\rho^1\right) \leqslant C_{\mathrm{tr}}^1 \rho. \tag{14.53}$$

Now we define the mapping $\mathcal{T} : Q^1(\mathcal{F}) \to \mathbb{R}^m$ as follows: $\mathcal{T}(\mathbf{t}) := (\mathcal{T}_1(\mathbf{t}), \ldots, \mathcal{T}_m(\mathbf{t}))$ where

$$\mathcal{T}_k(\mathbf{t}) := \begin{cases} \Delta(\mathbf{t}), & \text{if } k = 1; \\ t_k, & \text{otherwise,} \end{cases} \qquad \text{for } \mathbf{t} = (t_1, \ldots, t_m).$$

Then by Lemma 14.5.3,

$$\left|\det\left((\mathcal{T}')^{-1}\right)\right| \leqslant \frac{1}{1-r}. \tag{14.54}$$

Indeed,

$$\mathcal{T}'(\mathbf{t}) = \left(\begin{array}{c|cccc} \frac{\partial \Delta}{\partial t_1} & \frac{\partial \Delta}{\partial t_2} & \frac{\partial \Delta}{\partial t_3} & \cdots & \frac{\partial \Delta}{\partial t_m} \\ \hline \mathbf{0}_{m-1,1} & & & \mathbf{I}_{m-1} & \end{array}\right),$$

where $\mathbf{0}_{m-1,1}$ and $\mathbf{I}_{m-1}$ are the zero and identity matrices of indicated size respectively. Hence, $|\det(\mathcal{T}'(\mathbf{t}))| = \left|\frac{\partial \Delta}{\partial t_1}\right| \geqslant 1 - r$. This implies (14.54). Observe that by the definition of A_ρ^1 and by (14.46),

$$\mathcal{T}(A_\rho^1) \subset [-\rho, \rho] \times [-|I|, |I|]^{m-1}.$$

Hence,

$$\begin{aligned} \mathcal{L}^m(A_\rho^1) &= \mathcal{L}^m\left(\mathcal{T}^{-1}\left(\mathcal{T}(A_\rho^1)\right)\right) \\ &\leqslant \max_{\mathbf{y} \in \mathcal{T}(A_\rho^1)} |\det(\mathcal{T}^{-1})'(y)| \mathcal{L}^m\left(\mathcal{T}(A_\rho^1)\right) \\ &\leqslant \frac{2^m |I|^{m-1}}{1-r} \rho. \end{aligned}$$

That is, (14.53) holds with $C_{\mathrm{tr}}^1 = \frac{2^m |I|^{m-1}}{1-r}$. Using this and (14.52) we obtain that the transversality condition (14.41) holds with $C_{\mathrm{tr}} = 2 \cdot \frac{2^m |I|^{m-1}}{1-r}$. □

THEOREM 14.5.5. *For a residual subset of $\mathcal{F} \in \Theta_{0,\frac{1}{2}}(I)$ the attractor $\Lambda_\mathcal{F}$ satisfies*

$$\dim_{\mathrm{H}}(\Lambda_\mathcal{F}) = \min\{s(\mathcal{F}), 1\},$$

where $s(\mathcal{F})$ is the unique solution of Bowen's equation.

This theorem is deduced from a measure-theoretic result.

PROOF OF THEOREM 14.5.5. The proof is based on Theorem 14.5.2 and a theorem of Jonker and Veerman [**JV02**], which asserts that the Hausdorff dimension is lower semicontinuous on the space of conformal IFS. One-dimensional hyperbolic IFS in $\Theta(I)$ are conformal, so the theorem applies in our situation. We proved the claim on lower semicontinuity of Hausdorff dimension for self-similar IFS in Theorem 3.5.2, and the conformal case can be treated analogously. We leave the details as an exercise.

Define the functions $\alpha(\mathcal{F}) := \dim_{\mathrm{H}}(\Lambda_{\mathcal{F}})$ and $\beta(\mathcal{F}) = \min\{s(\mathcal{F}), 1\}$. First we prove that for any $0 < \gamma_1 < \gamma_2 < \frac{1}{2}$, we have the equality

$$\begin{aligned} &\{\mathcal{F} \in \Theta_{\gamma_1,\gamma_2}(I) : \ \alpha(\mathcal{F}) = \beta(\mathcal{F})\} = \\ &\{\mathcal{F} \in \Theta_{\gamma_1,\gamma_2}(I) : \ \alpha \text{ is continuous at } \mathcal{F}\}. \end{aligned} \tag{14.55}$$

Indeed, suppose that $\alpha(\mathcal{F}) = \beta(\mathcal{F})$. We know that the function α is lower semicontinuous, β is continuous on $\Theta_{\gamma_1,\gamma_2}(I)$ by Lemma 14.4.3(i), and $\alpha(\mathcal{F}) \leqslant \beta(\mathcal{F})$ by Lemma 14.2.8(ii), so it is immediate that α is continuous at $\mathcal{F}$. On the other hand, if α is continuous at $\mathcal{F}$ and $\alpha(\mathcal{F}) < \beta(\mathcal{F})$, then $\alpha(\mathcal{G}) < \beta(\mathcal{G})$ holds on a whole neighborhood of $\mathcal{F}$. However, this is impossible since by Theorem 14.5.2(i), $\alpha(\mathcal{F}^{\mathbf{t}}) = \beta(\mathcal{F}^{t})$ holds for arbitrarily small $\mathbf{t}$. Thus (14.55) is proved.

Using that continuity points of any function form a G_δ-set, we conclude that the set

$$\{\mathcal{F} \in \Theta_{\gamma_1,\gamma_2}(I) : \ \alpha(\mathcal{F}) = \beta(\mathcal{F})\}$$

is residual in each $\Theta_{\gamma_1,\gamma_2}(I)$ since it contains a dense G_δ-set. Finally, $\Theta_{0,\frac{1}{2}}(I) = \bigcup_{n\in\mathbb{N}} \Theta_{\frac{1}{n},\frac{1}{2}-\frac{1}{n}}(I)$, and this is an increasing family of subsets, so the theorem follows. □

Finally, we mention two "topologically generic" type results related to the SSP, WSP and Assouad dimension, due to Bárány, Kolossváry, Rams and Simon [**BSKR21**].

THEOREM 14.5.6. *For every $0 < \gamma_1 < \gamma_2 < 1$, there exists a residual subset $\Phi \subset \Theta_{\gamma_1,\gamma_2}(I)$ such that for $\mathcal{F} \in \Phi$ one of the two possibilities holds:*

(i) *either $\mathcal{F}$ has the Strong Separation Property (SSP) (there are no overlaps),*
(ii) *or $\mathcal{F}$ does not have the Weak Separation Property (WSP), and no exact overlaps either.*

Combining this with Theorems 14.3.5 and 14.3.6, we get

COROLLARY 14.5.7. *Fix arbitrary $0 < \gamma_1 < \gamma_2 < 1$. Consider the open subset of $\Gamma := \{\mathcal{F} \in \Theta_{\gamma_1,\gamma_2}(I) : s(\mathcal{F}) < 1\}$ of $\Theta_{\gamma_1,\gamma_2}(I)$. Then there is a residual subset of $\Phi \subset \Gamma$ such that for $\mathcal{F} \in \Phi$ one of the two possibilities holds:*

(i) *either $\mathcal{F}$ satisfies the SSP and then by Theorem 14.3.6, $\dim_{\mathrm{H}}(\Lambda_{\mathcal{F}}) = \dim_{\mathrm{A}}(\Lambda_{\mathcal{F}})$ and $0 < \mathcal{H}^{\dim_{\mathrm{H}}(\Lambda_{\mathcal{F}})}(\Lambda_{\mathcal{F}}) < \infty$,*
(ii) *or $\dim_{\mathrm{A}}(\Lambda_{\mathcal{F}}) = 1$ and $\mathcal{H}^{\dim_{\mathrm{H}}(\Lambda_{\mathcal{F}})}(\Lambda_{\mathcal{F}}) = 0$.*

APPENDIX A

Some elements of Linear algebra

A.1. Singular values and SVD

In this section we always work in $\mathbb{R}^d$ for $d > 1$. Let T be a nonsingular $d \times d$ matrix with real entries. It follows from the well known *singular value decomposition* (SVD), see e.g. [**AB02**], that there exist nonnegative reals arranged in nonincreasing order

$$\alpha_1 \geqslant \alpha_2 \geqslant \cdots \geqslant \alpha_d \geqslant 0 \tag{A.1}$$

and a diagonal matrix

$$D = \begin{pmatrix} \alpha_1 & 0 & \cdots & 0 \\ 0 & \alpha_2 & \ddots & \vdots \\ \vdots & \ddots & \ddots & 0 \\ 0 & \cdots & 0 & \alpha_d \end{pmatrix} \tag{A.2}$$

and $d \times d$ orthogonal matrices U, V such that

$$T = U \cdot D \cdot V^*. \tag{A.3}$$

The numbers in (A.1) are called the *singular values* of the matrix T. The meaning of the singular value decomposition is that if (the pairwise orthogonal unit vectors) $\mathbf{v}_1, \ldots, \mathbf{v}_d$ and $\mathbf{u}_1, \ldots, \mathbf{u}_d$ are the column vectors of the matrices U and V respectively, then

$$\alpha_j \cdot \mathbf{u}_j = T \cdot \mathbf{v}_j, \qquad j = 1, \ldots, d.$$

In particular, T maps the closed unit ball $\overline{B}(\mathbf{0}, 1)$ onto an ellipsoid with pairwise perpendicular semi-axes: $\alpha_j \cdot \mathbf{u}_j$, $j = 1, \ldots, d$. It is immediate from the SVD that the singular values of T are the positive square roots of the eigenvalues of the positive-definite symmetric matrix T^*T. It is also immediate from the SVD that

$$\alpha_1 = \|T\| \text{ and } \alpha_d = \|T^{-1}\|^{-1}. \tag{A.4}$$

We prove just the first equality since the second can be proved in the same way. In fact, from SVD: $T = UDV^*$ with U, V orthogonal. Thus,

$$\begin{aligned} \|T\|^2 &= \sup_{\|x\|=1} (Tx, Tx) = \sup_{\|x\|=1} (T^*Tx, x) \\ &= \sup_{\|x\|=1} (VD^2V^*x, x) = \sup_{\|x\|=1} (D^2V^*x, V^*x) \\ &= \sup_{\|y\|=1} (D^2y, y) = \alpha_1^2. \end{aligned}$$

A.2. The k-fold exterior product of $\mathbb{R}^d$.

Here we do not give the formal definition of the k-fold exterior product $\bigwedge^k(\mathbb{R}^d)$ of $\mathbb{R}^d$. Instead we list the most important properties of $\bigwedge^k(\mathbb{R}^d)$. Namely, $\bigwedge^k(\mathbb{R}^d)$ is a vector space over $\mathbb{R}$, where the vectors are formal expressions:

$$\bigwedge^k(\mathbb{R}^d) := \Big\{ \sum_{i=1}^{m} c_i u_{i1} \wedge u_{i2} \wedge \cdots \wedge u_{ik} : c_i \in \mathbb{R}, m \in \mathbb{N}, u_{ij} \in \mathbb{R}^d \Big\},$$

with addition and multiplication by scalar defined in such a way that the multilinear property and the antisymmetry hold:

(1) $u_1 \wedge \cdots \wedge (u_j + u_j') \wedge \cdots \wedge u_k = u_1 \wedge \cdots \wedge u_j \wedge \cdots \wedge u_k + u_1 \wedge \cdots \wedge u_j' \wedge \cdots \wedge u_k$.
(2) $c \cdot (u_1 \wedge \cdots \wedge u_j \wedge \cdots \wedge u_k) = u_1 \wedge \cdots \wedge c \cdot u_j \wedge \cdots \wedge u_k$.
(3) For a permutation π of $\{1, \dots, d\}$ we have

$$u_{\pi(1)} \wedge \cdots \wedge u_{\pi(k)} = \mathrm{sign}(\pi)(u_1 \wedge \cdots \wedge u_k).$$

A basis of the vector space $\bigwedge^k(\mathbb{R}^d)$ is

$$\{\mathrm{e}_{i_1} \wedge \cdots \wedge \mathrm{e}_{i_k}, \quad 1 \leqslant i_1 < \cdots < i_k \leqslant d\}.$$

Thus, $\bigwedge^k(\mathbb{R}^d)$ is an $\binom{d}{k}$-dimensional vector space.

The inner product is defined as follows: let $\alpha := u_1 \wedge \cdots \wedge u_k$, $\beta := v_1 \wedge \cdots \wedge v_k$. Then the inner product of α, β is defined as

$$\langle \alpha, \beta \rangle := \det \begin{pmatrix} \langle u_1, v_1 \rangle & \cdots & \langle u_1, v_d \rangle \\ \vdots & \ddots & \vdots \\ \langle u_d, v_1 \rangle & \cdots & \langle u_d, v_d \rangle \end{pmatrix}.$$

We extend the definition of the inner product to $\bigwedge^k(\mathbb{R}^d)$ in a bilinear way. Moreover, the norm an $\alpha \in \bigwedge^k(\mathbb{R}^d)$ is $\|\alpha\| = \sqrt{\langle \alpha, \alpha \rangle}$.

Note that if $\alpha = u_1 \wedge \cdots \wedge u_k$ then $\|\alpha\|$ is the volume of the parallelepiped spanned by the vectors $u_1, \dots, u_k$. In particular, $u_1 \wedge \cdots \wedge u_k = 0$ if and only if $\{u_1, \dots u_k\}$ is a linearly dependent set of vectors.

The **k-fold exterior product** $A^{\wedge k}$ of a matrix $A \in \mathbb{M}_d$ is the linear operator acting on $\bigwedge^k(\mathbb{R}^d)$ defined by:

$$u_1 \wedge \cdots \wedge u_k \mapsto Au_1 \wedge \cdots \wedge Au_k,$$

and extended to $\bigwedge^k(\mathbb{R}^d)$ in a linear way. Clearly,

$$(BA)^{\wedge k}(u_1 \wedge \cdots \wedge u_k) = B^{\wedge k} A^{\wedge k}(u_1 \wedge \cdots \wedge u_k). \tag{A.5}$$

It can be proved (see [**Kre85**, p. 44] and references there) that for any $A \in \mathbb{M}_d$,

$$\|A^{\wedge k}\| = \sup\Big\{ \|A^{\wedge k}\beta\| : \beta \in \bigwedge^k(\mathbb{R}^d), \ \|\beta\| = 1 \Big\} = \prod_{1 \leqslant i \leqslant k} \alpha_i, \tag{A.6}$$

where $\alpha_1 \geqslant \cdots \geqslant \alpha_d$ are the singular values of A. Therefore,

$$\|A^{\wedge k}\| \leqslant \alpha_1^k = \|A\|^k, \tag{A.7}$$

in view of (A.4), and

$$\alpha_k(A) = \frac{\|A^{\wedge k}\|}{\|A^{\wedge (k-1)}\|}, \quad k = 2, \dots, d. \tag{A.8}$$

APPENDIX B

Some elements of measure theory

B.1. Measure spaces

The material in this section is standard; we borrow our presentation from Einsiedler and Ward [**EW11**, A.1]. Let X be a set, and denote by 2^X the collection of all subsets of X.

DEFINITION B.1.1. A set $\mathscr{S} \subseteq 2^X$ is called a *semi-algebra* if

(1) $\varnothing \in \mathscr{S}$,
(2) $A, B \in \mathscr{S} \implies A \cap B \in \mathscr{S}$,
(3) if $A \in \mathscr{S}$ then the complement $X \backslash A$ is a finite union of pairwise disjoint elements of $\mathscr{S}$.

If in addition,

(4) $A \in \mathscr{S} \implies X \backslash A \in \mathscr{S}$, then $\mathscr{S}$ is called an *algebra*. Thus algebra is closed under finite unions and intersections.

If, furthermore,

(5) $A_n \in \mathscr{S}, n \in \mathbb{N} \implies \bigcup_{n=1}^{\infty} A_n \in \mathscr{S}$, then $\mathscr{S}$ is called an σ*-algebra.* A σ-algebra is closed under countable unions and intersections.

For any collection of sets $\mathscr{A} \subseteq 2^X$ denote by $\sigma(\mathscr{A})$ the smallest σ-algebra containing $\mathscr{A}$, called the σ-algebra *generated by* $\mathscr{A}$. This is well defined since the intersection of any number of σ-algebras is a σ-algebra.

A basic example of semi-algebra is the collection of subintervals in $[0,1]$. Another example, important for us, is the collection of cylinder sets in the symbolic space. That is, $\mathscr{S} := \{[\mathbf{i}] : \mathbf{i} \in \Sigma^*\}$ is a semi-algebra in Σ, and the set $\{\bigcup_{j=1}^{n} [\mathbf{i}_j] : n \in \mathbb{N} \text{ and } \mathbf{i}_j \in \Sigma^* \text{ for } j = 1, \ldots, n\}$ is an algebra.

Let $\mathscr{S} \subset 2^X$. A function $\mu : \mathscr{S} \to [0, \infty]$ is said to be *countably additive or* σ*-additive on* $\mathscr{S}$ if $\mu(\varnothing) = 0$ and for any finite or infinite sequence $\{A_n\}_{n=1}^{\infty}$ of pairwise disjoint elements of $\mathscr{S}$, such that $\bigcup_{n=1}^{\infty} A_N \in \mathscr{S}$, holds

$$\mu\Big(\bigcup_{n=1}^{\infty} A_n\Big) = \sum_{n=1}^{\infty} \mu(A_n).$$

DEFINITION B.1.2. The pair $(X, \mathscr{A})$, where $\mathscr{A}$ is a σ-algebra in X, is called a *measurable space.* A function $\mu : \mathscr{A} \to [0, \infty]$ is a *measure* if it is countably additive on $\mathscr{A}$. If $\mu(X) = 1$ then we say that $(X, \mathscr{A}, \mu)$ is a probability space. A measure space $(X, \mathscr{A}, \mu)$ is *complete* if for every $H \in \mathscr{A}$ with $\mu(H) = 0$, for all subsets $E \subset H$ we have $E \in \mathscr{A}$.

THEOREM B.1.3. *Let $\mathscr{S}$ be a semi-algebra on X. If $\mu : \mathscr{S} \to [0, \infty]$ is a σ-additive set-function, there is a unique measure defined on $\sigma(\mathscr{S})$ which extends μ.*

THEOREM B.1.4. *If $\mathscr{A}$ is an algebra and $(X, \sigma(\mathscr{A}), \mu)$ is a probability space then for any $B \in \sigma(\mathscr{A})$ and $\varepsilon > 0$ there is an $A \in \mathscr{A}$ with $\mu(A \backslash B \cup B \backslash A) < \varepsilon$.*

The measures are often defined via *outer measures.* An *outer measure* ν on X is defined on all subsets of X and takes values from $[0, \infty]$, so that

- $\nu(\varnothing) = 0$,
- $\nu(A) \leqslant \nu(B)$ if $A \subset B$,
- $\nu(\bigcup_{i=1}^{\infty} E_i) \leqslant \sum\limits_{i=1}^{\infty} \nu(E_i)$ for all sequence of sets $\{E_i\}_{i=1}^{\infty}$.

We say that a set E is *measurable* with respect to the outer measure ν if for every $A \subset X$ we have

$$\nu(A) = \nu(A \cap E) + \nu(A \backslash E).$$

THEOREM B.1.5. *Let $\mathcal{M}$ be the collection of all measurable sets for an outer measure ν. Then $\mathcal{M}$ is a σ-algebra and the restriction of ν to $\mathcal{M}$ is a measure.*

DEFINITION B.1.6.

(a) Let $(X, \mathscr{A}_X)$ and $(Y, \mathscr{A}_Y)$ be two measurable spaces. A map $T : X \to Y$ is called *measurable* if for every $A \in \mathscr{A}_Y$ we have $T^{-1}A \in \mathscr{A}_X$.

(b) Let $(X, \mathscr{A}_X, \mu_X)$ and $(Y, \mathscr{A}_Y, \mu_Y)$ be two measure spaces. A measurable map (or transformation) $T : X \to Y$ is called *measure-preserving* if or every $A \in \mathscr{A}_Y$ we have

$$\mu_X(T^{-1}A) = \mu_Y(A).$$

Measure-preserving transformations T_1 and T_2 are often identified if they agree on a set of full measure; in fact, a measure-preserving map may only be defined μ_X-a.e.

A convenient method for checking that the system is measure-preserving is based on the following

THEOREM B.1.7. *Let $(X, \mathcal{B}, \mu)$ be a probability measure space, and let $\mathscr{S}$ be a semi-algebra which generates the σ-algebra $\mathcal{B}$. A measurable map $T : X \to X$ is measure-preserving if and only if $\mu(T^{-1}B) = \mu(B)$ for all $B \in \mathscr{S}$.*

For a proof see [**EW11**, Theorem A.8].

B.2. Measures on metric spaces

Here we follow Mattila [**Mat95**, Section 1].

DEFINITION B.2.1. Let (X, ϱ) be a metric space. The *Borel σ-algebra $\mathcal{B}$* is the smallest σ-algebra in X containing all open (hence also closed) subsets.

Let $\mathcal{M}$ be a σ-algebra in X and μ a measure on $(X, \mathcal{M})$.

(1) μ is locally finite if for every $x \in X$ there is $r > 0$ such that $\mu(B(x, r)) < \infty$.
(2) μ is a Borel measure if all Borel sets are in $\mathcal{M}$.
(3) μ is Borel regular if it is a Borel measure and for every $A \subset X$ there is a Borel set $B \subset X$ such that $A \subset B$ and $\mu(A) = \mu(B)$.
(4) μ is a Radon measure if
   (a) μ is a Borel measure,
   (b) $\forall K \subset X$ compact, $\mu(K) < \infty$,
   (c) $\forall V \subset X$ open, $\mu(V) = \sup\{\mu(K) : K \subset V$ is compact $\}$
   (d) $\forall A \subset X$, $\mu(A) = \inf\{\mu(V) : A \subset V$ and V is open $\}$.

THEOREM B.2.2. *A measure μ on $\mathbb{R}^d$ is a Radon measure if and only if it is locally finite and Borel regular.*

We say that the outer measure ν is a *metric outer measure* if $\nu(A \cup B) = \nu(A) + \nu(B)$ holds for all $A, B \subset X$ with $\inf\{\varrho(a,b) : a \in A, b \in B\} > 0$.

THEOREM B.2.3. *If ν is a metric outer measure, then all Borel sets are ν-measurable.*

For the proof, see e.g., [**BP17**, Theorem 1.2.4].

EXAMPLES.

(1) The Lebesgue measure $\mathcal{L}^d$ on $\mathbb{R}^d$ is a Radon measure.
(2) The Dirac measure $\delta_a(A) := 1$ if $a \in A$ and $\delta_a(A) = 0$ if $a \notin A$ is a Radon measure.
(3) For every $s \geqslant 0$ the Hausdorff measure is a Borel regular measure but it need not be locally finite. So, in general the Hausdorff measure is not a Radon measure. However, for $A \subset \mathbb{R}^d$, with $\mathcal{H}^s(A) < \infty$, the restriction $\mathcal{H}^s|_A$ is a Radon measure. (See Mattila's book: [**Mat95**, p. 57].)

It is important to understand the relation between different measures, and in particular, the relation of a measure μ with respect to the Lebesgue measure. Let μ and ν be Radon measures on $\mathbb{R}^d$. We say that μ is *absolutely continuous* with respect to ν if

$$\nu(A) = 0 \text{ implies } \mu(A) = 0 \text{ for every Borel set } A \subset \mathbb{R}^d.$$

In this case, we write $\mu \ll \nu$. Further, we say that μ and ν are *mutually singular* if there exists a subset $A \subset \mathbb{R}^d$ such that $\mu(A) = \nu(\mathbb{R}^d \backslash A) = 0$. In this case, we write $\mu \perp \nu$.

DEFINITION B.2.4. Let μ and ν be Radon measures on $\mathbb{R}^d$. The upper and lower derivatives of μ with respect to ν at a point $x \in \mathbb{R}^d$ are defined by

$$\overline{D}(\mu, \nu, x) = \limsup_{r \to 0} \frac{\mu(B(x,r))}{\nu(B(x,r))},$$
$$\underline{D}(\mu, \nu, x) = \liminf_{r \to 0} \frac{\mu(B(x,r))}{\nu(B(x,r))}.$$

At the points x where the limit exists, we define the derivative of μ with respect to ν by

$$D(\mu, \nu, x) = \overline{D}(\mu, \nu, x) = \underline{D}(\mu, \nu, x).$$

For simplicity, if ν is the d-dimensional Lebesgue measure, we write $D(\mu, x) = \overline{D}(\mu, x) = \underline{D}(\mu, x)$.

THEOREM B.2.5. *Let μ and ν be Radon measures on $\mathbb{R}^d$.*

(i) *The derivative $D(\mu, \nu, x)$ exists and is finite for ν almost every $x \in \mathbb{R}^d$.*
(ii) *For all Borel sets $B \subset \mathbb{R}^d$,*

$$\int_B D(\mu, \nu, x)\, d\nu(x) \leqslant \mu(B)$$

with equality if $\mu \ll \nu$.

(iii) *$\mu \ll \nu$ if and only if $\underline{D}(\mu, \nu, x) < \infty$ for μ almost every $x \in \mathbb{R}^d$.*

For the proof of the theorem, see [**Mat95**, Theorem 2.12].

B.3. Covering Theorems

Here we mention three kinds of covering theorems on $\mathbb{R}^d$: the 4r-covering lemma, the Vitali covering theorems, and the Besicovitch Covering theorem.

B.3.1. 4r-covering lemma.

LEMMA B.3.1. *Let $\mathcal{C}$ be a family of closed balls contained in a bounded region of $\mathbb{R}^d$. For every i let $\widetilde{B}_i$ be the ball concentric with $\overline{B}_i$ and with the radius four times the radius of $\overline{B}_i$. Then there is a countable disjoint subcollection $\{\overline{B}_i\}$ of $\mathcal{C}$ such that $\bigcup_{\overline{B}\in\mathcal{C}} \overline{B} \subset \bigcup_i \widetilde{B}_i$.*

For the proof see [**Fal03**, Covering lemma 4.8]. We remark that if $\mathcal{C}$ is finite, then we can replace 4 with 3 in the lemma above. A version of this lemma is proved (as $5r$-covering theorem) in [**Mat95**, Theorem 2.1] for boundedly compact metric spaces (where all bounded closed sets are compact).

B.3.2. Besicovitch's covering theorem.

THEOREM B.3.2. *There exist integers $P(d)$ and $Q(d)$ with the following properties: Let $A \subset \mathbb{R}^d$ be a bounded set and let $\mathcal{B}$ be a family of closed balls of $\mathbb{R}^d$ such that for every $a \in A$ there exists a ball $B \in \mathcal{B}$ centered at a.*

(i) *There is a finite or countable sub-family $\{B_i\}$ of $\mathcal{B}$ such that*

$$\mathbb{1}_A \leqslant \sum_i \mathbb{1}_{B_i} \leqslant P(n).$$

That is, $\{B_i\}$ is a cover of A such that every element of A is covered by at most $P(n)$ elements of $\mathcal{B}$.

(ii) *There are $Q(d)$ sub-families $\mathcal{B}_1, \ldots, \mathcal{B}_d$ of $\mathcal{B}$ such that*
   (a) *$\mathcal{B}_i$ consists of pairwise disjoint balls for every i and*
   (b) $A \subset \bigcup_{i=1}^{Q(d)} \bigcup_{B\in\mathcal{B}_i} B$.

The proof can be found in [**Mat95**, p. 31].

B.3.3. Vitali's covering theorems. Let $\mathcal{V}$ be a collection of subsets of $\mathbb{R}^d$. We say that $\mathcal{V}$ is a Vitali class for the set $E \subset \mathbb{R}^d$ if

$$\inf\{|V| : V \in \mathcal{V} \text{ and } x \in V\} = 0, \quad \text{for all } x \in E.$$

First we state the Vitali covering theorem for the Lebesgue measure $\mathcal{L}^d$ in $\mathbb{R}^d$.

THEOREM B.3.3. *Let $A \subset \mathbb{R}^d$ let $\mathcal{B}$ be a family of closed balls in $\mathbb{R}^d$ which is a Vitali class for E. Then there is a disjoint subcollection $\{B_i\}$ of $\mathcal{B}$ such that*

$$\mathcal{L}^d\Big(A\setminus\bigcup_i B_i\Big) = 0. \tag{B.1}$$

Moreover, for every $\varepsilon > 0$ we can choose the disjoint subcollection $\{B_i\}$ such that besides (B.1) *we also have*

$$\sum_i \mathcal{L}^d(B_i) \leqslant \mathcal{L}^d(A) + \varepsilon. \tag{B.2}$$

The proof is presented in [**Mat95**, Theorem 2.2]. Now we state the Vitali covering theorem for Hausdorff measures. For the proof see [**Fal85**, p. 11].

THEOREM B.3.4. *Fix $s > 0$. Let $\mathcal{V}$ be a Vitali class of closed sets for a given $\mathcal{H}^s$-measurable set $E \subset \mathbb{R}^d$. Then we may select a finite or countable disjoint subcollection $\{V_i\}$ of $\mathcal{V}$ such that either*

(a) $\sum_i |V_i|^s = \infty$, *or*

(b) $\mathcal{H}^s\Big(E\backslash \bigcup_i V_i\Big) = 0$.

Moreover, if $\mathcal{H}^s(E) < \infty$ and $\varepsilon > 0$, then we can select the countable disjoint subcollection $\{V_i\}$ in such a way that besides (a) or (b) above, we also have

$$\mathcal{H}^s(E) \leqslant \sum_i |V_i|^s + \varepsilon.$$

Finally, we state the Vitali-type covering theorem for an arbitrary Radon measure on $\mathbb{R}^d$.

THEOREM B.3.5. *Let μ be a Radon measure on $\mathbb{R}^d$, $A \subset \mathbb{R}^d$, and $\mathcal{B}$ a collection of closed balls of $\mathbb{R}^d$ satisfying*

$$\inf\big\{r : \overline{B}(x,r) \in \mathcal{B}\big\} = 0 \quad \textit{for all } x \in A,$$

where $\overline{B}(x,r) = \big\{y \in \mathbb{R}^d : |x-y| \leqslant r\big\}$. Then we can select a finite or countable disjoint subcollection $\{B_i\}$ of $\mathcal{B}$ such that

$$\mu\Big(A\backslash \bigcup_i B_i\Big) = 0.$$

Note that in the last theorem $\mathcal{B}$ is not simply a Vitali class, since we required more. Namely, we assumed not just that every point of A is contained in an element of $\mathcal{B}$ of arbitrarily small diameter, but that every point of A is a **center** of an arbitrarily small ball from $\mathcal{B}$. The proof of the theorem can be found in [**Mat95**, p. 34].

B.4. Density theorems

We start with the classical

THEOREM B.4.1 (Lebesgue's density theorem). *Let A be a Borel set in $\mathbb{R}^d$. Then*

$$\lim_{r\to\infty} \frac{\mathcal{L}^d(B(x,r)\cap A)}{\mathcal{L}^d(B(x,r))} = 1 \textit{ for Lebesgue almost every } x.$$

It can be proved using Vitali's covering theorem. We will need a generalization of the Lebesgue density theorem for arbitrary measures.

THEOREM B.4.2.

(i) *Let ν be a Radon measure on $\mathbb{R}^d$ and let $A \subset \mathbb{R}^d$ be a Borel set. Then*

$$\lim_{r\to\infty} \frac{\nu(B(x,r)\cap A)}{\nu(B(x,r))} = 1 \textit{ for } \nu\textit{-almost every } x \in A. \tag{B.3}$$

(ii) *Let ν be a probability Borel measure on the symbolic space $\Sigma = \{1,\dots,m\}^{\mathbb{N}}$, equipped with the standard metric, or more generally, on an ultrametric Polish space, and let $A \subset \Sigma$ be a Borel set. Then* (B.3) *holds for ν-almost every $x \in A$.*

The proof of part (i) can be deduced from Theorem B.2.5 by taking $\mu = \nu|_A$, see [**Mat95**, Corollary 2.14]. The proof relies on Besicovitch's covering theorem B.3.2, and so Theorem B.4.2(i) is sometimes called "Besicovitch's density theorem". We give here a short proof of part (i) following [**KRS16**, Lemma 2.3].

For a proof of part (ii), see Miller [**Mil08**, Proposition 2.10]. It should be pointed out that the analog of (B.3) may fail for a general doubling metric space, see [**KRS16**, Example 5.6]. Note that, combining parts (i) and (ii) with the Fubini Theorem, we also obtain the analog of the Lebesgue density theorem for product measures in spaces like $\Sigma \times \mathbb{PR}^1$.

PROOF OF PART (i). Let

$$A_\varepsilon = \left\{ x \in A : \limsup_{r \to 0} \frac{\nu(B(x,r)\backslash A)}{\nu(B(x,r))} > \varepsilon \right\}.$$

It is enough to show that $\nu(A_\varepsilon) = 0$ for every $\varepsilon > 0$. Let $\delta > 0$ be arbitrary. Since ν is a Radon measure, there exists an open set G such that $\nu(G\backslash A) < \delta$. Then for every $x \in A_\varepsilon$ there exists $0 < r_x$ such that $B(x, r_x) \subset G$ and $\nu(B(x,r_x)\backslash A) > \varepsilon\nu(B(x,r_x))$. Consider the collection of balls $\mathcal{B} = \{B(x,r_x)\}_{x \in A_\varepsilon}$. Then by Besicovitch's covering theorem, there exist at most $Q(d)$ many subcollections $\mathcal{B}_i \subset \mathcal{B}$ of balls which are disjoint and $\bigcup_{i=1}^{Q(d)} \mathcal{B}_i$ is a cover of A_ε. Thus,

$$\begin{aligned}
\nu(A_\varepsilon) &\leqslant \nu\left(\bigcup_{i=1}^{Q(d)} \bigcup_{B \in \mathcal{B}_i} B \right) \leqslant \sum_{i=1}^{Q(d)} \sum_{B \in \mathcal{B}_i} \nu(B) \\
&\leqslant \varepsilon^{-1} \sum_{i=1}^{Q(d)} \sum_{B \in \mathcal{B}_i} \nu(B\backslash A) = \varepsilon^{-1} \sum_{i=1}^{Q(d)} \nu\left(\bigcup_{B \in \mathcal{B}_i} B\backslash A \right) \\
&\leqslant \varepsilon^{-1} Q(d) \nu(G\backslash A) < \delta \varepsilon^{-1} Q(d).
\end{aligned}$$

Since $\delta > 0$ was arbitrary we get that $\nu(A_\varepsilon) = 0$. □

B.5. Integral in terms of the distribution function

We cite the following useful formula for the record, which is a consequence of Fubini's Theorem in one of the forms. See Mattila's book [**Mat95**, Theorem 1.15] for an easy proof.

THEOREM B.5.1. *Let X be a separable metric space, μ a Borel measure on X, and f a nonnegative Borel function on X. Then*

$$\int_X f \, d\mu = \int_0^\infty \mu(\{x \in X : f(x) \geqslant t\}) \, dt.$$

APPENDIX C

Some elements of Harmonic Analysis

Here we collect some definitions and results needed in the book; the reader is referred to [**Kat76, Rud87, Fol89, Wol03, Mat15**] for details. We denote by $L^p(\mathbb{R}^d)$ the space of all (equivalence classes of) complex valued measurable functions for which $|f(x)|^p$ is Lebesgue integrable. For $f \in L^1(\mathbb{R})$ and for a finite Borel measure ν on the line, the Fourier transforms are defined by

$$\widehat{f}(t) = \int_{\mathbb{R}} e^{itx} f(x)\, dx \quad \text{and} \quad \widehat{\nu}(t) = \int_{\mathbb{R}} e^{itx}\, d\nu(x).$$

(Often, the formula $\int_{\mathbb{R}} e^{-itx} f(x)\, dx$ is used instead; our definition follows the convention of Probability Theory.) Note that $\widehat{\nu}$ is sometimes called the Fourier–Stieltjes transform, which is more historically accurate.

In the next theorem we collect the well-known and easy to check properties of the Fourier transform.

THEOREM C.0.1. *Let $f \in L^1(\mathbb{R})$. Then*

(i) $|\widehat{f}(t)| \leqslant \int |f(x)|\, dx = \|f\|_{L^1}$;

(ii) $\widehat{f}$ *is uniformly continuous on* $\mathbb{R}$;

(iii) *for $y \in \mathbb{R}$ denote $f_y(x) = f(x-y)$; then $\widehat{f_y}(t) = \widehat{f}(t)e^{ity}$;*

(iv) *for $r > 0$ denote $\phi(x) = f(rx)$; then $\widehat{\phi}(t) = r^{-1}\widehat{f}(t/r)$;*

(v) *denote $F(x) = \int_{-\infty}^{x} f(y)\, dy$; then $\widehat{F}(t) = -(it)^{-1}\widehat{f}(t)$.*

(vi) *if $f' \in L^1(\mathbb{R})$, then $\widehat{f'}(t) = it\widehat{f}(t)$.*

(vii) *If both $f, \widehat{f} \in L^1(\mathbb{R})$, then f is equivalent to a uniformly continuous function and*

$$f(x) = \frac{1}{2\pi} \int \widehat{f}(t) e^{-itx} dt \tag{C.1}$$

(viii) *If ν is a probability Borel measure and $\widehat{\nu} \in L^1(\mathbb{R})$, then ν has bounded continuous density: $f(y) = \frac{1}{2\pi}\int e^{-ity}\widehat{\nu}(t)dt$.*

(ix) *If ν is a probability Borel measure and $\widehat{\nu}(t) = O(|t|^{-(k+1+\varepsilon)})$ for some $\varepsilon > 0$, then ν has bounded k-times continuously differentiable density.*

LEMMA C.0.2 (Riemann–Lebesgue). *For $f \in L^1(\mathbb{R})$ we have*

$$\lim_{|t| \to \infty} \widehat{f}(t) = 0.$$

THEOREM C.0.3 (Plancherel). *There exists a unique bounded linear operator $\mathcal{F} : L^2(\mathbb{R}) \to L^2(\mathbb{R})$ such that*

$$\mathcal{F}f = \widehat{f} \quad \textit{for } f \in L^1(\mathbb{R}) \cap L^2(\mathbb{R}).$$

Moreover, $(2\pi)^{-1/2}\mathcal{F}$ *is a bijective isometry. In particular,*

$$\int f(x)\cdot\overline{g(x)}dx = \frac{1}{2\pi}\int \widehat{f}(t)\overline{\widehat{g}(t)}dt, \quad f \in L^2(\mathbb{R}),\ g \in L^2(\mathbb{R}). \tag{C.2}$$

In Fourier Analysis it is often convenient to work with the *Schwartz space* $\mathcal{S}$ consisting of those C^∞ complex functions which, together with all their derivatives, vanish at infinity faster than any power of $|x|$.

PROPOSITION C.0.4. *The Fourier transform is an isomorphism of the Schwartz space* $\mathcal{S}$ *onto itself.*

Similarly to the Fourier transform of an integrable function, if ν is a probability measure, then $\widehat{\nu}(t)$ is a bounded, uniformly continuous function on $\mathbb{R}$. If ν is absolutely continuous with respect to the Lebesgue measure, $d\nu = h(x)dx$, then $\widehat{\nu}(t) = \widehat{h}(t)$.

The Riemann–Lebesgue Lemma implies that if $\widehat{\nu}(t) \not\to 0$ as $|t| \to \infty$, then ν is not absolutely continuous. However, there exist singular measures whose Fourier transform tends to zero at infinity; such measures are sometimes called *Rajchman measures.* Among them is the *Cantor–Lebesgue measure* on a self-similar Cantor set of non-PV contraction rate, see Salem's book [**Sal63**] and Theorem 8.1.2. (See Appendix D for the definition of PV-numbers.)

THEOREM C.0.5 (Parseval formula). *(see* [**Kat76**, p.132]*) Let* ν *be a finite Borel measure on* $\mathbb{R}$ *and let* f *be a continuous function in* $L^1(\mathbb{R})$ *such that* $\widehat{f} \in L^1(\mathbb{R})$. *(This happens for example if* $f \in \mathcal{S}$.*) Then*

$$\int \overline{f(x)}\, d\nu(x) = (2\pi)^{-1}\int \overline{\widehat{f}(t)}\widehat{\nu}(t)\, dt. \tag{C.3}$$

PROOF. By Part (vi) of Theorem C.0.1, $f(x) = \frac{1}{2\pi}\int \widehat{f}(t)e^{-itx}\, dt$. Hence

$$\int f(x)d\nu(x) = \frac{1}{2\pi}\iint \widehat{f}(t)e^{-itx}d\nu(x)dt = \frac{1}{2\pi}\int \widehat{f}(t)\widehat{\nu}(-t)dt.$$

This immediately implies (C.3). □

COROLLARY C.0.6 (Uniqueness). *Let* ν *be a finite Borel measure on* $\mathbb{R}$ *such that* $\widehat{\nu}(t) \equiv 0$ *for all* $t \in \mathbb{R}$. *Then* ν *is the zero measure.*

LEMMA C.0.7. *(see* [**Wol03**, 3.12] *or* [**Mat15**, Theorem 3.3]*) Let* ν *be a finite Borel measure on* $\mathbb{R}$ *such that* $\widehat{\nu} \in L^2(\mathbb{R})$. *Then* $\nu \ll \mathcal{L}^1$ *and the density* $\frac{d\nu}{dx}$ *is in* $L^2(\mathbb{R})$.

PROOF. By Theorem C.0.3, there exists $h \in L^2(\mathbb{R})$ such that $\widehat{h} = \widehat{\nu}$. Now we prove that $h = \frac{d\nu}{dx}$. First observe that any $\phi \in \mathcal{S}$:

$$\begin{aligned}\int \overline{\phi(x)}h(x)\, dx = &\ (2\pi)^{-1}\textstyle\int \overline{\widehat{\phi}(\xi)}\widehat{h}(\xi)\, d\xi \\ = &\ (2\pi)^{-1}\textstyle\int \overline{\widehat{\phi}(\xi)}\widehat{\nu}(\xi)\, d\xi \ = \displaystyle\int \overline{\phi(x)}\, d\nu(x).\end{aligned}$$

Here we first used (C.2) and then (C.3) (Parseval formula). It follows that $\phi \mapsto \int \phi(x)\, d\nu(x)$ defines the same linear functional on $\mathcal{S}$ as $\phi \mapsto \int \phi(x)h(x)\, dx$, hence the latter extends by continuity to $C_0(\mathbb{R})$. This is because $C_0^\infty(\mathbb{R}) \subset \mathcal{S}$ is dense in $C_0(\mathbb{R})$. Thus $h \in L^1(\mathbb{R})$ and $d\nu(x) = h(x)\, dx$, as desired. □

The convolution of two measures μ, ν is defined by

$$(\mu * \nu)(E) = \int \mu(E - y)\, d\nu(y)$$

for all Borel sets E. Some of the basic properties of $\mu * \nu$ are the following.

THEOREM C.0.8.

(i) $\mu * \nu = \nu * \mu$;
(ii) $\widehat{\mu * \nu} = \widehat{\mu}\,\widehat{\nu}$;

DEFINITION C.0.9. Let ν_n, $n \geqslant 1$, be a sequence of finite Borel measures. The infinite convolution product $*_{n=1}^{\infty}\nu_n$ is said to converge to a measure μ if $*_{n=1}^{N}\nu_n \to \mu$ in the weak* topology.

We also need the following.

LEMMA C.0.10. *Let μ_n be a sequence of finite Borel measures such that $\mu_n \to \mu$ in the weak* topology, and $\|\mu_n\| \to \|\mu\|$. Then $\widehat{\mu}_n(t) \to \widehat{\mu}(t)$ pointwise.*

COROLLARY C.0.11. *If $\mu = *_{n=1}^{\infty}\nu_n$, then $\widehat{\mu}_n(t) = \prod_{n=1}^{\infty} \widehat{\nu}_n(t)$ for all $t \in \mathbb{R}$.*

APPENDIX D

Some facts about algebraic numbers

Recall that a number θ is *algebraic* if it is a zero of a polynomial equation with integer coefficients. It is an *algebraic integer* if the polynomial is monic, i.e. its leading coefficient is 1. There is a polynomial of minimal degree which has θ as its zero; it is called the *minimal polynomial* of θ. The other roots of the minimal polynomial are called the *conjugates* of θ. The following fact is immediate from definitions.

PROPOSITION D.0.1. *If θ is a rational number and an algebraic integer, then $\theta \in \mathbb{Z}$.*

Two important classes of algebraic integers are the PV (also called Pisot) numbers, and Salem numbers.

DEFINITION D.0.2. An algebraic integer $\theta > 1$ is a PV number if all its conjugates are less than one in modulus (i.e. lie inside the unit circle). PV numbers are often called Pisot numbers.

An algebraic integer $\theta > 1$ is a Salem number (after Raphael Salem) if all its conjugates are less than or equal to one in modulus, and there are some of modulus exactly one (i.e. it is not PV).

Both classes of numbers have many remarkable properties, and there are still many open questions. We list here some known results which are relevant for the study of self-similar measures, with selected (short, well-known) proofs, but mostly with references to the literature.

We use the notation $\|\cdot\|$ for the **distance to the nearest integer**, which is common in the theory of Diophantine approximation, that is,

$$\|x\| = \operatorname{dist}(x, \mathbb{Z}).$$

LEMMA D.0.3. *Let θ be a PV number. Then there exist $c > 0$ and $\rho \in (0,1)$ such that $\|\theta^n\| \leqslant c\rho^n$.*

PROOF. Let $\theta_1 = \theta, \theta_2, \ldots, \theta_d$ be all the roots of the minimal polynomial for the θ. Since $a_n = \sum_{j=1}^d \theta_j^n$ is a symmetric polynomial of d variables evaluated at the set of the roots of a polynomial of one variable with integer coefficients, one can show by using Newton's identities that $a_n = \sum_{j=1}^d \theta_j^n \in \mathbb{Z}$. It follows that

$$\|\theta^n\| \leqslant |\theta^n - a_n| \leqslant \sum_{j=2}^{d} |\theta_j|^n \leqslant (d-1)\rho^n,$$

where $\rho = \max_{j \geqslant 2} \|\theta_j\| < 1$, and the lemma is proved. $\square$

Essentially the same proof shows that $\|\alpha\theta^n\| \leqslant c\rho^n$ for any $\alpha \in \mathbb{Z}[\theta]$ (the ring of integer polynomials evaluated at θ). The following is a famous *open problem*:

PROBLEM D.0.4. Suppose that $\theta > 1$ and $\alpha > 0$ are such that $\|\alpha\theta^n\| \to 0$, as $n \to \infty$. Does it follow that θ is a PV number?

Classical results on this problem have been obtained by C. Pisot and R. Salem, but there has not been any progress since then. For the proofs, see [**Sal63**].

THEOREM D.0.5 (Pisot). *Let $\theta > 1$, $\alpha > 0$ be such that $\|\alpha\theta^n\| \to 0$.*
(i) *If $\sum_{n=1}^{\infty} \|\alpha\theta^n\|^2 < \infty$, then θ is a PV number.*
(ii) *If θ is algebraic, then θ is a PV number.*

THEOREM D.0.6 (Salem). *The set*

$$\{\theta > 1 :\ \exists\, \alpha > 0,\ \|\alpha\theta^n\| \to 0\}$$

is countable.

We sketched a proof of Theorem D.0.7 in Section 8.1.

The following may be found in [**Boy69**], [**Mey72**, 6.9], [**BDGGH**$^+$**92**, 5.5.1].

THEOREM D.0.7. *Let $\theta > 1$ be a Salem number. Then for any $\delta > 0$ there exists $t > 0$ such that $\|t\theta^n\| < \delta$ for all $n \in \mathbb{N}$.*

The structure of the set of PV numbers is well understood. The set is closed, with a smallest value of $1.324\cdots$, the real root of $x^3 - x - 1$, and the golden ratio $\frac{\sqrt{5}+1}{2}$ is the smallest limit point of PV numbers. Much less is known about Salem numbers. The following is an outstanding open problem.

PROBLEM D.0.8. Is there a smallest Salem number?

We needed yet another class of algebraic integers, introduced by Garsia [**Gar62**], which we call the set of *Garsia numbers*.

DEFINITION D.0.9. An algebraic integer $\theta > 1$ is a Garsia number if its minimal polynomial has constant term ± 2 and all its roots are greater than 1 in absolute value.

Note that θ and all its conjugates must have absolute values in $(1, 2]$, since their product equals 2. The roots of $x^m - 2$ are the simplest examples of Garsia numbers, but there are many more: for instance, the positive roots of $x^{p+n} - x^n - 2$, where $\max(p, n) \geqslant 2$, see [**Gar62**]. Garsia numbers have been studied in a paper by Hare and Panju [**HP13**], where it is shown, in particular, that $1.618\cdots$ and $1.465\cdots$, the roots of $x^2 - x - 1$ and $x^3 - x^2 - 1$ are the only limit points of Garsia numbers greater than $\sqrt{2}$.

We will need a lower bound on the absolute value of an integer polynomial evaluated at an algebraic integer. It is often referred to as *Garsia's Lemma.* Recall that the *height* of a polynomial is the maximum of the absolute values of its coefficients.

LEMMA D.0.10 ([**Gar62**]). *Let $\theta > 1$ be an algebraic integer, and $\theta_2, \ldots, \theta_d$ are the conjugates of $\theta = \theta_1$. Let σ be the number of j such that $|\theta_j| = 1$, If $P(x)$ is a polynomial with integer coefficients of height H of degree at most n for which $P(\theta) \neq 0$, then*

$$|P(\theta)| \geqslant \frac{\prod_{|\theta_j| \neq 1} |\theta_j| - 1|}{(n+1)^\sigma \left(\prod_{|\theta_j|>1} |\theta_j|\right)^{n+1} H^d}. \tag{D.1}$$

COROLLARY D.0.11.

(i) *If θ is a PV number and $H \in \mathbb{N}$, then there exists $c > 0$ such that for any polynomial $P(x)$ with integer coefficients of height at most H and degree n, we have*

$$P(1/\theta) \neq 0 \ \Rightarrow \ |P(1/\theta)| \geqslant c\theta^{-n},$$

independent of the degree of $P(x)$.

(ii) *Suppose that θ is a Garsia number.Then there exists $c > 0$ such that for any polynomial $P(x)$ of degree less than or equal to n, with integer coefficients and height 1, we have*

$$|P(1/\theta)| \geqslant c2^{-n}.$$

PROOF.

(i) Note that $P(1/x) = x^{-n}P^*(x)$, where $P^*(x)$ is a polynomial of degree n with integer coefficients and height at most H. Applying Lemma D.0.10 to $P^*(\theta)$ yields the result.

(ii) We again write $P(1/\theta) = \theta^{-n}P^*(\theta)$; now P^* has integer coefficients and height one. Note that $P^*(\theta) \neq 0$, since the minimal polynomial for θ has constant term ± 2, and P^* has height one. From Lemma D.0.10 we obtain

$$|P^*(\theta)| \geqslant c\Big(\prod_{j=2}^{d} |\theta_j|\Big)^{-n-1}.$$

By assumption, the minimal polynomial for θ has leading term x^d and the constant term ± 2, hence $\prod_{j=2}^{d} |\theta_j| = 2\theta^{-1}$, and the claim follows.

□

Bibliography

[AB02] Howard Anton and Robert C. Busby, *Contemporary Linear Algebra*, John Wiley & Sons, 2002. ↑413

[ABY10] Artur Avila, Jairo Bochi, and Jean-Christophe Yoccoz, *Uniformly hyperbolic finite-valued* $\mathrm{SL}(2,\mathbb{R})$*-cocycles*, Comment. Math. Helv. **85** (2010), no. 4, 813–884, DOI 10.4171/CMH/212. MR2718140 ↑297

[AF03] Robert A. Adams and John J. F. Fournier, *Sobolev spaces*, 2nd ed., Pure and Applied Mathematics (Amsterdam), vol. 140, Elsevier/Academic Press, Amsterdam, 2003. MR2424078 ↑184

[AFKP20] Shigeki Akiyama, De-Jun Feng, Tom Kempton, and Tomas Persson, *On the Hausdorff dimension of Bernoulli convolutions*, Int. Math. Res. Not. IMRN **19** (2020), 6569–6595, DOI 10.1093/imrn/rny209. MR4165484 ↑253

[AHW21] Amir Algom, Federico Rodriguez Hertz, and Zhiren Wang, *Pointwise normality and Fourier decay for self-conformal measures*, Adv. Math. **393** (2021), Paper No. 108096, 72, DOI 10.1016/j.aim.2021.108096. MR4340230 ↑238

[AHW22] Amir Algom, Federico Rodriguez Hertz, and Zhiren Wang, *Logarithmic Fourier decay for self conformal measures*, J. Lond. Math. Soc. (2) **106** (2022), no. 2, 1628–1661, DOI 10.1112/jlms.12608. MR4477227 ↑238

[AKT20] Jasmina Angelevska, Antti Käenmäki, and Sascha Troscheit, *Self-conformal sets with positive Hausdorff measure*, Bull. Lond. Math. Soc. **52** (2020), no. 1, 200–223, DOI 10.1112/blms.12320. MR4072045 ↑402, 403, 404

[AP96] Matthias Arbeiter and Norbert Patzschke, *Random self-similar multifractals*, Math. Nachr. **181** (1996), 5–42, DOI 10.1002/mana.3211810102. MR1409071 ↑140

[Arn98] Ludwig Arnold, *Random dynamical systems*, Springer Monographs in Mathematics, Springer-Verlag, Berlin, 1998, DOI 10.1007/978-3-662-12878-7. MR1723992 ↑341

[Ass77] Patrice Assouad, *Espaces métriques, plongements, facteurs* (French), Publications Mathématiques d'Orsay, No. 223-7769, Université Paris XI, U.E.R. Mathématique, Orsay, 1977. Thèse de doctorat. MR0644642 ↑215

[Ass79] Patrice Assouad, *Étude d'une dimension métrique liée à la possibilité de plongements dans* $\mathbb{R}^n$, C. R. Acad. Sci. Paris Sér. A-B **288** (1979), no. 15, A731–A734. MR532401 215

[AY84] James C. Alexander and James A. Yorke, *Fat baker's transformations*, Ergodic Theory Dynam. Systems **4** (1984), no. 1, 1–23, DOI 10.1017/S0143385700002236. MR758890 ↑41, 251

[AZ91] James C. Alexander and Don Zagier, *The entropy of a certain infinitely convolved Bernoulli measure*, J. London Math. Soc. (2) **44** (1991), no. 1, 121–134, DOI 10.1112/jlms/s2-44.1.121. MR1122974 ↑252

[Bak21] Simon Baker, *Iterated function systems with super-exponentially close cylinders*, Adv. Math. **379** (2021), Paper No. 107548, 13, DOI 10.1016/j.aim.2020.107548. MR4202041 ↑186, 187

[Ban18] Christoph Bandt, *Finite orbits in multivalued maps and Bernoulli convolutions*, Adv. Math. **324** (2018), 437–485, DOI 10.1016/j.aim.2017.11.011. MR3733892 ↑248

[Bár12] Balázs Bárány, *Dimension of the generalized 4-corner set and its projections*, Ergodic Theory Dynam. Systems **32** (2012), no. 4, 1190–1215, DOI 10.1017/S014338571100023X. MR2955310 ↑324

[Bár15] Balázs Bárány, *On the Ledrappier–Young formula for self-affine measures*, Math. Proc. Cambridge Philos. Soc. **159** (2015), no. 3, 405–432, DOI 10.1017/S0305004115000419. MR3413884 ↑343, 345, 367

[Bar93] Michael F. Barnsley, *Fractals everywhere*, 2nd ed., Academic Press Professional, Boston, MA, 1993. Revised with the assistance of and with a foreword by Hawley Rising, III. MR1231795 ↑ix, 82

[Bar96] Luis M. Barreira, *A nonadditive thermodynamic formalism and applications to dimension theory of hyperbolic dynamical systems*, Ergodic Theory Dynam. Systems **16** (1996), no. 5, 871–927, DOI 10.1017/S0143385700010117. MR1417767 ↑273, 290

[Bar07] Krzysztof Barański, *Hausdorff dimension of the limit sets of some planar geometric constructions*, Adv. Math. **210** (2007), no. 1, 215–245, DOI 10.1016/j.aim.2006.06.005. MR2298824 ↑54, 324, 339, 341

[Bar08] Krzysztof Barański, *Hausdorff dimension of self-affine limit sets with an invariant direction*, Discrete Contin. Dyn. Syst. **21** (2008), no. 4, 1015–1023, DOI 10.3934/dcds.2008.21.1015. MR2399447 ↑343

[BBR14] Krzysztof Barański, Balázs Bárány, and Julia Romanowska, *On the dimension of the graph of the classical Weierstrass function*, Adv. Math. **265** (2014), 32–59, DOI 10.1016/j.aim.2014.07.033. MR3255455 ↑156, 181

[BD85] Michael F. Barnsley and Stephen Demko, *Iterated function systems and the global construction of fractals*, Proc. Roy. Soc. London Ser. A **399** (1985), no. 1817, 243–275. MR799111 ↑82

[BDGGH+92] Marie-José Bertin, Annette Decomps-Guilloux, Marthe Grandet-Hugot, Martine Pathiaux-Delefosse, and Jean-Pierre Schreiber, *Pisot and Salem numbers*, Birkhäuser Verlag, Basel, 1992. With a preface by David W. Boyd, DOI 10.1007/978-3-0348-8632-1. MR1187044 ↑426

[Bed84] Tim Bedford, *Crinkly curves, Markov partitions and dimension*, Ph.D. thesis, University of Warwick, 1984. 53, 332

[Ber73] Claude Berge, *Graphs and hypergraphs*, North-Holland Mathematical Library, Vol. 6, North-Holland Publishing Co., Amsterdam-London; American Elsevier Publishing Co., Inc., New York, 1973. Translated from the French by Edward Minieka. MR0357172 ↑107

[Bes38] Abram S. Besicovitch, *On the fundamental geometrical properties of linearly measurable plane sets of points (II)*, Math. Ann. **115** (1938), no. 1, 296–329, DOI 10.1007/BF01448943. MR1513189 ↑69

[BG92] Christoph Bandt and Siegfried Graf, *Self-similar sets. VII. A characterization of self-similar fractals with positive Hausdorff measure*, Proc. Amer. Math. Soc. **114** (1992), no. 4, 995–1001, DOI 10.2307/2159618. MR1100644 ↑86, 97, 202

[BHR19] Balázs Bárány, Michael Hochman, and Ariel Rapaport, *Hausdorff dimension of planar self-affine sets and measures*, Invent. Math. **216** (2019), no. 3, 601–659, DOI 10.1007/s00222-018-00849-y. MR3955707 ↑58, 360, 361, 375, 386

[Bil60] Patrick Billingsley, *Hausdorff dimension in probability theory*, Illinois J. Math. **4** (1960), 187–209. MR131903 ↑80

[Bil61] Patrick Billingsley, *Hausdorff dimension in probability theory. II*, Illinois J. Math. **5** (1961), 291–298. MR120339 ↑80

[Bil99] Patrick Billingsley, *Convergence of probability measures*, second ed., Wiley Series in Probability and Statistics: Probability and Statistics, John Wiley & Sons, Inc., New York, 1999, A Wiley-Interscience Publication. MR1700749 77

[BK17] Balázs Bárány and Antti Käenmäki, *Ledrappier–Young formula and exact dimensionality of self-affine measures*, Adv. Math. **318** (2017), 88–129, DOI 10.1016/j.aim.2017.07.015. MR3689737 ↑57, 345, 358, 361

[BK21] Balázs Bárány, *Super-exponential condensation without exact overlaps*, Adv. Math. **379** (2021), 107549. MR4202042 186

[BKK18] Balázs Bárány, Antti Käenmäki, and Henna Koivusalo, *Dimension of self-affine sets for fixed translation vectors*, J. Lond. Math. Soc. (2) **98** (2018), no. 1, 223–252, DOI 10.1112/jlms.12132. MR3847239 ↑360

[BKM20] Balázs Bárány, Antti Käenmäki, and Ian D. Morris, *Domination, almost additivity, and thermodynamic formalism for planar matrix cocycles*, Israel J. Math. **239** (2020), no. 1, 173–214, DOI 10.1007/s11856-020-2054-4. MR4160885 ↑298, 301

[BL85] Philippe Bougerol and Jean Lacroix, *Products of random matrices with applications to Schrödinger operators*, Progress in Probability and Statistics, vol. 8, Birkhäuser Boston, Inc., Boston, MA, 1985, DOI 10.1007/978-1-4684-9172-2. MR886674 ↑347, 350, 351, 353

[Blu98] Christian Bluhm, *On a theorem of Kaufman: Cantor-type construction of linear fractal Salem sets*, Ark. Mat. **36** (1998), no. 2, 307–316, DOI 10.1007/BF02384771. MR1650442 ↑232

[BM09] Christoph Bandt and Mathias Mesing, *Self-affine fractals of finite type*, Convex and fractal geometry, Banach Center Publ., vol. 84, Polish Acad. Sci. Inst. Math., Warsaw, 2009, pp. 131–148, DOI 10.4064/bc84-0-9. MR2512375 ↑123

[BM15] Jairo Bochi and Ian D. Morris, *Continuity properties of the lower spectral radius*, Proc. Lond. Math. Soc. (3) **110** (2015), no. 2, 477–509, DOI 10.1112/plms/pdu058. MR3335285 ↑297

[BM18] Jairo Bochi and Ian D. Morris, *Equilibrium states of generalised singular value potentials and applications to affine iterated function systems*, Geom. Funct. Anal. **28** (2018), no. 4, 995–1028, DOI 10.1007/s00039-018-0447-x. MR3820437 ↑301

[BMS04] Dave Broomhead, James Montaldi, and Nikita Sidorov, *Golden gaskets: Variations on the Sierpiński sieve*, Nonlinearity **17** (2004), no. 4, 1455–1480, DOI 10.1088/0951-7715/17/4/017. MR2069714 ↑5, 6, 9, 27

[Bog07] Vladimir I. Bogachev, *Measure theory. Vol. I, II*, Springer-Verlag, Berlin, 2007, DOI 10.1007/978-3-540-34514-5. MR2267655 ↑77

[Bou28] Georges Bouligand, *Ensembles impropres et nombre dimensionnel*, Bull. Soc. Math. France **52** (1928), 320–334 and 361–376. 215

[Bow75] Rufus Bowen, *Equilibrium states and the ergodic theory of Anosov diffeomorphisms*, Lecture Notes in Mathematics, Vol. 470, Springer-Verlag, Berlin-New York, 1975. MR0442989 ↑46, 47, 293, 396, 400

[Boy69] David W. Boyd, *Transcendental numbers with badly distributed powers*, Proc. Amer. Math. Soc. **23** (1969), 424–427, DOI 10.2307/2037185. MR248094 ↑426

[BP17] Christopher J. Bishop and Yuval Peres, *Fractals in probability and analysis*, Cambridge Studies in Advanced Mathematics, vol. 162, Cambridge University Press, Cambridge, 2017, DOI 10.1017/9781316460238. MR3616046 ↑ix, xi, 13, 14, 23, 25, 37, 62, 67, 69, 97, 99, 107, 279, 333, 335, 417

[BR07] Christoph Bandt and Hui Rao, *Topology and separation of self-similar fractals in the plane*, Nonlinearity **20** (2007), no. 6, 1463–1474, DOI 10.1088/0951-7715/20/6/008. MR2327133 ↑101

[BR14] Balázs Bárány and Michał Rams, *Dimension of slices of Sierpiński-like carpets*, J. Fractal Geom. **1** (2014), no. 3, 273–294, DOI 10.4171/JFG/8. MR3276836 ↑131, 136

[BR18] Balázs Bárány and Michał Rams, *Dimension maximizing measures for self-affine systems*, Trans. Amer. Math. Soc. **370** (2018), no. 1, 553–576, DOI 10.1090/tran/7103. MR3717989 ↑360

[Bra51] Alfred Brauer, *On algebraic equations with all but one root in the interior of the unit circle*, Math. Nachr. **4** (1951), 250–257, DOI 10.1002/mana.3210040123. MR41975 ↑150

[Bré21] Julien Brémont, *Self-similar measures and the Rajchman property* (English, with English and French summaries), Ann. H. Lebesgue **4** (2021), 973–1004, DOI 10.5802/ahl.94. MR4315775 ↑238

[BRS19] Balázs Bárány, Michał Rams, and Károly Simon, *On the dimension of triangular self-affine sets*, Ergodic Theory Dynam. Systems **39** (2019), no. 7, 1751–1783, DOI 10.1017/etds.2017.110. MR3961673 ↑341, 342

[BS00] Luis Barreira and Jörg Schmeling, *Sets of "nontypical" points have full topological entropy and full Hausdorff dimension*, Israel J. Math. **116** (2000), 29–70, DOI 10.1007/BF02773211. MR1759398 ↑151

[BS14] Alexander I. Bufetov and Boris Solomyak, *On the modulus of continuity for spectral measures in substitution dynamics*, Adv. Math. **260** (2014), 84–129, DOI 10.1016/j.aim.2014.04.004. MR3209350 ↑237

[BSKR21] Balázs Bárány, Károly Simon, István Kolossváry, and Michał Rams, *Hausdorff measure and Assouad dimension of generic self-conformal IFS on the line*, Proc. Roy. Soc. Edinburgh Sect. A **151** (2021), no. 6, 2051–2081, DOI 10.1017/prm.2020.89. MR4349434 ↑412

[BU90] Tim Bedford and Mariusz Urbański, *The box and Hausdorff dimension of self-affine sets*, Ergodic Theory Dynam. Systems **10** (1990), no. 4, 627–644, DOI 10.1017/S0143385700005812. MR1091418 ↑329

[BV19] Emmanuel Breuillard and Péter P. Varjú, *On the dimension of Bernoulli convolutions*, Ann. Probab. **47** (2019), no. 4, 2582–2617, DOI 10.1214/18-AOP1324. MR3980929 ↑187

[BV20] Emmanuel Breuillard and Péter P. Varjú, *Entropy of Bernoulli convolutions and uniform exponential growth for linear groups*, J. Anal. Math. **140** (2020), no. 2, 443–481, DOI 10.1007/s11854-020-0100-0. MR4093919 ↑187, 252

[CJP$^+$06] Marianna Csörnyei, Thomas Jordan, Mark Pollicott, David Preiss, and Boris Solomyak, *Positive-measure self-similar sets without interior*, Ergodic Theory Dynam. Systems **26** (2006), no. 3, 755–758, DOI 10.1017/S0143385705000702. MR2237468 ↑212, 213

[CM92] Robert Cawley and R. Daniel Mauldin, *Multifractal decompositions of Moran fractals*, Adv. Math. **92** (1992), no. 2, 196–236, DOI 10.1016/0001-8708(92)90064-R. MR1155465 ↑140, 147

[CX99] Ercai Chen and Jincheng Xiong, *The pointwise dimension of self-similar measures*, Chinese Sci. Bull. **44** (1999), no. 23, 2136–2140, DOI 10.1007/BF03182695. MR1751239 ↑151

[dC92] Manfredo Perdigão do Carmo, *Riemannian geometry*, Mathematics: Theory & Applications, Birkhäuser Boston, Inc., Boston, MA, 1992. Translated from the second Portuguese edition by Francis Flaherty, DOI 10.1007/978-1-4757-2201-7. MR1138207 ↑393

[DEL63] Harold Davenport, Paul Erdős, and William J. LeVeque, *On Weyl's criterion for uniform distribution*, Michigan Math. J. **10** (1963), 311–314. MR153656 ↑232

[DF99] Persi Diaconis and David Freedman, *Iterated random functions*, SIAM Rev. **41** (1999), no. 1, 45–76, DOI 10.1137/S0036144598338446. MR1669737 ↑82

[DFW07] Xin-Rong Dai, De-Jun Feng, and Yang Wang, *Refinable functions with noninteger dilations*, J. Funct. Anal. **250** (2007), no. 1, 1–20, DOI 10.1016/j.jfa.2007.02.005. MR2345903 ↑237

[DNPV12] Eleonora Di Nezza, Giampiero Palatucci, and Enrico Valdinoci, *Hitchhiker's guide to the fractional Sobolev spaces*, Bull. Sci. Math. **136** (2012), no. 5, 521–573, DOI 10.1016/j.bulsci.2011.12.004. MR2944369 ↑184

[Dow11] Tomasz Downarowicz, *Entropy in dynamical systems*, New Mathematical Monographs, vol. 18, Cambridge University Press, Cambridge, 2011, DOI 10.1017/CBO9780511976155. MR2809170 ↑269

[DS86] Persi Diaconis and Mehrdad Shahshahani, *Products of random matrices and computer image generation*, Random matrices and their applications (Brunswick, Maine, 1984), Contemp. Math., vol. 50, Amer. Math. Soc., Providence, RI, 1986, pp. 173–182, DOI 10.1090/conm/050/841091. MR841091 ↑82

[DS17] Tushar Das and David Simmons, *The Hausdorff and dynamical dimensions of self-affine sponges: a dimension gap result*, Invent. Math. **210** (2017), no. 1, 85–134, DOI 10.1007/s00222-017-0725-5. MR3698340 ↑300

[Dur96] Richard Durrett, *Probability: Theory and examples*, 2nd ed., Duxbury Press, Belmont, CA, 1996. MR1609153 ↑46, 121

[Dur12] Richard Durrett, *Essentials of stochastic processes*, 2nd ed., Springer Texts in Statistics, Springer, New York, 2012, DOI 10.1007/978-1-4614-3615-7. MR2933766 ↑348

[dVK16] Martijn de Vries and Vilmos Komornik, *Expansions in noninteger bases*, Combinatorics, words and symbolic dynamics, Encyclopedia Math. Appl., vol. 159, Cambridge Univ. Press, Cambridge, 2016, pp. 18–58. MR3525481 ↑247

[DWX21] Juan Deng, Zhiying Wen, and Lifeng Xi, *Finite type in measure sense for self-similar sets with overlaps*, Math. Z. **298** (2021), no. 1-2, 821–837, DOI 10.1007/s00209-020-02632-3. MR4257110 ↑129

[Edg92] Gerald A. Edgar, *Fractal dimension of self-affine sets: Some examples*, Rend. Circ. Mat. Palermo (2) Suppl. **28** (1992), 341–358. Measure theory (Oberwolfach, 1990). MR1183060 ↑53, 55

[Edg98] Gerald A. Edgar, *Integral, probability, and fractal measures*, Springer-Verlag, New York, 1998, DOI 10.1007/978-1-4757-2958-0. MR1484412 ↑ix, 45, 80, 83

[EJK90] Paul Erdös, István Joó, and Vilmos Komornik, *Characterization of the unique expansions* $1 = \sum_{i=1}^{\infty} q^{-n_i}$ *and related problems* (English, with French summary), Bull. Soc. Math. France **118** (1990), no. 3, 377–390. MR1078082 ↑247

[Eks15] Fredrik Ekström, *The Fourier dimension is not finitely stable*, Real Anal. Exchange **40** (2014/15), no. 2, 397–402, DOI 10.14321/realanalexch.40.2.0397. MR3499772 ↑232

[EPS15] Fredrik Ekström, Tomas Persson, and Jörg Schmeling, *On the Fourier dimension and a modification*, J. Fractal Geom. **2** (2015), no. 3, 309–337, DOI 10.4171/JFG/23. MR3416842 ↑231, 232, 233

[Erd39] Paul Erdös, *On a family of symmetric Bernoulli convolutions*, Amer. J. Math. **61** (1939), 974–976, DOI 10.2307/2371641. MR311 ↑41, 228

[Erd40] Paul Erdös, *On the smoothness properties of a family of Bernoulli convolutions*, Amer. J. Math. **62** (1940), 180–186, DOI 10.2307/2371446. MR858 ↑41, 234, 235

[ES17] Fredrik Ekström and Jörg Schmeling, *A survey on the Fourier dimension*, Patterns of dynamics, Springer Proc. Math. Stat., vol. 205, Springer, Cham, 2017, pp. 67–87, DOI 10.1007/978-3-319-64173-7_5. MR3775405 ↑231

[EW11] Manfred Einsiedler and Thomas Ward, *Ergodic theory with a view towards number theory*, Graduate Texts in Mathematics, vol. 259, Springer-Verlag London, Ltd., London, 2011, DOI 10.1007/978-0-85729-021-2. MR2723325 ↑66, 255, 260, 261, 262, 263, 390, 415, 416

[Fal82] Kenneth J. Falconer, *Hausdorff dimension and the exceptional set of projections*, Mathematika **29** (1982), no. 1, 109–115, DOI 10.1112/S0025579300012201. MR673510 ↑184

[Fal85] Kenneth J. Falconer, *The geometry of fractal sets*, Cambridge University Press, 1985. ix, 15, 19, 23, 68, 69, 232, 418

[Fal87] Kenneth J. Falconer, *The Hausdorff dimension of some fractals and attractors of overlapping construction*, J. Statist. Phys. **47** (1987), no. 1-2, 123–132, DOI 10.1007/BF01009037. MR892926 ↑30, 155, 156, 157

[Fal88] Kenneth J. Falconer, *The Hausdorff dimension of self-affine fractals*, Math. Proc. Cambridge Philos. Soc. **103** (1988), no. 2, 339–350, DOI 10.1017/S0305004100064926. MR923687 ↑53, 155, 273, 277, 278, 287

[Fal89] Kenneth J. Falconer, *Dimensions and measures of quasi self-similar sets*, Proc. Amer. Math. Soc. **106** (1989), no. 2, 543–554, DOI 10.2307/2048840. MR969315 ↑84, 89, 113, 402

[Fal94] Kenneth J. Falconer, *Bounded distortion and dimension for nonconformal repellers*, Math. Proc. Cambridge Philos. Soc. **115** (1994), no. 2, 315–334, DOI 10.1017/S030500410007211X. MR1277063 ↑273, 290

[Fal97] Kenneth J. Falconer, *Techniques in fractal geometry*, John Wiley & Sons, Ltd., Chichester, 1997. MR1449135 ↑ix, xi, 33, 34, 37, 38, 86, 139, 402

[Fal03] Kenneth J. Falconer, *Fractal geometry*, 2nd ed., John Wiley & Sons, Inc., Hoboken, NJ, 2003. Mathematical foundations and applications, DOI 10.1002/0470013850. MR2118797 ↑ix, 17, 23, 26, 60, 62, 418

[Far] Ábel Farkas, personal communication. 101

[Far16] Ábel Farkas, *Projections of self-similar sets with no separation condition*, Israel J. Math. **214** (2016), no. 1, 67–107, DOI 10.1007/s11856-016-1345-2. MR3540606 ↑111

[Fed69] Herbert Federer, *Geometric measure theory*, Die Grundlehren der mathematischen Wissenschaften, Band 153, Springer-Verlag New York, Inc., New York, 1969. MR0257325 ↑70

[Fel71] William Feller, *An introduction to probability theory and its applications. Vol. II.*, 2nd ed., John Wiley & Sons, Inc., New York-London-Sydney, 1971. MR0270403 ↑261

[Fen] De-Jun Feng, personal communication. 331

[Fen99] *Proceedings of the International Mathematics Conference*, Mathematical Society of the Philippines, Quezon City, 1999. Held in honor of Professor Peng Yee Lee on his 60th birthday at Ateneo de Manila University, Manila, December 16–18, 1998; Matimyás Mat. **22** (1999), no. 2. MR1770160 ↑252

[Fen03] De-Jun Feng, *Smoothness of the L^q-spectrum of self-similar measures with overlaps*, J. London Math. Soc. (2) **68** (2003), no. 1, 102–118, DOI 10.1112/S002461070300437X. MR1980246 ↑124, 150

[Fen05] De-Jun Feng, *The limited Rademacher functions and Bernoulli convolutions associated with Pisot numbers*, Adv. Math. **195** (2005), no. 1, 24–101, DOI 10.1016/j.aim.2004.06.011. MR2145793 ↑150, 252

[Fen09] De-Jun Feng, *Lyapunov exponents for products of matrices and multifractal analysis. II. General matrices*, Israel J. Math. **170** (2009), 355–394, DOI 10.1007/s11856-009-0033-x. MR2506331 ↑273, 293

[Fen16] De-Jun Feng, *On the topology of polynomials with bounded integer coefficients*, J. Eur. Math. Soc. (JEMS) **18** (2016), no. 1, 181–193, DOI 10.4171/JEMS/587. MR3438383 ↑127

[Fen23] De-Jun Feng, *Dimension of invariant measures for affine iterated function systems*, Duke Math. J. **172** (2023), no. 4, 701–774, DOI 10.1215/00127094-2022-0014. MR4557759 ↑57, 321, 345, 361

[FF15] Ábel Farkas and Jonathan M. Fraser, *On the equality of Hausdorff measure and Hausdorff content*, J. Fractal Geom. **2** (2015), no. 4, 403–429, DOI 10.4171/JFG/27. MR3416846 ↑86, 97, 215, 216, 217

[FF22] De-Jun Feng and Zhou Feng, *Estimates on the dimension of self-similar measures with overlaps*, J. Lond. Math. Soc. (2) **105** (2022), no. 4, 2104–2135, DOI 10.1112/jlms.12555. MR4440532 ↑253

[FFJ15] Kenneth J. Falconer, Jonathan Fraser, and Xiong Jin, *Sixty years of fractal projections*, Fractal geometry and stochastics V, Progr. Probab., vol. 70, Birkhäuser/Springer, Cham, 2015, pp. 3–25, DOI 10.1007/978-3-319-18660-3_1. MR3558147 ↑184

[FH96] Kenneth J. Falconer and John D. Howroyd, *Projection theorems for box and packing dimensions*, Math. Proc. Cambridge Philos. Soc. **119** (1996), no. 2, 287–295, DOI 10.1017/S0305004100074168. MR1357045 ↑63

[FH97] Kenneth J. Falconer and John D. Howroyd, *Packing dimensions of projections and dimension profiles*, Math. Proc. Cambridge Philos. Soc. **121** (1997), no. 2, 269–286, DOI 10.1017/S0305004196001375. MR1426523 ↑63

[FH09] De-Jun Feng and Huyi Hu, *Dimension theory of iterated function systems*, Comm. Pure Appl. Math. **62** (2009), no. 11, 1435–1500, DOI 10.1002/cpa.20276. MR2560042 ↑46, 56, 242, 303, 305, 306, 313

[FH17] Jonathan M. Fraser and Douglas C. Howroyd, *Assouad type dimensions for self-affine sponges*, Ann. Acad. Sci. Fenn. Math. **42** (2017), no. 1, 149–174, DOI 10.5186/aasfm.2017.4213. MR3558522 ↑216

[FHOR15] Jonathan M. Fraser, A. M. Henderson, Eric J. Olson, and James C. Robinson, *On the Assouad dimension of self-similar sets with overlaps*, Adv. Math. **273** (2015), 188–214, DOI 10.1016/j.aim.2014.12.026. MR3311761 ↑215, 217, 218, 220

[FJ17] Jonathan M. Fraser and Thomas Jordan, *The Assouad dimension of self-affine carpets with no grid structure*, Proc. Amer. Math. Soc. **145** (2017), no. 11, 4905–4918, DOI 10.1090/proc/13629. MR3692005 ↑216

[FJS10] Andrew Ferguson, Thomas Jordan, and Pablo Shmerkin, *The Hausdorff dimension of the projections of self-affine carpets*, Fund. Math. **209** (2010), no. 3, 193–213, DOI 10.4064/fm209-3-1. MR2720210 ↑391, 392

[FK60] Hillel Furstenberg and Harry Kesten, *Products of random matrices*, Ann. Math. Statist. **31** (1960), 457–469, DOI 10.1214/aoms/1177705909. MR121828 ↑257

[FK11] De-Jun Feng and Antti Käenmäki, *Equilibrium states of the pressure function for products of matrices*, Discrete Contin. Dyn. Syst. **30** (2011), no. 3, 699–708, DOI 10.3934/dcds.2011.30.699. MR2784616 ↑273, 293, 301

[FK17] Kenneth J. Falconer and Tom Kempton, *The dimension of projections of self-affine sets and measures*, Ann. Acad. Sci. Fenn. Math. **42** (2017), no. 1, 473–486, DOI 10.5186/aasfm.2017.4232. MR3558544 ↑375, 380, 381, 384, 386, 389, 390

[FL02] De-Jun Feng and Ka-Sing Lau, *The pressure function for products of nonnegative matrices*, Math. Res. Lett. **9** (2002), no. 2-3, 363–378, DOI 10.4310/MRL.2002.v9.n3.a10. MR1909650 ↑273, 293

[FL09] De-Jun Feng and Ka-Sing Lau, *Multifractal formalism for self-similar measures with weak separation condition* (English, with English and French summaries), J. Math. Pures Appl. (9) **92** (2009), no. 4, 407–428, DOI 10.1016/j.matpur.2009.05.009. MR2569186 ↑76, 149, 293

[FLR02] Ai-Hua Fan, Ka-Sing Lau, and Hui Rao, *Relationships between different dimensions of a measure*, Monatsh. Math. **135** (2002), no. 3, 191–201, DOI 10.1007/s006050200016. MR1897575 ↑71

[Fol89] Gerald B. Folland, *Harmonic analysis in phase space*, Annals of Mathematics Studies, vol. 122, Princeton University Press, Princeton, NJ, 1989, DOI 10.1515/9781400882427. MR983366 ↑421

[Fra13] Jonathan M. Fraser, *Dimension theory and fractal constructions based on self-affine carpets*, ProQuest LLC, Ann Arbor, MI, 2013. Thesis (Ph.D.)–University of St. Andrews (United Kingdom). MR3271782 ↑26

[Fra14] Jonathan M. Fraser, *Assouad type dimensions and homogeneity of fractals*, Trans. Amer. Math. Soc. **366** (2014), no. 12, 6687–6733, DOI 10.1090/S0002-9947-2014-06202-8. MR3267023 ↑216

[Fra20] Jonathan M. Fraser, *Assouad dimension and fractal geometry*, Cambridge Tracts in Mathematics, vol. 222, Cambridge University Press, Cambridge, 2021, DOI 10.1017/9781108778459. MR4411274 ↑ix, 37, 216

[Fra21] Jonathan M. Fraser, *Fractal geometry of Bedford–McMullen carpets*, Thermodynamic formalism, Lecture Notes in Math., vol. 2290, Springer, Cham, [2021] ©2021, pp. 495–516, DOI 10.1007/978-3-030-74863-0_15. MR4436835 ↑54

[FS11] De-Jun Feng and Nikita Sidorov, *Growth rate for beta-expansions*, Monatsh. Math. **162** (2011), no. 1, 41–60, DOI 10.1007/s00605-010-0192-1. MR2747343 ↑246, 248

[FS14] De-Jun Feng and Pablo Shmerkin, *Nonconformal repellers and the continuity of pressure for matrix cocycles*, Geom. Funct. Anal. **24** (2014), no. 4, 1101–1128, DOI 10.1007/s00039-014-0274-7. MR3248481 ↑299

[Fur63] Harry Furstenberg, *Noncommuting random products*, Trans. Amer. Math. Soc. **108** (1963), 377–428, DOI 10.2307/1993589. MR163345 ↑346, 353

[Fur08] Hillel Furstenberg, *Ergodic fractal measures and dimension conservation*, Ergodic Theory Dynam. Systems **28** (2008), no. 2, 405–422, DOI 10.1017/S0143385708000084. MR2408385 ↑345, 358, 361

[FW05] De-Jun Feng and Yang Wang, *A class of self-affine sets and self-affine measures*, J. Fourier Anal. Appl. **11** (2005), no. 1, 107–124, DOI 10.1007/s00041-004-4031-4. MR2128947 ↑310, 311, 323, 324

[Gar62] Adriano M. Garsia, *Arithmetic properties of Bernoulli convolutions*, Trans. Amer. Math. Soc. **102** (1962), 409–432, DOI 10.2307/1993615. MR137961 ↑229, 230, 426

[Gar63] Adriano M. Garsia, *Entropy and singularity of infinite convolutions*, Pacific J. Math. **13** (1963), 1159–1169. MR156945 ↑83, 249, 251

[Gar20] Ignacio García, *Assouad dimension and local structure of self-similar sets with overlaps in* $\mathbb{R}^d$, Adv. Math. **370** (2020), 107244, 33, DOI 10.1016/j.aim.2020.107244. MR4104822 ↑220

[GKT02] Peter J. Grabner, Peter Kirschenhofer, and Robert F. Tichy, *Combinatorial and arithmetical properties of linear numeration systems*, Combinatorica **22** (2002), no. 2, 245–267, DOI 10.1007/s004930200011. Special issue: Paul Erdős and his mathematics. MR1909085 ↑252

[GP96] Dimitrios Gatzouras and Yuval Peres, *The variational principle for Hausdorff dimension: a survey*, Ergodic theory of $\mathbf{Z}^d$ actions (Warwick, 1993), London Math.

Soc. Lecture Note Ser., vol. 228, Cambridge Univ. Press, Cambridge, 1996, pp. 113–125, DOI 10.1017/CBO9780511662812.004. MR1411217 ↑300

[GP97] Dimitrios Gatzouras and Yuval Peres, *Invariant measures of full dimension for some expanding maps*, Ergodic Theory Dynam. Systems **17** (1997), no. 1, 147–167, DOI 10.1017/S0143385797060987. MR1440772 ↑300

[GS01] Paul Glendinning and Nikita Sidorov, *Unique representations of real numbers in noninteger bases*, Math. Res. Lett. **8** (2001), no. 4, 535–543, DOI 10.4310/MRL.2001.v8.n4.a12. MR1851269 ↑247

[GSSY16] Daniel Galicer, Santiago Saglietti, Pablo Shmerkin, and Alexia Yavicoli, *L^q dimensions and projections of random measures*, Nonlinearity **29** (2016), no. 9, 2609–2640, DOI 10.1088/0951-7715/29/9/2609. MR3544801 ↑191

[Hal13] Paul R. Halmos, *Measure Theory*, D. Van Nostrand Co., Inc., New York, N. Y., 1950. MR0033869 ↑66

[HHR21] Kathryn E. Hare, Kevin G. Hare, and Alex Rutar, *When the weak separation condition implies the generalized finite type condition*, Proc. Amer. Math. Soc. **149** (2021), no. 4, 1555–1568, DOI 10.1090/proc/15307. MR4242311 ↑123, 127

[HJJL12] Risto Hovila, Esa Järvenpää, Maarit Järvenpää, and François Ledrappier, *Besicovitch–Federer projection theorem and geodesic flows on Riemann surfaces*, Geom. Dedicata **161** (2012), 51–61, DOI 10.1007/s10711-012-9693-5. MR2994030 ↑70

[HK97] Brian R. Hunt and Vadim Yu. Kaloshin, *How projections affect the dimension spectrum of fractal measures*, Nonlinearity **10** (1997), no. 5, 1031–1046, DOI 10.1088/0951-7715/10/5/002. MR1473372 ↑70, 176

[HL28] Godfrey H. Hardy and John E. Littlewood, *Some properties of fractional integrals. I*, Math. Z. **27** (1928), no. 1, 565–606, DOI 10.1007/BF01171116. MR1544927 ↑72

[HL93] Tian You Hu and Ka-Sing Lau, *Fractal dimensions and singularities of the Weierstrass type functions*, Trans. Amer. Math. Soc. **335** (1993), no. 2, 649–665, DOI 10.2307/2154398. MR1076614 ↑181

[HL95] Irene Hueter and Steven P. Lalley, *Falconer's formula for the Hausdorff dimension of a self-affine set in $\boldsymbol{R}^2$*, Ergodic Theory Dynam. Systems **15** (1995), no. 1, 77–97, DOI 10.1017/S0143385700008257. MR1314970 ↑57, 299, 359

[Hoc14a] Michael Hochman, *Lectures on dynamics, fractal geometry, and metric number theory*, J. Mod. Dyn. **8** (2014), no. 3-4, 437–497, DOI 10.3934/jmd.2014.8.437. MR3345838 ↑380

[Hoc14b] Michael Hochman, *On self-similar sets with overlaps and inverse theorems for entropy*, Ann. of Math. (2) **180** (2014), no. 2, 773–822, DOI 10.4007/annals.2014.180.2.7. MR3224722 ↑32, 41, 56, 156, 185, 186, 187, 189, 191, 221, 223, 252, 360, 361, 375

[Hoc15] Michael Hochman, *On self-similar sets with overlaps and inverse theorems for entropy*, Ann. of Math. (2) **180** (2014), no. 2, 773–822, DOI 10.4007/annals.2014.180.2.7. MR3224722 ↑156, 187, 188

[HP13] Kevin G. Hare and Maysum Panju, *Some comments on Garsia numbers*, Math. Comp. **82** (2013), no. 282, 1197–1221, DOI 10.1090/S0025-5718-2012-02636-6. MR3008855 ↑230, 426

[HR22] Michael Hochman and Ariel Rapaport, *Hausdorff dimension of planar self-affine sets and measures with overlaps*, J. Eur. Math. Soc. (JEMS) **24** (2022), no. 7, 2361–2441, DOI 10.4171/jems/1127. MR4413769 ↑58, 361

[HS12] Michael Hochman and Pablo Shmerkin, *Local entropy averages and projections of fractal measures*, Ann. of Math. (2) **175** (2012), no. 3, 1001–1059, DOI 10.4007/annals.2012.175.3.1. MR2912701 ↑92, 375, 378, 380, 381, 386

[HS18] Kevin G. Hare and Nikita Sidorov, *A lower bound for the dimension of Bernoulli convolutions*, Exp. Math. **27** (2018), no. 4, 414–418, DOI 10.1080/10586458.2017.1301841. MR3894719 ↑252, 253

[HT94] Xiaoyu Hu and S. James Taylor, *Fractal properties of products and projections of measures in $\boldsymbol{R}^d$*, Math. Proc. Cambridge Philos. Soc. **115** (1994), no. 3, 527–544, DOI 10.1017/S0305004100072285. MR1269937 ↑65

[Hut81] John E. Hutchinson, *Fractals and self-similarity*, Indiana Univ. Math. J. **30** (1981), no. 5, 713–747, DOI 10.1512/iumj.1981.30.30055. MR625600 ↑12, 15, 27, 28, 37, 38, 40

[Jär94] Maarit Järvenpää, *On the upper Minkowski dimension, the packing dimension, and orthogonal projections*, Ann. Acad. Sci. Fenn. Ser. A I Math. Dissertationes **99** (1994), 34. MR1316487 ↑63

[JJ14] Thomas Jordan and Natalia Jurga, *Self-affine sets with noncompactly supported random perturbations*, Ann. Acad. Sci. Fenn. Math. **39** (2014), no. 2, 771–785, DOI 10.5186/aasfm.2014.3948. MR3237050 ↑276

[JM20] Natalia Jurga and Ian D. Morris, *Analyticity of the affinity dimension for planar iterated function systems with matrices which preserve a cone*, Nonlinearity **33** (2020), no. 4, 1572–1593, DOI 10.1088/1361-6544/ab60d6. MR4072386 ↑299

[JP06] Thomas Jordan and Mark Pollicott, *Properties of measures supported on fat Sierpinski carpets*, Ergodic Theory Dynam. Systems **26** (2006), no. 3, 739–754, DOI 10.1017/S0143385705000696. MR2237467 ↑212

[JPS07] Thomas Jordan, Mark Pollicott, and Károly Simon, *Hausdorff dimension for randomly perturbed self affine attractors*, Comm. Math. Phys. **270** (2007), no. 2, 519–544, DOI 10.1007/s00220-006-0161-7. MR2276454 ↑273, 275, 276, 280, 282, 285, 289

[JR21] Thomas Jordan and Ariel Rapaport, *Dimension of ergodic measures projected onto self-similar sets with overlaps*, Proc. Lond. Math. Soc. (3) **122** (2021), no. 2, 191–206, DOI 10.1112/plms.12337. MR4218958 ↑192

[JSS11] Thomas Jordan, Pablo Shmerkin, and Boris Solomyak, *Multifractal structure of Bernoulli convolutions*, Math. Proc. Cambridge Philos. Soc. **151** (2011), no. 3, 521–539, DOI 10.1017/S0305004111000466. MR2838343 ↑248

[JV02] Leo B. Jonker and Jan J. P. Veerman, *Semicontinuity of dimension and measure for locally scaling fractals*, Fund. Math. **173** (2002), no. 2, 113–131, DOI 10.4064/fm173-2-2. MR1924810 ↑89, 402, 412

[JW35] Børge Jessen and Aurel Wintner, *Distribution functions and the Riemann zeta function*, Trans. Amer. Math. Soc. **38** (1935), no. 1, 48–88, DOI 10.2307/1989728. MR1501802 ↑78

[JY05] Ning Jin and Stephen S. T. Yau, *General finite type IFS and M-matrix*, Comm. Anal. Geom. **13** (2005), no. 4, 821–843. MR2191909 ↑123

[K04] Antti Käenmäki, *On natural invariant measures on generalised iterated function systems*, Ann. Acad. Sci. Fenn. Math. **29** (2004), no. 2, 419–458. MR2097242 ↑273, 279, 291

[Kah71] Jean-Pierre Kahane, *Sur la distribution de certaines séries aléatoires* (French), Colloque de Théorie des Nombres (Univ. Bordeaux, Bordeaux, 1969), Bull. Soc. Math. France, Mém. No. 25, Soc. Math. France, Paris, 1971, pp. 119–122, DOI 10.24033/msmf.42. MR0360498 ↑234, 238

[Kat76] Yitzhak Katznelson, *An introduction to harmonic analysis*, Second corrected edition, Dover Publications, Inc., New York, 1976. MR0422992 ↑421, 422

[Kau68] Robert Kaufman, *On Hausdorff dimension of projections*, Mathematika **15** (1968), 153–155, DOI 10.1112/S0025579300002503. MR248779 ↑63, 172

[Kau81] Robert Kaufman, *On the theorem of Jarník and Besicovitch*, Acta Arith. **39** (1981), no. 3, 265–267, DOI 10.4064/aa-39-3-265-267. MR640914 ↑232

[Kem13] Tom Kempton, *Counting β-expansions and the absolute continuity of Bernoulli convolutions*, Monatsh. Math. **171** (2013), no. 2, 189–203, DOI 10.1007/s00605-013-0512-3. MR3077931 ↑245, 246

[Kem16] Tom Kempton, *Sets of β-expansions and the Hausdorff measure of slices through fractals*, J. Eur. Math. Soc. (JEMS) **18** (2016), no. 2, 327–351, DOI 10.4171/JEMS/591. MR3459953 ↑246

[Ken97] Richard Kenyon, *Projecting the one-dimensional Sierpinski gasket*, Israel J. Math. **97** (1997), 221–238, DOI 10.1007/BF02774038. MR1441250 ↑101, 202, 221, 222

[Ker36] Richard Kershner, *On Singular Fourier–Stieltjes Transforms*, Amer. J. Math. **58** (1936), no. 2, 450–452, DOI 10.2307/2371053. MR1507167 ↑237

[Kit21] Samuel Kittle, *Absolute continuity of self-similar measures*, `arXiv:2103.12684` (2021). 190

[KK19] Vilmos Komornik and Derong Kong, *Bases in which some numbers have exactly two expansions*, J. Number Theory **195** (2019), 226–268, DOI 10.1016/j.jnt.2018.06.004. MR3867441 ↑247

[KL98] Vilmos Komornik and Paola Loreti, *Unique developments in noninteger bases*, Amer. Math. Monthly **105** (1998), no. 7, 636–639, DOI 10.2307/2589246. MR1633077 ↑247

[KM18] Antti Käenmäki and Ian D. Morris, *Structure of equilibrium states on self-affine sets and strict monotonicity of affinity dimension*, Proc. Lond. Math. Soc. (3) **116** (2018), no. 4, 929–956, DOI 10.1112/plms.12089. MR3789836 ↑293, 301

[KN74] Lauwerens Kuipers and Harald Niederreiter, *Uniform distribution of sequences*, Pure and Applied Mathematics, Wiley-Interscience [John Wiley & Sons], New York-London-Sydney, 1974. MR0419394 ↑239

[KO20] Antti Käenmäki and Tuomas Orponen, *Absolute continuity in families of parametrised nonhomogeneous self-similar measures*, (2020), Preprint, `arXiv:1812.05006`. 159, 191

[KOR18] Antti Käenmäki, Tuomo Ojala, and Eino Rossi, *Rigidity of quasisymmetric mappings on self-affine carpets*, Int. Math. Res. Not. IMRN **12** (2018), 3769–3799, DOI 10.1093/imrn/rnw336. MR3815187 ↑217, 218

[KP96] Richard Kenyon and Yuval Peres, *Measures of full dimension on affine-invariant sets*, Ergodic Theory Dynam. Systems **16** (1996), no. 2, 307–323, DOI 10.1017/S0143385700008828. MR1389626 ↑304

[KPV22] Victor Kleptsyn, Mark Pollicott, and Polina Vytnova, *Uniform lower bounds on the dimension of Bernoulli convolutions*, Adv. Math. **395** (2022), Paper No. 108090, 55, DOI 10.1016/j.aim.2021.108090. MR4363586 ↑253

[Kra06] Aleksey S. Kravchenko, *Completeness of the space of separable measures in the Kantorovich–Rubinshteĭn metric* (Russian, with Russian summary), Sibirsk. Mat. Zh. **47** (2006), no. 1, 85–96, DOI 10.1007/s11202-006-0009-6; English transl., Siberian Math. J. **47** (2006), no. 1, 68–76. MR2215298 ↑37, 38

[Kre85] Ulrich Krengel, *Ergodic theorems*, De Gruyter Studies in Mathematics, vol. 6, Walter de Gruyter & Co., Berlin, 1985. With a supplement by Antoine Brunel, DOI 10.1515/9783110844641. MR797411 ↑258, 259, 414

[KRS16] Antti Käenmäki, Tapio Rajala, and Ville Suomala, *Local homogeneity and dimensions of measures*, Ann. Sc. Norm. Super. Pisa Cl. Sci. (5) **16** (2016), no. 4, 1315–1351. MR3616336 ↑420

[KS19] István Kolossváry and Károly Simon, *Triangular Gatzouras–Lalley-type planar carpets with overlaps*, Nonlinearity **32** (2019), no. 9, 3294–3341, DOI 10.1088/1361-6544/ab1757. MR3987800 ↑321, 336, 341, 343, 344

[KSS95] Mike Keane, Meir Smorodinsky, and Boris Solomyak, *On the morphology of γ-expansions with deleted digits*, Trans. Amer. Math. Soc. **347** (1995), no. 3, 955–966, DOI 10.2307/2154880. MR1290723 ↑32, 225

[KSS03] Michael Keane, Károly Simon, and Boris Solomyak, *The dimension of graph directed attractors with overlaps on the line, with an application to a problem in fractal image recognition*, Fund. Math. **180** (2003), no. 3, 279–292, DOI 10.4064/fm180-3-5. MR2063630 ↑36

[KT20] Kirill Kamalutdinov and Andrei Tetenov, *On one-point intersection property for self-similar fractals*, Nonlinearity **33** (2020), no. 1, 408–416, DOI 10.1088/1361-6544/ab4e0e. MR4039777 ↑101, 206, 208

[KV10] Antti Käenmäki and Markku Vilppolainen, *Dimension and measures on sub-self-affine sets*, Monatsh. Math. **161** (2010), no. 3, 271–293, DOI 10.1007/s00605-009-0144-9. MR2726214 ↑292

[Lal97] Steven P. Lalley, *β-expansions with deleted digits for Pisot numbers β*, Trans. Amer. Math. Soc. **349** (1997), no. 11, 4355–4365, DOI 10.1090/S0002-9947-97-02069-2. MR1451608 ↑54, 122, 125

[Lal98] Steven P. Lalley, *Random series in powers of algebraic integers: Hausdorff dimension of the limit distribution*, J. London Math. Soc. (2) **57** (1998), no. 3, 629–654, DOI 10.1112/S0024610798005948. MR1659849 ↑249, 250, 251, 252

[Lau95] Ka-Sing Lau, *Self-similarity, L^p-spectrum and multifractal formalism*, Fractal geometry and stochastics (Finsterbergen, 1994), Progr. Probab., vol. 37, Birkhäuser, Basel, 1995, pp. 55–90. MR1391971 ↑149

[Led92] François Ledrappier, *On the dimension of some graphs*, Symbolic dynamics and its applications (New Haven, CT, 1991), Contemp. Math., vol. 135, Amer. Math. Soc., Providence, RI, 1992, pp. 285–293, DOI 10.1090/conm/135/1185095. MR1185095 ↑41, 181

[LG92] Steven P. Lalley and Dimitrios Gatzouras, *Hausdorff and box dimensions of certain self-affine fractals*, Indiana Univ. Math. J. **41** (1992), no. 2, 533–568, DOI 10.1512/iumj.1992.41.41031. MR1183358 ↑54, 335, 336, 338

[LN98] Ka-Sing Lau and Sze-Man Ngai, *L^q-spectrum of the Bernoulli convolution associated with the golden ratio*, Studia Math. **131** (1998), no. 3, 225–251. MR1644468 ↑150

[LN99] Ka-Sing Lau and Sze-Man Ngai, *Multifractal measures and a weak separation condition*, Adv. Math. **141** (1999), no. 1, 45–96, DOI 10.1006/aima.1998.1773. MR1667146 ↑103, 104, 147, 150

[LN00] Ka-Sing Lau and Sze-Man Ngai, *Second-order self-similar identities and multifractal decompositions*, Indiana Univ. Math. J. **49** (2000), no. 3, 925–972, DOI 10.1512/iumj.2000.49.1789. MR1803217 ↑131

[LN07] Ka-Sing Lau and Sze-Man Ngai, *A generalized finite type condition for iterated function systems*, Adv. Math. **208** (2007), no. 2, 647–671, DOI 10.1016/j.aim.2006.03.007. MR2304331 ↑115, 123

[LNW09] Ka-Sing Lau, Sze-Man Ngai, and Xiang-Yang Wang, *Separation conditions for conformal iterated function systems*, Monatsh. Math. **156** (2009), no. 4, 325–355, DOI 10.1007/s00605-008-0052-4. MR2486602 ↑402, 403

[LPS02] Elon Lindenstrauss, Yuval Peres, and Wilhelm Schlag, *Bernoulli convolutions and an intermediate value theorem for entropies of K-partitions*, J. Anal. Math. **87** (2002), 337–367, DOI 10.1007/BF02868480. Dedicated to the memory of Thomas H. Wolff. MR1945288 ↑185

[LS22] Jialun Li and Tuomas Sahlsten, *Trigonometric series and self-similar sets*, J. Eur. Math. Soc. (JEMS) **24** (2022), no. 1, 341–368, DOI 10.4171/jems/1102. MR4375453 ↑238

[LW04] Ka-Sing Lau and Xiang-Yang Wang, *Iterated function systems with a weak separation condition*, Studia Math. **161** (2004), no. 3, 249–268, DOI 10.4064/sm161-3-3. MR2033017 ↑109, 116, 131

[LY85a] François Ledrappier and Lai-Sang Young, *The metric entropy of diffeomorphisms. I. Characterization of measures satisfying Pesin's entropy formula*, Ann. of Math. (2) **122** (1985), no. 3, 509–539, DOI 10.2307/1971328. MR819556 ↑303, 305, 361

[LY85b] François Ledrappier and Lai-Sang Young, *The metric entropy of diffeomorphisms. II. Relations between entropy, exponents and dimension*, Ann. of Math. (2) **122** (1985), no. 3, 540–574, DOI 10.2307/1971329. MR819557 ↑305, 355, 361

[Mac11] John M. Mackay, *Assouad dimension of self-affine carpets*, Conform. Geom. Dyn. **15** (2011), 177–187, DOI 10.1090/S1088-4173-2011-00232-3. MR2846307 ↑216

[Mañ12] Ricardo Mañé, *Ergodic theory and differentiable dynamics*, Ergebnisse der Mathematik und ihrer Grenzgebiete (3) [Results in Mathematics and Related Areas (3)], vol. 8, Springer-Verlag, Berlin, 1987. Translated from the Portuguese by Silvio Levy, DOI 10.1007/978-3-642-70335-5. MR889254 ↑257

[Mar54] John M. Marstrand, *Some fundamental geometrical properties of plane sets of fractional dimensions*, Proc. London Math. Soc. (3) **4** (1954), 257–302, DOI 10.1112/plms/s3-4.1.257. MR63439 ↑62, 63, 67, 68

[Mat75] Pertti Mattila, *Hausdorff dimension, orthogonal projections and intersections with planes*, Ann. Acad. Sci. Fenn. Ser. A I Math. **1** (1975), no. 2, 227–244. MR0409774 ↑63, 66, 67

[Mat90] Pertti Mattila, *Orthogonal projections, Riesz capacities, and Minkowski content*, Indiana Univ. Math. J. **39** (1990), no. 1, 185–198, DOI 10.1512/iumj.1990.39.39011. MR1052016 ↑66, 67, 166, 172

[Mat95] Pertti Mattila, *Geometry of sets and measures in Euclidean spaces*, Cambridge Studies in Advanced Mathematics, vol. 44, Cambridge University Press,

Cambridge, 1995. Fractals and rectifiability, DOI 10.1017/CBO9780511623813. MR1333890 ↑ix, xi, 23, 24, 59, 60, 63, 65, 67, 68, 69, 74, 166, 175, 202, 277, 319, 356, 416, 417, 418, 419, 420

[Mat15] Pertti Mattila, *Fourier analysis and Hausdorff dimension*, Cambridge Studies in Advanced Mathematics, vol. 150, Cambridge University Press, Cambridge, 2015, DOI 10.1017/CBO9781316227619. MR3617376 ↑ix, xi, 185, 231, 421, 422

[McM84] Curt McMullen, *The Hausdorff dimension of general Sierpiński carpets*, Nagoya Math. J. **96** (1984), 1–9, DOI 10.1017/S0027763000021085. MR771063 ↑53, 332, 335

[Mey72] Yves Meyer, *Algebraic numbers and harmonic analysis*, North-Holland Mathematical Library, Vol. 2, North-Holland Publishing Co., Amsterdam-London; American Elsevier Publishing Co., Inc., New York, 1972. MR0485769 ↑426

[Mil08] Benjamin Miller, *The existence of measures of a given cocycle. I. Atomless, ergodic σ-finite measures*, Ergodic Theory Dynam. Systems **28** (2008), no. 5, 1599–1613, DOI 10.1017/S0143385707001113. MR2449546 ↑420

[MM97] Pertti Mattila and R. Daniel Mauldin, *Measure and dimension functions: measurability and densities*, Math. Proc. Cambridge Philos. Soc. **121** (1997), no. 1, 81–100, DOI 10.1017/S0305004196001089. MR1418362 ↑26

[MMR00] Pertti Mattila, Manuel Morán, and José-Manuel Rey, *Dimension of a measure*, Studia Math. **142** (2000), no. 3, 219–233, DOI 10.4064/sm-142-3-219-233. MR1792606 ↑43, 44

[Mor46] P. A. P. Moran, *Additive functions of intervals and Hausdorff measure*, Proc. Cambridge Philos. Soc. **42** (1946), 15–23, DOI 10.1017/s0305004100022684. MR14397 ↑27

[Mor22] Ian D. Morris, *Fast approximation of the affinity dimension for dominated affine iterated function systems* (English, with English and Finnish summaries), Ann. Fenn. Math. **47** (2022), no. 2, 645–694, DOI 10.54330/afm.116153. MR4407233 ↑299

[MS98] R. Daniel Mauldin and Károly Simon, *The equivalence of some Bernoulli convolutions to Lebesgue measure*, Proc. Amer. Math. Soc. **126** (1998), no. 9, 2733–2736, DOI 10.1090/S0002-9939-98-04460-8. MR1458276 ↑79

[MS13] Anthony Manning and Károly Simon, *Dimension of slices through the Sierpinski carpet*, Trans. Amer. Math. Soc. **365** (2013), no. 1, 213–250, DOI 10.1090/S0002-9947-2012-05586-3. MR2984058 ↑211, 244, 369

[MS19a] Ian D. Morris and Cagri Sert, *A converse statement to Hutchinson's theorem and a dimension gap for self-affine measures*, arXiv e-prints (2019), `arXiv:1909.08532`. 300

[MS19b] Ian D. Morris and Pablo Shmerkin, *On equality of Hausdorff and affinity dimensions, via self-affine measures on positive subsystems*, Trans. Amer. Math. Soc. **371** (2019), no. 3, 1547–1582, DOI 10.1090/tran/7334. MR3894027 ↑360, 361

[MS21] Ian D. Morris and Cagri Sert, *A strongly irreducible affine iterated function system with two invariant measures of maximal dimension*, Ergodic Theory Dynam. Systems **41** (2021), no. 11, 3417–3438, DOI 10.1017/etds.2020.107. MR4321717 ↑293

[MU03] R. Daniel Mauldin and Mariusz Urbański, *Graph directed Markov systems*, Cambridge Tracts in Mathematics, vol. 148, Cambridge University Press, Cambridge, 2003. Geometry and dynamics of limit sets, DOI 10.1017/CBO9780511543050. MR2003772 ↑x

[MW88] R. Daniel Mauldin and Stanley C. Williams, *Hausdorff dimension in graph directed constructions*, Trans. Amer. Math. Soc. **309** (1988), no. 2, 811–829, DOI 10.2307/2000940. MR961615 ↑33, 34

[Neu01] Jörg Neunhäuserer, *Properties of some overlapping self-similar and some self-affine measures*, Acta Math. Hungar. **92** (2001), no. 1-2, 143–161, DOI 10.1023/A:1013716430425. MR1924256 ↑156, 180, 329

[Nga97] Sze-Man Ngai, *A dimension result arising from the L^q-spectrum of a measure*, Proc. Amer. Math. Soc. **125** (1997), no. 10, 2943–2951, DOI 10.1090/S0002-9939-97-03974-9. MR1402878 ↑150

[Ngu02] Nhu Nguyen, *Iterated function systems of finite type and the weak separation property*, Proc. Amer. Math. Soc. **130** (2002), no. 2, 483–487, DOI 10.1090/S0002-9939-01-06063-4. MR1862129 ↑127

[NN03] Sze-Man Ngai and Nhu Nguyen, *The Heighway dragon revisited*, Discrete Comput. Geom. **29** (2003), no. 4, 603–623, DOI 10.1007/s00454-003-0778-7. MR1976609 ↑7

[NW01] Sze-Man Ngai and Yang Wang, *Hausdorff dimension of self-similar sets with overlaps*, J. London Math. Soc. (2) **63** (2001), no. 3, 655–672, DOI 10.1017/S0024610701001946. MR1825981 ↑122, 123, 124, 125, 126, 127

[NW05] Sze-Man Ngai and Yang Wang, *Self-similar measures associated to IFS with nonuniform contraction ratios*, Asian J. Math. **9** (2005), no. 2, 227–244, DOI 10.4310/AJM.2005.v9.n2.a7. MR2176605 ↑156, 179, 180

[OW03] Lars Olsen and Steffen Winter, *Normal and nonnormal points of self-similar sets and divergence points of self-similar measures*, J. London Math. Soc. (2) **67** (2003), no. 1, 103–122, DOI 10.1112/S0024610702003630. MR1942414 ↑151

[Par97] William Parry, *Skew products of shifts with a compact Lie group*, J. London Math. Soc. (2) **56** (1997), no. 2, 395–404, DOI 10.1112/S0024610797005462. MR1489145 ↑260, 348

[Par04] William Parry, *Topics in ergodic theory*, Cambridge Tracts in Mathematics, vol. 75, Cambridge University Press, Cambridge-New York, 1981. MR614142 ↑271

[Per94] Yuval Peres, *The self-affine carpets of McMullen and Bedford have infinite Hausdorff measure*, Math. Proc. Cambridge Philos. Soc. **116** (1994), no. 3, 513–526, DOI 10.1017/S0305004100072789. MR1291757 ↑335

[Pet83] Karl Petersen, *Ergodic theory*, Cambridge Studies in Advanced Mathematics, vol. 2, Cambridge University Press, Cambridge, 1983, DOI 10.1017/CBO9780511608728. MR833286 ↑46, 255, 269, 271, 348

[PRSS01] Yuval Peres, Michał Rams, Károly Simon, and Boris Solomyak, *Equivalence of positive Hausdorff measure and the open set condition for self-conformal sets*, Proc. Amer. Math. Soc. **129** (2001), no. 9, 2689–2699, DOI 10.1090/S0002-9939-01-05969-X. MR1838793 ↑98, 402, 403

[PS95] Mark Pollicott and Károly Simon, *The Hausdorff dimension of λ-expansions with deleted digits*, Trans. Amer. Math. Soc. **347** (1995), no. 3, 967–983, DOI 10.2307/2154881. MR1290729 ↑32, 155, 156, 163, 164, 167, 226

[PS96] Yuval Peres and Boris Solomyak, *Absolute continuity of Bernoulli convolutions, a simple proof*, Math. Res. Lett. **3** (1996), no. 2, 231–239, DOI 10.4310/MRL.1996.v3.n2.a8. MR1386842 ↑168

[PS98] Yuval Peres and Boris Solomyak, *Self-similar measures and intersections of Cantor sets*, Trans. Amer. Math. Soc. **350** (1998), no. 10, 4065–4087, DOI 10.1090/S0002-9947-98-02292-2. MR1491873 ↑156, 238

[PS00a] Yuval Peres and Wilhelm Schlag, *Smoothness of projections, Bernoulli convolutions, and the dimension of exceptions*, Duke Math. J. **102** (2000), no. 2, 193–251, DOI 10.1215/S0012-7094-00-10222-0. MR1749437 ↑156, 167, 172, 184, 185

[PS00b] Yuval Peres and Boris Solomyak, *Existence of L^q dimensions and entropy dimension for self-conformal measures*, Indiana Univ. Math. J. **49** (2000), no. 4, 1603–1621, DOI 10.1512/iumj.2000.49.1851. MR1838304 ↑71, 72, 73, 89, 92, 93

[PSS00a] Yuval Peres, Wilhelm Schlag, and Boris Solomyak, *Sixty years of Bernoulli convolutions*, Fractal geometry and stochastics, II (Greifswald/Koserow, 1998), Progr. Probab., vol. 46, Birkhäuser, Basel, 2000, pp. 39–65. MR1785620 ↑79, 211, 233, 235, 239

[PSS00b] Yuval Peres, Károly Simon, and Boris Solomyak, *Self-similar sets of zero Hausdorff measure and positive packing measure*, Israel J. Math. **117** (2000), 353–379, DOI 10.1007/BF02773577. MR1760599 ↑194, 195, 201, 206, 207, 223

[PSS03] Yuval Peres, Károly Simon, and Boris Solomyak, *Fractals with positive length and zero Buffon needle probability*, Amer. Math. Monthly **110** (2003), no. 4, 314–325, DOI 10.2307/3647880. MR1984572 ↑69

[PSS06] Yuval Peres, Károly Simon, and Boris Solomyak, *Absolute continuity for random iterated function systems with overlaps*, J. London Math. Soc. (2) **74** (2006), no. 3, 739–756, DOI 10.1112/S0024610706023258. MR2286443 ↑276

[PT93] Jacob Palis and Floris Takens, *Hyperbolicity and sensitive chaotic dynamics at homoclinic bifurcations*, Cambridge Studies in Advanced Mathematics, vol. 35, Cambridge University Press, Cambridge, 1993. Fractal dimensions and infinitely many attractors. MR1237641 ↑31, 199

[PU89] Feliks Przytycki and Mariusz Urbański, *On the Hausdorff dimension of some fractal sets*, Studia Math. **93** (1989), no. 2, 155–186, DOI 10.4064/sm-93-2-155-186. MR1002918 ↑41, 54, 249, 335

[PU10] Feliks Przytycki and Mariusz Urbański, *Conformal fractals: ergodic theory methods*, London Mathematical Society Lecture Note Series, vol. 371, Cambridge University Press, Cambridge, 2010, DOI 10.1017/CBO9781139193184. MR2656475 ↑25, 42, 43, 46, 255, 261, 266, 306, 393

[PV15] Mark Pollicott and Polina Vytnova, *Estimating singularity dimension*, Math. Proc. Cambridge Philos. Soc. **158** (2015), no. 2, 223–238, DOI 10.1017/S030500411400053X. MR3310242 ↑299

[PW94] Mark Pollicott and Howard Weiss, *The dimensions of some self-affine limit sets in the plane and hyperbolic sets*, J. Statist. Phys. **77** (1994), no. 3-4, 841–866, DOI 10.1007/BF02179463. MR1301464 ↑246, 329

[QR03] Martine Queffélec and Olivier Ramaré, *Analyse de Fourier des fractions continues à quotients restreints* (French, with English summary), Enseign. Math. (2) **49** (2003), no. 3-4, 335–356. MR2028020 ↑232

[Ram02] Michał Rams, *Packing dimension estimation for exceptional parameters*, Israel J. Math. **130** (2002), 125–144, DOI 10.1007/BF02764074. MR1919375 ↑194, 195, 197, 198

[Ram05] Michał Rams, *Generic behavior of iterated function systems with overlaps*, Pacific J. Math. **218** (2005), no. 1, 173–186, DOI 10.2140/pjm.2005.218.173. MR2224595 ↑194, 195, 198, 201, 202

[Ran89] David A. Rand, *The singularity spectrum $f(\alpha)$ for cookie-cutters*, Ergodic Theory Dynam. Systems **9** (1989), no. 3, 527–541, DOI 10.1017/S0143385700005162. MR1016670 ↑140

[Rap17] Ariel Rapaport, *A self-similar measure with dense rotations, singular projections and discrete slices*, Adv. Math. **321** (2017), 529–546, DOI 10.1016/j.aim.2017.10.007. MR3715720 ↑358

[Rap18] Ariel Rapaport, *On self-affine measures with equal Hausdorff and Lyapunov dimensions*, Trans. Amer. Math. Soc. **370** (2018), no. 7, 4759–4783, DOI 10.1090/tran/7099. MR3812095 ↑360

[Rap22a] Ariel Rapaport, *On self-affine measures associated to strongly irreducible and proximal systems*, (2022), Preprint, `arXiv:2212.07215`. 361

[Rap22b] Ariel Rapaport, *Proof of the exact overlaps conjecture for systems with algebraic contractions* (English, with English and French summaries), Ann. Sci. Éc. Norm. Supér. (4) **55** (2022), no. 5, 1357–1377. MR4517689 ↑33, 187

[Rén70] Alfréd Rényi, *Probability theory*, North-Holland Series in Applied Mathematics and Mechanics, Vol. 10, North-Holland Publishing Co., Amsterdam-London; American Elsevier Publishing Co., Inc., New York, 1970. Translated by László Vekerdi. MR0315747 ↑70

[Rie95] Rolf Riedi, *An improved multifractal formalism and self-similar measures*, J. Math. Anal. Appl. **189** (1995), no. 2, 462–490, DOI 10.1006/jmaa.1995.1030. MR1312056 ↑72, 145, 147, 149

[Rog70] C. Ambrose Rogers, *Hausdorff measures*, Cambridge University Press, London-New York, 1970. MR0281862 ↑ix, 279

[Roh52] Vladimir A. Rohlin, *On the fundamental ideas of measure theory*, Amer. Math. Soc. Translation **1952** (1952), no. 71, 55. MR0047744 ↑261, 265

[Ros14] Eino Rossi, *Local dimensions of measures on infinitely generated self-affine sets*, J. Math. Anal. Appl. **413** (2014), no. 2, 1030–1039, DOI 10.1016/j.jmaa.2013.12.030. MR3159819 ↑282

[Roy68] Halsey L. Royden, *Real analysis*, The Macmillan Company, New York; Collier Macmillan Ltd., London, 1963. MR0151555 ↑265

[RS21] Haojie Ren and Weixiao Shen, *A dichotomy for the Weierstrass-type functions*, Invent. Math. **226** (2021), no. 3, 1057–1100, DOI 10.1007/s00222-021-01060-2. MR4337976 ↑182

[Rud76] Walter Rudin, *Principles of mathematical analysis*, 3rd ed., International Series in Pure and Applied Mathematics, McGraw-Hill Book Co., New York-Auckland-Düsseldorf, 1976. MR0385023 ↑3

[Rud87] Walter Rudin, *Real and complex analysis*, 3rd ed., McGraw-Hill Book Co., New York, 1987. MR924157 ↑421

[Rue04] David Ruelle, *Thermodynamic formalism*, 2nd ed., Cambridge Mathematical Library, Cambridge University Press, Cambridge, 2004. The mathematical structures of equilibrium statistical mechanics, DOI 10.1017/CBO9780511617546. MR2129258 ↑396

[Rui09] Víctor Ruiz, *Dimension of homogeneous rational self-similar measures with overlaps*, J. Math. Anal. Appl. **353** (2009), no. 1, 350–361, DOI 10.1016/j.jmaa.2008.12.008. MR2508873 ↑130, 131, 136, 211

[Rut] Alex Rutar, personal communication. 218

[Rut21a] Alex Rutar, *Geometric and combinatorial properties of self-similar multifractal measures*, (2021), Preprint, `arXiv:2008.00197`. 150

[Rut21b] Alex Rutar, *A multifractal decomposition for self-similar measures with exact overlaps*, (2021), Preprint, `arXiv:2104.06997`. 150

[RV20] Ariel Rapaport and Péter P. Varjú, *Self-similar measures associated to a homogeneous system of three maps*, (2020), Preprint, `arXiv:2010.01022`. 32, 33, 187

[Sal63] Raphaël Salem, *Algebraic numbers and Fourier analysis*, D. C. Heath and Company, Boston, Mass., 1963. MR0157941 ↑228, 229, 232, 238, 422, 426

[Sch94] Andreas Schief, *Separation properties for self-similar sets*, Proc. Amer. Math. Soc. **122** (1994), no. 1, 111–115, DOI 10.2307/2160849. MR1191872 ↑97

[She18] Weixiao Shen, *Hausdorff dimension of the graphs of the classical Weierstrass functions*, Math. Z. **289** (2018), no. 1-2, 223–266, DOI 10.1007/s00209-017-1949-1. MR3803788 ↑181

[Shm14] Pablo Shmerkin, *On the exceptional set for absolute continuity of Bernoulli convolutions*, Geom. Funct. Anal. **24** (2014), no. 3, 946–958, DOI 10.1007/s00039-014-0285-4. MR3213835 ↑41, 156, 188, 191, 227, 240

[Shm19] Pablo Shmerkin, *On Furstenberg's intersection conjecture, self-similar measures, and the L^q norms of convolutions*, Ann. of Math. (2) **189** (2019), no. 2, 319–391, DOI 10.4007/annals.2019.189.2.1. MR3919361 ↑41, 156, 189, 190, 192, 210, 241

[Sid03] Nikita Sidorov, *Almost every number has a continuum of β-expansions*, Amer. Math. Monthly **110** (2003), no. 9, 838–842, DOI 10.2307/3647804. MR2024754 ↑244

[Sid09] Nikita Sidorov, *Expansions in noninteger bases: lower, middle and top orders*, J. Number Theory **129** (2009), no. 4, 741–754, DOI 10.1016/j.jnt.2008.11.003. MR2499403 ↑247

[Sim02] Károly Simon, *Multifractals and the dimension of exceptions*, Real Anal. Exchange **27** (2001/02), no. 1, 191–207. MR1887691 ↑199

[Sim93] Károly Simon, *Hausdorff dimension for noninvertible maps*, Ergodic Theory Dynam. Systems **13** (1993), no. 1, 199–212, DOI 10.1017/S014338570000729X. MR1213088 ↑155, 163

[Sim12] David Simmons, *Conditional measures and conditional expectation; Rohlin's disintegration theorem*, Discrete Contin. Dyn. Syst. **32** (2012), no. 7, 2565–2582, DOI 10.3934/dcds.2012.32.2565. MR2900561 ↑269

[Sol95] Boris Solomyak, *On the random series $\sum \pm\lambda^n$ (an Erdős problem)*, Ann. of Math. (2) **142** (1995), no. 3, 611–625, DOI 10.2307/2118556. MR1356783 ↑32, 41, 156, 168, 185

[Sol98] Boris Solomyak, *Measure and dimension for some fractal families*, Math. Proc. Cambridge Philos. Soc. **124** (1998), no. 3, 531–546, DOI 10.1017/S0305004198002680. MR1636589 ↑53, 156, 170, 172, 226, 273, 275, 287

[Sol21] Boris Solomyak, *Fourier decay for self-similar measures*, Proc. Amer. Math. Soc. **149** (2021), no. 8, 3277–3291, DOI 10.1090/proc/15515. MR4273134 ↑238

[SS98] Károly Simon and Boris Solomyak, *Correlation dimension for self-similar Cantor sets with overlaps*, Fund. Math. **155** (1998), no. 3, 293–300. MR1607450 ↑199, 372

[SS02] Károly Simon and Boris Solomyak, *On the dimension of self-similar sets*, Fractals **10** (2002), no. 1, 59–65, DOI 10.1142/S0218348X02000963. MR1894903 ↑30, 53, 157

[SS06] Pablo Shmerkin and Boris Solomyak, *Zeros of $\{-1,0,1\}$ power series and connectedness loci for self-affine sets*, Experiment. Math. **15** (2006), no. 4, 499–511. MR2293600 ↑169, 170

[SS16] Pablo Shmerkin and Boris Solomyak, *Absolute continuity of self-similar measures, their projections and convolutions*, Trans. Amer. Math. Soc. **368** (2016), no. 7, 5125–5151, DOI 10.1090/tran6696. MR3456174 ↑72, 159, 188, 190, 191, 224, 241, 242

[SSS18] Santiago Saglietti, Pablo Shmerkin, and Boris Solomyak, *Absolute continuity of nonhomogeneous self-similar measures*, Adv. Math. **335** (2018), 60–110, DOI 10.1016/j.aim.2018.06.015. MR3836658 ↑179, 181, 191

[Ste89] J. Michael Steele, *Kingman's subadditive ergodic theorem* (English, with French summary), Ann. Inst. H. Poincaré Probab. Statist. **25** (1989), no. 1, 93–98. MR995293 ↑257

[Str23] Lauritz Streck, *On absolute continuity and maximal Garsia entropy for self-similar measures with algebraic contraction ratio*, (2023), Preprint, `arXiv:2303.07785`. 237

[SV98] Nikita Sidorov and Anatoly Vershik, *Ergodic properties of the Erdős measure, the entropy of the golden shift, and related problems*, Monatsh. Math. **126** (1998), no. 3, 215–261, DOI 10.1007/BF01367764. MR1651776 ↑252

[SV02] Grzegorz Świątek and Jan J. P. Veerman, *On a conjecture of Furstenberg*, Israel J. Math. **130** (2002), 145–155, DOI 10.1007/BF02764075. MR1919376 ↑221

[SV19] Károly Simon and Lajos Vágó, *Singularity versus exact overlaps for self-similar measures*, Proc. Amer. Math. Soc. **147** (2019), no. 5, 1971–1986, DOI 10.1090/proc/14042. MR3937675 ↑210, 211

[SY97] Timothy D. Sauer and James A. Yorke, *Are the dimensions of a set and its image equal under typical smooth functions?*, Ergodic Theory Dynam. Systems **17** (1997), no. 4, 941–956, DOI 10.1017/S0143385797086252. MR1468109 ↑43

[TKV15] Andrey Tetenov, Kirill Kamalutdinov, and Dmitry Vaulin, *Self-Similar Jordan Arcs Which Do Not Satisfy OSC*, (2015), Preprint, `arXiv:1512.00290`. 101

[Tor] Lidia Torma, *The dimension theory of some special families of self-similar fractals of overlapping construction satisfying the weak separation property. master's thesis, institute of mathematics, budapest university of technology and economics, 2015*, `http://math.bme.hu/~lidit/Thesi/MSC_BME_TLB.pdf`. 131

[Tri82] Claude Tricot Jr., *Two definitions of fractional dimension*, Math. Proc. Cambridge Philos. Soc. **91** (1982), no. 1, 57–74, DOI 10.1017/S0305004100059119. MR633256 ↑25, 62

[Tsu01] Masato Tsujii, *Fat solenoidal attractors*, Nonlinearity **14** (2001), no. 5, 1011–1027, DOI 10.1088/0951-7715/14/5/306. MR1862809 ↑182

[TT85] S. James Taylor and Claude Tricot Jr., *Packing measure, and its evaluation for a Brownian path*, Trans. Amer. Math. Soc. **288** (1985), no. 2, 679–699, DOI 10.2307/1999958. MR776398 ↑25, 197

[Urb16] Mariusz Urbański, *Random countable alphabet conformal iterated function systems satisfying the transversality condition*, Nonlinearity **29** (2016), no. 3, 1096–1123, DOI 10.1088/0951-7715/29/3/1096. MR3465996 ↑276

[Var19a] Péter P. Varjú, *Absolute continuity of Bernoulli convolutions for algebraic parameters*, J. Amer. Math. Soc. **32** (2019), no. 2, 351–397, DOI 10.1090/jams/916. MR3904156 ↑41, 156, 190, 230

[Var19b] Péter P. Varjú, *On the dimension of Bernoulli convolutions for all transcendental parameters*, Ann. of Math. (2) **189** (2019), no. 3, 1001–1011, DOI 10.4007/annals.2019.189.3.9. MR3961088 ↑54, 156, 187, 248

[Via14] Marcelo Viana, *Lectures on Lyapunov exponents*, Cambridge Studies in Advanced Mathematics, vol. 145, Cambridge University Press, Cambridge, 2014, DOI 10.1017/CBO9781139976602. MR3289050 ↑346

[VY22] Péter P. Varjú and Han Yu, *Fourier decay of self-similar measures and self-similar sets of uniqueness*, Anal. PDE **15** (2022), no. 3, 843–858, DOI 10.2140/apde.2022.15.843. MR4442842 ↑238

[Wal82] Peter Walters, *An introduction to ergodic theory*, Graduate Texts in Mathematics, vol. 79, Springer-Verlag, New York-Berlin, 1982. MR648108 ↑46, 47, 48, 121, 122, 130, 133, 134, 137, 255, 269, 271

[Wol03] Thomas H. Wolff, *Lectures on harmonic analysis*, University Lecture Series, vol. 29, American Mathematical Society, Providence, RI, 2003. With a foreword by Charles Fefferman and a preface by Izabella Łaba; Edited by Łaba and Carol Shubin, DOI 10.1090/ulect/029. MR2003254 ↑421, 422

[You82] Lai-Sang Young, *Dimension, entropy and Lyapunov exponents*, Ergodic Theory Dynam. Systems **2** (1982), no. 1, 109–124, DOI 10.1017/s0143385700009615. MR684248 ↑71

[Yu22] Han Yu, *Bernoulli convolutions with Garsia parameters in* $(1, \sqrt{2}]$ *have continuous density functions*, Proc. Amer. Math. Soc. **150** (2022), no. 10, 4359–4368, DOI 10.1090/proc/15971. MR4470180 ↑230

[Zer96] Martin P. W. Zerner, *Weak separation properties for self-similar sets*, Proc. Amer. Math. Soc. **124** (1996), no. 11, 3529–3539, DOI 10.1090/S0002-9939-96-03527-7. MR1343732 ↑97, 103, 112, 113

Index

Basic notation

$\mathbb{Z}, \mathbb{R}, \mathbb{C}$	the sets of integer, real and complex numbers.
$\mathbb{R}^d$	the d dimensional Euclidean space.
$\mathbb{S}^{d-1}$	the unit sphere in $\mathbb{R}^d$.
$\mathbb{PR}^1$	the real projective space of $\mathbb{R}^d$.
$G(d,m)$	the Grassmannian manifold of m-dimensional linear subspaces of $\mathbb{R}^d$.
$\bigwedge^k \mathbb{R}^d$	k-th exterior product of $\mathbb{R}^d$ vector spaces
$(\cdot,\cdot); \langle\cdot,\cdot\rangle$	standard inner product $\mathbb{R}^d$.
$B(\mathbf{x},r), B_r(\mathbf{x})$	open ball centered at $\mathbf{x}$ of radius r.
$\overline{B}(\mathbf{x},r), \overline{B}_r(\mathbf{x})$	closed ball centered at $\mathbf{x}$ of radius r.
$B(H,\delta)$	open δ-neighborhood of the set H, that is $B(H,\delta) := \{\mathbf{x} : \exists \mathbf{h} \in H,\ \|\mathbf{x}-\mathbf{h}\| < \delta\}$.
$\bar{A} = \mathrm{clos}(A)$	closure of the set A.
∂A	boundary of the set A.
$|A|$	diameter of the set A.
A^c	complement of the set A.
$\mathrm{int}(A)$	interior of the set A.
$\mathbb{1}_A$	characteristic (indicator) function of the set A.
$\#A$	cardinality of the set A.
$\mathrm{dist}(A,B)$	distance between the sets A, B in a metric space (X,ρ), that is, $\mathrm{dist}(A,B) := \inf\{\rho(a,b) : a \in A, b \in B\}$.
$\mathrm{dist}_H(A,B)$	Hausdorff distance between the sets A, B in a metric space (X,ρ), see (1.37).
Σ	symbolic space $\{1,\ldots,m\}^{\mathbb{N}}$.
Σ_n	the set $\{1,\ldots,m\}^n$.
Σ^*	the set of finite words in the alphabet $\{1,\ldots,m\}$, i.e., $\bigcup_{n=0}^{\infty} \Sigma_n$.
σ	the left shift operator on Σ.
$\mathbf{i}, \mathbf{j}, \mathbf{k}, \boldsymbol{\ell}$	elements of $\Sigma^* \cup \Sigma$.
$\mathbf{i}|_n$	the finite word formed by the first n symbols of $\mathbf{i} \in \Sigma \cup \Sigma^*$.
$\mathbf{i}^-$	the finite word obtained by deleting the last symbol of the finite word $\mathbf{i}$.
$|\mathbf{i}|$	length of the finite word $\mathbf{i} \in \Sigma^*$.
$[\mathbf{i}], [i_1 \ldots i_n]$	cylinder set on Σ, i.e. $\{\mathbf{j} \in \Sigma : j_k = i_k \text{ for } k = 1,\ldots,n\}$.
Π	the natural projection from the symbolic space Σ to $\mathbb{R}^d$, see (1.42).

$\mathbf{i} \wedge \mathbf{j}$	the common prefix of $\mathbf{i}$ and $\mathbf{j}$ for $\mathbf{i}, \mathbf{j} \in \bigcup_{n=1}^{\infty} \{1, \ldots, m\}^n \cup \{1, \ldots, m\}^{\mathbb{N}}$.
$S_{i_1 \ldots i_n}, S_{\mathbf{i}}$	n-fold composition: $S_{i_1}(S_{i_2}(\cdots(S_{i_n}(x)))) = S_{i_1} \circ \cdots \circ S_{i_n}(x)$ corresponding to the finite word $\mathbf{i} = (i_1, \ldots, i_n)$.
$\dim_{\mathrm{S}}(\Lambda), \dim_{\mathrm{S}}(\nu)$	similarity dimension of the self-similar set and measure.
$s(A_1, \ldots, A_m)$, $\dim_{\mathrm{aff}}(\mathcal{F})$	affinity dimension of the self-affine IFS $\mathcal{F} = (A_i x + t_i)_{i=1}^m$
$r_{\mathbf{i}}$	the product $r_{i_1} \cdots r_{i_n}$ for a finite word $\mathbf{i} = (i_1 \ldots i_n)$.
$p_{\mathbf{i}}$	the product of probabilities $p_{i_1} \cdots p_{i_n}$ for a finite word $\mathbf{i} = (i_1 \ldots i_n)$.
T^n	for a map $T : X \to X$ we write T^n for the n-fold composition of T, that is, $T^n := \underbrace{T \circ \cdots \circ T}_{n}$.
$\Lambda_{i_1 \ldots i_n}$	geometric cylinder set $S_{i_1 \ldots i_n}(\Lambda) = \Pi([i_1 \ldots i_n])$.
$\widetilde{\mathcal{M}}_r, \mathcal{M}_r$	symbolic Moran minimal cut-set $\{\mathbf{i} \in \Sigma : \lvert\Lambda_{\mathbf{i}}\rvert \leqslant r < \lvert\Lambda_{\mathbf{i}^-}\rvert\}$ at scale r , and the geometric Moran cover $\{\Pi([\mathbf{i}])\}_{\mathbf{i} \in \widetilde{\mathcal{M}}_r}$
Proj_{θ}	orthogonal projection along the line in the direction $\theta \in \mathbb{PR}^{d-1}$.
$\mathrm{proj}_{\mathbf{v}}$	orthogonal projection onto the line determined by the vector $\mathbf{v}$.
π_i	projection to the i-th axis in $\mathbb{R}^d$ for $d \geqslant 2$, that is, $\pi_i(x_1, \ldots, x_d) := x_i$ for $(x_1, \ldots, x_d) \in \mathbb{R}^d$.
$\mu\vert_A$	restriction of the measure μ to the set A.
$\mu_x^{\xi}, \mu_x^{\mathbf{v}}$	the family of conditional measures of μ with respect to the measurable partition ξ and, in particular, for the partition defined by $\mathrm{proj}_{\mathbf{v}}^{-1}$.
$f_*\mu$	push-forward of the measure μ with respect to f, that is, $f_*\mu(A) = \mu(f^{-1}(A))$.
$\mathcal{L}^d$	Lebesgue measure on $\mathbb{R}^d$.
$\mathcal{H}^s$	s-dimensional Hausdorff measure.
$\mathcal{H}^s_{\infty}$	s-dimensional Hausdorff content.
$\mathcal{P}^s, \widetilde{\mathcal{P}}^s$	s-dimensional packing measure and s-dimensional outer packing measure.
$\dim_{\mathrm{H}}(A)$	Hausdorff dimension of the set A.
$\dim_{\mathrm{P}}(A)$	packing dimension of the set A.
$N_{\delta}(A)$	the smallest number of sets of diameter δ needed to cover the bounded set A.
$\overline{\dim}_{\mathrm{B}}(A)$, $\underline{\dim}_{\mathrm{B}}(A)$	upper and lower box dimensions of the set A; if it exists it is denoted by $\dim_{\mathrm{B}}(A)$.
$\dim_{\mathrm{A}}(A)$	Assouad dimension of the set A.
$\dim_{\mathrm{F}}(\mu), \dim_{\mathrm{F}}(A)$	Fourier dimension of the measure μ and the set A.
$\widehat{f}, \widehat{\mu}$	Fourier transform of the function f and the measure μ.

$\mu \ll \nu$	the measure μ is absolutely continuous with respect to ν.
$\mu \perp \nu$	the probability measures μ and ν are mutually singular.
$L^q(\mathbb{R}^d)$	space of (equivalence classes of) functions f for which $\int \lvert f\rvert^q d\mathcal{L}^d < \infty$.
$\mathcal{E}_t(\mu)$	t-energy of the measure μ.
$\mathrm{spt}(\mu)$	support of the measure μ.
$\underline{D}(\nu, x)$	lower derivative of the measure ν at x.
$\frac{d\nu}{dx}$	Radon-Nikodym derivative of the measure ν with respect to the Lebesgue measure.
$\Theta^s_*(\nu, x), \Theta^{*s}(\nu, x)$	lower and upper s-densities of the measure ν at x.
$\overline{d}_\mu(x), \underline{d}_\mu(x)$	upper and lower local dimension of the measure μ at x; if it exists it is denoted by $d_\mu(x)$.
$\chi^{\mathbf{p}}_{\mathbf{r}}, \chi_\mu$	Lyapunov exponent of the measure $\mathbf{p}^{\mathbb{N}} = (p_1, \dots, p_m)^{\mathbb{N}}$ and of the measure μ in self-similar case.
$\chi_i, \chi_i(\mu)$	The ith Lyapunov exponent of the measure μ in self-affine case.
$h_{\mathbf{p}}$	entropy of the Bernoulli measure $\mathbf{p}^{\mathbb{N}}$
$D_q(\mu)$	L^q dimension of the measure μ.
$\overline{\dim}_e(\mu), \underline{\dim}_e(\mu)$	upper and lower entropy dimension of the measure μ.
$\overline{\dim}_{\mathrm{H}}(\mu), \underline{\dim}_{\mathrm{H}}(\mu)$	upper and lower Hausdorff dimension of the measure μ.
$\tau(\mu, q)$	L^q spectrum of the measure μ.
$\mathcal{D}_n$ and $\mathscr{D}_n$	partition of $\mathbb{R}^d$ into dyadic cubes.
$H_\mu(\mathcal{P}), H(\mu, \mathcal{P})$	entropy of the partition $\mathcal{P}$ with respect to μ.
$I_\mu(\xi\vert\mathcal{B}), H_\mu(\xi\vert\mathcal{B})$	conditional information and conditional entropy of the countable partition ξ given a partition or sub-σ-algebra $\mathcal{B}$.
$h_n(\mu)$	entropy of the partition $\mathcal{D}_n$.
$H_r(\mu)$	r-scale entropy of μ.
$\lVert\nu\rVert_{2,\gamma}$	homogeneous Sobolev γ-norm: $\lVert\nu\rVert^2_{2,\gamma} = \int_{\mathbb{R}^n} \lvert\widehat{\nu}(\xi)\rvert^2 \lvert\xi\rvert^{2\gamma}\, d\xi$.
$M_d(\mathbb{R}), M_d(\mathbb{Z})$	the set of $d \times d$ matrices with elements from $\mathbb{R}$ and $\mathbb{Z}$.
$GL(d, \mathbb{R})$	the set of invertible $d \times d$ matrices with elements from $\mathbb{R}$.
$SL(d, \mathbb{R})$	the set of invertible $d \times d$ matrices with elements from $\mathbb{R}$ of determinant 1.
$O(d, \mathbb{R}), SO(d, \mathbb{R})$	orthogonal subgroup and special orthogonal subgroup.
A^*	transpose of the matrix A.
$\alpha_i(A)$	ith singular value of the matrix A.
$\varphi^s(A)$	singular value function $\varphi^s(A) = \alpha_1(A) \cdots \alpha_{\lceil s\rceil}(A)^{s-(\lceil s\rceil-1)}$.
$A^{\wedge k}$	kth exterior product of the matrix A.
$A\vert_V$	linear operator of A restricted to the subspace V.

Selected Published Titles in This Series

276 **Balázs Bárány, Károly Simon, and Boris Solomyak,** Self-similar and Self-affine Sets and Measures, 2023

274 **Willi Freeden and M. Zuhair Nashed,** Recovery Methodologies: Regularization and Sampling, 2023

273 **Gennadiy Feldman,** Characterization of Probability Distributions on Locally Compact Abelian Groups, 2023

272 **Tadashi Ochiai,** Iwasawa Theory and Its Perspective, Volume 1, 2023

271 **Hideto Asashiba,** Categories and Representation Theory, 2022

270 **Leslie Hogben, Jephian C.-H. Lin, and Bryan L. Shader,** Inverse Problems and Zero Forcing for Graphs, 2022

269 **Ralph S. Freese, Ralph N. McKenzie, George F. McNulty, and Walter F. Taylor,** Algebras, Lattices, Varieties, 2022

268 **Ralph S. Freese, Ralph N. McKenzie, George F. McNulty, and Walter F. Taylor,** Algebras, Lattices, Varieties, 2022

267 **Dragana S. Cvetković Ilić,** Completion Problems on Operator Matrices, 2022

266 **Kate Juschenko,** Amenability of Discrete Groups by Examples, 2022

265 **Nabil H. Mustafa,** Sampling in Combinatorial and Geometric Set Systems, 2022

264 **J. Scott Carter and Seiichi Kamada,** Diagrammatic Algebra, 2021

263 **Vugar E. Ismailov,** Ridge Functions and Applications in Neural Networks, 2021

262 **Ragnar-Olaf Buchweitz,** Maximal Cohen–Macaulay Modules and Tate Cohomology, 2021

261 **Shiri Artstein-Avidan, Apostolos Giannopoulos, and Vitali D. Milman,** Asymptotic Geometric Analysis, Part II, 2021

260 **Lindsay N. Childs, Cornelius Greither, Kevin P. Keating, Alan Koch, Timothy Kohl, Paul J. Truman, and Robert G. Underwood,** Hopf Algebras and Galois Module Theory, 2021

259 **William Heinzer, Christel Rotthaus, and Sylvia Wiegand,** Integral Domains Inside Noetherian Power Series Rings, 2021

258 **Pramod N. Achar,** Perverse Sheaves and Applications to Representation Theory, 2021

257 **Juha Kinnunen, Juha Lehrbäck, and Antti Vähäkangas,** Maximal Function Methods for Sobolev Spaces, 2021

256 **Michio Jimbo, Tetsuji Miwa, and Fedor Smirnov,** Local Operators in Integrable Models I, 2021

255 **Alexandre Boritchev and Sergei Kuksin,** One-Dimensional Turbulence and the Stochastic Burgers Equation, 2021

254 **Karim Belabas and Henri Cohen,** Numerical Algorithms for Number Theory, 2021

253 **Robert R. Bruner and John Rognes,** The Adams Spectral Sequence for Topological Modular Forms, 2021

252 **Julie Déserti,** The Cremona Group and Its Subgroups, 2021

251 **David Hoff,** Linear and Quasilinear Parabolic Systems, 2020

250 **Bachir Bekka and Pierre de la Harpe,** Unitary Representations of Groups, Duals, and Characters, 2020

249 **Nikolai M. Adrianov, Fedor Pakovich, and Alexander K. Zvonkin,** Davenport–Zannier Polynomials and Dessins d'Enfants, 2020

248 **Paul B. Larson and Jindrich Zapletal,** Geometric Set Theory, 2020

247 **István Heckenberger and Hans-Jürgen Schneider,** Hopf Algebras and Root Systems, 2020

For a complete list of titles in this series, visit the
AMS Bookstore at **www.ams.org/bookstore/survseries/**.